Value Distribution of Meromorphic Functions

Translations of
MATHEMATICAL MONOGRAPHS

Volume 236

Value Distribution of Meromorphic Functions

Anatoly A. Goldberg
Iossif V. Ostrovskii

Translated by Mikhail Ostrovskii

With an Appendix
by Alexandre Eremenko and James K. Langley

American Mathematical Society
Providence, Rhode Island

А. А. Гольдберг, И. В. Островский

РАСПРЕДЕЛЕНИЕ ЗНАЧЕНИЙ МЕРОМОРФНЫХ ФУНКЦИЙ

"Наука", Москва, 1970

This work was originally published in Russian by NAUKA under the title "Распределение значений мероморфных функций", 1970. The present translation was created under license for the American Mathematical Society and is published by permission.

Translated from the Russian by Mikhail Ostrovskii

2000 *Mathematics Subject Classification.* Primary 30D30, 30D35;
Secondary 30D20, 30D15.

For additional information and updates on this book, visit
www.ams.org/bookpages/mmono-236

Library of Congress Cataloging-in-Publication Data

Gol′dberg, A. A. (Anatolii Asirovich)
[Raspredelenie znachenii meromorfnykh funktsii. English]
Value distribution of meromorphic functions / Anatoly A. Goldberg, Iossif V. Ostrovskii ; translated by Mikhail Ostrovskii.
p. cm. — (Translations of mathematical monographs ; v. 236)
Includes bibliographical references and indexes.
ISBN 978-0-8218-4265-2 (alk. paper)
1. Functions, Meromorphic. I. Ostrovskii, I. V. (Iosif Vladimirovich) II. Title.

QA331.G57413 2008
515′.982—dc22 2008060001

∞ The paper used in this book is acid-free and falls within the guidelines
established to ensure permanence and durability.
Visit the AMS home page at `http://www.ams.org/`

10 9 8 7 6 5 4 3 2 1 13 12 11 10 09 08

To our teachers:

Lev Izrailevich Volkovyskiĭ
and Boris Yakovlevich Levin

Contents

Preface

In this edition we corrected some errors and misprints and added references and footnotes related to recent achievements in the topics considered in the original edition. A survey of the results obtained since 1970 is in Appendix B, after the authors' Notes. There is also a separate list of references added to the English edition, which follows the original Bibliography. These references are numbered consecutively, to distinguish them from those in the main Bibliography.

Preface to the Russian Edition

The origins of the value distribution theory of meromorphic functions go back to the classical theorems of Sokhotskii–Casorati (1868), Weierstrass (1876), and Picard (1879). In the last decade of the 19th and the first two decades of the 20th centuries, these theorems underwent further development through investigations on the zero distribution of entire functions carried out mainly by the French school (Hadamard, Borel, Valiron, and others). The analytical machinery intrinsically connected with meromorphic functions was built in the 1920s by the Finnish mathematician Rolf Nevanlinna. After his work, the value distribution theory acquired, in some way, a complete form. The main classical results of the theory of entire functions have been included in Nevanlinna's theory in a natural way.

The results of Nevanlinna's theory, which, at the present time, can be regarded as classical, are considered in the first part of this book, consisting of Chapters 1–3 and Sections 1–3 of Chapter 4. Our aim here is to introduce beginners to the subject, and therefore our exposition does not skip details. We foresee reproaches of more qualified readers in this respect.

While the first part of this monograph treats mostly results obtained before the 1950s, the remainder is dedicated to modern research.

In spite of a certain completeness of the value distribution theory, the study of even most classical problems has not been brought to an end. On the contrary, it becomes more and more extensive. Many important problems remain open, and new problems arise in the process of further investigation.

The achievements are so essential and diverse that they admit the free coexistence of several monographs on the value distribution theory. In any case, the reader who is acquainted with Hayman's book [**Hay64**], will not find any substantial intersection with our book except for the classical theorems which had been known as early as the 1930s.

The value distribution theory of meromorphic functions occupies one of the central places in Complex Analysis. Extensive research is devoted to its connections with other areas of mathematics (topology, differential geometry, measure theory, potential theory, and others), to extention of its results to larger classes of functions (meromorphic functions in arbitrary plane regions and Riemann surfaces, algebroid functions, functions of several variables, meromorphic curves), and also to its applications, mainly in the analytic theory of differential equations.

In this book, the related topics are left aside, and the main attention is concentrated on the internal problems of the value distribution theory, which include the following:

(i) To what extent are the main results of Nevanlinna's theory definitive and cannot be improved further?
(ii) What properties of Picard's exceptional values are preserved for a wider class of exceptional values considered in the value distribution theory?
(iii) What are the connections between Nevanlinna's characteristics and other quantities that describe asymptotic properties of entire and meromorphic functions?
(iv) Also, we study asymptotic properties and value distribution of meromorphic functions that belong to some special classes which are, on one hand, sufficiently narrow to give new information not implied by general theorems and, on the other hand, sufficiently wide to be of interest for the general theory.
(v) Moreover, we study the value distribution with respect to arguments (not only with respect to moduli as in the classical Nevanlinna theory).

We pay much attention to the examples of functions with "pathological" properties. Without them, the reader would get a restricted image of the theory. Examples of functions with unusual properties play in the value distribution theory an important role, the same way counterexamples do in Real Analysis.

In conclusion, we indicate terminology, conventions, and notation used throughout the book without further notice.

Meromorphic functions are functions meromorphic in the complex plane, unless explicitly stated otherwise. Uniform (absolute or usual) convergence of a series or of an infinite product is understood as the corresponding convergence of its remainder. Thus, the behavior of finitely many terms does not affect the convergence or its type.

If we consider a continuous function with a removable discontinuity, we assume that this discontinuity is removed. For example, $f(0) = 1$ for $f(\lambda) = \pi\lambda \cot \pi\lambda$.

By $[\alpha]$ we denote the integer part of the number α. Square brackets are also used to separate expressions in formulas if this cannot cause any confusion. We also use square brackets for references. We denote by α^+ the number $(|\alpha| + \alpha)/2$.

The closure of the angle $\{\alpha < \arg z < \beta\}$ in the complex plane is denoted by $\{\alpha \leq \arg z \leq \beta\}$. Thus $0 \in \{\alpha \leq \arg z \leq \beta\}$.

Writing $\sum_k a_k$ or $\prod_k a_k$ we do not exclude the case where the sequence $\{a_k\}$ is finite or empty. In the latter case we assume that $\sum_k a_k = 0$ and $\prod_k a_k = 0$.

In each chapter, the formulas are numbered within sections. Thus, the reference (4.20) means formula (4.20) of the same chapter. Similar numeration is used for theorems. When we refer to a formula or a theorem from another chapter, we include the chapter number.

The facts related to the history of the problems under consideration, along with bibliographical references, are placed in the Notes at the end of the book. References to the Bibliography in the text are given only to results that lie beyond the scope of standard graduate courses in Real and Complex Analysis.

We note that the results of Sections 3 and 5 of Chapter 1 and of Section 3 of Chapter 3 are used in Chapter 6 only, and can be skipped in the first reading.

Preliminary versions of Sections 7 and 8 of Chapter 1; Chapter 2; Sections 1 and 2 of Chapter 3; Sections 1–4 and 6 of Chapter 4; Sections 1 and 2 of Chapter 5; Chapter 7; and Appendix A have been written by A. Goldberg. Sections 1–5 of Chapter 1; Section 3 of Chapter 3; Section 5 of Chapter 4; Sections 3–6 of Chapter 5; and Chapter 6 have been written by I. Ostrovskii.

We express our deep gratitude to Nikolai Govorov, Larisa Kudina, Victor Petrenko, David Potyagailo, Elena Sergienko, and Stanislav Tushkanov for their help with the book at various stages. Our special thanks to our editor Naum Landkof.

Anatoly A. Goldberg and Iossif V. Ostrovskii

CHAPTER 1

Characteristics of the Behavior of a Meromorphic Function and the First Fundamental Theorem

The basis of the value distribution theory of meromorphic functions consists of several formulas connecting the behavior of a meromorphic function with the distribution of its zeros and poles. At the beginning of this chapter we derive these formulas. Later, we use them to introduce the main characteristics of the behavior of a meromorphic function and study connections between them. One of the most important connections is the celebrated theorem which R. Nevanlinna called "the first fundamental theorem of the value distribution theory".

1. Auxiliary results

Let D be a bounded domain in the complex plane, whose boundary Γ consists of finitely many piecewise-analytic curves; by $\frac{\partial}{\partial n}$ we denote the operator of differentiation in the direction of the inward normal vector to Γ, and by Δ, the operator $\frac{\partial^2}{\partial x^2} + \frac{\partial^2}{\partial y^2}$.

We use the following formula:

$$\iint_D (u\Delta v - v\Delta u)d\sigma = -\int_\Gamma \left(u\frac{\partial v}{\partial n} - v\frac{\partial u}{\partial n}\right) ds, \tag{1.1}$$

where $u = u(x,y)$ and $v = v(x,y)$ are twice continuously differentiable in some domain containing the closure $\bar{D}$ of the domain D, σ is Lebesgue measure in the plane, and s is one-dimensional Lebesgue measure (length) on Γ. This formula is usually called the *second Green formula*. It can be obtained from the well-known Green formula

$$\iint_D \left(\frac{\partial Q}{\partial x} - \frac{\partial P}{\partial y}\right) d\sigma = \int_\Gamma P dx + Q dy$$

in the following way. Let

$$P = -u\frac{\partial v}{\partial y}, \qquad Q = u\frac{\partial v}{\partial x}.$$

We get the formula

$$\iint_D \left(u\Delta v + \frac{\partial u}{\partial x}\frac{\partial v}{\partial x} + \frac{\partial u}{\partial y}\frac{\partial v}{\partial y}\right) d\sigma = \int_\Gamma -u\frac{\partial v}{\partial y}dx + u\frac{\partial v}{\partial x}dy,$$

which can be rewritten as

$$\iint_D \left(u\Delta v + \frac{\partial u}{\partial x}\frac{\partial v}{\partial x} + \frac{\partial u}{\partial y}\frac{\partial v}{\partial y}\right) d\sigma = -\int_\Gamma u\frac{\partial v}{\partial n}ds. \tag{1.2}$$

Subtracting from (1.2) the formula which is obtained from (1.2) if we interchange u and v, we get (1.1).

We also need the notion of a *Green function.* The *Green function* of a domain D is a function $G(\zeta, z)$, defined for $\zeta \in \bar{D},\ z \in \bar{D},\ \zeta \neq z$, and satisfying the following conditions:

(i) For each fixed $z \in D$,

$$G(\zeta, z) = \ln \frac{1}{|\zeta - z|} + h_z(\zeta),$$

where $h_z(\zeta)$ is a function that is harmonic in D and continuous in $\bar{D}$.

(ii) If $\zeta \in \Gamma,\ z \in \bar{D},\ \zeta \in \bar{D}$, and $z \in \Gamma$, then

$$G(\zeta, z) = 0.$$

The Green function exists and is unique. The uniqueness can be proved in the following way. Assuming that $G_1(\zeta, z)$ and $G_2(\zeta, z)$ are two Green functions of D, we choose $z \in D$ and consider the function $v(\zeta) = G_1(\zeta, z) - G_2(\zeta, z)$. The condition (i) implies that this function is harmonic in D and continuous in $\bar{D}$, and the condition (ii) implies that it is equal to zero on Γ. By the maximum and the minimum principles for harmonic functions we get $v(\zeta) = 0$, and hence $G_1(\zeta, z) \equiv G_2(\zeta, z)$.

We would like to observe that condition (i) implies that there exists a disc K_z centered at z and such that $G(\zeta, z) > 0$ for $\zeta \in K_z$. Using the minimum principle for $D \backslash K_z$, we conclude that $G(\zeta, z) > 0$ for all $\zeta \in D,\ z \in D$.

We will not use the existence of the Green functions for multiply connected domains; for this reason, we do not discuss this matter.

For a simply connected domain, the existence of the Green function follows from the classical results on conformal mappings. In fact, by the Riemann theorem, there exists a conformal mapping $w = f_z(\zeta)$ which maps the domain D onto the disc $\{|w| < 1\}$ in such a way that $f_z(z) = 0$. By the theorem on the boundary behavior of conformal mappings, the function f_z is continuous on $\bar{D}$, and $|f_z(\zeta)| = 1$ for $\zeta \in \Gamma$. Let

$$G(\zeta, z) = \ln \frac{1}{|f_z(\zeta)|}; \tag{1.3}$$

it is easy to check that this function satisfies (i) and (ii).

EXAMPLE 1. Let $D = \{|\zeta| < R\}$. Then

$$f_z(\zeta) = \frac{R(\zeta - z)}{R^2 - \zeta\bar{z}}, \qquad G(\zeta, z) = \ln \left| \frac{R^2 - \zeta\bar{z}}{R(\zeta - z)} \right|. \tag{1.4}$$

EXAMPLE 2. Let $D = \{|\zeta| < R,\ \operatorname{Im}\zeta > 0\}$. Then

$$f_z(\zeta) = \frac{R(\zeta - z)}{R^2 - \zeta\bar{z}} \frac{R^2 - \zeta z}{R(\zeta - \bar{z})}, \qquad G(\zeta, z) = \ln \left| \frac{R^2 - \zeta\bar{z}}{R(\zeta - z)} \frac{R(\zeta - \bar{z})}{R^2 - \zeta z} \right|. \tag{1.5}$$

Suppose that D is simply connected. Let $\gamma_1, \dots, \gamma_p$ be analytic curves forming Γ, let $A_l,\ l = 1, \dots, p$, be the common endpoint of γ_l and γ_{l+1} (we use the convention: $\gamma_{p+1} \equiv \gamma_1$), and let $\pi\alpha_l$ $(0 < \alpha_l \leq 2)$, $l = 1, \dots, p$, be the angles between γ_l and γ_{l+1}. By the Riemann–Schwarz symmetry principle (see Privalov [**Pri56**],

Chapter XII, §2.4) the function f_z has analytic continuation to a certain domain containing $\bar{D}\backslash\{A_1, \dots, A_p\}$.

We assume (without mentioning this) that in the intersection of a sufficiently small neighborhood U_l of A_l with D the function f_z has a representation

$$f_z(\zeta) = (\zeta - A_l)^{1/\alpha_l}\varphi_l(\zeta) + w_l, \tag{1.6}$$

where $\varphi_l(\zeta)$ is an analytic function in U_l, $\varphi_l(A_l) \neq 0$, $|w_l| = 1$, $l = 1, \dots, p$. We omit the discussion of the conditions on D under which such a representation holds. For our purposes it is enough to observe that the domain $D = \{|\zeta| < R\}$ satisfies this condition since it is bounded by an analytic curve and there are no points A_l. The domain $D = \{|\zeta| < R, \operatorname{Im}\zeta > 0\}$ also belongs to this class. In fact, in this case the points A_l, $l = 1, 2$, are the points $\zeta = \pm R$, $\alpha_l = \frac{1}{2}$, $l = 1, 2$. Equation (1.5) implies that $f_z(\zeta)$ is analytic at $\zeta = \pm R$ and $\frac{d}{d\zeta} f_z(\zeta)$ has simple zeros at $\zeta = \pm R$; hence the representation (1.6) holds.

Further we regard the derivative $\frac{\partial G}{\partial n}$ with respect to the variable ζ as a function of z. This derivative exists for all $\zeta \neq A_l$ $(l = 1, \dots, p)$. Since $G(\zeta, z) > 0$ for $\zeta \in D$, $z \in D$ and $G(\zeta, z) = 0$ for $\zeta \in \Gamma$, $z \in D$, we have $\frac{\partial G}{\partial n} \geq 0$. Along the curve Γ the following equation is satisfied:

$$\frac{f_z'(\zeta)}{f_z(\zeta)} d\zeta = i\frac{\partial G}{\partial n} ds \qquad (\zeta \neq A_l,\ l = 1, \dots, p). \tag{1.7}$$

In fact, computing the differential $d\ln f_z(\zeta)$ along the curve Γ and using the Cauchy–Riemann conditions, we get

$$\begin{aligned} d\ln f_z(\zeta) &= \frac{\partial}{\partial s} \ln f_z(\zeta) ds = \left(\frac{\partial}{\partial s} \ln|f_z(\zeta)| + i\frac{\partial}{\partial s} \arg f_z(\zeta)\right) ds \\ &= i\frac{\partial}{\partial s} \arg f_z(\zeta) ds = -i\frac{\partial}{\partial n} \ln|f_z(\zeta)| ds = i\frac{\partial G}{\partial n} ds. \end{aligned}$$

The equation (1.7) can sometimes be used to find $\frac{\partial G}{\partial n}$.

EXAMPLE 1. Let $D = \{\zeta : \ |\zeta| < R\}$; then

$$\frac{\partial G}{\partial n} ds = -i\left[\ln\frac{R(\zeta - z)}{R^2 - \zeta\bar{z}}\right]' d\zeta = \frac{R^2 - |z|^2}{|\zeta - z|^2} d\theta, \qquad \zeta = Re^{i\theta}. \tag{1.8}$$

The function $\frac{R^2 - |z|^2}{|\zeta - z|^2}$ is called the *Poisson kernel.* Observe that

$$\frac{R^2 - |z|^2}{|\zeta - z|^2} = \frac{R^2 - r^2}{R^2 + r^2 - 2Rr\cos(\varphi - \theta)} = \operatorname{Re}\frac{\zeta + z}{\zeta - z}, \qquad \zeta = Re^{i\theta},\ z = re^{i\varphi}. \tag{1.9}$$

EXAMPLE 2. Let $D = \{\zeta : \ |\zeta| < R, \ \operatorname{Im}\zeta > 0\}$. Then

$$\frac{\partial G}{\partial n} ds = -i\left[\ln\left(\frac{R(\zeta - z)}{R^2 - \zeta\bar{z}}\,\frac{R^2 - \zeta z}{R(\zeta - \bar{z})}\right)\right]' d\zeta.$$

On the semicircle $\{\zeta = Re^{i\theta},\ 0 < \theta < \pi\}$ we have

$$\frac{\partial G}{\partial n} ds = \left\{\frac{R^2 - |z|^2}{|\zeta - z|^2} - \frac{R^2 - |z|^2}{|\zeta - \bar{z}|^2}\right\} d\theta; \tag{1.10}$$

on the interval $\{\zeta = t, \ -R < t < R\}$ we have

$$\frac{\partial G}{\partial n} ds = 2\left\{\frac{r\sin\varphi}{|z-t|^2} - \frac{R^2 r\sin\varphi}{|R^2 - zt|^2}\right\} dt, \qquad z = re^{i\varphi}. \tag{1.11}$$

Since the function $f_z(\zeta)$ is analytic on Γ, except at the points A_l, $l = 1, \ldots, p$, (1.7) implies that, with the exception of these points, the function $\frac{\partial G}{\partial n}$ is infinitely differentiable on Γ. By (1.6), the singularities of $\frac{\partial G}{\partial n}$ at the points A_l are such that the product $|\zeta - A_l|^{\frac{1}{2}} \frac{\partial G}{\partial n} \left(\leq |\zeta - A_l|^{1-\frac{1}{\alpha_l}} \frac{\partial G}{\partial n}\right)$ is bounded. This observation shows that all integrals containing $\frac{\partial G}{\partial n}$ (considered later) are convergent.

THEOREM 1.1. *Let D be a simply connected domain with a piecewise analytic boundary Γ, and let $u(z)$ be a twice continuously differentiable*[1] *function in some domain containing $\bar{D}$, excluding a finite set $\{c_1, \ldots, c_q\} \subset \bar{D}$. In a neighborhood of these points, let $u(z)$ have the form*

$$u(z) = d_k \ln|z - c_k| + u_k(z), \tag{1.12}$$

where d_k are constants and $u_k(z)$ are twice continuously differentiable functions in a neighborhood of points c_k $(k = 1, \ldots, q)$. Then

$$u(z) + \frac{1}{2\pi}\iint_D G(\zeta, z)\Delta u(\zeta) d\sigma = \frac{1}{2\pi}\int_\Gamma u(\zeta)\frac{\partial G}{\partial n} ds - \sum_{c_k \in D} d_k G(c_k, z) \tag{1.13}$$

for $z \in D$, $z \neq c_1, \ldots, c_q$.

PROOF. We exclude from D discs of radius ε centered at $c_1, \ldots, c_q$, z, and $A_1, \ldots, A_p$. We assume that ε is so small that the discs do not intersect, and that if the center of such a disc is inside D, then the whole disc is inside D. We denote by D_ε the domain so obtained, and by Γ_ε the part of Γ which is not in any of the discs. By $C(\varepsilon, a)$ we denote the intersection of $\{z : \ |z - a| = \varepsilon\}$ and D.

Letting $u = u(\zeta)$ and $v = G(\zeta, z)$, we write the second Green formula for D_ε. Since $\Delta v = 0$, we get[2]

$$\begin{aligned}\iint_{D_\varepsilon} & v\Delta u d\sigma \\ &= \left(\int_{\Gamma_\varepsilon} + \int_{C(\varepsilon,z)} + \sum_{k=1}^{q}\int_{C(\varepsilon,c_k)} + \sum_{l=1}^{p}\int_{C(\varepsilon,A_l)}\right)\left(u\frac{\partial v}{\partial n} - v\frac{\partial u}{\partial n}\right) ds.\end{aligned} \tag{1.14}$$

Let us find the limits of each of the integrals as $\varepsilon \to 0$. Since $v = 0$ for $\zeta \in \Gamma$, we have

$$\lim_{\varepsilon\to 0}\int_{\Gamma_\varepsilon}\left(u\frac{\partial v}{\partial n} - v\frac{\partial u}{\partial n}\right) ds = \int_\Gamma u(\zeta)\frac{\partial G}{\partial n} ds.$$

Furthermore, by the mean value theorem, for each $a \in \bar{D}$ we have

$$\int_{C(\varepsilon,a)}\left(u\frac{\partial v}{\partial n} - v\frac{\partial u}{\partial n}\right) ds = (\text{the length of } C(\varepsilon, a))\left(u\frac{\partial v}{\partial n} - v\frac{\partial u}{\partial n}\right)\Big|_{\zeta*},$$

[1] With respect to the variables x and y, $z = x + iy$.

[2] If some points are repeated twice among $\{c_1, \ldots, c_q\}$ and $\{A_1, \ldots, A_p\}$, we write the corresponding integrals only once.

where ζ^* is a point on $C(\varepsilon, a)$. Since the length of $C(\varepsilon, z)$ is $2\pi\varepsilon$ and on $C(\varepsilon, z)$ the following asymptotic estimates:

$$v = \ln\frac{1}{\varepsilon} + O(1), \qquad \frac{\partial v}{\partial n} = \frac{\partial v}{\partial \varepsilon} = -\frac{1}{\varepsilon} + O(1), \qquad u = u(z) + o(1), \qquad \frac{\partial u}{\partial n} = O(1),$$

hold as $\varepsilon \to 0$, we see that

$$\lim_{\varepsilon\to 0} \int_{C(\varepsilon,z)} \left(u\frac{\partial v}{\partial n} - v\frac{\partial u}{\partial n}\right) ds = -2\pi u(z).$$

If $c_k \in D$, then on $C(\varepsilon, c_k)$ the following asymptotic estimates hold as $\varepsilon \to 0$:

$$u = d_k \ln\varepsilon + O(1), \qquad \frac{\partial u}{\partial n} = \frac{\partial u}{\partial \varepsilon} = \frac{d_k}{\varepsilon} + O(1),$$

$$v = G(c_k, z) + o(1), \qquad \frac{\partial v}{\partial n} = O(1).$$

Therefore, for $c_k \in D$ we have

$$\lim_{\varepsilon\to 0} \int_{C(\varepsilon,c_k)} \left(u\frac{\partial v}{\partial n} - v\frac{\partial u}{\partial n}\right) ds = -2\pi d_k G(c_k, z).$$

If a is either one of the points c_k belonging to Γ or one of the points A_l, $l = 1, \dots, p$, then, in any case, the length of $C(\varepsilon, a)$ is $O(\varepsilon)$, and the following asymptotic estimates are valid on $C(\varepsilon, a)$:

$$u = O(|\ln\varepsilon|), \qquad \frac{\partial u}{\partial n} = O\left(\frac{1}{\varepsilon}\right), \qquad v = o(1), \qquad \frac{\partial v}{\partial n} = O\left(\frac{1}{\sqrt{\varepsilon}}\right)$$

(the last estimate follows from (1.6) and (1.3)). Therefore we have

$$\lim_{\varepsilon\to 0} \int_{C(\varepsilon,a)} \left(u\frac{\partial v}{\partial n} - v\frac{\partial u}{\partial n}\right) ds = 0.$$

Taking the limit as $\varepsilon \to 0$ in (1.14), we get the statement of the theorem. □

Observe that the integral in the left-hand side of (1.13) is, by definition, assumed to be equal to

$$\lim_{\varepsilon\to 0} \iint_{D_\varepsilon} G(\zeta, z)\Delta u(\zeta) d\sigma.$$

REMARK. The statement of Theorem 1.1 remains true even if the function $u(z)$, in addition to singularities c_k on the boundary Γ of the region D whose behavior is determined by (1.12), has a finite set of singularities μ_k, $k = 1, \dots, q$, such that on the arcs $C(\varepsilon, \mu_k)$ the following asymptotic estimates hold (as $\varepsilon \to 0$):

$$u = o\left(\frac{1}{\sqrt{\varepsilon}}\right), \qquad \frac{\partial u}{\partial n} = O\left(\frac{1}{\varepsilon}\right).$$

2. The Nevanlinna, Poisson–Jensen, and Shimizu–Ahlfors formulas

We are going to use several formulas connecting the behavior of a meromorphic function with the distribution of its zeros and poles. Most of the formulas we need are special cases of (1.13).

THEOREM 2.1. *Let D be a simply connected region with a piecewise analytic boundary Γ and $f(z) \not\equiv 0$ be a meromorphic function in $\bar{D}$. Then*

$$\ln|f(z)| = \frac{1}{2\pi}\int_\Gamma \ln|f(\zeta)|\frac{\partial G}{\partial n}ds - \sum_{a_m\in D} G(a_m, z) + \sum_{b_n\in D} G(b_n, z), \tag{2.1}$$

where the a_m are the zeros of $f(z)$, and the b_n are its poles.[3]

PROOF. We apply Theorem 1.1 with $u(z) = \ln|f(z)|$. If z is neither a pole nor a zero of $f(z)$, then $\Delta u(z) = 0$. Therefore the integral in the left-hand side of (1.13) is equal to zero. If the point c_k is a zero (pole) of $f(z)$ of order χ_k, then equality (1.12) with $d_k = \chi_k$ ($d_k = -\chi_k$) is valid in the neighborhood of c_k. Therefore the right-hand side of (1.13) is equal to the right-hand side of (2.1). □

REMARK. The remark after Theorem 1.1 implies that formula (2.1) remains valid if $f(z)$ is meromorphic everywhere in $\bar{D}$ except possibly finitely many singular points $\mu_k \in \Gamma$, $k = 1, \dots, q$, such that in the intersection of a small neighborhood of μ_k with $\bar{D}$ the function $f(z)$ admits a representation

$$f(z) = g_k((z-\mu_k)^{\alpha_k}),$$

where $g_k(\zeta)$ are functions meromorphic in the neighborhood of $\zeta = 0$, and $\alpha_k > 0$ are some constants ($k = 1, \dots, q$).

One of the most important special cases of formula (2.1) is the case where D is a disc $\{|z| < R\}$.

THEOREM 2.2. *Let $f(z) \not\equiv 0$ be a meromorphic function in $\{|z| \le R\}$. Then the following formula, called the Poisson–Jensen formula, holds:*

$$\begin{aligned}\ln|f(z)| = {} & \frac{1}{2\pi}\int_0^{2\pi} \ln|f(Re^{i\theta})| \operatorname{Re}\frac{Re^{i\theta}+z}{Re^{i\theta}-z}d\theta \\ & - \sum_{|a_m|<R} \ln\left|\frac{R^2-\bar{a}_m z}{R(z-a_m)}\right| + \sum_{|b_n|<R} \ln\left|\frac{R^2-\bar{b}_n z}{R(z-b_n)}\right|,\end{aligned} \tag{2.2}$$

where the a_m are the zeros of $f(z)$, and the b_n are its poles.

The proof immediately follows from formulas (1.4), (1.8), and (1.9).

Applying (2.1) to the half-disc $D = \{|z| < R,\ \operatorname{Im} z > 0\}$, and using the identities (1.5), (1.6), and (1.11), we get the following result.

---

[3]The summands related to the multiple zeros and poles are repeated in the right-hand side of (2.1) an appropriate number of times.

THEOREM 2.3. *Let $f(z) \not\equiv 0$ be a meromorphic function in the half-disc $\{|z| \le R,\ \operatorname{Im} z \ge 0\}$. Then*

$$
\begin{aligned}
\ln|f(z)| &= \frac{1}{\pi}\int_{-R}^{R} \ln|f(t)| \left\{ \frac{r\sin\varphi}{|z-t|^2} - \frac{R^2 r \sin\varphi}{|R^2 - zt|^2} \right\} dt \\
&\quad + \frac{1}{2\pi}\int_0^{\pi} \ln|f(Re^{i\theta})| \left\{ \frac{R^2 - r^2}{|Re^{i\theta} - z|^2} - \frac{R^2 - r^2}{|Re^{-i\theta} - z|^2} \right\} d\theta \\
&\quad - \sum_{|a_m|<R,\ \operatorname{Im} a_m > 0} \ln \left| \frac{R^2 - \bar a_m z}{R(z - a_m)} \cdot \frac{R(z - \bar a_m)}{R^2 - a_m z} \right| \\
&\quad + \sum_{|b_n|<R,\ \operatorname{Im} b_n > 0} \ln \left| \frac{R^2 - \bar b_n z}{R(z - b_n)} \cdot \frac{R(z - \bar b_n)}{R^2 - b_n z} \right|,
\end{aligned} \tag{2.3}
$$

where the a_m are the zeros of the function $f(z)$, and the b_n are its poles.

REMARK. Formulas (2.2) and (2.3) are valid for functions $f(z)$ satisfying the conditions of the remark after Theorem 2.1 in the disc $\{|z| \le R\}$ and the half-disc $\{|z| \le R,\ \operatorname{Im} z \ge 0\}$, respectively.

THEOREM 2.4. *Let $f(z)$ be a non-identically zero meromorphic function in the disc $\{|z| \le R\}$. The following formulas hold (the last of them under the assumption $f(0) \neq 0,\ \infty$):*

$$
\begin{aligned}
\ln f(z) &= \frac{1}{2\pi}\int_0^{2\pi} \ln|f(Re^{i\theta})| \frac{Re^{i\theta} + z}{Re^{i\theta} - z} d\theta \\
&\quad - \sum_{|a_m|<R} \ln \frac{R^2 - \bar a_m z}{R(z - a_m)} + \sum_{|b_n|<R} \ln \frac{R^2 - \bar b_n z}{R(z - b_n)} + iC;
\end{aligned} \tag{2.4}
$$

$$
\begin{aligned}
\frac{d^p}{dz^p} \ln f(z) &= \frac{1}{2\pi}\int_0^{2\pi} \ln|f(Re^{i\theta})| \frac{p!\,2Re^{i\theta}}{(Re^{i\theta} - z)^{p+1}} d\theta \\
&\quad + (p-1)! \sum_{|a_m|<R} \left(\frac{\bar a_m^p}{(R^2 - \bar a_m z)^p} - \frac{(-1)^p}{(z - a_m)^p} \right) \\
&\quad - (p-1)! \sum_{|b_n|<R} \left(\frac{\bar b_n^p}{(R^2 - \bar b_n z)^p} - \frac{(-1)^p}{(z - b_n)^p} \right);
\end{aligned} \tag{2.5}
$$

$$
\begin{aligned}
\left.\frac{d^p}{dz^p} \ln f(z)\right|_{z=0} &= \frac{2p!}{R^p}\frac{1}{2\pi}\int_0^{2\pi} \ln|f(Re^{i\theta})| e^{-ip\theta} d\theta \\
&\quad + (p-1)! \sum_{|a_m|<R} \left(\frac{\bar a_m^p}{R^{2p}} - \frac{1}{a_m^p} \right) - (p-1)! \sum_{|b_n|<R} \left(\frac{\bar b_n^p}{R^{2p}} - \frac{1}{b_n^p} \right),
\end{aligned} \tag{2.6}
$$

where C is a real constant, $p = 1, 2, \dots$, the a_m are the zeros of the function $f(z)$, and the b_n are its poles.

PROOF. The functions in both sides of (2.4) are analytic as functions of z. By Theorem 2.2 the real parts of these functions coincide. In view of the Cauchy–Riemann equations it follows that the functions are equal up to a purely imaginary additive constant. The validity of (2.4) follows. Formula (2.5) follows from (2.4) by differentiation. Finally, (2.6) follows from (2.5) if we let $z = 0$. □

Observe that in the case where $f(z)$ has neither poles nor zeros in $\{|z| \le R\}$, formula (2.4) coincides with the classical Schwarz formula written for $\ln f(z)$. We refer to (2.1) and (2.3)–(2.6) as the *Nevanlinna formulas.*

EXERCISE. Obtain analogs of (2.4) and (2.5) for meromorphic functions in the half-disc $\{|z| \le R,\ \operatorname{Im} z \ge 0\}$.

THEOREM 2.5. *Let $f(z) \not\equiv 0$ be a meromorphic function in the disc $\{|z| \le R\}$ and let*

$$f(z) = c_\lambda z^\lambda + c_{\lambda+1} z^{\lambda+1} + \cdots, \quad c_\lambda \ne 0, \tag{2.7}$$

be its Laurent expansion in a neighborhood of $z = 0$. Then the following formula (called the Jensen formula) holds:

$$\begin{aligned} \ln |c_\lambda| = \frac{1}{2\pi} \int_0^{2\pi} \ln |f(Re^{i\theta})| d\theta - \sum_{0<|a_m|<R} \ln \frac{R}{|a_m|} \\ + \sum_{0<|b_n|<R} \ln \frac{R}{|b_n|} - \lambda \ln R, \end{aligned} \tag{2.8}$$

where the a_m are the zeros of $f(z)$, and the b_n are its poles.

PROOF. Let $z \to 0$ in the Poisson–Jensen formula. For the sums in the right-hand side we have

$$\begin{aligned} -\sum_{|a_m|<R} + \sum_{|b_n|<R} &= -\sum_{0<|a_m|<R} + \sum_{0<|b_n|<R} + \lambda \ln \frac{|z|}{R} \\ &= -\sum_{0<|a_m|<R} \ln \frac{R}{|a_m|} + \sum_{0<|b_n|<R} \ln \frac{R}{|b_n|} + o(1) + \lambda \ln \frac{|z|}{R}. \end{aligned}$$

Since in the left-hand side we get

$$\ln |f(z)| = \lambda \ln |z| + \ln |c_\lambda| + o(1),$$

formula (2.8) follows. □

THEOREM 2.6. *Let $f(z) \not\equiv 0$ be a meromorphic function in the disc $\{|z| \le R\}$. If $f(0) \ne \infty$, then the following formula (called the Shimizu–Ahlfors formula) holds:*

$$\begin{aligned} &\frac{1}{\pi} \iint_{|z|\le R} \left(\ln \frac{R}{|z|} \right) \frac{|f'(z)|^2}{(1+|f(z)|^2)^2} d\sigma(z) \\ &= \frac{1}{2\pi} \int_0^{2\pi} \ln \sqrt{1+|f(Re^{i\theta})|^2} d\theta - \ln \sqrt{1+|f(0)|^2} + \sum_{|b_n|<R} \ln \frac{R}{|b_n|}, \end{aligned} \tag{2.9}$$

where the b_n are the poles of $f(z)$. In the case $f(0) = \infty$, the right-hand side must be replaced by

$$\frac{1}{2\pi} \int_0^{2\pi} \ln \sqrt{1+|f(Re^{i\theta})|^2} d\theta - \ln |c_\lambda| - \lambda \ln R + \sum_{0<|b_n|<R} \ln \frac{R}{|b_n|},$$

where c_λ is the same as in (2.7).

PROOF. We use Theorem 1.1 with $u(z) = \ln\sqrt{1+|f(z)|^2}$, $D = \{|z| \le R\}$. To compute Δu we let $A(z) = \operatorname{Re} f(z)$, $B(z) = \operatorname{Im} f(z)$. Then

$$u = \frac{1}{2}\ln(1 + A^2 + B^2), \qquad \frac{\partial u}{\partial x} = \frac{A\frac{\partial A}{\partial x} + B\frac{\partial B}{\partial x}}{1 + A^2 + B^2},$$

$$\frac{\partial^2 u}{\partial x^2} = \frac{1}{(1 + A^2 + B^2)^2}\left\{\left[\left(\frac{\partial A}{\partial x}\right)^2 + \left(\frac{\partial B}{\partial x}\right)^2 + A\frac{\partial^2 A}{\partial x^2} + B\frac{\partial^2 B}{\partial x^2}\right] \times (1 + A^2 + B^2) - 2\left(A\frac{\partial A}{\partial x} + B\frac{\partial B}{\partial x}\right)^2\right\},$$

$$\frac{\partial^2 u}{\partial y^2} = \frac{1}{(1 + A^2 + B^2)^2}\left\{\left[\left(\frac{\partial A}{\partial y}\right)^2 + \left(\frac{\partial B}{\partial y}\right)^2 + A\frac{\partial^2 A}{\partial y^2} + B\frac{\partial^2 B}{\partial y^2}\right] \times (1 + A^2 + B^2) - 2\left(A\frac{\partial A}{\partial y} + B\frac{\partial B}{\partial y}\right)^2\right\}.$$

Since

$$\Delta A = \Delta B = 0,$$

$$\left(\frac{\partial A}{\partial x}\right)^2 + \left(\frac{\partial B}{\partial x}\right)^2 = \left(\frac{\partial A}{\partial y}\right)^2 + \left(\frac{\partial B}{\partial y}\right)^2 = |f'(z)|^2,$$

$$\frac{\partial A}{\partial x}\frac{\partial B}{\partial x} + \frac{\partial A}{\partial y}\frac{\partial B}{\partial y} = 0,$$

we get

$$\Delta u = \frac{2|f'(z)|^2}{(1 + |f(z)|^2)^2}. \tag{2.10}$$

Observe that in a neighborhood of a pole b_k of order χ_k of the function $f(z)$, relation (1.12) is valid with $c_k = b_k$ and $d_k = -\chi_k$. Taking into account (1.4) and (1.8), we get

$$\begin{aligned} &\ln\sum\sqrt{1 + |f(z)|^2} + \frac{1}{\pi}\iint_{|\zeta|\le R}\ln\left|\frac{R^2 - \zeta z}{R(z - \zeta)}\right|\frac{|f'(\zeta)|^2}{(1 + |f(\zeta)|^2)^2}d\sigma(\zeta) \\ &\quad= \frac{1}{2\pi}\int_0^{2\pi}\frac{R^2 - |z|^2}{|Re^{i\theta} - z|^2}\ln\sqrt{1 + |f(Re^{i\theta})|^2}d\theta + \sum_{|b_n|<R}\ln\left|\frac{R^2 - \bar{b}_n z}{R(z - b_n)}\right|. \end{aligned} \tag{2.11}$$

If $f(0) \ne \infty$, then, letting $z = 0$, we get (2.9). If $f(0) = \infty$, we take into account that

$$\ln\sqrt{1 + |f(z)|^2} = \lambda\ln|z| + \ln|c_\lambda| + o(1),$$

$$\sum_{|b_n|<R}\ln\left|\frac{R^2 - b_n z}{R(z - b_n)}\right| = \lambda\ln\frac{|z|}{R} + \sum_{0<|b_n|<R}\ln\frac{R}{|b_n|} + o(1),$$

as $z \to 0$. It remains to let $z \to 0$ in (2.11). □

3. The Carleman and Levin formulas

THEOREM 3.1. *Let $f(z)$ be a non-identically-zero function in the semi-annulus $\{0 < R_0 \le |z| \le R,\ \operatorname{Im} z \ge 0\}$. Then the following formula (called the Carleman formula) holds:*

$$\begin{aligned}(3.1)\qquad &\sum_m \left(\frac{1}{r_m} - \frac{r_m}{R^2}\right) \sin\varphi_m - \sum_n \left(\frac{1}{\rho_n} - \frac{\rho_n}{R^2}\right) \sin\psi_n \\ &= \frac{1}{2\pi}\int_{R_0}^{R} \left(\frac{1}{t^2} - \frac{1}{R^2}\right) \ln|f(t)f(-t)|dt + \frac{1}{\pi R}\int_0^{\pi} \ln|f(Re^{i\theta})|\sin\theta d\theta \\ &+ Q(R, R_0; f),\end{aligned}$$

where the $r_m e^{i\varphi_m}$ are the zeros and the $\rho_n e^{i\psi_n}$ are the poles of the function $f(z)$ in $\{R_0 < |z| < R,\ \operatorname{Im} z > 0\}$, and $Q(R, R_0; f) = O(1)$ as $R \to \infty$.

We start by proving the following analog of Theorem 1.1.

THEOREM 3.2. *Let $D = \{R_0 < |z| < R,\ \operatorname{Im} z > 0\}$, and let the function $u(z)$ satisfy the hypothesis of Theorem 1.1. Then*

$$\begin{aligned}&-\frac{1}{2\pi}\iint_D \left[\operatorname{Im}\left(\frac{1}{\zeta} + \frac{\zeta}{R^2}\right)\right] \Delta u(\zeta) d\sigma(\zeta) \\ (3.2)\qquad &= \frac{1}{2\pi}\int_{R_0}^{R} \left(\frac{1}{t^2} - \frac{1}{R^2}\right) [u(t) + u(-t)]dt + \frac{1}{\pi R}\int_0^{\pi} u(Re^{i\theta})\sin\theta d\theta \\ &+ \sum_{c_k \in D} d_k \operatorname{Im}\left(\frac{1}{c_k} + \frac{c_k}{R^2}\right) + Q^{(1)}(R, R_0; u),\end{aligned}$$

where $Q^{(1)}(R, R_0; u) = O(1)$.

PROOF. First we consider the case where none of the points c_k belongs to $\{|z| = R_0,\ \operatorname{Im} z \ge 0\}$. We exclude from D discs of sufficiently small radius ε centered at $c_1, \dots, c_q$. We use the second Green formula for the obtained domain with

$$u = u(\zeta), \qquad v = -\operatorname{Im}\left(\frac{1}{\zeta} + \frac{\zeta}{R^2}\right).$$

It is easy to see that the function $v(\zeta)$ satisfies the following conditions:

a) $\Delta v = 0$.

b) For ζ in the semicircle $\{|z| = R,\ \operatorname{Im} z \ge 0\}$,

$$v = 0, \qquad \frac{\partial v}{\partial n} = \frac{2}{R^2}\sin\theta, \qquad \zeta = Re^{i\theta}.$$

c) For ζ in the line segments $\{z = t,\ R_0 < t < R\}$ and $\{z = t,\ -R < t < -R_0\}$,

$$v = 0, \qquad \frac{\partial v}{\partial n} = \frac{1}{t^2} - \frac{1}{R^2}.$$

Let $\varepsilon \to 0$. We compute the integrals over the boundaries of the excluded discs using an argument similar to the one used in the proof of Theorem 1.1; we get formula (3.2) with

$$\begin{aligned}Q^{(1)}(R,R_0;u) &= \frac{1}{2\pi}\int_{|\zeta|=R_0,\ 0\le\arg\zeta\le\pi}\left(u\frac{\partial v}{\partial n}-v\frac{\partial u}{\partial n}\right)ds\\ &= -\frac{1}{2\pi}\int_0^{\pi}\left[u(R_0e^{i\theta})\left(\frac{1}{R_0^2}+\frac{1}{R^2}\right)\sin\theta+\left(\frac{1}{R_0}-\frac{R_0}{R^2}\right)\sin\theta\frac{\partial}{\partial R_0}u(R_0e^{i\theta})\right]R_0\,d\theta\\ &= O(1)\quad\text{as}\quad R\to\infty.\end{aligned}$$

We get rid of the condition that none of the points c_k belongs to $\{|z|=R_0,\ \operatorname{Im} z\ge 0\}$ in the following way. Choose $R_0'>R_0$ so that the semicircle $\{|z|=R_0',\ \operatorname{Im} z\ge 0\}$ will not contain the points c_k. Write formula (3.2) for the domain $D'=\{R_0<|z|<R_0',\ \operatorname{Im} z>0\}$. Then set

$$\begin{aligned}Q^{(1)}(R,R_0;u) &= Q^{(1)}(R,R_0';u)-\frac{1}{2\pi}\int_{R_0}^{R_0'}\left(\frac{1}{t^2}-\frac{1}{R^2}\right)[u(t)+u(-t)]dt\\ &\quad-\sum_{c_k\in D\setminus D'}d_k\operatorname{Im}\left(\frac{1}{c_k}+\frac{c_k}{R^2}\right)+\frac{1}{2\pi}\iint_{D\setminus D'}\left[-\operatorname{Im}\left(\frac{1}{\zeta}+\frac{\zeta}{R^2}\right)\right]\Delta u(\zeta)d\sigma(\zeta).\end{aligned}$$

It is easy to see that $Q^{(1)}(R,R_0;u)=O(1)$ as $R\to\infty$, and that the equation (3.2) is satisfied. □

Theorem 3.1 can be derived from Theorem 3.2 if we let $u(\zeta)=\ln|f(\zeta)|$ (cf. the proof of Theorem 2.1); also, we take the equalities

$$-\operatorname{Im}\left(\frac{1}{\zeta}+\frac{\zeta}{R^2}\right)=\left(\frac{1}{r}-\frac{r}{R^2}\right)\sin\theta,\qquad \zeta=re^{i\theta}$$

into account. Letting $u(z)=\ln\sqrt{1+|f(z)|^2}$ in (3.2), and using (2.10), we get an analog of the Shimizu–Ahlfors formula.

THEOREM 3.3. *Let $f(z)\not\equiv 0$ be a meromorphic function in the semi-annulus $\{0<R_0\le|z|\le R,\ \operatorname{Im} z\ge 0\}$. Then*

$$\begin{aligned}\frac{1}{\pi}\int_{R_0}^{R}\int_0^{\pi}\left(\frac{1}{r}-\frac{r}{R^2}\right)\sin\theta\frac{|f'(re^{i\theta})|^2}{(1+|f(re^{i\theta})|^2)^2}r\,dr\,d\theta\\ =\frac{1}{2\pi}\int_{R_0}^{R}\left(\frac{1}{t^2}-\frac{1}{R^2}\right)\ln\sqrt{(1+|f(t)|^2)(1+|f(-t)|^2)}\,dt\\ +\frac{1}{\pi R}\int_0^{\pi}\ln\sqrt{1+|f(Re^{i\theta})|^2}\sin\theta\,d\theta\\ +\sum_n\left(\frac{1}{\rho_n}-\frac{\rho_n}{R^2}\right)\sin\psi_n+Q^{(2)}(R,R_0;f),\end{aligned}$$

where the $\rho_ne^{i\psi_n}$ are the poles of $f(z)$, and $Q^{(2)}(R,R_0;f)=O(1)$ as $R\to\infty$.

We now turn to the Levin formula.

THEOREM 3.4. *Let $f(z)$ be a non-identically-zero meromorphic function in the domain $\{|z - i\frac{R}{2}| \le \frac{R}{2}\} \cup \{|z| \le R_0\}$, $0 < R_0 < R$. Then*

$$\begin{aligned} &\sum_m \left(\frac{\sin\varphi_m}{r_m} - \frac{1}{R}\right) - \sum_n \left(\frac{\sin\psi_n}{\rho_n} - \frac{1}{R}\right) \\ &\quad = \frac{1}{2\pi}\int_{\arcsin(R_0R^{-1})}^{\pi-\arcsin(R_0R^{-1})} \ln|f(R\sin\theta e^{i\theta})| \frac{d\theta}{R\sin^2\theta} + \tilde{Q}(R, R_0; f), \end{aligned} \tag{3.3}$$

where the $r_m e^{i\theta}$ are the zeros and the $\rho_n e^{i\psi_n}$ are the poles of the function $f(z)$ that belong to the domain $\{|z - i\frac{R}{2}| < \frac{R}{2},\ |z| > R_0\}$, and $\tilde{Q}(R, R_0; f) = O(1)$ as $R \to \infty$.

We derive this theorem from a suitable analog of Theorem 1.1.

THEOREM 3.5. *Let $D = \{|z - i\frac{R}{2}| < \frac{R}{2},\ |z| > R_0\}$, and let $u(z)$ satisfy the hypothesis of Theorem 1.1 in the domain $D \cup \{|z| < R_1\}$, $R_0 < R_1 < R$. Then*

$$\begin{aligned} &-\frac{1}{2\pi}\iint_D \operatorname{Im}\left(\frac{1}{\zeta} + \frac{i}{R}\right) \Delta u(\zeta) d\sigma(\zeta) \\ &\quad = \frac{1}{2\pi}\int_{\arcsin(R_0R^{-1})}^{\pi-\arcsin(R_0R^{-1})} u(R\sin\theta e^{i\theta}) \frac{d\theta}{R\sin^2\theta} \\ &\quad\quad + \sum_{c_k \in D} d_k \operatorname{Im}\left(\frac{1}{c_k} + \frac{i}{R}\right) + \tilde{Q}^{(1)}(R, R_0; u), \end{aligned} \tag{3.4}$$

where $\tilde{Q}^{(1)}(R, R_0; u) = O(1)$ as $R \to \infty$.

PROOF. The argument is similar to the one used in the proof of Theorems 1.1 and 3.2. First we assume that none of the points c_k belongs to the semicircle $\{|z| = R_0,\ \operatorname{Im} z > 0\}$. We delete from the domain D the discs of radius ε centered at $c_1, \dots, c_q$, and use the second Green formula for the obtained domain with

$$u = u(\zeta), \qquad v = -\operatorname{Im}\left(\frac{1}{\zeta} + \frac{i}{R}\right).$$

Letting $\varepsilon \to 0$ and observing that for z from the set $\{|z - i\frac{R}{2}| = \frac{R}{2},\ |z| \ge R_0\} = \{z = R\sin\theta e^{i\theta}, \arcsin(R_0R^{-1}) \le \theta \le \pi - \arcsin(R_0R^{-1})\}$, we have

$$v = 0, \qquad \frac{\partial v}{\partial n} = \frac{1}{R^2\sin^2\theta},$$

we get formula (3.4) and, furthermore,

$$\begin{aligned} \tilde{Q}^{(1)}(R, R_0; u) &= \frac{1}{2\pi}\int_{\arcsin(R_0R^{-1})}^{\pi-\arcsin(R_0R^{-1})} \Big\{u(R_0e^{i\theta})\Big(-\frac{-\sin\theta}{R^2}\Big) \\ &\qquad - \left(\frac{\sin\theta}{R_0} - \frac{1}{R}\right)\frac{\partial}{\partial R_0}u(R_0e^{i\theta})\Big\} R_0\, d\theta \\ &= O(1) \quad \text{as } R \to \infty. \end{aligned}$$

If the semicircle $\{|z| = R_0,\ \operatorname{Im} z > 0\}$ contains some of the points c_k, then we first use formula (3.4) for the domain $D' = \{|z - i\frac{R}{2}| < \frac{R}{2},\ |z| > R_0'\}$, where R_0', $R_0 < R_0' < R$, is chosen in such a way that $\{|z| = R_0',\ \operatorname{Im} z > 0\}$ does not contain points c_k. We get the result for the domain D that is in a way similar to the one used at the end of the proof of Theorem 3.2. □

We get Theorem 3.4 from Theorem 3.5 by letting $u(\zeta) = \ln|f(\zeta)|$. If we let $u(\zeta) = \ln\sqrt{1+|f(\zeta)|^2}$, we get the following analog of the Shimizu–Ahlfors formula.

THEOREM 3.6. *Let $f(z)$ be a non-identically-zero meromorphic function in the domain $\{|z - i\frac{R}{2}| \le \frac{R}{2}\} \cup \{|z| \le R_0\}$, $0 < R_0 < R$. Then*

$$\frac{1}{\pi}\iint_{|z-i\frac{R}{2}|\le\frac{R}{2},\ |z|\ge R_0}\left(\frac{\sin\theta}{r} - \frac{1}{R}\right)\frac{|f'(re^{i\theta})|^2}{(1+|f(re^{i\theta})|^2)^2}r\,dr\,d\theta$$
$$= \frac{1}{2\pi}\int_{\arcsin(R_0R^{-1})}^{\pi-\arcsin(R_0R^{-1})}\ln\sqrt{1+|f(R\sin\theta e^{i\theta})|^2}\frac{d\theta}{R\sin^2\theta}$$
$$+\sum_n\left(\frac{\sin\psi_n}{\rho_n} - \frac{1}{R}\right) + \tilde{Q}^{(2)}(R, R_0; f),$$

where the $\rho_n e^{i\psi_n}$ are the poles of $f(z)$ that belong to the domain $\{|z - i\frac{R}{2}| < \frac{R}{2}$, $|z| > R_0\}$, and $\tilde{Q}^{(2)}(R, R_0; f) = O(1)$ as $R \to \infty$.

4. The Nevanlinna characteristics. The first fundamental theorem of the value distribution theory

Let $f(z)$ be a meromorphic function. We introduce several real functions defined on $[0, \infty)$ which characterize the behavior of $f(z)$. These functions will be called the *Nevanlinna characteristics* of $f(z)$.

The number of the poles of $f(z)$ in the disc $\{|z| \le r\}$ will be denoted by $n(r, f)$; we assume that a pole of order m contributes m to the value of $n(r, f)$. It is clear that $n(r, f)$ is integer-valued, non-decreasing, and right semicontinuous on $[0, \infty)$. A point $r_0 \in (0, \infty)$ is a point of discontinuity for $n(r, f)$ if and only if the circle $\{|z| = r_0\}$ contains poles of $f(z)$; the value of the jump at r_0 is equal to the number of such poles. The set of points of discontinuity for $n(r, f)$ cannot have limit points in $[0, \infty)$; therefore on each interval $[a, b] \subset [0, \infty)$ the function $n(r, f)$ is piecewise constant.

Let

$$N(r, f) = \int_0^r \frac{n(t, f) - n(0, f)}{t}\,dt + n(0, f)\ln r.$$

This characteristic also describes the location of the poles of $f(z)$. The function $N(r, f)$ is a continuous non-decreasing function on $(0, \infty)$. Furthermore, as a function of $\ln r$, this function is convex; that is, $N(e^u, f)$ is a convex function of u for $-\infty < u < \infty$. In fact, since

$$N(e^u, f) = \int_{-\infty}^{u}[n(e^v, f) - n(0, f)]\,dv + n(0, f)u,$$

$N(e^u, f)$ is an integral of a non-decreasing function. It is worth mentioning that the characteristics $n(r, f)$ and $N(r, f)$ take into account only the absolute values of the poles of $f(z)$.

The characteristic $N(r, f)$ is convenient because it is related, in a natural way, to the Jensen formula (2.8). The Jensen formula can be written as

$$N\left(r, \frac{1}{f}\right) - N(r, f) = \frac{1}{2\pi}\int_0^{2\pi}\ln|f(re^{i\theta})|\,d\theta - \ln|c_\lambda|. \tag{4.1}$$

This statement follows from the equalities

$$\sum_{0<|b_n|<r} \ln \frac{r}{|b_n|} = N(r,f) - n(0,f)\ln r,$$

$$\sum_{0<|a_m|<r} \ln \frac{r}{|a_m|} = N\left(r,\frac{1}{f}\right) - n\left(0,\frac{1}{f}\right)\ln r,$$

$$-\lambda = n(0,f) - n\left(0,\frac{1}{f}\right),$$

where the a_m are the zeros of $f(z)$, the b_n are its poles, and the number λ is determined by (2.7). The third equality is obvious, and the other two can be obtained from the properties of the Stieltjes integral. In fact,

$$\begin{aligned}\sum_{0<|b_n|<r} \ln \frac{r}{|b_n|} &= \int_{+0}^{r} \ln \frac{r}{t}\, dn(t,f) = \int_{0}^{r} \ln \frac{r}{t}\, d[n(t,f) - n(0,f)] \\ &= \ln \frac{r}{t}[n(t,f) - n(0,f)]\Big|_{+0}^{r} + \int_0^r \frac{n(t,f)-n(0,f)}{t}\,dt \\ &= \int_0^r \frac{n(t,f)-n(0,f)}{t}\,dt.\end{aligned}$$

In order to introduce other characteristics we need the function $\ln^+ x$, defined on the semi-axis $\{x \ge 0\}$ by the equality

$$\ln^+ x = \max(\ln x, 0).$$

The following relations are obvious:

$$\ln x = \ln^+ x - \ln^+ \frac{1}{x}, \quad |\ln x| = \ln^+ x + \ln^+ \frac{1}{x}, \quad \ln^+ x = \ln x^\wedge, \tag{4.2}$$

where $x \ge 0$, $x^\wedge = \max(x,1)$. We also need the following inequalities:

$$\ln^+ \left|\prod_{\nu=1}^{n} x_\nu\right| \le \sum_{\nu=1}^{n} \ln^+ |x_\nu|, \tag{4.3}$$

$$\ln^+ \left|\sum_{\nu=1}^{n} x_\nu\right| \le \sum_{\nu=1}^{n} \ln^+ |x_\nu| + \ln n, \tag{4.4}$$

$$|\ln^+ |x_1| - \ln^+ |x_2|| \le \left|\ln\left|\frac{x_1}{x_2}\right|\right|, \tag{4.5}$$

$$|\ln^+ |x_1| - \ln^+ |x_2|| \le \ln^+ |x_1 - x_2| + \ln 2, \tag{4.6}$$

where $x_1, x_2, \ldots, x_n$ are arbitrary complex numbers. Inequalities (4.3) and (4.4) can be obtained in the following way:

$$\begin{aligned}\ln^+ \left|\prod_{\nu=1}^{n} x_\nu\right| &= \ln^+ \left(\prod_{\nu=1}^{n} |x_\nu|\right) \le \ln^+ \left(\prod_{\nu=1}^{n} |x_\nu|^\wedge\right) \\ &= \ln \left(\prod_{\nu=1}^{n} |x_\nu|^\wedge\right) = \sum_{\nu=1}^{n} \ln |x_\nu|^\wedge = \sum_{\nu=1}^{n} \ln^+ |x_\nu|;\end{aligned}$$

$$\begin{aligned}\ln^+\left|\sum_{\nu=1}^n x_\nu\right| &\le \ln^+\left(\sum_{\nu=1}^n |x_\nu|\right) \le \ln^+\left(\sum_{\nu=1}^n |x_\nu|^\wedge\right)\\ &= \ln\left(\sum_{\nu=1}^n |x_\nu|^\wedge\right) \le \ln\left(n \max_{1\le\nu\le n} |x_\nu|^\wedge\right)\\ &= \ln\left(\max_{1\le\nu\le n}|x_\nu|^\wedge\right) + \ln n \le \sum_{\nu=1}^n \ln|x_\nu|^\wedge + \ln n = \sum_{\nu=1}^n \ln^+|x_\nu| + \ln n.\end{aligned}$$

Inequalities (4.5) and (4.6) follow from (4.3) and (4.4), respectively. In fact, inequality (4.3) implies

$$\ln^+|x_1| = \ln^+\left|\frac{x_1}{x_2}x_2\right| \le \ln^+\left|\frac{x_1}{x_2}\right| + \ln^+|x_2|,$$

therefore, using the second of relations (4.2), we get

$$\ln^+|x_1| - \ln^+|x_2| \le \left|\ln\left|\frac{x_1}{x_2}\right|\right|. \tag{4.5$'$}$$

Similarly, in view of (4.4),

$$\ln^+|x_1| \le \ln^+(|x_1-x_2|+|x_2|) \le \ln^+|x_2| + \ln^+|x_1-x_2| + \ln 2;$$

hence

$$\ln^+|x_1| - \ln^+|x_2| \le \ln^+|x_1-x_2| + \ln 2. \tag{4.6$'$}$$

Since in (4.5′) and (4.6′), the variables x_1 and x_2 are interchangeable, we get inequalities (4.5) and (4.6).

Let

$$m(r,f) = \frac{1}{2\pi}\int_0^{2\pi} \ln^+|f(re^{i\varphi})|d\varphi.$$

This function characterizes the growth of the function $f(z)$. Later we will see that $m(r,f)$ is a continuous function of r. Now we only mention that the behavior of this function can be less regular than that of the function $N(r,f)$, which, as we have already remarked, is a non-decreasing convex function of $\ln r$.

EXAMPLE. Let

$$f(z) = \frac{3}{\pi^2}\sum_{n=1}^\infty \frac{1}{n^2(z-n)}.$$

Then

$$|f(z)| \le \frac{3}{\pi^2}\sum_{n=1}^\infty \frac{2}{n^2} = 1$$

outside the union of the discs $\{|z-n|<\frac{1}{2}\}$; therefore $m\left(n+\frac{1}{2},f\right) = 0$, $n = 1,2\dots$. At the same time, it is clear that $m(n,f)>0$, $n=1,2,\dots$.

Suppose $a\ne\infty$ is a complex number. The function $m\big(r,\frac{1}{f-a}\big)$ characterizes the proximity of $f(z)$ to a on the circle $\{|z|=r\}$. It is clear that the functions $n\big(r,\frac{1}{f-a}\big)$ and $N\big(r,\frac{1}{f-a}\big)$ characterize the moduli of the *a-points* of the function $f(z)$, that is, the moduli of the roots of the equation

$$f(z) = a.$$

In the cases where the function $f(z)$ under cosideration is clear from the context, we write $m(r,a)$, $n(r,a)$, and $N(r,a)$ instead of $m\big(r,\frac{1}{f-a}\big)$, $n\big(r,\frac{1}{f-a}\big)$, and

$N(r, \frac{1}{f-a})$. Also, we write $m(r,\infty)$, $n(r,\infty)$, and $N(r,\infty)$ instead of $m(r,f)$, $n(r,f)$, and $N(r,f)$. We also use the notation $m(r,a,f)$, $n(r,a,f)$, and $N(r,a,f)$. We introduce the characteristic

$$T(r,f) = m(r,f) + N(r,f).$$

It turns out that for each non-constant function $f(z)$, the sum $m(r,a)+N(r,a)$ is "almost independent" of a. Roughly speaking, this means that if $f(z)$ takes the value a more often than the value b, then it "approaches" a more slowly than it "approaches" b. The precise statement of this result is given in the following theorem, which Nevanlinna named "the first fundamental theorem of the value distribution theory".

THEOREM 4.1. *Let $f(z)$ be a non-constant meromorphic function. Then*

$$m(r,a) + N(r,a) = T(r,f) + \varepsilon(r,a) \tag{4.7}$$

for each complex number $a \neq \infty$, where $\varepsilon(r,a) = O(1)$ as $r \to \infty$.

PROOF. We use the Jensen formula in the form (4.1) for $f(z) - a$ (we can do this because $f(z)$ is non-constant). Since the poles of this function coincide with the poles of $f(z)$, we get

$$N\left(r, \frac{1}{f-a}\right) - N(r,f) = \frac{1}{2\pi}\int_0^{2\pi} \ln|f(re^{i\varphi}) - a|\, d\varphi - \ln|c_\lambda(a)|, \tag{4.1$'$}$$

where $c_\lambda(a)$ is the first non-zero coefficient in the Laurent series of $f(z) - a$ in a neighborhood of $z = 0$.

Since by the first of the equalities in (4.2) we have

$$\frac{1}{2\pi}\int_0^{2\pi} \ln|f(re^{i\varphi}) - a|\, d\varphi = m(r, f-a) - m\left(r, \frac{1}{f-a}\right),$$

(4.1′) can be rewritten in the form

$$N\left(r, \frac{1}{f-a}\right) + m\left(r, \frac{1}{f-a}\right) = N(r,f) + m(r, f-a) - \ln|c_\lambda(a)|.$$

Inequality (4.6) implies that

$$|\ln^+|f-a| - \ln^+|f|| \leq \ln^+|a| + \ln 2;$$

therefore $|m(r, f-a) - m(r,f)| \leq \ln^+|a| + \ln 2$. Letting

$$\varepsilon(r,a) = m(r, f-a) - m(r,f) - \ln|c_\lambda(a)|,$$

we get (4.7). To finish the proof, it remains to observe that

$$|\varepsilon(r,a)| \leq \ln^+|a| + \ln 2 + |\ln|c_\lambda(a)|| = O(1) \quad \text{as} \quad r \to \infty. \qquad \square$$

Let us show that for a non-constant meromorphic function $f(z)$, the characteristic $T(r,f)$ tends to ∞ as $r \to \infty$, and hence the term $\varepsilon(r,a)$ in (4.7) can be regarded as unessential. Let $a = f(0)$; then $n(0,a) > 0$, and we have

$$N(r,a) = \int_0^r \frac{n(t,a) - n(0,a)}{t}\, dt + n(0,a)\ln r \geq n(0,a)\ln r \to \infty.$$

Since $m(r,a)$ is non-negative, (4.7) implies

$$T(r,f) \geq N(r,a) + O(1) \to \infty \quad \text{as} \quad r \to \infty.$$

Our next purpose is to find integral representations of the characteristic $T(r,f)$. We write $f(z)$ as a quotient

$$f(z) = \frac{f_1(z)}{f_2(z)},$$

where $f_1(z)$ and $f_2(z) = z^\lambda + c_{\lambda+1}z^{\lambda+1} + \cdots$ $(\lambda \geq 0)$ are entire functions without common zeros. Let

$$u(z) = \max(|f_1(z)|, |f_2(z)|).$$

Then, despite the non-uniqueness of the choice of $f_1(z)$ and $f_2(z)$, we have

$$T(r,f) = \frac{1}{2\pi}\int_0^{2\pi} \ln u(re^{i\varphi})\,d\varphi. \tag{4.8}$$

In fact, it is easy to see that

$$\ln^+|f(z)| = \ln u(z) - \ln|f_2(z)|.$$

On the other hand, by the Jensen formula (4.1), we have

$$N\left(r, \frac{1}{f_2}\right) = \frac{1}{2\pi}\int_0^{2\pi} \ln|f_2(re^{i\varphi})|\,d\varphi.$$

Since $N(r,f) = N\left(r, \frac{1}{f_2}\right)$, it follows that

$$\begin{aligned} T(r,f) &= m(r,f) + N(r,f) \\ &= \frac{1}{2\pi}\int_0^{2\pi} [\ln u(re^{i\varphi}) - \ln|f_2(re^{i\varphi})|]\,d\varphi + \frac{1}{2\pi}\int_0^{2\pi} \ln|f_2(re^{i\varphi})|\,d\varphi \\ &= \frac{1}{2\pi}\int_0^{2\pi} \ln u(re^{i\varphi})\,d\varphi. \end{aligned}$$

Shimizu and Ahlfors proved a more refined theorem than Theorem 4.1. Also, their theorem admits an interesting geometric interpretation.

Denote by S_0 the Riemann sphere, and by $\omega = p(w)$, the map inverse to the stereographic projection of S_0 onto the extended complex w-plane. Let w and a be two different points in the extended complex plane. Denote by $[w,a]$ the length of the line segment joining the points $p(w)$ and $p(a)$ on the sphere S_0. Since the diameter of S_0 is equal to 1, we see that $[w,a] \leq 1$ for each pair of points. It is easy to check that

$$[w,a] = \frac{|w-a|}{\sqrt{1+|w|^2}\sqrt{1+|a|^2}}, \qquad w, a \neq \infty,$$

$$[w,\infty] = \frac{1}{\sqrt{1+|w|^2}}.$$

The ratio of the length elements on S_0 and on the w-plane is equal to

$$\frac{|dp(w)|}{|dw|} = \frac{1}{1+|w|^2},$$

and the ratio of the area elements $d\omega(w)$ on S_0 and $d\sigma(w)$ on the w-plane is equal to

$$\frac{d\omega(w)}{d\sigma(w)} = \left(\frac{|dp(w)|}{|dw|}\right)^2 = \frac{1}{(1+|w|^2)^2}.$$

We introduce the *spherical derivative* of a meromorphic function $f(z)$ by

$$\overset{\circ}{f}(z) = \frac{|f'(z)|}{1+|f(z)|^2} = \frac{|dp(f(z))|}{|df(z)|}\frac{|df(z)|}{|dz|} = \frac{|dp(f(z))|}{|dz|}.$$

It is easy to see that the function $\overset{\circ}{f}(z)$ is continuous at each finite point of the z-plane.

Suppose that the function $w = f(z)$ maps the disc $\{|z| \le r\}$ onto the Riemann surface F_r. The average number of sheets of the surface F_r can be determined as the quotient of the area of F_r in the spherical metric and the area of the sphere S_0, which is π. If we denote the average number of sheets by $\overset{\circ}{A}(r, f)$, we get

$$\overset{\circ}{A}(r, f) = \frac{1}{\pi}\operatorname{area}(F_r) = \frac{1}{\pi}\iint_{F_r} d\omega(w) = \frac{1}{\pi}\iint_{|w|\le\infty} n(r, w)\, d\omega(w).$$

On the other hand,

$$\overset{\circ}{A}(r, f) = \frac{1}{\pi}\iint_{F_r} d\omega(w) = \frac{1}{\pi}\iint_{|z|\le r}\left(\frac{d\omega(w)}{d\sigma(z)}\right) d\sigma(z) = \frac{1}{\pi}\iint_{|z|\le r} (\overset{\circ}{f}(z))^2 d\sigma(z).$$

Let $f(z)$ be a non-constant meromorphic function. If $f(0) \ne a$, we let

$$\overset{\circ}{m}(r, a) = \frac{1}{2\pi}\int_0^{2\pi} \ln\frac{1}{[f(re^{i\varphi}), a]}\, d\varphi - \ln\frac{1}{[f(0), a]},$$

and if $f(0) = a$, we let

$$\overset{\circ}{m}(r, a) = \frac{1}{2\pi}\int_0^{2\pi} \ln\frac{1}{[f(re^{i\varphi}), a]}\, d\varphi - \ln\frac{\sqrt{1+|a|^2}}{|c_\lambda(a)|}, \qquad a \ne \infty,$$

$$\overset{\circ}{m}(r, \infty) = \frac{1}{2\pi}\int_0^{2\pi} \ln\frac{1}{[f(re^{i\varphi}), \infty]}\, d\varphi - \ln|c_\lambda(0)|,$$

where $c_\lambda(a)$ is the first non-zero coefficient in the Laurent series of $f(z) - a$ in a neighborhood of $z = 0$. Observe that the integral from the definition of $\overset{\circ}{m}(r, a)$ is different from the one in the definition of $m(r, a)$: now the distance between $f(re^{i\varphi})$ and a is not the plane distance, but the length of a chord of the Riemann sphere.

The function

$$\overset{\circ}{T}(r, f) = \int_0^r \frac{\overset{\circ}{A}(t, f)}{t}\, dt$$

is called the *Shimizu–Ahlfors characteristic* of $f(z)$. By the continuity of $\overset{\circ}{f}(z)$, the function $\overset{\circ}{A}(r, f)$ is a continuous function of r, and $\overset{\circ}{A}(r, f) = O(r^2)$ as $r \to 0$; therefore the function $\overset{\circ}{T}(r, f)$ is finite and continuous for $0 \le r < \infty$. Since

$$\frac{d\overset{\circ}{A}(r, f)}{dr} = \frac{r}{\pi}\int_0^{2\pi} (\overset{\circ}{f}(re^{i\varphi}))^2 d\varphi > 0 \quad \text{for} \quad r > 0,$$

we see that the function $\overset{\circ}{A}(r, f)$ is strictly increasing on $[0, \infty)$, and the function $\overset{\circ}{T}(r, f)$ is twice continuously differentiable and strictly increasing on $[0, \infty)$, and it is convex as a function of $\ln r$. Also, the second derivative of $\overset{\circ}{T}(r, f)$ with respect to $\ln r$ is continuous for $r \ge 0$ and positive for $r > 0$.

THEOREM 4.2. *Let $f(z)$ be a non-constant meromorphic function. Then*

$$\overset{\circ}{m}(r, a) + N(r, a) = \overset{\circ}{T}(r, f) \tag{4.9}$$

for each complex number a from the extended plane.

PROOF. We need the equality

$$\mathring{T}(r,f) = \frac{1}{\pi}\iint_{|z|\le r}\left(\ln\frac{r}{|z|}\right)(\mathring{f}(z))^2 d\sigma(z), \tag{4.10}$$

which can be obtained in the following way:

$$\frac{1}{\pi}\iint_{|z|\le r}\left(\ln\frac{r}{|z|}\right)(\mathring{f}(z))^2 d\sigma(z) = \int_0^r \ln\frac{r}{t}\left\{\frac{1}{\pi}\int_0^{2\pi}(\mathring{f}(te^{i\varphi}))^2 t\,d\varphi\right\}dt$$

$$= \int_0^r \ln\frac{r}{t}\,d\mathring{A}(t,f) = \ln\frac{r}{t}\mathring{A}(t,f)\Big|_0^r + \int_0^r \frac{\mathring{A}(t,f)}{t}dt = \mathring{T}(r,f).$$

Using the introduced notation, we can rewrite (2.9) from Theorem 2.6 in the form

$$\mathring{T}(r,f) = \mathring{m}(r,\infty) + N(r,\infty). \tag{4.9$'$}$$

Thus (4.9) has been proved for $a=\infty$; observe that a similar equality for the Nevanlinna characteristic $T(r,f)$ is valid by definition.

Now assume that $a\ne\infty$. Let

$$F(z,a) = \frac{1+\bar{a}f(z)}{f(z)-a}. \tag{4.11}$$

We show that (a) $\mathring{T}(r,F(z,a)) = \mathring{T}(r,f)$, and (b) the value $\mathring{m}(r,\infty)$ for the function $F(z,a)$ coincides with the value $\mathring{m}(r,a)$ for the function $f(z)$. Therefore the result will follow if we write equality (4.9$'$) for $F(z,a)$.

Using the identities

$$F'(z,a) = -\frac{1+|a|^2}{(f(z)-a)^2}f'(z)$$

and

$$|f-a|^2 + |1+\bar{a}f|^2 = (1+|a|^2)(1+|f|^2),$$

we obtain

$$\mathring{F}(z,a) = \frac{|F'(z,a)|}{1+|F(z,a)|^2} = \frac{(1+|a|^2)|f'(z)|}{|f(z)-a|^2+|1+\bar{a}f(z)|^2} = \mathring{f}(z).$$

As a consequence we get $\mathring{A}(r,F(z,a)) = \mathring{A}(r,f)$, and hence the statement (a).

Equation (4.11) implies

$$[F,\infty] = \frac{1}{\sqrt{1+|F|^2}} = \frac{|f-a|}{\sqrt{|f-a|^2+|1+\bar{a}f|^2}} = \frac{|f-a|}{\sqrt{1+|a|^2}\sqrt{1+|f|^2}} = [f,a].$$

If $f(0)\ne a$, this equality immediately implies (b). If $f(0)=a$, we must take into account the fact that the Laurent series for $F(z,a)$ in a neighborhood of $z=0$ has the form

$$F(z,a) = \frac{1+|a|^2}{c_\lambda(a)}z^{-\lambda} + \cdots. \qquad \square$$

REMARK. The basic idea of the argument used to derive (4.9) from its special case $a=\infty$ is the following. The Möbius transformation

$$w_1 = \frac{1+\bar{a}w}{w-a}$$

can be geometrically interpreted as the rotation of the Riemann sphere S_0 under which the image $p(a)$ of the point a is mapped onto the north pole $p(\infty)$ of S_0. If we use this idea, the argument becomes intuitively obvious.

Let us prove that Theorem 4.1 is a corollary of Theorem 4.2. To do this it is enough to show that

$$|m(r,a) - \overset{\circ}{m}(r,a)| < d(a), \tag{4.12}$$

where $d(a)$, $0 < d(a) < \infty$, depends on a only.

For $a \neq \infty$ we have

$$\frac{1}{[f,a]} = \frac{\sqrt{1+|f|^2}\sqrt{1+|a|^2}}{|f-a|} \geq \left(\frac{1}{|f-a|}\right)^\wedge,$$

$$\begin{aligned}\frac{1}{|f,a|} &\leq \frac{\sqrt{1+(|f-a|+|a|)^2}\sqrt{1+|a|^2}}{|f-a|}\\ &\leq \frac{1+|f-a|+|a|}{|f-a|}\sqrt{1+|a|^2} \leq \left(\frac{1+|a|}{|f-a|}+1\right)(1+|a|)\\ &\leq \left[\left(\frac{1}{|f-a|}\right)^\wedge (1+|a|) + \left(\frac{1}{|f-a|}\right)^\wedge\right](1+|a|) \leq \left(\frac{1}{|f-a|}\right)^\wedge (2+|a|)^2.\end{aligned}$$

Hence

$$\ln^+ \frac{1}{|f-a|} \leq \ln \frac{1}{[f,a]} \leq \ln^+ \frac{1}{|f-a|} + 2\ln(2+|a|),$$

and we get (4.12). It is obvious that

$$|f|^\wedge \leq \frac{1}{[f,\infty]} \leq |f|^\wedge \sqrt{2};$$

therefore

$$\ln^+|f| \leq \ln \frac{1}{[f,\infty]} \leq \ln^+|f| + \frac{1}{2}\ln 2,$$

and (4.12) remains true for $a = \infty$.

Observe that (4.12) implies

$$|T(r,f) - \overset{\circ}{T}(r,f)| \leq d_1(f),$$

where $d_1(f) = \frac{1}{2}\ln(2+2|f(0)|^2)$ if $f(0) \neq \infty$, and $d_1(f) = \frac{1}{2}\ln 2 + |\ln|c_\lambda(0)||$ if $f(0) = \infty$; that is, the characteristics $T(r,f)$ and $\overset{\circ}{T}(r,f)$ have a bounded difference as $r \to \infty$.

EXERCISE. Prove the following formula, similar to (4.8):

$$\overset{\circ}{T}(r,f) = \frac{1}{2\pi}\int_0^{2\pi} \ln v(re^{i\varphi})\,d\varphi - \ln v(0),$$

where

$$v(z) = \sqrt{|f_1(z)|^2 + |f_2(z)|^2},$$

and $f_1(z)$, $f_2(z)$ have the same meaning as in the definition of the function $u(z)$ used in (4.8). (Hint. Use the equality $\ln v(z) = \ln\sqrt{1+|f(z)|^2} - \ln|f_2(z)|$, the Jensen formula for $f_2(z)$, and (4.9′).)

As we have already mentioned, the characteristic $\overset{\circ}{T}(r,f)$ is a non-decreasing convex function of $\ln r$. It turns out that the same statement is valid for $T(r,f)$.

This fact is not trivial because the behavior of the term $m(r, f)$ in the sum $m(r, f) + N(r, f)$ can be quite irregular (see the example on p. 15). We will use the following theorem of H. Cartan.

THEOREM 4.3. *Let $f(z)$ be a non-constant meromorphic function. Then*

$$T(r, f) = \frac{1}{2\pi} \int_0^{2\pi} N(r, e^{i\theta})\, d\theta + d(f), \tag{4.13}$$

where the constant $d(f)$ does not depend on r.

PROOF. Observe that in passing we will prove measurability of $N(r, e^{i\theta})$ as a function of θ.

Since $N(r, a)$ is non-decreasing and convex in $\ln r$, we see that (4.13) immediately implies that $T(r, f)$ is also non-decreasing and convex in $\ln r$.

We need the formula

$$\frac{1}{2\pi} \int_0^{2\pi} \ln |w - e^{i\theta}|\, d\theta = \ln^+ |w| \tag{4.14}$$

and the inequality

$$\frac{1}{2\pi} \int_0^{2\pi} |\ln |w - e^{i\theta}||\, d\theta \le \ln^+ |w| + 2\ln 2. \tag{4.15}$$

Formula (4.14) can be proved as follows. We write down (2.8) for $f(z) = w - z$ and $R = 1$. If $|w| \ge 1$, then this function has neither zeros nor poles in the disc $\{|z| < 1\}$, and we get

$$\ln |w| = \frac{1}{2\pi} \int_0^{2\pi} \ln |w - e^{i\theta}|\, d\theta.$$

If $0 < |w| < 1$, then $f(z)$ has a zero at w in $\{|z| < 1\}$, and (2.8) implies

$$\ln |w| = \frac{1}{2\pi} \int_0^{2\pi} \ln |w - e^{i\theta}|\, d\theta - \ln \frac{1}{|w|}.$$

Thus (4.14) has been proved for $w \neq 0$. For $w = 0$ the formula is obvious.

Furthermore, using (4.2), we get

$$\frac{1}{2\pi} \int_0^{2\pi} |\ln |w - e^{i\theta}||\, d\theta = -\frac{1}{2\pi} \int_0^{2\pi} \ln |w - e^{i\theta}|\, d\theta + 2 \cdot \frac{1}{2\pi} \int_0^{2\pi} \ln^+ |w - e^{i\theta}|\, d\theta.$$

Observing that $\ln^+ |w - e^{i\theta}| \le \ln^+ |w| + \ln 2$ and using (4.14), we get (4.15).

We now turn to the proof of Theorem 4.3. First we assume that $f(0) \neq \infty$. We apply the Jensen formula (4.1) to $f(z) - e^{i\theta}$. If $f(0) \neq e^{i\theta}$, we obtain

$$N(r, e^{i\theta}) - N(r, \infty) = \frac{1}{2\pi} \int_0^{2\pi} \ln |f(re^{i\varphi}) - e^{i\theta}|\, d\varphi - \ln |f(0) - e^{i\theta}|. \tag{4.16}$$

The right-hand side of this equality is measurable in θ. Since the equality holds for all $\theta \in [0, 2\pi)$ except possibly one (for which $e^{i\theta} = f(0)$), it follows that $N(r, e^{i\theta})$ is also measurable in θ on $[0, 2\pi)$. We integrate equality (4.16) over $[0, 2\pi]$ and divide the result by 2π. We get

$$\begin{aligned}\frac{1}{2\pi}\int_0^{2\pi} & N(r,e^{i\theta})\,d\theta - N(r,\infty)\\ &= \frac{1}{2\pi}\int_0^{2\pi}\left\{\frac{1}{2\pi}\int_0^{2\pi}\ln|f(re^{i\varphi})-e^{i\theta}|\,d\varphi\right\}d\theta\\ &\quad -\frac{1}{2\pi}\int_0^{2\pi}\ln|f(0)-e^{i\theta}|\,d\theta,\end{aligned}\tag{4.17}$$

provided that the first integral in the right-hand side is well defined. We now show this and also justify the change of the order of integration in this integral. By the Fubini theorem, it suffices to establish that

$$I=\frac{1}{2\pi}\int_0^{2\pi}\left\{\frac{1}{2\pi}\int_0^{2\pi}|\ln|f(re^{i\varphi})-e^{i\theta}||\,d\theta\right\}d\varphi<\infty.$$

This follows from (4.15):

$$I\le\frac{1}{2\pi}\int_0^{2\pi}\{\ln^+|f(re^{i\varphi})|+2\ln 2\}\,d\varphi=m(r,f)+2\ln 2<\infty.$$

Changing the order of integration in (4.17) and using (4.14), we get

$$\frac{1}{2\pi}\int_0^{2\pi}N(r,e^{i\theta})\,d\theta-N(r,\infty)=\frac{1}{2\pi}\int_0^{2\pi}\ln^+|f(re^{i\varphi})|\,d\varphi-\ln^+|f(0)|.$$

The obtained equality is equivalent to (4.13) with $d(f)=\ln^+|f(0)|$.

If $f(0)=\infty$, then the right-hand side of (4.16) will contain $\ln|c_\lambda|$ instead of $\ln|f(0)-e^{i\theta}|$, where c_λ is defined in (2.7). Therefore, using the same argument as above, we get (4.13) with $d(f)=\ln|c_\lambda|$. □

The integral in the right-hand side of (4.13) can be regarded as an integral average of $N(r,a)$ over the circle $|a|=1$ or over the equator of the sphere S_0. One can get similar formulas for average values of $N(r,a)$ over certain other sufficiently large subsets of the sphere S_0 (see R. Nevanlinna [**NevR74**, Chapter VI, §4]). We give here only one identity for $\overset{\circ}{T}(r,f)$, which will be used in Chapter 4.

THEOREM 4.4. *Let $f(z)$ be a non-constant meromorphic function. Then*

$$\overset{\circ}{T}(r,f)=\frac{1}{\pi}\iint_{|a|\le\infty}N(r,a)\,d\omega(a).\tag{4.18}$$

PROOF. Identity (4.18) could be obtained by a change of the order of integration:

$$\begin{aligned}\overset{\circ}{T}(r,f)&=\int_0^r\frac{\overset{\circ}{A}(t,f)}{t}\,dt=\frac{1}{\pi}\int_0^r\frac{dt}{t}\iint_{|a|\le\infty}n(t,a)\,d\omega(a)\\ &=\frac{1}{\pi}\iint_{|a|\le\infty}d\omega(a)\int_0^r\frac{n(t,a)}{t}\,dt=\frac{1}{\pi}\iint_{|a|\le\infty}N(r,a)\,d\omega(a).\end{aligned}$$

However, to justify the change, we need to show that $n(t,a)$ is measurable as a function of two variables t and a on $\{0\le t\le r,\ |a|\le\infty\}$. This would require a lengthy argument. We pass over the difficulty in the following way. We use the equality

$$\iint_{|a|\le\infty} \ln\frac{1}{[w,a]}\,d\omega(a) = \frac{\pi}{2}, \tag{4.19}$$

which is valid for an arbitrary complex number w, $|w| \le \infty$. In fact, after the change of variables

$$a' = \frac{1+a\bar{w}}{a-w},$$

which is equivalent to a rotation of the sphere S_0 under which the point $p(w)$ is mapped onto $p(\infty)$, we get

$$\iint_{|a|\le\infty} \ln\frac{1}{[w,a]}\,d\omega(a) = \iint_{|a'|\le\infty} \ln\frac{1}{[\infty,a']}\,d\omega(a')$$
$$= \iint_{|a'|\le\infty} \frac{\ln\sqrt{1+|a'|^2}}{(1+|a'|^2)^2}\,d\sigma(a') = 2\pi\int_0^\infty \frac{\ln\sqrt{1+r^2}}{(1+r^2)^2}\,r\,dr = \frac{\pi}{2}.$$

Equation (4.9) implies that

$$N(r,a) = \overset{\circ}{T}(r,f) - \frac{1}{2\pi}\int_0^{2\pi} \ln\frac{1}{[f(re^{i\varphi}),a]}\,d\varphi + \ln\frac{1}{[f(0),a]} \tag{4.20}$$

provided $a \ne f(0)$. We integrate this equation with respect to $d\omega(a)$ over $|a| \le \infty$. Measurability of $N(r,a)$ in a follows from measurability of the right-hand side of (4.20). Using (4.19) and changing the order of integration, we obtain

$$\frac{1}{\pi}\iint_{|a|\le\infty} N(r,a)\,d\omega(a)$$
$$= \overset{\circ}{T}(r,f) - \frac{1}{2\pi}\int_0^{2\pi} d\varphi\frac{1}{\pi}\iint_{|a|\le\infty} \ln\frac{1}{[f(re^{i\varphi}),a]}\,d\omega(a)$$
$$+ \frac{1}{\pi}\iint_{|a|\le\infty} \ln\frac{1}{[f(0),a]}\,d\omega(a)$$
$$= \overset{\circ}{T}(r,f) - \frac{1}{2\pi}\int_0^{2\pi} \frac{1}{2}\,d\varphi + \frac{1}{2} = \overset{\circ}{T}(r,f),$$

that is, (4.18). The legitimacy of the change of the order of integration follows immediately from Fubini's theorem, since the integrand is non-negative and measurable in φ and a. □

5. The angular Nevanlinna characteristics, Tsuji characteristics, and related analogs of the first fundamental theorem

The characteristics introduced in the previous section are fundamental for the study of functions meromorphic in the whole complex plane. These characteristics do not take into account, at least explicitly, the arguments of a-points. For this reason, in the studies devoted to the distribution of arguments of a-points, an essential role is played by other characteristics, which we now introduce.

Let $f(z)$ be meromorphic in the closed half-plane $\{\operatorname{Im} z \ge 0\}$. Let the $\rho_n e^{i\psi_n}$ be its poles; we count each pole according to its multiplicity. Let

$$c(r,f) = \sum_{1<\rho_n\le r} \sin\psi_n.$$

The value of the characteristic $c(r,f)$ heavily depends on the arguments of the poles of $f(z)$. Roughly speaking, the closer a pole $\rho_n e^{i\psi_n}$ to the real line $\{\operatorname{Im} z = 0\}$ (in the sense that ψ_n is close to either 0 or π), the smaller its contribution to $c(r,f)$. The poles lying on the line $\{\operatorname{Im} z = 0\}$ do not contribute to $c(r,f)$ at all. It is clear that $c(r,f)$ is non-decreasing, right-continuous in r on $[1,\infty)$, and piecewise constant on each line segment $[a,b] \subset [1,\infty)$.

Let

$$C(r,f) = 2\int_1^r c(t,f)\left(\frac{1}{t^2}+\frac{1}{r^2}\right)dt.$$

This characteristic is naturally connected with the Carleman formula (3.1), which can be written in the form

$$\begin{aligned} C\left(r,\frac{1}{f}\right) - C(r,f) &= \frac{1}{\pi}\int_1^r \left(\frac{1}{t^2}-\frac{1}{r^2}\right)\ln|f(t)f(-t)|\,dt \\ &\quad + \frac{2}{\pi r}\int_0^\pi \ln|f(re^{i\theta})|\sin\theta\,d\theta + Q(r,1;f). \end{aligned} \tag{5.1}$$

This statement follows from the equation

$$C(r,f) = 2\sum_{1<\rho_n<r}\left(\frac{1}{\rho_n}-\frac{\rho_n}{r^2}\right)\sin\psi_n, \tag{5.1'}$$

which can be verified as follows:

$$\begin{aligned} 2\sum_{1\le\rho_n\le r}\left(\frac{1}{\rho_n}-\frac{\rho_n}{r^2}\right)\sin\psi_n &= 2\int_1^r\left(\frac{1}{t}-\frac{t}{r^2}\right)dc(t,f) \\ &= 2\left(\frac{1}{t}-\frac{t}{r^2}\right)c(t,f)\Big|_1^r + 2\int_1^r c(t,f)\left(\frac{1}{t^2}+\frac{1}{r^2}\right)dt = C(r,f). \end{aligned}$$

Since each summand in (5.1′) is non-negative and non-decreasing, $C(r,f)$ is non-decreasing in r.

We also introduce the characteristics

$$A(r,f) = \frac{1}{\pi}\int_1^r\left(\frac{1}{t^2}-\frac{1}{r^2}\right)[\ln^+|f(t)| + \ln^+|f(-t)|]\,dt,$$

$$B(r,f) = \frac{2}{\pi r}\int_0^\pi \ln^+|f(re^{i\varphi})|\sin\varphi\,d\varphi, \quad 1\le r\le\infty.$$

These functions describe the growth of $f(z)$ as $|z|\to\infty$; the function A specifies the growth of f along the real line $\{\operatorname{Im} z = 0\}$, and the function B, in the open half-plane $\{\operatorname{Im} z > 0\}$.

When there is no danger of confusion, the functions

$$A\left(r,\frac{1}{f-a}\right),\ B\left(r,\frac{1}{f-a}\right),\ C\left(r,\frac{1}{f-a}\right),\quad a\ne\infty,$$

will be denoted by

$$A(r,a),\ B(r,a),\ C(r,a),$$

respectively. These functions characterize the proximity of $f(z)$ to a and the distribution of a-points. We also denote

$$A(r,\infty) = A(r,f), \qquad B(r,\infty) = B(r,f), \qquad C(r,\infty) = C(r,f).$$

Finally, we introduce

$$S(r,f) = A(r,f) + B(r,f) + C(r,f).$$

The characteristics A, B, C, c, and S will be called the *Nevanlinna characteristics for a half-plane.*

The following result is an analog of the first fundamental theorem.

THEOREM 5.1. *Let $f(z)$ be a non-constant meromorphic function in the upper half-plane $\{\operatorname{Im} z \geq 0\}$. Then*

$$A(r,a) + B(r,a) + C(r,a) = S(r,f) + \varepsilon_1(r,a) \tag{5.2}$$

for each complex $a \neq \infty$, where $\varepsilon_1(r,a) = O(1)$ as $r \to \infty$.

PROOF. Following the analogy with Theorem 4.1, we apply the Carleman formula in the form (5.1) to $f(z) - a$. Taking into consideration the first of the equations in (4.2), we get

$$\begin{aligned} &A\left(r, \frac{1}{f-a}\right) + B\left(r, \frac{1}{f-a}\right) + C\left(r, \frac{1}{f-a}\right) \\ &\quad = A(r, f-a) + B(r, f-a) + C(r,f) + Q(r,1; f-a). \end{aligned}$$

Now we use (4.6) and obtain

$$\begin{aligned} |A(r,f-a) - A(r,f)| &\leq 2(\ln^+|a| + \ln 2)\frac{1}{\pi}\int_1^r \left(\frac{1}{t^2} - \frac{1}{r^2}\right) dt \\ &\leq \frac{2}{\pi}(\ln^+|a| + \ln 2), \\ |B(r,f-a) - B(r,f)| &\leq (\ln^+|a| + \ln 2)\frac{2}{\pi r}\int_0^\pi \sin\varphi\, d\varphi \\ &\leq \frac{4}{\pi}(\ln^+|a| + \ln 2). \end{aligned}$$

Hence we get (5.2) with

$$|\varepsilon_1(r,a)| \leq \frac{6}{\pi}(\ln^+|a| + \ln 2) + |Q(r,1;f-a)| = O(1)$$

as $r \to \infty$. □

Later we will need a more general result, dealing with functions meromorphic in an angle $\{\alpha \leq \arg z \leq \beta\}$, $0 < \beta - \alpha \leq 2\pi$. Letting $k = \pi(\beta-\alpha)^{-1}$, $1 \leq r < \infty$, we introduce the functions that we call the *angular Nevanlinna characteristics*:

$$A_{\alpha\beta}(r,f) = \frac{k}{\pi}\int_1^r \left(\frac{1}{t^k} - \frac{t^k}{r^{2k}}\right)[\ln^+|f(te^{i\alpha})| + \ln^+|f(te^{i\beta})|]\frac{dt}{t};$$

$$B_{\alpha\beta}(r,f) = \frac{2k}{\pi r^k}\int_\alpha^\beta \ln^+|f(re^{i\varphi})| \sin k(\varphi-\alpha)\, d\varphi;$$

$$C_{\alpha\beta}(r,f) = 2k\int_1^r c_{\alpha\beta}(t,f)\left(\frac{1}{t^k} + \frac{t^k}{r^{2k}}\right)\frac{dt}{t},$$

where

$$c_{\alpha\beta}(r,f) = \sum_{1<\rho_n\leq r,\quad \alpha\leq\psi_n\leq\beta} \sin k(\psi_n - \alpha),$$

and the $\rho_n e^{i\psi_n}$ are the poles of $f(z)$ counted according to their multiplicity;

$$S_{\alpha\beta}(r,f) = A_{\alpha\beta}(r,f) + B_{\alpha\beta}(r,f) + C_{\alpha\beta}(r,f).$$

Note that the characteristics $A(r,f)$, $B(r,f)$, $C(r,f)$, and $S(r,f)$ are obtained from the angular characteristics if $\alpha=0$, $\beta=\pi$. For functions $A_{\alpha\beta}(r,a)$, $B_{\alpha\beta}(r,a)$, and $C_{\alpha\beta}(r,a)$ we use the same convention as for their half-plane analogs.

THEOREM 5.2. *Let $f(z)$ be a non-constant meromorphic function in the angle $\{\alpha\le\arg z\le\beta\}$, $0<\beta-\alpha\le 2\pi$. Then*

$$A_{\alpha\beta}(r,a)+B_{\alpha\beta}(r,a)+C_{\alpha\beta}(r,a)=S_{\alpha\beta}(r,f)+\varepsilon_2(r,a)$$

for each complex number $a\ne\infty$, where $\varepsilon_2(r,a)=O(1)$ as $r\to\infty$.

PROOF. This theorem is a corollary of Theorem 5.1. In fact, consider in the half-plane $\{\operatorname{Im} z\ge 0\}$ the branch of $w=z^{\frac{1}{k}}$ that is positive for $z>0$; we introduce the function $f_1(z)=f(z^{\frac{1}{k}}e^{i\alpha})$. This function is meromorphic in the half-plane $\{\operatorname{Im} z\ge 0\}$, except at the origin. Straightforward verification shows that

$$A(r,f_1)=A_{\alpha\beta}(r^{\frac{1}{k}},f),\qquad B(r,f_1)=B_{\alpha\beta}(r^{\frac{1}{k}},f),\qquad C(r,f_1)=C_{\alpha\beta}(r^{\frac{1}{k}},f);$$

hence

$$S(r,f_1)=S_{\alpha\beta}(r^{\frac{1}{k}},f).$$

Applying Theorem 5.1 to $f_1(z)$, we get the desired result. □

Now we turn to the characteristics generated by the Levin formula (3.3). These characteristics will be called the *Tsuji characteristics.*

Let $f(z)$ be a non-constant meromorphic function in the half-plane $\{\operatorname{Im} z\ge 0\}$. Denote by $\mathfrak{n}(r,f)$ the number of poles of $f(z)$ contained in the set

$$\left\{\left|z-i\frac{r}{2}\right|\le\frac{r}{2},\ |z|>1\right\}$$

(each pole is counted according to its multiplicity). The characteristic $\mathfrak{n}(r,f)$ takes into account the arguments of the poles. The characteristic $\mathfrak{n}(r,f)$ can only decrease if we fix the moduli of the poles and change their arguments in such a way that they will become closer to the real line $\{\operatorname{Im} z=0\}$.

Let

$$\mathfrak{N}(r,f)=\int_1^r\frac{\mathfrak{n}(t,f)}{t^2}\,dt,\qquad 1<r<\infty.$$

In this notation, the Levin formula (3.3) can be written as

$$\mathfrak{N}\left(r,\frac{1}{f}\right)-\mathfrak{N}(r,f)=\frac{1}{2\pi}\int_{\kappa(r)}^{\pi-\kappa(r)}\ln|f(r\sin\varphi e^{i\varphi})|\frac{d\varphi}{r\sin^2\varphi}+\tilde{Q}(r,1;f),\tag{5.3}$$

where

$$\kappa(r)=\arcsin\left(\frac{1}{r}\right).$$

This statement follows from the equality

$$\sum_{1<\rho_n\le r\sin\psi_n}\left(\frac{\sin\psi_n}{\rho_n}-\frac{1}{r}\right)=\mathfrak{N}(r,f),$$

where the $\rho_n e^{i\psi_n}$ are the poles of the function $f(z)$. Indeed, observe that the function $\mathfrak{n}(t,f)$ has jumps only at the points $t=\rho_n/\sin\psi_n$, and the values of the jumps are equal to the numbers of poles of $f(z)$ lying on the circular arc

$$\left\{\left|z-i\frac{t}{2}\right|=\frac{t}{2},\ |z|>1\right\}.$$

Hence

$$\sum_{1<\rho_n\le\sin\psi_n}\left(\frac{\sin\psi_n}{\rho_n}-\frac{1}{r}\right)=\int_1^r\left(\frac{1}{t}-\frac{1}{r}\right)d\mathfrak{n}(t,f)$$
$$=\left(\frac{1}{t}-\frac{1}{r}\right)\mathfrak{n}(t,f)\Big|_1^r+\int_1^r\frac{\mathfrak{n}(t,f)}{t^2}dt=\mathfrak{N}(r,f).$$

We also introduce

$$\mathfrak{m}(r,f)=\frac{1}{2\pi}\int_{\kappa(r)}^{\pi-\kappa(r)}\ln^+|f(r\sin\varphi e^{i\varphi})|\frac{d\varphi}{r\sin^2\varphi}$$

and

$$\mathfrak{T}(r,f)=\mathfrak{m}(r,f)+\mathfrak{N}(r,f).$$

With the customary convention for the meaning of the symbols $\mathfrak{m}(r,a)$, $\mathfrak{n}(r,a)$, and $\mathfrak{N}(r,a)$, we get the following analog of the first fundamental theorem.

THEOREM 5.3. *Let $f(z)$ be a non-constant meromorphic function in the half-plane $\{\operatorname{Im} z\ge 0\}$. Then*

$$\mathfrak{m}(r,a)+\mathfrak{N}(r,a)=\mathfrak{T}(r,f)+\varepsilon_3(r,a),\qquad 1<r<\infty,$$

for each complex number $a\ne\infty$, where $\varepsilon_3(r,a)=O(1)$ as $r\to\infty$.

PROOF. This is proved by an argument similar to that in the proofs of Theorems 5.1 and 5.2. We apply the Levin formula in the form (5.3) to $f(z)-a$. Combining this result with the observation that (4.6) implies

$$|\mathfrak{m}(r,f-a)-\mathfrak{m}(r,f)|\le(\ln^+|a|+\ln 2)\frac{1}{2\pi}\int_{\kappa(r)}^{\pi-\kappa(r)}\frac{d\varphi}{r\sin^2\varphi}$$
$$=(\ln^+|a|+\ln 2)\frac{1}{2\pi r}2\cot\kappa(r)$$
$$=(\ln^+|a|+\ln 2)\frac{\sqrt{r^2-1}}{\pi r}\le\frac{1}{\pi}(\ln^+|a|+\ln 2),$$

we get the desired result. □

Using Theorems 3.3 and 3.6 it is possible to prove analogs of the Shimizu–Ahlfors theorem for angular Nevanlinna characteristics and for Tsuji characteristics. Since these analogs will not be used in this book, we are not going to do this. Among the results of this type we prove only the following fact.

THEOREM 5.4. *There exist continuous non-decreasing functions $\overset{\circ}{S}_{\alpha\beta}(r,f)$ and $\overset{\circ}{\mathfrak{T}}(r,f)$ such that*

$$S_{\alpha\beta}(r,f)=\overset{\circ}{S}_{\alpha\beta}(r,f)+O(1),$$
$$\mathfrak{T}(r,f)=\overset{\circ}{\mathfrak{T}}(r,f)+O(1).$$

PROOF. Let

$$\overset{\circ}{S}(r,f)=\frac{2}{\pi}\int_1^r\int_1^\pi\left(\frac{1}{t}-\frac{t}{r^2}\right)(\overset{\circ}{f}(te^{i\theta}))^2\sin\theta t\,dt\,d\theta,$$

$$\overset{\circ}{\mathfrak{T}}(r,f)=\frac{1}{\pi}\iint_{|z-\frac{ir}{2}|\le\frac{r}{2},\ |z|\ge 1}\left(\frac{\sin\theta}{t}-\frac{1}{r}\right)(\overset{\circ}{f}(te^{i\theta}))^2t\,dt\,d\theta.$$

Since the integrands in both integrals are non-negative and are non-decreasing as functions of r, it follows that $\overset{\circ}{S}(r,f)$ and $\overset{\circ}{\mathfrak{T}}(r,f)$ are non-decreasing functions of r. Using the inequality

$$\left|\ln^+|f| - \ln\sqrt{1+|f|^2}\right| \leq \frac{1}{2}\ln 2$$

and Theorems 3.3 and 3.6, we get the estimates

$$|S(r,f) - \overset{\circ}{S}(r,f)| \leq \frac{2}{\pi}\cdot\frac{1}{2}\ln 2 + \frac{4}{\pi}\cdot\frac{1}{2}\ln 2 + |Q(r,1;f)| + |Q^{(2)}(r,1;f)|$$

and

$$|\mathfrak{T}(r,f) - \overset{\circ}{\mathfrak{T}}(r,f)| \leq \frac{1}{\pi}\cdot\frac{1}{2}\ln 2 + |\tilde{Q}(r,1;f)| + |\tilde{Q}^{(2)}(r,1;f)|.$$

These estimates imply the statement of the theorem for $S = S_{0\pi}$ and $\mathfrak{T}$. Generalization for $S_{\alpha\beta}(r,f)$ can be achieved in the same way as in Theorem 5.2. □

In many situations the characteristics $\overset{\circ}{S}_{\alpha\beta}(r,f)$ and $\overset{\circ}{\mathfrak{T}}(r,f)$ are more convenient than $S_{\alpha\beta}(r,f)$ and $\mathfrak{T}(r,f)$.

6. Relations between the characteristics

The relations which we derive in this section will be used very often in this book, usually without reference.

Let $f_1(z),\dots,f_q(z)$ be meromorphic functions. Inequalities (4.3) and (4.4) immediately imply

$$m\left(r,\prod_{j=1}^{q} f_j\right) \leq \sum_{j=1}^{q} m(r,f_j) \tag{6.1}$$

and

$$m\left(r,\sum_{j=1}^{q} f_j\right) \leq \sum_{j=1}^{q} m(r,f_j) + \ln q. \tag{6.2}$$

If $f_j(z)$ has a pole of order $\lambda_j \geq 0$ at z_0, then the order of the pole of $\prod_{j=1}^q f_j(z)$ at z_0 is at most $\sum_{j=1}^q \lambda_j$, and the order of the pole of $\sum_{j=1}^q f_j(z)$ at z_0 is at most $\max_{1\leq j\leq q}\lambda_j \leq \sum_{j=1}^q \lambda_j$. Hence, if we denote by $n_j(D)$ the number of the poles of $f_j(z)$ in a set $D \subset \mathbf{C}$, and by $n(D)$ the number of the poles of $\prod_{j=1}^q f_j$ (or $\sum_{j=1}^q f_j$) in D by $n(D)$, then

$$n(D) \leq \sum_{j=1}^{q} n_j(D).$$

In particular, for $D = \{0 < |z| \leq t\}$ we get

$$n\left(t,\prod_{j=1}^{q} f_j\right) - n\left(0,\prod_{j=1}^{q} f_j\right) \leq \sum_{j=1}^{q}[n(t,f_j) - n(0,f_j)],$$

$$n\left(t,\sum_{j=1}^{q} f_j\right) - n\left(0,\sum_{j=1}^{q} f_j\right) \leq \sum_{j=1}^{q}[n(t,f_j) - n(0,f_j)],$$

and for $D = \{z = 0\}$ we get

$$n\left(0,\prod_{j=1}^{q} f_j\right) \leq \sum_{j=1}^{q} n(0,f_j), \qquad n\left(0,\sum_{j=1}^{q} f_j\right) \leq \sum_{j=1}^{q} n(0,f_j).$$

If $r \geq 1$, these inequalities imply

$$N\left(r, \prod_{j=1}^{q} f_j\right) \leq \sum_{j=1}^{q} N(r, f_j) \tag{6.3}$$

and

$$N\left(r, \sum_{j=1}^{q} f_j\right) \leq \sum_{j=1}^{q} N(r, f_j). \tag{6.4}$$

Adding (6.1) to (6.3), and (6.2) to (6.4), we get for $r \geq 1$:

$$T\left(r, \prod_{j=1}^{q} f_j\right) \leq \sum_{j=1}^{q} T(r, f_j), \tag{6.5}$$

$$T\left(r, \sum_{j=1}^{q} f_j\right) \leq \sum_{j=1}^{q} T(r, f_j) + \ln q. \tag{6.6}$$

Theorem 4.1 can be regarded as the following property of the characteristic $T(r, f)$:

$$T\left(r, \frac{1}{f-a}\right) = T(r, f) + O(1). \tag{6.7}$$

Let us prove the following more general fact:

$$T(r, L(f)) = T(r, f) + O(1), \tag{6.8}$$

where

$$L(w) = \frac{\alpha w + \beta}{\gamma w + \delta}, \quad \alpha\delta - \beta\gamma \neq 0,$$

is a Möbius transformation. Properties of Möbius transformations imply that it suffices to consider the following cases: 1) $L(w) = aw$, 2) $L(w) = w - a$, 3) $L(w) = \frac{1}{w}$, where a is a complex number, $a \neq 0, \infty$. By (4.5) and (4.6) we have

$$|m(r, af) - m(r, f)| \leq |\ln |a||,$$

$$|m(r, f - a) - m(r, f)| \leq \ln^+ |a| + \ln 2.$$

Since it is clear that $N(r, af) = N(r, f)$ and $N(r, f - a) = N(r, f)$, we get (6.8) in cases 1) and 2). In case 3), relation (6.8) coincides with (6.7) for $a = 0$.

Observe that if $|f(0)| = 1$, then (4.1) immediately implies

$$T\left(r, \frac{1}{f}\right) = T(r, f). \tag{6.8$'$}$$

In a similar manner, for meromorphic functions in an angle $\{\alpha \leq \arg z \leq \beta\}$, $0 < \beta - \alpha \leq 2\pi$, we get $(r \geq 1)$

$$D_{\alpha\beta}\left(r, \prod_{j=1}^{q} f_j\right) \leq \sum_{j=1}^{q} D_{\alpha\beta}(r, f_j), \tag{6.9}$$

$$D_{\alpha\beta}\left(r, \sum_{j=1}^{q} f_j\right) \leq \sum_{j=1}^{q} D_{\alpha\beta}(r, f_j) + \ln q, \tag{6.10}$$

where $D_{\alpha\beta}(r, f)$ is any of the characteristics $A_{\alpha\beta}(r, f)$, $B_{\alpha\beta}(r, f)$, and $C_{\alpha\beta}(r, f)$;

$$S_{\alpha\beta}(r, L(f)) = S_{\alpha\beta}(r, f) + O(1). \tag{6.11}$$

For functions analytic in the half-plane

$$\{\operatorname{Im} z \geq 0\}$$

we get

$$D\left(r, \prod_{j=1}^{q} f_j\right) \leq \sum_{j=1}^{q} D(r, f_j), \tag{6.12}$$

$$D\left(r, \sum_{j=1}^{q} f_j\right) \leq \sum_{j=1}^{q} D(r, f_j) + O(1), \tag{6.13}$$

where $D(r, f)$ is $\mathfrak{m}(r, f)$ or $\mathfrak{N}(r, f)$; and also

$$\mathfrak{T}(r, L(f)) = \mathfrak{T}(r, f) + O(1) \tag{6.14}$$

From (6.5), (6.6), and (6.7) we immediately get

$$T(r, R(f_1, \ldots, f_q)) = O\left(\sum_{j=1}^{q} T(r, f_j)\right), \tag{6.15}$$

where $R(w_1, \ldots, w_q)$ is a rational function in $w_1, \ldots, w_q$. Equality (6.15) can be made more precise in several important cases. Let

$$F(z) = f_0(z) + f_1(z)f(z) + \cdots + f_q(z)[f(z)]^q,$$

where $f_0(z), \ldots, f_q(z)$, $f(z)$ are meromorphic functions. Then

$$T(r, F) \leq qT(r, f) + \sum_{j=0}^{q} T(r, f_j) + O(1). \tag{6.16}$$

This statement can be proved using induction on q. For $q = 1$, inequality (6.16) immediately follows from (6.5) and (6.6). Assume that (6.16) has been proved for $q - 1$. Then

$$T(r, F_1) \leq (q-1)T(r, f) + \sum_{j=1}^{q} T(r, f_j) + O(1),$$

where $F_1(z) = f_1(z) + f_2(z)f(z) + \cdots + f_q(z)[f(z)]^{q-1}$. Since $F(z) = f_0(z) + f(z)F_1(z)$, we see that

$$T(r, F) \leq T(r, f_0) + T(r, f) + T(r, F_1) + O(1) \leq qT(r, f) + \sum_{j=0}^{q} T(r, f_j) + O(1).$$

The following inequalities can be proved using the same argument that was used to prove (6.16):

$$m(r, F) \leq qm(r, f) + \sum_{j=0}^{q} m(r, f_j) + O(1), \tag{6.17}$$

$$N(r, F) \leq qN(r, f) + \sum_{j=1}^{q} N(r, f_j). \tag{6.18}$$

(Inequality (6.18) can also be easily proved directly.)

DEFINITION. A meromorphic function $f(z)$ is called a *rational function of degree* d if it can be written as a quotient of two polynomials, $P(z)$ and $Q(z)$, which do not have common zeros, and the maximum of the degrees of $P(z)$ and $Q(z)$ is equal to d.

THEOREM 6.1. *If $R(u)$ is a rational function of degree d and $f(z)$ is a meromorphic function, then*

$$T(r, R(f)) = dT(r, f) + O(1). \tag{6.19}$$

We start by proving the following result.

THEOREM 6.2. *If $P(u)$ is a polynomial of degree p and $f(z)$ is a meromorphic function, then*

$$T(r, P(f(z))) = pT(r, f(z)) + O(1), \tag{6.20}$$

$$m(r, P(f(z))) = pm(r, f(z)) + O(1), \tag{6.21}$$

$$N(r, P(f(z))) = pN(r, f(z)). \tag{6.22}$$

The equation (6.22) is obvious, because if $f(z)$ has a pole of order $\lambda \geq 0$ at z_0, then $P(f(z))$ has a pole of order λp at z_0.

To prove (6.21) we need the following lemma, which will be used many times later.

LEMMA 6.1. *Let K be a positive number and let E be the set of those φ satisfying $0 \leq \varphi < 2\pi$ for which $|f(re^{i\varphi})| > K$. Then*

$$\frac{1}{2\pi}\int_E \ln|f(re^{i\varphi})|\, d\varphi = m(r, f) + O(1).$$

This lemma immediately follows from the equality

$$\frac{1}{2\pi}\int_E \ln|f(re^{i\varphi})|\, d\varphi = m\left(r, \frac{f}{K}\right) + \frac{\text{meas}\, E}{2\pi}\ln K = m(r, f) + O(1).$$

To prove (6.21) we observe that (6.17) implies

$$m(r, P(f)) \leq pm(r, f) + O(1). \tag{6.23}$$

Let $P(u) = a_p u^p + a_{p-1}u^{p-1} + \cdots + a_0$, $a_p \neq 0$. Let $K > 1$ be so large that

$$\left|a_p + \frac{a_{p-1}}{u} + \cdots + \frac{a_0}{u^p}\right| \geq \frac{1}{2}|a_p|$$

for $|u| > K$. Then for $|f(z)| > K$ we have $|P(f(z))| \geq |f(z)|^p \left|\frac{a_p}{2}\right|$. Denote by E the set of those arcs of $\{|z| = r\}$ for which $|f(z)| > K$. Using the lemma, we get

$$\begin{aligned} m(r, P(f)) &\geq \frac{1}{2\pi}\int_E \ln|P(f(re^{i\varphi}))|\, d\varphi \geq \frac{1}{2\pi}\int_E \ln\left\{|f(re^{i\varphi})|^p\left|\frac{a_p}{2}\right|\right\} d\varphi \\ &= \frac{p}{2\pi}\int_E \ln|f(re^{i\varphi})|\, d\varphi + O(1) = pm(r, f) + O(1). \end{aligned} \tag{6.24}$$

The assertion (6.21) follows from (6.23) and (6.24). The assertion (6.20) follows from (6.21) and (6.22).

PROOF OF THEOREM 6.1. Let $R(u) = P(u)/Q(u)$. Without loss of generality we may assume that the degree of $Q(u)$ is equal to d. If this was not the case, we would use

$$T\left(r, \frac{P(f)}{Q(f)}\right) = T\left(r, \frac{Q(f)}{P(f)}\right) + O(1).$$

Also, we may assume that the degree of $P(u)$ is strictly less than d, because, if the degree of $P(u)$ is d, we may use the equation

$$T\left(r, \frac{P(f)}{Q(f)}\right) = T\left(r, \frac{P(f)}{Q(f)} - R(\infty)\right) + O(1) = T\left(r, \frac{P_1(f)}{Q(f)}\right) + O(1),$$

where $P_1(u) = P(u) - Q(u)R(\infty)$ is a polynomial whose degree is strictly less than d. So we assume that the degree of $Q(u)$ is d and the degree of $P(u)$ is strictly less than d. We cover the zeros of $Q(u)$ by a collection of open discs Γ in such a way that none of the zeros of $P(u)$ is contained in the closure of Γ. Denote by Ω the complement of Γ in the extended complex plane. Since $R(u) \to 0$ and $1/Q(u) \to 0$ as $u \to \infty$, the functions $R(u)$ and $1/Q(u)$ are bounded on Ω. There exist real numbers $0 < M_1, M_2 < \infty$ such that $M_1 < |P(u)| < M_2$ for each $u \in \Gamma$. Hence for $u \in \Gamma$ we have

$$\frac{M_1}{|Q(u)|} \le |R(u)| \le \frac{M_2}{|Q(u)|}.$$

Let E_1 and E_2 be the subsets of $\{|z| = r\}$ where $f(z) \in \Gamma$ and $f(z) \in \Omega$, respectively. Then

$$\begin{aligned} m(r, R(f)) &= \frac{1}{2\pi}\int_{z\in E_1} \ln^+ |R(f(z))|\, d\varphi + \frac{1}{2\pi}\int_{z\in E_2} \ln^+ |R(f(z))|\, d\varphi \\ &= \frac{1}{2\pi}\int_{z\in E_1} \ln^+ |R(f(z))|\, d\varphi + O(1). \end{aligned}$$

Furthermore,

$$\begin{aligned} \frac{1}{2\pi}\int_{z\in E_1} \ln^+ |R(f(z))|\, d\varphi &\le \frac{1}{2\pi}\int_{z\in E_1} \ln^+ \frac{M_2}{|Q(f(z))|}\, d\varphi \\ &\le \frac{1}{2\pi}\int_0^{2\pi} \ln^+ \frac{M_2}{|Q(f(z))|}\, d\varphi = m\left(r, \frac{M_2}{Q(f)}\right) = m\left(r, \frac{1}{Q(f)}\right) + O(1), \end{aligned}$$

$$\begin{aligned} \frac{1}{2\pi}\int_{z\in E_1} \ln^+ |R(f(z))|\, d\varphi &\ge \frac{1}{2\pi}\int_{z\in E_1} \ln^+ \frac{M_1}{|Q(f(z))|}\, d\varphi \\ &= m\left(r, \frac{M_1}{Q(f)}\right) - \frac{1}{2\pi}\int_{z\in E_2} \ln^+ \frac{M_1}{|Q(f(z))|}\, d\varphi = m\left(r, \frac{1}{Q(f)}\right) + O(1). \end{aligned}$$

Combining these estimates we get

$$m(r, R(f)) = m\left(r, \frac{1}{Q(f)}\right) + O(1). \tag{6.25}$$

On the other hand, the poles of the functions $R(f(z))$ and $1/Q(f(z))$ coincide. In fact, if the function $f(z)$ has a pole at z_0, then the order of the pole of $Q(f(z))$ at z_0 is greater than the order of the pole of $P(f(z))$ at z_0, and $R(f(z_0)) = 0$. Hence the function $R(f(z))$ can have poles only at the zeros of the function $Q(f(z))$. The function $Q(f(z))$ has zeros only at those points at which $f(z)$ is equal to one of the zeros of $Q(u)$. Since the zeros of $P(u)$ are different from the zeros of $Q(u)$, we have $|P(f(z))| + |Q(f(z))| > 0$. Therefore the functions $R(f(z))$ and $1/Q(f(z))$ have the same poles. Hence $N(r, R(f)) = N(r, 1/Q(f))$. Adding this equality to (6.25) we obtain

$$T(r, R(f)) = T(r, 1/Q(f)) + O(1).$$

By (6.20) we have

$$T(r, 1/Q(f)) = T(r, Q(f)) + O(1) = dT(r, f) + O(1).$$

Theorem 6.1 has been proved. □

Relations (6.14) and (6.20) are special cases of Theorem 6.1. We would like to mention also the following.

COROLLARY. *If $R(z)$ is a rational function of degree d, then*

$$T(r, R(z)) = d \ln r + O(1). \tag{6.26}$$

In fact, $T(r, z) = \ln^+ r$, and (6.26) follows from (6.19).

EXERCISES. (i) Let p be a natural number. Prove the relation $T(r, [f(z)]^p) = pT(r, f(z))$.

(ii) Prove that $T(r, \cosh z) = 2T(r, e^z) + O(1) = \frac{2}{\pi} r + O(1)$.

A theorem similar to Theorem 6.1 holds for the characteristics $S_{\alpha\beta}$ and $\mathfrak{T}$.

THEOREM 6.3. *Let $R(u)$ be a rational function of degree d. If $f(z)$ is meromorphic in an angle $\{\alpha \le \arg z \le \beta\}$, $0 < \beta - \alpha \le 2\pi$, then*

$$S_{\alpha\beta}(r, R(f)) = dS_{\alpha\beta}(r, f) + O(1).$$

If $f(z)$ is meromorphic in the half-plane $\{\operatorname{Im} z \ge 0\}$, then

$$\mathfrak{T}(r, R(f)) = d\mathfrak{T}(r, f) + O(1).$$

This theorem can be proved by minor modifications of the argument used in the proof of Theorem 6.1. We leave it to the reader as an exercise.

DEFINITION. A meromorphic function is called *transcendental* if it is not rational.

THEOREM 6.4. *A meromorphic function $f(z)$ is transcendental if and only if*

$$\liminf_{r \to \infty} \frac{T(r, f)}{\ln r} = \infty. \tag{6.27}$$

Sufficiency of condition (6.27) follows from (6.26). Let us prove the necessity. Assume that (6.27) does not hold, that is,

$$\liminf_{r \to \infty} \frac{T(r, f)}{\ln r} = K < \infty. \tag{6.28}$$

Then $f(z)$ has at most $p = [K]$ poles. In fact, let us enumerate them according to their moduli, the pole with the smallest modulus coming first. If there is a $(p+1)$st pole, denote its modulus by r_0. We have

$$\begin{aligned} N(r, f) &\ge \int_{r_0}^{r} \frac{n(t, f) - n(0, f)}{t} \, dt + n(0, f) \ln r \\ &\ge \int_{r_0}^{r} \frac{p + 1 - n(0, f)}{t} \, dt + n(0, f) \ln r \\ &= [p + 1 - n(0, f)] \ln \frac{r}{r_0} + n(0, f) \ln r = (p + 1) \ln r + O(1). \end{aligned}$$

Hence

$$\liminf_{r\to\infty}\frac{T(r,f)}{\ln r}\ge\liminf_{r\to\infty}\frac{N(r,f)}{\ln r}\ge p+1>K,$$

and we get a contradiction with (6.28). In a similar way, we prove that $f(z)$ has at most p zeros. Therefore there exists a rational function $R(z)$ such that

$$g(z)=f(z)R(z)$$

has neither zeros nor poles. Since

$$T(r,g)\le T(r,f)+T(r,R)=T(r,f)+O(\ln r),$$

it follows that

$$\liminf_{r\to\infty}\frac{T(r,g)}{\ln r}\le K_1<\infty.$$

We need to show that $g(z)\equiv$ const. This statement can be derived from the following lemma.

LEMMA 6.2. *Let $g(z)$ be an entire function without zeros that satisfies*

$$\liminf_{r\to\infty}\frac{T(r,g)}{r^\lambda}=0$$

for some $\lambda>0$. Then

$$g(z)=e^{P(z)},$$

where $P(z)$ is a polynomial of degree less than λ.

PROOF. Observe that the function

$$P(z)=\ln g(z)=\int_0^z\frac{g'(\zeta)}{g(\zeta)}\,d\zeta+\ln g(0)$$

is an entire function. Let

$$P(z)=\sum_{p=0}^{\infty}c_pz^p.$$

By (2.6), for all $r>0$ and $p=1,2,\dots$, we have

$$\frac{d^p}{dz^p}\ln g(z)\bigg|_{z=0}=\frac{2p!}{r^p}\frac{1}{2\pi}\int_0^{2\pi}\ln|g(re^{i\theta})|e^{-ip\theta}\,d\theta;$$

hence

$$\begin{aligned}|c_p|&=\frac{1}{p!}\left|\frac{d^p}{dz^p}\ln g(z)\bigg|_{z=0}\right|\le\frac{2}{r^p}\frac{1}{2\pi}\int_0^{2\pi}|\ln|g(re^{i\theta})||\,d\theta\\&=\frac{2}{r^p}\left[m(r,g)+m\left(r,\frac{1}{g}\right)\right].\end{aligned}$$

Since $m(r,g)\le T(r,g)$, and (by Theorem 4.1) $m(r,\frac{1}{g})\le T(r,g)+O(1)$, we have

$$|c_p|\le\frac{2}{r^p}[2T(r,g)+O(1)].$$

Letting $r\to\infty$, we see that $c_p=0$ for $p\ge\lambda$. □

We are mostly interested in transcendental meromorphic functions. For this reason, in our estimates for $T(r,f)$, terms which are $O(\ln r)$ can be disregarded.

THEOREM 6.5. *Let $f(z)$ be a meromorphic function, let $R_j(z)$, $0 \le j \le p$, be rational functions, $R_p(z) \not\equiv 0$, and let*

$$F(z) = R_p(z)[f(z)]^p + R_{p-1}(z)[f(z)]^{p-1} + \cdots + R_0(z).$$

Then

$$T(r, F) = pT(r, f) + O(\ln r), \tag{6.29}$$

$$m(r, F) = pm(r, f) + O(\ln r), \tag{6.30}$$

$$N(r, F) = pN(r, f) + O(\ln r). \tag{6.31}$$

PROOF. If $f(z)$ is rational, the result is trivial. For this reason we assume that $f(z)$ is transcendental. Let M be a finite set containing the poles of all functions $R_j(z)$, $0 \le j \le p$, and the zeros of $R_p(z)$. If the function $f(z)$ has a pole of order λ (≥ 0) at some $z_0 \notin M$, then $F(z)$ has a pole of order λp at z_0. Hence $n(r, F) = pn(r, f) + O(1)$, and we get equality (6.31). From (6.17) we get $m(r, F) \le pm(r, f) + O(\ln r)$. Let us show that the inverse inequality is also true. Let $R_j(z) = \beta_j z^{\alpha_j}(1 + o(1))$ as $z \to \infty$, where $\beta_j \neq 0$ and α_j is an integer. Denote by λ the largest of the differences $\alpha_j - \alpha_p$, $0 \le j \le p-1$. Let $\chi = \max(1, \lambda + 1, \alpha_p)$. Denote by E the set of those z for which

$$|f(z)| \ge |z|^{\chi}, \qquad \Omega = \{|z| < \infty\} \backslash E.$$

It is clear that for $z \in E$, $|z| \ge r_0$ we have

$$\begin{aligned}
&\left| 1 + \frac{R_{p-1}(z)}{R_0(z)} \frac{1}{f(z)} + \cdots + \frac{R_0(z)}{R_p(z)} \frac{1}{[f(z)]^p} \right| \\
&\quad = \left| 1 + \frac{\beta_{p-1}}{\beta_p} \frac{z^{\alpha_{p-1} - \alpha_p}(1 + o(1))}{f(z)} + \cdots + \frac{\beta_0}{\beta_p} \frac{z^{\alpha_0 - \alpha_p}(1 + o(1))}{[f(z)]^p} \right| \\
&\quad \ge 1 - \left| \frac{\beta_{p-1}}{\beta_p} \right| |z|^{\lambda - \chi}(1 + o(1)) - \cdots - \left| \frac{\beta_0}{\beta_p} \right| |z|^{\lambda - p\chi}(1 + o(1)) > \frac{1}{2},
\end{aligned}$$

$$|F(z)| \ge \frac{|\beta_p|}{2}(1 + o(1))|z|^{\alpha_p}|f(z)|^p.$$

Let $E(r) = E \cap \{|z| = r\}$, and $\Omega(r) = \Omega \cap \{|z| = r\}$. Then ($z = re^{i\varphi}$)

$$\begin{aligned}
m(r, F) &\ge \frac{1}{2\pi} \int_{E(r)} \ln^+ |F(z)|\, d\varphi \ge O(\ln r) + \frac{1}{2\pi} \int_{E(r)} \ln^+ \{|f(z)|\}^p \, d\varphi \\
&= O(\ln r) + pm(r, f) - \frac{1}{2\pi} \int_{\Omega(r)} \ln^+ \{|f(z)|^p\} \, d\varphi \\
&\ge O(\ln r) + pm(r, f) - p\chi \ln r = pm(r, f) + O(\ln r).
\end{aligned}$$

We have proved (6.30). Equality (6.29) follows from (6.30) and (6.31). □

COROLLARY. *Let $f(z)$ be a meromorphic function with expansion $f(z) = c_\lambda z^\lambda + c_{\lambda+1} z^{\lambda+1} + \cdots$, $c_\lambda \neq 0$, in a neighborhood of $z = 0$. Let $f_1(z) = f(z)/c_\lambda z^\lambda$. Then*

$$T(r, f_1) = T(r, f) + O(\ln r), \qquad m(r, f_1) = m(r, f) + O(\ln r),$$

$$N(r, f_1) = N(r, f) + O(\ln r), \qquad m\left(r, \frac{1}{f_1}\right) = m\left(r, \frac{1}{f}\right) + O(\ln r),$$

$$N\left(r, \frac{1}{f_1}\right) = N\left(r, \frac{1}{f}\right) + O(\ln r).$$

This corollary allows us, in many cases, to assume that the transcendental function $f(z)$ under consideration is normalized by the condition $f(0) = 1$.

EXERCISE. Let $f(z)$ be a meromorphic function, and $R_j(z)$, $1 \le j \le 4$, rational functions that satisfy the condition $R_1(z)R_4(z) - R_2(z)R_3(z) \not\equiv 0$. Set

$$F(z) = \frac{R_1(z)f(z) + R_2(z)}{R_3(z)f(z) + R_4(z)}.$$

Show that $T(r, F) = T(r, f) + O(\ln r)$.

Proofs of the following two theorems are left to the reader as exercises.

THEOREM 6.6. *Let $f(z)$ be a meromorphic function, p a natural number, and $f_1(z) = f(z^p)$. Then*

$$m(r, f_1) = m(r^p, f), \qquad N(r, f_1) = N(r^p, f),$$
$$T(r, f_1) = T(r^p, f).$$

COROLLARY. *Let p be a natural number, and $f(z)$ a meromorphic function such that $f_1(z) = f(z^{1/p})$ is single-valued. Then*

$$m(r, f_1) = m(r^{1/p}, f), \qquad N(r, f_1) = N(r^{1/p}, f),$$
$$T(r, f_1) = T(r^{1/p}, f).$$

THEOREM 6.7. *Let $f(z)$ be a meromorphic function, $f_1(z) = f(Az)$, where $A \ne 0$. Then*

$$m(r, f_1) = m(|A|r, f), \qquad N(r, f_1) = N(|A|r, f) - n(0, f)\ln|A|,$$
$$T(r, f_1) = T(|A|r, f) - n(0, f)\ln|A|.$$

7. Connection between the growth of the modulus of a meromorphic function and the growth of its characteristic $T(r, f)$

Let $f(z)$ be a meromorphic function. Let

$$M(r, f) = \max_{|z|=r} |f(z)|.$$

We study relations between $M(r, f)$ and $T(r, f)$. First we consider the case where f is an entire function.

THEOREM 7.1. *Let $f(z)$ be an entire function, and let $0 < r < R$. Then*

$$T(r, f) \le \ln^+ M(r, f) \le \frac{R + r}{R - r} T(R, f). \tag{7.1}$$

PROOF. Since $\ln^+ |f(z)| \le \ln^+ M(r, f)$ for $|z| = r$, we have

$$T(r, f) = m(r, f) \le \ln^+ M(r, f).$$

Next we suppose that $|f(z_0)| = M(r, f)$, $|z_0| = r$. We use the Poisson–Jensen formula (2.2) for the disc $\{|z| \le R\}$. We get

$$\begin{aligned}\ln M(r, f) &= \ln|f(z_0)| \\ &= \frac{1}{2\pi}\int_0^{2\pi} \ln|f(Re^{i\theta})| \operatorname{Re}\frac{Re^{i\theta} + z_0}{Re^{i\theta} - z_0}\, d\theta - \sum_{|a_\mu|<R} \ln\left|\frac{R^2 - \bar{a}_\mu z_0}{R(r_0 - a_\mu)}\right|. \end{aligned} \tag{7.2}$$

Since the summands in the right-hand side of (7.2) represent the values of the Green function for a disc, they are positive. Also

$$0 < \operatorname{Re}\frac{Re^{i\theta}+z_0}{Re^{i\theta}-z_0} = \frac{R^2-|z_0|^2}{|Re^{i\theta}-z_0|^2} \le \frac{R+r}{R-r}. \tag{7.3}$$

Therefore

$$\begin{aligned} &\frac{1}{2\pi}\int_0^{2\pi} \ln|f(Re^{i\theta})| \operatorname{Re}\frac{Re^{i\theta}+z_0}{Re^{i\theta}-z_0}\,d\theta \\ &\le \frac{1}{2\pi}\int_0^{2\pi} \ln^+|f(Re^{i\theta})| \operatorname{Re}\frac{Re^{i\theta}+z_0}{Re^{i\theta}-z_0}\,d\theta \\ &\le \frac{R+r}{R-r}\frac{1}{2\pi}\int_0^{2\pi} \ln^+|f(Re^{i\theta})|\,d\theta = \frac{R+r}{R-r}m(R,f) \end{aligned} \tag{7.4}$$

and

$$\ln M(r,f) \le \frac{R+r}{R-r}m(R,f) = \frac{R+r}{R-r}T(R,f).$$

Since the right-hand side of this inequality is positive, we have proved the right part of (7.1). □

Now we consider the connection between $T(r,f)$ and $\ln M(r,f)$ for meromorphic functions. Unlike an entire f, for a non-constant meromorphic f, it is possible that $\liminf_{r\to\infty} M(r,f) < \infty$: in the example on page 15 we have

$$M(n+1/2,f) \le 1, \quad n = 0,1,2,\dots .$$

Hence, in general, the left inequality in (7.1) does not hold for meromorphic functions. If $f(z)$ has a pole on the circle $\{|z| = r\}$, then $M(r,f) = +\infty$. Hence, in general, an upper estimate of $M(r,f)$ in terms of $T(R,f)$, $R > r$, is impossible. Nevertheless, there is an analog of the right inequality in (7.1) for meromorphic functions, though it is significantly weaker. We mean the following theorem due to R. Nevanlinna.

THEOREM 7.2. *Let $f(z)$ be a meromorphic function, and $k > 1$ a real number. Then*

$$\frac{1}{r}\int_0^r \ln^+ M(t,f)\,dt \le C(k)T(kr,f), \tag{7.5}$$

where the constant $C(k) > 1$ depends only on k.

First we prove two lemmas, which will be used not only in the proof of Theorem 7.2.

LEMMA 7.1. *Let $R' > R > 1$. Then*

$$n(R,f) \le \frac{N(R',f)}{\ln\frac{R'}{R}} \le \frac{R'}{R'-R}N(R',f).$$

PROOF. Indeed,

$$\begin{aligned}
N(R', f) &\geq \int_R^{R'} \frac{n(t, f) - n(0, f)}{t} dt + n(0, f) \ln R' \\
&\geq [n(R, f) - n(0, f)] \ln \frac{R'}{R} + n(0, f) \ln R' \\
&= n(R, f) \ln \frac{R'}{R} + n(0, f) \ln R \geq n(R, f) \ln \frac{R'}{R} \\
&= n(R, f) \int_R^{R'} \frac{dt}{t} \geq n(R, f) \frac{R' - R}{R'}. \qquad \square
\end{aligned}$$

LEMMA 7.2. *Let $\varphi(x)$ be an integrable even real-valued function on $(-a, a)$, non-increasing on $(0, a)$. (The case $\varphi(0) = +\infty$ is allowed.) Let $E \subset (-a, a)$ be a measurable subset,* $\operatorname{meas} E = 2b$. *Then*

$$\int_E \varphi(x)\, dx \leq \int_{-b}^{b} \varphi(x)\, dx.$$

PROOF. Let $E_1 = E \cap (-b, b)$, $E_2 = E \backslash E_1$, $E_3 = (-b, b) \backslash E_1$. It is clear that $\operatorname{meas} E_2 = \operatorname{meas} E_3 = 2b - \operatorname{meas} E_1$. For $x \in (-b, b)$ we have $\varphi(x) \geq \varphi(b)$, and for $x \notin (-b, b)$ we have $\varphi(x) \leq \varphi(b)$. Hence

$$\int_{E_2} \varphi(x)\, dx \leq \varphi(b) \operatorname{meas} E_2 = \varphi(b) \operatorname{meas} E_3 \leq \int_{E_3} \varphi(x)\, dx.$$

Adding the integral

$$\int_{E_1} \varphi(x)\, dx$$

to both sides of the inequality, we get the desired inequality. $\square$

PROOF OF THEOREM 7.2. Denote by a_μ the zeros, and by b_ν the poles of $f(z)$. Let $0 \leq t \leq r$, $R = \sqrt{k}r$. Using the Poisson–Jensen formula for the disc $\{|z| \leq R\}$, we get

$$\begin{aligned}
\ln |f(te^{i\varphi})| &= \frac{1}{2\pi} \int_0^{2\pi} \ln |f(Re^{i\theta})| \operatorname{Re} \left(\frac{Re^{i\theta} + te^{i\varphi}}{Re^{i\theta} - te^{i\varphi}} \right) d\theta \\
&\quad + \sum_{|b_\nu| < R} \ln \left| \frac{R^2 - \bar{b}_\nu te^{i\varphi}}{R(te^{i\varphi} - b_\nu)} \right| - \sum_{|a_\mu| < R} \ln \left| \frac{R^2 - \bar{a}_\mu te^{i\varphi}}{R(te^{i\varphi} - a_\mu)} \right| \\
&\leq \frac{1}{2\pi} \int_0^{2\pi} \ln |f(Re^{i\theta})| \operatorname{Re} \left(\frac{Re^{i\theta} + te^{i\varphi}}{Re^{i\theta} - te^{i\varphi}} \right) d\theta \\
&\quad + \sum_{|b_\nu| < R} \ln \left| \frac{R^2 - \bar{b}_\nu te^{i\varphi}}{R(te^{i\varphi} - b_\nu)} \right|.
\end{aligned} \tag{7.6}$$

We estimate the integral in the right-hand side using (7.4), and estimate the sum using

$$\left| \frac{R^2 - b_\nu te^{i\varphi}}{R(te^{i\varphi} - b_\nu)} \right| \leq \frac{R^2 + |b_\nu| t}{R|t - |b_\nu||} \leq \frac{2R}{|t - |b_\nu||}.$$

Then we get in succession the following inequalities:

$$\ln|f(te^{i\varphi})| \leq \frac{R+t}{R-t} m(R,f) + \sum_{|b_\nu|<R} \ln \frac{2R}{|t-|b_\nu||};$$

$$\ln^+ M(t,f) \leq \frac{R+t}{R-t} m(R,f) + \sum_{|b_\nu|\leq R} \ln \frac{2R}{|t-|b_\nu||}$$

$$\leq \frac{R+r}{R-r} m(R,f) + \sum_{|b_\nu|\leq R} \ln \frac{2R}{|t-|b_\nu||};$$

$$\frac{1}{r}\int_0^r \ln^+ M(t,f)\,dt \leq \frac{R+r}{R-r} m(R,f) + \sum_{|b_\nu|\leq R} \frac{1}{r}\int_0^r \ln \frac{2R}{|t-|b_\nu||}\,dt. \tag{7.7}$$

Using Lemma 7.2 and straightforward transformations, we get

$$\int_0^r \ln \frac{2R}{|t-|b_\nu||}\,dt = \int_{-|b_\nu|}^{r-|b_\nu|} \ln \frac{2R}{|\tau|}\,d\tau \leq \int_{-r/2}^{r/2} \ln \frac{2R}{|\tau|}\,d\tau$$

$$= 2\int_0^{r/2} \ln \frac{2R}{\tau}\,d\tau = r \ln \frac{4eR}{r} = r\ln(4e\sqrt{k}).$$

Hence (7.7) implies

$$\frac{1}{r}\int_0^r \ln^+ M(t,f)\,dt \leq \frac{\sqrt{k}+1}{\sqrt{k}-1} m(\sqrt{k}r, f) + n(R,f)\ln(4e\sqrt{k}). \tag{7.8}$$

Taking $R' = \sqrt{k}R$, we estimate $n(R,f)$ from above using Lemma 7.1; also, we take into account the inequalities

$$m(\sqrt{k}r,f) \leq T(\sqrt{k}r,f) \leq T(kr,f), \qquad N(R',f) \leq T(R',f) \leq T(kr,f).$$

Hence (7.8) implies (7.5) with

$$C(k) = \frac{\sqrt{k}+1}{\sqrt{k}-1} + \frac{\ln(4e\sqrt{k})}{\ln\sqrt{k}}. \qquad \square$$

COROLLARY. *Under the hypothesis of Theorem 7.2, the inequality*

$$\frac{1}{r}\int_0^r \ln^+ M\left(t, \frac{1}{f-a}\right) dt \leq 2C(k)T(kr,f)$$

holds for $r > r_0$, $r_0 = r_0(k,a)$.

To prove this corollary, it suffices to apply Theorem 7.2 to the function $1/(f(z)-a)$, and to use the first fundamental theorem (Theorem 4.1).

We shall prove two more theorems connecting the behavior of the modulus of a meromorphic function f with the behavior of $T(r,f)$.

THEOREM 7.3. *Let $f(z)$ be a meromorphic function, and let k, δ, and r be real numbers satisfying $k > 1$, $0 < \delta \leq 2\pi$, $r > 1$. There exists a constant $C_1(k,\delta)$ such that each measurable subset $E_r \subset [-\pi,\pi]$ with* $\operatorname{meas} E_r = \delta$ *satisfies*

$$\int_{E_r} \ln^+ |f(re^{i\varphi})|\,d\varphi \leq C_1(k,\delta)T(kr,f), \tag{7.9}$$

where $C_1(k,\delta) = \frac{6k}{k-1}\delta \ln \frac{2\pi e}{\delta} \to 0$ as $\delta \to 0$.

PROOF. First we assume that $\delta < \pi$. We write inequality (7.6) with $t = r$ for the disc $\{|z| \le R\}$, $R = \sqrt{k}r$:

$$\begin{aligned}\ln|f(re^{i\varphi})| &\le \frac{1}{2\pi}\int_0^{2\pi} \ln|f(Re^{i\theta})| \operatorname{Re}\left(\frac{Re^{i\theta}+re^{i\varphi}}{Re^{i\theta}-re^{i\varphi}}\right) d\theta \\ &\quad + \sum_{|b_\nu|<R} \ln\left|\frac{R^2-\bar{b}_\nu re^{i\varphi}}{R(re^{i\varphi}-b_\nu)}\right| \\ &\le \frac{R+r}{R-r} m(R,f) + \sum_{|b_\nu|<R} \ln \frac{2R}{|re^{i\varphi}-b_\nu|} \\ &\le \frac{R+r}{R-r} T(R,f) + n(0,f)\ln\frac{2R}{r} + \sum_{0<|b_\nu|<R} \ln\frac{2R}{|re^{i(\varphi-\beta_\nu)}-|b_\nu||},\end{aligned}$$

$$\beta_\nu = \arg b_\nu.$$

The right-hand side of the inequality is non-negative; therefore we may replace $\ln|f(re^{i\varphi})|$ by $\ln^+|f(re^{i\varphi})|$ in the left-hand side. Let us integrate the obtained inequality over the set E_r. We get

$$\begin{aligned}\int_{E_r} \ln^+|f(re^{i\varphi})|\,d\varphi &\le \delta\frac{R+r}{R-r}T(R,f) + \delta n(0,f)\ln\frac{2R}{r} \\ &\quad + \sum_{0<|b_\nu|<R}\int_{E_r} \ln\frac{2R}{|re^{i(\varphi-\beta_\nu)}-|b_\nu||}\,d\varphi.\end{aligned}$$

We apply Lemma 7.2 and get

$$\begin{aligned}\int_{E_r} \ln\frac{2R}{|re^{i(\varphi-\beta_\nu)}-|b_\nu||}\,d\varphi &\le \int_{-\delta/2}^{\delta/2} \ln\frac{2R}{|re^{i\varphi}-|b_\nu||}\,d\varphi \\ &\le \int_{-\delta/2}^{\delta/2} \ln\frac{2R}{r|\sin\varphi|}\,d\varphi \le 2\int_0^{\delta/2} \ln\frac{\sqrt{k}\pi}{\varphi}\,d\varphi = \delta\ln\frac{2e\sqrt{k}\pi}{\delta}.\end{aligned}$$

Hence

$$\begin{aligned}\int_{E_r} \ln^+|f(re^{i\varphi})|\,d\varphi &\le \delta\frac{\sqrt{k}+1}{\sqrt{k}-1}T(\sqrt{k}r,f) + \delta n(0,f)\ln 2\sqrt{k} \\ &\quad + \left\{n(\sqrt{k}r,f) - n(0,f)\right\}\delta\ln\frac{2\sqrt{k}e\pi}{\delta} \\ &\le \delta\frac{\sqrt{k}+1}{\sqrt{k}-1}T(kr,f) + n(\sqrt{k}r,f)\delta\ln\frac{2\sqrt{k}e\pi}{\delta}.\end{aligned} \tag{7.10}$$

We estimate $n(\sqrt{k}r,f)$ using Lemma 7.1 with $R' = kr$. We get

$$n(\sqrt{k}r,f) \le \frac{N(kr,f)}{\ln\sqrt{k}} \le \frac{T(kr,f)}{\ln\sqrt{k}}. \tag{7.11}$$

Using (7.10) and (7.11), we obtain

$$\int_{E_r} \ln^+|f(re^{i\varphi})|\,d\varphi \le C(k,\delta)T(kr,f),$$

where

$$C(k,\delta)=\delta\left(\frac{\sqrt{k}+1}{\sqrt{k}-1}+\frac{\ln\frac{2\sqrt{k}e\pi}{\delta}}{\ln\sqrt{k}}\right)=\delta\left(\frac{2\sqrt{k}}{\sqrt{k}-1}+\frac{2}{\ln k}\ln\left(\frac{2e\pi}{\delta}\right)\right)$$
$$<\left(\frac{4k}{k-1}+\frac{2}{\ln k}\right)\delta\ln\frac{2e\pi}{\delta}<\frac{6k}{k-1}\delta\ln\frac{2e\pi}{\delta}=C_1(k,\delta),$$

and (7.9) is proved.

For $\pi\le\delta\le 2\pi$ we use the trivial estimate

$$\int_{E_r}\ln^+|f(re^{i\varphi})|\,d\varphi\le 2\pi m(r,f)\le 2\pi T(r,f).$$

Since $T(r,f)\le T(kr,f)$ and $2\pi\le 2\delta\le C_1(k,\delta)$, inequality (7.9) has been proved in this case also. □

REMARK. It is clear that we can replace $C_1(k,\delta)$ by $C(k,\delta)<C_1(k,\delta)$ in inequality (7.9), but $C_1(k,\delta)$ is simpler than $C(k,\delta)$.

THEOREM 7.4. *Let E be a measurable subset of $[-\pi,\pi]$, $\Phi(\varphi)$ an integrable non-negative function on E, and $R(r)$ a positive function on $\{r\ge r_0>0\}$ that satisfies $R(r)\to\infty$ as $r\to\infty$, and is such that $R(kr)=O\{R(r)\}$ for some real $k>1$. Suppose that $f(z)$ is a meromorphic function, such that $T(r,f)=O\{R(r)\}$, and for each $\delta>0$ there exists a subset $E_\delta\subset E,\operatorname{meas}E_\delta=\delta$ such that one of the following three conditions:*

$$\text{(a)}\qquad \ln^+|f(re^{i\varphi})|\le\Phi(\varphi)R(r)+o(R(r)),$$
$$\text{(b)}\qquad \ln^+|f(re^{i\varphi})|\ge\Phi(\varphi)R(r)+o(R(r)),$$
$$\text{(c)}\qquad \ln^+|f(re^{i\varphi})|=\Phi(\varphi)R(r)+o(R(r))$$

is satisfied uniformly in φ for $\varphi\in E\backslash E_\delta$. Then

$$\text{(7.12)}\qquad \text{(a}'\text{)}\qquad \limsup_{r\to\infty}\frac{1}{R(r)}\int_E\ln^+|f(re^{i\varphi})|\,d\varphi\le\int_E\Phi(\varphi)\,d\varphi,$$

$$\text{(7.13)}\qquad \text{(b}'\text{)}\qquad \liminf_{r\to\infty}\frac{1}{R(r)}\int_E\ln^+|f(re^{i\varphi})|\,d\varphi\ge\int_E\Phi(\varphi)\,d\varphi,$$

$$\text{(7.14)}\qquad \text{(c}'\text{)}\qquad \lim_{r\to\infty}\frac{1}{R(r)}\int_E\ln^+|f(re^{i\varphi})|\,d\varphi=\int_E\Phi(\varphi)\,d\varphi,$$

respectively.

PROOF. Let us prove (a′). Using (7.9) we get

$$\int_E\ln^+|f(re^{i\varphi})|\,d\varphi=\int_{E\backslash E_\delta}\ln^+|f(re^{i\varphi})|\,d\varphi+\int_{E_\delta}\ln^+|f(re^{i\varphi})|\,d\varphi$$
$$\le R(r)\int_{E\backslash E_\delta}\Phi(\varphi)\,d\varphi+o(R(r))+C_1(k,\delta)T(kr,f)$$
$$\le R(r)\int_E\Phi(\varphi)\,d\varphi+o(R(r))+C_1(k,\delta)O\{R(r)\}.$$

Hence

$$\limsup_{r\to\infty}\frac{1}{R(r)}\int_E\ln^+|f(re^{i\varphi})|\,d\varphi\le\int_E\Phi(\varphi)\,d\varphi+C_1(k,\delta)A,$$

where A is a constant. Letting $\delta \to 0$, we get (a′). The proof of (b′) is trivial after we observe that

$$\lim_{\delta\to 0} \int_{E_\delta} \Phi(\varphi)\, d\varphi = 0.$$

The equation (c′) follows from (a′) and (b′). □

REMARK 1. If we replace the function $\Phi(\varphi)$ in Theorem 7.4 by a bounded function $\Phi(\varphi, r)$ depending on a parameter r, and allow the set E_δ to be dependent on r also, then, under condition (c), we have

$$\int_E \ln^+ |f(re^{i\varphi})|\, d\varphi = R(r) \int_E \Phi(\varphi, r)\, d\varphi + o(R(r)). \tag{7.15}$$

The proof follows the same lines as the proof of Theorem 7.4. In the case where conditions (a) and (b) are satisfied, the equality in (7.15) will be replaced by the corresponding inequality.

REMARK 2. In conditions (a), (b), and (c) of Theorem 7.4, the interval $[r_0, \infty)$ can be replaced by a set unbounded from above. Then (a′), (b′), and (c′) will take place for the same set of values of r.

CHAPTER 2

Meromorphic Functions of Finite Order

1. The growth scale for meromorphic functions

We classify meromorphic functions according to the rate of growth of $T(r, f)$ as $r \to \infty$.

By Theorem 4.3 of Chapter 1, the characteristic $T(r, f)$ is a non-decreasing function of r. Now we introduce the basic notions related to the growth scale of non-decreasing functions.

Let $\alpha(r)$ be a function defined for $r > 0$ and non-negative and non-decreasing for sufficiently large r (if $\alpha(r)$ satisfies this condition, we write $\alpha(r) \in \Lambda$). The number

$$\rho = \rho[\alpha] = \limsup_{r \to \infty} \frac{\ln^+ \alpha(r)}{\ln r}$$

is called the *order* of $\alpha(r)$. If we replace the upper limit by the lower limit in this definition, we get the definition of the *lower order*, which will be denoted by $\lambda = \lambda[\alpha]$. The function $\alpha(r)$ is said to have *finite* (*infinite*) order if $\rho < \infty$ ($\rho = \infty$). Suppose that $\alpha(r)$ has finite order ρ. The number

$$\sigma = \sigma[\alpha] = \limsup_{r \to \infty} \frac{\alpha(r)}{r^\rho}$$

is called the *magnitude of type* of the function $\alpha(r)$. If $\sigma = 0$, we say that $\alpha(r)$ has *minimal type*; if $0 < \sigma < \infty$, we say that $\alpha(r)$ has *normal* (or *mean*) *type*; if $\sigma = \infty$, we say that $\alpha(r)$ has *maximal type*.

Let ρ be the order of $\alpha(r)$, $\rho < \infty$, and μ an arbitrary number satisfying $\mu > \rho$. Then, for sufficiently large r, we have $\alpha(r) < r^{(\rho+\mu)/2}$, and the integral

$$\int_1^\infty \frac{\alpha(r)}{r^{\mu+1}} dr \tag{1.1}$$

converges. On the other hand, if the integral (1.1) converges ($\mu > 0$), then the order of $\alpha(r)$ is not greater than μ; furthermore, if the order of $\alpha(r)$ is equal to μ, then $\alpha(r)$ is of minimal type. In fact, for $r > r_0(\varepsilon)$ we have

$$\varepsilon > \int_r^\infty \frac{\alpha(t)}{t^{\mu+1}} dt \geq \alpha(r) \int_r^\infty \frac{dt}{t^{\mu+1}} = \frac{\alpha(r)}{\mu r^\mu},$$

from which we get

$$\alpha(r) < \varepsilon \mu r^\mu.$$

Thus, the order of $\alpha(r)$ is equal to the infimum of those μ for which the integral (1.1) converges (if the set is empty, then $\rho[\alpha] = \infty$).

Let $\alpha(r)$ be a function of finite order ρ. We say that $\alpha(r)$ belongs to the *convergence class* or to the *divergence class* depending on whether the integral

$$\int_1^\infty \frac{\alpha(r)}{r^{\rho+1}}dr$$

converges or diverges.

The reasoning presented above implies that $\alpha(r)$ has minimal type if it belongs to the convergence class. Functions of normal and maximal type belong to the divergence class.

EXAMPLE. Let $\alpha(r) = r^\beta(\ln r)^\gamma$ for $r > 2$, $0 < \beta < \infty$, $-\infty < \gamma < \infty$. Then $\alpha(r)$ is of order β; $\alpha(r)$ is of maximal, normal, or minimal type depending on whether $\gamma > 0$, $\gamma = 0$, or $\gamma < 0$. The function $\alpha(r)$ belongs to the divergence class if $\gamma \geq -1$, and to the convergence class if $\gamma < -1$.

DEFINITION. We say that functions $\alpha_1(r), \alpha_2(r) \in \Lambda$ are of the same *growth category* if they have the same order and, if the order is finite, have the same type and either both belong to the convergence class, or both belong to the divergence class. We say that $\alpha_2(r)$ is of *higher growth category* than $\alpha_1(r)$ if one of the following conditions is satisfied:

1) $\rho[\alpha_2] > \rho[\alpha_1]$.
2) $\rho[\alpha_1] = \rho[\alpha_2] < \infty$, $\alpha_1(r)$ is of minimal type, and $\alpha_2(r)$ is of normal or maximal type.
3) $\rho[\alpha_1] = \rho[\alpha_2] < \infty$, $\alpha_1(r)$ is of normal type, and $\alpha_2(r)$ is of maximal type.
4) $\rho[\alpha_1] = \rho[\alpha_2] < \infty$, $\alpha_1(r)$ and $\alpha_2(r)$ are of minimal type, $\alpha_1(r)$ belongs to the convergence class, and $\alpha_2(r)$ belongs to the divergence class.

It is easy to see that

$$\liminf_{r\to\infty} \frac{\alpha_1(r)}{\alpha_2(r)} = 0$$

if $\alpha_2(r)$ is of higher growth category than $\alpha_1(r)$, but the condition $\alpha_1(r) = o(\alpha_2(r))$ does not have to be satisfied. Indeed, consider $\alpha_2(r) = r^{2+\sin(\ln^+ \ln^+ r)}$ and $\alpha_1(r) = r^2$. It is clear that $\alpha_1, \alpha_2 \in \Lambda$, $\rho[\alpha_2] = 3$, and $\rho[\alpha_1] = 2$. Therefore $\alpha_2(r)$ is of higher growth category than $\alpha_1(r)$. However, $\alpha_2(r_n) = o(\alpha_1(r_n))$ for

$$r_n = \exp\exp\left(\frac{3}{2}\pi + 2\pi n\right).$$

On the other hand, $\alpha_1(r)$ and $\alpha_2(r)$ may be of the same growth category, but $\alpha_1(r) = o(\alpha_2(r))$. It happens, for example, if $\alpha_1(r) = r\ln^+ r$ and $\alpha_2(r) = r(\ln^+ r)^2$.

It is easy to verify that the functions $\alpha(r) \in \Lambda$, $\alpha_1(r) = A\alpha(r)+B(A > 0)$,[1] and $\alpha_2(r) = \alpha(ar+b)$, where $a > 0$, are of the same growth category (here and later, we define functions $\alpha_j(r)$ for sufficiently large r; the values of $\alpha_j(r)$ on finite intervals do not affect the growth category). The magnitudes of the types of these functions are related in the following way: $\sigma[\alpha_1] = A\sigma[\alpha]$, $\sigma[\alpha_2] = a^\rho\sigma[\alpha]$. If $\alpha_3(r) = \alpha(r^\beta)$, $\beta > 0$, then $\rho[\alpha_3] = \beta\rho[\alpha]$ and $\sigma[\alpha_3] = \sigma[\alpha]$. We leave the verification of these statements to interested readers.

---

[1] If $\alpha(r) = O(1)$ and $B < 0$, we also assume that $\alpha_1(r) \in \Lambda$.

THEOREM 1.1. *Let $\alpha(r) \in \Lambda$, $\alpha(r) \to \infty$ $(r \to \infty)$, and*

$$\alpha_1(r) = C + \int_{r_0}^{r} \frac{\alpha(t)}{t} dt, \qquad r_0 > 0.$$

Then the functions $\alpha(r)$ and $\alpha_1(r)$ are of the same growth category.

PROOF. We have

$$\alpha_1(er) - C \geq \int_r^{er} \frac{\alpha(t)}{t} dt \geq \alpha(r) \int_r^{er} \frac{dt}{t} = \alpha(r). \tag{1.2}$$

Hence the growth category of $\alpha(r)$ cannot be higher than the growth category of $\alpha_1(r)$. On the other hand,

$$\alpha_1(r) \leq C + \alpha(r) \int_{r_0}^{r} d \ln t = C + \alpha(r) \ln \frac{r}{r_0},$$

$$\frac{\ln^+ \alpha_1(r)}{\ln r} \leq \frac{\ln^+ \alpha(r)}{\ln r} + \frac{\ln^+ \ln \frac{r}{r_0}}{\ln r} + O\left(\frac{1}{\ln r}\right) = \frac{\ln^+ \alpha(r)}{\ln r} + o(1);$$

hence $\rho[\alpha_1] \leq \rho[\alpha]$. Since the inverse inequality has already been verified, $\rho[\alpha_1] = \rho[\alpha] = \rho$. If $\rho = 0$, then both functions $\alpha(r)$ and $\alpha_1(r)$ are of maximal type and thus are of the same growth category. Now let $0 < \rho < \infty$. If $\sigma[\alpha] < \infty$, then for $r > r_1(\varepsilon)$ we have $\alpha(r) < (\sigma[\alpha] + \varepsilon) r^\rho$ and

$$\alpha_1(r) \leq O(1) + \int_{r_1}^{r} \frac{(\sigma[\alpha] + \varepsilon) t^\rho}{t} dt \leq O(1) + \frac{\sigma[\alpha] + \varepsilon}{\rho} r^\rho.$$

Therefore $\sigma[\alpha_1] \leq \rho^{-1} \sigma[\alpha]$. This inequality is trivial if $\sigma[\alpha] = \infty$. The inequality (1.2) implies that $\sigma[\alpha] \leq e^\rho \sigma[\alpha_1]$. Combining these inequalities, we conclude that $\alpha_1(r)$ and $\alpha(r)$ are of the same type. If $\alpha_1(r)$ belongs to the convergence class, then $\alpha(r)$ also belongs to the convergence class (see (1.2)). If $\alpha(r)$ belongs to the convergence class, then $\alpha(r)$ and thus also $\alpha_1(r)$ have minimal type. On the other hand,

$$\int_1^r \frac{\alpha(t)}{t^{\rho+1}} dt = \int_1^r \frac{d\alpha_1(t)}{t^\rho} = \frac{\alpha_1(r)}{r^\rho} - \alpha_1(1) + \rho \int_1^r \frac{\alpha_1(t)}{t^{\rho+1}} dt.$$

Letting $r \to \infty$, we get

$$\rho \int_1^\infty \frac{\alpha_1(t)}{t^{\rho+1}} dt = \alpha_1(1) + \int_1^\infty \frac{\alpha(t)}{t^{\rho+1}} dt < \infty. \tag{1.3}$$

Hence $\alpha_1(r)$ also belongs to the convergence class. □

COROLLARY. *Let $f(z)$ be a meromorphic function. Then*

1) *$T(r, f)$, $\overset{\circ}{T}(r, f)$, and $\overset{\circ}{A}(r, f)$ have the same growth category, except in the case where $\overset{\circ}{A}(r, f) = O(1)$ as $r \to \infty$ and $\overset{\circ}{A}(r, f)$ is not identically 0.*
2) *$N(r, a)$ and $n(r, a)$ have the same growth category, except in the case where $n(r, a) = O(1)$ and $n(r, a)$ is not identically 0. For any a, the category of growth of $N(r, a)$ cannot exceed the category of growth of $T(r, f)$.*

In the case where $\overset{\circ}{A}(r, f) = O(1)$ and $\overset{\circ}{A}(r, f)$ is not identically 0, it is clear that $\overset{\circ}{T}(r, f) \sim \text{const} \cdot \ln r$. In the case where $n(r, a) = O(1)$ and $n(r, a)$ is not identically

0, we have $N(r,a) \sim \text{const} \cdot \ln r$. Thus, in these cases the categories of $\overset{\circ}{A}(r,f)$ and $n(r,a)$ are lower than the categories of $\overset{\circ}{T}(r,f)$ and $N(r,a)$, respectively.

To prove the corollary, it is enough to use the definitions of $\overset{\circ}{T}(r,f)$ and $N(r,a)$. We write the definition of $N(r,a)$ in the following way. Let r_1 be the least absolute value of a non-zero a-point of $f(z)$, $r > r_1$. Then

$$\begin{aligned} N(r,a) &= \int_{r_1}^{r} \frac{n(t,a)-n(0,a)}{t}\,dt + n(0,a)\ln r \\ &= \int_{r_1}^{r} \frac{n(t,a)}{t}\,dt + n(0,a)\ln r_1. \end{aligned}$$

The first fundamental theorem implies that $N(r,a) \le T(r,f) + O(1)$; hence the growth category of $N(r,a)$ does not exceed the growth category of $T(r,f)$.

The proof of the following theorem is left to interested readers.

THEOREM 1.2. *Suppose $\alpha_1(r) \in \Lambda$, $\alpha_2(r) \in \Lambda$ are such that $\alpha_2(r)$ is not identically 0, and the growth category of $\alpha_1(r)$ does not exceed the growth category of $\alpha_2(r)$. Then the growth category of $\alpha(r) = \alpha_1(r) + \alpha_2(r) + O(1)$ coincides with the growth category of $\alpha_2(r)$.*

DEFINITION. The *order, lower order, type, magnitude of type class*, and the *growth category* of a meromorphic function f are defined as those of the characteristic $T(r,f)$.

THEOREM 1.3. *If $f(z)$ is an entire function, then it has the same growth category as $\ln M(r,f)$.*

PROOF. Letting $R = kr$, $k > 1$, in formula (7.1) of Chapter 1, we get $T(r,f) \le \ln^+ M(r,f) \le \frac{k+1}{k-1}T(kr,f)$. Since $T(r,f)$ and $\frac{k+1}{k-1}T(kr,f)$ have the same growth category, we get the desired result. □

REMARK. If $0 < \rho[f] = \rho[T(r,f)] = \rho < \infty$, then the equation (7.1) of Chapter 1 implies

$$\sigma[f] \le \sigma[\ln M(r,f)] \le \frac{k+1}{k-1}k^{\rho}\sigma[f].$$

If $k > 1$, the minimum of $\frac{k+1}{k-1}k^{\rho}$ is attained at $k = k(\rho) = \frac{1}{\rho}(1 + \sqrt{1+\rho^2})$. The estimate

$$\sigma[\ln M(r,f)] \le \frac{k(\rho)+1}{k(\rho)-1}[k(\rho)]^{\rho}\sigma[f]$$

is not sharp for the class of entire functions. The following estimate holds:

$$\sigma[\ln M(r,f)] \le \begin{cases} \pi\rho\csc\pi\rho\sigma[f] & \text{if } 0 \le \rho \le \frac{1}{2}, \\ \pi\rho\sigma[f] & \text{if } \frac{1}{2} \le \rho < \infty. \end{cases}$$

Proof of this estimate will be given in Appendix A, where we also show that this estimate is sharp.

Relations obtained in Section 6 of Chapter 1 imply several results on growth of meromorphic functions.

THEOREM 1.4. *Let $f(z)$ be a meromorphic function, $f_1(z) = R_1[f(z)]$, and $f_2(z) = R_2[z, f(z)]$, where $R_1(u)$ is a rational function of degree d, and $R_2(u,v)$ is a rational function of u and a polynomial of degree p in v. Then $f(z)$, $f_1(z)$, and $f_2(z)$ are of the same growth category, and the magnitudes of their types are related by the equalities $\sigma[f_1] = d\sigma[f]$ and $\sigma[f_2] = p\sigma[f]$.*

THEOREM 1.5. *Let $f(z)$ be a meromorphic function, and let $f_1(z) = f(z^n)$, where n is a positive integer. Then $\rho[f_1] = n\rho[f]$, $\sigma[f_1] = \sigma[f]$, and the functions $f(z)$ and $f_1(z)$ are in the same class.*

THEOREM 1.6. *Let $f(z)$ be a meromorphic function, and let $f_1(z) = f(az+b)$, $a \neq 0$. Then $f(z)$ and $f_1(z)$ as well as $N(r,f)$ and $N(r,f_1)$ are of the same growth category; furthermore $\sigma[f_1] = |a|^\rho \sigma[f]$ and $\rho = \rho[f] = \rho[f_1]$.*

PROOF. It suffices to prove Theorem 1.6 in the cases where either $b = 0$ or $a = 1$. In the former case the assertion follows from Theorem 6.7 of Chapter 1. In the latter case, the obvious (for $r > |b|$) inequalities

$$\mathring{A}(r-|b|, f(z)) \leq \mathring{A}(r, f(z+b)) \leq \mathring{A}(r+|b|, f(z)),$$
$$n(r-|b|, f(z)) \leq n(r, f(z+b)) \leq n(r+|b|, f(z)),$$

and the corollary of Theorem 1.1 imply that $f(z)$ and $f(z+b)$ (as well as $N(r, f(z))$ and $N(r, f(z+b))$) have the same growth category.

Let $0 < \rho < \infty$. We show that $\sigma[f(z)] = \sigma[f(z+b)]$. It is clear that $\mathring{A}(r, f(z))$ and $\mathring{A}(r, f(z+b))$ tend to ∞ as $r \to \infty$ (otherwise we would have $\mathring{T}(r, f(z)) = O(\ln r)$ and $\mathring{T}(r, f(z+b)) = O(\ln r)$). Let $r > 2|b|$. Then

$$\begin{aligned}\mathring{T}(r, f(z+b)) &= \int_0^r \frac{\mathring{A}(t, f(z+b))}{t}dt \\ &\geq \int_{2|b|}^r \frac{\mathring{A}(t, f(z+b))}{t}dt \geq \int_{2|b|}^r \frac{\mathring{A}(t-|b|, f(z))}{t}dt \\ &= \int_{|b|}^{r-|b|} \frac{\mathring{A}(t, f(z))}{t+|b|}dt = (1+o(1)) \int_0^{r-|b|} \frac{\mathring{A}(t, f(z))}{t}dt \\ &= (1+o(1))\mathring{T}(r-|b|, f)\end{aligned}$$

(we used l'Hôpital's rule). We get

$$\mathring{T}(r, f(z)) = \mathring{T}(r, f[(z+b)-b]) \geq (1+o(1))\mathring{T}(r-|b|, f(z+b)).$$

Hence

$$\begin{aligned}\mathring{T}(r, f(z+b)) &\geq (1+o(1))\mathring{T}(r-|b|, f(z)) \\ &\geq (1+o(1))\mathring{T}(r-2|b|, f(z+b)).\end{aligned} \tag{1.4}$$

Therefore $\sigma[f(z+b)] = \sigma[f(z)]$. □

REMARK. The following inequality (for an arbitrary a) can be proved in the same way as (1.4):

$$\begin{aligned}N(r, a, f(z+b)) &\geq (1+o(1))N(r-|b|, a, f(z)) \\ &\geq (1+o(1))N(r-2|b|, a, f(z+b)).\end{aligned} \tag{1.5}$$

THEOREM 1.7. *Let $f_1(z)$ and $f_2(z)$ be meromorphic functions such that the growth category of $f_1(z)$ does not exceed the growth category of $f_2(z)$. Then the growth categories of $f_1(z) \pm f_2(z)$, $f_1(z)f_2(z)$, $f_1(z)/f_2(z)$, and $f_2(z)/f_1(z)$ do not exceed the growth category of $f_2(z)$. Furthermore, they coincide with the growth category of $f_2(z)$ if $f_1(z)$ is not identically 0 and the growth category of $f_1(z)$ is lower than the growth category of $f_2(z)$.*

PROOF. Since

$$T(r, f_1 \pm f_2) \leq T(r, f_1) + T(r, f_2) + O(1),$$
$$T(r, f_1 f_2) \leq T(r, f_1) + T(r, f_2) + O(1),$$
$$T\left(r, \frac{f_1}{f_2}\right) \leq T(r, f_1) + T(r, f_2) + O(1),$$

Theorem 1.2 implies the first statement. We prove the second statement for the product $f_1(z)f_2(z)$ only (proofs for $f_1 \pm f_2$, f_1/f_2, and f_2/f_1 are similar). Assume that the growth category of $f_1(z)f_2(z)$ is lower than the growth category of $f_2(z)$. Since $f_2(z) = \{f_1(z)f_2(z)\}/f_1(z)$, by the first part of the theorem, the growth category of $f_2(z)$ does not exceed the maximum of the growth categories of $f_1(z)f_2(z)$ and $f_1(z)$, which are lower than the growth category of $f_2(z)$. We get a contradiction. □

Let $\{r_\nu\}$ be a non-decreasing sequence of positive numbers tending to ∞. Denote by $\nu(r)$ the number of the members of this sequence satisfying $0 < r_\nu \leq r$. Let

$$\nu_1(r) = \int_0^r \frac{\nu(t)}{t}\, dt.$$

It is clear that $\nu(r) \in \Lambda$, $\nu_1(r) \in \Lambda$, and by Theorem 1.1, these functions have the same growth category.

THEOREM 1.8. *Let $\nu(r)$ be of order ρ, $0 < \rho < \infty$. Then the series*

$$\sum_{\nu=1}^{\infty} \left(\frac{1}{r_\nu}\right)^{\mu} \tag{1.6}$$

converges if $\mu > \rho$, and diverges if $\mu < \rho$. For $\mu = \rho$, the series converges or diverges depending on whether $\nu(r)$ belongs to the convergence class or to the divergence class.

PROOF. Obviously (see above) it suffices to show that the series (1.6) and the integral

$$\int_{r_1}^{\infty} \frac{\nu(r)}{r^{1+\mu}}\, dr \tag{1.7}$$

are either both convergent of both divergent. For $0 < r_0 < r_1$ we have

$$\sum_{r_\nu \leq r} \left(\frac{1}{r_\nu}\right)^{\mu} = \int_{r_0}^{r} \frac{d\nu(t)}{t^\mu} = \frac{\nu(r)}{r^\mu} + \mu \int_{r_0}^{r} \frac{\nu(t)}{t^{\mu+1}}\, dt. \tag{1.8}$$

If the integral (1.7) diverges, then the inequality

$$\sum_{r_\nu \leq r} \left(\frac{1}{r_\nu}\right)^{\mu} \geq \mu \int_{r_0}^{r} \frac{\nu(t)}{t^{\mu+1}}\, dt$$

implies that the series (1.6) also diverges. If the integral (1.7) converges, then, as we have seen, $\nu(r) = o(r^\mu)$; therefore, letting $r \to \infty$ in (1.8), we get

$$\sum_{\nu=1}^{\infty} \left(\frac{1}{r_\nu}\right)^\mu = \mu \int_{r_0}^{\infty} \frac{\nu(t)}{t^{\mu+1}} dt. \qquad \square$$

COROLLARY. *Let $f(z)$ be a meromorphic function, and the $z_\nu(a)$ be its non-zero a-points. Let $\rho = \rho[f]$ and $\rho_1 = \rho[n(r,a)]$. Then the series*

$$\sum_\nu \frac{1}{|z_\nu(a)|^\mu} \tag{1.9}$$

converges for $\mu > \rho_1$, and hence for $\mu > \rho$. The series (1.9) converges if $\mu = \rho_1$ and $n(r,a)$ belongs to the convergence class. If $\mu < \rho_1$, the series (1.9) diverges.

DEFINITION. Let $f(z)$ be a transcendental meromorphic function. A complex number a is called a *Picard exceptional value* of the function $f(z)$ if the number of its a-points in $\{|z| < \infty\}$ is finite, that is, if $N(r,a) = O(\ln r)$. A complex number a is called a *Borel exceptional value* if the growth category of $n(r,a)$ is lower than the growth category of $f(z)$.

In the cases where the order of $f(z)$ is non-zero, or where $n(r,a) \to \infty$ as $r \to \infty$, we may replace $n(r,a)$ by $N(r,a)$ in the definition of a Borel exceptional value.

It is clear that a Picard exceptional value is a Borel exceptional value. The converse is not true.

EXAMPLE: $f(z) = e^{z^2} \sin z$. The function e^{z^2} has order 2, and the function $\sin z$ has order 1. Hence $f(z)$ has order 2. On the other hand, $n(r,0) = 1 + 2[r/\pi]$. Hence, the order of $n(r,0)$ is 1. Thus 0 is a Borel exceptional value, but it is not a Picard exceptional value.

2. Proximate order

A function $\rho(r)$ defined on $[0,\infty)$ is called a *proximate order* if it satisfies the conditions

1) $\rho(r) \geq 0$;
2) $\lim_{r\to\infty} \rho(r) = \rho, \ 0 \leq \rho < \infty$;
3) $\rho(r)$ is continuously differentiable on $[0,\infty)$;
4) $\lim_{r\to\infty} r\rho'(r) \ln r = 0$.

Proximate orders $\rho_1(r)$ and $\rho_2(r)$ are called *equivalent* if $\rho_1(r)\rho_2(r) = o(\ln^{-1} r)$. Let $\alpha(r) \in \Lambda$ be of finite order $\rho[\alpha]$. A proximate order $\rho(r)$ is called a *proximate order of the function* $\alpha(r)$ if $0 < \sigma^*[\alpha] < \infty$, where

$$\sigma^*[\alpha] = \limsup_{r\to\infty} \frac{\alpha(r)}{r^{\rho(r)}}.$$

The number $\sigma^*[\alpha]$ is called the *magnitude of type* of $\alpha(r)$ with respect to the proximate order $\rho(r)$. It is clear that if $\rho(r)$ is a proximate order of $\alpha(r)$, then $\rho(r) \to \rho = \rho[\alpha]$ as $r \to \infty$. The magnitude of type $\sigma^*[\alpha]$ does not change if the proximate order $\rho_1(r)$ is replaced by a proximate order $\rho_2(r)$, since $r^{\rho_1(r)} \sim r^{\rho_2(r)}$. Using a proximate order $\rho(r)$ and the magnitude of type $\sigma^*[\alpha]$, we can introduce the notions of type, convergence class, and growth category of $\alpha(r)$ with respect to $\rho(r)$. It can be verified that all theorems of the previous section (except for

Theorem 1.1 in the case $\rho = 0$) remain true if we understand the category of $\alpha(r)$ as the category with respect to a proximate order $\rho(r)$. We are not going to use this.

If a function $\alpha(r) \in \Lambda$ has finite lower order $\lambda[\alpha]$, then its *lower proximate order* is defined as a proximate order $\lambda(r)$ such that $0 < \sigma_*[\alpha] < \infty$, where

$$\sigma_*[\alpha] = \liminf_{r\to\infty} \frac{\alpha(r)}{r^{\lambda(r)}}.$$

It is easy to see, that if $\lambda(r)$ is a lower proximate order of $\alpha(r)$, then $\lambda(r) \to \lambda = \lambda[\alpha]$.

Applications of proximate orders are based on the following theorem.

THEOREM 2.1. *Each positive continuous function $\alpha(r) \in \Lambda$ has a proximate order. Each positive continuous function $\alpha(r) \in \Lambda$ of finite lower order has a lower proximate order.*

PROOF. As is easy to see, it suffices to find $\rho(r)$ and $\lambda(r)$ for which condition 3) in the definition of a proximate order is replaced by the following weaker condition: $\rho(r)$ and $\lambda(r)$ are continuous on $[0, \infty)$ and continuously differentiable everywhere on $[0, \infty)$ except for a set of isolated points at which they have one-sided derivatives. In fact, in sufficiently small neighborhoods of the points where the derivatives do not exist, the functions $\rho(r)$ and $\lambda(r)$ can be corrected by at most $o\{(\ln r)^{-1}\}$ in such a way that the resulting functions be continuously differentiable everywhere, and still satisfy conditions 1), 2), and 4).

First we prove the existence of a proximate order $\rho(r)$. Suppose that $\alpha(r)$ is defined on $[0, \infty)$ and $\alpha(r) \to \infty$ as $r \to \infty$. (If $\alpha(r) = O(1)$, we put $\rho(r) \equiv 0$.)

Let ρ be the order of $\alpha(r)$. Denote by $d(r)$ the function

$$d(r) = \frac{\ln^+ \alpha(r)}{\ln r}.$$

Then

$$\limsup_{r\to\infty} d(r) = \rho.$$

There are two possibilities:

A) there exists a sequence $r_k \to \infty$, $1 \le k \to \infty$, such that $d(r_k) > \rho$;

B) $d(r) \le \rho$ for all $r \ge r_0 > e$.

Assume that A) is satisfied. Let $\varphi(r) = \max_{x\ge r} d(x)$. It is clear that $\varphi(r)$ is well defined and is a non-increasing, positive function, and $\varphi(r) \to \rho$ as $r \to \infty$. It is also clear that the set M of those r for which $\varphi(r) = d(r)$ is unbounded.

Let $r_1 > e^{e^e}$ and $r_1 \in M$. Set $\rho(r) = \varphi(r_1)$ for $0 \le r \le r_1$. Denote by t_1 the least integer exceeding $1 + r_1$ and such that $\varphi(r_1) > \varphi(t_1)$. Set $\rho(r) = \varphi(r_1)$ for $r_1 < r \le t_1$. Let u_1 be the x-coordinate of the first point of intersection of the curves $y = \rho(r_1) - \ln_3 x + \ln_3 t_1$ and $y = \varphi(x)$ for $x > t_1$.[2] Set $\rho(r) = \rho(r_1) - \ln_3 r + \ln_3 t_1$ for $t_1 \le r \le u_1$. It is clear that on this interval, $\rho(r) \ge \varphi(r)$, and the equality holds at $r = u_1$ only. Let $r_2 = \min(M \cap \{r \ge u_1\})$. If $r_2 > u_1$, set $\rho(r) = \varphi(r)$ for $u_1 \le r \le r_2$. Since $\varphi(r)$ is constant for $u_1 \le r \le r_2$, it follows that $\rho(r)$ is also constant on this interval. We repeat the same construction with r_1 replaced

[2]We use the notation $\ln_j x$ for the j-th iteration of the logarithm ($j \ge 0$): $\ln_0 x = x$, $\ln_j x = \ln(\ln_{j-1} x)$. We use the notation $\exp_j x$ for the j-th iteration of the exponential function: $\exp_0 x = x$, $\exp_j x = \exp(\exp_{j-1} x)$.

by r_2. Continuing in such a manner, we define $\rho(r)$ on $[0,\infty)$, since $r_n - r_{n-1} \geq u_{n-1} - r_{n-1} \geq t_{n-1} - r_{n-1} \geq 1$ and $r_n \to \infty$. The function $\rho(r)$ is continuous and has a continuous derivative everywhere except for the points t_n and u_n, $n \geq 1$, where it has one-sided derivatives. Since

$$\rho'(r) = 0 \quad \text{or} \quad \rho'(r) = -\frac{1}{r \ln r \ln_2 r},$$

condition 4) is satisfied. In view of our construction, $\rho(r) \geq \varphi(r) \geq d(r)$ if $r \geq r_1$, $\rho(r_n) = \varphi(r_n) = d(r_n)$, $n \geq 1$, and $\rho(r)$ is a non-increasing function. Hence

$$\lim_{r\to\infty} \rho(r) = \lim_{r\to\infty} \varphi(r) = \rho,$$

and condition 2) is satisfied. Furthermore,

$$\alpha(r) \leq r^{d(r)} \leq r^{\rho(r)} \quad \text{and} \quad \alpha(r_n) = r_n^{d(r_n)} = r_n^{\rho(r_n)}$$

if $r \geq r_1$. Therefore

$$\sigma^*[\alpha] = \limsup_{r\to\infty} \frac{\alpha(r)}{r^{\rho(r)}} = 1.$$

Now we consider case B). If $d(r_n) = \rho$ for a sequence $\{r_n\}$ satisfying $r_n \to \infty$, then we may set $\rho(r) \equiv \rho$. Now, suppose $0 < d(r) < \rho$ for all $r \geq r_0 > \exp_3 1$. Set $\psi(r) = \max_{r_0 \leq x \leq r} d(x)$. It is clear that $\psi(r)$ is a continuous non-decreasing function, and the set L of those $r \geq r_0$ for which $d(r) = \psi(r)$ is unbounded. Suppose $r_1 > r_0$ and let s_1 be the largest x-coordinate of the intersection points of the curves $y = \rho + \ln_3 x - \ln_3 r_1$ and $y = \psi(x)$. It is clear that if r_1 is sufficiently large, then there exists an intersection point such that $r_0 < s_1 < r_1$ and $L \cap [r_0, s_1] \neq \emptyset$. Let $t_1 = \max\{L \cap [r_0, s_1]\}$. Set $\rho(r) \equiv \psi(t_1) = d(t_1)$ for $0 \leq r \leq s_1$. Choose $r_2 > r_1 + 1$ to be so large that the largest x-coordinate of intersection points of the curves $y = \rho + \ln_3 x - \ln_3 r_2$ and $y = \psi(x)$, denoted by s_2, is greater than r_1, and $L \cap [r_1, s_2] \neq \emptyset$. Let $t_2 = \max\{L \cap [r_1, s_2]\}$. Denote by u_1, $s_1 \leq u_1 \leq r_1$, the point at which $\rho + \ln_3 x - \ln_3 r_1 = \psi(t_2)$. Set $\rho(r) = \rho + \ln_3 r - \ln_3 r_1$ for $s_1 \leq r \leq u_1$ and $\rho(r) \equiv \psi(t_2) = d(t_2)$ for $u_1 \leq r \leq s_2$. We continue in an obvious way. Since $r_n > r_{n-1} + 1$, the function $\rho(r)$ is defined for all $r \geq 0$. It is clear from the construction that $\rho(r)$ is continuous for $r \geq 0$, and the inequality $\rho \geq \rho(r) \geq \psi(r) \geq d(r)$ is satisfied; furthermore $\rho(t_n) = \psi(t_n) = d(t_n)$, $t_n \to \infty$. Therefore

$$\sigma^*[\alpha] = \limsup_{r\to\infty} \frac{\alpha(r)}{r^{\rho(r)}} = 1.$$

Since $\psi(r) \to \rho$ as $r \to \infty$, we have $\rho(r) \to \rho$. Finally, the derivatives $\rho'(r)$ exist everywhere except for the points s_n, u_n, and are equal to either 0 or $(r \ln r \ln_2 r)^{-1}$, and thus condition 4) is satisfied.

The desired function is constructed in all cases.

To construct a lower proximate order $\lambda(r)$, we introduce the function $d_1(r)$ by

$$d_1(r) = \frac{\ln r}{\ln r + \ln^+ \alpha(r)}.$$

Then

$$\limsup_{r\to\infty} d_1(r) = \frac{1}{1+\lambda},$$

and using the same argument as above, we can construct a proximate order $\rho_1(r) \to (1+\lambda)^{-1}$ such that $0 < \rho_1(r) < 1$ and $d_1(r) \leq \rho_1(r)$ for all r, $r_0 \leq r < \infty$, and, furthermore, $d_1(r_n) = \rho_1(r_n)$ for a sequence $\{r_n\}$ satisfying $\lim_{n\to\infty} r_n = \infty$.

Set $\lambda(r) = \frac{1}{\rho_1(r)} - 1$. It is easy to see that $\lambda(r)$ is a proximate order, $\lambda(r) \to \lambda = \lambda[\alpha]$, and, moreover, $\alpha(r) \geq r^{\lambda(r)}$ for all r satisfying $r_0 \leq r < \infty$, and $\alpha(r_n) = r_n^{\lambda(r_n)}$, $n = 1, 2, \ldots$. Hence $\lambda(r)$ is a lower proximate order of $\alpha(r)$. □

Observe that the proximate order $\rho(r)$ constructed in the proof is a monotone function, $\alpha(r) \leq r^{\rho(r)}$ for $r \geq r_0$, and there exists a sequence $\{r_k\}$, $r_k \to \infty$, such that $\alpha(r) = r^{\rho(r)}$ for $r = r_k$.

A proximate order $\rho(r)$ is convenient because the function $r^{\rho(r)}$ has many properties of the power function r^ρ.

DEFINITION. A positive continuous function $L(r)$, defined for $r \geq r_0$, is called *slowly varying* if

$$\lim_{r\to\infty} \frac{L(kr)}{L(r)} = 1$$

uniformly on each interval $0 < a \leq k \leq b < \infty$.

THEOREM 2.2. *If $\rho(r)$ is a proximate order, then $L(r) = r^{\rho(r)-\rho}$ is a slowly varying function.*

PROOF. Assume, without loss of generality, that $0 < a < 1$ and $b = 1/a$. Then

$$\ln \frac{L(kr)}{L(r)} = [\rho(kr) - \rho] \ln k + [\rho(kr) - \rho(r)] \ln r.$$

For each $\varepsilon > 0$ there exists r_0 such that for $ar > r_0 > 1$, the conditions

$$|\rho(r) - \rho| < \varepsilon \quad \text{and} \quad |\rho'(r)| r \ln r < \varepsilon$$

hold. By Lagrange's mean value theorem, applied to $\rho(e^t)$, $t = \ln r$ (for $r > r_0$), there exists $k^* = k^*(r)$, $a \leq k^* \leq b$, such that

$$|\rho(kr) - \rho(r)| = k^* r |\rho'(k^* r)| |\ln k| < \varepsilon \frac{\ln b}{\ln(k^* r)} < \frac{\varepsilon \ln b}{\ln(ar)}.$$

Then

$$\left| \ln \frac{L(kr)}{L(r)} \right| \leq \varepsilon \ln b + \varepsilon \ln b \frac{\ln r}{\ln(ar)} \leq \varepsilon \ln b \left(1 + \frac{\ln r_0}{\ln(ar_0)}\right).$$

Hence $\ln(L(kr)/L(r))$ converges uniformly to 0, and $L(kr)L(r))$ converges uniformly to 1. □

COROLLARY. *If $\rho(r)$ is a proximate order and $0 < a < b < \infty$, then for each $\varepsilon > 0$ there exists r_0 such that for all $r > r_0$ and $a \leq k \leq b$, the inequality*

$$(1 - \varepsilon) k^\rho r^{\rho(r)} < (kr)^{\rho(kr)} < (1 + \varepsilon) k^\rho r^{\rho(r)}$$

holds.

In fact, if $L(r)$ is the function from Theorem 2.2, then

$$\frac{L(kr)}{L(r)} = \frac{(kr)^{\rho(kr)}}{k^\rho r^{\rho(r)}}.$$

THEOREM 2.3. *If $\rho(r)$ is a proximate order and $\rho > 0$, then the function $V(r) = r^{\rho(r)}$ is increasing for sufficiently large r.*

In fact,

$$V'(r) = r^{\rho(r)-1}\{\rho(r) + \rho'(r) r \ln r\}.$$

The function in braces approaches $\rho > 0$ as $r \to \infty$; hence, for sufficiently large r, we have $V'(r) > 0$.

THEOREM 2.4. *Let $\rho(r)$ be a proximate order, $\rho(r) \to \rho$ as $r \to \infty$. Then*

$$\int_1^r t^{\rho(t)-q} dt \sim \frac{r^{\rho(r)+1-q}}{\rho + 1 - q}$$

if $q < \rho + 1$, and

$$\int_r^\infty t^{\rho(t)-q} dt \sim \frac{r^{\rho(r)+1-q}}{q - \rho - 1}$$

if $q > \rho + 1$.

The statements can be easily verified using l'Hôpital's rule.

EXERCISE 1. Let $0 \le p < \rho < p+1$, and let $\rho(r)$ be a proximate order satisfying $\rho(r) \to \rho$. Prove that

$$\lim_{r\to\infty} \int_{c/r}^\infty \frac{(r\tau)^{\rho(r\tau)}}{r^{\rho(r)}} \frac{d\tau}{\tau^{p+1}(1+\tau)} = \int_0^\infty \frac{\tau^{\rho-p-1}}{1+\tau} d\tau = \frac{\pi}{\sin \pi(\rho - p)};$$

that is, it is possible to pass to the limit under the integral sign.

HINT. Split the interval $[c/r, \infty)$ into three subintervals: $[c/r, \varepsilon]$, $[\varepsilon, \varepsilon^{-1}]$, and $[\varepsilon^{-1}, \infty)$, where $\varepsilon > 0$ is a small positive number. The first integral is estimated as

$$\int_{c/r}^{\varepsilon} \frac{(r\tau)^{\rho(r\tau)}}{r^{\rho(r)}} \frac{d\tau}{\tau^{p+1}(1+\tau)} \le r^{p-\rho(r)} \int_c^{\varepsilon r} t^{\rho(t)-p-1} dt \underset{r\to\infty}{\longrightarrow} \frac{\varepsilon^{\rho-p}}{\rho - p}.$$

The last relation can be obtained by combining Theorem 2.4 and Theorem 2.2 (instead of Theorem 2.4, one can use l'Hôpital's rule). The third integral can be estimated in a similar way. To estimate the second integral, one can use the corollary of Theorem 2.2.

EXERCISE 2. Let $\rho(r)$ be a proximate order satisfying $\rho(r) \to \rho$. Then for each fixed $k > 0$ we have

$$\lim_{R\to\infty} \int_1^{kR} r^{\rho(r)-\rho(R)-1} dr = \infty.$$

HINT. Let $L(r) = r^{\rho(r)-\rho}$ and fix $0 < \varepsilon < k$. Then

$$\int_1^{kR} r^{\rho(r)-\rho(R)-1} dr = \int_1^{kR} \frac{L(r)}{L(R)} \left(\frac{R}{r}\right)^{\rho(R)-\rho} \frac{dr}{r}$$

$$\ge \int_{\varepsilon R}^{kR} \frac{L(r)}{L(R)} \left(\frac{R}{r}\right)^{\rho(R)-\rho} \frac{dr}{r} \ge \left(\min_{\varepsilon R \le r \le kR} \frac{L(r)}{L(R)}\right) \varepsilon^{|\rho(R)-\rho|} \ln \frac{k}{\varepsilon}.$$

Applying Theorem 2.2 we get

$$\liminf_{R\to\infty} \int_1^{kR} r^{\rho(r)-\rho(R)-1} dr \ge \ln \frac{k}{\varepsilon}.$$

It remains to use the possibility of choosing ε arbitrarily small.

3. Infinite products

We assume that the reader is familiar with the basic properties of infinite products of analytic functions (see, for example, Privalov [**Pri56**, Chapter IX]). It will be convenient for us to deviate from the standard definitions and to consider an infinite product $\prod_{k=1}^{\infty} f_k(z)$ to be *convergent at* z_0 if for some N there exists a finite non-zero limit

$$\lim_{n\to\infty} \prod_{k=N}^{n} f_k(z_0).$$

We consider $\prod_{k=1}^{n} f_k(z)$ to be *uniformly convergent* on a set M if for some N the products $\prod_{k=N}^{n} f_k(z)$ converge uniformly on $z \in M$ to some function $R_N(z) \neq 0, \infty$ as $n \to \infty$.

In a similar way, we modify the standard definition of the absolute convergence of an infinite product.

This modification of the standard definitions is convenient for us, because, with this modification, the fact that some terms of the product are equal to 0 or ∞ at some points does not affect the convergence of the product and the type of this convergence.

The following theorem is presented in many texts.

THEOREM 3.1. *Let $f_k(z)$, $k = 1, 2, \ldots$, be analytic functions in a domain G and $|f_k(z)| \leq a_k$, $k \geq k_0 \geq 1$ for $z \in G$. If the series $\sum_k a_k$ is convergent, then the product $\prod_{k=1}^{\infty}[1 + f_k(z)]$ converges uniformly and absolutely on G to an analytic function.*

Theorem 3.1 implies the following result.

THEOREM 3.2. *Let a_k, $k = 1, 2, \ldots$, be a sequence of complex numbers satisfying $0 < |a_1| \leq |a_2| \leq \cdots$ and*

$$\sum_{k=1}^{\infty} \frac{1}{|a_k|^q} < \infty, \tag{3.1}$$

where q is a positive integer. Suppose that an entire function $g(z)$ has the power series representation of the form

$$g(z) = 1 + c_q z^q + c_{q+1} z^{q+1} + \cdots . \tag{3.2}$$

Then the product

$$f(z) = \prod_{k=1}^{\infty} g\left(\frac{z}{a_k}\right) \tag{3.3}$$

converges absolutely and uniformly on each bounded disc to an entire function $f(z)$.

PROOF. By (3.2), $|g(z) - 1| \leq A|z|^q$ for $|z| < r_0$, where A is some constant. Consider a disc $\{|z| \leq R\}$ of an arbitrarily large radius R. The inequality $R < |a_k| r_0$ holds for $k \geq k_0 = k_0(R)$. We have

$$\prod_{k=k_0}^{\infty} g\left(\frac{z}{a_k}\right) = \prod_{k=k_0}^{\infty} \left\{1 + \left[g\left(\frac{z}{a_k}\right) - 1\right]\right\}.$$

Since for $k \geq k_0$ and $|z| \geq R$ we have

$$\left|g\left(\frac{z}{a_k}\right) - 1\right| \leq A\left|\frac{z}{a_k}\right|^q \leq AR^q \frac{1}{|a_k|^q},$$

by (3.1) and Theorem 3.1, the product (3.3) converges absolutely and uniformly in $\{|z| < R\}$ to an analytic function $f(z)$. □

EXAMPLE 1. Denote by $E(z,p)$ the entire function given by

$$E(z,p) = \begin{cases} 1 - z & \text{if } p = 0, \\ (1-z)e^{z+\frac{z^2}{2}+\cdots+\frac{z^p}{p}} & \text{if } p \geq 1. \end{cases}$$

The function $E(z,p)$ is called a *Weierstrass primary factor.* It is clear that $E(z,p) - 1$ has a zero of order $(p+1)$ at $z = 0$. If $p \geq 1$, we have

$$E'(z,p) = -z^p \exp\left(z + \frac{z^2}{2} + \cdots + \frac{z^p}{p}\right).$$

Thus, if the sequence $\{a_k\}$ is such that the series (3.1) converges for $q = p+1$, then, by Theorem 3.2, the product

$$f(z) = \prod_{k=1}^{\infty} E\left(\frac{z}{a_k}, p\right) \tag{3.4}$$

converges absolutely and uniformly on each bounded disc, and hence it is an entire function, called the *Weierstrass canonical product of genus p.* It is clear that $f(z)$ has zeroes at the points a_k, and only there.

EXAMPLE 2. Let $g(z)$ be an arbitrary entire function satisfying $g(0) = 1$. Suppose that in a neighborhood of $z = 0$ the function $1/g$ admits a representation of the form $1/g(z) = 1 + c_1 z + c_2 z^2 + \cdots$. Let $P_{n-1}(z) = 1 + c_1 z + \cdots + c_{n-1} z^{n-1}$, $R_n(z) = \sum_{j=n}^{\infty} c_j z^j$. Then the entire function

$$P_{n-1}(z)g(z) = \left(\frac{1}{g(z)} - R_n(z)\right) g(z) = 1 - R_n(z)g(z) = 1 - c_n z^n + \cdots$$

satisfies the hypothesis of Theorem 3.2 with $q = n$.

EXAMPLE 3. If a sequence $\{a_k\}$ is such that the series (3.1) is convergent with $q = 2$, then the product

$$\prod_{k=1}^{\infty} \cos \frac{z}{a_k}$$

converges absolutely and uniformly in each bounded disc, and represents an entire function.

Suppose that $g(z)$ and $f(z)$ satisfy the hypothesis of Theorem 3.2. Our next purpose is to find an upper estimate for $\ln M(r,f)$ provided an upper estimate for $\ln M(r,g)$ is given.

THEOREM 3.3. *Suppose that the hypothesis of Theorem 3.2 is satisfied and* $\ln M(r,g) = O(r^{\kappa})$ *as* $r \to \infty$, *where* $0 < \kappa < q$. *Denote by* $\nu(r)$ *the number of points* a_k *in the disc* $\{|z| \leq r\}$. *Then there exists a constant* C, *depending on* $g(z)$, *but not on the sequence* $\{a_k\}$, *such that*

$$\ln M(r,f) < C \int_0^{\infty} \frac{\nu(r\tau)}{\tau^{1+\kappa}} \frac{d\tau}{1+\tau^{q-\kappa}}, \qquad 0 < r < \infty.$$

PROOF. By C with different indices we denote constants which depend on $g(z)$ only. Since for $r \to 0$ we have $\ln M(r,g) = O(r^q)$, for all r, $0 < r < \infty$, the following condition holds:

$$\ln M(r, g(z)) \le C_1 \Psi(r), \quad \text{where} \quad \Psi(r) = \frac{r^q}{1 + r^{q-\kappa}}.$$

By (3.3), we get

$$\begin{aligned} \ln M(r,f) &\le \sum_{k=1}^{\infty} \ln M\left(r, g\left(\frac{z}{a_k}\right)\right) \le C_1 \sum_{k=1}^{\infty} \Psi\left(\frac{r}{|a_k|}\right) \\ &= C_1 \int_0^{\infty} \Psi\left(\frac{r}{t}\right) d\nu(t) = C_1 \Phi(r). \end{aligned}$$

Convergence of the series (3.1) implies that $\nu(r) = o(r^q)$ as $r \to \infty$. In addition, $\nu(r) = 0$ for $0 \le r < |a_1|$. Taking this into account and integrating by parts we get

$$\Phi(r) = \int_0^{\infty} \nu(t) \Psi'\left(\frac{r}{t}\right) \frac{r\,dt}{t^2} = \int_0^{\infty} \nu(r\tau) \Psi'\left(\frac{1}{\tau}\right) \frac{d\tau}{\tau^2}.$$

It is easy to check that

$$\frac{\kappa \tau^{1-\kappa}}{1 + \tau^{q-\kappa}} < \Psi'\left(\frac{1}{\tau}\right) = \tau^{1-\kappa} \frac{q\tau^{q-\kappa} + \kappa}{(1 + \tau^{q-\kappa})^2} < \frac{q\tau^{1-\kappa}}{1 + \tau^{q-\kappa}}.$$

This immediately implies the conclusion of the theorem with $C = C_1 q$.

We will need also the following inequality:

$$\Phi(r) \ge C_1 \kappa \int_0^{\infty} \frac{\nu(r\tau)}{\tau^{1+\kappa}} \frac{d\tau}{1 + \tau^{q-\kappa}}, \quad 0 < r < \infty.$$

COROLLARY. *If $f(z)$ is a Weierstrass canonical product (3.4) of genus p, then*

$$\ln M(r,f) < C(p) \int_0^{\infty} \frac{n(r\tau, 0)}{\tau^{1+p}} \frac{d\tau}{1+\tau} = C(p) r^{p+1} \int_0^{\infty} \frac{n(t,0)\,dt}{t^{p+1}(r+t)}. \tag{3.5}$$

In fact, if $p \ge 1$, we can apply Theorem 3.3 with $g(z) = E(z,p)$, $\kappa = p$, $q = p+1$. If $p = 0$, we use the same argument as above with $\Psi(r) = \ln(1+r)$. We get

$$\ln M(r,f) \le C_1 \int_0^{\infty} n(r\tau, 0) \Psi'\left(\frac{1}{\tau}\right) \frac{d\tau}{\tau^2}.$$

But in this case $\Psi'\left(\frac{1}{\tau}\right) = \tau/(1+\tau)$, and we immediately get the desired inequality. Observe that, as is easy to see, if $p = 0$ we can set $C(0) = 1$. It can be shown that $C(p)$ can be chosen as $4(p+1)(2 + \ln p)$ for $p \ge 1$ (see Hayman [**Hay64**]).

REMARK 1. If $f(z)$ is a Weierstrass canonical product of genus p, then

$$\ln M(r,f) \le C(p) \left\{ r^p \int_0^r \frac{n(t,0)}{t^{p+1}} dt + r^{p+1} \int_r^{\infty} \frac{n(t,0)}{t^{p+2}} dt \right\}. \tag{3.6}$$

In fact, for each R, $0 < R < \infty$, we have

$$\begin{aligned} r^{p+1} &\int_0^\infty \frac{n(t,0)}{t^{p+1}(r+t)}dt \\ &\leq r^{p+1}\int_0^R \frac{n(t,0)}{t^{p+1}(t+r)}dt + r^{p+1}\int_R^\infty \frac{n(t,0)}{t^{p+1}(t+r)}dt \\ &\leq r^p \int_0^R \frac{n(t,0)}{t^{p+1}}dt + r^{p+1}\int_R^\infty \frac{n(t,0)}{t^{p+2}}dt. \end{aligned} \tag{3.7}$$

If $R = r$, from (3.5) and (3.7) we get (3.6).

REMARK 2. If $f(z)$ is a Weierstrass canonical product of genus p, then

$$\ln M(r,f) = o(r^{p+1}). \tag{3.8}$$

This relation can be derived from (3.5) since

$$\int_0^\infty \frac{n(t,0)\,dt}{t^{p+1}(r+t)} = o(1) \quad \text{as} \quad r\to\infty.$$

The legitimacy of passing to the limit under the integral sign can be easily justified, since the integrand tends to 0 monotonically as $r \to \infty$.

We establish the following convention: for the value of p in the Weierstrass canonical products of genus p we take, unless explicitly stated otherwise, the least non-negative integer such that the series (3.1) converges for $q = p+1$. This non-negative integer will be called the *genus of the sequence* $\{a_k\}$.

THEOREM 3.4 (Borel). *The order of the Weierstrass canonical product $f(z)$ is equal to the order of $n(r,0)$.*

PROOF. Let ρ_1 be the order of $n(r,0)$. This number is equal to the order of $N(r,0)$ (see the corollary of Theorem 1.1); hence $\rho_1 \leq \rho$, where ρ is the order of the canonical product. Since p is the least integer for which the series $\sum_k |a_k|^{-p-1}$ is convergent, we have $p \leq \rho_1 \leq p+1$. By Remark 2, we get $\rho \leq p+1$; therefore, if $\rho_1 = p+1$, then $\rho_1 \geq \rho$. Now assume that $\rho_1 < p+1$. Then for each ρ', $\rho_1 < \rho' < p+1$, we have $n(r,0) < C_2 r^{\rho'}$ for all $r > 0$, where C_2 is some constant. From (3.5) we get

$$\ln M(r,f) < C(p)C_2 r^{\rho'} \int_0^\infty \frac{\tau^{\rho'-p-1}}{1+\tau}d\tau = C_3 r^{\rho'}.$$

Hence $\rho \leq \rho'$ and $\rho \leq \rho_1$. Therefore $\rho \leq \rho_1$ is satisfied in all cases, and together with $\rho_1 \leq \rho$ this proves the theorem. □

4. Hadamard's and Lindelöf's theorems

Canonical products can be used to find convenient representations of meromorphic functions of finite order.

THEOREM 4.1 (Hadamard). *Let $f(z)$ be a meromorphic function of finite order ρ, and let $\{a_\nu\}$ and $\{b_\mu\}$ be the sequences of zeros and poles of the function $f(z)$, which are different from $z = 0$. Let p_1 be the genus of $\{a_\nu\}$, and p_2 the genus of*

$\{b_\mu\}$. *Suppose that in a neighborhood of* $z = 0$ *the function* $f(z)$ *has a representation* $f(z) = c_\lambda z^\lambda + c_{\lambda+1} z^{\lambda+1} + \cdots$ *with* $c_\lambda \neq 0$. *Then*

$$f(z) = z^\lambda e^{P(z)} \frac{\prod_{a_\nu} E\left(\frac{z}{a_\nu}, p_1\right)}{\prod_{b_\mu} E\left(\frac{z}{b_\mu}, p_2\right)}, \tag{4.1}$$

where $P(z)$ *is a polynomial of degree* $q \leq [\rho]$.

PROOF. Consider an auxiliary function

$$F(z) = f(z) z^{-\lambda} \frac{\prod_{b_\mu} E\left(\frac{z}{b_\mu}, p_2\right)}{\prod_{a_\nu} E\left(\frac{z}{a_\nu}, p_1\right)}.$$

It is clear that $F(z)$ is an entire function without zeros. By Theorems 1.7 and 3.4, we have

$$\rho_1 = \rho[F] \leq \max(\rho[f], \rho[n(r,0,f)], \rho[n(r,\infty,f)]) = \rho.$$

Using Lemma 6.2 of Chapter 1, we get $F(z) = \exp P(z)$, where $P(z)$ is a polynomial whose degree does not exceed $[\rho_1] \leq [\rho]$. □

DEFINITION. Let $f(z)$ be a meromorphic function of finite order ρ. Then it has a representation of the form (4.1). The number $p = \max(q, p_1, p_2)$ is called the *genus of the function* $f(z)$.

EXERCISE. Prove the inequality $p \leq \rho \leq p + 1$.

REMARK. Suppose that the hypothesis of Theorem 4.1 is satisfied. Denote the sequence of zeros and poles of the function $f(z)$ by $\{c_k\}$. Then the representation (4.1) remains valid if we replace p_1 and p_2 by p' in the Weierstrass canonical products. The same representation remains valid if we replace p_1 and p_2 by the genus p of the function $f(z)$. In fact, suppose, for example, that $p_1 < p$, $p_2 < p$. Then

$$f(z) = z^\lambda e^{P(z)+P_1(z)+P_2(z)} \frac{\prod_{a_\nu} E\left(\frac{z}{a_\nu}, p\right)}{\prod_{b_\mu} E\left(\frac{z}{b_\mu}, p\right)},$$

where

$$P_1(z) = -\frac{z^{p_1+1}}{p_1+1} \sum_\nu \frac{1}{a_\nu^{p_1+1}} - \cdots - \frac{z^p}{p} \sum_\nu \frac{1}{a_\nu^p},$$

$$P_2(z) = \frac{z^{p_2+1}}{p_2+1} \sum_\mu \frac{1}{b_\mu^{p_2+1}} + \cdots + \frac{z^p}{p} \sum_\mu \frac{1}{b_\mu^p}.$$

The series forming the coefficients of polynomials $P_1(z)$ and $P_2(z)$ are absolutely convergent by the definitions of the numbers p_1 and p_2. Since the degree of the polynomial $P(z) + P_1(z) + P_2(z)$ does not exceed p, and hence does not exceed $[\rho]$, the statement is proved in this case. All other cases can be considered in a similar manner.

Denote by $N(r; a, b)$, $a \neq b$, the sum $N(r, a) + N(r, b)$; also let $n(r; a, b) = n(r, a) + n(r, b)$. We are going to establish connections between the orders and types of the functions $f(z)$ and $N(r; a, b)$. It is clear that without loss of generality we may assume that $a = 0$, $b = \infty$, and $f(0) = 1$.

The easiest case is where the order ρ of the function $f(z)$ is non-integer.

THEOREM 4.2. *If the order ρ of a meromorphic function $f(z)$ is non-integer, then the order of $N(r;0,\infty)$ is equal to ρ. The types of $T(r,f)$ and $N(r;0,\infty)$ with respect to an arbitrary proximate order $\rho(r)$, $\rho(r) \to \rho$, coincide.*

PROOF. We represent $f(z)$ in the form (4.1). It is clear that $\rho[N(r;0,\infty)] \leq \rho[f] = \rho$. Suppose that $\rho_1 = \rho[N(r,0)] \geq \rho[N(r,\infty)]$. Then $\rho[N(r;0,\infty)] = \rho_1$.

By Theorem 3.4, the order of the function

$$\pi(z) = \frac{\prod_{a_\nu} E\left(\frac{z}{a_\nu}, p_1\right)}{\prod_{b_\mu} E\left(\frac{z}{b_\mu}, p_2\right)}$$

is equal to ρ_1, since $\rho_1 = \rho[N(r,0)] \leq \rho[\pi(z)] \leq \max(\rho[N(r,0)], \rho[N(r,\infty)]) = \rho_1$. Since $\rho[e^{-P(z)}] \leq p < \rho$ (ρ is non-integer!), we have

$$\rho_1 = \rho[\pi(z)] = \max(\rho[f(z)], \rho[e^{-P(z)}]) = \rho.$$

Suppose that

$$\limsup_{r\to\infty} \frac{N(r;0,\infty)}{r^{\rho(r)}} = \Delta_N, \qquad \limsup_{r\to\infty} \frac{T(r,f)}{r^{\rho(r)}} = \Delta_f. \tag{4.2}$$

By $N(r;0,\infty) \leq 2T(r,f) + O(1)$, we get $\Delta_N \leq 2\Delta_f$. Represent $f(z)$ in the form (4.1) with $p_1 = p_2 = p$ (see the remark to Theorem 4.1). Using formulas (6.5) and (6.7) from Chapter 1 and the corollary of Theorem 3.3, we get

(4.3)

$$\begin{aligned}
T(r,f) &\leq O(r^p) + \ln M\left(r, \prod_{a\nu} E\left(\frac{z}{a_\nu}, p\right)\right) + \ln M\left(r, \prod_{b_\mu} E\left(\frac{z}{b_\mu}, p\right)\right) \\
&\leq O(r^p) + C(p)\int_0^\infty \frac{n(r\tau,0)}{\tau^{1+p}}\frac{d\tau}{1+\tau} + C(p)\int_0^\infty \frac{n(r\tau,\infty)}{\tau^{1+p}}\frac{d\tau}{1+\tau} \\
&= O(r^p) + C(p)\int_0^\infty \frac{n(r\tau;0,\infty)}{\tau^{1+p}}\frac{d\tau}{1+\tau} \\
&= O(r^p) + C(p)\int_0^\infty \frac{dN(r\tau;0,\infty)}{\tau^p(1+\tau)} \\
&= O(r^p) + C(p)\frac{N(r\tau;0,\infty)}{\tau^{p(1+\tau)}}\Bigg|_{\tau=0}^{\tau=\infty} + C(p)\int_0^\infty N(r\tau;0,\infty)\frac{(p+1)\tau+p}{\tau^{p+1}(1+\tau)^2}d\tau \\
&\leq O(r^p) + C(p)(p+1)\int_0^\infty \frac{N(r\tau;0,\infty)}{\tau^{p+1}}\frac{d\tau}{1+\tau}.
\end{aligned}$$

We have used the facts that $\rho < p+1$ and $N(r\tau;0,\infty) = o(\tau^{p+1})$. Assume that $\Delta_N < \infty$. Then the condition $N(r;0,\infty) \leq (\Delta_N + \varepsilon)r^{\rho(r)}$ is satisfied for $r > c(\varepsilon), \varepsilon > 0$. Therefore (4.3) implies

$$\frac{T(r,f)}{r^{\rho(r)}} \leq o(1) + C(p)(p+1)(\Delta_N + \varepsilon)\int_{c/r}^\infty \frac{(r\tau)^{\rho(r\tau)}}{r^{\rho(r)}}\frac{d\tau}{\tau^{p+1}(1+\tau)}.$$

Taking limits in this inequality as $r \to \infty$, and using Exercise 1 from Section 2 of this chapter, we get

$$\Delta_f \leq C(p)(p+1)(\Delta_N + \varepsilon)\frac{\pi}{\sin \pi(\rho - p)}.$$

Since ε can be arbitrarily small, we get

$$\Delta_f \le C(p)(p+1)\frac{\pi}{\sin \pi(\rho - p)}\Delta_N. \tag{4.4}$$

This inequality is obvious if $\Delta_N = \infty$. Together with the inequality $\Delta_f \ge 2^{-1}\Delta_N$ it concludes the proof. □

COROLLARY. *If a meromorphic function $f(z)$ has non-integer order ρ, $[\rho] = p$, then*

$$\limsup_{r\to\infty} \frac{N(r;0,\infty)}{T(r,f)} \ge \frac{\sin \pi(\rho - p)}{C(p)(p+1)\pi} > 0. \tag{4.4$'$}$$

Let $\rho(r)$ be a proximate order of $T(r,f)$. Then, by (4.4) we get

$$\limsup_{r\to\infty} \frac{N(r;0,\infty)}{T(r,f)} \ge \limsup_{r\to\infty} \frac{N(r;0,\infty)}{\Delta_f r^{\rho(r)}} = \frac{\Delta_N}{\Delta_f} \ge \frac{\sin \pi(\rho - p)}{C(p)(p+1)\pi}.$$

We now consider the case where a meromorphic function $f(z)$ has integer order $\rho \ge 1$. Then the genus of $f(z)$ is equal to $\rho - 1$ or ρ. Using (4.1), the function $f(z)$ can be represented in the form

$$f(z) = e^{P(z)} \frac{\prod_{a_\nu} E\left(\frac{z}{a_\nu}, \rho\right)}{\prod_{b_\mu} E\left(\frac{z}{b_\mu}, \rho\right)}, \tag{4.1$'$}$$

where

$$P(z) = c_\rho z^\rho + c_{\rho-1} z^{\rho-1} + \cdots + c_1 z.$$

First we prove the following theorem.

THEOREM 4.3. *Let $f(z)$ be a meromorphic function of integer order $\rho \ge 1$, and let $\rho(r)$ be a proximate order, $\rho(r) \to \rho$ as $r \to \infty$. Suppose that*

$$\Delta_n = \limsup_{r\to\infty} \frac{n(r;0,\infty)}{r^{\rho(r)}} < \infty.$$

Then ($\varepsilon > 0$, $r \ge r(\varepsilon)$)

$$T(r,f) = \frac{r^\rho}{\pi}|K(r)| + (\Delta_n + \varepsilon)\omega_\varepsilon(r) r^{\rho(r)}, \tag{4.5}$$

where the value of $|\omega_\varepsilon(r)|$ is bounded by a constant M_ρ, which depends on ρ only, and

$$K(r) = c_\rho + \frac{1}{\rho}\sum_{|a_\nu|\le r} a_\nu^{-\rho} - \frac{1}{\rho}\sum_{|b_\mu|\le r} b_\mu^{-\rho}.$$

PROOF. Using (4.1$'$), we represent the function $f(z)$ as a product, $f(z) = f_1(z)f_2(z)$, where

$$f_1(z) = \exp\{z^\rho K(r)\},$$

$$f_2(z) = e^{c_{\rho-1}z^{\rho-1}+\cdots+c_1 z} \cdot \frac{\prod_{|a_\nu|\le r} E\left(\frac{z}{a_\nu}, \rho - 1\right) \prod_{|a_\nu|>r} E\left(\frac{z}{a_\nu}, \rho\right)}{\prod_{|b_\mu|\le r} E\left(\frac{z}{b_\mu}, \rho - 1\right) \prod_{|b_\mu|>r} E\left(\frac{z}{b_\mu}, \rho\right)}.$$

It is easy to see that

$$T(r, f_1(z)) = \frac{r^\rho}{\pi}|K(r)|.$$

Since

$$|T(r,f) - T(r,f_1)| \le T(r,f_2) + O(1),$$

it remains to estimate $T(r,f_2)$. Using inequality (3.6) and equality (6.8′) from Chapter 1, we get

$$
\begin{aligned}
T(r,f_2) &\le T(r, e^{c_{\rho-1}z^{\rho-1}+\dots+c_1 z}) + T\left(r, \prod_{|a_\nu|\le r} E\left(\frac{z}{a_\nu}, \rho-1\right)\right) \\
&\quad + T\left(r, \prod_{|a_\nu|>r} E\left(\frac{z}{a_\nu}, \rho\right)\right) + T\left(r, \prod_{|b_\mu|\le r} E\left(\frac{z}{b_\mu}, \rho-1\right)\right) \\
&\quad + T\left(r, \prod_{|b_\mu|>r} E\left(\frac{z}{b_\mu}, \rho\right)\right) \\
&\le O(r^{\rho-1}) + \ln M\left(r, \prod_{|a_\nu|\le r} E\left(\frac{z}{a_\nu}, \rho-1\right)\right) + \ln M\left(r, \prod_{|a_\nu|>r} E\left(\frac{z}{a_\nu}, \rho\right)\right) \\
&\quad + \ln M\left(r, \prod_{|b_\mu|\le r} E\left(\frac{z}{b_\mu}, \rho-1\right)\right) + \ln M\left(r, \prod_{|b_\mu|>r} E\left(\frac{z}{b_\mu}, \rho\right)\right) \\
&\le O(r^{\rho-1}) + C(\rho-1)\left\{r^{\rho-1}\int_{r_0}^{r} \frac{n(t;0,\infty)}{t^\rho}dt + r^\rho n(r;0,\infty)\int_r^\infty \frac{dt}{t^{\rho+1}}\right\} \\
&\quad + C(\rho) r^{\rho+1}\int_r^\infty \frac{n(t;0,\infty)}{t^{\rho+2}}dt = O(r^{\rho-1}) \\
&\quad + C(\rho-1)\left\{r^{\rho-1}\int_{r_0}^{r} \frac{n(t;0,\infty)}{t^\rho}dt + \frac{1}{\rho}n(r;0,\infty)\right\} \\
&\quad + C(\rho) r^{\rho+1}\int_r^\infty \frac{n(t;0,\infty)}{t^{\rho+2}}dt, \qquad r_0>0.
\end{aligned}
$$

We may assume that the inequality $n(r;0,\infty) \le (\Delta_n+\varepsilon) r^{\rho(r)}$ is satisfied for all $r \ge r_0$. Taking into account Theorem 2.4, we get

$$
\begin{aligned}
T(r,f_2) &\le O(r^{\rho-1}) + (\Delta_n+\varepsilon)\left\{C(\rho-1)r^{\rho-1}\int_{r_0}^{r} t^{\rho(t)-\rho}dt\right. \\
&\qquad\qquad \left. + \frac{C(\rho-1)}{\rho} r^{\rho(r)} + C(\rho) r^{\rho+1}\int_r^\infty t^{\rho(t)-\rho-2}dt\right\} \\
&= O(r^{\rho-1}) + (\Delta_n+\varepsilon)(1+o(1))r^{\rho(r)}\left\{C(\rho-1)\left(1+\frac{1}{\rho}\right) + C(\rho)\right\} \\
&= (\Delta_n+\varepsilon)(1+o(1))r^{\rho(r)}\left\{C(\rho-1)\left(1+\frac{1}{\rho}\right) + C(\rho)\right\},
\end{aligned}
$$

and the proof is complete. □

THEOREM 4.4 (Lindelöf). *Suppose that the hypothesis of Theorem 4.3 is satisfied. Let*

$$
L(r) = r^{\rho(r)-\rho}, \qquad \Delta_K = \limsup_{r\to\infty} \frac{|K(r)|}{L(r)}, \qquad \Omega = \max(\Delta_K, \Delta_n).
$$

Then the function $f(z)$ has minimal, normal, or maximal type with respect to $\rho(r)$ depending on whether $\Omega = 0$, $0 < \Omega < \infty$, or $\Omega = \infty$, respectively.

PROOF. We denote by A_j various positive constants depending on ρ only. The relation (4.5) implies that, if $\Delta_n < \infty$, then $\Delta_f \leq A_1(\Delta_K + \Delta_n + \varepsilon)$, and hence $\Delta_f \leq A_2\Omega$. If $\Delta_n = \infty$, the inequality is trivial. Relation (4.5) implies, also, that $\Delta_K \leq A_3(\Delta_f + \Delta_n + \varepsilon)$ if $\Delta_n < \infty$, and hence $\Delta_K \leq A_3(\Delta_f + \Delta_n)$. It is clear that this inequality is valid also if $\Delta_n = \infty$. Lemma 7.1 of Chapter 1 implies that $n(r; 0, \infty) \leq N(er; 0, \infty) \leq 2T(er, f) + O(1)$. Therefore $\Delta_n \leq 2e^\rho \Delta_f$. Hence $\Delta_K \leq A_3(1+2e^\rho)\Delta_f$. We get $\Omega \leq A_4\Delta_f$. Using the inequality $A_4^{-1}\Omega \leq \Delta_f \leq A_2\Omega$ we complete the proof. □

REMARK. It is easy to show (cf. the proof of Theorem 1.1) that $e^{-\rho}\Delta_n \leq \Delta_N \leq \rho^{-1}\Delta_n$. Therefore Theorem 4.4 remains true if we replace Ω in its statement by $\Omega' = \max(\Delta_K, \Delta_N)$.

Alternatively, Theorem 4.4 can be proved without referring to Theorem 4.3 but using formula (2.6) from Chapter 1 instead. The idea of this proof is due to N. I. Akhiezer [**Akh27**] and L. A. Rubel [**Rub63**]. For simplicity, we restrict our attention to the case $\rho(r) \equiv \rho$, $L(r) \equiv 1$, leaving the general case to interested readers.

Without loss of generality we may assume that $f(0) = 1$. Suppose that in a neighborhood of $z = 0$ we have

$$\ln f(z) = \sum_{k=1}^{\infty} \gamma_k z^k.$$

It is easy to derive from (4.1′) that

$$\gamma_k = \begin{cases} c_k, & k = 1, \dots, \rho, \\ -\frac{1}{k}\sum_\nu a_\nu^{-k} + \frac{1}{k}\sum_\mu b_\mu^{-k}, & k = \rho+1, \rho+2, \dots. \end{cases} \tag{4.6}$$

Denote by $F_k(r)$ the Fourier coefficient of the function $\ln|f(re^{i\varphi})|$, $0 \leq \varphi \leq 2\pi$:

$$F_k(r) = \frac{1}{2\pi}\int_0^{2\pi} \ln|f(re^{i\varphi})| e^{-ik\varphi}\, d\varphi, \qquad k = 0, \pm 1, \pm 2, \dots.$$

It is clear that

$$|F_k(r)| \leq \frac{1}{2\pi}\int_0^{2\pi} |\ln|f(re^{i\varphi})||\, d\varphi = m(r, f) + m\left(r, \frac{1}{f}\right) \leq 2T(r, f). \tag{4.7}$$

Formula (2.6) from Chapter 1 can be rewritten as ($k = 1, 2, \dots$)

$$\gamma_k = \frac{2}{r^k}F_k(r) + \frac{1}{k}\sum_{|a_\nu| \leq r}\left(\frac{\bar{a}_\nu^k}{r^{2k}} - \frac{1}{a_\nu^k}\right) - \frac{1}{k}\sum_{|b_\mu| \leq r}\left(\frac{\bar{b}_\mu^k}{r^{2k}} - \frac{1}{b_\mu^k}\right). \tag{4.8}$$

If $k = \rho$ and $\gamma_k = c_\rho$, equality (4.8) can be rewritten as

$$K(r) = \frac{2}{r^\rho}F_\rho(r) + \frac{1}{\rho}\sum_{|a_\nu| \leq r}\frac{\bar{a}_\nu^\rho}{r^{2\rho}} - \frac{1}{\rho}\sum_{|b_\mu| \leq r}\frac{\bar{b}_\mu^\rho}{r^{2\rho}}. \tag{4.9}$$

Using (4.7) we get

$$\begin{aligned}|K(r)| &\leq \frac{4T(r,f)}{r^\rho} + \frac{1}{\rho r^\rho}\left\{\sum_{|a_\nu|\leq r}\left(\frac{|\bar{a}_\nu|}{r}\right)^\rho + \sum_{|b_\mu|\leq r}\left(\frac{|\bar{b}_\mu|}{r}\right)^\rho\right\}\\ &\leq \frac{4T(r,f)}{r^\rho} + \frac{n(r;0,\infty)}{\rho r^\rho}.\end{aligned}$$

Hence

$$\Delta_K \leq 4\Delta_f + \rho^{-1}\Delta_n.$$

Using this inequality we get $\Omega \leq A_5\Delta_f$, as in our first proof of Theorem 4.4.

Assume now that $\Omega < \infty$. If $r > r_0(\varepsilon)$, then $|K(r)| < \Delta_K + \varepsilon$ and $n(r;0,\infty) < (\Delta_n + \varepsilon)r^\rho$, $\varepsilon > 0$. We estimate the Fourier coefficients $F_k(r)$. If $k > \rho$, from (4.8) we get

$$-\frac{1}{k}\sum_{|a_\nu|>r} a_\nu^{-k} + \frac{1}{k}\sum_{|b_\mu|>r} b_\mu^{-k} = \frac{2}{r^k}F_k(r) + \frac{1}{k}\sum_{|a_\nu|\leq r}\frac{\bar{a}_\nu^k}{r^{2k}} - \frac{1}{k}\sum_{|b_\mu|\leq r}\frac{\bar{b}_\mu^k}{r^{2k}};$$

hence

$$\begin{aligned}|F_k(r)| &\leq \frac{1}{2k}n(r;0,\infty) + \frac{r^k}{2k}\int_r^\infty \frac{dn(t;0,\infty)}{t^k}\\ &= \frac{r^k}{2}\int_r^\infty \frac{n(t;0,\infty)}{t^{k+1}}dt \leq \frac{\Delta_n + \varepsilon}{2(k-\rho)}r^\rho, \qquad k > \rho.\end{aligned}\tag{4.10}$$

If $1 \leq k \leq \rho - 1$, then in view of (4.8) we obtain

$$\begin{aligned}|F_k(r)| &\leq \frac{|c_k|}{2}r^k + \frac{1}{2k}n(r;0,\infty) + \frac{r^k}{2k}\left\{\sum_{|a_\nu|\leq r}|a_\nu|^{-k} + \sum_{|b_\mu|\leq r}|b_\mu|^{-k}\right\}\\ &\leq \frac{|c_k|}{2}r^k + \frac{n(r;0,\infty)}{2k} + \frac{r^k}{2k}\int_0^r \frac{dn(t;0,\infty)}{t^k}dt\\ &= \frac{|c_k|}{2}r^k + \frac{n(r;0,\infty)}{k} + \frac{r^k}{2}\int_0^r \frac{n(t;0,\infty)}{t^{k+1}}dt\\ &\leq O(r^{\rho-1}) + (\Delta_n + \varepsilon)\left(\frac{1}{k} + \frac{1}{2(\rho-k)}\right)r^\rho.\end{aligned}\tag{4.11}$$

From (4.9) we get

$$|F_\rho(r)| \leq \frac{|K(r)|}{2}r^\rho + \frac{n(r;0,\infty)}{2\rho} \leq \frac{1}{2}\left(\Delta_K + \varepsilon + \frac{\Delta_n + \varepsilon}{\rho}\right)r^\rho.\tag{4.12}$$

Finally, by Jensen's formula,

$$F_0(r) = N(r,0) - N(r,\infty)$$

and

$$|F_0(r)| \leq N(r;0,\infty) \leq \frac{\Delta_n + \varepsilon}{\rho}r^\rho + O(1).\tag{4.13}$$

By Parseval's identity and the identity $F_{-k}(r) = \overline{F_k(r)}$, we have

$$\frac{1}{2\pi}\int_0^{2\pi} \ln^2|f(re^{i\varphi})|d\varphi = \sum_{k=-\infty}^{\infty}|F_k(r)|^2 = |F_0(r)|^2 + 2\sum_{k=1}^{\infty}|F_k(r)|^2.\tag{4.14}$$

Using inequalities (4.10)–(4.13), and the elementary inequality $(|a|+|b|)^2 \le 2(|a|^2+|b|^2)$, we get

$$\begin{aligned}\frac{1}{2\pi}&\int_0^{2\pi} \ln^2|f(re^{i\varphi})|\,d\varphi\\ &\le O(r^{2\rho-2}) + 2\Bigg\{\left(\frac{\Delta_n+\varepsilon}{\rho}\right)^2 + (\Delta_n+\varepsilon)^2\sum_{k=1}^{\rho-1}\left(\frac{1}{k}+\frac{1}{2(\rho-k)}\right)^2\\ &\qquad + \frac{(\Delta_K+\varepsilon)^2}{2} + \frac{(\Delta_n+\varepsilon)^2}{2\rho^2} + \frac{(\Delta_n+\varepsilon)^2}{4}\sum_{k=\rho+1}^{\infty}\frac{1}{(k-\rho)^2}\Bigg\} r^{2\rho}\\ &= O(r^{2\rho-2}) + \{(\Delta_K+\varepsilon)^2 + A_6(\Delta_n+\varepsilon)^2\} r^{2\rho},\end{aligned}$$

where

$$A_6 = \frac{3}{\rho^2} + 2\sum_{k=1}^{\rho-1}\left(\frac{1}{k}+\frac{1}{2(\rho-k)}\right)^2 + \frac{1}{2}\sum_{k=\rho+1}^{\infty}\frac{1}{(k-\rho)^2} < \infty.$$

By virtue of the Cauchy–Bunyakovskii inequality,

$$\begin{aligned}m(r,f) + m\left(r,\frac{1}{f}\right) &= \frac{1}{2\pi}\int_0^{2\pi} |\ln|f(re^{i\varphi})||\,d\varphi \le \left(\frac{1}{2\pi}\int_0^{2\pi}\ln^2|f(re^{i\varphi})|\,d\varphi\right)^{1/2}\\ &\le r^\rho\{(\Delta_K+\varepsilon)^2 + A_6(\Delta_n+\varepsilon)^2 + O(r^{-2})\}^{1/2},\end{aligned}$$

$$\begin{aligned}2T(r,f) &= m(r,f) + m\left(r,\frac{1}{f}\right) + N(r;0,\infty)\\ &\le r^\rho\{(\Delta_K+\varepsilon)^2 + A_6(\Delta_n+\varepsilon)^2 + O(r^{-2})\}^{1/2} + \frac{\Delta_n+\varepsilon}{\rho}r^\rho + O(1).\end{aligned}$$

Hence

$$2\Delta_f \le \{(\Delta_K+\varepsilon)^2 + A_6(\Delta_n+\varepsilon)^2\}^{1/2} + \frac{\Delta_n+\varepsilon}{\rho},$$

$$\Delta_f \le \frac{1}{2}\{\Delta_K^2 + A_6\Delta_n^2\}^{1/2}\frac{\Delta_n}{\rho} + A_7\Omega.$$

Therefore, we have proved the inequality $A_5\Omega \le \Delta_f \le A_7\Omega$, thus completing the proof of Theorem 4.4.

EXERCISE. Prove Theorem 4.2 using formula (4.8), estimates (4.7), (4.10), (4.11), (4.13), and equality (4.14).

Now we consider the case of order $\rho = 0$.

THEOREM 4.5. *Let $f(z)$ be a transcendental meromorphic function of order $\rho = 0$. Then the types of $T(r,f)$ and $N(r;0,\infty)$ with respect to an arbitrary proximate order $\rho(r)$, $\rho(r)\to 0$, coincide.*

PROOF. Let

$$f(z) = \frac{\prod_{a_\nu}\left(1-\frac{z}{a_\nu}\right)}{\prod_{b_\mu}\left(1-\frac{z}{b_\mu}\right)} = \frac{\pi_1(z)}{\pi_2(z)}. \tag{4.15}$$

From inequality (3.6) we get

$$
\begin{aligned}
T(r,f) &\le T(r,\pi_1)+T(r,\pi_2) \le \ln M(r,\pi_1)+\ln M(r,\pi_2)\\
&\le \int_0^r \frac{n(t;0,\infty)}{t}dt + r\int_r^\infty \frac{n(t;0,\infty)\,dt}{t^2}\\
&= N(r;0,\infty) + r\int_r^\infty \frac{dN(t;0,\infty)}{t}\\
&= N(r;0,\infty) - N(r;0,\infty) + r\int_r^\infty \frac{N(t;0,\infty)}{t^2}\,dt\\
&= r\int_r^\infty \frac{N(t;0,\infty)}{t^2}\,dt.
\end{aligned}
\tag{4.16}
$$

Suppose that $\ln r = o(r^{\rho(r)})$. Then, using Theorem 2.4, it is easy to see that $\Delta_f \le \Delta_N$. On the other hand, it is clear that $\Delta_N \le 2\Delta_f$. These two inequalities imply the statement of the theorem in this case.

If the condition $\ln r = o(r^{\rho(r)})$ is not satisfied, then $T(r,f)$ has maximal type. If $N(r;0,\infty)$ has normal or minimal type, then $N(r;0,\infty) = O(\ln r)$ for an infinite set of values of r. In this case the function $f(z)$ must have finitely many zeros and poles, and hence by (4.15) is a rational function, a contradiction. □

REMARK 1. Under the hypothesis of Theorem 4.5, if $f(z)$ is an entire function, then $\Delta_N \le \Delta_f$, and hence $\Delta_f = \Delta_N$.

REMARK 2. In the case where the proximate order $\rho(r)$ approaches 0, the types of the functions $n(r;0,\infty)$ and $N(r;0,\infty)$ can differ. For example, if $n(r;0,\infty) \sim Kr^{\rho(r)}$, then $N(r;0,\infty) \sim K\int_{r_0}^r t^{\rho(t)-1}dt$ and, as is easy to check,

$$
r^{\rho(r)} = o\left(\int_{r_0}^r t^{\rho(t)-1}dt\right).
$$

Denote by $\widehat{N}(r;a,b)$ the function $\max(N(r,a),N(r,b))$. Suppose that the hypothesis of Theorem 4.5 is satisfied. We estimate $T(r,f)$ from above using, instead of

$$
T\left(r,\frac{\pi_1}{\pi_2}\right) \le \ln M(r,\pi_1) + \ln M(r,\pi_2) + O(1),
$$

a sharper inequality which can be obtained from (4.8) in Chapter 1:

$$
T\left(r,\frac{\pi_1}{\pi_2}\right) \le \max\{\ln M(r,\pi_1), \ln M(r,\pi_2)\}.
$$

Then, in the same way as we proved (4.16), we get

$$
\begin{aligned}
T(r,f) &\le \max\{\ln M(r,\pi_1), \ln M(r,\pi_2)\}\\
&\le \max\left\{r\int_r^\infty \frac{N(t,0)}{t^2}\,dt,\quad r\int_r^\infty \frac{N(t,\infty)}{t^2}\,dt\right\}\\
&\le r\int_r^\infty \frac{\widehat{N}(t;0,\infty)}{t^2}\,dt.
\end{aligned}
\tag{4.17}
$$

This inequality implies $\Delta_f \le \Delta_{\widehat{N}}$. On the other hand, $\widehat{N}(r;0,\infty) \le T(r,f) + O(1)$ and $\Delta_{\widehat{N}} \le \Delta_f$. Hence $\Delta_f = \Delta_{\widehat{N}}$, that is,

$$\limsup_{r\to\infty} \frac{T(r,f)}{r^{\rho(r)}} = \limsup_{r\to\infty} \frac{\widehat{N}(r;0,\infty)}{r^{\rho(r)}}.$$

It is clear that 0 and ∞ in the last equality can be replaced by arbitrary a and b. Let $\rho(r)$ be a proximate order of $T(r,f)$. Then

$$1 \ge \limsup_{r\to\infty} \frac{\widehat{N}(r;a,b)}{T(r,f)} \ge \limsup_{r\to\infty} \frac{\widehat{N}(r;a,b)}{\Delta_f r^{\rho(r)}} = \frac{\Delta_{\widehat{N}}}{\Delta_f} = 1,$$

that is,

$$\limsup_{r\to\infty} \frac{\widehat{N}(r;a,b)}{T(r,f)} = 1. \tag{4.18}$$

In the case where $f(z)$ is a rational function, (4.18) can be easily verified, using equality (6.26) from Chapter 1. This implies the following theorem.

THEOREM 4.6 (Valiron). *Let $f(z)$ be a meromorphic function of order $\rho = 0$. If the inequality*

$$\limsup_{r\to\infty} \frac{N(r,a)}{T(r,f)} < 1$$

holds for some a, then

$$\limsup_{r\to\infty} \frac{N(r,b)}{T(r,f)} = 1$$

for all $b \ne a$.

5. Examples

We consider examples of entire transcendental functions and find some relations for them. We will use these relations on many occasions.[3]

1°. Let p be a non-negative integer, $p < \eta_1 \le \eta_2 < p+1$. Let $l(r)$ be a function, continuously differentiable on $[0,\infty)$; and such that

1) $l'(r) r \ln r \to 0$ as $r \to \infty$;

2) $\liminf_{r\to\infty} l(r) = \eta_1$, $\limsup_{r\to\infty} l(r) = \eta_2$.

If $\eta_1 = \eta_2$, the function $l(r)$ is a proximate order. The function $l(r)$ preserves many properties of a proximate order, sometimes with minor modifications. Among these properties we mention those that will be used later. 1) The function $r^{l(r)}$ is increasing for sufficiently large r (this can be proved in the same way as Theorem 2.3). 2) If $0 < a \le k \le b < \infty$, then

$$l(kr) - l(r) = o\left(\frac{1}{\ln r}\right) \quad \text{and} \quad \frac{(kr)^{l(kr)}}{(kr)^{l(r)}} \to 1 \qquad (r \to \infty)$$

uniformly in k. The proof is based on the same tricksas our proof of Theorem 2.2,

[3]This section may be skipped on the first reading and be consulted when we refer to it in later sections.

and we leave it to interested readers. 3) The following inequalities hold:

$$\limsup_{r\to\infty} \frac{\int_1^r t^{l(t)-p-1}dt}{r^{l(r)-p}} \le \frac{1}{\eta_1 - p},$$

$$\limsup_{r\to\infty} \frac{\int_r^\infty t^{l(t)-p-2}dt}{r^{l(r)-p-1}} \le \frac{1}{p+1-\eta_2}.$$

Let us prove, for example, the latter inequality. Using a well-known result from calculus, we get

$$\limsup_{r\to\infty} \frac{\int_r^\infty t^{l(t)-p-2}dt}{r^{l(r)-p-1}} \le \limsup_{r\to\infty} \frac{\frac{d}{dr}\int_r^\infty t^{l(t)-p-2}dt}{\frac{d}{dr}r^{l(r)-p-1}}$$
$$= \limsup_{r\to\infty} \frac{1}{p+1-l(r)-l'(r)r\ln r} = \frac{1}{p+1-\eta_2}.$$

Now we consider the Weierstrass canonical products of genus p with positive zeros $1 \le a_1 \le \cdots$, such that $n(r,0) \sim \Delta r^{l(r)}$ $(0 < \Delta < \infty)$:

$$f(z) = \prod_{k=1}^\infty E\left(\frac{z}{a_k}, p\right). \tag{5.1}$$

Let us find an asymptotic formula for $\ln f(z)$ in the angle $\delta < \arg z < 2\pi - \delta$ for the branch of $\ln f(z)$ for which $\ln f(0) = 0$. We represent $\ln f(z)$ as follows:

$$\begin{aligned} \ln f(z) &= \sum_{k=1}^\infty \ln E\left(\frac{z}{a_k}, p\right) = \int_0^\infty \ln E\left(\frac{z}{t}, p\right) dn(t,0) \\ &= n(t,0)\ln E\left(\frac{z}{t}, p\right)\Big|_{t=0}^\infty - \int_0^\infty n(t,0)\frac{d}{dt}\ln E\left(\frac{z}{t}, p\right) dt \\ &= -z^{p+1}\int_0^\infty n(t,0)\frac{dt}{t^{p+1}(t-z)}. \end{aligned} \tag{5.1$'$}$$

Consider an arbitrarily small number ε, $0 < \varepsilon < 1$. If $r > r_0(\varepsilon)$, then $|n(r,0) - \Delta r^{l(r)}| < \varepsilon\Delta r^{l(r)}$. Let $z = re^{i\varphi}$. Observe that for $\delta < \varphi < 2\pi - \delta$ we have $|t - z| \ge (t + r)\sin\frac{\delta}{2}$ (this statement can be easily proved by minimizing $|x - e^{i\varphi}|/(x+1)$ over $0 \le x < \infty$).

There exists a constant $K \ge \Delta > 0$ such that the inequality $n(r,0) < Kr^{l(r)}$ holds for $0 < r < \infty$. Let $\delta_1 > 0$ be such that $1 > K_1 = \eta_1 - p - \delta_1 > 0$ and $K_2 = p + 1 - \eta_2 - \delta_1 > 0$. Select numbers a and b, $0 < a < 1 < b < \infty$, such that $2KK_1^{-1}b^{-K_1} < \varepsilon$, $2KK_2^{-1}b^{-K_2} < \varepsilon$ and find $r_1(\varepsilon) > a^{-1}r_0(\varepsilon)$ such that for $r \ge r_1(\varepsilon)$ the following inequalities hold:

1) $(1-\varepsilon)(kr)^{l(r)} \le (kr)^{l(kr)} \le (1+\varepsilon)(kr)^{l(r)}$ for $a \le k \le b$;
2) $\int_1^{ar} t^{l(t)-p-1}dt \le K_1^{-1}(ar)^{l(ar)-p}$;
3) $\int_{br}^\infty t^{l(t)-p-2}dt \le K_2^{-1}(br)^{l(br)-p-1}$;
4) $\eta_1 - \delta_1 \le l(r) \le \eta_2 + \delta_1$.

Now, for $r \geq r_1(\varepsilon)$, we estimate the difference:

$$
\begin{aligned}
&\left| z^{p+1} \int_0^\infty \frac{n(t,0)dt}{t^{p+1}(t-z)} - z^{p+1}\Delta \int_0^\infty \frac{t^{l(r)}dt}{t^{p+1}(t-z)} \right| \\
&\leq r^{p+1} \left| \int_0^{ar} \frac{n(t,0)dt}{t^{p+1}(t-z)} + \int_{br}^\infty \frac{n(t,0)dt}{t^{p+1}(t-z)} + \int_{ar}^{br} \frac{n(t,0)-\Delta t^{l(t)}}{t^{p+1}(t-z)} dt \right. \\
&\qquad \left. + \Delta \int_{ar}^{br} \frac{t^{l(t)} - t^{l(r)}}{t^{p+1}(t-z)} dt - \Delta \int_0^{ar} \frac{t^{l(r)}dt}{t^{p+1}(t-z)} - \Delta \int_{br}^\infty \frac{t^{l(r)}dt}{t^{p+1}(t-z)} \right| \\
&\leq \frac{r^{p+1}}{\sin\frac{\delta}{2}} \left\{ \frac{1}{r} \int_1^{ar} \frac{n(t,0)}{t^{p+1}} dt + \int_{br}^\infty \frac{n(t,0)}{t^{p+2}} dt + \int_{ar}^{br} \frac{|n(t,0)-\Delta t^{l(t)}|}{t^{p+1}(t+r)} dt \right. \\
&\qquad \left. + \Delta \int_{ar}^{br} \frac{|t^{l(t)} - t^{l(r)}|}{t^{p+1}(t+r)} dt + \frac{\Delta}{r} \int_0^{ar} \frac{t^{l(r)}dt}{t^{p+1}} + \Delta \int_{br}^\infty \frac{t^{l(r)}}{t^{p+2}} dt \right\} \\
&\leq \frac{r^{p+1}}{\sin\frac{\delta}{2}} \left\{ \frac{K}{r} \int_1^{ar} t^{l(t)-p-1} dt + K \int_{br}^\infty t^{l(t)-p-2} dt + \varepsilon\Delta \int_{ar}^{br} \frac{t^{l(t)}dt}{t^{p+1}(t+r)} \right. \\
&\qquad + \frac{\Delta}{r^{p+1}} \int_a^b \frac{|(\tau r)^{l(\tau r)} - (\tau r)^{l(r)}|}{\tau^{p+1}(\tau+1)} d\tau \\
&\qquad \left. + \frac{\Delta}{l(r)-p} (ar)^{l(r)-p} \frac{1}{r} + \frac{\Delta}{p+1-l(r)} (br)^{l(r)-p-1} \right\} \\
&\leq \frac{r^{p+1}}{\sin\frac{\delta}{2}} \left\{ \frac{K}{r} K_1^{-1} (ar)^{l(ar)-p} + K K_2^{-1} (br)^{l(br)-p-1} \right. \\
&\qquad + \varepsilon\Delta r^{-p-1} \int_a^b \frac{(\tau r)^{l(\tau r)} d\tau}{\tau^{p+1}(1+\tau)} + \varepsilon\Delta r^{-p-1} \int_a^b \frac{(\tau r)^{l(r)} d\tau}{\tau^{p+1}(1+\tau)} \\
&\qquad \left. + \Delta K_1^{-1} r^{-p-1} a^{K_1} r^{l(r)} + \Delta K_2^{-1} r^{-p-1} b^{-K_2} r^{l(r)} \right\} \\
&\leq \csc\frac{\delta}{2} \left\{ K K_1^{-1} a^{-p} (1+\varepsilon)(ar)^{l(r)} + K K_2^{-1} b^{-p-1} (1+\varepsilon)(br)^{l(r)} \right. \\
&\qquad \left. + \varepsilon\Delta(2+\varepsilon) \int_a^b \frac{(\tau r)^{l(r)} d\tau}{\tau^{p+1}(1+\tau)} + \varepsilon r^{l(r)} + \varepsilon r^{l(r)} \right\} \\
&\leq r^{l(r)} \csc\frac{\delta}{2} \left\{ K K_1^{-1} a^{K_1} (1+\varepsilon) + K K_2^{-1} b^{-K_2} (1+\varepsilon) \right. \\
&\qquad \left. + \varepsilon\Delta(2+\varepsilon) \int_0^\infty \frac{\tau^{l(r)-p-1} d\tau}{1+\tau} + 2\varepsilon \right\} \\
&\leq \varepsilon r^{l(r)} \csc\frac{\delta}{2} \left\{ 2(1+\varepsilon) + \Delta(2+\varepsilon) \frac{\pi}{\sin\pi(l(r)-p)} + 2\varepsilon \right\} \\
&\leq \varepsilon r^{l(r)} \csc\frac{\delta}{2} \{ 6 + 3\Delta\pi \max[\csc\pi K_1, \csc\pi K_2] \} \\
&= \varepsilon A r^{l(r)} \csc\frac{\delta}{2}.
\end{aligned}
$$

The integral

$$I(z) = \int_0^\infty \frac{t^{l(r)}\,dt}{t^{p+1}(t-z)}$$

can be easily evaluated using the residue calculus:

$$I(z) = -\frac{\pi e^{-i\pi(l(r)-p)}}{\sin \pi(l(r)-p)} \cdot \operatorname{res}_{t=z} \frac{t^{l(r)-p-1}}{t-z} = -\frac{\pi e^{-i\pi(l(r)-p)}}{\sin \pi(l(r)-p)} z^{l(r)-p-1}.$$

Thus we get for $\{\delta < \arg z < 2\pi < 2\pi - \delta,\ |z| > r_1(\varepsilon)\}$ the inequality

$$\left| \ln f(z) - \frac{\pi\Delta}{\sin \pi(l(r)-p)} e^{-i\pi(l(r)-p)} z^{l(r)} \right| \le \varepsilon A r^{l(r)} \csc \frac{\delta}{2};$$

that is, for $\delta < \varphi < 2\pi - \delta$, the following formulas hold uniformly:

$$\ln f(re^{i\varphi}) = \frac{\pi\Delta}{\sin \pi l(r)} e^{il(r)(\varphi-\pi)} r^{l(r)} + o(r^{l(r)}), \tag{5.2$_1$}$$

$$\ln |f(re^{i\varphi})| = \pi\Delta \frac{\cos l(r)(\varphi-\pi)}{\sin \pi l(r)} r^{l(r)} + o(r^{l(r)}), \tag{5.2$_2$}$$

$$\arg f(re^{i\varphi}) = \pi\Delta \frac{\sin l(r)(\varphi-\pi)}{\sin \pi l(r)} r^{l(r)} + o(r^{l(r)}). \tag{5.2$_3$}$$

In the case where $l(r) = \rho(r) \to \rho$, $p < \rho < p+1$, that is, $l(r)$ is a proximate order, these formulas can be rewritten as

$$\ln f(re^{i\varphi}) = \frac{\pi\Delta}{\sin \pi\rho} e^{i\rho(\varphi-\pi)} r^{\rho(r)} + o(r^{\rho(r)}), \tag{5.3$_1$}$$

$$\ln |f(re^{i\varphi})| = \pi\Delta \frac{\cos \rho(\varphi-\pi)}{\sin \pi\rho} r^{\rho(r)} + o(r^{\rho(r)}), \tag{5.3$_2$}$$

$$\arg f(re^{i\varphi}) = \pi\Delta \frac{\sin \rho(\varphi-\pi)}{\sin \pi\rho} r^{\rho(r)} + o(r^{\rho(r)}). \tag{5.3$_3$}$$

Now we estimate $\ln M(r, f)$. We cannot use estimate (5.2$_2$) for this purpose because (5.2$_2$) is known for $\delta < \varphi < 2\pi - \delta$ only. For this reason, we use a rougher estimate. Inequality (3.5) and formula (5.1$'$) imply

$$\ln M(r, f) < C(p)(-1)^p \ln f(-r).$$

Using (5.2$_2$) we get

$$\frac{\pi\Delta}{|\sin \pi l(r)|} r^{l(r)} + o(r^{l(r)}) < \ln M(r, f) < C(p) \frac{\pi\Delta}{|\sin \pi l(r)|} r^{l(r)} + o(r^{l(r)}).$$

It is clear that for $p = 0$ we have

$$\ln M(r, f) = \ln f(-r) = \frac{\pi\Delta}{\sin \pi l(r)} r^{l(r)} + o(r^{l(r)}). \tag{5.2$'_2$}$$

Let us compute the Nevanlinna characteristic for the function $f(z)$ defined by (5.1). Note that if $\psi(r) \to \infty$ as $r \to \infty$, then $\{\psi(r) + o(\psi(r))\}^+ = \{\psi(r)\}^+ + o(\psi(r))$. If, in addition, an estimate of $o(\psi(r))$ in the left-hand side is uniform with respect to some parameter, then the same is true for $o(\psi(r))$ in the right-hand side. The function $r^{l(r)} = R(r)$ satisfies the assumptions of Theorem 7.4 (Chapter 1) for an arbitrary $k > 1$. In fact, $R(kr) \le (1 + o(1))k^{l(r)} R(r) \le (1 + o(1))k^{\eta_2} R(r)$ and

$T(r, f) \le \ln M(r, f) = O(R(r))$. Using formula (7.15) from Chapter 1 and (5.2_2), we get

$$T(r,f) = m(r,f) = \frac{1}{2\pi}\int_0^{2\pi} \ln^+ |f(re^{i\varphi})|\, d\varphi$$
$$= r^{l(r)}\frac{\Delta}{2}\int_0^{2\pi}\left\{\frac{\cos l(r)(\varphi-\pi)}{\sin \pi l(r)}\right\}^+ d\varphi + o(r^{l(r)}).$$

An elementary computation shows that for $p < \lambda < p+1$ we have

$$I_1(\lambda) = \frac{1}{2}\int_0^{2\pi}\left\{\frac{\cos\lambda(\varphi-\pi)}{\sin\pi\lambda}\right\}^+ d\varphi = \begin{cases} \frac{p+|\sin\pi\lambda|}{\lambda|\sin\pi\lambda|} & \text{if } p<\lambda<p+\frac{1}{2},\\ \frac{p+1}{\lambda|\sin\pi\lambda|} & \text{if } p+\frac{1}{2}\le\lambda<p+1.\end{cases}$$

Hence

$$T(r,f) = \Delta I_1(l(r))r^{l(r)} + o(r^{l(r)}). \tag{5.4}$$

Taking into account condition 2) for $l(r)$, we get that for $r \ge r_0$ the value of $I_1(l(r))$ is bounded below and bounded above by positive constants. Therefore the function $f(z)$ has order η_2. The lower order of $f(z)$ is equal to η_1. If $l(r) = \rho(r) \to \rho$ as $r \to \infty$, then relation (5.4) can be written as

$$T(r,f) = \Delta I_1(\rho)r^{\rho(r)} + o(r^{\rho(r)}). \tag{5.4′}$$

It is clear that $n(r, f) \equiv 0$ and $N(r, f) \equiv 0$. We find $m\left(r, \frac{1}{f}\right)$ using a similar argument:

$$\begin{aligned} m\left(r,\frac{1}{f}\right) &= r^{l(r)}\frac{\Delta}{2}\int_0^{2\pi}\left\{-\frac{\cos l(r)(\varphi-\pi)}{\sin\pi l(r)}\right\}^+ d\varphi + o(r^{l(r)})\\ &= \Delta I_2(l(r))r^{l(r)} + o(r^{l(r)}),\end{aligned} \tag{5.5}$$

where

$$I_2(\lambda) = \frac{1}{2}\int_0^{2\pi}\left\{-\frac{\cos\lambda(\varphi-\pi)}{\sin\pi\lambda}\right\}^+ d\varphi$$
$$= \begin{cases} \frac{p}{\lambda|\sin\pi\lambda|} & \text{if } p<\lambda<p+\frac{1}{2},\\ \frac{p+1-|\sin\pi\lambda|}{\lambda|\sin\pi\lambda|} & \text{if } p+\frac{1}{2}\le\lambda<p+1.\end{cases}$$

Let us compute $N\left(r, \frac{1}{f}\right)$. Choose an arbitrary $\varepsilon > 0$. Let $r_0(\varepsilon)$ be such that $|n(r, 0) - \Delta r^{l(r)}| < \varepsilon r^{l(r)}$ for $r > r_0(\varepsilon)$. Select a, $0 < a < 1$, to be so small that $a^{\eta_1} < \eta_1\varepsilon$. For $r > r_1(\varepsilon, a) \ge a^{-1}r_0(\varepsilon)$ and $a \le k \le 1$, we have the inequality

$|(kr)^{l(r)} - (kr)^{l(r)}| < \varepsilon(kr)^{l(r)}$. The following estimate holds:

$$\begin{aligned}
&\left|N\left(r, \frac{1}{f}\right) - \Delta \int_{ar}^{r} t^{l(r)-1}dt\right| \\
&\leq \int_{1}^{ar} \frac{n\left(t, \frac{1}{f}\right)}{t}dt + \left|\int_{ar}^{r} \frac{n\left(t, \frac{1}{f}\right) - \Delta t^{l(t)}}{t}dt\right| + \left|\Delta \int_{ar}^{r} \frac{t^{l(t)} - t^{l(r)}}{t}dt\right| \\
&\leq K \int_{1}^{ar} t^{l(t)-1}dt + \varepsilon \int_{ar}^{r} t^{l(t)-1}dt + \Delta \int_{a}^{r} \frac{|(\tau r)^{l(\tau r)} - (\tau r)^{l(r)}|}{\tau}d\tau \\
&\leq K \int_{1}^{ar} t^{l(t)-1}dt + \varepsilon(1+\varepsilon) \int_{a}^{1} \frac{(\tau r)^{l(r)}}{\tau}d\tau + \Delta\varepsilon \int_{a}^{1} \frac{(\tau r)^{l(r)}}{\tau}d\tau \\
&= K \int_{1}^{ar} t^{l(t)-1}dt + \varepsilon(1+\Delta+\varepsilon)r^{l(r)} \int_{a}^{1} \tau^{l(r)-1}d\tau \\
&\leq K \int_{1}^{ar} t^{l(t)-1}dt + \varepsilon\frac{1+\Delta+\varepsilon}{l(r)}r^{l(r)}.
\end{aligned}$$

On the other hand,

$$\begin{aligned}
\limsup_{r\to\infty} \frac{\int_1^{ar} t^{l(t)-1}dt}{r^{l(r)}} &\leq \limsup_{r\to\infty} \frac{\frac{d}{dr}\int_1^{ar} t^{l(t)-1}dt}{\frac{d}{dr}r^{l(r)}} \\
&= \limsup_{r\to\infty} a^{l(r)} \frac{(ar)^{l(ar)}}{(ar)^{l(r)}(l(r) + l'(r)r\ln r)} = \frac{a^{\eta_1}}{\eta_1} < \varepsilon.
\end{aligned}$$

Since

$$\int_{ar}^{r} t^{l(r)-1}dt = \frac{r^{l(r)}}{l(r)}(1 - a^{l(r)}),$$

for $r \geq r_2 \geq r_1(\varepsilon, a)$ we have

$$\left|N\left(r, \frac{1}{f}\right) - \frac{\Delta r^{l(r)}}{l(r)}\right| \leq \varepsilon r^{l(r)} \left(K + \Delta + \frac{1+\Delta+\varepsilon}{\eta_1}\right).$$

Thus

$$N\left(r, \frac{1}{f}\right) = \frac{\Delta}{l(r)}r^{l(r)} + o(r^{l(r)}). \tag{5.6}$$

Suppose $a \neq 0, \infty$, and let $E_\delta(r)$ be the set

$$\bigcup_k \{|l(r)(\varphi - \pi) - \pi(k + 1/2)| < \delta\} \cup [0, \delta] \cup [2\pi - \delta, 2\pi],$$

where the union is taken over those integers k for which $-l(r)-1/2 \leq k \leq l(r)-1/2$ (if $p = 0$, there may be no such k). It is easy to see that $\operatorname{meas} E_\delta(r) \leq A\delta$, where A is a constant independent of r. From (5.2_2) we see that for $r \geq r_0$ and $\varphi \in [0, 2\pi]\backslash E_\delta(r)$,

$$\ln^+ \frac{1}{|f(re^{i\varphi}) - a|} = O(1).$$

Hence by Theorem 7.4 of Chapter 1 and Remark 1 following it, we get

$$m(r, a, f) = o(r^{l(r)}). \tag{5.7}$$

The First Fundamental Theorem together with (5.7) and (5.4) implies

$$N(r, a, f) = T(r, f) + o(r^{l(r)}) = \Delta I_1(l(r))r^{l(r)} + o(r^{l(r)}). \tag{5.8}$$

We have already proved an estimate for $\ln M(r,f)$, but it is sharp for $p=0$ only. Now we shall get a sharp estimate in the case where $l(r)=\rho(r)\to\rho$ as $r\to\infty$, that is, where $l(r)$ is a proximate order. We shall complement (5.3_2) by the following asymptotic inequality:

$$\ln|f(re^{i\varphi})|\le\pi\Delta\frac{\cos\rho(\varphi-\pi)}{\sin\pi\rho}r^{\rho(r)}+o(r^{\rho(r)}) \tag{5.9}$$

as $r\to\infty$, uniformly with respect to φ, $0\le\varphi\le 2\pi$.

Denote by D_δ the domain $\{|z|>1,|\arg z|<\delta\}$, where $0<\delta<\frac{\pi}{2}\min\left(1,\frac{1}{\rho}\right)$. Choose an arbitrarily small $\omega>0$ and denote by μ the number

$$\mu=\frac{\cos\rho(\delta-\pi)}{\cos\rho\delta}+\omega\frac{\sin\pi\rho}{\pi\Delta\cos\rho\delta}.$$

In the domain D_δ, consider the function

$$\psi_\delta(z)=f(z)[f(-z)]^{-\mu}.$$

For the ray $\arg z=\delta$ we have the following estimate:

$$\begin{aligned}\ln|\psi_\delta(re^{i\delta})|&=\ln|f(re^{i\delta})|-\mu\ln|f(re^{i(\delta+\pi)})|\\&=\pi\Delta\frac{\cos\rho(\delta-\pi)}{\sin\pi\rho}r^{\rho(r)}-\mu\pi\Delta\frac{\cos\rho\delta}{\sin\pi\rho}r^{\rho(r)}+o(r^{\rho(r)})=-\omega r^{\rho(r)}+o(r^{\rho(r)}).\end{aligned}$$

A similar estimate holds for the ray $\arg z=-\delta$. Therefore the function $\psi_\delta(z)$ is bounded on the boundary D_δ with ∞ excluded:

$$|\psi_\delta(z)|\le M,\quad z\in\partial D_\delta\backslash\{\infty\}.$$

Consider ρ' satisfying $\rho<\rho'<\frac{\pi}{2\delta}$, and an arbitrary number $\xi>0$. The function $\psi_\delta(z)\exp\{-\xi z^{\rho'}\}$ is continuous in the closure of D_δ and in the closure of D_δ in the extended complex plane, if we define its value at ∞ as 0. In fact, for $|\varphi|\le\delta$ we have

$$\begin{aligned}\ln\left|\psi_\delta(re^{i\varphi})e^{-\xi r^{\rho'}e^{i\varphi\rho'}}\right|&\le\ln M(r,f)-\mu\ln|f(re^{i(|\varphi|+\pi)})|-\xi r^{\rho'}\cos\varphi\rho'\\&\le O(r^{\rho(r)})-\mu\pi\Delta\frac{\cos\rho|\varphi|}{\sin\pi\rho}r^{\rho(r)}+o(r^{\rho(r)})-\xi r^{\rho'}\cos\delta\rho'\\&=O(r^{\rho(r)})-\xi r^{\rho}\cos\delta\rho'\to-\infty\quad\text{as}\quad r\to\infty.\end{aligned}$$

Hence we may use the maximum modulus principle for $\psi_\delta(z)\exp(-\xi z^{\rho'})$. On the other hand, for $z\in D_\delta$ we have $|\exp(-\xi z^{\rho'})|<1$. Hence

$$|\varphi_\delta(z)\exp(-\xi z^{\rho'})|\le M$$

for $z\in D_\delta$. Passing in this inequality to the limit as $\xi\to 0$, we see that for $z\in D_\delta$ the inequality $|\psi_\delta(z)|\le M$ holds. For $|\varphi|\le\delta$ we have

$$\begin{aligned}\ln|f(re^{i\varphi})|&=\mu\ln|f(re^{i(|\varphi|+\pi)})|+\ln|\psi_\delta(re^{i\varphi})|\\&\le\mu\pi\Delta\frac{\cos\rho|\varphi|}{\sin\pi\rho}r^{\rho(r)}+o(r^{\rho(r)})+\ln M\\&\le\frac{\mu\pi\Delta}{\sin\pi\rho}r^{\rho(r)}+o(r^{\rho(r)}).\end{aligned}\tag{5.10}$$

Let $\varepsilon > 0$ be an arbitrarily small positive number. Choose $\delta > 0$ and $\omega > 0$ in such a way that for $0 \le \varphi \le \delta$ and $2\pi - \delta \le \varphi \le 2\pi$ the inequality

$$\varepsilon + \Delta\pi \frac{\cos \rho(\varphi - \pi)}{\sin \pi\rho} > \Delta\pi \frac{\cos \rho\pi}{\sin \pi\rho} > \frac{\mu\pi\Delta}{\sin \pi\rho} - \varepsilon$$

holds. Then by (5.10) we get that for $0 \le \varphi \le \delta$ and $2\pi - \delta \le \varphi \le 2\pi$ the inequality

$$\ln |f(re^{i\varphi})| \le \Delta\pi \frac{\cos \rho(\varphi - \pi)}{\sin \pi\rho} r^{\rho(r)} + 2\varepsilon r^{\rho(r)} + o(r^{\rho(r)})$$

holds uniformly with respect to φ.

By (5.3_2) this inequality holds also for $\delta \le \varphi \le 2\pi - \delta$. Thus, we have proved inequality (5.9).

Relation (5.3_2) implies that

$$\ln |f(re^{i\varphi_0})| = \frac{\pi\Delta}{\sin \pi\rho} r^{\rho(r)} + o(r^{\rho(r)}),$$

where $\varphi_0 = \pi$ for even p and $\varphi_0 = \pi + \frac{\pi}{\rho}$ for odd p. Together with (5.9) it implies

$$\ln M(r, f) = \frac{\pi\Delta}{\sin \pi\rho} r^{\rho(r)} + o(r^{\rho(r)}). \tag{5.11}$$

The function $f(z)$ given by (5.1) will be used in several constructions. First, let us mention the following theorem.

THEOREM 5.1. *Let $\rho(r)$ be a proximate order, $\rho(r) \to \rho$ as $r \to \infty$, $L(r) = r^{\rho(r)-\rho}$. There exists a function $F(z)$ analytic in $\{|\arg z| < \pi\}$, taking positive values on the positive semi-axis, and such that for every δ satisfying $0 < \delta < \pi$ the condition*

$$F(re^{i\varphi}) = (1 + o(1))L(r), \qquad r \to \infty,$$

holds uniformly with respect to φ in $\{|\varphi| < \pi - \delta\}$.

Indeed, let $f(z)$ be a canonical product with positive zeros, such that $n(r, 0) \sim \pi^{-1} r^{1/2+\rho(r)-\rho}$. The desired function $F(z)$ can be chosen in $\{|\arg z| < \pi\}$ as

$$F(z) = \frac{\ln f(-z)}{\sqrt{z}},$$

where the branch of $\sqrt{z}$ is chosen in such a way that $\sqrt{r} > 0$ for $r > 0$. Since $f(-r) > 1$, we have $F(r) > 0$. Taking into account inequality (5.3_1) we obtain (for $|\varphi| < \pi - \delta$)

$$F(re^{i\varphi}) = \frac{e^{i\frac{\varphi}{2}} r^{\frac{1}{2}+\rho(r)-\rho} + o(r^{\frac{1}{2}+\rho(r)-\rho})}{e^{i\frac{\varphi}{2}} r^{\frac{1}{2}}} = L(r) + o(L(r)).$$

Let $f(z)$ be an entire function of order ρ, $0 < \rho < \infty$. The inequality $T(r, f) \le \ln^+ M(r, f) \le 3T(2r, f)$ (see (7.1) in Chapter 1) implies that if

$$0 < \liminf_{r\to\infty} \frac{\ln M(r, f)}{r^\rho} \le \limsup_{r\to\infty} \frac{\ln M(r, f)}{r^\rho} < \infty,$$

then

$$0 < \liminf_{r\to\infty} \frac{T(r, f)}{r^\rho} \le \limsup_{r\to\infty} \frac{T(r, f)}{r^\rho} < \infty,$$

and vice versa. It is natural to ask whether the existence of the limit

$$\lim_{r\to\infty} \frac{\ln M(r, f)}{r^\rho} \ne 0, \infty \tag{5.12}$$

implies the existence of the limit

$$\lim_{r\to\infty}\frac{T(r,f)}{r^\rho}\neq 0,\infty, \tag{5.13}$$

and vice versa. It turns out that in both cases the answer to this question is negative. Let us give the corresponding examples.

EXAMPLE 1. We give an example showing that the existence of the limit in (5.12) does not imply the existence of the limit in (5.13). First we let $0<\rho<1/2$. We construct two proximate orders, $\rho_1(r)$ and $\rho_2(r)$, satisfying for $r>r_0$ the inequality

$$\rho-\frac{\ln\ln r}{\ln r}\le\rho_j(r)\le\rho,\qquad j=1,2.$$

Observe that $r^{\rho-\frac{\ln\ln r}{\ln r}}=\frac{r^\rho}{\ln r}$. Choose $\rho_j(r)$ such that

$$\max(\rho_1(r),\rho_2(r))\equiv\rho,$$

the equality $\rho_1(r_n)=\rho-\frac{\ln\ln r_n}{\ln r_n}$ hold for some sequence $r_n\to\infty$, and $\rho_1(r_n')=\rho_2(r_n')=\rho$ for some sequence $r_n'\to\infty$. Let $V_j(z)$ be canonical products of genus 0 with positive zeros satisfying $n(r,0,V_j)\sim r^{\rho_j(r)}$. Consider the function

$$f(z)=V_1(z)+V_2(-z).$$

It is clear that $|f(\bar z)|=|f(z)|$; therefore it suffices to get suitable estimates for $|f(z)|$ in the upper half-plane $\{\operatorname{Im}z\ge0\}$. We use relations ($5.3_2$) and (5.11). Since

$$M(r,f)\le2\max_{j=1,2}M(r,V_j),$$

we have

$$\ln M(r,f)\le\frac{\pi}{\sin\pi\rho}r^\rho+o(r^\rho).$$

On the other hand,

$$V_j(r)\le\exp(\pi\cot\rho\pi r^{\rho_j(r)}+o(r^{\rho_j(r)}))$$

and

$$V_j(-r)=\exp\left(\frac{\pi}{\sin\rho\pi}r^{\rho_j(r)}+o(r^{\rho_j(r)})\right).$$

Denote by $s(r)$ the function that is equal to 1 at those values of r for which $\rho_1(r)=\rho$, and is equal to -1 if $\rho_1(r)<\rho$ (at such points, $\rho_2(r)=\rho$). It is easy to see that

$$|f(-s(r)r)|=\exp\left\{\frac{\pi}{\sin\pi\rho}r^\rho+o(r^\rho)\right\};$$

therefore

$$\ln M(r,f)=\frac{\pi}{\sin\pi\rho}r^\rho+o(r^\rho),$$

and (5.12) is satisfied. For $r=r_n$ we have

$$T(r_n,V_1)\le\ln M(r_n,V_1)=O(r_n^{\rho_1(r_n)})=O\left(\frac{r_n^\rho}{\ln r_n}\right).$$

Taking into account (5.4′), we get

$$
\begin{aligned}
T(r_n,f) &= T(r_n,V_2) + O\left(\frac{r_n^\rho}{\ln r_n}\right) \\
&= I_1(\rho) r_n^{\rho_2(r_n)} + o(r_n^{\rho_2(r_n)}) + O\left(\frac{r_n^\rho}{\ln r_n}\right) \\
&= I_1(\rho) r_n^\rho + o(r_n^\rho) = \frac{1}{\rho} r_n^\rho + o(r_n^\rho).
\end{aligned}
$$

For $r = r_n'$ and $\delta \le \varphi \le \pi/2 - \delta$ or $\pi/2 + \delta \le \varphi \le \pi - \delta$, $\delta > 0$, by (5.3_2) we have

$$
\ln |f(r_n' e^{i\varphi})| = \pi \frac{\max\{\cos\rho(\pi-\varphi), \cos\rho\varphi\}}{\sin \pi\rho} r_{n'}^\rho + o(r_{n'}^\rho).
$$

In view of Theorem 7.4 from Chapter 1 and Remark 2 following it, we get

$$
\begin{aligned}
T(r_n', f) = m(r_n', f) &= \frac{r_n'^{\,\rho}}{\sin\pi\rho} \int_0^\pi \max\{\cos\rho(\pi-\varphi), \cos\rho\varphi\}\, d\varphi + o(r_n'^{\,\rho}) \\
&= \frac{2 r_n'^{\,\rho}}{\sin\pi\rho} \int_0^{\pi/2} \cos\rho\varphi \, d\varphi + o(r_n'^{\,\rho}) = \frac{2\sin\frac{\pi\rho}{2}}{\rho\sin\pi\rho} r_n'^{\,\rho} + o(r_n'^{\,\rho}) \\
&= \frac{1}{\rho}\sec\frac{\pi\rho}{2} r_n'^{\,\rho} + o(r_n'^{\,\rho}).
\end{aligned}
$$

Hence

$$
\liminf_{r\to\infty} \frac{T(r,f)}{r^\rho} \le \frac{1}{\rho} < \frac{1}{\rho}\sec\frac{\pi\rho}{2} \le \limsup_{r\to\infty}\frac{T(r,f)}{r^\rho};
$$

that is, the limit in (5.13) does not exist. If instead of $f(z)$ we consider the functions $f(z^p)$, $p = 2, 3, 4, \dots$, then by Theorem 6.6 of Chapter 1 and the obvious equality $\ln M(r, f(z^p)) = \ln M(r^p, f(z))$, we get analogous examples for each ρ, $0 < \rho < \frac{p}{2}$, that is, for each ρ, $0 < \rho < \infty$.

EXAMPLE 2. We now show that the existence of the limit in (5.13) does not imply the existence of the limit in (5.12). The idea of this example is similar to the idea of Example 1, but the construction is somewhat more technical. Let $0 < \rho < 1/2$. Consider the equation

$$
\begin{aligned}
F(x,y) &= y\cot\rho\pi\left\{\frac{y}{x\cos\rho\pi} - 1\right\} \\
&\quad + \sqrt{x^2\left[1 + \cot^2\rho\pi\left\{\frac{y}{x\cos\rho\pi}-1\right\}^2\right] - y^2} \\
&\quad - 2\sin\frac{\rho\pi}{2}\sqrt{1+\cot^2\rho\pi\left\{\frac{y}{x\cos\rho\pi}-1\right\}^2} = 0,
\end{aligned}
\tag{5.14}
$$

where the arithmetical values of the square roots are taken, and x and y are real variables. It is easy to check that $F(1,1) = 0$, $F_x'(1,1) = \tan\frac{\rho\pi}{2}$, and $F_y'(1,1) = \tan\frac{\rho\pi}{2}$. The well-known theorems on implicit functions imply that in a sufficiently small neighborhood of $(1,1)$ equation (5.14) defines a univalent continuously differentiable decreasing function $y = y(x)$, $y(1) = 1$, $|y'(x) - (-1)| < 1/2$.

Let $L_1(r) = 1 + \eta \sin\ln\ln(e+r)$, $r \ge 0$, where $\eta > 0$ is chosen to be so small that the interval $[1-\eta, 1+\eta]$ be contained in the interval where the function $y(x)$

with described properties is defined. Let $L_2(r) = y(L_1(r))$. We may assume that η is so small that the inequality

$$\cos \pi\rho < y(x)/x < \sec \pi\rho$$

is satisfied for $1 - \eta \le x \le +\eta$. Hence, for all $r > 0$ the inequality

$$\cos \pi\rho < L_2(r)/L_1(r) < \sec \pi\rho \tag{5.15}$$

holds. Denote by $V_j(r)$ the canonical product of genus 0 with positive zeros, for which $n(r, 0, V_j) \sim r^\rho L_j(r)$, $j = 1, 2$. It is easy to verify that the functions $\rho_j(r)$ defined by the equations $r^{\rho_j(r)} = r^\rho L_j(r)$ are proximate orders, and $\rho_j(r) \to \rho$ as $r \to \infty$. As in Example 1 we let

$$f(z) = V_1(z) + V_2(-z).$$

Using the same argument as in Example 1, we get

$$\begin{aligned}\ln M(r, f) &= \max_{j=1,2} \ln M(r, V_j) + o(r^\rho) \\ &= \frac{\pi}{\sin \pi\rho} r^\rho \max\{L_1(r), L_2(r)\} + o(r^\rho).\end{aligned}$$

Hence

$$\begin{aligned}\limsup_{r\to\infty} \frac{\ln M(r, f)}{r^\rho} &= \frac{\pi}{\sin \pi\rho} \limsup_{r\to\infty} \max\{L_1(r), L_2(r)\} \\ &= \frac{\pi}{\sin \pi\rho} \max\{(1 + \eta), y(1 - \eta)\} \\ &\ge (1 + \eta)\frac{\pi}{\sin \pi\rho}, \\ \liminf_{r\to\infty} \frac{\ln M(r, f)}{r^\rho} &= \frac{\pi}{\sin \pi\rho} \liminf_{r\to\infty} \max\{L_1(r), L_2(r)\} < \frac{\pi}{\sin \pi\rho}.\end{aligned}$$

Thus the limit in (5.12) does not exist. Denote by $\varphi(r)$, $0 < \varphi(r) < \pi$, the (unique) solution of the equation

$$L_1(r) \cos \rho(\pi - \varphi) = L_2(r) \cos \rho\varphi. \tag{5.16}$$

The existence of the solution follows from (5.15). From (5.3_2) it is easy to get that for each positive δ we have

$$\ln |f(re^{i\varphi})| = \begin{cases} \pi r^\rho L_2(r) \frac{\cos \rho\varphi}{\sin \rho\pi} + o(r^\rho), & \delta \le \varphi \le \varphi(r) - \delta, \\ \pi r^\rho L_1(r) \frac{\cos \rho(\pi - \varphi)}{\sin \rho\pi} + o(r^\rho), & \varphi(r) + \delta \le \varphi \le \pi - \delta. \end{cases}$$

Taking into account Theorem 7.4 of Chapter 1 and Remark 1 following it, we obtain

$$\begin{aligned} &T(r, f) \\ &= \frac{r^\rho}{\sin \pi\rho} \left\{ L_2(r) \int_0^{\varphi(r)} \cos \rho\varphi + L_1(r) \int_{\varphi(r)}^{\pi} \cos \rho(\pi - \varphi) d\varphi \right\} + o(r^\rho) \\ &= \frac{r^\rho}{\rho \sin \pi\rho} \{L_2(r) \sin \rho\varphi(r) + L_1(r) \sin \rho(\pi - \varphi(r))\} + o(r^\rho).\end{aligned} \tag{5.17}$$

In view of (5.16),

$$\tan\rho\varphi(r) = \cot\rho\pi\left(\frac{L_2(r)}{L_1(r)\cos\rho\pi} - 1\right),$$

$$\sin\rho\varphi(r) = \frac{\cot\rho\pi\left(\frac{L_2(r)}{L_1(r)\cos\rho\pi} - 1\right)}{\left\{1 + \cot^2\rho\pi\left(\frac{L_2(r)}{L_1(r)\cos\rho\pi} - 1\right)^2\right\}^{\frac{1}{2}}},$$

$$\sin\rho[\pi - \varphi(r)] = \left\{1 - \left[\frac{L_2(r)}{L_1(r)}\cos\rho\varphi(r)\right]^2\right\}^{\frac{1}{2}}$$

$$= \frac{\left\{1 + \cot^2\rho\pi\left\{\frac{L_2(r)}{L_1(r)\cos\rho\pi} - 1\right\}^2 - \frac{L_2^2(r)}{L_1^2(r)}\right\}^{\frac{1}{2}}}{\left\{1 + \cot^2\rho\pi\left\{\frac{L_2(r)}{L_1(r)\cos\rho\pi} - 1\right\}^2\right\}^{\frac{1}{2}}}.$$

From (5.14) and the definition of the function $L_2(r) = y(L_1(r))$, we see that

$$L_2(r)\sin\rho\varphi(r) + L_1(r)\sin\rho(\pi - \varphi(r)) \equiv 2\sin\frac{\rho\pi}{2}.$$

Now we can rewrite (5.17) as

$$T(r,f) = \frac{r^\rho}{\rho\cos\frac{\rho\pi}{2}} + o(r^\rho);$$

that is, the limit in (5.13) is defined. Considering the functions $f(z^p)$, as in Example 1, we remove the restriction $\rho < 1/2$.

It is worth mentioning that examples of the same type as Example 1 can be constructed using another idea. We restrict our attention to the case of $\rho = 1$ because in this case the construction is simpler.

EXAMPLE 1′. We define the integer sequence $\{r_k\}$ by the following recurrent relation: $r_1 = 1$, $r_{k+1} = 3r_k^2$. Take $p_k = r_k^2$ and $\eta_{2k} = 1$, $k = 0, 1, 2, \ldots$. Let $\eta_{2k+1} = 1$ if $r_{2n-1} \leq 2k+1 < r_{2n}$, and $\eta_{2k+1} = 0$ if $r_{2n} \leq 2k+1 < r_{2n+1}$. Let $\eta'_{2k} = 0$ and $\eta'_{2k+1} = \eta_{2k+1}$. Note that for an integer $s \geq r_{k+1}$ we have

$$\text{(5.18)} \qquad \frac{p_k^s}{s!} < \frac{r_k^{2s}e^s}{s^s} \leq \left(\frac{r_k^2 e}{r_{k+1}}\right)^s = \left(\frac{e}{3}\right)^s.$$

The function

$$f(z) = \sum_{s=0}^{\infty}\eta_s\frac{z^s}{s!}$$

has the desired properties. It is clear that

$$\cosh r \leq M(r,f) = f(r) \leq e^r.$$

Hence

$$\lim_{r\to\infty}\frac{\ln M(r,f)}{r} = 1.$$

If $|z| = p_{2n-1}$, then (see (5.18)) the following inequality holds:

$$|e^z - f(z)| = \left|\sum_{s=0}^{\infty}(1-\eta_s)\frac{z^s}{s!}\right| \le \sum_{s=0}^{r_{2n-1}} \frac{|z|^s}{s!} + \sum_{s=r_{2n}}^{\infty} \frac{|z|^s}{s!}$$
$$\le p_{2n-1}^{r_{2n-1}} \sum_{s=0}^{\infty} \frac{1}{s!} + \sum_{s=r_{2n}}^{\infty} \left(\frac{e}{3}\right)^s \le e p_{2n-1}^{\sqrt{p_{2n-1}}} + o(1).$$

Hence, for $r = p_{2n-1}$ we have

$$f(re^{i\varphi}) = \exp(re^{i\varphi}) + O(\exp(\sqrt{r}\ln r))$$

and

$$T(r,f) = T(r,e^z) + O(\sqrt{r}\ln r) = \frac{r}{\pi} + O(\sqrt{r}\ln r).$$

Thus

$$\liminf_{r\to\infty} \frac{T(r,f)}{r} \le \frac{1}{\pi}. \tag{5.19}$$

In a similar manner, for $|z| = p_{2n}$ we get

$$|f(z) - \cosh z| = \left|\sum_{s=0}^{\infty} \eta_s' \frac{z^s}{s!}\right| \le \sum_{s=0}^{r_{2n}} \frac{|z|^s}{s!} + \sum_{s=r_{2n+1}}^{\infty} \frac{|z|^s}{s!}$$
$$\le p_{2n}^{r_{2n}} e + o(1) = e p_{2n}^{\sqrt{p_{2n}}} + o(1).$$

Therefore for $r = p_{2n}$ we have

$$f(re^{i\varphi}) = \cosh(re^{i\varphi}) + O(\exp(\sqrt{r}\ln r)),$$
$$T(r,f) = T(r,\cosh z) + O(\sqrt{r}\ln r) = \frac{2r}{\pi} + O(\sqrt{r}\ln r).$$

(Here we have used the result of Exercise 2 to Theorem 6.1 in Chapter 1.) It follows that

$$\limsup_{r\to\infty} \frac{T(r,f)}{r} \ge \frac{2}{\pi}. \tag{5.20}$$

Combining (5.19) and (5.20) we conclude that the limit of $T(r,f)/r$ as $r \to \infty$ does not exist.

2°. We now consider the entire functions of integer order $\rho > 0$ with positive zeros such that $n(r,0) \sim \Delta r^{\rho(r)}$, where $0 < \Delta < \infty$ and $\rho(r)$ is a proximate order, $\rho(r) \to \rho$ as $r \to \infty$. We use the notation from the proof of Theorem 4.3. In this case $f(z) = f_1(z)f_2(z)$, where

$$f_1(z) = \exp\left\{ z^\rho \left(c_\rho + \frac{1}{\rho} \sum_{0<a_\nu\le r} a_\nu^{-\rho} \right) \right\},$$

$$f_2(z) = e^{c_{\rho-1}z^{\rho-1}+\dots+c_1 z} \prod_{0<a_\nu\le r} E\left(\frac{z}{a_\nu}, \rho-1\right) \prod_{a_\nu>r} E\left(\frac{z}{a_\nu}, \rho\right)$$
$$= e^{c_{\rho-1}z^{\rho-1}+\dots+c_1 z} f_3(z).$$

Consider the function $\ln f(z)$ in the angle $\{0 < \delta \le \arg z \le 2\pi - \delta\}$, $\ln f(0) = \ln f_2(0) = \ln f_1(0) = \ln f_3(0) = 0$. We have

$$|\ln f(z) - \ln f_1(z)| = |\ln f_2(z)| \le O(|z|^{\rho-1}) + |\ln f_3(z)|.$$

On the other hand, the function $\ln f_3(z)$ can be represented in the form (cf. the computation in part 1° of this section)

$$\begin{aligned}\ln f_3(z) &= -z^\rho \int_0^r n(t,0)\frac{dt}{t^\rho(t-z)} - z^\rho n(r,0)\int_r^\infty \frac{dt}{t^\rho(t-z)} \\ &\quad - z^{\rho+1}\int_r^\infty \{n(t,0)-n(r,0)\}\frac{dt}{t^{\rho+1}(t-z)} \\ &= -z^\rho \int_0^r n(t,0)\frac{dt}{t^\rho(t-z)} - \frac{z^\rho}{\rho r^\rho}n(r,0) - z^{\rho+1}\int_r^\infty n(t,0)\frac{dt}{t^{\rho+1}(t-z)}.\end{aligned}$$

For $\delta < \varphi < 2\pi - \delta$ we have $|t - re^{i\varphi}| \geq \sin\frac{\delta}{2}(t+r)$; hence

$$|\ln f_3(re^{i\varphi})| \leq \frac{1}{\sin\frac{\delta}{2}}\left\{r^\rho \int_0^r \frac{n(t,0)}{t^\rho(t+r)}dt + \frac{1}{\rho}n(r,0) + r^{\rho+1}\int_r^\infty \frac{n(t,0)}{t^{\rho+1}(t+r)}dt\right\}.$$

Estimating the sum in braces, we derive, as in the proof of Theorem 4.3, that

$$|\ln f_3(re^{i\varphi})| = O(r^{\rho(r)}) \tag{5.21}$$

uniformly with respect to φ, $\delta < \varphi < 2\pi - \delta$. Hence

$$\ln f(z) = z^\rho\left(c_\rho + \frac{1}{\rho}\sum_{0<a_\nu\leq r} a_\nu^{-\rho}\right) + O(r^{\rho(r)}), \tag{5.22$_1$}$$

$$\ln|f(z)| = \operatorname{Re}\left\{z^\rho\left(c_\rho + \frac{1}{\rho}\sum_{0<a_\nu\leq r} a_\nu^{-\rho}\right)\right\} + O(r^{\rho(r)}), \tag{5.22$_2$}$$

$$\arg f(z) = \operatorname{Im}\left\{z^\rho\left(c_\rho + \frac{1}{\rho}\sum_{0<a_\nu\leq r} a_\nu^{-\rho}\right)\right\} + O(r^{\rho(r)}). \tag{5.22$_3$}$$

Note that in order to prove these relations, we used the relation $n(r,0) = O(r^{\rho(r)})$ only. Consider first the case where the integral

$$\int_1^\infty t^{\rho(t)-\rho-1}dt \tag{5.23}$$

is divergent. In this case we introduce the notation ($r \geq 2$)

$$L_1(r) = \int_1^r t^{\rho(t)-\rho-1}dt, \qquad r^{\rho_1(r)} = r^\rho L_1(r).$$

It is easy to check that $\rho_1(r)$ is a proximate order,

$$\rho_1(r) \to \rho, \qquad r^\rho = o(r^{\rho_1(r)}), \qquad r^{\rho(r)} = o(r^{\rho_1(r)}) \quad \text{as} \quad r \to \infty.$$

Using (5.22$_1$) we get

$$\ln f(z) = z^\rho \frac{1}{\rho}\sum_{0<a_\nu\leq r} a_\nu^{-\rho} + o(r^{\rho_1(r)}).$$

By l'Hôpital's rule,

$$\frac{1}{\rho}\sum_{0<a_\nu\le r} a_\nu^{-\rho} = \frac{1}{\rho}\int_0^r \frac{dn(t,0)}{t^\rho} = \frac{n(r,0)}{\rho r^\rho} + \int_0^r \frac{n(t,0)}{t^{\rho+1}}dt$$

$$= O(r^{\rho(r)-\rho}) + \Delta(1+o(1))\int_1^r t^{\rho(t)-\rho-1}dt = \Delta(1+o(1))L_1(r).$$

Hence

$$\ln f(re^{i\varphi}) = \Delta e^{i\rho\varphi} r^{\rho_1(r)} + o(r^{\rho_1(r)}), \tag{5.24$_1$}$$

$$\ln|f(re^{i\varphi})| = \Delta r^{\rho_1(r)}\cos\rho\varphi + o(r^{\rho_1(r)}), \tag{5.24$_2$}$$

and these estimates are uniform with respect to $\delta < \varphi < 2\pi - \delta$. From Theorem 4.3 we get

$$T(r,f) = m(r,f) = \frac{\Delta}{\pi} r^{\rho_1(r)} + o(r^{\rho_1(r)}).$$

By Theorem 7.4 of Chapter 1 we have

$$N(r,f) = 0, \qquad m\left(r,\frac{1}{f}\right) = \frac{\Delta}{\pi} r^{\rho_1(r)} + o(r^{\rho_1(r)}),$$

$$N\left(r,\frac{1}{f}\right) \sim \frac{\Delta}{\rho} r^{\rho(r)} = o(r^{\rho_1(r)}),$$

and for $a \neq 0, \infty$,

$$m\left(r,\frac{1}{f-a}\right) = o(r^{\rho_1(r)}).$$

Hence, by the First Fundamental Theorem,

$$N\left(r,\frac{1}{f-a}\right) = \frac{\Delta}{\pi} r^{\rho_1(r)} + o(r^{\rho_1(r)}).$$

It is worth mentioning that it can be shown, using the method of part 1°, that if in (5.24$_2$) we replace the equality sign by $\le$, the obtained inequality will hold everywhere in the complex plane.

EXERCISE. Using the infinite product representation of the entire function $1/\Gamma(z)$, where $\Gamma(z)$ is the Euler gamma-function, namely

$$\frac{1}{\Gamma(z)} = ze^{\gamma z}\sum_{n=1}^{\infty} E\left(-\frac{z}{n},1\right)$$

($\gamma = 0,577\ldots$ is the Euler–Mascheroni constant), prove the asymptotic formula for $\Gamma(z)$ in the domain $\{|\arg z| < \pi - \delta\}$, $\delta > 0$:

$$\ln\Gamma(z) = z\ln z + o(r\ln r).$$

Now we suppose that the integral in (5.23) converges. Let

$$\rho_2(r) = \rho + (\ln r)^{-1}\ln\left\{\int_r^\infty t^{\rho(t)-\rho-1}dt\right\}, \qquad r>1,$$

$$r^{\rho_2(r)} = r^\rho \int_r^\infty t^{\rho(t)-\rho-1}dt = r^\rho L_2(r).$$

As before, we check that $r^{\rho(r)} = o(r^{\rho_2(r)})$, $r^{\rho_2(r)} = o(r^{\rho})$. In the case under consideration the function $n(r,0)$ belongs to the convergence class, and the function $f(z)$ can be represented as

$$f(z) = e^{c'_\rho z^\rho + \cdots} \prod_\nu E\left(\frac{z}{a_\nu}, \rho - 1\right) = f_4(z) f_2(z),$$

where $f_2(z)$ is defined as before and

$$f_4(z) = \exp\left\{ z^\rho \left(c'_\rho - \frac{1}{\rho} \sum_{|a_\nu|>r} a_\nu^{-\rho} \right) \right\}.$$

The series $\sum_\nu a_\nu^{-\rho}$ converges since $n(r,0)$ belongs to the convergence class.

Relation (5.21) implies that the relation

$$|\ln f(re^{i\varphi}) - \ln f_4(re^{i\varphi})| \le O(r^{\rho(r)}) + O(r^{\rho-1}) = o(r^{\rho_2(r)}) \tag{5.25}$$

holds uniformly in $\varphi = \arg z$ within the angle $0 < \delta < \arg z < 2\pi - \delta$. In the same way as before we get

$$-\frac{1}{\rho} \sum_{a_\nu > r} a_\nu^{-\rho} = -\frac{1}{\rho} \int_r^\infty \frac{dn(t,0)}{t^\rho} = \frac{1}{\rho} \frac{n(r,0)}{r^\rho} - \int_r^\infty \frac{n(t,0)}{t^{1+\rho}} dt$$

$$= O(r^{\rho(r)-\rho}) - \Delta(1+o(1)) \int_r^\infty t^{\rho(t)-\rho-1} dt = -\Delta(1+o(1)) L_2(r).$$

In view of (5.25),

$$\ln f(re^{i\varphi}) = e^{i\rho\varphi} c'_\rho r^\rho - \Delta e^{i\rho\varphi} r^{\rho_2(r)} + o(r^{\rho_2(r)}).$$

If $c'_\rho = 0$, then

$$\ln f(re^{i\varphi}) = -\Delta e^{i\rho\varphi} r^{\rho_2(r)} + o(r^{\rho_2(r)}), \tag{5.26$_1$}$$

$$\ln |f(re^{i\varphi})| = -\Delta \cos \rho\varphi r^{\rho_2(r)} + o(r^{\rho_2(r)}). \tag{5.26$_2$}$$

If $c'_\rho \ne 0$, then

$$\ln f(re^{i\varphi}) = e^{i\rho\varphi} c'_\rho r^\rho + o(r^\rho), \tag{5.27$_1$}$$

$$\ln |f(re^{i\varphi})| = |c'_\rho| \cos(\rho\varphi + \arg c'_\rho) r^\rho + o(r^\rho). \tag{5.27$_2$}$$

As in the previous example, it can be shown that if we replace in (5.26$_2$) and (5.27$_2$) the equality sign by $\le$, then the obtained inequality will be valid everywhere in the complex plane.

For the Nevanlinna characteristics we get the same formulas as in the case where the integral (5.23) diverges, but with $\rho_1(r)$ replaced by $\rho_2(r)$ if $c'_\rho = 0$, and with $\rho_1(r)$ replaced by ρ and Δ replaced by $|c'_\rho|$ if $c'_\rho \ne 0$. To apply (4.5), we take into account that

$$c_\rho = c'_\rho - \frac{1}{\rho} \sum_\nu a_\nu^{-\rho}.$$

Finally, consider the case $\rho = 0$. Let $\rho(r)$ be a proximate order satisfying $\rho(r) \to 0$ as $r \to \infty$, and let $r^{\rho(r)}$ be monotone for $r \ge r_0$ and go to ∞ as $r \to \infty$. Consider a canonical product of genus 0 with positive zeros such that

$n(r,0) \sim \Delta r^{\rho(r)}$, $0<\Delta<\infty$. Without loss of generality we may assume that all the zeros are greater than 1. It is easy to check that

$$N(r,0)=\int_1^r \frac{n(t,0)}{t}dt=\Delta(1+o(1))\int_1^r t^{\rho(t)-1}dt=\Delta(1+o(1))L_1(r),$$

$$L_1\left(\frac{r}{2}\right)\sim L_1(r), \quad r^{\rho(r)}=o(L_1(r)) \text{ as } r\to\infty.$$

Let $\delta<\varphi<2\pi-\delta$. We apply (5.1′) to $f(z)$. We get

$$\begin{aligned}\ln f(re^{i\varphi})-N(r,0)&=-z\int_1^\infty \frac{n(t,0)dt}{t(t-z)}-\int_1^r\frac{n(t,0)}{t}dt\\&=-\int_1^r\frac{n(t,0)}{t-z}dt-z\int_r^\infty\frac{n(t,0)}{t(t-z)}dt,\end{aligned}$$

$$\begin{aligned}|\ln f(re^{i\varphi})-N(r,0)|&\le \csc\frac{\delta}{2}\left\{\int_1^r\frac{n(t,0)}{t+r}dt+r\int_r^\infty\frac{n(t,0)}{t(t+r)}dt\right\}\\&\le\csc\frac{\delta}{2}\left\{n(r,0)+r\int_r^\infty\frac{n(t,0)}{t^2}dt\right\}\\&\le\Delta(1+o(1))\csc\frac{\delta}{2}\left\{r^{\rho(r)}+r\int_r^\infty t^{\rho(t)-2}dt\right\}\\&=2\Delta(1+o(1))\csc\frac{\delta}{2}r^{\rho(r)}=o(N(r,0)).\end{aligned}$$

Therefore, uniformly with respect to φ satisfying $\delta<\varphi<2\pi-\delta$ the relations

$$\ln f(re^{i\varphi})=N(r,0)+o(N(r,0)), \tag{5.28$_1$}$$

$$\ln|f(re^{i\varphi})|=N(r,0)+o(N(r,0)), \tag{5.28$_2$}$$

$$\arg f(re^{i\varphi})=o(N(r,0)) \tag{5.28$_3$}$$

hold. It is clear that $\ln M(r,f)=\ln f(-r)=N(r,0)+o(N(r,0))$. Since

$$N(r,0)\le T(r,f)\le \ln M(r,f), \tag{5.29}$$

we get

$$T(r,f)=N(r,0)+o(N(r,0)), \qquad m\left(r,\frac{1}{f}\right)=o(N(r,0)). \tag{5.30}$$

It is easy to check that relations (5.30) hold even without the requirement that the zeros are positive. In fact, let

$$f(z)=\prod_\nu\left(1-\frac{z}{a_\nu}\right), \qquad \check f(z)=\prod_\nu\left(1-\frac{z}{|a_\nu|}\right).$$

Obviously, $M(r,f)\le M(r,\check f)=\check f(-r)$ and $N(r,0,f)=N(r,0,\check f)$. From (5.29) and (5.30) we get

$$\begin{aligned}N(r,0,f)\le T(r,f)\le \ln M(r,\check f)&=N(r,0,\check f)+o(N(r,0,\check f))\\&=N(r,0,f)+o(N(r,0,f));\end{aligned}$$

that is, (5.30) remains true. Using the positivity of the zeros of $f(z)$, relation (5.28_2), the First Fundamental Theorem, and Theorem 7.4 of Chapter 1, we conclude that

$$m(r,a,f) = o(N(r,0)), \qquad N(r,a) = (1+o(1))N(r,0), \quad a \neq 0, \infty.$$

3°. Finally we consider the Mittag-Leffler entire function

$$E_\rho(z) = \sum_{k=0}^{\infty} \frac{z^k}{\Gamma(1+\frac{k}{\rho})}, \qquad 0 < \rho < \infty. \tag{5.31}$$

Using Stirling's formula, it is easy to verify that this power series has an infinite radius of convergence. The function $E_\rho(z)$ is a generalization of the exponential function e^z, which corresponds to the case $\rho = 1$. For $\rho = 1/2$ we have $E_{1/2}(z) = \cosh\sqrt{z}$. We are going to find some asymptotic relations for $E_\rho(z)$. To do so we need the Hankel integral representation for the Γ-function. Let $s = \sigma + i\tau$ be an arbitrary complex number. Denote by $L(\theta, a)$, $\frac{\pi}{2} < \theta \le \pi$, $0 < a < \infty$, the boundary of the domain $\{|\arg z| < \theta, |z| > a\}$, parameterized in the negative direction.

Consider the integral

$$F(s) = \frac{1}{2\pi i}\int_{L(\theta,a)} e^t t^{-s}\,dt, \tag{5.32}$$

where by t^{-s} we mean the principal value of the power function. It is easy to check that the integral (5.32) converges absolutely and uniformly with respect to s, for $|s| \le R$, where R is an arbitrarily large number. Using the Morera theorem for the function $F(s)$ and changing the order of integration, we conclude that $F(s)$ is an entire function. The integral (5.32) does not depend on a, since the integral of $e^t t^{-s}$ over the boundary of the region $\{|\arg z| < \theta,\ a_1 < |z| < a_2\}$ is equal to zero by the Cauchy theorem. The integral (5.32) does not depend on θ, $\frac{\pi}{2} < \theta \le \pi$, either. In fact, let $\frac{\pi}{2} < \theta_1 < \theta_2 \le \pi$. The difference

$$\int_{L(\theta_1,a)} e^t t^{-s}\,dt - \int_{L(\theta_2,a)} e^t t^{-s}\,dt \tag{5.33}$$

does not depend on a, and is equal to the integral of $e^t t^{-s}$ taken over the boundary of the set $\{\theta_1 < |\arg z| < \theta_2, |z| > a\}$. On the other hand, on the arcs $\{\theta_1 < |\arg z| < \theta_2, |z| = a\}$ the integrand satisfies the estimate $|e^t t^{-s}| \le e^{a\cos\theta_1} a^{-\sigma} e^{\theta_2 \tau} = o(a^{-1})$ as $a \to +\infty$. Therefore the difference (5.33) approaches zero as $a \to +\infty$. Since the difference (5.33) does not depend on a, it should be equal to 0.

Thus, in (5.32) we do not need to indicate the dependence of $F(s)$ on θ and a. Now we let $0 < s = \sigma < 1$. We have

$$\begin{aligned} F(\sigma) &= \frac{1}{2\pi i}\int_{L(\pi,a)} e^t t^{-\sigma}\,dt \\ &= \frac{e^{\pi i\sigma}}{2\pi i}\int_{-\infty}^{-a} e^x |x|^{-\sigma}\,dx \\ &\quad + \frac{1}{2\pi}\int_{-\pi}^{\pi} e^{ae^{i\varphi}} a^{1-\sigma} e^{i(1-\sigma)\varphi}\,d\varphi + \frac{e^{-\pi i\sigma}}{2\pi i}\int_{-a}^{-\infty} e^x |x|^{-\sigma}\,dx \\ &= \frac{\sin\pi\sigma}{\pi}\int_a^{\infty} e^{-x} x^{-\sigma}\,dx + a^{1-\sigma}\frac{1}{2\pi}\int_{-\pi}^{\pi} e^{ae^{i\varphi} + i(1-\sigma)\varphi}\,d\varphi. \end{aligned}$$

Letting $a \to 0$ and using the formula for the complement of the Γ-function, we get

$$F(\sigma) = \frac{\sin \pi\sigma}{\pi} \int_0^\infty e^{-x} x^{-\sigma} dx = \frac{\sin \pi\sigma}{\pi} \Gamma(1-\sigma) = \frac{1}{\Gamma(\sigma)}.$$

Since the entire functions $F(s)$ and $1/\Gamma(s)$ coincide on the interval $0 < \sigma = s < 1$, they are identically equal. From (5.32) we get the Hankel formula

$$(5.34)\qquad \frac{1}{\Gamma(s)} = \frac{1}{2\pi i} \int_{L(\theta,a)} e^t t^{-s} dt, \qquad \frac{\pi}{2} < \theta \le \pi,\ 0 < a < \infty.$$

We introduce the notation $\alpha = 1/\rho$. Let $|z| < 1$, $a > 1$, and $\frac{\pi}{2} < \theta \le \pi$. Using (5.31) and (5.34) we get

$$(5.35)\qquad \begin{aligned} E_\rho(z) &= \sum_{k=0}^\infty \frac{z^k}{2\pi i} \int_{L(\theta,a)} e^t t^{-\alpha k - 1} dt \\ &= \frac{1}{2\pi i} \int_{L(\theta,a)} \frac{e^t}{t} \left\{ \sum_{k=0}^\infty \left(\frac{z}{t^\alpha} \right)^k \right\} dt = \frac{1}{2\pi i} \int_{L(\theta,a)} \frac{e^t t^{\alpha-1}}{t^\alpha - z} dt. \end{aligned}$$

The integral in the right-hand side of (5.35) defines, for $|z| < a^\alpha$, an analytic function of z, which coincides with $E_\rho(z)$. Since a can be chosen to be arbitrarily large, we get the following integral representation for $E_\rho(z)$:

$$(5.36)\qquad E_\rho(z) = \frac{1}{2\pi i} \int_{L(\theta,a)} \frac{e^t t^{\alpha-1}}{t^\alpha - z} dt, \qquad \frac{\pi}{2} < \theta \le \pi,\ a > |z|^\rho.$$

Suppose first that $1/2 < \rho < \infty$. Assume that $|z| > 1$. Let $z = re^{i\varphi}$. Suppose that $|\varphi| \le \frac{\pi}{2\rho} + \delta$, $0 < \delta < \pi - \frac{\pi}{2\rho}$. Then

$$(5.37)\qquad \begin{aligned} E_\rho(re^{i\varphi}) &= \frac{1}{2\pi i} \int_{L(\pi,(2r)^\rho)} \frac{e^t t^{\alpha-1}}{t^\alpha - z} dt \\ &= \frac{1}{2\pi i} \int_{L(\pi,1)} \frac{e^t t^{\alpha-1}}{t^\alpha - z} dt + \operatorname{res}_{t=z^\rho} \frac{e^t t^{\alpha-1}}{t^\alpha - z}. \end{aligned}$$

In writing this representation we take into account the fact that the function

$$e^t t^{\alpha-1}(t^\alpha - z)^{-1}$$

has only one pole in the domain $\{|\arg t| < \pi,\ 1 < |t| < (2r)^\rho\}$ at $t = r^\rho e^{i\rho\varphi}$. Since for $|\varphi| \le \frac{\pi}{2\rho} + \sigma$, $t^\alpha = |t|^\alpha e^{\pm i\pi\alpha}$, $r \ge 1$, $|t| \ge 1$, the absolute value $\left|\frac{t^\alpha}{z} - 1\right|$ is bounded below by a positive constant m, it follows that

$$\left| \frac{1}{2\pi i} \int_{L(\pi,1)} \frac{e^t t^{\alpha-1}}{t^\alpha - z} dt \right| \le \frac{1}{|z|} \frac{1}{2\pi m} \int_1^\infty e^{-x} x^{\alpha-1} dx < \frac{1}{|z|} \frac{\Gamma(\alpha)}{2\pi m}.$$

Therefore, in view of (5.37),

$$(5.38)\qquad E_\rho(z) = O\left(\frac{1}{|z|}\right) + \rho e^{z^\rho}, \qquad |\arg z| \le \frac{\pi}{2\rho} + \delta < \frac{\pi}{\rho}.$$

If $\frac{\pi}{2\rho} + \delta \le |\varphi| \le \pi$, we choose θ satisfying $\frac{\pi}{2} < \theta < \frac{\pi}{2} + \rho\delta$. Then

$$(5.39)\qquad E_\rho(re^{i\varphi}) = \frac{1}{2\pi i} \int_{L(\theta,(2r)^\rho)} \frac{e^t t^{\alpha-1} dt}{t^\alpha - z} = \frac{1}{2\pi i} \int_{L(\theta,1)} \frac{e^t t^{\alpha-1}}{t^\alpha - z} dt,$$

since under the conditions listed above, $|\rho\varphi| > \theta$, $\rho\varphi - 2k\pi\rho \le \rho\pi - 2k\pi\rho < -\theta$, and $\rho\varphi + 2k\pi \ge -\pi\rho + 2k\pi > \theta$ for $k = 1, 2, \dots$, and the integrand does not have poles in $\{|\arg t|\leq\theta,|t|>1\}$. In the same way as before, we show that the integral in the right-hand side of (5.39) is $O(|z|^{-1})$. Thus we get the following uniform estimate in the complex plane:

$$E_\rho(z)=\begin{cases}\rho e^{z^\rho}+O\left(\frac{1}{|z|}\right), & |\arg z|\leq\frac{\pi}{2\rho},\\ O\left(\frac{1}{|z|}\right), & \pi\geq|\arg z|\geq\frac{\pi}{2\rho}.\end{cases}\tag{5.40}$$

In some cases, for $\pi\geq|\arg z|\geq\frac{\pi}{2\rho}+\delta$, $\delta>0$, a sharper estimate than $O(\frac{1}{|z|})$ is needed for $E_\rho(z)$. To get it, we transform the integral (5.39) as follows:

$$\begin{aligned}\frac{1}{2\pi i}\int_{L(\theta,1)}\frac{e^t t^{\alpha-1}}{t^\alpha-z}dt&=-\frac{1}{z}\frac{1}{2\pi i}\int_{L(\theta,1)}e^t t^{\alpha-1}dt+\frac{1}{z}\frac{1}{2\pi i}\int_{L(\theta,1)}\frac{e^t t^{2\alpha-1}}{t^\alpha-z}dt\\&=-\frac{1}{z}\frac{1}{\Gamma(1-\alpha)}+O\left(\frac{1}{|z|^2}\right).\end{aligned}$$

(We used (5.34) here.) Hence

$$E_\rho(z)=-\frac{1}{z}\frac{1}{\Gamma(1-\alpha)}+O\left(\frac{1}{|z|^2}\right)\text{ for }\pi\geq|\arg z|\geq\frac{\pi}{2\rho}+\delta.\tag{5.40$'$}$$

If $\rho>1/2$ and $\rho\neq1$, then $1/\Gamma(1-\alpha)\neq0$. If $\rho=1$, formula (5.40′) gives a rougher estimate, because $E_1(z)=e^z$. If $\rho=1/2$, then, as noted above, $E_\rho(z)=\cosh\sqrt{z}$, and we easily get

$$E_{1/2}(z)=\frac{1}{2}e^{z^{1/2}}+\omega(z),\qquad|\omega(z)|\leq\frac{1}{2}.\tag{5.41}$$

For $0<\rho<1/2$, our investigation is somewhat more complicated. Let $|\varphi_0|\leq\pi$. We find $\theta=\theta(\varphi_0)$, $\pi\geq\theta>\frac{\pi}{2}$, and a small number $\varepsilon(\varphi_0)>0$ such that for $|\varphi-\varphi_0|\leq\varepsilon(\varphi_0)$ and an arbitrary integer k, the numbers $\rho\varphi+2k\pi\rho$ differ from θ and $-\theta$. Then, as before,

$$E_\rho(z)=\frac{1}{2\pi i}\int_{L(\theta,1)}\frac{e^t t^{\alpha-1}dt}{t^\alpha-z}+\sum\operatorname{res}\frac{e^t t^{\alpha-1}}{t^\alpha-z},\tag{5.42}$$

where the sum of residues is taken over all poles of the function $e^t t^{\alpha-1}(t^\alpha-z)^{-1}$ in the domain $\{|\arg t|<\theta,|t|>1\}$, that is, at the points $t=r^\rho e^{i\rho(\varphi+2k\pi)}$, where the integers k are such that $-\theta<\rho\varphi+2k\pi\rho<\theta$. The integral in (5.42) can be estimated using the same method as before. Therefore

$$E_\rho(z)=O\left(\frac{1}{|z|}\right)+\sum_{|\varphi+2k\pi|<\frac{\theta}{\rho}}\rho e^{z^\rho e^{i2k\pi\rho}},\qquad|\arg z|\leq\pi.\tag{5.43}$$

This estimate is uniform with respect to $|\varphi-\varphi_0|\leq\varepsilon(\varphi_0)$. Using the Heine-Borel lemma, we cover the interval $[-\pi,\pi]$ by finitely many intervals $\{|\varphi-\varphi_\mu|\leq\varepsilon(\varphi_\mu)\}$, $\mu=1,2,\ldots,\mu_0$, and conclude that the estimate (5.43) is uniform in the complex plane. It is necessary to take into account that θ in (5.43) will take, depending on $\varphi=\arg z$, μ_0 different values. If we restrict our attention to the main term of the decomposition, we get from (5.43) that

$$E_\rho(z)=(1+o(1))\rho e^{z^\rho},\tag{5.44}$$

uniformly with respect to $\arg z$ for $|\arg z|\leq\pi-\delta$, $\delta>0$.

Since the coefficients in (5.31) are positive, it follows that ($\rho>0$):

$$M(r,E_\rho(z))=E_\rho(r)=(1+o(1))\rho e^{r^\rho},\qquad T(r,E_\rho)\leq r^\rho+O(1).$$

Applying Theorem 7.4 of Chapter 1, we get

$$T(r, E_\rho) = \begin{cases} \frac{1}{\pi\rho} r^\rho + o(r^\rho), & \frac{1}{2} \le \rho < \infty, \\ \frac{\sin \pi\rho}{\pi\rho} r^\rho + o(r^\rho), & 0 < \rho \le \frac{1}{2}. \end{cases}$$

Taking into account (5.40′), we conclude that

$$m\left(r, \frac{1}{E_\rho}\right) = \begin{cases} o(r^\rho), & \rho \ne 1, \\ \frac{r}{\pi}, & \rho = 1, \end{cases}$$

$$m\left(r, \frac{1}{E_\rho - a}\right) = o(r^\rho), \qquad 0 < \rho < \infty,\ a \ne 0, \infty.$$

By the First Fundamental Theorem we get

$$N(r, 0, E_\rho) = \begin{cases} T(r, E_\rho) + o(r^\rho), & \rho \ne 1, \\ 0, & \rho = 1, \end{cases}$$

$$N(r, a, E_\rho) = T(r, E_\rho) + o(r^\rho), \quad a \ne 0, \infty; \quad 0 < \rho < \infty.$$

CHAPTER 3

The Second Fundamental Theorem

1. Lemma on the logarithmic derivative

The second fundamental theorem is the deepest and the most important result of the value distribution theory. As we shall see later, the classical Picard and Borel theorems are its corollaries. To prove the second fundamental theorem we need some auxiliary results, which we are going to present now.

THEOREM 1.1. *Let $f(z)$ be a meromorphic function satisfying $f(0) = 1$. For all r and R satisfying $1 < r < R < \infty$ and all α, $0 < \alpha < 1$, the following inequality holds:*

$$m\left(r, \frac{f'}{f}\right) \leq \frac{1}{\alpha} \ln^+ T(R, f) + \max\left(2, \frac{1}{\alpha}\right) \left\{\ln^+ \frac{1}{R-r} + \ln R\right\} + \frac{1}{\alpha} \ln \frac{64}{1-\alpha}. \tag{1.1}$$

We need the following lemma.

LEMMA 1.1. *Let $f(x)$ and $g(x)$ be non-negative measurable functions on $[a, b]$, and let*

$$A = \int_a^b g(x)\, dx > 0.$$

Then

$$\frac{1}{A} \int_a^b \{\ln^+ f(x)\} g(x)\, dx \leq \ln^+ \left\{\frac{1}{A} \int_a^b f(x) g(x)\, dx\right\} + \ln 2. \tag{1.2}$$

PROOF. Note that in the case $g(x) \equiv 1$ inequality (1.2) becomes

$$\frac{1}{b-a} \int_a^b \ln^+ f(x)\, dx \leq \ln^+ \left\{\frac{1}{b-a} \int_a^b f(x)\, dx\right\} + \ln 2. \tag{1.2'}$$

To prove the lemma, we set

$$m = \frac{1}{A} \int_a^b [f(x)]^\wedge g(x)\, dx, \qquad \varphi(x) = [f(x)]^\wedge - m.$$

Note that $m \geq 1$. It is clear that

$$\int_a^b \varphi(x) g(x)\, dx = 0.$$

We have

$$\frac{1}{A}\int_a^b \{\ln^+ f(x)\}g(x)dx = \frac{1}{A}\int_a^b \{\ln[f(x)]^\wedge\}g(x)dx$$
$$= \ln m + \frac{1}{A}\int_a^b \ln\left(1+\frac{\varphi(x)}{m}\right) g(x)dx \le \ln m + \frac{1}{A}\int_a^b \frac{\varphi(x)}{m} g(x)dx = \ln^+ m.$$

Thus

$$\frac{1}{A}\int_a^b \{\ln[f(x)]^\wedge\}g(x)dx \le \ln^+\left\{\frac{1}{A}\int_a^b [f(x)]^\wedge g(x)dx\right\}$$
$$\le \ln^+\left\{\frac{1}{A}\int_a^b [f(x)+1]g(x)dx\right\} = \ln^+\left\{\frac{1}{A}\int_a^b f(x)g(x)dx + 1\right\}$$
$$\le \ln^+\left\{\frac{1}{A}\int_a^b f(x)g(x)dx\right\} + \ln 2. \qquad \square$$

PROOF OF THEOREM 1.1. Note that the condition $f(0) = 1$ implies (see (6.8′) in Chapter 1) that $T(r, \frac{1}{f}) = T(r, f)$; therefore we have

$$m\left(r, \frac{1}{f}\right) + m(r, f) \le 2T(r, f), \qquad N(r; 0, \infty) \le 2T(r, f).$$

Using formula (2.5) of Chapter 1 for the disc $\{|z| < s\}$, $s > 0$, and letting $p = 1$ in this formula, we get

$$\begin{aligned}\frac{f'(z)}{f(z)} &= \frac{1}{2\pi}\int_0^{2\pi} \ln|f(se^{i\theta})| \frac{2se^{i\theta}}{(se^{i\theta}-z)^2} d\theta \\ &+ \sum_{|a_m|<s} \frac{s^2-|a_m|^2}{(s^2-\bar{a}_m z)(z-a_m)} - \sum_{|b_n|<s} \frac{s^2-|b_n|^2}{(s^2-\bar{b}_n z)(z-b_n)}, \quad |z|<s.\end{aligned} \tag{1.3}$$

Denote by $\{c_q\}$ the set-theoretic sum of the sequences $\{a_m\}$ and $\{b_n\}$. Observe that

$$\frac{1}{2\pi}\int_0^{2\pi} |\ln|f(se^{i\theta})||d\theta = m\left(s, \frac{1}{f}\right) + m(s, f) \le 2T(s, f),$$

and for $|c_q| < s$,

$$\left|\frac{s^2-|c_q|^2}{(s^2-\bar{c}_q z)(z-c_q)}\right| \le \frac{s^2-|c_q|^2}{(s^2-|c_q|s)|z-c_q|} = \frac{s+|c_q|}{s}\cdot\frac{1}{|z-c_q|} \le \frac{2}{|z-c_q|}.$$

Hence

$$\left|\frac{f'(z)}{f(z)}\right| \le \frac{4sT(s,f)}{(s-|z|)^2} + 2\sum_{|c_q|<s}\frac{1}{|z-c_q|}, \qquad |z|<s. \tag{1.3′}$$

Using the well-known inequality

$$\left(\sum_k d_k\right)^\alpha \le \sum_k d_k^\alpha, \qquad d_k \ge 0,\ 0<\alpha<1,$$

we get

$$\int_0^{2\pi}\left|\frac{f'(re^{i\varphi})}{f(re^{i\varphi})}\right|^\alpha d\varphi \le \int_0^{2\pi}\left\{\frac{4sT(s,f)}{(s-r)^2}+2\sum_{|c_q|<s}\frac{1}{|re^{i\varphi}-c_q|}\right\}^\alpha d\varphi$$

$$\le \int_0^{2\pi}\left\{\frac{4^\alpha s^\alpha T^\alpha(s,f)}{(s-r)^{2\alpha}}+2^\alpha\sum_{|c_q|<s}\frac{1}{|re^{i\varphi}-c_q|^\alpha}\right\}d\varphi$$

$$= 2\pi\frac{4^\alpha s^\alpha T^\alpha(s,f)}{(s-r)^{2\alpha}}+2^\alpha\sum_{|c_q|<s}\int_0^{2\pi}\frac{d\varphi}{|re^{i\varphi}-c_q|^\alpha}.$$

Since

$$\int_0^{2\pi}\frac{d\varphi}{|re^{i\varphi}-c_q|^\alpha}=\int_0^{2\pi}\frac{d\varphi}{|re^{i\varphi}-|c_q||^\alpha}\le\int_0^{2\pi}\frac{d\varphi}{|r\sin\varphi|^\alpha}$$

$$=\frac{4}{r^\alpha}\int_0^{\pi/2}\frac{d\varphi}{\sin^\alpha\varphi}\le\frac{4}{r^\alpha}\left(\frac{\pi}{2}\right)^\alpha\int_0^{\pi/2}\frac{d\varphi}{\varphi^\alpha}=\frac{2\pi}{r^\alpha(1-\alpha)},$$

we get the inequality

$$\int_0^{2\pi}\left|\frac{f'(re^{i\varphi})}{f(re^{i\varphi})}\right|^\alpha d\varphi\le 2\pi\frac{4^\alpha s^\alpha T^\alpha(s,f)}{(s-r)^{2\alpha}}+2\pi\frac{2^\alpha n(s;0,\infty)}{(1-\alpha)r^\alpha}.$$

Now let $s=\frac{1}{2}(R+r)$. Observe that then

$$r<s<R,\quad s-r=R-s=\frac{1}{2}(R-r),$$

and since the function $T(s,f)$ is non-decreasing, we get

$$T(s,f)\le T(R,f).$$

By Lemma 7.1 of Chapter 1,

$$n(s;0,\infty)\le\frac{R}{R-s}N(R;0,\infty)=\frac{2R}{R-r}N(R;0,\infty)\le\frac{4R}{R-r}T(R,f).$$

Thus $(R>r\ge 1)$,

$$\frac{1}{2\pi}\int_0^{2\pi}\left|\frac{f'(re^{i\varphi})}{f(re^{i\varphi})}\right|^\alpha d\varphi\le\frac{2^{4\alpha}R^\alpha T^\alpha(R,f)}{(R-r)^{2\alpha}}+\frac{2^{2+2\alpha}RT(R,f)}{(1-\alpha)(R-r)r^\alpha}$$

$$\le\frac{16}{1-\alpha}T^{\wedge}(R,f)\left\{\frac{1}{R^\alpha}\left(1-\frac{r}{R}\right)^{-2\alpha}+\frac{1}{r^\alpha}\left(1-\frac{r}{R}\right)^{-1}\right\}\tag{1.4}$$

$$\le\frac{32}{1-\alpha}T^{\wedge}(R,f)r^{-\alpha}\left(1-\frac{r}{R}\right)^{-(2\alpha)^{\wedge}}.$$

By Lemma 1.1,

$$m\left(r,\frac{f'}{f}\right)=\frac{1}{\alpha}\frac{1}{2\pi}\int_0^{2\pi}\ln^+\left|\frac{f'(re^{i\varphi})}{f(re^{i\varphi})}\right|^\alpha d\varphi$$

$$\le\frac{1}{\alpha}\left\{\ln^+\left[\frac{1}{2\pi}\int_0^{2\pi}\left|\frac{f'(re^{i\varphi})}{f(re^{i\varphi})}\right|^\alpha d\varphi\right]+\ln 2\right\};$$

therefore

$$m\left(r,\frac{f'}{f}\right)\le\frac{1}{\alpha}\ln^+\left\{\frac{32}{1-\alpha}T^{\wedge}(R,f)r^{-\alpha}\left(1-\frac{r}{R}\right)^{-(2\alpha)^{\wedge}}\right\}+\frac{1}{\alpha}\ln 2.\tag{1.5}$$

Using inequality (4.3) of Chapter 1, we easily obtain the conclusion of the theorem. □

Later we will need an estimate for $m(r, \frac{f'}{f})$ in terms of $T(r, f)$ rather than in terms of $T(R, f)$, $R > r$. It is clear that we cannot get such an estimate taking the limit as $R \to r$ in (1.1). We shall use for this matter the following theorem of Borel–Nevanlinna.

THEOREM 1.2 (Borel–Nevanlinna). *Let $u(r)$ be a continuous, non-decreasing function on $[r_0, \infty)$, tending to $+\infty$ as $r \to \infty$. Let $\varphi(u)$ be a continuous positive non-increasing function on $[u_0, \infty)$, $u_0 = u(r_0)$, having zero limit as $u \to \infty$ and satisfying*

$$\int_{u_0}^{\infty} \varphi(u)du < \infty. \tag{1.6}$$

Then for all $r \geq r_0$ except, possibly, a set of finite measure,

$$u\{r + \varphi(u(r))\} < u(r) + 1. \tag{1.7}$$

PROOF. Denote by E the closed subset of $[r_0, \infty)$ on which inequality (1.7) is not satisfied, that is,

$$u\{r + \varphi(u(r))\} \geq u(r) + 1. \tag{1.8}$$

Denote by $E(r)$ the set $E \cap [r, \infty)$. We assume that for all $r \geq r_0$ the sets $E(r)$ are non-empty; otherwise the theorem is trivial. Let $r_1 = \min r$, $r \in E$; $r_1 \geq r_0$. Let r_1' be the least value of r satisfying $u(r) = u(r_1) + 1$; it is clear that $r_1' > r_1$. On the other hand, $u\{r_1 + \varphi(u(r_1))\} \geq u(r_1) + 1 = u(r_1')$. Since the function $u(r)$ is non-decreasing, we have $r_1 + \varphi(u(r_1)) \geq r_1'$, $r_1' - r_1 \leq \varphi(u(r_1))$. Suppose that we have already found the values $r_1, \dots, r_n, r_1', \dots, r_n'$. Set $r_{n+1} = \min r$, $r \in E(r_n')$, and let r_{n+1}' be the least number r satisfying $u(r) = u(r_{n+1}) + 1$ and $r_{n+1}' > r_{n+1}$. We have $u\{r_{n+1} + \varphi(u(r_{n+1}))\} \geq u(r_{n+1}) + 1 = u(r_{n+1}')$ and $r_{n+1}' - r_{n+1} \leq \varphi(u(r_{n+1}))$. Since $u(r_{n+1}) \geq u(r_n') = u(r_n) + 1$, we get $u(r_{n+1}) - u(r_n) \geq 1$ and $u(r_{n+1}) \geq u(r_1) + n \geq u_0 + n$. Therefore $u(r_n) \to \infty$ and $r_n \to \infty$ as $n \to \infty$. The definition of the sequence r_n implies that (1.7) holds on (r_{n-1}', r_n), $n = 1, 2, \dots, r_0' = r_0$. Hence the set E is covered by the sets $[r_n, r_n']$, $n = 1, 2, \dots$. On the other hand,

$$\sum_{n=1}^{\infty}(r_n' - r_n) \leq \sum_{n=1}^{\infty} \varphi(u(r_n)) \leq \sum_{n=1}^{\infty} \varphi(u_0 + n - 1)$$
$$\leq \varphi(u_0) + \int_{u_0}^{\infty} \varphi(u)du < \infty.$$

Observe that we have also proved that the set E can be covered by pairwise disjoint closed intervals of finite total length with endpoints going to infinity.

R. Nevanlinna showed that condition (1.6) cannot be weakened. In fact, let u_0 be a real number, and $\varphi(u)$ a function satisfying the assumptions of Theorem 1.2, except that

$$\int_{u_0}^{\infty} \varphi(u)du = \infty. \tag{1.9}$$

Set

$$r(u) = \int_{u_0}^{u} \varphi(t)dt, \qquad u \geq u_0.$$

Then the function $u(r)$ inverse to $r(u)$ satisfies all assumptions of Theorem 1.2. Condition (1.9) implies that $u(r)$ is defined everywhere on $[0, \infty)$. Let us show that (1.8) is satisfied everywhere on $[0, \infty)$. Indeed, by the mean value theorem we get

$$u\{r + \varphi(u(r))\} - u(r) = u'_r\{r + \theta\varphi(u(r))\}\varphi(u(r)),$$

where $0 < \theta = \theta(r) < 1$. On the other hand,

$$\begin{aligned} u'_r\{r + \theta\varphi(u(r))\} &= \frac{1}{r'_u\{u(r + \theta\varphi(u(r)))\}} \\ &= \frac{1}{\varphi\{u(r + \theta\varphi(u(r)))\}} \geq \frac{1}{\varphi(u(r))}; \end{aligned}$$

hence (1.8) is satisfied. □

EXERCISE 1. Let $u(r)$ be a function satisfying the assumptions of Theorem 1.2, $\varepsilon > 0$. Then for all $r \geq r_0$ except, possibly, a set of finite measure, the inequality

$$u\left(r + \frac{1}{\ln u(r)}\right) < \{u(r)\}^{1+\varepsilon}$$

holds.

HINT. Use the argument of Theorem 1.2 with $u_1(r) = \sqrt{\ln u(r)}$ instead of $u(r)$ and with $\varphi(u) = u^{-2}$.

EXERCISE 2. A set $E \subset [1, \infty)$ is called a set of *finite logarithmic measure* if the integral $\int_E r^{-1}dr$ converges. For all $r \geq r_0$ except, possibly, a set of finite logarithmic measure, the inequality

$$u\left(r + \frac{r}{\ln u(r)}\right) < \{u(r)\}^{1+\varepsilon}$$

holds.

HINT. Use the result of Exercise 1 for $u(e^r)$, and observe that

$$\exp\left(r + \frac{1}{\ln u(e^r)}\right) \geq e^r\left(1 + \frac{1}{\ln u(e^r)}\right).$$

Let $f(z)$ be a meromorphic function. Denote by $Q(r, f)$ an arbitrary function defined on $\{0 \leq r < \infty\}$ and such that:

1) If $f(z)$ is of finite order, then $Q(r, f) = O(\ln r)$ as $r \to \infty$.

2) If $f(z)$ is of infinite order, then $Q(r, f) = O(\ln T(r, f) + \ln r)$ as $r \to \infty$, excluding, possibly, a set of closed intervals in $[0, \infty)$ with finite total length.

It is clear that $Q(r, f) \pm Q(r, f) = Q(r, f)$, and $AQ(r, f) = Q(r, f)$; if $T(r, f_1) = o(T(r, f))$, then

$$Q(r, ff_1) = Q(r, f), \qquad Q(r, f_1 + f) = Q(r, f).$$

THEOREM 1.3 (Lemma on the logarithmic derivative). *Each meromorphic function $f(z)$ satisfies*

$$m\left(r, \frac{f'}{f}\right) = Q(r, f). \tag{1.10}$$

PROOF. Observe that it is enough to prove the theorem for functions $f(z)$ satisfying $f(0) = 1$. In fact, in the general case, $f(z) = Az^p\varphi(z)$, where A is a constant, p is an integer, and $\varphi(0) = 1$. If we assume that for $\varphi(z)$ the theorem has already been proved, we get

$$m\left(r, \frac{f'}{f}\right) = m\left(r, \frac{p}{z} + \frac{\varphi'}{\varphi}\right) = m\left(r, \frac{\varphi'}{\varphi}\right) + O(1)$$
$$= Q(r, \varphi) + O(1) = Q(r, f).$$

So let $f(0) = 1$. Assume that $T(r, f) > 1$ for $r \geq r_0 > 1$. Taking $R > r$, $\alpha = 1/2$ in (1.1), we get for $r \geq r_0$

$$m\left(r, \frac{f'}{f}\right) \leq 2\ln T(R, f) + 2\ln^+ \frac{1}{R-r} + 2\ln R + 2\ln 128. \tag{1.11}$$

If the function $f(z)$ is of finite order, then $\ln T(r, f) = O(\ln r)$ and taking $R = 2r$ in (1.11), we get

$$m\left(r, \frac{f'}{f}\right) = O(\ln 2r) = O(\ln r).$$

Now suppose that $f(z)$ is of infinite order. Let R in (1.11) be equal to $R = r + \{T(r, f)\}^{-2} \leq 2r$. By Theorem 1.2 everywhere except, possibly, a set of finite measure the inequality

$$T(R, f) = T(r + \{T(r, f)\}^{-2}, f) < T(r, f) + 1 \tag{1.7'}$$

holds. Here we use (1.7) with $u(r) = T(r, f)$ and $\varphi(u) = u^{-2}$. Now it follows from (1.11) that outside the exceptional set of values of r, the inequality

$$m\left(r, \frac{f'}{f}\right) \leq 6\ln T(r, f) + 2\ln r + O(1)$$

holds; that is, (1.10) is satisfied. □

The question arises whether Theorem 1.3 is valid in the general case for functions of infinite order if in the definition of $Q(r, f)$ we do not mention the exceptional set; that is, whether the following estimate holds:

$$m\left(r, \frac{f'}{f}\right) = O(\ln T(r, f) + \ln r), \qquad r \to \infty. \tag{1.12}$$

Observe that the estimate

$$m\left(r, \frac{f'}{f}\right) = O(\ln T(kr, f) + \ln r), \qquad r \to \infty \tag{1.13}$$

holds for each $k > 1$; to get this estimate it is enough to let $R = kr$ in (1.11).

Following Hayman [**Hay65**] we show, however, that the estimate (1.12) is not true in general. If (1.12) were valid, then for each transcendental entire function $f(z)$ we would have

$$T(r, f') = m(r, f') \leq m(r, f) + m\left(r, \frac{f'}{f}\right)$$
$$= T(r, f) + O(\ln T(r, f) + \ln r) = (1 + o(1))T(r, f).$$

It turns out that not only this estimate is not always true (as R. Nevanlinna supposed), but for each positive arbitrarily rapidly increasing function $\Psi(T)$, $\Psi(T) \to \infty$ as $T \to \infty$, there exists an entire function $f(z)$ such that

$$\limsup_{r\to\infty} \frac{T(r,f')}{\Psi(T(r,f))} = \infty. \tag{1.14}$$

Let $r_n = 2^{n-1}$, $n = 1,2,\dots$, and let $\lambda_1, \lambda_2, \dots$ be a a strictly increasing sequence of positive integers satisfying $n = o(\ln \lambda_n)$, $n\lambda_n 2^{n\lambda_n} = o(\lambda_{n+1})$, and $\Psi(n\lambda_{n-1}) = o(\ln \lambda_n)$ as $n \to \infty$. Set

$$f(z) = \sum_{n=1}^{\infty} \left(\frac{z}{r_n}\right)^{\lambda_n}.$$

Then

$$f'(z) = \sum_{n=1}^{\infty} \frac{\lambda_n}{r_n} \left(\frac{z}{r_n}\right)^{\lambda_n - 1}.$$

For $|z| = r_\nu$, $\nu \geq 2$ we have

$$|f(z)| \leq \sum_{n=1}^{\infty} \left(\frac{r_\nu}{r_n}\right)^{\lambda_n} \leq \sum_{n=1}^{\nu-1} r_\nu^{\lambda_n} + 1 + \sum_{n=\nu+1}^{\infty} 2^{-\lambda_n} \leq (\nu-1) r_\nu^{\lambda_{\nu-1}} + 2,$$

$$T(r_\nu, f) \leq \lambda_{\nu-1} \ln r_\nu + \ln(\nu - 1) + O(1) \sim \nu\lambda_{\nu-1} \ln 2. \tag{1.15}$$

If $|z| = r_\nu$, $\nu \geq 2$, then

$$\begin{aligned}
|f'(z)| &\geq \frac{\lambda_\nu}{r_\nu} - \sum_{n=1}^{\nu-1} \frac{\lambda_n}{r_n}\left(\frac{r_\nu}{r_n}\right)^{\lambda_n-1} - \sum_{n=\nu+1}^{\infty} \frac{\lambda_n}{r_n}\left(\frac{r_\nu}{r_n}\right)^{\lambda_n-1} \\
&\geq \frac{\lambda_\nu}{r_\nu} - (\nu-1)\lambda_{\nu-1} r_\nu^{\lambda_{\nu-1}-1} - \frac{1}{r_\nu} \sum_{n=\nu+1}^{\infty} \lambda_n \left(\frac{r_\nu}{r_n}\right)^{\lambda_n} \\
&\geq \frac{1}{r_\nu}\left\{\lambda_\nu - (\nu-1)\lambda_{\nu-1} r_\nu^{\lambda_{\nu-1}} - \sum_{n=\nu+1}^{\infty} \lambda_n 2^{-\lambda_n}\right\} \\
&\geq \frac{1}{r_\nu}\{\lambda_\nu - (\nu-1)\lambda_{\nu-1} 2^{(\nu-1)\lambda_{\nu-1}} - 2\} = (1+o(1))\frac{\lambda_\nu}{r_\nu}.
\end{aligned}$$

Hence

$$T(r_\nu, f') = m(r_\nu, f') \geq \ln \lambda_\nu - \ln r_\nu + o(1) = (1+o(1)) \ln \lambda_\nu. \tag{1.16}$$

Using (1.15) and (1.16) we see that for $\nu \geq \nu_0$,

$$\frac{T(r_\nu, f')}{\Psi(T(r_\nu, f))} \geq \frac{(1+o(1))\ln \lambda_\nu}{\Psi(\nu\lambda_{\nu-1})} \to \infty \quad \text{as} \quad \nu \to \infty,$$

that is, (1.14) holds.

If $T = o(\Psi(T))$ as $T \to \infty$, then

$$\frac{m(r_\nu, f'/f)}{\Psi(T(r_\nu, f))} \geq \frac{T(r_\nu, f')}{\Psi(T(r_\nu, f))} - \frac{T(r_\nu, f)}{\Psi(T(r_\nu, f))} \to \infty \quad \text{as} \quad \nu \to \infty.$$

Thus, not only the estimate $m(r, f'/f) = o(T(r,f))$ is not always true for functions of infinite order (this is weaker than (1.12)), but no estimate of the form $m(r, f'/f) = O(\Psi(T(r,f)))$ holds unless we exclude some values of r. □

For the functions of finite order the estimate $m(r, f'/f) = O(\ln r)$ can be made sharper. We show that if $f(z)$ has order ρ, then

$$\limsup_{r\to\infty} (\ln r)^{-1} m\left(r, \frac{f'}{f}\right) \le (\rho - 1)^+. \tag{1.17}$$

From the argument at the beginning of the proof of Theorem 1.3 we see that without loss of generality we may assume that $f(0) = 1$.

Let $R = 2r$ in (1.5). We have

$$m\left(r, \frac{f'}{f}\right) \le \frac{1}{\alpha} \ln^+ \left\{\frac{32}{1-\alpha} T^\wedge(2r, f) r^{-\alpha} 2^2\right\} + \frac{1}{\alpha} \ln 2.$$

Since for each $\varepsilon > 0$ the relation $T(2r, f) = O(r^{\rho+\varepsilon})$ holds, we get

$$m\left(r, \frac{f'}{f}\right) \le \frac{1}{\alpha} \ln^+ \{O(r^{\rho+\varepsilon-\alpha})\} + O(1) \le \frac{1}{\alpha}(\rho + \varepsilon - \alpha)^+ \ln r + O(1);$$

therefore

$$\limsup_{r\to\infty} (\ln r)^{-1} m\left(r, \frac{f'}{f}\right) \le \frac{1}{\alpha}(\rho + \varepsilon - \alpha)^+.$$

Letting $\varepsilon \to 0$ and $\alpha \to 1$ in this inequality, we get (1.17).

Observe that the function $f(z) = e^{z^p}$ (p is a positive integer) satisfies $\rho = p$, $m(r, f'/f) = (\rho - 1) \ln r + \ln \rho$, $r \ge 1$, and hence the two sides of (1.17) are equal.

EXERCISE 1. Prove that for functions of order $\rho < 1$,

$$m\left(r, \frac{f'}{f}\right) = o(1), \qquad r \to \infty.$$

(We have

$$m\left(r, \frac{f'}{f}\right) \le \frac{1}{\alpha} \frac{1}{2\pi} \int_0^{2\pi} \left|\frac{f'(re^{i\varphi})}{f(re^{i\varphi})}\right|^\alpha d\varphi;$$

we use estimate (1.4) with $R = 2r$ for the integral and choose the number α such that $\rho < \alpha < 1$.)

EXERCISE 2. Prove that for the functions of finite lower order λ,

$$\liminf_{r\to\infty} (\ln r)^{-1} m\left(r, \frac{f'}{f}\right) \le (\lambda - 1)^+.$$

(Let $r = r_n$, $R = 2r_n$ in (1.5), where the sequence $\{r_n\}$ is such that $T(2r_n, f) = O(r_n^{\lambda+\varepsilon})$.)

EXERCISE 3. Prove that

$$\int_1^s m\left(r, \frac{f'}{f}\right) dr = O\left\{\int_{s/2}^s [\ln T(r, f) + \ln r] dr\right\}, \qquad s \to \infty.$$

(Let $R = (s + r)/2$ in (1.11) and integrate with respect to r from 1 to s.)

EXERCISE 4. Prove that

$$m\left(r, \frac{f'}{f}\right) = O(\ln T(r, f))$$

everywhere outside a set of finite logarithmic measure. (Use inequality (1.5) and Exercise 2 after Theorem 1.2.)

EXERCISE 5. Show that for non-integer ρ and for $f(z) = \prod_{k=1}^{\infty} E\left(\frac{z}{k^{1/\rho}}, [\rho]\right)$ the two sides of (1.17) are equal.

(First we need to prove that for every $\delta > 0$, the relation

$$\frac{f'(re^{i\varphi})}{f(re^{i\varphi})} = \frac{\pi\rho}{\sin\pi\rho} e^{-i\rho\pi} e^{i(\rho-1)\varphi} r^{\rho-1} + o(r^{\rho-1})$$

holds uniformly in φ, $\delta \le \varphi \le 2\pi - \delta$. This can be done using the asymptotic formula (5.2_1) of Chapter 2. In fact, letting $z = re^{i\varphi}$, $z^\rho = r^\rho e^{i\rho\varphi}$ and using the above-mentioned formula, we get that the function

$$g(z) = \ln f(z) - \frac{\pi}{\sin\pi\rho} e^{-i\rho\pi} z^\rho,$$

which is analytic for $0 < \varphi < 2\pi$, admits an estimate

$$g(z) = o(r^\rho),$$

uniform with respect to φ for $\frac{\delta}{2} \le \varphi \le 2\pi - \frac{\delta}{2}$.

Assuming that z is in the angle $\{\delta \le \varphi \le 2\pi - \delta\}$ and observing that then the circle $\left\{|\zeta - z| = |z|\sin\frac{\delta}{2}\right\}$ is contained in $\left\{\frac{\delta}{2} \le \varphi \le 2\pi - \frac{\delta}{2}\right\}$, by the Cauchy formula

$$g'(z) = \frac{1}{2\pi i}\int_{|\zeta - z| = |z|\sin\frac{\delta}{2}} \frac{g(\zeta)}{(\zeta - z)^2} d\zeta$$

we get

$$g'(z) = o(r^{\rho-1})$$

uniformly for $\delta \le \varphi \le 2\pi - \delta$.)

2. The Nevanlinna second fundamental theorem

We need some auxiliary functions. Let $f(z)$ be a meromorphic function. Denote by $n_1(r, a, f) = n_1(r, a)$ the number of a-points of the function $f(z)$ in the disc $\{|z| \le r\}$, where an a-point of order m is counted $m - 1$ times. If the orders of a-points are disregarded, and each a-point is counted once, we get the function $\bar{n}(r, a)$. It is clear that $n(r, a) = n_1(r, a) + \bar{n}(r, a)$. Using the functions $n_1(r, a)$ and $\bar{n}(r, a)$ we define the functions $N_1(r, a)$ and $\bar{N}(r, a)$ in the same way as we defined $N(r, a)$ using $n(r, a)$.

For example,

$$N_1(r, a) = \int_0^r \frac{n_1(t, a) - n_1(0, a)}{t} dt + n_1(0, a)\ln r.$$

It is clear that the function $n_1(r, a)$ does not count simple a-points at all. Since for $a \ne \infty$ each a-point of order m of $f(z)$ is a zero of order $m - 1$ of $f'(z)$, we have

$$\sum_{a \ne \infty} n_1(r, a, f) = n(r, 0, f'),$$

where the sum in the left-hand side has at most finitely many non-zero terms. Each pole of order m of the function $f(z)$ is a pole of order $m + 1$ of the function $f'(z)$.

Therefore $n(r,\infty,f') = n(r,\infty,f) + \bar{n}(r,\infty,f)$ and $n_1(r,\infty,f) = 2n(r,\infty,f) - n(r,\infty,f')$. We set $n_1(r,f) = n_1(r,\infty,f)$, $N_1(r,f) = N_1(r,\infty,f)$, etc. From the above equalities we get

$$n_1(r) = \sum_a n_1(r,a,f) = n\left(r,\frac{1}{f'}\right) + 2n(r,f) - n(r,f'),$$

and hence

$$N_1(r) = \sum_a N_1(r,a,f) = N\left(r,\frac{1}{f'}\right) + 2N(r,f) - N(r,f').$$

One drawback of the introduced notation $n_1(r)$ and $N_1(r)$ is that it does not reflect the function $f(z)$ used in the definition, but this will not cause any inconvenience.

We present one more equality which will be needed on many occasions:

$$N\left(r,\frac{f'}{f}\right) = \bar{N}(r,f) + \bar{N}\left(r,\frac{1}{f}\right).$$

This equality immediately follows from the easily verifiable fact that f'/f has simple poles at the zeros and poles of $f(z)$ and has no other poles.

Now we are ready to state the second fundamental theorem of R. Nevanlinna.

THEOREM 2.1. *Suppose $f(z)$ is a meromorphic function that is not identically constant, and let $a_1,\dots,a_q$ be distinct complex numbers, one of which may be equal to ∞. Then*

$$\sum_{\nu=1}^{q} m(r,a_\nu) \le 2T(r,f) - N_1(r) + Q(r,f). \tag{2.1}$$

PROOF. It is clear that it suffices to consider the case where all numbers a_ν, $\nu = 1,\dots,q$, are finite, and to prove the inequality

$$m(r,\infty) + \sum_{\nu=1}^{q} m(r,a_\nu) + N_1(r) \le 2T(r,f) + Q(r,f). \tag{2.2}$$

Consider the rational function

$$h(w) = \sum_{\nu=1}^{q} \frac{1}{w - a_\nu}$$

of degree q. Let $H(z) = h(f(z))$. By Theorem 6.1 of Chapter 1 we get

$$T(r,H) = qT(r,f) + O(1).$$

It is easy to see that

$$N(r,H) = \sum_{\nu=1}^{q} N(r,a_\nu,f).$$

Taking into account the first fundamental theorem, we get

$$\begin{aligned} m(r,H) &= T(r,H) - N(r,H) \\ &= qT(r,f) - \sum_{\nu=1}^{q} N(r,a_\nu,f) + O(1) = \sum_{\nu=1}^{q} m(r,a_\nu,f) + O(1). \end{aligned} \tag{2.3}$$

Note that by the lemma on the logarithmic derivative,

$$
\begin{aligned}
m(r,H) &\le m\left(r,\frac{1}{f'}\right)+m(r,f'H)\\
&\le m\left(r,\frac{1}{f'}\right)+\sum_{\nu=1}^{q} m\left(r,\frac{f'}{f-a_\nu}\right)+\ln q = m\left(r,\frac{1}{f'}\right)+Q(r,f).
\end{aligned}
\tag{2.4}
$$

From (2.3) and (2.4) we get

$$
\sum_{\nu=1}^{q} m(r,a_\nu) \le m\left(r,\frac{1}{f'}\right)+Q(r,f). \tag{2.5}
$$

Thus

$$
\begin{aligned}
&m(r,\infty)+\sum_{\nu=1}^{q} m(r,a_\nu)+N_1(r)\\
&\quad\le m(r,f)+m\left(r,\frac{1}{f'}\right)+N\left(r,\frac{1}{f'}\right)+2N(r,f)-N(r,f')+Q(r,f)\\
&\quad= T(r,f)+T\left(r,\frac{1}{f'}\right)+N(r,f)-N(r,f')+Q(r,f)\\
&\quad= T(r,f)+\{T(r,f')-N(r,f')\}+N(r,f)+Q(r,f)\\
&\quad= T(r,f)+N(r,f)+m(r,f')+Q(r,f).
\end{aligned}
\tag{2.6}
$$

Since

$$
m(r,f') \le m(r,f)+m\left(r,\frac{f'}{f}\right)=m(r,f)+Q(r,f), \tag{2.7}
$$

inequality (2.6) implies (2.2). □

REMARK 1. It is easy to check that the quantity $2T(r,f)-N_1(r)$ from the right-hand side of (2.1) can be replaced by

$$
-\frac{1}{2\pi}\int_0^{2\pi} \ln \mathring{f}(re^{i\varphi})\,d\varphi,
$$

where $\mathring{f}(z)$ is the spherical derivative of $f(z)$ (see Section 4 of Chapter 1). In fact,

$$
\begin{aligned}
&-\frac{1}{2\pi}\int_0^{2\pi} \ln\frac{|f'(re^{i\varphi})|}{1+|f(re^{i\varphi})|^2}d\varphi+N_1(r)\\
&\quad= -\frac{1}{2\pi}\int_0^{2\pi} \ln|f'(re^{i\varphi})|d\varphi+2\frac{1}{2\pi}\int_0^{2\pi} \ln\sqrt{1+|f(re^{i\varphi})|^2}d\varphi+N_1(r)\\
&\quad= -m(r,f')+m\left(r,\frac{1}{f'}\right)+2\mathring{m}(r,f)+N\left(r,\frac{1}{f'}\right)+2N(r,f)-N(r,f')\\
&\quad= 2T(r,f)+O(1)-T(r,f')+T\left(r,\frac{1}{f'}\right)=2T(r,f)+O(1).
\end{aligned}
$$

REMARK 2. The result of Exercise 1 on p. 94 implies that for functions of order less than 1, we may assume that $Q(r,f)=O(1)$ in (2.1).

REMARK 3. Inequality (2.5) implies the following lemma, which will play an important role in Chapter 6.

LEMMA 2.1. *Let $a_1, \dots, a_p$ be distinct complex numbers, different from 0 and ∞. Then*

$$\sum_{\nu=1}^{p} m(r, a_\nu) \leq m\left(r, \frac{f}{f'}\right) + Q(r, f).$$

In particular, for each $a \neq 0, \infty$,

$$m(r, a) \leq m\left(r, \frac{f}{f'}\right) + Q(r, f).$$

In fact, writing (2.5) for the numbers $0, a_1, \dots, a_p$ $(q = p + 1)$, and observing that

$$m\left(r, \frac{1}{f'}\right) \leq m\left(r, \frac{f}{f'}\right) + m\left(r, \frac{1}{f}\right) = m\left(r, \frac{f}{f'}\right) + m(r, 0),$$

we get the desired inequality.

We write some inequalities, immediately following from (2.1). Letting $m(r, a_\nu) = T(r, f) - N(r, a_\nu) + O(1)$ in (2.1), we get

$$(q-2)T(r, f) \leq \sum_{\nu=1}^{q} N(r, a_\nu) - N_1(r) + Q(r, f). \tag{2.8}$$

Replacing $N_1(r)$ by $\sum_{\nu=1}^{q} N_1(r, a_\nu) \leq N_1(r)$, we get from (2.1) and (2.8), respectively, the following inequalities:

$$\sum_{\nu=1}^{q} \{m(r, a_\nu) + N_1(r, a_\nu)\} \leq 2T(r, f) + Q(r, f), \tag{2.9}$$

$$(q-2)T(r, f) \leq \sum_{\nu=1}^{q} \bar{N}(r, a_\nu) + Q(r, f). \tag{2.10}$$

Theorem 2.1 immediately implies the following theorem, which goes back to Borel.

THEOREM 2.2. *A transcendental meromorphic function $f(z)$ cannot have more than two Borel exceptional values.*

PROOF. Assume that $f(z)$ has three Borel exceptional values a_1, a_2, a_3. From (2.10) with $q = 3$ we get

$$T(r, f) \leq \sum_{\nu=1}^{3} N(r, a_\nu) + Q(r, f). \tag{2.11}$$

If the function $f(z)$ is of infinite order, then the $N(r, a_\nu)$, $\nu = 1, 2, 3$, are of finite order; hence, there is a constant $K < \infty$ such that for $r \geq r_0$ the inequality $\sum_{\nu=1}^{3} N(r, a_\nu) \leq r^K$ holds. Then outside a set E of finite measure L,

$$T(r, f) \leq r^K + O(\ln T(r, f) + \ln r) \leq r^K + \frac{1}{2}T(r, f)$$

and

$$T(r, f) \leq 2r^K.$$

If $r \in E$, then there exists r', $r \leq r' \leq r + L + 1$, such that $r' \notin E$. Then

$$T(r, f) \leq T(r', f) \leq 2r'^K \leq 2(r + L + 1)^K \leq 3r^K$$

if r is sufficiently large. Hence $T(r, f)$ is of finite order, and we obtain a contradiction.

If $f(z)$ is of finite order, then $Q(r, f) = o(T(r, f))$, and (2.11) can be rewritten as

$$(1 + o(1))T(r, f) \leq \sum_{\nu=1}^{3} N(r, a_\nu). \tag{2.12}$$

This immediately implies that the growth category of at least one of the functions $N(r, a_\nu)$, $\nu = 1, 2, 3$, is not lower than the growth category of $T(r, f)$. If $f(z)$ has positive order, we already have a contradiction. If $f(z)$ has zero order, we need to consider also the case where $n(r, a_\nu) = O(1)$, that is, $N(r, a_\nu) = O(\ln r)$, $\nu = 1, 2, 3$. From (2.12) we get $T(r, f) = O(\ln r)$; that is, $f(z)$ is a rational function. □

COROLLARY (Picard). *A transcendental meromorphic function $f(z)$ cannot have more than two Picard exceptional values.*

EXERCISE. Prove the corollary using (2.11) directly, without using Theorem 2.2.

THEOREM 2.3. *If $f(z)$ is a meromorphic function, then* ($l = 1, 2, 3, \dots$)

$$T(r, f^{(l)}) \leq (l + 1)T(r, f) + Q(r, f), \tag{2.13}$$

$$Q(r, f^{(l)}) = Q(r, f). \tag{2.14}$$

PROOF. Combining inequality (2.7) with $N(r, f') = N(r, f) + \bar{N}(r, f) \leq 2N(r, f)$, we get $T(r, f') \leq 2T(r, f) + Q(r, f)$. From this it is easy to derive that $Q(r, f') = Q(r, f)$, and prove (2.14) for an arbitrary $l \geq 1$ using induction. Then by the lemma on the logarithmic derivative,

$$\begin{aligned} m(r, f^{(l)}) &\leq m\left(r, \frac{f^{(l)}}{f^{(l-1)}}\right) + \cdots + m\left(r, \frac{f'}{f}\right) + m(r, f) \\ &= Q(r, f^{(l-1)}) + \cdots + Q(r, f) + m(r, f) = m(r, f) + Q(r, f). \end{aligned} \tag{2.15}$$

On the other hand,

$$N(r, f^{(l)}) = N(r, f) + l\bar{N}(r, f) \leq (l + 1)N(r, f). \tag{2.16}$$

Inequality (2.13) follows from (2.15) and (2.16). □

COROLLARY. *If $f(z)$ is a meromorphic function, then*

$$m\left(r, \frac{f^{(l)}}{f}\right) = Q(r, f).$$

The second fundamental theorem can be generalized in many directions if we consider inequalities involving not only the characteristics of the meromorphic function itself but also the characteristics of its derivatives. We consider here only the simplest generalizations of inequality (2.11), which we will need later.

THEOREM 2.4 (Milloux). *Let $f(z)$ be a meromorphic function such that $f^{(k)}(z)$ is not identically constant for $k \geq 1$. Let a and b be (finite) complex numbers, $b \neq 0$.*

Then

$$
\begin{aligned}
T(r,f) \leq \bar{N}(r,f) &+ N\left(r, \frac{1}{f-a}\right) \\
&+ \bar{N}\left(r, \frac{1}{f^{(k)}-b}\right) - N\left(r, \frac{f^{(k)}-b}{f^{(k+1)}}\right) + Q(r,f).
\end{aligned} \tag{2.17}
$$

PROOF. It is clear that it is enough to prove (2.17) for $a = 0$, $b = 1$, that is, to prove the inequality

$$
\begin{aligned}
T(r,f) \leq \bar{N}(r,f) &+ N\left(r, \frac{1}{f}\right) \\
&+ \bar{N}\left(r, \frac{1}{f^{(k)}-1}\right) - N\left(r, \frac{f^{(k)}-1}{f^{(k+1)}}\right) + Q(r,f),
\end{aligned} \tag{2.17$'$}
$$

since (2.17) follows from (2.17′) if we pass to the function $F(z) = (f(z) - a)/b$.

By the first fundamental theorem,

$$T(r,f) = m\left(r, \frac{1}{f}\right) + N\left(r, \frac{1}{f}\right) + O(1).$$

On the other hand, by the Corollary to Theorem 2.3,

$$
\begin{aligned}
m\left(r, \frac{1}{f}\right) &= m\left(r, \frac{f^{(k)}}{f} - \frac{f^{(k)}-1}{f^{(k+1)}}\frac{f^{(k+1)}}{f}\right) \\
&\leq m\left(r, \frac{f^{(k)}}{f}\right) + m\left(r, \frac{f^{(k)}-1}{f^{(k+1)}}\right) + m\left(r, \frac{f^{(k+1)}}{f}\right) + \ln 2 \\
&= m\left(r, \frac{f^{(k)}-1}{f^{(k+1)}}\right) + Q(r,f).
\end{aligned}
$$

Again by the first fundamental theorem,

$$m\left(r, \frac{f^{(k)}-1}{f^{(k+1)}}\right) = m\left(r, \frac{f^{(k+1)}}{f^{(k)}-1}\right) + N\left(r, \frac{f^{(k+1)}}{f^{(k)}-1}\right) - N\left(r, \frac{f^{(k)}-1}{f^{(k+1)}}\right) + O(1).$$

Since

$$\frac{f^{(k+1)}}{f^{(k)}-1} = \frac{(f^{(k)}-1)'}{f^{(k)}-1},$$

we have

$$m\left(r, \frac{f^{(k+1)}}{f^{(k)}-1}\right) = Q(r, f^{(k)}-1) = Q(r,f)$$

and

$$
\begin{aligned}
N\left(r, \frac{f^{(k+1)}}{f^{(k)}-1}\right) &= \bar{N}\left(r, \frac{1}{f^{(k)}-1}\right) + \bar{N}(r, f^{(k)}-1) \\
&= \bar{N}\left(r, \frac{1}{f^{(k)}-1}\right) + \bar{N}(r,f).
\end{aligned}
$$

Combining these estimates we get (2.17′). □

It is not difficult to prove an inequality similar to (2.17) in which the characteristics $T(r,f)$ would be estimated from above in terms of the three functions N, but without taking into account the number of poles.

THEOREM 2.5 (Hiong King-Lai). *Let $f(z)$ be a meromorphic function such that $f^{(k)}(z)$ is not identically constant for some $k \geq 1$. Let a, b, c be (finite) complex numbers, $b \neq 0$, $c \neq 0$. Then*

$$T(r,f) \leq N\left(r, \frac{1}{f-a}\right) + N\left(r, \frac{1}{f^{(k)}-b}\right) + N\left(r, \frac{1}{f^{(k)}-c}\right) + Q(r,f). \tag{2.18}$$

PROOF. By the first fundamental theorem,

$$T(r,f) = m(r, f-a) + N(r, f-a) + O(1) \tag{2.19}$$

and

$$\begin{aligned} m(r, f-a) &\leq m(r, f^{(k)}) + m\left(r, \frac{f-a}{f^{(k)}}\right) \\ &= m(r, f^{(k)}) + m\left(r, \frac{f^{(k)}}{f-a}\right) + N\left(r, \frac{f^{(k)}}{f-a}\right) - N\left(r, \frac{f-a}{f^{(k)}}\right) + O(1). \end{aligned} \tag{2.20}$$

Later we shall use the equality

$$\begin{aligned} N&\left(r, \frac{f^{(k)}}{f-a}\right) - N\left(r, \frac{f-a}{f^{(k)}}\right) \\ &= N(r, f^{(k)}) - N\left(r, \frac{1}{f^{(k)}}\right) + N\left(r, \frac{1}{f-a}\right) - N(r, f-a), \end{aligned}$$

which easily follows from the Jensen formula (4.1) of Chapter 1. Since

$$m\left(r, \frac{f^{(k)}}{f-a}\right) = Q(r,f),$$

using (2.19) and (2.20) we get

$$T(r,f) \leq T(r, f^{(k)}) - N\left(r, \frac{1}{f^{(k)}}\right) + N\left(r, \frac{1}{f-a}\right) + Q(r,f). \tag{2.21}$$

Applying to $f^{(k)}(z)$ inequality (2.8) with $q = 3$, $a_1 = 0$, $a_2 = b$, $a_3 = c$, we get

$$\begin{aligned} T(r, f^{(k)}) &\leq N\left(r, \frac{1}{f^{(k)}}\right) + N\left(r, \frac{1}{f^{(k)}-b}\right) + N\left(r, \frac{1}{f^{(k)}-c}\right) \\ &\quad - N_1(r, f^{(k)}) - N\left(r, \frac{1}{f^{(k+1)}}\right) + Q(r, f^{(k)}). \end{aligned} \tag{2.22}$$

By (2.14), (2.21), (2.22), and the obvious equality

$$N_1(r, f^{(k)}) = N(r,f) + (k-1)\bar{N}(r,f),$$

we get the inequality

$$\begin{aligned} T(r,f) &\leq N\left(r, \frac{1}{f-a}\right) + N\left(r, \frac{1}{f^{(k)}-b}\right) + N\left(r, \frac{1}{f^{(k)}-c}\right) \\ &\quad - \left\{N(r,f) + (k-1)\bar{N}(r,f) + N\left(r, \frac{1}{f^{(k+1)}}\right)\right\} + Q(r,f). \end{aligned}$$

Leaving out the expression in braces, we get (2.18). □

We show that we can delete $N(r, 1/(f^{(k)}-c))$ in (2.18) if we multiply the $N(r, 1/(f-a))$ and $N(r, 1/(f^{(k)}-b))$ by 3 and 4, respectively. More precisely, we prove the following theorem.

THEOREM 2.6 (Hayman). *Under the assumptions of Theorem 2.4, the inequality*

$$T(r,f) \le \left(2+\frac{1}{k}\right) N\left(r,\frac{1}{f-a}\right) + \left(2+\frac{2}{k}\right) \bar{N}\left(r,\frac{1}{f^{(k)}-b}\right) + Q(r,f) \tag{2.23}$$

holds.

PROOF. As in the proof of Theorem 2.4, it suffices to consider the case $a=0$, $b=1$. We introduce the following notation:

$$\varphi(z) = f^{(k)}(z) - 1, \qquad g(z) = \left(\frac{\varphi'}{\varphi}\right)^{k+1} \frac{1}{\varphi}.$$

If the function $\varphi(z)$ has a zero of order μ at z_0, then $g(z)$ has a pole of order $\mu+k+1$ at z_0. If the function $f(z)$ has a pole of order $m \ge 1$ at z_0, then $\varphi(z)$ has a pole of order $m+k$ at z_0, $(\varphi'/\varphi)^{k+1}$ has a pole of order $k+1$, and $g(z)$ has a zero of order $m-1$ if $m \ge 2$, and $g(z_0) \ne 0, \infty$ if $m=1$. The function $g(z)$ also has zeros at the zeros of φ'/φ, and $g(z) \ne 0, \infty$ at all points except those listed above. Let $N_s(r,f)$ take into account simple poles of the function $f(z)$ in the disc $\{|z| \le r\}$ only. The previous argument implies that

$$\bar{N}(r,g) = \bar{N}\left(r,\frac{1}{\varphi}\right),$$

$$\bar{N}\left(r,\frac{1}{g}\right) = \bar{N}(r,f) - N_s(r,f) + \bar{N}\left(r,\frac{\varphi}{\varphi'}\right).$$

Hence

$$\begin{aligned} N\left(r,\frac{g'}{g}\right) &= \bar{N}(r,g) + \bar{N}\left(r,\frac{1}{g}\right) \\ &= \bar{N}\left(r,\frac{1}{\varphi}\right) + \bar{N}(r,f) - N_s(r,f) + \bar{N}\left(r,\frac{\varphi}{\varphi'}\right). \end{aligned} \tag{2.24}$$

Suppose that $f(z)$ has pole of order one at z_0. Then in a neighborhood of z_0, the function $\varphi(z)$ and its derivatives can be represented as

$$\varphi(z) = \frac{A}{(z-z_0)^{k+1}} + h(z), \qquad \varphi'(z) = \frac{-(k+1)A}{(z-z_0)^{k+2}} + h'(z),$$

$$\varphi''(z) = \frac{(k+1)(k+2)A}{(z-z_0)^{k+3}} + h''(z),$$

where $A \ne 0$, and $h(z)$ is a function holomorphic in a neighborhood of the point z_0. We have

$$\frac{g'(z)}{g(z)} = \frac{(k+1)\varphi\varphi'' - (k+2)\varphi'^2}{\varphi\varphi'} = (k+1)\frac{\varphi''}{\varphi'} - (k+2)\frac{\varphi'}{\varphi}.$$

It is easy to see that

$$(k+1)\varphi\varphi'' - (k+2)\varphi'^2 = \frac{(k+1)^2(k+2)Ah(z_0)}{(z-z_0)^{k+3}} + P(z),$$

where $P(z)$ has a pole of order at most $k+2$ at z_0, and hence g'/g has a zero of order at least k at z_0. Thus[1]

$$kN_s(r,f) \le N\left(r,\frac{g}{g'}\right). \tag{2.25}$$

Using the first fundamental theorem, the lemma on the logarithmic derivative, and (2.14), we get from (2.24) and (2.25):

$$\begin{aligned} kN_s(r,f) &\le N\left(r,\frac{g'}{g}\right) + m\left(r,\frac{g'}{g}\right) - m\left(r,\frac{g}{g'}\right) + O(1) \\ &\le N\left(r,\frac{g'}{g}\right) + Q(r,f) \\ &= \bar{N}\left(r,\frac{1}{\varphi}\right) + \bar{N}(r,f) - N_s(r,f) + \bar{N}\left(r,\frac{\varphi}{\varphi'}\right) + Q(r,f). \end{aligned} \tag{2.26}$$

Since the multiple poles are counted at least twice in $N(r,f)$, we get, using (2.17′),

$$\begin{aligned} N_s(r,f) + 2\{\bar{N}(r,f) - N_s(r,f)\} &\le N(r,f) \le T(r,f) \\ &\le \bar{N}(r,f) + N\left(r,\frac{1}{f}\right) + \bar{N}\left(r,\frac{1}{\varphi}\right) - N\left(r,\frac{\varphi}{\varphi'}\right) + Q(r,f) \end{aligned}$$

and

$$\bar{N}(r,f) \le N_s(r,f) + N\left(r,\frac{1}{f}\right) + \bar{N}\left(r,\frac{1}{\varphi}\right) - N\left(r,\frac{\varphi}{\varphi'}\right) + Q(r,f).$$

Multiplying this inequality by $k+1$ and adding inequality (2.26) to it we get

$$\begin{aligned} k\bar{N}(r,f) \le (k+1)N\left(r,\frac{1}{f}\right) &+ (k+2)\bar{N}\left(r,\frac{1}{\varphi}\right) \\ &- kN\left(r,\frac{\varphi}{\varphi'}\right) - N_1\left(r,\frac{\varphi}{\varphi'}\right) + Q(r,f). \end{aligned}$$

Substituting this inequality in (2.17′) and leaving out the negative summands, we get

$$T(r,f) \le \left(2+\frac{1}{k}\right)N\left(r,\frac{1}{f}\right) + \left(2+\frac{2}{k}\right)\bar{N}\left(r,\frac{1}{\varphi}\right) + Q(r,f),$$

that is, inequality (2.23) with $a=0$, $b=1$. □

Using Theorems 2.4, 2.5, and 2.6 it is possible to prove analogs of Theorem 2.2 and the Corollary of Theorem 2.2. For example, from Theorem 2.6 we get that if a meromorphic function $f(z)$ has a Borel (Picard) exceptional value $a \ne \infty$, then $f^{(k)}(z)$, $k \ge 1$, cannot have a Borel (Picard) exceptional value $b \ne 0, \infty$.

In fact, if $f(z)$ is of infinite order, then $N(r,a,f)$ and $N(r,b,f^{(k)})$ are of finite order, and there is a constant $K>0$ such that

$$N(r,a,f) + N(r,b,f^{(k)}) < r^K$$

for $r \ge r_0$. Arguing in the same way as in the proof of Theorem 2.2, but using (2.23) instead of (2.11), we conclude that $T(r,f) < 3r^K$ for sufficiently large r. Hence $f(z)$ is of finite order. For transcendental functions of finite order, (2.23) implies

$$(1+o(1))T(r,f) \le \left(2+\frac{1}{k}\right)N(r,a,f) + \left(2+\frac{2}{k}\right)N(r,b,f^{(k)}). \tag{2.27}$$

[1] If $g' \equiv 0$, then f is rational with $f(\infty)=\infty$, and (2.23) holds trivially in this case.

The category of $N(r, a, f)$ is lower than the category of $T(r, f)$, and the category of $N(r, b, f^{(k)})$ is lower than the category of $T(r, f^{(k)})$, by the definition of a Borel exceptional value. On the other hand, by Theorem 2.3 the category of $T(r, f^{(k)})$ does not exceed the category of $T(r, f)$. Therefore the right-hand side of (2.27) has lower category than the left-hand side, and we get a contradiction.

Theorem 2.6 implies the following result.

THEOREM 2.7 (Hayman). *If $g(z)$ is an entire function and $g(z)g''(z)$ does not have zeros, then $g(z) = e^{Az+B}$, where A and B are constants, $A \neq 0$.*

PROOF. Let

$$f(z) = g(z)/g'(z).$$

It is clear that $N(r, 0, f) \equiv 0$ and that $1/f(z)$ is an entire function. Since

$$f'(z) = 1 - \frac{g(z)g''(z)}{[g'(z)]^2},$$

we also have $N(r, 1, f') \equiv 0$. Applying (2.23) with $k = 1$, $a = 0$, $b = 1$, we get

$$T(r, f) = Q(r, f);$$

therefore

$$T(r, f) = O(\ln r),$$

and hence the function $f(z)$ is rational, and $1/f(z)$ is a polynomial. It follows that $g(z) = \exp P(z)$, where $P(z)$ is a polynomial. Since $g''(z) = (P'^2 + P'')e^P$, it follows that $g''(z)$ has no zeros if and only if $P'^2 + P'' = \text{const} \neq 0$, that is, if $P(z) = Az + B$, $A \neq 0$. □

REMARK. The statement of Theorem 2.7 remains true if we replace the condition $g(z)g''(z) \neq 0$ by the condition $g(z)g^{(n)}(z) \neq 0$ for some $n \geq 2$ (Tumura, Clunie). Proof of this result is rather deep, and we do not include it (it can be found in Hayman's book [**Hay64**, §3.5]).

REMARK. Langley [**104**] generalized Theorem 2.7 to meromorphic functions: the only meromorphic functions with the property $f(z)f''(z) \neq 0$ are the exponential function and $f(z) = c(z - a)^{-n}$. The same result for the product $ff^{(k)}$, $k \geq 3$, was earlier proved by Frank [**52**].

3. Analogs of the second fundamental theorem for angular Nevanlinna characteristics and for Tsuji characteristics

First we prove the corresponding analogs of the lemma on the logarithmic derivative.

It is natural to expect that for the function $f(z)$ meromorphic in the angle $\alpha \leq \arg z \leq \beta$, $0 < \beta - \alpha \leq 2\pi$, the relation

$$A_{\alpha\beta}\left(r, \frac{f'}{f}\right) + B_{\alpha\beta}\left(r, \frac{f'}{f}\right) = o(S_{\alpha\beta}(r, f))$$

holds as $r \to \infty$ outside an exceptional set. However, this statement conjectured by R. Nevanlinna [**NevR25b**] in 1925 has not been proved or disproved yet.[2] We

[2] A counterexample was constructed by Goldberg [**63**].

estimate the sum $A_{\alpha\beta}(r, f'/f) + B_{\alpha\beta}(r, f'/f)$ only for functions $f(z)$ meromorphic everywhere in the complex plane, and we give it in terms of $T(r, f)$, and not $S_{\alpha\beta}(r, f)$.

Let $f(z)$ be a meromorphic function, let $\{\alpha \leq \arg z \leq \beta\}$ be an angle, and let $k = \pi/(\beta - \alpha)$. Denote by $Q_{\alpha\beta}(r, f)$ an arbitrary function defined for $\{1 \leq r < \infty\}$ and such that

1) If the function $f(z)$ satisfies the condition

$$\int_1^{\infty} r^{-k-1} \ln^+ T(r, f)\, dr < \infty, \tag{3.1}$$

then $Q_{\alpha\beta}(r, f) = O(1)$.

2) If condition (3.1) is not satisfied, then $Q_{\alpha\beta}(r, f) = O(\ln T(r, f) + \ln r)$ as $r \to \infty$, possibly omitting a set of finite measure.

THEOREM 3.1. *Each meromorphic function $f(z)$ satisfies*

$$A_{\alpha\beta}(r, f'/f) + B_{\alpha\beta}(r, f'/f) = Q_{\alpha\beta}(r, f).$$

PROOF. It is easy to see that

$$B_{\alpha\beta}\left(r, \frac{f'}{f}\right) = \frac{2k}{\pi r^k} \int_{\alpha}^{\beta} \ln^+ \left| \frac{f'(re^{i\varphi})}{f(re^{i\varphi})} \right| \sin k(\varphi - \alpha) d\varphi \leq \frac{4k}{r^k} m\left(r, \frac{f'}{f}\right). \tag{3.2}$$

If the function $f(z)$ satisfies condition (3.1), then $\ln^+ T(r, f) = o(r^k)$ because

$$\frac{\ln^+ T(r, f)}{2^{k+1} r^k} \leq \int_r^{2r} \frac{\ln^+ T(t, f)}{t^{k+1}} dt = o(1).$$

Using estimate (1.13) for $m(r, f'/f)$, we get

$$B_{\alpha\beta}\left(r, \frac{f'}{f}\right) \leq \frac{4k}{r^k}(o[(2r)^k] + O(\ln r)) = o(1) = Q_{\alpha\beta}(r, f).$$

If condition (3.1) is not satisfied, then the estimate $B_{\alpha\beta}(r, f'/f) = Q_{\alpha\beta}(r, f)$ follows from (3.2) and the lemma on the logarithmic derivative.

Thus, everything is reduced to a proof of the relation

$$A_{\alpha\beta}\left(r, \frac{f'}{f}\right) = Q_{\alpha\beta}(r, f). \tag{3.3}$$

We prove the following lemma which plays the same role as Theorem 1.1 in the proof of the lemma on the logarithmic derivative.

LEMMA 3.1. *Let $f(z)$ be a meromorphic function satisfying $f(0) = 1$. Then the following inequality holds:*

$$A_{\alpha\beta}\left(r, \frac{f'}{f}\right) \leq K \left\{ \left(\frac{R}{r}\right)^k \int_1^R \frac{\ln^+ T(t, f)}{t^{k+1}} dt + \ln^+ \frac{r}{R - r} + \ln \frac{R}{r} + 1 \right\}, \tag{3.4}$$

where $1 \leq r < R < \infty$, and K is a constant which is independent of r and R.

PROOF. Our argument is based on inequality (1.3′) (we use the same notation as in the proof of Theorem 1.1). Let $z = te^{i\varphi}$ and $s = \sqrt{\lambda} t$, where $\lambda = Rr^{-1}$ in inequality (1.3′). We get

$$\left| \frac{f'(te^{i\varphi})}{f(te^{i\varphi})} \right| \leq U_1(t) + U_2(t),$$

where

$$U_1(t) = \frac{4\sqrt{\lambda}}{t(\sqrt{\lambda}-1)^2} T(\sqrt{\lambda}t, f) \le \frac{16\lambda^{3/2}}{(\lambda-1)^2} T(\lambda t, f), \qquad t \ge 1,$$

and

$$U_2(t) = \frac{\sqrt{\lambda}}{\sqrt{\lambda}-1} \sum_{|c_q|<\sqrt{\lambda}t} \frac{1}{|t-|c_q||} \le \frac{2\lambda}{\lambda-1} \sum_{|c_q|<\sqrt{\lambda}t} \frac{1}{|t-|c_q||}.$$

Therefore

$$\begin{aligned}
A_{\alpha\beta}\left(r, \frac{f'}{f}\right) &= \frac{k}{\pi}\int_1^r \left(\frac{1}{t^k} - \frac{t^k}{r^{2k}}\right)\left[\ln^+\left|\frac{f'(te^{i\alpha})}{f(te^{i\alpha})}\right| + \ln^+\left|\frac{f'(te^{i\beta})}{f(te^{i\beta})}\right|\right]\frac{dt}{t} \\
&\le \frac{k}{\pi}\int_1^r \left(\frac{1}{t^k} - \frac{t^k}{r^{2k}}\right) 2\ln^+[U_1(t) + U_2(t)]\frac{dt}{t} \\
&\le \frac{2k}{\pi}\int_1^r \left(\frac{1}{t^k} - \frac{t^k}{r^{2k}}\right)[\ln^+ U_1(t) + \ln^+ U_2(t) + \ln 2]\frac{dt}{t} \\
&\le \frac{2k}{\pi}\left\{\int_1^r t^{-k-1}\ln^+ U_1(t)dt + \int_1^r \left(\frac{1}{t^k} - \frac{t^k}{r^{2k}}\right)\ln^+ U_2(t)\frac{dt}{t} + \frac{\ln 2}{k}\right\} \\
&= \frac{2k}{\pi}\left\{I_1 + I_2 + \frac{\ln 2}{k}\right\}.
\end{aligned}$$

Now we estimate the integrals I_1 and I_2. By K we denote positive constants independent of t, r, λ, and hence, of $R = \lambda r$.

We have

$$\begin{aligned}
I_1 &\le \int_1^r \ln^+\left\{\frac{16\lambda^{3/2}}{(\lambda-1)^2}T(\lambda t, f)\right\}\frac{dt}{t^{k+1}} \\
&\le \int_1^r \left\{\ln 16 + \frac{3}{2}\ln\lambda + 2\ln^+\frac{1}{\lambda-1} + \ln^+ T(\lambda t, f)\right\}\frac{dt}{t^{k+1}} \\
&\le \int_1^r \frac{\ln^+ T(\lambda t, f)}{t^{k+1}}dt + K\left(\ln^+\frac{1}{\lambda-1} + \ln\lambda + 1\right).
\end{aligned}$$

In a similar way we get

$$\begin{aligned}
I_2 &\le \int_1^r \ln^+\left\{\frac{2\lambda}{\lambda-1}\sum_{|c_q|<\sqrt{\lambda}t}\frac{1}{|t-|c_q||}\right\}\left(\frac{1}{t^k} - \frac{t^k}{r^{2k}}\right)\frac{dt}{t} \\
&\le \int_1^r \ln^+\left\{\sum_{|c_q|<\sqrt{\lambda}t}\frac{1}{|t-|c_q||}\right\}\left(\frac{1}{t^k} - \frac{t^k}{r^{2k}}\right)\frac{dt}{t} + K\left(\ln^+\frac{1}{\lambda-1} + \ln\lambda + 1\right) \\
&= \tilde{I}_2 + K\left(\ln^+\frac{1}{\lambda-1} + \ln\lambda + 1\right).
\end{aligned}$$

Consider the integral $\tilde{I}_2$. Integrating by parts we get

$$\begin{aligned}\tilde{I}_2 &= \int_1^r \left\{ \int_1^t \ln^+ \left[\sum_{|c_q|<\sqrt{\lambda}\tau} \frac{1}{|\tau - |c_q||} \right] d\tau \right\} \left\{ \frac{k+1}{t^{k+2}} + \frac{k-1}{r^{2k}} t^{k-2} \right\} dt \\
&\le 2(k+1) \int_1^r \left\{ \int_1^t \ln^+ \left[\sum_{|c_q|<\sqrt{\lambda}\tau} \frac{1}{|\tau - |c_q||} \right] d\tau \right\} \frac{dt}{t^{k+2}} \\
&= 4(k+1) \int_1^r \left\{ \int_1^t \ln^+ \left[\sum_{|c_q|<\sqrt{\lambda}\tau} \frac{1}{|\tau - |c_q||} \right]^{1/2} d\tau \right\} \frac{dt}{t^{k+2}} \\
&\le 4(k+1) \int_1^r \left\{ \frac{1}{t-1} \int_1^t \ln^+ \left[\sum_{|c_q|<\sqrt{\lambda}\tau} \frac{1}{|\tau - |c_q||^{1/2}} \right] d\tau \right\} \frac{dt}{t^{k+1}}.\end{aligned}$$

To estimate the expression in braces, we use inequality (1.2′). We get[3]

$$\begin{aligned}&\frac{1}{t-1} \int_1^t \ln^+ \left[\sum_{|c_q|<\sqrt{\lambda}\tau} \frac{1}{|\tau - |c_q||^{1/2}} \right] d\tau \\
&\qquad\le \ln^+ \left\{ \frac{1}{t-1} \int_0^t \left[\sum_{|c_q|<\sqrt{\lambda}t} \frac{1}{|\tau - |c_q||^{1/2}} \right] d\tau \right\} + \ln 2 \\
&\qquad\le \ln^+ \left\{ \frac{1}{t-1} \sum_{|c_q|<\sqrt{\lambda}t} \int_0^t \frac{d\tau}{|\tau - |c_q||^{1/2}} \right\} + \ln 2 \\
&\qquad\le \ln^+ \left\{ \frac{1}{t-1} n(\sqrt{\lambda}t; 0, \infty) \int_{-t}^t \frac{d\tau}{|\tau|^{1/2}} \right\} + \ln 2 \\
&\qquad= \ln^+ \left\{ \frac{1}{t-1} 4\sqrt{t} n(\sqrt{\lambda}t; 0, \infty) \right\} + \ln 2.\end{aligned}$$

By Lemma 7.1 of Chapter 1,

$$n(\sqrt{\lambda}t; 0, \infty) \le \frac{\lambda t}{\lambda t - \sqrt{\lambda}t} N(\lambda t; 0, \infty) \le \frac{\sqrt{\lambda}}{\sqrt{\lambda} - 1} 2T(\lambda t, f) \le \frac{4\lambda}{\lambda - 1} T(\lambda t, f);$$

thus, we get the following estimate for I_2:

$$I_2 \le K \left\{ \int_1^r t^{-k-1} \ln^+ T(\lambda t, f)\, dt + \ln^+ \frac{1}{\lambda - 1} + \ln \lambda + 1 \right\}.$$

Therefore

$$\begin{aligned}A_{\alpha\beta}\left(r, \frac{f'}{f}\right) &\le K \left\{ \int_1^r t^{-k-1} \ln^+ T(\lambda t, f)\, dt + \ln^+ \frac{1}{\lambda - 1} + \ln \lambda + 1 \right\} \\
&\le K \left\{ \lambda^k \int_1^{\lambda r} t^{-k-1} \ln^+ T(t, f)\, dt + \ln^+ \frac{1}{\lambda - 1} + \ln \lambda + 1 \right\}.\end{aligned}$$

Recalling that $\lambda = Rr^{-1}$, we get the estimate (3.4). □

[3] Below we use Lemma 7.2 of Chapter 1.

Now we prove (3.3). As in the proof of Theorem 1.3, we can assume without loss of generality that $f(0) = 1$. If the function $f(z)$ satisfies condition (3.1), then, letting $R = 2r$ in (3.4), we get $A_{\alpha\beta}(r, f'/f) = O(1)$. Now suppose that condition (3.1) is not satisfied. Inequality (3.4) implies that

$$A_{\alpha\beta}\left(r, \frac{f'}{f}\right) \leq K\left\{\left(\frac{R}{r}\right)^k \ln^+ T(R, f) + \ln^+ \frac{R}{R-r} + \ln\frac{R}{r} + 1\right\}.$$

We set

$$R = r + \{T(r, f)\}^{-2}$$

and observe that then condition (1.7′) is satisfied everywhere except for a set of finite measure. Therefore, outside a set of finite measure, $A_{\alpha\beta}(r, f'/f) = O(\ln T(r, f) + \ln r)$. This proves (3.3) and concludes the proof of the theorem. □

It turns out to be possible to prove a complete analog of the lemma on the logarithmic derivative for the Tsuji characteristics.

Let $f(z)$ be a meromorphic function in the half-plane $\{\operatorname{Im} z \geq 0\}$. Denote by $\mathfrak{Q}(r, f)$ an arbitrary function defined for $\{1 \leq r < \infty\}$ and such that

1) If $\overset{\circ}{\mathfrak{T}}(r, f)$ is of finite order, then $\mathfrak{Q}(r, f) = O(\ln r)$.

2) If $\overset{\circ}{\mathfrak{T}}(r, f)$ is of infinite order, then

$$\mathfrak{Q}(r, f) = O(\ln \overset{\circ}{\mathfrak{T}}(r, f) + \ln r)$$

as $r \to \infty$, except, possibly, for a set of finite measure.

THEOREM 3.2. *For each function $f(z)$ that is meromorphic in the half-plane $\{\operatorname{Im} z \geq 0\}$, the following relation holds:*

$$\mathfrak{m}\left(r, \frac{f'}{f}\right) = \mathfrak{Q}(r, f).$$

PROOF. First we obtain the following estimate, similar to estimate (1.1):

$$\mathfrak{m}\left(r, \frac{f'}{f}\right) \leq K\left(\ln^+ \overset{\circ}{\mathfrak{T}}(R, f) + \ln^+ \frac{1}{R-r} + \ln R + 1\right), \tag{3.5}$$

where R and r are arbitrary numbers satisfying the condition $2 \leq r < R < \infty$, and K is a constant which does not depend on r and R.

We write formula (1.3) for the disc $\{|z| \leq \frac{1}{2}s\}$ and the function $f\left(z + \frac{1}{2}is\right)$, and after that set $z + \frac{1}{2}is = \zeta$. We get

$$\begin{aligned}\frac{f'(\zeta)}{f(\zeta)} &= \frac{1}{2\pi}\int_0^{2\pi} \ln\left|f\left(\frac{s}{2}e^{i\theta} + \frac{is}{2}\right)\right| \frac{se^{i\theta}}{\left(\frac{s}{2}e^{i\theta} - \zeta + \frac{is}{2}\right)^2}\,d\theta \\ &+ \sum_{\left|a_m - \frac{is}{2}\right| < \frac{s}{2}} \frac{\frac{s^2}{4} - \left|a_m - \frac{is}{2}\right|^2}{\left(\frac{s^2}{4} - \overline{\left[a_m - \frac{is}{2}\right]}\left[\zeta - \frac{is}{2}\right]\right)(\zeta - a_m)} \\ &- \sum_{\left|b_n - \frac{is}{2}\right| < \frac{s}{2}} \frac{\frac{s^2}{4} - \left|b_n - \frac{is}{2}\right|^2}{\left(\frac{s^2}{4} - \overline{\left[b_n - \frac{is}{2}\right]}\left[\zeta - \frac{is}{2}\right]\right)(\zeta - b_n)}, \qquad \left|\zeta - i\frac{s}{2}\right| < \frac{s}{2}.\end{aligned} \tag{3.6}$$

In what follows, we use the notation $\zeta = r\sin\vartheta e^{i\vartheta}$, $\kappa(r) \leq \vartheta \leq \pi - \kappa(r)$, $\kappa(r) = \arcsin\frac{1}{r}$, $s = \frac{1}{2}(R + r)$, and denote by K positive constants which do not depend on R and r. Observe that

$$\left|\zeta - \frac{is}{2}\right| \le \left|r\sin\kappa(r)e^{i\kappa(r)} - \frac{is}{2}\right| = \sqrt{1 - \frac{s}{r} + \frac{s^2}{4}}$$

implies

$$\frac{s}{2}\left|\zeta - \frac{is}{2}\right| \ge \frac{s}{2} - \sqrt{1 - \frac{s}{r} + \frac{s^2}{4}} \ge \frac{s-r}{st}.$$

Formula (3.6) yields the inequality

$$\left|\frac{f'(\zeta)}{f(\zeta)}\right| \le \Phi_1(\zeta) + \Phi_2(\zeta), \tag{3.7}$$

where

$$\Phi_1(\zeta) = \frac{s^3r^2}{(s-r)^2}\frac{1}{2\pi}\int_0^{2\pi}\left|\ln\left|f\left(\frac{s}{2}e^{i\theta} + \frac{is}{2}\right)\right|\right|d\theta, \tag{3.8}$$

$$\Phi_2(\zeta) = \frac{s^2r}{s-r}\sum_{|c_q - \frac{is}{2}| < \frac{s}{2}} \frac{1}{|\zeta - c_q|},$$

and $\{c_q\} = \{a_m\} \cup \{b_n\}$. By (3.7),

$$\begin{aligned}
\mathfrak{m}\left(r, \frac{f'}{f}\right) &\le \frac{1}{2\pi}\int_{\kappa(r)}^{\pi-\kappa(r)} \ln^+[\Phi_1(\zeta) + \Phi_2(\zeta)]\frac{d\vartheta}{r\sin^2\vartheta} \\
&\le \frac{1}{2\pi}\int_{\kappa(r)}^{\pi-\kappa(r)} \ln^+\Phi_1(\zeta)\frac{d\vartheta}{r\sin^2\vartheta} + \frac{1}{2\pi}\int_{\kappa(r)}^{\pi-\kappa(r)} \ln^+\Phi_2(\zeta)\frac{d\vartheta}{r\sin^2\vartheta} \\
&\quad + \frac{\ln 2}{2\pi}\int_{\kappa(r)}^{\pi-\kappa(r)}\frac{d\vartheta}{r\sin^2\vartheta} \\
&= I_1 + I_2 + \frac{\ln 2}{2\pi}\frac{2}{r}\cot\kappa(r).
\end{aligned}$$

Let us estimate the integrals I_1 and I_2. For I_1 we note that taking $\left[-\frac{\pi}{2}, \frac{3\pi}{2}\right]$ as the integration interval instead of $[0, 2\pi]$ and making the change of variable $\theta = 2\tau - \frac{\pi}{2}$, we get

$$\begin{aligned}
\Phi_1(\zeta) &= \frac{s^3r^2}{(s-r)^2}\frac{1}{\pi}\int_0^{\pi}|\ln|f(s\sin\tau e^{i\tau})||d\tau \\
&= \frac{s^3r^2}{(s-r)^2}\left\{\frac{1}{\pi}\int_{\kappa(s)}^{\pi-\kappa(s)} s\sin^2\tau|\ln|f(s\sin\tau e^{i\tau})||\frac{d\tau}{s\sin^2\tau}\right. \\
&\qquad \left. + \frac{1}{\pi}\left[\int_0^{\kappa(s)} + \int_{\pi-\kappa(s)}^{\pi}\right]|\ln|f(s\sin\tau e^{i\tau})||d\tau\right\} \\
&\le \frac{s^3r^2}{(s-r)^2}\{2s(\mathfrak{m}(s,0) + \mathfrak{m}(s,\infty)) + K\}.
\end{aligned}$$

In view of Theorem 5.3 in Chapter 1, $\mathfrak{m}(s,a) \le \mathfrak{T}(s,f) + O(1)$, and by Theorem 5.4 of Chapter 1, $\mathfrak{T}(s,f) = \mathring{\mathfrak{T}}(s,f) + O(1)$. Therefore

$$\Phi_1(\zeta) \le K\frac{R^6}{(R-r)^2}[\mathring{\mathfrak{T}}(R,f) + 1],$$

whence

$$I_1 \leq K\left\{\ln^+ \mathring{\mathfrak{T}}(R,f) + \ln^+ \frac{1}{R-r} + \ln R + 1\right\} \frac{2}{\pi r} \cot \kappa(r).$$

Furthermore, since $\frac{2}{\pi r} \cot \kappa(r) \leq 1$, we get

$$I_1 \leq K\left\{\ln^+ \mathring{\mathfrak{T}}(R,f) + \ln^+ \frac{1}{R-r} + \ln R + 1\right\}. \tag{3.9}$$

We now estimate I_2. We have

$$\begin{aligned} I_2 &\leq \frac{1}{2\pi}\int_{\kappa(r)}^{\pi-\kappa(r)} \ln^+\left(\sum_{\left|c_q-\frac{is}{2}\right|<\frac{s}{2}} \frac{1}{|\zeta-c_q|}\right)\frac{d\vartheta}{r\sin^2\vartheta} \\ &\quad + K\left(\ln^+\frac{1}{R-r} + \ln R\right) \\ &= \tilde{I}_2 + K\left(\ln^+\frac{1}{R-r} + \ln R\right). \end{aligned} \tag{3.10}$$

Moreover,

$$\begin{aligned} \tilde{I}_2 &= \frac{2}{\pi}\int_{\kappa(r)}^{\pi-\kappa(r)} \ln^+\left(\sum_{\left|c_q-i\frac{s}{2}\right|<\frac{s}{2}} \frac{1}{|\zeta-c_q|}\right)^{1/4}\frac{d\vartheta}{r\sin^2\vartheta} \\ &\leq \frac{2}{\pi}\int_{\kappa(r)}^{\pi-\kappa(r)} \ln^+\left(\sum_{\left|c_q-\frac{is}{2}\right|<\frac{s}{2}} \frac{1}{|\zeta-c_q|^{1/4}}\right)\frac{d\vartheta}{r\sin^2\vartheta}. \end{aligned}$$

From Lemma 1.1 we get

$$\begin{aligned} \tilde{I}_2 &\leq \frac{2}{\pi}\frac{2}{r}\cot\kappa(r) \\ &\quad\times\left\{\ln^+\left[\frac{r}{2\cot\kappa(r)}\int_{\kappa(r)}^{\pi-\kappa(r)}\left(\sum_{\left|c_q-\frac{is}{2}\right|<\frac{s}{2}} \frac{1}{|\zeta-c_q|^{1/4}}\right)\frac{d\vartheta}{r\sin^2\vartheta}\right] + \ln 2\right\} \\ &\leq K\left\{\ln^+\left[\sum_{\left|c_q-\frac{is}{2}\right|<\frac{s}{2}} \int_{\kappa(r)}^{\pi-\kappa(r)} \frac{1}{|\zeta-c_q|^{1/4}}\frac{d\vartheta}{r\sin^2\vartheta}\right] + 1\right\} \\ &\leq K\left\{\ln^+\left[r\sum_{\left|c_q-\frac{is}{2}\right|<\frac{s}{2}} \int_{\kappa(r)}^{\pi-\kappa(r)} \frac{d\vartheta}{|\zeta-c_q|^{1/4}}\right] + 1\right\}. \end{aligned}$$

Using the notation $\beta_q = \arg c_q$ we obtain

$$\begin{aligned}\int_{\kappa(r)}^{\pi-\kappa(r)} \frac{d\vartheta}{|\zeta - c_q|^{1/4}} &= \int_{\kappa(r)}^{\pi-\kappa(r)} \frac{d\vartheta}{|r\sin\vartheta e^{i(\vartheta-\beta_q)} - |c_q||^{1/4}} \\ &\le \int_0^\pi \frac{d\vartheta}{|r\sin\vartheta\sin(\vartheta-\beta_q)|^{1/4}} \\ &\le \left(\int_0^\pi \frac{d\vartheta}{(\sin\vartheta)^{1/2}} \int_0^\pi \frac{d\vartheta}{|\sin(\vartheta-\beta_q)|^{1/2}}\right)^{1/2} \\ &= \int_0^\pi \frac{d\vartheta}{(\sin\vartheta)^{1/2}}.\end{aligned}$$

Therefore[4]

$$\tilde{I}_2 \le K\{\ln r + \ln^+ \mathfrak{n}(s;0,\infty) + 1\}.$$

Since

$$\begin{aligned}\mathfrak{n}(s;0,\infty) &\le \frac{R^2}{R-s}\int_s^R \frac{\mathfrak{n}(t;0,\infty)}{t^2}dt \le \frac{R^2}{R-s}\mathfrak{N}(R;0,\infty) \\ &\le \frac{R^2}{R-s}(2\mathring{\mathfrak{T}}(R,f) + O(1)) \le \frac{KR^2}{R-r}(\mathring{\mathfrak{T}}(R,f) + O(1)),\end{aligned}$$

we get

$$\tilde{I}_2 \le K\left\{\ln^+ \mathring{\mathfrak{T}}(R,f) + \ln^+ \frac{1}{R-r} + \ln R + 1\right\}. \tag{3.11}$$

Now (3.9), (3.10), and (3.11) yield (3.5).

To prove Theorem 3.2 it is enough to let $R = r + \{\mathring{\mathfrak{T}}(r,f)\}^{-2}$ in (3.5). □

We thus turn to analogs of the second fundamental theorem. We say that the characteristics $C_{\alpha\beta}(r,a)$ and $\mathfrak{N}(r,a)$ correspond to the characteristic $N(r,a)$, the characteristics $A_{\alpha\beta}(r,a) + B_{\alpha\beta}(r,a)$ and $\mathfrak{m}(r,a)$ correspond to the characteristic $m(r,a)$, and the characteristics $S_{\alpha\beta}(r,f)$ and $\mathring{\mathfrak{T}}(r,f)$ correspond to the characteristic $T(r,f)$. Corresponding to $\bar{N}(r,a)$ are the characteristics $\bar{C}_{\alpha\beta}(r,a)$ and $\bar{\mathfrak{N}}(r,a)$ defined as follows:

$$\bar{C}_{\alpha\beta}(r,a) = 2\sum_{\substack{1<r_m<r\\ \alpha<\varphi_m<\beta}} \left(\frac{1}{r_m^k} - \frac{r_m^k}{r^{2k}}\right)\sin k(\varphi_m - \alpha),$$

$$\bar{\mathfrak{N}}(r,a) = \sum_{1<r_m<r\sin\varphi_m} \left(\frac{\sin\varphi_m}{r_m} - \frac{1}{r}\right),$$

where $r_m e^{i\varphi_m}$ are the a-points of the function $f(z)$ counted without multiplicity. We also define in a natural way the characteristics $C_{1\alpha\beta}(r)$ and $\mathfrak{N}_1(r)$ corresponding to the characteristic $N_1(r)$, and the characteristics $C_{1\beta}(r,a)$ and $\mathfrak{N}_1(r,a)$ corresponding to the characteristic $N_1(r,a)$.

Analyzing the proofs of Theorems 2.1, 2.4, 2.5, and 2.6, we see that they use only those properties of the Nevanlinna characteristic that were described in Theorems 4.1 and 6.1 of Chapter 1, relations (6.1), (6.2), (6.3), (6.4), and (4.1) of Chapter 1, and in Theorem 1.3 of Chapter 3. On the other hand, similar theorems and relations are valid also for the angular Nevanlinna characteristics (Theorems

[4]We set $\mathfrak{n}(r;a,b) = \mathfrak{n}(r,a) + \mathfrak{n}(r,b)$; $\mathfrak{N}(r;a,b) = \mathfrak{N}(r,a) + \mathfrak{N}(r,b)$.

5.1 and 6.3 of Chapter 1, relations (6.9), (6.10), (6.12), and (6.13) of Chapter 1, and Theorem 3.1 of Chapter 3), and for the Tsuji characteristics (Theorem 5.3 and 6.3 of Chapter 1, relations (6.9), (6.10), (6.12), and (6.13) of Chapter 1, and Theorem 3.2 of Chapter 3).

Therefore we have the following theorem.

THEOREM 3.3. *Theorems* 2.1, 2.3, 2.4, 2.5, 2.6 *and Lemma* 2.1 *remain true if the Nevanlinna characteristics are replaced by the corresponding angular Nevanlinna characteristics, and* $Q(r, f)$ *is replaced by* $Q_{\alpha\beta}(r, f)$.

Theorems 2.1, 2.3, 2.4, 2.5, 2.6 *and Lemma* 2.1 *remain true if the Nevanlinna characteristics are replaced by the corresponding Tsuji characteristics, and* $Q(r, f)$ *is replaced by* $\mathfrak{Q}(r, f)$.

CHAPTER 4

Deficient Values

1. Exceptional values

By the first fundamental theorem the relation $m(r,a)+N(r,a) = T(r,f)+O(1)$ holds for each a, so the dependence of the sum $m(r,a)+N(r,a)$ on a is inessential. On the other hand, the contribution to this sum of each of the summands may depend heavily on a, although, as we shall see later, for the "majority" of the values of a, the summand $N(r,a)$ is dominant. We introduce the following notation:

$$\delta(a,f) = \delta(a) = \liminf_{r\to\infty} \frac{m(r,a)}{T(r,f)} = 1 - \limsup_{r\to\infty} \frac{N(r,a)}{T(r,f)},$$

$$\Delta(a,f) = \Delta(a) = \limsup_{r\to\infty} \frac{m(r,a)}{T(r,f)} = 1 - \liminf_{r\to\infty} \frac{N(r,a)}{T(r,f)}.$$

The quantity $\delta(a)$ is called the *Nevanlinna deficiency* (or simply *deficiency*) of the function $f(z)$ at the point a; the quantity $\Delta(a)$ is called the *Valiron deficiency*. It is clear that

$$0 \le \delta(a) \le \Delta(a) \le 1. \tag{1.1}$$

If $\Delta(a) > 0$, then a is called a *Valiron exceptional value*, and if $\delta(a) > 0$, then a is called a *Nevanlinna exceptional value*, or *deficient value*. Denote by $E_P(f)$, $E_B(f)$, $E_N(f)$, and $E_V(f)$ the sets of Picard, Borel, Nevanlinna, and Valiron exceptional values, respectively.

Let $f(z)$ be a transcendental meromorphic function. We already observed in Section 1 of Chapter 2 that $E_P(f) \subset E_B(f)$. Inequality (1.1) immediately implies that $E_N(f) \subset E_V(f)$. Since for $a \in E_P(f)$ we have $N(r,a) = O(\ln r)$, it is easy to see that $a \in E_P(f)$ implies $\delta(a) = \Delta(a) = 1$.

Thus

$$E_P(f) \subset E_N(f) \subset E_V(f). \tag{1.2}$$

Let us show that $E_B(f) \subset E_V(f)$. In fact, if $a \notin E_V(f)$, then $\Delta(a) = 0$ and $N(r,a) \sim T(r,f)$. If $n(r,a) \to \infty$ as $r \to \infty$, then, by Theorem 1.1 of Chapter 2, the growth category of $n(r,a)$ coincides with the growth category of $N(r,a)$, and thus $a \notin E_B(f)$. On the other hand, if $n(r,a) = O(1)$, then $a \in E_P(f)$, and hence a belongs to both $E_B(f)$ and $E_V(f)$. Thus we obtain also

$$E_P(f) \subset E_B(f) \subset E_V(f). \tag{1.3}$$

The question arises: is it possible to find an inclusion between the sets $E_B(f)$ and $E_N(f)$, and thus to write (1.2) and (1.3) as one inclusion? Examples given below (see Remark 3 at the end of Section 4) and Examples 3 and 4 in Section 2 of Chapter 5 show that there is no general set-theoretical dependence between the sets $E_B(f)$

and $E_N(f)$. Namely, for some a and f, we may have 1) $a \in E_B(f) \cap E_N(f)$; 2) $a \in E_B(f)\backslash E_N(f)$; 3) $a \in E_N(f)\backslash E_B(f)$.

The problem of the structure of the sets $E_P(f)$ and $E_B(f)$ is relatively simple. In Theorem 2.2 of Chapter 3 and its Corollary it was shown that for transcendental meromorphic functions $f(z)$ the sets $E_P(f)$ and $E_B(f)$ consist of at most two points. For meromorphic functions of non-integer order ρ and for meromorphic functions of zero order, this statement can be made more precise.

THEOREM 1.1. *Transcendent meromorphic function of non-integer or zero order cannot have more than one Borel (and hence Picard) exceptional value.*

PROOF. Assume the contrary; that is, the function $f(z)$ has two Borel exceptional values. Without loss of generality we may assume that 0 and ∞ are the Borel exceptional values; otherwise we can get into such a situation by using a linear-fractional transformation. First we consider the case $\rho = 0$. Then, if $n(r,0)$ and $n(r,\infty)$ have lower category than $T(r,f)$, we have $n(r,0) = O(1)$ and $n(r,\infty) = O(1)$. By the Hadamard Theorem (Theorem 4.1 of Chapter 2), the function $f(z)$ is rational, contrary to the assumption. Now we consider the case when ρ (the order of $f(z)$) is non-integer. If both $N(r,0)$ and $N(r,\infty)$ have lower category than $T(r,f)$, then the same property is shared by $N(r;0,\infty)$. From Theorem 4.2 (Chapter 2) it follows that the orders and types of the functions $T(r,f)$ and $N(r;0,\infty)$ coincide.

Let us prove that if $N(r;0,\infty)$ is of convergence class, then $T(r,f)$ also is of convergence class. We assume, without loss of generality, that $f(0) \neq 0, \infty$. Let

$$\int_0^\infty \frac{N(r;0,\infty)}{r^{\rho+1}}dr = M < \infty.$$

Using inequality (4.3) of Chapter 2, we get

$$\begin{aligned}\int_1^\infty \frac{T(r,f)}{r^{\rho+1}}dr &\le O(1)\int_1^\infty r^{p-\rho-1}dr \\ &\quad + C(p)(p+1)\int_1^\infty \frac{dr}{r^{\rho+1}}\int_0^\infty \frac{N(r\tau;0,\infty)}{\tau^{p+1}}\frac{d\tau}{1+\tau} \\ &= O(1) + C(p)(p+1)\int_0^\infty \frac{d\tau}{\tau^{p+1}(1+\tau)}\int_1^\infty \frac{N(r\tau;0,\infty)}{r^{\rho+1}}dr \\ &\le O(1) + C(p)(p+1)\int_0^\infty \frac{\tau^{\rho-p-1}d\tau}{1+\tau}\int_0^\infty \frac{N(t;0,\infty)}{t^{\rho+1}}dt \\ &= O(1) + C(p)(p+1)M\frac{\pi}{\sin\pi(\rho-p)} < \infty.\end{aligned}$$

Legitimacy of the change of order of integration follows from the standard tests. We get a contradiction with the assumption that the category of $N(r;0,\infty)$ is lower than the category of $T(r,f)$. □

Now we give examples showing that the statements about $E_P(f)$ and $E_B(f)$ cannot be made more precise. Detailed analysis of these examples is left as an exercise for interested readers.

A. Order $\rho = 0$. Each transcendental meromorphic function of zero order with infinitely many zeros and poles and such that $n(r,\infty) = o(n(r,0))$ has neither Picard nor Borel exceptional values. Each transcendental entire function of zero order has ∞ as its only Picard and Borel exceptional value.

B. Order $\rho = n = 1, 2, \dots$. The function $\exp(z^n)$ has two Picard exceptional values: 0 and ∞. For $f(z) = \sin z^n$, the sets $E_P(f) = E_B(f)$ consist of one point ∞. The function $(\sin z^n)/\sinh z^n$ does not have Borel exceptional values.

C. The order ρ is non-integer. Let

$$f(z) = \prod_{n=1}^{\infty} E\left(\frac{z}{n^{1/\rho}}, p\right), \qquad p = [\rho].$$

Then ∞ is the only Borel and Picard exceptional value of $f(z)$. The function $f(z)/f(-z)$ does not have Borel (and Picard) exceptional values.

D. Order $\rho = \infty$. The function $\exp(e^z)$ has two Picard exceptional values: 0 and ∞; the function $\sin e^z$ has ∞ as its only Borel and Picard exceptional value; the function $(\sin e^z)/\sinh e^z$ does not have Borel exceptional values.

In all of the examples, Picard exceptional values were 0 and ∞. Using linear-fractional transformations they can be replaced by arbitrary numbers a and b.

The problem of the characterization of the structure of $E_V(f)$ and $E_N(f)$ is much more complicated. We study this problem in the forthcoming sections.

Note that by Theorems 6.6 and 6.7 of Chapter 1, the sets of exceptional values in all senses under consideration (Picard, Borel, Nevanlinna, and Valiron) of the functions $f(z)$, $f(z^n)$, and $f(Az)$, $n = 1, 2, \dots, A \neq 0$, coincide; they also have the same Valiron and Nevanlinna deficiencies. This observation will be repeatedly used later on.

EXERCISE 1. Let $f(z)$ be a canonical product of genus $p \geq 0$ with positive zeros such that $n(r, 0) \sim Ar^{\rho(r)}$, where $0 < A < \infty$, $\rho(r)$ is a proximate order, $\rho(r) \to \rho$. The results of Section 5 in Chapter 2 show that

(1) If $\rho = 0$, then $\Delta(a) = 0$ for all $a \neq \infty$.
(2) If $\rho > 0$ and $\rho = p$ or $\rho = p + 1$, then $\delta(\infty) = \delta(0) = 1$ and $\Delta(a) = 0$ for all $a \neq 0, \infty$.
(3) If $p < \rho < p + 1$, then $\Delta(a) = 0$ for all $a \neq 0, \infty$, and

$$\delta(0, f) = \Delta(0, f) = \omega(\rho) = \begin{cases} \frac{p}{p + |\sin \rho\pi|}, & p < \rho \leq p + \frac{1}{2}, \\ \frac{p + 1 - |\sin \rho\pi|}{p + 1}, & p + \frac{1}{2} \leq \rho < p + 1. \end{cases} \tag{1.4}$$

EXERCISE 2. Let $f(z)$ be a canonical product of genus $p \geq 0$ with positive zeros such that $n(r, 0) \sim Ar^{l(r)}$, where $0 < A < \infty$, $l(r)$ is a differentiable function on $[r_0, \infty)$, $rl'(r) \ln r \to 0$ as $r \to \infty$, and

$$\limsup_{r \to \infty} l(r) = l_2 < p + 1,$$

$$\liminf_{r \to \infty} l(r) = l_1 > p.$$

Prove that $\Delta(a, f) = 0$ for all $a \neq 0, \infty$, and

$$\delta(0, f) = \min_{l_1 \leq t \leq l_2} \omega(t),$$

$$\Delta(0, f) = \max_{l_1 \leq t \leq l_2} \omega(t),$$

where $\omega(t)$ is defined by (1.4).

EXERCISE 3. Prove that for the Mittag-Leffler function $E_\rho(z)$, where $0 < \rho < \infty$, $\rho \neq 1$, the deficiency $\Delta(a) = 0$ for all $a \neq \infty$.

EXERCISE 4. Let $f(z)$ be a meromorphic function of normal type of order ρ, $0 < \rho < \infty$. Prove that if $\delta(a) = 1$, then a is a Borel exceptional value.

2. The set of Valiron exceptional values

We start with the following Valiron theorem.

THEOREM 2.1. *For each meromorphic function $f(z)$ that is not identically constant, the two-dimensional Lebesgue measure of $E_V(f)$ is zero.*

PROOF. This theorem explains why the values a for which $\Delta(a) > 0$ are called *exceptional*. More precise estimates of "smallness" of the set $E_V(f)$ are known; they are due to Valiron and Ahlfors. We are not going to present them because their statements and proofs require additional background in potential theory.[1]

We use the notation of Section 4 in Chapter 1. The measure of a measurable set D in an a-plane will be denoted by $|D|$, and the two-dimensional measure of the corresponding set $\tilde{D} = p(D) \subset S_0$ will be denoted by $|D|_s$.

We use the following equality:

$$\iint_{|a|\leq\infty} \ln \frac{1}{[\omega, a]} d\omega(a) = \frac{\pi}{2}, \tag{2.1}$$

which was proved in Section 4 of Chapter 1 (see (4.19)).

Equality (4.20) of Chapter 1 implies the inequality

$$N(r, a) \leq \overset{\circ}{T}(r, f) + \ln \frac{1}{[f(0), a]}. \tag{2.2}$$

Let D be a measurable subset of an extended a-plane. Integrating inequality (2.2) over D and using (2.1), we get

$$\begin{aligned} \iint_D N(r, a) d\omega(a) &\leq \overset{\circ}{T}(r, f)|D|_s + \iint_D \ln \frac{1}{[f(0), a]} d\omega(a) \\ &\leq \overset{\circ}{T}(r, f)|D|_s + \iint_{|a|\leq\infty} \ln \frac{1}{[f(0), a]} d\omega(a) = \overset{\circ}{T}(r, f)|D|_s + \frac{\pi}{2}. \end{aligned} \tag{2.3}$$

Let $0 < \eta < 1/2$. Denote by $B(r)$ the set of those values of a for which

$$N(r, a) \leq \overset{\circ}{T}(r, f) - \{\overset{\circ}{T}(r, f)\}^{\frac{1}{2}+\eta}. \tag{2.4}$$

Since $N(r, a)$ is measurable as a function of a, the set $B(r)$ is measurable. We show that

$$|B(r)|_s < \frac{\pi}{\{\overset{\circ}{T}(r, f)\}^{\frac{1}{2}+\eta}}. \tag{2.5}$$

Assume the contrary; that is, assume that

$$|B(r)|_s \geq \frac{\pi}{\{\overset{\circ}{T}(r, f)\}^{\frac{1}{2}+\eta}}. \tag{2.6}$$

---

[1]Nevanlinna proved that $E_V(f)$ always has zero logarithmic capacity.

Denote by $B_1(r)$ a subset of $B(r)$ satisfying

$$|B_1(r)|_s = \frac{\pi}{\{\overset{\circ}{T}(r,f)\}^{\frac{1}{2}+\eta}}, \tag{2.7}$$

and let $C(r) = \{|a| \le \infty\} \backslash B_1(r)$. Then, combining (4.18) of Chapter 1 with (2.3), (2.4), and (2.7), we get

$$\begin{aligned}
\overset{\circ}{T}(r,f) &= \frac{1}{\pi}\iint_{B_1(r)} N(r,a)\,d\omega(a) + \frac{1}{\pi}\iint_{C(r)} N(r,a)\,d\omega(a) \\
&\le \frac{1}{\pi}\left\{\overset{\circ}{T}(r,f) - \{\overset{\circ}{T}(r,f)\}^{\frac{1}{2}+\eta}\right\}|B_1(r)|_s + \frac{1}{\pi}\overset{\circ}{T}(r,f)|C(r)|_s + \frac{1}{2} \\
&= \overset{\circ}{T}(r,f)\frac{|S_0|_s}{\pi} - \{\overset{\circ}{T}(r,f)\}^{\frac{1}{2}+\eta}\frac{|B_1(r)|_s}{\pi} + \frac{1}{2} = \overset{\circ}{T}(r,f) - \frac{1}{2},
\end{aligned}$$

a contradiction.

Note that for $x > 1$ the function $x - x^{\frac{1}{2}+\eta}$ is strictly increasing. Choose r_1 such that $\overset{\circ}{T}(r_1,f) > 1$, and define a sequence $\{r_n\}$ using the following recurrence: r_n is the (uniquely determined) solution of the equation

$$\overset{\circ}{T}(r,f) - \{\overset{\circ}{T}(r,f)\}^{\frac{1}{2}+\eta} = \overset{\circ}{T}(r_{n-1},f), \qquad n = 2,3,\ldots.$$

Since

$$\overset{\circ}{T}(r_n,f) - \overset{\circ}{T}(r_{n-1},f) = \{\overset{\circ}{T}(r_n,f)\}^{\frac{1}{2}+\eta} > 1, \tag{2.8}$$

we have $\overset{\circ}{T}(r_n,f) \to \infty$ as $n \to \infty$. It is clear that the sequence $\{r_n\}$ is increasing and $\lim_{n\to\infty} r_n = \infty$. If $a \notin B(r_n)$, then the inequality

$$N(r,a) \ge \overset{\circ}{T}(r,f) - 2\{\overset{\circ}{T}(r,f)\}^{\frac{1}{2}+\eta} \tag{2.9}$$

holds for $r_n \le r \le r_{n+1}$. In fact,

$$\begin{aligned}
N(r,a) \ge N(r_n,a) &> \overset{\circ}{T}(r_n,f) - \{\overset{\circ}{T}(r_n,f)\}^{\frac{1}{2}+\eta} \\
&\ge \overset{\circ}{T}(r_n,f) - \{\overset{\circ}{T}(r_n,f)\}^{\frac{1}{2}+\eta} + \left\{\overset{\circ}{T}(r,f) - \{\overset{\circ}{T}(r,f)\}^{\frac{1}{2}+\eta}\right\} \\
&\quad - \left\{\overset{\circ}{T}(r_{n+1},f) - \{\overset{\circ}{T}(r_{n+1},f)\}^{\frac{1}{2}+\eta}\right\} \\
&= \overset{\circ}{T}(r,f) - \{\overset{\circ}{T}(r,f)\}^{\frac{1}{2}+\eta} - \{\overset{\circ}{T}(r_n,f)\}^{\frac{1}{2}+\eta} \ge \overset{\circ}{T}(r,f) - 2\{\overset{\circ}{T}(r,f)\}^{\frac{1}{2}+\eta}.
\end{aligned}$$

Hence, if $a \notin \bigcup_{k=n}^{\infty} B(r_k)$, then inequality (2.9) holds for all $r \ge r_n$; thus $\Delta(a) = 0$. Therefore

$$E_V(f) \subset \bigcup_{k=n}^{\infty} B(r_k)$$

for each n. On the other hand, by (2.5) and (2.8), we get

$$\begin{aligned}
\left|\bigcup_{k=n}^{\infty} B(r_k)\right|_s &\le \sum_{k=n}^{\infty} |B(r_k)|_s \le \pi\sum_{k=n}^{\infty}\{\overset{\circ}{T}(r_k,f)\}^{-\frac{1}{2}-\eta} \\
&= \pi\sum_{k=n}^{\infty}\frac{\overset{\circ}{T}(r_k,f) - \overset{\circ}{T}(r_{k-1},f)}{\{\overset{\circ}{T}(r_k,f)\}^{1+2\eta}} = \pi\sum_{k=n}^{\infty}\int_{r_{k-1}}^{r_k}\frac{d\overset{\circ}{T}(r,f)}{\{\overset{\circ}{T}(r_k,f)\}^{1+2\eta}} \\
&\le \pi\sum_{k=n}^{\infty}\int_{r_{k-1}}^{r_k}\frac{d\overset{\circ}{T}(r,f)}{\{\overset{\circ}{T}(r,f)\}^{1+2\eta}} = \pi\int_{r_{n-1}}^{\infty}\frac{d\overset{\circ}{T}(r,f)}{\{\overset{\circ}{T}(r,f)\}^{1+2\eta}} = \frac{\pi}{2\eta}\frac{1}{\{\overset{\circ}{T}(r_{n-1},f)\}^{2\eta}}.
\end{aligned}$$

Hence

$$\lim_{n\to\infty}\left|\bigcup_{k=n}^{\infty} B(r_k)\right|_s = 0$$

and

$$|E_V(f)|_s = 0.$$

Denote by $E_V^n(f)$ the intersection $E_V(f)\cap\{|a|<n\}$, $n=1,2,\dots$. Since

$$0 = |E_V^n(f)|_s = \iint_{E_V^n(f)} \frac{d\sigma(a)}{(1+|a|^2)^2} \geq \frac{1}{(1+n^2)^2}\iint_{E_V^n(f)} d\sigma(a) = \frac{|E_V^n(f)|}{(1+n^2)^2},$$

we get $|E_V^n(f)| = 0$. The set $E_V(f)$ may differ from $\bigcup_{n=1}^{\infty} E_V^n(f)$ by one point $a=\infty$ only and has zero two-dimensional measure. □

EXERCISE. Using Cartan's identity (4.13) of Chapter 1, prove that the intersection of $E_V(f)$ with the unit circle has one-dimensional measure zero. Use this result to show that the intersection of $E_V(f)$ with an arbitrary circle or line has one-dimensional measure zero.

EXAMPLE. The set $E_V(f)$ can have the cardinality of a continuum.[2]

Denote $N_p = 2^{3^p}$, $p = 0,1,2,\dots$, $q(p) = [\log_2 p]+1$, $p\geq 1$ ($\log_2 p$ is the logarithm with base 2). We select $\{\theta(n)\}$, $n=1,2,\dots$, to be an integer-valued sequence, so rapidly increasing that $\theta(n+1) > \theta(n)+1$ and

$$2^{-\theta(q(p))} < \frac{N_p!}{(2N_p)!}. \tag{2.10}$$

It is clear that for $n \leq N_p$ the inequality

$$2^{-\theta(q(p))} < \frac{n!}{(2n)!} \tag{2.11}$$

holds.

Denote by M the set of those real numbers $x\in\left(\frac{1}{2},1\right)$ that can be represented in the form

$$x = \frac{1}{2}+\frac{1}{2}\left\{\frac{\alpha_1}{2^{\theta(1)}}+\frac{\alpha_2}{2^{\theta(2)}}+\cdots+\frac{\alpha_n}{2^{\theta(n)}}+\cdots\right\}, \tag{2.12}$$

where $\alpha_1,\alpha_2,\dots,\alpha_n,\dots$ is an arbitrary sequence of 0s and 1s with infinitely many 1s. The cardinality of the set of such sequences coincides with the cardinality of the interval $(0,1)$ of real numbers. Since the representation of a real number x in the form (2.12) is unique, the set M has cardinality of the continuum.

If $p=\gamma_1+\gamma_2 2+\cdots+\gamma_q 2^{q-1}$, $q=q(p)$, $\gamma_j = 0,1$, we let

$$a_p = \frac{1}{2}+\frac{1}{2}\left\{\frac{\gamma_1}{2^{\theta(1)}}+\frac{\gamma_2}{2^{\theta(2)}}+\cdots+\frac{\gamma_q}{2^{\theta(q)}}\right\}. \tag{2.13}$$

Now we let $b_0=b_1=b_2=3/4$ and

$$b_n = a_p \quad\text{for}\quad N_{p-1} < n \leq N_p,\quad p=1,2,\dots. \tag{2.14}$$

[2] Hayman [**72**] constructed an example of an entire function such that $E_V(f)$ contains an arbitrary given F_σ set of zero capacity.

It is clear that $1/2 < b_j < 1$, $j \geq 0$. Denote by $\varphi(z)$ the entire function

$$\varphi(z) = \sum_{j=0}^{\infty} \frac{b_j}{j!} z^j.$$

Obviously, $M(r, \varphi) = \varphi(r)$, $\frac{1}{2}e^r < \varphi(r) < e^r$. Consider the entire function $\psi(z) = 2e^z - \varphi(z)$. It is easy to see that $M(r, \psi) = \psi(r)$ and $e^r < \psi(r) < 2e^r$. Using inequality (7.1) of Chapter 1, we get

$$T(r, \psi) \geq \frac{1}{3} \ln \psi\left(\frac{r}{2}\right) > \frac{1}{6} r.$$

Let a_0 be a complex number satisfying $\Delta(a_0, \psi) = 0$, that is, $N(r, a_0, \psi) = (1 + o(1))T(r, \psi) > \frac{1}{6}r(1 + o(1))$. The existence of such a number a_0 follows from Theorem 2.1. The function

$$f(z) = (\varphi(z) + a_0)e^{-z}$$

has the desired properties.

First of all, we have

$$T(r, f) \leq \ln M(r, f) \leq \ln M(r, e^{-z}) + \ln M(r, \varphi) + O(1) \leq 2r + O(1).$$

Let us estimate $N(r, 2, f)$ from below. The solutions of the equation $f(z) = 2$ coincide with solutions of the equation $2e^z - \varphi(z) = a_0$; therefore $N(r, 2, f) \equiv N(r, a_0, \psi) > \frac{1}{6}(1 + o(1))$. Hence

$$T(r, f) \geq N(r, 2, f) + O(1) \geq \frac{1}{6} r(1 + o(1)). \tag{2.15}$$

Now we show that each $x \in M$ satisfies $\Delta(x, f) = 1$. Choose $x \in M$,

$$x = \frac{1}{2} + \frac{1}{2}\left\{\frac{\gamma_1}{2^{\theta(1)}} + \frac{\gamma_2}{2^{\theta(2)}} + \cdots + \frac{\gamma_n}{2^{\theta(n)}} + \cdots\right\}, \qquad \gamma_j = 0, 1. \tag{2.16}$$

Let $\{\gamma_{n_k}\}_{k=1}^{\infty}$ be the subsequence of the sequence $\{\gamma_n\}$ consisting of the members that are equal to 1. We introduce the correspondence that assigns to x the sequence

$$p_k = \gamma_1 + \gamma_2 2 + \cdots + \gamma_{n_k} 2^{n_k - 1}, \qquad k = 1, 2, \ldots,$$

of positive integers. It is clear that $n_k = q(p_k)$ and that $p_k \to \infty$ as $k \to \infty$. Let $r_k = \frac{1}{3} N_{p_k}$. We estimate $N(r_k, x, f)$ from above. Since the solutions of the equation

$$(\varphi(z) + a_0)e^{-z} = x$$

coincide with the solutions of

$$xe^z - \varphi(z) - a_0 = 0,$$

we have $N(r_k, x, f) \equiv N(r_k, 0, \Phi(z))$, where

$$\Phi(z) = xe^z - \varphi(z) - a_0 = -a_0 + \sum_{j=0}^{\infty} \frac{x - b_j}{j!} z^j.$$

Since $x \in \left(\frac{1}{2}, 1\right)$ and $b_j \in \left(\frac{1}{2}, 1\right)$, it follows that $|x - b_j| < \frac{1}{2}$ for all $j \geq 0$. By (2.13), (2.16), and (2.11), for $N_{p_{k-1}} < j \leq N_{p_k}$ we have

$$\begin{aligned} x - b_j = x - a_{p_k} &= \frac{1}{2} \sum_{n=n_k+1}^{\infty} \frac{\gamma_n}{2^{\theta(n)}} \leq \frac{1}{2} \sum_{n=n_k+1}^{\infty} 2^{-\theta(n)} \\ &< \frac{1}{2} \sum_{n=\theta(n_k)}^{\infty} 2^{-n} = 2^{-\theta(n_k)} = 2^{-\theta(q(p_k))} < \frac{j!}{(2j)!}. \end{aligned}$$

Then for $|z| = r_k$ the inequality

$$\begin{aligned} |\Phi(z)| &\leq |a_0| + \sum_{j=0}^{\infty} \frac{|x - b_j|}{j!} r_k^j \\ &\leq |a_0| + \frac{1}{2} \sum_{j=0}^{N_{p_k-1}} \frac{r_k^j}{j!} + \sum_{j=N_{p_k-1}+1}^{N_{p_k}} \frac{j!}{(2j)!} \frac{1}{j!} r_k^j + \frac{1}{2} \sum_{j=N_{p_k}+1}^{\infty} \frac{r_k^j}{j!} \end{aligned} \tag{2.17}$$

holds. Now we estimate each sum in the right-hand side of (2.17). We get

$$\begin{aligned} \sum_{j=0}^{N_{p_k-1}} \frac{r_k^j}{j!} \leq r_k^{N_{p_k-1}} \sum_{j=0}^{\infty} \frac{1}{j!} &= e \exp\left\{N_{p_k-1} \ln r_k\right\} \\ &= e \exp\{(3r_k)^{1/3} \ln r_k\} = o\left(e^{\sqrt{r_k}}\right), \end{aligned}$$

since $N_{p-1} = N_p^{1/3}$. We also have

$$\sum_{j=N_{p_k}+1}^{\infty} \frac{r_k^j}{j!} \leq \sum_{j=N_{p_k}+1}^{\infty} \left(\frac{er_k}{j}\right)^j < \sum_{j=N_{p_k}+1}^{\infty} \left(\frac{er_k}{N_{p_k}}\right)^j < \sum_{j=1}^{\infty} \left(\frac{e}{3}\right)^j = \frac{e}{3-e},$$

$$\sum_{j=N_{p_k-1}+1}^{N_{p_k}} \frac{r_k^j}{(2j)!} < \sum_{j=0}^{\infty} \frac{r_k^j}{(2j)!} = \cosh \sqrt{r_k} = \left\{\frac{1}{2} + o(1)\right\} e^{\sqrt{r_k}}.$$

Combining these estimates, we get

$$M(r_k, \Phi) \leq \left(\frac{1}{2} + o(1)\right) e^{\sqrt{r_k}},$$

$$T(r_k, \Phi) \leq \ln M(r_k, \Phi) = \sqrt{r_k} + O(1).$$

Now,

$$N(r_k, x, f) = N(r_k, 0, \Phi) \leq T(r_k, \Phi) + O(1) \leq \sqrt{r_k} + O(1). \tag{2.18}$$

From (2.15) and (2.18) we get

$$\liminf_{r\to\infty} \frac{N(r, x, f)}{T(r, f)} \leq \liminf_{k\to\infty} \frac{N(r_k, x, f)}{T(r_k, f)} \leq \lim_{k\to\infty} \frac{\sqrt{r_k} + O(1)}{\frac{1}{6} r_k} = 0,$$

that is, $\Delta(x, f) = 1$.

Thus the function $f(z)$ has all the desired properties.

3. Deficiency relation

In Section 2 of Chapter 3 (see (2.9)) we proved the inequality

$$\sum_{\nu=1}^{q}\{m(r,a_\nu)+N_1(r,a_\nu)\}\le 2T(r,f)+Q(r,f), \tag{3.1}$$

where $a_1,\dots,a_q$ are arbitrary pairwise distinct complex numbers. Denote by $\varepsilon(a)$ the quantity

$$\varepsilon(a)=\liminf_{r\to\infty}\frac{N_1(r,a)}{T(r,f)},$$

and call it the *ramification index* (or simply the *index*) of the function $f(z)$ at the point a. If $\varepsilon(a)>0$, then a is called an *index value.* We choose a sequence $r_k\to\infty$ such that $Q(r_k,f)=o(T(r_k,f))$. Such a sequence exists by the definition of $Q(r,f)$ (for rational functions $f(z)$, we use Remark 2 on page 97). Let $r=r_k$ in (3.1), divide the obtained inequality by $T(r_k,f)$, and pass to the lower limit. We get

$$\sum_{\nu=1}^{q}\left\{\liminf_{k\to\infty}\frac{m(r_k,a_\nu)}{T(r_k,f)}+\liminf_{k\to\infty}\frac{N_1(r_k,a_\nu)}{T(r_k,f)}\right\}\le 2,$$

from where we obtain

$$\sum_{\nu=1}^{q}\{\delta(a_\nu)+\varepsilon(a_\nu)\}\le 2. \tag{3.2}$$

Since $m(r,a)+N_1(r,a)\le m(r,a)+N(r,a)=T(r,f)+O(1)$, we have

$$0\le\delta(a)+\varepsilon(a)\le 1. \tag{3.3}$$

We show that the set of those points a for which $\delta(a)+\varepsilon(a)>0$ is at most countable. In fact, let E be the set of those numbers a for which $\delta(a)+\varepsilon(a)>0$, and E_k be the set of those numbers a for which $\delta(a)+\varepsilon(a)>\frac{1}{k}$. It is clear that $E=\bigcup_{k=1}^{\infty}E_k$; therefore it is enough to show that the sets E_k are finite. To do this, suppose that for some k the set E_k contains more than $2k$ points a_ν, $\nu=1,\dots,n_k$, $n_k>2k$. Choosing $q=n_k$ in (3.2) we get a contradiction.

If the set of points a for which $\delta(a)+\varepsilon(a)>0$ is finite, then we may let $\{a_\nu\}$ in (3.2) be the set of all these points. If this set can be listed as a sequence $\{a_1,a_2,a_3,\dots\}$, then, taking the limit in (3.2) as $q\to\infty$, we get

$$\sum_{\nu-1}^{\infty}\{\delta(a_\nu)+\varepsilon(a_\nu)\}\le 2.$$

We combine these cases (of a finite and a countable set of values a) writing

$$\sum_{a}\{\delta(a)+\varepsilon(a)\}\le 2. \tag{3.4}$$

Despite the fact that the sum is taken over all complex numbers a, its terms are positive for a finite or countable set of values of a only, and the sum is to be understood as a sum of finitely many numbers or a sum of a series with positive terms. Inequality (3.4) is called the *deficiency relation*; it is one of the main results of the Nevanlinna theory. Since for the Picard exceptional values $\delta(a)=1$, the Picard theorem follows from (3.4): the set $E_P(f)$ contains at most two points. Before (see Theorem 2.2 in Chapter 3) this theorem was derived as a corollary of the Borel theorem. Note that we have also proved the following result, which we state as a separate theorem.

THEOREM 3.1. *The set $E_N(f)$ is at most countable.*

EXERCISE. Prove that for $f(z) = \sin z$ we have $\Delta(a) = 0$ for all $a \neq \infty$, $\delta(\infty) = 1$, and $\varepsilon(1) = \varepsilon(-1) = 1/2$.

4. The structure of the set of deficient values

For the meromorphic functions of zero order, the statement of Theorem 3.1 can be made more precise. In fact, Valiron's theorem (Theorem 4.6, Chapter 2) can be restated in the following way: a meromorphic function of zero order cannot have more than one deficient value.

For the meromorphic functions of order ρ, $0 < \rho \leq \infty$, Theorem 3.1 cannot be improved (unless we impose other restrictions on a meromorphic function in addition to prescribing its magnitude of order).

THEOREM 4.1. *Let $0 < \rho \leq \infty$, and let M be an arbitrary at most countable subset of the extended complex plane (M may be empty). Then there exists a meromorphic function of order ρ whose set of deficient values coincides with M.*

PROOF. First we prove that for each ρ, $0 \leq \rho \leq \infty$, there exists a meromorphic function of order ρ that has no deficient values. Let $\{p_n\}$ be an arbitrary sequence of non-negative integers.

Consider the function

$$f(z) = \frac{1}{z} + \sum_{n=1}^{\infty} \frac{1}{2^n}\left(\frac{1}{z-2^n}\right)^{p_n+2} = \frac{1}{z} + \psi(z). \tag{4.1}$$

It is clear that the series in (4.1) converges absolutely and uniformly on bounded sets in the complex plane. We show that for any choice of the sequence $\{p_n\}$, the set $E_N(f)$ is empty. Let $r_\nu = 2^{\nu-2}3 = 2^\nu\frac{3}{4}$. For $|z| = r_\nu$, $\nu \geq 3$, we have

$$\begin{aligned}
|\psi(z)| &\leq \sum_{n=1}^{\infty} \frac{1}{2^n}\left(\frac{1}{|z-2^n|}\right)^{p_n+2} \leq \sum_{n=1}^{\infty} \frac{1}{2^n}\left(\frac{1}{|r_\nu-2^n|}\right)^{p_n+2} \\
&\leq \sum_{n=1}^{\infty} \frac{1}{2^n}\frac{1}{|r_\nu-2^n|^2} = \sum_{n=1}^{\nu-1} \frac{1}{r_\nu^2 2^n}\frac{1}{\left(1-\frac{2^n}{r_\nu}\right)^2} + \sum_{n=\nu}^{\infty} \frac{1}{2^{3n}}\frac{1}{(1-r_\nu 2^{-n})^2} \\
&< \frac{1}{r_\nu^2}\sum_{n=1}^{\nu-1} \frac{1}{2^n}\frac{1}{\left(1-\frac{2^{\nu-1}}{r_\nu}\right)^2} + \sum_{n=\nu}^{\infty} \frac{1}{2^{3n}}\frac{1}{(1-r_\nu 2^{-\nu})^2} \\
&= \frac{9}{r_\nu^2}\sum_{n=1}^{\nu-1} \frac{1}{2^n} + 16\sum_{n=\nu}^{\infty} \frac{1}{2^{3n}} < \frac{9}{r_\nu^2} + \frac{128}{7}\frac{1}{2^{3\nu}} = \frac{9}{r_\nu^2} + \frac{54}{7}\frac{1}{r_\nu^3}.
\end{aligned}$$

Equality (4.1) implies that

$$|f(z)| = \frac{1}{r_\nu} + O\left(\frac{1}{r_\nu^2}\right)$$

for $|z| = r_\nu$. Hence for $0 < |a| \leq \infty$ we have $m(r_\nu, a, f) = O(1)$, $r_\nu \to \infty$, and $m(r_\nu, 0, f) = \ln r_\nu + o(1)$. Therefore $\delta(a) = 0$ for each a and $E_N(f) = \emptyset$.

Now we show that choosing a suitable sequence $\{p_n\}$ we get a function $f(z)$ with a prescribed order ρ. Note that if $|z| = r$, where $|r - 2^n| \geq 1$, $n = 1, 2, \dots$, $r \geq 3$, then

$$|f(z)| \leq \frac{1}{r} + \sum_{n=1}^{\infty} \frac{1}{2^n} \left(\frac{1}{|z - 2^n|} \right)^{p_n + 2} \leq \frac{1}{r} + \sum_{n=1}^{\infty} \frac{1}{2^n} \leq \frac{4}{3}$$

and

$$m(r, f) = O(1). \tag{4.2}$$

Now, for $2^{n-1} \leq r < 2^n$ we have

$$n(r, f) = 2n - 1 + p_1 + p_2 + \cdots + p_{n-1}. \tag{4.3}$$

If all $p_n \equiv 0$, then $n(r, f) \sim \frac{2}{\ln 2} \ln r$ and $N(r, f) \sim \frac{\ln^2 r}{\ln 2}$. Denote by I the set $[3, \infty)$ with the intervals $|r - 2^n| < 1$, $n = 2, 3, \dots$, removed. By (4.2), $T(r, f) \sim \frac{\ln^2 r}{\ln 2}$ on I, and by monotonicity of $T(r, f)$, this relation remains true for all r. For $p_n \equiv 0$ the function $f(z)$ has zero order.

Now let $0 < \rho < \infty$. Set $p_n = [2^{n\rho}]$. Then (4.3) implies ($2^{n-1} \leq r < 2^n$):

$$n(r, f) \sim \frac{2^{n\rho}}{2^\rho - 1},$$

$$(1 + o(1)) \frac{r^\rho}{2^\rho - 1} \leq n(r, f) \leq (1 + o(1)) \frac{2^\rho r^\rho}{2^\rho - 1},$$

$$(1 + o(1)) \frac{r^\rho}{\rho(2^\rho - 1)} \leq N(r, f) \leq (1 + o(1)) \frac{2^\rho r^\rho}{\rho(2^\rho - 1)}.$$

Using (4.2) as before, we show that

$$\ln T(r, f) \sim \rho \ln r,$$

that is, the function $f(z)$ has order ρ.

If we set $p_n = 2^{(n+1)^2}$, we get (for $2^{n-1} \leq r < 2^n$)

$$n(r, f) \geq p_{n-1} = 2^{n^2} \geq 2^{\left(\frac{\ln r}{\ln 2}\right)^2},$$

$$\ln T(r, f) \geq \ln N(r, f) \geq \ln n\left(\frac{r}{e}, f\right) \geq \frac{(\ln r - 1)^2}{\ln 2};$$

that is, the function $f(z)$ has order $\rho = \infty$.

It is easy to give an example of a meromorphic function $f(z)$ of order zero, such that $E_N(f)$ consists of a single point a. If $a = \infty$, we let $f(z)$ be an arbitrary entire function $g(z)$ of zero order; if $a \neq \infty$, we set $f(z) = a + 1/g(z)$.

Now let $0 < \rho \leq \infty$, and let M be an arbitrary nonempty at most countable set. We show that there exists a meromorphic function $f(z)$ of order ρ, for which $E_N(f) = M$. It is clear that without loss of generality we may assume that $\infty \notin M$.

We start with the case $\rho = 1$. Let θ_j, $j = 0, \pm 1, \pm 2, \pm 3, \dots$, be a sequence of numbers from the interval $(-\pi, \pi)$, $\pi > \theta_{j+1} - \theta_j > 0$, $\theta_j \to \pi$ as $j \to +\infty$, and $\theta_j \to -\pi$ as $j \to -\infty$. Let a_j, $j = 0, \pm 1, \pm 2, \dots$, be a sequence of complex numbers whose set of terms coincides with M.

Let $\{c_j\}$ be a sequence of positive numbers such that

$$\sum_{j=-\infty}^{\infty} c_j |a_j| = S_1 < \infty, \qquad \sum_{j=-\infty}^{\infty} c_j = S_2 < \infty.$$

Set

$$f_1(z) = \sum_{j=-\infty}^{\infty} c_j a_j \exp(ze^{-i\theta_j}), \tag{4.4}$$

$$f_2(z) = \sum_{j=-\infty}^{\infty} c_j \exp(ze^{-i\theta_j}). \tag{4.5}$$

It is clear that these series converge absolutely and uniformly on bounded sets and represent entire functions. Let

$$f(z) = \frac{f_1(z)}{f_2(z)}.$$

We show that the meromorphic function $f(z)$ has the desired properties.

For $|z| = r$ we have $|\exp(ze^{-i\theta_j})| \le e^r$, hence

$$|f_1(z)| \le \sum_{j=-\infty}^{\infty} c_j |a_j| e^r = S_1 e^r,$$

$$|f_2(z)| \le \sum_{j=-\infty}^{\infty} c_j e^r = S_2 e^r.$$

Therefore

$$\begin{aligned} T(r,f) &\le T(r,f_1) + T(r,f_2) + O(1) \\ &\le \ln M(r,f_1) + \ln M(r,f_2) + O(1) \le 2r + O(1). \end{aligned} \tag{4.6}$$

We fix an integer ν. Let $\theta_j' \equiv \theta_j (\operatorname{mod} 2\pi)$, $\theta_\nu - \pi < \theta_j' \le \theta_\nu + \pi$. It is clear that $\theta'_{\nu\pm 1} = \theta_{\nu\pm 1}$. Let $\eta = \eta(\nu) > 0$. Consider the function $f(z)$ in the angle

$$\frac{\theta_\nu + \theta_{\nu-1}}{2} + \eta < \varphi = \arg z < \frac{\theta_{\nu+1} + \theta_\nu}{2} - \eta, \quad 0 < \eta < \frac{\theta_{\nu+1} - \theta_{\nu-1}}{4}. \tag{4.7}$$

We have

$$\begin{aligned} f(z) - a_\nu &= \frac{f_1(z) - a_\nu f_2(z)}{f_2(z)} = \frac{\sum_{j\ne\nu} c_j (a_j - a_\nu) \exp(ze^{-i\theta_j'})}{\sum_j c_j \exp(ze^{-i\theta_j'})} \\ &= \frac{\sum_{j\ne\nu} c_j (a_j - a_\nu) \exp\{z(e^{-i\theta_j'} - e^{-i\theta_\nu})\}}{c_\nu + \sum_{j\ne\nu} c_j \exp\{z(e^{-i\theta_j'} - e^{-i\theta_\nu})\}}. \end{aligned} \tag{4.8}$$

Observe that

$$\begin{aligned} |\exp\{z(e^{-i\theta_j'} - e^{-i\theta_\nu})\}| &= \exp\{r(\cos(\varphi - \theta_j') - \cos(\varphi - \theta_\nu))\} \\ &= \exp\left\{2r \sin\left(\varphi - \frac{\theta_j' + \theta_\nu}{2}\right) \sin\frac{\theta_j' - \theta_\nu}{2}\right\}. \end{aligned} \tag{4.9}$$

Let $\theta_{\nu+1} \le \theta_j' \le \theta_\nu + \pi$. Then

$$\eta \le \frac{\theta_j' - \theta_{\nu+1}}{2} + \eta < \frac{\theta_j' + \theta_\nu}{2} - \varphi < \frac{\theta_j' - \theta_{\nu-1}}{2} - \eta < \pi - \eta$$

and

$$\sin\left(\varphi - \frac{\theta_j' + \theta_\nu}{2}\right) \sin\frac{\theta_j' - \theta_\nu}{2} \le -\sin\eta \sin\frac{\theta_{\nu+1} - \theta_\nu}{2} = -A_\nu' \sin\eta.$$

If $\theta_\nu - \pi < \theta'_j \le \theta_{\nu-1}$, then

$$\eta \le \frac{\theta_{\nu-1} - \theta'_j}{2} + \eta < \varphi - \frac{\theta'_j + \theta_\nu}{2} < \frac{\theta_{\nu+1} - \theta'_j}{2} - \eta < \pi - \eta$$

and

$$\sin\left(\varphi - \frac{\theta'_j + \theta_\nu}{2}\right) \sin\frac{\theta'_j - \theta_\nu}{2} \le -\sin\eta \sin\frac{\theta_\nu - \theta_{\nu-1}}{2} = -A''_\nu \sin\eta.$$

If we denote $A_\nu = \min(A'_\nu, A''_\nu)$, then by (4.9) we get the inequality

$$|\exp\{z(e^{-i\theta'_j} - e^{-i\theta_\nu})\}| \le \exp\{-2rA_\nu \sin\eta\},$$

valid for all $j \ne \nu$. Now,

$$\left|\sum_{j\ne\nu} c_j \exp\{z(e^{-i\theta'_j} - e^{-i\theta_\nu})\}\right| \le S_2 \exp\{-2rA_\nu \sin\eta\}, \tag{4.10}$$

$$\left|\sum_{j\ne\nu} c_j(a_j - a_\nu) \exp\{z(e^{-i\theta'_j} - e^{-i\theta_\nu})\}\right| \le (S_1 + |a_\nu|S_2) \exp\{-2rA_\nu \sin\eta\}. \tag{4.11}$$

If $r \ge r_\nu(\eta)$, the right-hand side of (4.10) is less than $c_\nu/2$, and by (4.8) and (4.11) we get

$$|f(z) - a_\nu| \le \frac{S_1 + |a_\nu|S_2}{c_\nu - \frac{c_\nu}{2}} \exp\{-2rA_\nu \sin\eta\}, \tag{4.12}$$

$$\ln^+ \frac{1}{|f(z) - a_\nu|} \ge 2rA_\nu \sin\eta + O(1) \tag{4.12$_1$}$$

uniformly in the angle (4.7). Now

$$\begin{aligned} m(r, a_\nu) &\ge \frac{1}{2\pi} \int_{\frac{\theta_\nu + \theta_{\nu-1}}{2} + \eta}^{\frac{\theta_{\nu+1} + \theta_\nu}{2} - \eta} \ln^+ \frac{1}{|f(re^{i\varphi}) - a_\nu|} d\varphi \\ &\ge \frac{1}{\pi}\left(\frac{\theta_{\nu+1} - \theta_{\nu-1}}{2} - 2\eta\right) rA_\nu \sin\eta + O(1) = B_\nu(\eta)r + O(1), \end{aligned} \tag{4.13}$$

where $B_\nu(\eta) > 0$. From (4.6) and (4.13) we get

$$\delta(a_\nu) = \liminf_{r\to\infty} \frac{m(r, a_\nu)}{T(r, f)} \ge \frac{B_\nu(\eta)}{2} > 0.$$

Thus we have shown that $M \subset E_N(f)$. Inequalities (4.6) and (4.13) imply also

$$B_0(\eta_0)r + O(1) \le T(r, f) \le 2r + O(1),$$

where

$$0 < \eta_0 < \frac{\theta_1 - \theta_{-1}}{4}.$$

Hence $f(z)$ is of order 1.

Let δ be an arbitrary positive number. The interval $[-\pi + \delta, \pi - \delta]$ contains finitely many points of the form $(\theta_j + \theta_{j+1})/2$. We cover these points by intervals of length 2η centered at them, where $\eta > 0$ is so small that the total length of the intervals is less than δ. Denote by I the union of the intervals. If $\varphi \in [-\pi + \delta, \pi - \delta]\backslash I$, then by (4.12) the functions $|f(re^{i\varphi})|$ and $|f(re^{i\varphi}) - a|^{-1}$, $a \notin M$, $a \ne \infty$, converge uniformly to some bounded functions as $r \to \infty$.

By Theorem 7.4 of Chapter 1 we conclude that

$$m(r,a) = o(T(r,f)) \quad \text{and} \quad \delta(a) = 0$$

for $a \notin M$. Thus we have proved that $M = E_N(f)$.

Now we show how to change the above construction in order to get a similar example in the case $0 < \rho < 1/2$. As before, we choose the sequences θ_j, a_j, c_j. Now the role of e^z will be played by the function

$$V(z) = \prod_{n=1}^{\infty} \left(1 + \frac{z}{n^{1/\rho}}\right).$$

The results of Section 5 in Chapter 2 imply that

$$\ln |V(re^{i\varphi})| = \frac{\pi \cos \rho\varphi}{\sin \pi\rho} r^\rho + o(r^\rho), \qquad r \to \infty, \tag{4.14}$$

uniformly in $\varphi \in [-\pi + \delta, \pi - \delta]$. Furthermore, if we replace $=$ in (4.14) by $\le$, the obtained inequality is satisfied uniformly in $\varphi \in [-\pi, \pi]$. In addition,

$$\ln M(r,V) = \frac{\pi}{\sin \pi\rho} r^\rho + o(r^\rho). \tag{4.15}$$

Now we set

$$f_1(z) = \sum_{j=-\infty}^{\infty} c_j a_j V(ze^{-i\theta_j}),$$

$$f_2(z) = \sum_{j=-\infty}^{\infty} c_j V(ze^{-i\theta_j}),$$

and, as before, $f(z) = f_1(z)/f_2(z)$. Instead of (4.6) we get

$$T(r,f) \le \ln M(r,f_1) + \ln M(r,f_2) + O(1) \le \frac{2\pi}{\sin \pi\rho} r^\rho + o(r^\rho). \tag{4.16}$$

Fix θ_ν and define θ_j' as before. Then the equality

$$f(z) - a_\nu = \frac{\sum_{j \ne \nu} c_j (a_j - a_\nu) \frac{V(ze^{-i\theta_j'})}{V(ze^{-i\theta_\nu})}}{c_\nu + \sum_{j \ne \nu} c_j \frac{V(ze^{-i\theta_j'})}{V(ze^{-i\theta_\nu})}} \tag{4.17}$$

holds in the angle (4.7). Denote by $c_\rho(\varphi)$ the periodic (with period 2π) extension of the function $\cos \rho\varphi$ from $[-\pi, \pi]$ to the whole real line. Then the role of the equality (4.9) will be played by the inequality

$$\ln \left| \frac{V\left(re^{i(\varphi - \theta_j')}\right)}{V\left(re^{i(\varphi - \theta_\nu)}\right)} \right| \le r^\rho \frac{\pi}{\sin \pi\rho} \{c_\rho(\varphi - \theta_j') - c_\rho(\varphi - \theta_\nu)\} + o(r^\rho). \tag{4.18}$$

To derive (4.18), we use (4.14) with the equality sign for $\ln |V(re^{i(\varphi - \theta_\nu)})|$ and with the inequality sign for $\ln |V(re^{i(\varphi - \theta_j')})|$. To continue the proof along the same lines as before, we need to show that there exists a positive constant $K_\nu(\eta)$ such that for φ satisfying the inequality (4.7) and for all $j \ne \nu$,

$$c_\rho(\varphi - \theta_j') - c_\rho(\varphi - \theta_\nu) < -K_\nu(\eta). \tag{4.19}$$

The proof of (4.19) is based on simple trigonometric transformations, although often requires tedious computations. We leave the detailed proof of (4.19) to interested readers, providing only a short sketch of one of the possible routes.

a) Let $0 < \theta \le \pi$; then $c_\rho(\psi) > c_\rho(\psi - \theta)$ for $\theta/2 > \psi > \theta/2 - \pi$.
b) Let $-\pi \le \theta < 0$; then $c_\rho(\psi) > c_\rho(\psi - \theta)$ for $\theta/2 < \psi < \theta/2 + \pi$.
c) Let $0 < \theta'' \le \theta \le \pi$, $-\pi \le \psi \le \theta''/2$. Then

$$\max\{c_\rho(\psi - \pi),\ c_\rho(\psi - \theta'')\}c_\rho\} \ge c_\rho(\psi - \theta).$$

d) Let $-\pi \le \theta \le \theta' < 0$, $\theta'/2 \le \psi \le \pi$. Then

$$\max\{c_\rho(\psi - \pi),\ c_\rho(\psi - \theta')\} \ge c_\rho(\psi - \theta).$$

e) If $-\pi \le \theta' < 0 < \theta'' \le \pi$, $\theta'/2 < \psi < \theta''/2$, and $\theta \in [-\pi, \pi]\backslash[\theta', \theta'']$, then

$$c_\rho(\psi) > \max\{c_\rho(\psi - \theta'),\ c_\rho(\psi - \theta'')\} \ge c_\rho(\psi - \theta).$$

f) If $-\pi \le \theta' < \theta'/2 + \eta < 0 < \theta''/2 - \eta < \theta'' \le \pi$ and $\theta'/2 + \eta \le \psi \le \theta''/2 - \eta$, then, by the Weierstrass theorem,

$$c_\rho(\psi) - \max\{c_\rho(\psi - \theta'),\ c_\rho(\psi - \theta'')\} \ge A_\rho(\eta, \theta', \theta'') > 0$$

and

$$c_\rho(\psi) - c_\rho(\psi - \theta) \ge A_\rho(\eta, \theta', \theta'').$$

for $\theta \in [-\pi, \pi]\backslash[\theta', \theta'']$.

Now it is easy to see that in (4.19) we can set

$$K_\nu(\eta) = A_\rho(\eta, \theta_{\nu-1} - \theta_\nu, \theta_{\nu+1} - \theta_\nu).$$

To verify this, it is enough to let $\psi = \varphi - \theta_\nu$, $\theta' = \theta_{\nu-1} - \theta_\nu$, and $\theta'' = \theta_{\nu+1} - \theta_\nu$ in (4.19).

Now, after proving inequality (4.19), we get, by (4.17), (4.18), and (4.19),

$$|f(z) - a_\nu| \le \frac{S_1 + |a_\nu| S_2}{c_\nu - \frac{c_\nu}{2}} \exp\left\{-r^\rho \frac{\pi K_\nu(\eta)}{\sin \pi\rho} + o(r^\rho)\right\} \tag{4.20}$$

for sufficiently large r and for φ satisfying (4.7).

Using inequalities (4.16) and (4.20) as we used inequalities (4.6) and (4.12) before, we see that $E_N(f) = M$ and that $f(z)$ has order ρ, $0 < \rho < 1/2$.

Considering the functions $f(z^n), n = 1, 2, \ldots$, we get examples of meromorphic functions with $E_N(f) = M$ and an arbitrary order ρ, $0 < \rho < n/2$. Since n can be chosen to be arbitrarily large, it follows that the order ρ can be an arbitrary positive finite number. Of course, there is no need to consider the case $\rho = 1$ separately. We started with the case $\rho = 1$ because in this case the function $f(z)$ is particularly simple, and it will be convenient to use it in our construction of an example for $\rho = \infty$.

Let $f_1(z)$ and $f_2(z)$ be given by (4.4) and (4.5), respectively. Set

$$F_1(z) = f_1(e^z), \quad F_2(z) = f_2(e^z), \quad F(z) = F_1(z)/F_2(z).$$

Let

$$\psi(y) = \sup_{-\infty < j < \infty} \cos(y - \theta_j), \qquad -\infty < y < \infty.$$

It is easy to see that $\psi(y) = \cos(y - \theta_j)$ for $(\theta_{j-1} + \theta_j)/2 \le y \le (\theta_j + \theta_{j+1})/2$, $\psi(\pm\pi) = 1$. The function $\psi(y)$ is continuous for $-\infty < y < \infty$, periodic with period 2π, and satisfies the inequality $0 < \psi(y) \le 1$. We easily get

$$
\begin{aligned}
|F_2(x+iy)| &\le \sum_{j=-\infty}^{\infty} c_j \exp\{e^x \cos(y-\theta_j)\} \\
&\le \sum_{j=-\infty}^{\infty} c_j \exp\{e^x \psi(y)\} = S_2 \exp\{e^x \psi(y)\}
\end{aligned} \tag{4.21}
$$

and

$$
|F_1(x+iy)| \le S_1 \exp\{e^x \psi(y)\}. \tag{4.22}
$$

Later on we shall need the following

LEMMA 4.1 (Fuchs, Hayman). *Let $\varphi(t)$ be a bounded, non-negative function, integrable on each segment $0 \le t \le r$, and such that there exists a finite limit*

$$
\lim_{r\to\infty} \frac{1}{r} \int_0^r \varphi(t)\, dt = l.
$$

Then

$$
I(r) = \frac{1}{2\pi} \int_0^{\pi/2} e^{r\cos\theta} \varphi(r\sin\theta)\, d\theta = \frac{\{l+o(1)\}e^r}{2\sqrt{2\pi r}}, \qquad r \to \infty.
$$

PROOF. First we observe that

$$
\frac{1}{2\pi} \int_0^{\pi/2} e^{r\cos\theta} d\theta = (1+o(1)) \frac{e^r}{2\sqrt{2\pi r}}, \qquad r \to \infty. \tag{4.23}
$$

This follows from the equalities

$$
\frac{1}{2\pi} \int_0^{\pi/2} e^{r\cos\theta} d\theta = \frac{1}{4\pi} \int_{-\pi/2}^{\pi/2} e^{r\cos\theta} d\theta = \frac{1}{4\pi} \int_{-\pi}^{\pi} e^{r\cos\theta} d\theta + O(1),
$$

$$
\frac{1}{2\pi} \int_{-\pi}^{\pi} e^{r\cos\theta} d\theta = I_0(r) = (1+o(1)) \frac{e^r}{\sqrt{2\pi r}}, \tag{4.24}
$$

where $I_0(r) = J_0(ir)$ is a cylindric function of imaginary parameter.

For convenience of the readers who are not familiar with formula (4.24), we include an independent proof of formula (4.23).

We start with

$$
\int_0^{\pi/2} e^{r\cos\theta} d\theta = e^r \int_0^{\pi/2} e^{r(\cos\theta-1)} d\theta = e^r \int_0^{\pi/2} e^{-2r\sin^2\frac{\theta}{2}} d\theta. \tag{4.25}
$$

On the other hand,

$$
\begin{aligned}
\int_0^{\pi/2} e^{-2r\sin^2\frac{\theta}{2}} d\theta &= \int_0^{r^{-1/4}} e^{-2r\sin^2\frac{\theta}{2}} d\theta + \int_{r^{-1/4}}^{\pi/2} e^{-2r\sin^2\frac{\theta}{2}} d\theta \\
&= \int_0^{r^{-1/4}} e^{-2r\sin^2\frac{\theta}{2}} d\theta + O\left(e^{-\frac{1}{3}\sqrt{r}}\right), \qquad r \to \infty.
\end{aligned} \tag{4.26}
$$

But for $0 \le \theta \le r^{-1/4}$ we have

$$
h(r)\left(\frac{\theta}{2}\right)^2 \le \sin^2\frac{\theta}{2} \le \left(\frac{\theta}{2}\right)^2,
$$

where $h(r) \to 1$ as $r \to \infty$. Therefore

$$\int_0^{r^{-1/4}} e^{-2r\sin^2\frac{\theta}{2}}d\theta \le \int_0^{r^{-1/4}} e^{-\frac{r}{2}h(r)\theta^2}d\theta < \int_0^{\infty} e^{-\frac{r}{2}h(r)\theta^2}d\theta$$
$$= \sqrt{\frac{\pi}{2}}\frac{1}{\sqrt{rh(r)}} = (1+o(1))\sqrt{\frac{\pi}{2r}},$$

$$\int_0^{r^{-1/4}} e^{-2r\sin^2\frac{\theta}{2}}d\theta \ge \int_0^{r^{-1/4}} e^{-\frac{r}{2}\theta^2}d\theta = \int_0^{\infty} e^{-\frac{r}{2}\theta^2}d\theta - \int_{r^{-1/4}}^{\infty} e^{-\frac{r}{2}\theta^2}d\theta$$
$$\ge \sqrt{\frac{\pi}{2r}} - \int_{r^{-1/4}}^{1} e^{-\frac{r}{2}\theta^2}d\theta - \int_1^{\infty} e^{-\frac{r}{2}\theta}d\theta$$
$$> \sqrt{\frac{\pi}{2r}} - e^{-\frac{\sqrt{r}}{2}} - \frac{2}{r}e^{-\frac{r}{2}} = (1+o(1))\sqrt{\frac{\pi}{2r}}.$$

Hence

$$\int_0^{r^{-1/4}} e^{-2r\sin^2\frac{\theta}{2}}d\theta = (1+o(1))\sqrt{\frac{\pi}{2r}}.$$

Using (4.26) and (4.25), we get (4.23).

To complete the proof of the lemma, it is enough to show that

$$I(r) - \frac{l}{2\pi}\int_0^{\pi/2} e^{r\cos\theta}d\theta = o\left(\frac{e^r}{\sqrt{r}}\right), \quad r \to \infty. \tag{4.27}$$

Since $\varphi(r\sin\theta) = O(1)$,

$$\int_0^{r^{-2/3}} e^{r\cos\theta}d\theta = O\left(e^r r^{-\frac{2}{3}}\right) = o\left(\frac{e^r}{\sqrt{r}}\right),$$

$$\int_{\pi/3}^{\pi/2} e^{r\cos\theta}d\theta = O\left(e^{\frac{r}{2}}\right) = o\left(\frac{e^r}{\sqrt{r}}\right),$$

in order to prove (4.27), it is enough to check that

$$\begin{aligned}\int_{r^{-2/3}}^{\pi/3} e^{r\cos\theta}\varphi(r\sin\theta)d\theta - l\int_{r^{-2/3}}^{\pi/3} e^{r\cos\theta}d\theta \\ = \int_{r^{-2/3}}^{\pi/3} e^{r\cos\theta}\{\varphi(r\sin\theta) - l\}d\theta = o\left(\frac{e^r}{\sqrt{r}}\right).\end{aligned} \tag{4.28}$$

We introduce

$$\chi(\theta) = \frac{e^{r\cos\theta}}{r\cos\theta}, \qquad \Phi(\theta) = \int_0^{r\sin\theta} \varphi(t)dt - lr\sin\theta.$$

It is easy to see that for $r^{-2/3} \le \theta \le r^{-1/4}$ and for sufficiently large r we have $\chi'(\theta) < 0$, and also that $\Phi(\theta) = o(r\sin\theta)$ by the condition of the lemma and the fact that $r\sin\theta \geq (1+o(1))r^{1/3}$ uniformly in θ as $r\to\infty$. Now,

$$\begin{aligned}\int_{r^{-2/3}}^{\pi/3} e^{r\cos\theta}\{\varphi(r\sin\theta)-l\}\,d\theta &= \int_{r^{-2/3}}^{\pi/3}\chi(\theta)\,d\Phi(\theta)\\ &= \chi(\theta)\Phi(\theta)|_{r^{-2/3}}^{\pi/3} - \int_{r^{-2/3}}^{\pi/3}\chi'(\theta)\Phi(\theta)\,d\theta\\ &= o(\chi(\theta)r\sin\theta)|_{r^{-2/3}}^{\pi/3} - o(1)\int_{r^{-2/3}}^{\pi/3}\chi'(\theta)r\sin\theta\,d\theta\\ &= o(\chi(\theta)r\sin\theta)|_{r^{-2/3}}^{\pi/3} + o(1)\int_{r^{-2/3}}^{\pi/3}\chi(\theta)r\cos\theta\,d\theta\\ &= o\left(e^{\frac{r}{2}}\right)+o\left(e^{r}r^{-\frac{2}{3}}\right)+o(1)\int_{r^{-2/3}}^{\pi/3}e^{r\cos\theta}d\theta = o\left(\frac{e^r}{\sqrt{r}}\right).\end{aligned}$$

Here we used (4.23). Hence relation (4.28) and thus also the lemma have been proved. □

We continue our construction of the example. By (4.21) and (4.22),

$$\begin{aligned}T(r,F) &\leq T(r,F_1)+T(r,F_2)+O(1) = m(r,F_1)+m(r,F_2)+O(1)\\ &\leq 2\cdot\frac{1}{2\pi}\int_{-\pi}^{\pi}e^{r\cos\theta}\psi(r\sin\theta)\,d\theta+O(1)\\ &= \frac{1}{\pi}\int_{-\pi/2}^{\pi/2}e^{r\cos\theta}\psi(r\sin\theta)\,d\theta+O(1)\\ &= \frac{1}{\pi}\int_{0}^{\pi/2}e^{r\cos\theta}\{\psi(r\sin\theta)+\psi(-r\sin\theta)\}\,d\theta+O(1).\end{aligned}\tag{4.29}$$

Now we apply Lemma 4.1 with $\varphi(t)=\psi(t)+\psi(-t)$. Since the function $\psi(y)$ is periodic with period 2π, we have

$$\lim_{r\to\infty}\frac{1}{r}\int_0^r\varphi(t)dt = \lim_{r\to\infty}\frac{1}{r}\int_{-r}^{r}\psi(t)dt = \frac{1}{\pi}\int_{-\pi}^{\pi}\psi(t)dt = A,\quad 0<A<\infty.$$

Inequality (4.29) and Lemma 4.1 imply that

$$T(r,F)\leq(1+o(1))\frac{Ae^r}{\sqrt{2\pi r}},\quad r\to\infty.\tag{4.30}$$

Now we fix ν. Define the function $\psi_\nu(y)$ by the equation $\psi_\nu(y)=2A_\nu\sin\eta$ for

$$\frac{\theta_\nu+\theta_{\nu-1}}{2}+\eta<y<\frac{\theta_{\nu+1}+\theta_\nu}{2}-\eta,$$

where η and A_ν are defined in the same way as in the construction of the example for $\rho=1$. If

$$-\pi\leq y\leq\frac{\theta_\nu+\theta_{\nu-1}}{2}+\eta\quad\text{and}\quad\frac{\theta_{\nu+1}+\theta_\nu}{2}-\eta\leq y\leq\pi,$$

we set $\psi_\nu(y)=0$. We extend the function $\psi_\nu(y)$ as periodic with period 2π from the interval $[-\pi,\pi]$ to the whole real line $\{-\infty<y<\infty\}$. By (4.12$_1$) we have

$$\ln^+\frac{1}{|F(x+iy)-a_\nu|}=\ln^+\frac{1}{|f(e^{x+iy})-a_\nu|}\geq e^x\psi_\nu(y)+O(1).$$

Hence

$$m(r, a_\nu, F) \geq \frac{1}{2\pi} \int_{-\pi}^{\pi} e^{r\cos\theta} \psi_\nu(r\sin\theta)d\theta + O(1). \tag{4.31}$$

As in the above argument we show that

$$\begin{aligned} &\frac{1}{2\pi} \int_{-\pi}^{\pi} e^{r\cos\theta} \psi_\nu(r\sin\theta)d\theta \\ &\quad = \frac{1}{\pi} \int_{0}^{\pi/2} e^{r\cos\theta} \{\psi_\nu(r\sin\theta) + \psi_\nu(-r\sin\theta)\}d\theta + O(1) \\ &\quad = (1 + o(1)) \frac{B_\nu e^r}{\sqrt{2\pi r}}, \qquad r \to \infty, \end{aligned} \tag{4.32}$$

where

$$B_\nu = \lim_{r\to\infty} \frac{1}{r} \int_{-r}^{r} \psi_\nu(t)dt = \frac{1}{\pi} \int_{-\pi}^{\pi} \psi_\nu(t)dt = \frac{1}{\pi} \left(\frac{\theta_{\nu+1} - \theta_\nu}{2} - 2\eta \right) 2A_\nu \sin\eta.$$

Relations (4.31) and (4.32) imply that

$$m(r, a_\nu, F) \geq (1 + o(1)) \frac{B_\nu e^r}{\sqrt{2\pi r}},$$

which, together with (4.30), implies

$$\delta(a_\nu, F) \geq \frac{B_\nu}{A} > 0,$$

that is, $M \subset E_N(F)$. The proof of the fact that $a \notin M$ implies $a \notin E_N(F)$ requires much more complicated estimates and tedious computations. We do not include it here. □

REMARK 1. In the case where the set M is finite, the construction of a meromorphic function $f(z)$ for which $E_N(f) = M$ becomes much simpler. When constructing the functions $f_1(z)$ and $f_2(z)$, it is enough to consider finite sums.

REMARK 2. Investigation of the constructed examples would be easier and more complete if we use some deep theorems from the theory of entire functions given in B. Ya. Levin's book [**Lev80**, Chapters II and VI]. The readers acquainted with the material of these chapters are invited to prove, as an exercise, that $a \notin M$ implies $a \notin E_N(F)$, where $F(z)$ is the function of infinite order that we constructed at the end of the proof.

REMARK 3. It is easy to see that there exists a meromorphic function of any order for which there is a deficient value that is not a Borel exceptional value. If $\rho > 0$, it is enough to consider a meromorphic function with three deficient values. At least one of them is not a Borel exceptional value by the Borel theorem. If $\rho = 0$, an example can be constructed as follows. Let $g_1(z)$ and $g_2(z)$ be two Weierstrass canonical products of genus zero with positive zeros such that

$$n(r, 0, g_1) \sim \ln r, \qquad n(r, 0, g_2) \sim \ln^2 r.$$

Then ∞ is a deficient value with $\delta(\infty) = 1$ for the function $f(z) = g_2(z)/g_1(-z)$, but $\infty \notin E_B(f)$. We leave verification of this statement to interested readers. (Hint: Use the asymptotic formulas in Section 5 of Chapter 2.)

5. Entire functions with infinitely many deficient values

Since $\delta(\infty, f) = 1$ for each entire function $f(z)$, we mean finite deficient values when speaking about deficient values of entire functions. The problem of the structure of this set for entire functions of finite order was elusive until recently. Although the complete solution is not known yet,[3] the following result is available.

THEOREM 5.1 (N. U. Arakelyan). *Let $M = \{a_k\}_{k=1}^{\infty}$ be an arbitrary countable set, and ρ an arbitrary number satisfying the condition $1/2 < \rho < \infty$. Then there exists an entire function of order ρ whose set of deficient values contains the set M.*

The restriction $\rho > 1/2$ in this theorem is necessary. We will show in Chapter 5, page 207, that the entire functions of order $\rho \leq 1/2$ cannot have deficient values. We note that in proving Theorem 5.1, Arakelyan disproved R. Nevanlinna's conjecture ([**NevR74**, Chapter 10, Section 222]) that an entire function of order $\rho > 1/2$ cannot have more than $[2\rho]$ deficient values.

The idea of the proof of Theorem 5.1 can be described as follows. Let $M = \{a_k\}_{k=1}^{\infty}$, and let $\{\Omega_k\}_{k=1}^{\infty}$ be a sequence of pairwise disjoint sets such that for $r \geq r(k)$, the circle $\{|z| = r\}$ intersects the set Ω_k along an arc of length $\geq s_k r$, where $s_k > 0, k = 1, 2, \ldots$. On the set $\bigcup_{k=1}^{\infty} \Omega_k$, define a piecewise constant function $\varphi(z)$ by the equalities

$$\varphi(z) = a_k, \qquad z \in \Omega_k.$$

In view of the theorems from approximation theory that we prove below, it turns out that for a special choice of the sets Ω_k, it is possible to construct an entire function $G(z)$ of normal type and of order ρ, such that

$$|\varphi(z) - G(z)| < e^{-q_k|z|^{\rho}}, \qquad z \in \Omega_k, \quad |z| \geq r(k),$$

where $q_k > 0$ $(k = 1, 2, \ldots)$. Since the set of values θ for which $|G(re^{i\theta}) - a_k| < e^{-q_k r^{\rho}}$ for $r \geq r(k)$ has measure $\geq s_k$, we have $m(r, a_k, G) \geq \frac{1}{2\pi} s_k q_k r^{\rho}$ $(r \geq r(k))$, and hence $\delta(a_k, G) > 0$, $k = 1, 2, \ldots$.

Now we present the necessary results from approximation theory.

THEOREM 5.2 (M. V. Keldysh). *Let L be a rectifiable curve joining points a and b. For each ε and d, $0 < \varepsilon < 1$, $0 < d < 1$, there exists a polynomial $P(z)$ such that*

a) *For all z that are not in the d-neighborhood of L,*

$$\left|\frac{1}{z-a} - P\left(\frac{1}{z-b}\right)\right| < \varepsilon.$$

b) *For all z satisfying $\{|z - b| \geq d\}$,*

$$\left|P\left(\frac{1}{z-b}\right)\right| < \exp\left\{C\left(1 + \ln\frac{1}{\varepsilon d}\right)e^{\frac{C\cdot\text{length}(L)}{d}}\right\},$$

where $C > 1$ is an absolute constant.

PROOF. Let l be the length of the curve L, $m = \left[\frac{4l}{d}\right] + 1$. We choose points $z_0 = a, z_1, z_2, \ldots, z_{m-1}, z_m = b$ on L such that $\max_{0\leq k\leq m-1} |z_{k+1} - z_k| \leq \frac{1}{4}d$. Now we perform a construction in which we assign a polynomial $P_k(z)$ to each point

[3]Now this question is completely solved by Eremenko [**43**]: For every at most countable set E in the plane and every $\rho > 1/2$, there exists an entire function f of order ρ for which $E_N(f) = E$.

$z_k, k = 0, 1, \ldots, m$. The polynomial $P(z)$ whose existence is claimed in the theorem will be equal to $P_m(z)$.

The polynomials $P_k(z)$ are defined inductively. Let $P_0(z) \equiv z$. Suppose that we have already constructed the polynomial $P_k(z)$ of degree $n_k > k$. We now construct the polynomial $P_{k+1}(z)$ as follows.

The function $P_k(\frac{1}{z-z_k})$ is holomorphic in the domain $\{|z - z_{k+1}| > |z_k - z_{k+1}|\}$ and is bounded as $z \to \infty$. Therefore, in the indicated domain it is represented by the Laurent series

$$P_k\left(\frac{1}{z-z_k}\right) = \sum_{q=0}^{\infty} \frac{A_q^{(k)}}{(z-z_{k+1})^q}.$$

The polynomial $P_{k+1}(z)$ will be of the form $\sum_{q=0}^{n_{k+1}} A_q^{(k)} z^q$, with a suitable number $n_{k+1} > k+1$.

At the moment we use the notation n_{k+1} for an arbitrary positive integer. Let us estimate the quantity

$$U(n_{k+1}) = \max_{|z-z_{k+1}|\geq d} \left| P_k\left(\frac{1}{z-z_k}\right) - \sum_{q=0}^{n_{k+1}} \frac{A_q^{(k)}}{(z-z_{k+1})^q} \right|.$$

For $|z - z_{k+1}| > r > |z_k - z_{k+1}|$ we have

$$\begin{aligned} P_k\left(\frac{1}{z-z_k}\right) - \sum_{q=0}^{n_{k+1}} \frac{A_q^{(k)}}{(z-z_{k+1})^q} &= \sum_{q=n_{k+1}+1}^{\infty} \frac{A_q^{(k)}}{(z-z_{k+1})^q} \\ &= \sum_{q=n_{k+1}+1}^{\infty} \frac{1}{(z-z_{k+1})^q} \frac{1}{2\pi i} \int_{|t-z_{k+1}|=r} P_k\left(\frac{1}{t-z_k}\right)(t-z_{k+1})^{q-1}dt \\ &= \frac{1}{2\pi i} \int_{|t-z_{k+1}|=r} P_k\left(\frac{1}{t-z_k}\right)\left\{ \sum_{q=n_{k+1}+1}^{\infty} \frac{(t-z_{k+1})^{q-1}}{(z-z_{k+1})^q} \right\} dt \\ &= \frac{1}{2\pi i} \int_{|t-z_{k+1}|=r} P_k\left(\frac{1}{t-z_k}\right)\left(\frac{t-z_{k+1}}{z-z_{k+1}}\right)^{n_{k+1}} \frac{dt}{z-t}. \end{aligned} \tag{5.1}$$

Set

$$M_k = \max_{|z-z_k|\geq d} \left| P_k\left(\frac{1}{z-z_k}\right) \right|.$$

Note that the function $(t-z_k)^{n_k} P_k(\frac{1}{t-z_k})$ is holomorphic in the disc $\{|t - z_k| < d\}$, and its modulus does not exceed $d^{n_k} M_k$ on the circle. By the maximum modulus principle, its modulus does not exceed $d^{n_k} M_k$ inside the circle also. Hence we have the estimate

$$\left| P_k\left(\frac{1}{t-z_k}\right) \right| \leq \frac{d^{n_k} M_k}{|t-z_k|^{n_k}}, \qquad |t - z_k| < d. \tag{5.2}$$

For $|t - z_{k+1}| = \frac{1}{2}d$ we have $|t - z_k| \geq |t - z_{k+1}| - |z_{k+1} - z_k| \geq \frac{1}{2}d - \frac{1}{4}d = \frac{1}{4}d$; therefore

$$\left| P_k\left(\frac{1}{t-z_k}\right) \right| \leq 4^{n_k} M_k, \qquad |t - z_{k+1}| = \frac{1}{2}d.$$

Letting $r = \frac{1}{2}d$, $|z - z_{k+1}| \geq d$ in (5.1) and estimating the integral, we get the inequality

$$U(n_{k+1}) \leq 4^{n_k} M_k \left(\frac{\frac{1}{2}d}{d}\right)^{n_{k+1}} = M_k 2^{2n_k - n_{k+1}}.$$

If we set

$$n_{k+1} = 2n_k + k + 2 + \left[\frac{\ln^+(M_k/\varepsilon)}{\ln 2}\right], \tag{5.3}$$

then, as is easy to see, we have

$$U(n_{n+1}) < \frac{\varepsilon}{2^{k+1}}. \tag{5.4}$$

Defining n_{k+1} by formula (5.3), we obtain the desired polynomial $P_{k+1}(z)$.

Now we check that the polynomial $P(z) = P_m(z)$ satisfies conditions a) and b).

Outside the d-neighborhood of the curve L we have

$$\begin{aligned}\left|\frac{1}{z-a} - P\left(\frac{1}{z-b}\right)\right| &= \left|P_0\left(\frac{1}{z-z_0}\right) - P_m\left(\frac{1}{z-z_m}\right)\right| \\ &\leq \sum_{k=0}^{m-1}\left|P_k\left(\frac{1}{z-z_k}\right) - P_{k+1}\left(\frac{1}{z-z_{k+1}}\right)\right| \\ &\leq \sum_{k=0}^{m-1} U(n_{k+1}) < \sum_{k=0}^{m-1}\frac{\varepsilon}{2^{k+1}} < \varepsilon.\end{aligned}$$

Hence condition a) is satisfied.

To show that condition b) is satisfied, we must estimate M_m. To do this we need to consider the behavior of the quantities M_k and n_k.

Since the domain $\{|z - z_{k+1}| > d\}$ is contained in the domain $\{|z - z_k| \geq \frac{3}{4}d\}$, we have

$$\begin{aligned}M_{k+1} &= \max_{|z-z_{k+1}|\geq d}\left|P_{k+1}\left(\frac{1}{z-z_{k+1}}\right)\right| \\ &\leq \max_{|z-z_{k+1}|\geq d}\left|P_{k+1}\left(\frac{1}{z-z_{k+1}}\right) - P_k\left(\frac{1}{z-z_k}\right)\right| + \max_{|z-z_k|\geq \frac{3}{4}d}\left|P_k\left(\frac{1}{z-z_k}\right)\right|.\end{aligned}$$

Using inequalities (5.4) and (5.2), we get the relation

$$M_{k+1} < \frac{\varepsilon}{2^{k+1}} + \left(\frac{4}{3}\right)^{n_k} M_k.$$

Letting $\tilde{M}_k = \max(M_k, \varepsilon)$, we have

$$\tilde{M}_{k+1} < 2^{n_k}\tilde{M}_k. \tag{5.5}$$

Observe that by (5.3) the inequality $n_k < \frac{1}{2}n_{k+1}$ holds; therefore from (5.5) we get that

$$\tilde{M}_{k+1} < \tilde{M}_0 2^{n_k + n_{k-1} + \cdots + n_0} < \tilde{M}_0 2^{n_k(1 + \frac{1}{2} + \frac{1}{2^2} + \cdots + \frac{1}{2^k})} < \tilde{M}_0 2^{2n_k}. \tag{5.6}$$

Further, by (5.3) we have

$$n_{k+1} < 2\left(n_k + k + 1 + \ln\frac{\tilde{M}_0 2^{2n_k}}{\varepsilon}\right).$$

Taking into account that $n_k > k$, $n_0 = 1$, and $n_{k-1} < \frac{1}{2}n_k$, we have

$$n_{k+1} < 2\left(4n_k + \ln\frac{\tilde{M}_0}{\varepsilon}\right),$$

$$n_{k+1} + \ln\frac{\tilde{M}_0}{\varepsilon} < 8\left(n_k + \ln\frac{\tilde{M}_0}{\varepsilon}\right),$$

from which we get

$$n_{k+1} + \ln\frac{\tilde{M}_0}{\varepsilon} < 8^{k+1}\left(1 + \ln\frac{\tilde{M}_0}{\varepsilon}\right),$$

$$n_{k+1} < 2^{3(k+1)}\left(1 + \ln^+\frac{M_0}{\varepsilon}\right).$$

Since $M_0 = \frac{1}{d}$ and hence $\ln^+\frac{M_0}{\varepsilon} = \ln\frac{1}{\varepsilon d}$ and $m < \frac{4l}{d} + 1$, we have

$$n_{m-1} < 2^{\frac{12l}{d}}\left(1 + \ln\frac{1}{\varepsilon d}\right) < e^{\frac{9l}{d}}\left(1 + \ln\frac{1}{\varepsilon d}\right).$$

This inequality together with (5.6) imply

$$\begin{aligned} M_m &< \tilde{M}_0 2^{2n_{m-1}} < \exp\left\{2n_{m-1} + \ln\frac{1}{\varepsilon d}\right\} \\ &< \exp\left\{2e^{\frac{9l}{d}}\left(1 + \ln\frac{1}{\varepsilon d}\right) + \ln\frac{1}{\varepsilon d}\right\} < \exp\left\{3e^{\frac{9l}{d}}\left(1 + \ln\frac{1}{\varepsilon d}\right)\right\}. \end{aligned}$$

The proof of the theorem is complete. □

We shall need the following lemma, which is a corollary of Theorem 5.2.

LEMMA 5.1. *Let γ be a closed rectifiable curve, and l a rectifiable curve joining a point of the curve γ to a point b which does not belong to γ. Let $0 < \varepsilon < 1$ and $0 < d < 1$ be given. Then there exists a function $Q(\zeta, z)$, rational in z and having a single pole at b, that is defined for $\{\zeta \in \gamma\} \times \{|z| < \infty\}$, piecewise constant[4] in ζ and is such that the following conditions are satisfied:*

a′) *For all z outside the d-neighborhood of the curve $\gamma \cup l$ and for all $\zeta \in \gamma$,*

$$\left|Q(\zeta, z) - \frac{1}{\zeta - z}\right| < \varepsilon.$$

b′) *For all z in $\{|z - b| \geq d\}$ and all $\zeta \in \gamma$,*

$$|Q(\zeta, z)| < \exp\left\{C\left(1 + \ln\frac{1}{\varepsilon d}\right)e^{\frac{C}{d}\{\text{length}(\gamma) + \text{length}(l)\}}\right\},$$

where $C > 1$ is an absolute constant.

PROOF. Let ζ be an arbitrary point on the curve γ. Taking $a = \zeta$ and choosing as L the curve joining a to b and consisting of the curve l and a part of the curve γ, we apply Theorem 5.2. We denote the obtained function $P\left(\frac{1}{z-b}\right)$ by $R(\zeta, z)$. It is clear that conditions a′) and b′) will be satisfied for the function $R(\zeta, z)$, but it may or may not be piecewise constant in ζ.

---

[4]Piecewise constant means that the curve γ can be cut into finitely many pieces γ_j in such a way that the restriction of $Q(\zeta, z)$ to the set $\{\zeta \in \gamma_j\} \times \{|z| < \infty\}$ does not depend on ζ.

Denote by U the complement of the d-neighborhood of the curve $\gamma \cup l$. Since U is a closed set and $R(\zeta, z) \to 0$ as $z \to \infty$, condition a$'$) for $R(\zeta, z)$ implies that

$$\max_{z\in U}\left|R(\zeta,z)-\frac{1}{\zeta-z}\right|<\varepsilon.$$

Choose $\varepsilon_1(\zeta)$, $0<\varepsilon_1(\zeta)<\varepsilon$, such that the last inequality remains true if we replace ε by $\varepsilon_1(\zeta)$. Then for $\zeta'\in\gamma$, $|\zeta-\zeta'|<\varepsilon_2(\zeta)=d^2(\varepsilon-\varepsilon_1(\zeta))$, the inequality

$$\begin{aligned}\max_{z\in U}\left|R(\zeta,z)-\frac{1}{\zeta'-z}\right| &\le \max_{z\in U}\left(\left|R(\zeta,z)-\frac{1}{\zeta-z}\right|+\left|\frac{1}{\zeta-z}-\frac{1}{\zeta'-z}\right|\right)\\ &\le \varepsilon_1(\zeta)+\frac{|\zeta-\zeta'|}{d^2}<\varepsilon\end{aligned}$$

holds. By the Heine–Borel lemma, there is a finite subset $\{a_j\}$ of γ such that for each $\zeta\in\gamma$ there is $a_j\in\{a_j\}$ satisfying the condition

$$|a_j-\zeta|<\varepsilon_2(a_j).$$

Denoting by $a_{j(\zeta)}$ one of the points satisfying this condition, we can easily check that the function $Q(\zeta,z)=R(a_{j(\zeta)},z)$ satisfies the conditions of the lemma. □

THEOREM 5.3 (M. V. Keldysh). *Let $f(z)$ be a function analytic inside the angle $\{|\arg z|<\beta\}$, $0<\beta<\pi$, and satisfying there the condition*

$$\ln|f(z)|<K(1+|z|^\rho), \tag{5.7}$$

where $K>0$, $\rho>1/2$, are constants. For each α, $0<\alpha<\min(\beta,\pi-\frac{\pi}{2\rho})$, there exists an entire function $G(z)$ whose growth does not exceed the normal type of order ρ, such that

$$G(z)=f(z)+g(z),\qquad |\arg z|\le\alpha,$$

where $g(z)$ is bounded for $|\arg z|\le\alpha$.

PROOF. Let $\alpha<\gamma<\min(\beta,\pi-\frac{\pi}{2\rho})$, $\kappa=\frac{\pi}{2\pi-2\gamma}$. Note that $\kappa<\rho$. We construct an auxiliary function $G(z,\zeta,A)$, which is an entire function in the variable z, jointly continuous in the variables z,ζ,A on the set $\{|z|<\infty\}\times\{|\arg\zeta|\ge\beta\}\times\{A>0\}$, and such that the following conditions are satisfied:

a) $|\frac{1}{\zeta-z}-G(z,\zeta,A)|<e^{-AC_1|\zeta|^\kappa}$ for $z\in\{|\arg z|\le\alpha\}\cup\{1<|z|<C_2|\zeta|\}$, $C_2<1$;

b) $|G(z,\zeta,A)|<e^{AC_3|z|^\kappa}$ for $1<|\zeta|<C_2^{-1}|z|+1$, $|z|>1$, where here and after $C_1,C_2,\dots$ denote positive constants depending on α, β, and γ only.

Denote by $L(R,\theta)$ the boundary of the domain $D(R,\theta)=\{|\pi-\arg z|<\theta,$ $|z|>R\}$ parameterized in the negative direction. We consider the branch of the function $(-z)^\kappa$ in $D(R,\theta)$ taking positive values on the ray $\{\arg z=\pi\}$. Observe that the values of the function $(-z)^\kappa$ on the rays $\{\arg z=\pi\pm(\pi-\gamma)\}$ are pure imaginary.

If z is outside the closed domain $\bar D(R,\pi-\gamma)$ and $\zeta\in\{|\arg\zeta-\pi|\le\pi-\beta\}$, then by the Cauchy theorem,

$$\begin{aligned}&\frac{1}{2\pi i}\int_{L(R',\pi-\gamma)}\frac{e^{A(-t)^\kappa}\,dt}{(t-\zeta)(t-z)}-\frac{1}{2\pi i}\int_{L(R,\pi-\gamma)}\frac{e^{A(-t)^\kappa}\,dt}{(t-\zeta)(t-z)}\\ &\qquad=\begin{cases}0 & \text{if } |\zeta|<R<R',\\ \frac{e^{A(-\zeta)^\kappa}}{\zeta-z} & \text{if } R<|\zeta|<R'.\end{cases}\end{aligned}\tag{5.8}$$

For $R > |\zeta|$ and $z \notin \bar{D}(R, \pi - \gamma)$, we set

$$G(z,\zeta,A) = \frac{e^{-A(-\zeta)^\kappa}}{2\pi i} \int_{L(R,\pi-\gamma)} \frac{e^{A(-t)^\kappa}\, dt}{(t-\zeta)(t-z)}. \tag{5.9}$$

It is clear that the integral in (5.9) is an analytic function of z outside the domain $D(R, \pi - \gamma)$. By (5.8) the right-hand side of (5.9) does not change if we replace the integration over $L(R < \pi - \gamma)$ by the integration over $L(R', \pi - \gamma)$ with $R' > R$. We see that the function $G(z, \zeta, A)$ has an analytic continuation into the whole plane $\{|z| < \infty\}$ and thus determines an entire function. We keep the notation $G(z, \zeta, A)$ for this entire function and show that it satisfies conditions a) and b).

First we verify that a) is satisfied. If z is not in $D\left(\frac{1}{2}R, \pi - \alpha\right)$ ($\supset D(R, \pi - \gamma)$), and $\zeta \in D(R, \pi - \beta)$, then, by (5.8), we have

$$G(z,\zeta,A) = \frac{1}{\zeta - z} + \frac{e^{-A(-\zeta)^\kappa}}{2\pi i} \int_{L(R,\pi-\gamma)} \frac{e^{A(-t)^\kappa}\, dt}{(t-\zeta)(t-z)}.$$

Since for $|\arg \zeta - \pi| \le \pi - \beta$ the condition $\operatorname{Re}(-\zeta)^\kappa \ge \cos\{\kappa(\pi - \beta)\}|\zeta|^\kappa > 0$ holds, whereas for $|\arg t - \pi| = \pi - \gamma$ we have $\operatorname{Re}(-t)^\kappa = 0$, we get the estimate

$$\begin{aligned} \left| G(z,\zeta,A) - \frac{1}{\zeta - z} \right| &\le \frac{e^{-AC_5|\zeta|^\kappa}}{2\pi} \\ &\times \left\{ \frac{e^{AR^\kappa}}{|\zeta| - R} \int_{\substack{|t|=R \\ |\pi - \arg t| \le \pi - \gamma}} \frac{|dt|}{|t-z|} + \int_{\substack{R<|t|<\infty \\ |\pi - \arg t| = \pi-\gamma}} \frac{|dt|}{|t-\zeta||t-z|} \right\} \\ &= \frac{e^{-AC_5|\zeta|^\kappa}}{2\pi} \left\{ \frac{e^{AR^\kappa}}{|\zeta| - R} I_1 + I_2 \right\}, \qquad 0 < C_5 < 1. \end{aligned} \tag{5.10}$$

Since the distance between the arc $\{|t| = R, |\arg t - \pi| \le \pi - \gamma\}$ and the exterior of the domain $D\left(\frac{1}{2}R, \pi - \alpha\right)$ is at least $C_6 R$, we have $I_1 \le C_7$. Let us estimate I_2. Observe that for $|\arg t - \pi| = \pi - \gamma$ and $|\arg \zeta - \pi| \le \pi - \beta$, the inequality $|t - \zeta| \ge |t| \sin(\beta - \gamma)$ holds. Therefore for $|z| \le \frac{1}{2}R$ we have

$$I_2 \le 2 \int_R^\infty \frac{du}{u \sin(\beta - \gamma)\left(u - \frac{1}{2}R\right)} \le \frac{4}{R \sin(\beta - \gamma)}.$$

If $|\pi - \arg z| \ge \pi - \alpha$, taking into account $|t - z| \ge |t| \sin(\gamma - \alpha)$, we obtain

$$I_2 \le 2 \int_R^\infty \frac{du}{(u \sin(\beta - \gamma))(u \sin(\gamma - \alpha))} \le \frac{2}{R \sin(\beta - \gamma) \sin(\gamma - \alpha)}.$$

Hence, if z is outside the domain $D\left(\frac{1}{2}R, \pi - \alpha\right)$, then $I_2 \le C_8 R^{-1}$.

Now we set $R = C_9|\zeta|$ in (5.10), where $C_9 = \left(\frac{1}{2}C_5\right)^{1/\kappa} < 1$, and assume that $|z| > 1$. Then $R > 2|z| > 2$, and we get the estimate

$$\begin{aligned} \left| G(z,\zeta,A) - \frac{1}{\zeta - z} \right| &\le \frac{e^{-AC_5|\zeta|^\kappa}}{2\pi} \left\{ \frac{e^{\frac{1}{2}AC_5|\zeta|^\kappa}}{(C_9^{-1} - 1)R} C_7 + C_8 R^{-1} \right\} \\ &\le C_{10} e^{-\frac{1}{2}AC_5|\zeta|^\kappa}, \qquad 1 < |z| < \frac{1}{2}C_9|\zeta|. \end{aligned}$$

To verify condition b), we set $R = 2C_2^{-1}|z| + 2$ in (5.9). We get $(1 < |\zeta| < C_2^{-1}|z| + 1 = \frac{1}{2}R)$

$$
\begin{aligned}
&|G(z,\zeta, A)| \\
&\quad \le \frac{e^{-AC_5|\zeta|^\kappa}}{2\pi}\left\{\frac{e^{C_2^{-\kappa}A4^\kappa|z|^\kappa}}{R-|\zeta|}4(\pi-\gamma) + \int_{\substack{R<|t|<\infty \\ |\pi-\arg t|=\pi-\gamma}} \frac{|dt|}{(|t|-|\zeta|)(|t|-|z|)}\right\} \\
&\quad \le \frac{1}{2\pi}\left\{e^{AC_{11}|z|^\kappa}4\pi + 2\int_R^\infty \frac{du}{(u-|z|)^2}\right\} \\
&\quad \le \frac{1}{2\pi}\{e^{AC_{11}|z|^\kappa}4\pi + 2\} \le 3e^{AC_{11}|z|^\kappa}.
\end{aligned}
$$

Thus, we have constructed the auxiliary function. Set

$$G(z,\zeta) = G(z,\zeta, 2KC_1^{-1}|\zeta|^{\rho-\kappa}),$$

where K is the same constant as in (5.7). The function $G(z,\zeta)$ is entire in the variable z, jointly continuous on the set $\{|z| < \infty\} \times \{|\arg\zeta| \ge \beta\}$, and satisfies the conditions

c) For $|\arg z| \le \alpha$ and for $1 < |z| < C_2|\zeta|$, $C_2 < 1$,

$$\left|\frac{1}{\zeta - z} - G(z,\zeta)\right| < e^{-2K|\zeta|^\rho}. \tag{5.11}$$

d) For $1 < |\zeta| < |z|$,

$$|G(z,\zeta)| < e^{KC_{12}|z|^\rho} \tag{5.12}$$

(to derive this condition from b) we take into account that $\kappa < \rho$).

Now we turn to the construction of the desired entire function $G(z)$. Set

$$G_n(z) = \frac{1}{2\pi i}\int_{\substack{C_2^{-1}<|\zeta|<n \\ \arg\zeta=\pm\beta}} f(\zeta)G(z,\zeta)d\zeta + \frac{1}{2\pi i}\int_{\substack{|\zeta|=n \\ |\arg\zeta|\le\beta}} f(\zeta)\frac{d\zeta}{\zeta - z} \tag{5.13}$$

(here and after the paths are traversed as follows: the ray $\{\arg\zeta = \beta\}$ is traversed from ∞ to 0, the ray $\{\arg\zeta = -\beta\}$ is traversed from 0 to ∞, the circle $\{|\zeta| = \text{const}\}$ is traversed counterclockwise).

The function $G_n(z)$ is analytic if $|z| < n$. It is clear that for $m > n > C_2^{-1}$ we have

$$
\begin{aligned}
G_m(z) - G_n(z) &= \frac{1}{2\pi i}\int_{\substack{n<|\zeta|<m \\ \arg\zeta=\pm\beta}} f(\zeta)G(z,\zeta)d\zeta \\
&\quad + \frac{1}{2\pi i}\left\{\int_{\substack{|\zeta|=m \\ |\arg\zeta|\le\beta}} - \int_{\substack{|\zeta|=n \\ |\arg\zeta|\le\beta}}\right\} f(\zeta)\frac{d\zeta}{\zeta - z}.
\end{aligned} \tag{5.14}
$$

By the Cauchy theorem,

$$0 = \frac{1}{2\pi i}\left\{\int_{\substack{n<|\zeta|<m \\ \arg\zeta=\pm\beta}} + \int_{\substack{|\zeta|=m \\ |\arg\zeta|\le\beta}} - \int_{\substack{|\zeta|=n \\ |\arg\zeta|\le\beta}}\right\} f(\zeta)\frac{d\zeta}{\zeta - z} \tag{5.15}$$

for $|z| < n$. Subtracting (5.15) from (5.14), we get

$$G_m(z) - G_n(z) = \frac{1}{2\pi i}\int_{\substack{n<|\zeta|<m \\ \arg\zeta=\pm\beta}} f(\zeta)\left\{G(z,\zeta) - \frac{1}{\zeta - z}\right\}d\zeta, \qquad |z| < n.$$

From this, using (5.11), we have, for $1 < |z| < C_2 n$,

$$|G_m(z) - G_n(z)| \le \frac{1}{2\pi} \int_{\substack{n<|\zeta|<m \\ \arg\zeta=\pm\beta}} \exp\{K(1+|\zeta|^\rho) - 2K|\zeta|^\rho\}|d\zeta|. \tag{5.16}$$

Since the right-hand side goes to 0 as $n, m \to \infty$, the sequence $\{G_n(z)\}$ converges to some entire function $G(z)$ uniformly in each disc of finite radius. Let us show that this function has the desired properties.

By the Cauchy theorem, for $z \in \{C_2^{-1} < |z| < n, |\arg z| < \beta\}$, we have

$$f(z) = \frac{1}{2\pi i}\left\{ -\int_{\substack{|\zeta|=C_2^{-1} \\ |\arg\zeta|\le\beta}} + \int_{\substack{C_2^{-1}<|\zeta|<n \\ \arg\zeta=\pm\beta}} + \int_{\substack{|\zeta|=n \\ |\arg\zeta|=\beta}} \right\} f(\zeta)\frac{d\zeta}{\zeta - z}. \tag{5.17}$$

Subtracting (5.17) from (5.13), we get

$$\begin{aligned} G_n(z) - f(z) &= \frac{1}{2\pi i}\int_{\substack{C_2^{-1}<|\zeta|<n \\ \arg\zeta=\pm\beta}} f(\zeta)\left\{G(z,\zeta) - \frac{1}{\zeta - z}\right\} d\zeta \\ &\quad - \frac{1}{2\pi i}\int_{\substack{|\zeta|=C_2^{-1} \\ |\arg\zeta|\le\beta}} f(\zeta)\frac{d\zeta}{\zeta - z}. \end{aligned}$$

For $|\arg z| \le \alpha$, $2C_2^{-1} < |z| < n$, by (5.11), we get the estimate

$$|G_n(z) - f(z)| \le \frac{1}{2\pi}\int_{\substack{C_2^{-1}<|\zeta|<n \\ \arg\zeta=\pm\beta}} \exp\{K(1+|\zeta|^\rho) - 2K|\zeta|^\rho\}|d\zeta| + e^{KC_{13}},$$

from which, letting $n \to \infty$, we infer the relation ($|\arg z| \le \alpha$, $|z| > C_2^{-1}$)

$$|G(z) - f(z)| \le \frac{1}{2\pi}\int_{\substack{C_2^{-1}<|\zeta|<\infty \\ \arg\zeta=\pm\beta}} \exp\{K(1+|\zeta|^\rho) - 2K|\zeta|^\rho\}|d\zeta| + e^{KC_{13}}.$$

Thus we have proved that the difference $G(z) - f(z)$ is bounded for $|\arg z| \le \alpha$.

Now we estimate the growth of the function $G(z)$. Letting $m \to \infty$ in (5.16), we get ($1 < |z| < C_2 n$)

$$|G(z) - G_n(z)| \le \frac{1}{2\pi}\int_{\substack{n<|\zeta|<\infty \\ \arg\zeta=\pm\beta}} \exp\{K(1+|\zeta|^\rho) - 2K|\zeta|^\rho\}|d\zeta|.$$

Hence, for sufficiently large n and for $1 < |z| < C_2 n$,

$$|G(z) - G_n(z)| < 1;$$

therefore

$$\left|G(z) - G_{[C_2^{-1}|z|]+1}(z)\right| < 1 \tag{5.18}$$

for sufficiently large $|z|$. Furthermore, in view of (5.7), (5.12), and (5.13), for $n = [C_2^{-1}|z|] + 1$ we have the inequality

$$\begin{aligned} |G_n(z)| &< \frac{1}{2\pi}\int_{\substack{C_2^{-1}<|\zeta|<n \\ \arg\zeta=\pm\beta}} \exp\{K(1+|\zeta|^\rho) + KC_{12}|z|^\rho\}|d\zeta| \\ &\quad + \frac{1}{2\pi}\exp\{K(1+n^\rho)\}2\beta n \\ &\le e^{KC_{15}n^\rho} \le e^{KC_{16}|z|^\rho}. \end{aligned} \tag{5.19}$$

By (5.18) and (5.19), for sufficiently large $|z|$ we have $|G(z)| < \exp\{KC_{17}|z|^\rho\}$. The proof of the theorem is complete. □

We need the following lemma, which can be derived from Theorem 5.3.

LEMMA 5.2. *Let $f(z)$ be a function satisfying the assumptions of Theorem 5.3 for $\frac{1}{2} < \rho \le 1$. Then, for each α, $0 < \alpha < \min(\beta, \frac{1}{2}(\pi - \frac{\pi}{2\rho}))$, there exists an entire function $G(z)$ whose growth does not exceed the normal type of order ρ, such that*

$$G(z) = f(z) + g(z)$$

for $|\arg z| \le \alpha$, where $g(z)$ is a function satisfying the estimate

$$|g(z)| < e^{-|z|^\rho} \tag{5.20}$$

in the angle $\{|\arg z| \le \alpha\}$ for sufficiently large $|z|$.

PROOF. For each δ, $0 < \delta < \frac{1}{2}(\pi - \frac{\pi}{2\rho})$, we can find an entire function $\omega_\rho(z;\delta)$ not exceeding the normal type of order ρ and satisfying the estimate

$$e^{-K(\delta)|z|^\rho} \le |\omega_\rho(z;\delta)| \le e^{-2|z|^\rho}, \qquad 2 < K(\delta) < \infty,$$

in the angle $\{|\arg z| \le \frac{1}{2}(\pi - \frac{\pi}{2\rho}) - \delta\}$ for sufficiently large $|z|$. If $1/2 < \rho < 1$, this can be done in the following way. Let

$$\omega(z) = \prod_{k=1}^{\infty} \left(1 - \frac{z}{k^{1/\rho}}\right).$$

The asymptotic formula (5.3_2) in Chapter 2 implies that for $0 < \delta \le \arg z \le \pi - \frac{\pi}{2\rho} - \delta$ and for sufficiently large $|z|$, the estimate

$$e^{-K_1(\delta)|z|^\rho} \le |\omega(z)| \le e^{-K_2(\delta)|z|^\rho}$$

holds, where $0 < K_2(\delta) < K_1(\delta) < \infty$. Therefore we can take

$$\omega_\rho(z;\delta) = \omega\left(\lambda_\delta z e^{\frac{i}{2}(\pi - \frac{\pi}{2\rho})}\right), \qquad \lambda_\delta = \left(\frac{2}{K_2(\delta)}\right)^{1/\rho}.$$

If $\rho = 1$, set

$$\omega_1(z;\delta) = e^{-4z}.$$

Choose δ in such a way that $\{|\arg z| \le \alpha\} \subset \{|\arg z| < \frac{1}{2}(\pi - \frac{\pi}{2\rho}) - \delta\} \subset \{|\arg z| \le \beta\}$. Consider the function

$$f_1(z) = f(z)/\omega_\rho(z;\delta).$$

It is clear that this function is analytic in the angle $\{|\arg z| \le \frac{1}{2}(\pi - \frac{\pi}{2\rho}) - \delta\}$ and can be estimated there as

$$|f_1(z)| \le e^{K(1+|z|^\rho)}$$

($0 < K < \infty$ does not depend on z). We apply Theorem 5.3 to $f_1(z)$ (the role of the angle $\{|\arg z| < \beta\}$ is played by the angle $\{|\arg z| < \frac{1}{2}(\pi - \frac{\pi}{2\rho}) - \delta\}$). We get an entire function $G_1(z)$ not exceeding the normal type of order ρ, such that

$$G_1(z) = f_1(z) + g_1(z),$$

where $g_1(z)$ is bounded for $|\arg z| < \alpha$. Letting $G(z) = G_1(z)\omega_\rho(z;\delta)$, we see that we get an entire function with the desired properties. □

Now we turn to the proof of Theorem 5.1.

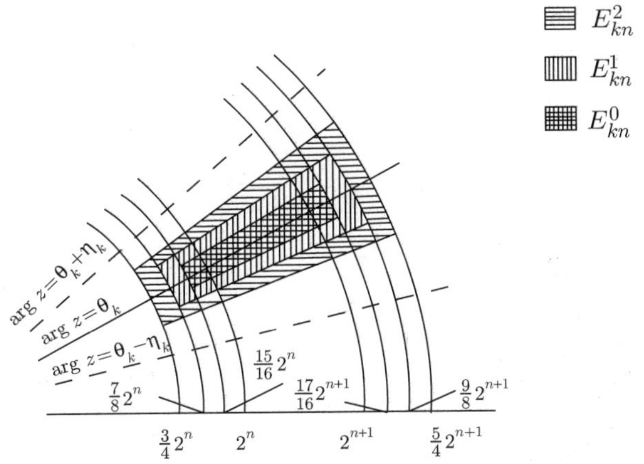

FIGURE 1

PROOF. Observe that it suffices to consider the case $1/2 < \rho \le 1$. In fact, if $\rho > 1$, we choose a positive integer N such that $N < \rho \le 2N$. We construct an entire function $G(z)$ of order $\frac{\rho}{2N}$ whose set of deficient values contains $\{a_k\}_{k=1}^{\infty}$. The function $G(z^{2N})$ has order ρ and, by the Remark in Section 1 (p. 115), its set of deficient values also contains $\{a_k\}_{k=1}^{\infty}$.

We now introduce some notation to be used here.

α: a number satisfying $0 < \alpha < \frac{1}{2}(\pi - \frac{\pi}{2\rho})$.

$\{\theta_k\}_{k=1}^{\infty}$: a sequence of positive numbers such that $\theta_k \uparrow \alpha$ as $k \uparrow +\infty$.

$\theta_{-k} = -\theta_k$, $k = 1, 2, \ldots$, $\theta_0 = 0$.

$\eta_k = \frac{1}{2}\min(|\theta_{k+1} - \theta_k|, |\theta_k - \theta_{k-1}|)$.

$\varepsilon_k = \exp(-2^{10}C\eta_k^{-1})$, where C is the constant in Lemma 5.1.

$\delta_{kn} = \exp(-\varepsilon_k 2^{n\rho})$.

$\Delta_k(\kappa) = \{|\arg z - \theta_k| \le \kappa\eta_k\}$, $0 < \kappa < 1$.

$\sigma_n(\lambda) = \{(1-\lambda)2^n \le |z| \le (1+\lambda)2^{n+1}\}$, $0 < \lambda < 1$.

$E^i_{kn} = \Delta_k\left(\frac{i+1}{4}\right) \cap \sigma_n(2^{i-4})$, $i = 0, 1, 2$ (for $k > 0$; see Figure 1).

γ^i_{kn} is the boundary of the set E^i_{kn}.

$l^i_{kn} = \{|z| = (1+2^{i-4})2^{n+1},\ (-1)^n\theta_k \le (-1)^n \arg z \le \pi\}$.

Let $\{n_k\}_{k=1}^{\infty}$ be a sequence of even positive integers, $n_k \uparrow \infty$ as $k \uparrow \infty$; define the sequence $\{n_{-k}\}_{k=1}^{\infty}$ by $n_{-k} = n_k + 1$, $k = 1, 2, \ldots$. Set (see Figure 2)

$$E_i = \bigcup_{|k|=1}^{\infty} \bigcup_{m=0}^{\infty} E^i_{k, n_k+2m}.$$

Denote by Γ_i the boundary of the set E_i. It is easy to see that

$$\Gamma_i = \bigcup_{|k|=1}^{\infty} \bigcup_{m=0}^{\infty} \gamma^i_{k, n_k+2m}.$$

We choose the sequence $\{n_k\}_{k=1}^{\infty}$ to be rapidly increasing so that the following

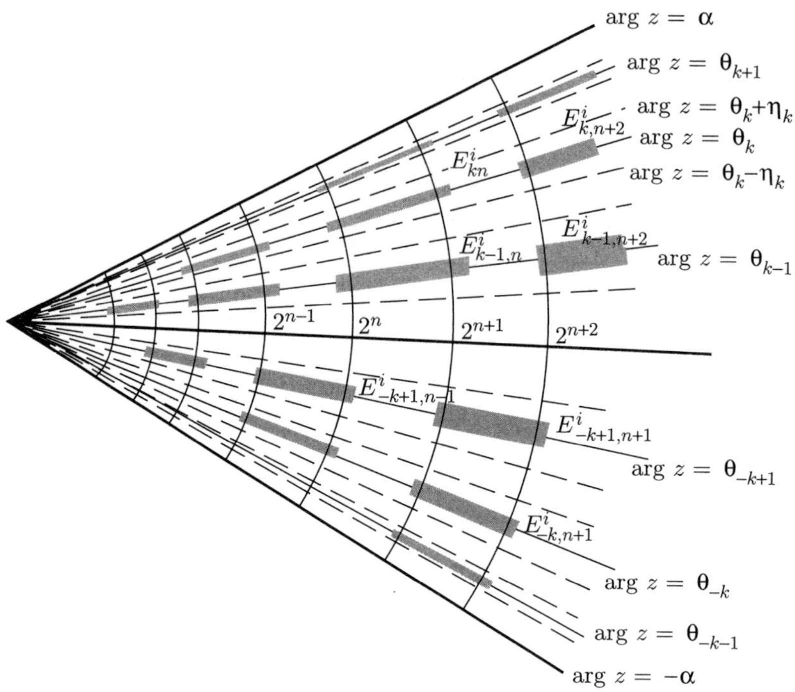

FIGURE 2. n is even

conditions hold:

$$\prod_{|k|=1}^{\infty}\prod_{n=n_k}^{\infty}\frac{1+\delta_{kn}}{1-\delta_{kn}}<\frac{16}{15}, \tag{5.21}$$

$$\varepsilon_k^{-1}<2^{n_k\rho}, \tag{5.22}$$

$$\max(2^{n+8},|a_{|k|}|)<\exp(\varepsilon_k 2^{n\rho-1}) \quad \text{for} \quad n\geq n_k, \tag{5.23}$$

$$\int_{\Gamma_1}|\zeta^{-2}||d\zeta|<1. \tag{5.24}$$

Note that $E^0_{kn}\subset E^1_{kn}\subset E^2_{kn}$, and hence $E_0\subset E_1\subset E_2$. Later on, we use the following convention: we consider the sets E^i_{kn}, γ^i_{kn}, l^i_{kn} only for those pairs (k,n) that participate in forming the sets E_i (that is, $k\neq 0$, $n=n_k+2m$, $m=0,1,2,\dots$).

Our proof is organized as follows. We wish to approximate the function $\varphi(z)$ defined by the equalities

$$\varphi(z)=a_{|k|}, \qquad z\in E^1_{kn},$$

by an entire function $G(z)$ having normal type of order ρ and such that

$$|\varphi(z)-G(z)|\leq\exp\left\{-\frac{1}{2}\varepsilon_k|z|^{\rho}\right\}, \qquad z\in E^0_{kn}.$$

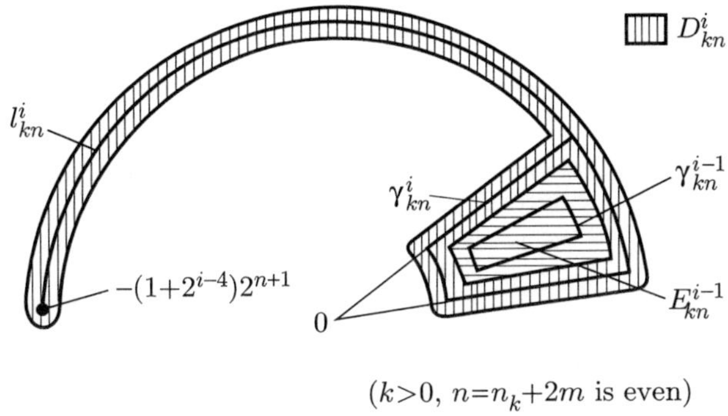

FIGURE 3

To this end, we first construct (Lemma 5.3) a function $\omega(z)$, analytic in the half-plane $\{\operatorname{Re} z > -1\}$ and such that

$$|e^{-\varepsilon_k z^\rho}| \le |\omega(z)| \le 2|e^{-\varepsilon_k z^\rho}| \quad \text{for} \quad z \in E^1_{kn},$$
$$|\omega(z)| \le \exp\{(|z|+1)^\rho\} \quad \text{for} \quad \operatorname{Re} z > -1.$$

Next, with the help of the function $\omega(z)$ we construct (Lemma 5.4) a function $F(z)$, analytic in the half-plane $\{\operatorname{Re} z > 0\}$ and such that

$$|\varphi(z) - F(z)| < \frac{1}{2}\left|\exp\left\{-\frac{1}{2}\varepsilon_k z^\rho\right\}\right|, \quad z \in E^0_{kn},$$
$$\ln|F(z)| \le O\left((|z|+1)^\rho\right), \quad \operatorname{Re} z > 0.$$

Finally, using Lemma 5.2, we approximate the function $F(z)$ in the angle $\{|\arg z| \le \alpha\}$ by an entire function $G(z)$ and get the desired function.

In our proofs of Lemmas 5.3 and 5.4 we use an auxiliary function $Q_i(\zeta, z)$, defined on the set $\{\zeta \in \Gamma_i\} \times \{|z| < \infty\}$, $i = 0, 1, 2$. We construct this function first. It is easy to check that the curves γ^i_{kn} and l^i_{kn} ($i = 1, 2$) do not intersect the set E_{i-1}, and their distance from this set is at least $d_{kn} = \eta_k 2^{n-4}$. Denote by D^i_{kn} the d_{kn}-neighborhood of the curve $\gamma^i_{kn} \cup l^i_{kn}$ (see Figure 3). We apply Lemma 5.1 with $\gamma = \gamma^i_{kn}$, $l = l^i_{kn}$, $b = -(1+2^{i-4})2^{n+1}$, $\varepsilon = 4^{-n-2}\exp\{-5\varepsilon_k 2^{n\rho}\}$, and $d = d_{kn}$. We get a function $Q^i_{kn}(\zeta, z)$, defined on the set $\{\zeta \in \gamma^i_{kn}\} \times \{|z| < \infty\}$, holomorphic with respect to z in the half-plane $\{\operatorname{Re} z > -1\}$, piecewise constant with respect to ζ, and satisfying the conditions:[5]

$$\left|Q^i_{kn}(\zeta, z) - \frac{1}{\zeta - z}\right| < 4^{-n-2}\exp(-5\varepsilon_k 2^{n\rho}), \quad \zeta \in \gamma^i_{kn}, \quad z \in CD^i_{kn}, \tag{5.25}$$

$$|Q^i_{kn}(\zeta, z)| < \exp(\varepsilon_k^{1/2} 2^{n\rho - 1}), \quad \zeta \in \gamma^i_{kn}, \quad \operatorname{Re} z > -1. \tag{5.26}$$

We derive inequality (5.26) from statement b$'$) of Lemma 5.1. We have

$$\frac{1}{\varepsilon d} = \frac{1}{4^{-n-2}\exp(-5\varepsilon_k 2^{n\rho})\eta_k 2^{n-4}} = \frac{1}{\eta_k} 2^{n+8}\exp(5\varepsilon_k 2^{n\rho}).$$

By (5.22),

$$\frac{1}{\eta_k} = \frac{1}{2^{10}C}\ln\frac{1}{\varepsilon_k} \le \frac{1}{\varepsilon_k} \le 2^{n_k\rho} \le 2^n < 2^{n+8}$$

[5]In this section we denote by CD the complement of the set D in the complex plane.

for $n \geq n_k$, and by (5.23),

$$2^{n+8} < \exp(\varepsilon_k 2^{n\rho-1}) < \exp(\varepsilon_k 2^{n\rho}).$$

Therefore

$$\frac{1}{\varepsilon d} < \exp(7\varepsilon_k 2^{n\rho}), \qquad \ln\frac{1}{\varepsilon d} < 7\varepsilon_k 2^{n\rho},$$

and, taking into account (5.22) again, we get

$$1 + \ln\frac{1}{\varepsilon d} < 1 + 7\varepsilon_k 2^{n\rho} < 8\varepsilon_k 2^{n\rho} = \varepsilon_k 2^{n\rho+3}.$$

Furthermore, it is easy to check that $\text{length}(\gamma^i_{kn}) < 2^{n+3}$, $\text{length}(l^i_{kn}) < 2^{n+3}$, and therefore

$$\frac{1}{d}(\text{length}(\gamma) + \text{length}(l)) \leq \frac{2^{n+4}}{\eta_k 2^{n-4}} = \frac{2^8}{\eta_k}.$$

Thus

$$\exp\left\{C\left(1 + \ln\frac{1}{\varepsilon d}\right) e^{\frac{C}{d}(\text{length}(\gamma)+\text{length}(l))}\right\}$$
$$< \exp\left\{C\varepsilon_k 2^{n\rho+3} e^{\frac{2^8 C}{\eta_k}}\right\} = \exp\left\{C\varepsilon_k^{3/4} 2^{n\rho+3}\right\}.$$

Observing that[6]

$$C\varepsilon_k^{1/4} = C\exp\left\{-\frac{2^8 C}{\eta_k}\right\} \leq \frac{\eta_k}{2^8}e^{-1} < \frac{\pi}{2^{10}e} < 2^{-9},$$

we get inequality (5.26).

The function $Q^i_{kn}(\zeta, z)$ is defined on the set $\{\zeta \in \gamma^i_{kn}\} \times \{|z| < \infty\}$. Since for fixed i the sets γ^i_{kn} are pairwise disjoint and their union is equal to Γ_i, we can define a function $Q_i(\zeta, z)$ on the set $\{\zeta \in \Gamma_i\} \times \{|z| < \infty\}$ by letting

$$Q_i(\zeta, z) = Q^i_{kn}(\zeta, z) \quad \text{for} \quad \zeta \in \gamma^i_{kn}, \quad |z| < \infty.$$

As is easy to see, inequalities (5.25) and (5.26) will remain valid if we replace $Q^i_{kn}(\zeta, z)$ by $Q_i(\zeta, z)$.

We define a function $\psi(z)$ on the set E_2 by

$$\psi(z) = e^{-\varepsilon_k z^\rho}, \qquad z \in E^2_{kn}.$$

Observe that for $z \in E^i_{kn}$,

$$\exp\left\{-\frac{5}{2}\varepsilon_k 2^{n\rho}\right\} \leq |\psi(z)| \leq \exp\left\{-\varepsilon_k 2^{n\rho-1}\right\}. \tag{5.27}$$

LEMMA 5.3. *There exists a function $\omega(z)$ that is analytic in the half-plane $\{\operatorname{Re} z > -1\}$ and satisfies*

$$1 < |\omega(z)/\psi(z)| < 2, \quad z \in E_1, \tag{5.28}$$

$$|\omega(z)| < \exp\{(|z|+1)^\rho\}, \quad \operatorname{Re} z > -1. \tag{5.29}$$

---

[6]It is easy to see that the function e^x/x takes its minimum value on $(0, \infty)$ at $x = 1$; hence $e^{-x} \leq (ex)^{-1}$ for $x > 0$.

PROOF. We set

$$\omega_{kn}(z) = 1 + \frac{1}{2\pi i}\int_{\gamma^2_{kn}} \{\psi(\zeta) - 1\} Q_2(\zeta, z)\, d\zeta. \tag{5.30}$$

Since $Q_i(\zeta, z)$ is a piecewise constant function of ζ, and it is analytic with respect to z in the half-plane $\{\operatorname{Re} z > -1\}$, the integral in (5.30) makes sense, and the function $\omega_{kn}(z)$ is analytic in the half-plane $\{\operatorname{Re} z > -1\}$. Let us show that the following inequalities hold:

$$|\omega_{kn}(z)| < \exp(\varepsilon_k^{1/2} 2^{n\rho+1}), \quad \operatorname{Re} z > -1, \tag{5.31}$$

$$3/4 < |\omega_{kn}(z)/\psi(z)| < 5/4, \quad z \in E^1_{kn}, \tag{5.32}$$

$$|\omega_{kn}(z) - 1| < \delta_{kn}, \quad z \in CD^2_{kn} \backslash E^2_{kn}. \tag{5.33}$$

Since $\operatorname{length}(\gamma^2_{kn}) < 2^{n+3}$ and $|\psi(\zeta)| < 1$, using (5.26) we get

$$|\omega_{kn}(z)| \le 1 + \frac{1}{2\pi} \cdot 2 \cdot 2^{n+3} \exp(\varepsilon_k^{1/2} 2^{n\rho}) < 2^{n+3} \exp(\varepsilon_k^{1/2} 2^{n\rho}).$$

On the other hand, by (5.23), we have $2^{n+3} < \exp(\varepsilon_k 2^{n\rho-1}) < \exp(\varepsilon_k^{1/2} 2^{n\rho})$, and (5.31) is proved.

Further, we set

$$\psi_{kn}(z) = 1 + \frac{1}{2\pi i}\int_{\gamma^2_{kn}} \{\psi(\zeta) - 1\} \frac{d\zeta}{\zeta - z}.$$

By the Cauchy formula,

$$\psi_{kn}(z) = \begin{cases} \psi(z), & z \in E^1_{kn}, \\ 1, & z \in CE^2_{kn}. \end{cases} \tag{5.34}$$

Since

$$\omega_{kn}(z) - \psi_{kn}(z) = \frac{1}{2\pi i}\int_{\gamma^2_{kn}} \{\psi(\zeta) - 1\} \left\{ Q_2(\zeta, z) - \frac{1}{\zeta - z} \right\} d\zeta,$$

from (5.25) we get that for $z \in CD^2_{kn}$ the inequality

$$\begin{aligned} |\omega_{kn}(z) - \psi_{kn}(z)| &\le \frac{1}{2\pi} \cdot 2 \cdot 2^{n+3} \cdot 4^{-n-2} \exp(-5\varepsilon_k 2^{n\rho}) \\ &< \frac{1}{4} \exp(-5\varepsilon_k 2^{n\rho}) \end{aligned} \tag{5.35}$$

holds, from which, by (5.34) and (5.27), we obtain

$$|\omega_{kn}(z) - \psi(z)| < \frac{1}{4}|\psi(z)|, \qquad z \in E^1_{kn} \cap CD^2_{kn} = E^1_{kn}.$$

Thus (5.32) is proved. Relation (5.33) immediately follows from (5.35) and (5.34).

We define the desired function $\omega(z)$ as

$$\omega(z) = \frac{3}{2} \prod_{|k|=1}^{\infty} \prod_{m=0}^{\infty} \omega_{k, n_k + 2m}(z).$$

The infinite product converges absolutely and uniformly in each bounded region Ω in $\{\operatorname{Re} z > -1\}$, and hence the function $\omega(z)$ is analytic in this half-plane. In fact, for sufficiently large n we have $\Omega \subset CD^2_{kn} \backslash E^2_{kn}$; hence, all $\omega_{k,n_k+2m}(z)$ except possibly finitely many satisfy inequality (5.33). Since (5.21) implies the convergence of the series

$$\sum_{|k|=1}^{\infty}\sum_{m=0}^{\infty}\delta_{k,n_k+2m},$$

we can apply Theorem 3.1 of Chapter 2.

Now let $z\in E_1$. Since the sets E^1_{kn} are pairwise disjoint, z belongs to only one of them; for the corresponding multiplier $\omega_{kn}(z)$, inequality (5.32) holds. The other sets E^1_{kn} do not contain z, even $z\in CD^2_{kn}\backslash E^2_{kn}$ holds, and the corresponding multipliers satisfy (5.33). Taking into account (5.21) we have (the product sign with a prime means that one of the terms is omitted)

$$\begin{aligned}|\omega(z)|&\le\frac{3}{2}\cdot\frac{5}{4}|\psi(z)|\prod_{|k|=1}^{\infty}{\prod_{m=0}^{\infty}}'(1+\delta_{k,n_k+2m})\\&\le\frac{15}{8}|\psi(z)|\prod_{|k|=1}^{\infty}\prod_{n=n_k}^{\infty}\frac{1+\delta_{kn}}{1-\delta_{kn}}\le 2|\psi(z)|,\end{aligned}$$

$$\begin{aligned}|\omega(z)|&\ge\frac{3}{2}\cdot\frac{3}{4}|\psi(z)|\prod_{|k|=1}^{\infty}{\prod_{m=0}^{\infty}}'(1-\delta_{k,nk+2m})\\&\ge\frac{9}{8}|\psi(z)|\prod_{|k|=1}^{\infty}\prod_{n=n_k}^{\infty}\frac{1-\delta_{kn}}{1+\delta_{kn}}\ge|\psi(z)|.\end{aligned}$$

Let us prove inequality (5.29). For each $z\in\{\operatorname{Re}z>-1\}\cap\{|z|\ge 2\}$ we choose a positive integer N such that $2^N\le|z|<2^{N+1}$. Then $z\in CD^2_{kn}\backslash E^2_{kn}$ for all sets of indices k and n, except possibly those for which $n=N-1,N,N+1$.

We divide the set of all pairs (k,n) of indices used in the construction of the set E_2 into two subsets: the set V_1 of those pairs for which $z\in CD^2_{kn}\backslash E^2_{kn}$, and the set V_2 of the remaining pairs. It is clear that by (5.33) and (5.21) we have

$$\prod_{(k,n)\in V_1}|\omega_{kn}(z)|\le\prod_{(k,n)\in V_1}(1+\delta_{kn})\le\prod_{|k|=1}^{\infty}\prod_{n=n_k}^{\infty}\frac{1+\delta_{kn}}{1-\delta_{kn}}\le\frac{16}{15}.$$

Now, if V_2 is non-empty, by (5.31) we get

$$\begin{aligned}\prod_{(k,n)\in V_2}|\omega_{kn}(z)|&\le\prod_{(k,n)\in V_2}\exp(\varepsilon_k^{1/2}2^{n\rho})\le\prod_{|k|=1}^{\infty}\prod_{n=N-1}^{N+1}\exp(\varepsilon_k^{1/2}2^{n\rho})\\&\le\prod_{|k|=1}^{\infty}\exp(3\varepsilon_k^{1/2}2^{(N+1)\rho})=\exp\left(3\cdot 2^{(N+1)\rho}\sum_{|k|=1}^{\infty}\varepsilon_k^{1/2}\right).\end{aligned}$$

Since

$$\sum_{|k|=1}^{\infty}\varepsilon_k^{1/2}=\sum_{|k|=1}^{\infty}\exp\left(-\frac{2^9C}{\eta_k}\right)\le\sum_{|k|=1}^{\infty}\frac{\eta_k}{2^9C}<\frac{\pi}{2^{10}}<\frac{1}{2^8},$$

we get the estimate

$$\begin{aligned}|\omega(z)|&\le\frac{3}{2}\cdot\frac{16}{15}\exp(2^{N\rho-5})<\exp(2^{N\rho-5}+1)\\&<\exp\{(2^N+1)^\rho\}<\exp\{(|z|+1)^\rho\}.\end{aligned}$$

The lemma is proved. □

Define a function $\varphi(z)$ on E_1 by the equality

$$\varphi(z) = a_{|k|}, \quad z \in E^1_{kn}.$$

LEMMA 5.4. *There exists a function $F(z)$ analytic in the half-plane $\{\operatorname{Re} z > 0\}$, such that*

$$|\varphi(z) - F(z)| < \frac{1}{2}|\psi(z)|, \quad z \in E_0, \tag{5.36}$$

$$|F(z)| \leq \exp\{D_1(|z|+1)^\rho\}, \quad \operatorname{Re} z > 0, \tag{5.37}$$

where $D_1 > 0$ is a constant.

PROOF. In view of (5.27), inequality (5.25) implies

$$\left|Q_1(\zeta, z) - \frac{1}{\zeta - z}\right| < \left|\frac{\psi(\zeta)}{\zeta}\right|^2, \quad \zeta \in \gamma^1_{kn}, \quad z \in CD^1_{kn}. \tag{5.38}$$

Observe that this inequality is valid, in particular, for $\zeta \in \Gamma_1$, $z \in E_0$, and also for $\zeta \in \Gamma_1$, $|z| \leq \frac{1}{4}|\zeta|$. Since for $\zeta \in \gamma^1_{kn}$ the inequality $|\zeta|^\rho \geq 2^{n\rho-1} \geq \varepsilon_k^{1/2} 2^{n\rho-1}$ holds, (5.26) implies

$$|Q_1(\zeta, z)| \leq \exp(|\zeta|^\rho), \quad \zeta \in \Gamma_1, \quad z \in \{\operatorname{Re} z > -1\}. \tag{5.39}$$

We set

$$\Gamma_{1n} = \bigcup_{n_k + 2m \leq n} \gamma^1_{k, n_k+2m}, \quad n \geq n_1.$$

Observe that Γ_{1n} is contained in the disc $\{|z| \leq \frac{9}{4}2^n\}$, $\Gamma_1 = \bigcup_{n=n_1}^{\infty} \Gamma_{1n}$, and $\Gamma_1 \backslash \Gamma_{1n}$ is contained in the complement of the disc $\{|z| < \frac{7}{8}2^n\}$.

Consider a sequence of functions analytic in the half-plane $\{\operatorname{Re} z > -1\}$:

$$H_n(z) = \frac{1}{2\pi i} \int_{\Gamma_{1n}} \frac{\varphi(\zeta)}{\omega(\zeta)} Q_1(\zeta, z)\, d\zeta.$$

We show that this sequence converges uniformly in each bounded region contained in $\{\operatorname{Re} z > 0\}$, and that the function $H(z) = \lim_{n\to\infty} H_n(z)$ satisfies the conditions

$$\left|H(z) - \frac{\varphi(z)}{\omega(z)}\right| < \frac{1}{4}, \quad z \in E_0; \tag{5.40}$$

$$|H(z)| \leq \exp\{D_2(|z|+1)^\rho\}, \quad \operatorname{Re} z > 0, \tag{5.41}$$

where $D_2 > 0$ is a constant.

Let N be such that $|z| \leq 2^N$ and $m > n > N + 2$. Since, by the Cauchy theorem,

$$0 = \frac{1}{2\pi i} \int_{\Gamma_{1m} \backslash \Gamma_{1n}} \frac{\varphi(\zeta)}{\omega(\zeta)} \frac{d\zeta}{\zeta - z},$$

we have

$$H_m(z) - H_n(z) = \frac{1}{2\pi i} \int_{\Gamma_{1m} \backslash \Gamma_{1n}} \frac{\varphi(\zeta)}{\omega(\zeta)} \left\{Q_1(\zeta, z) - \frac{1}{\zeta - z}\right\} d\zeta.$$

Using (5.28), (5.23), and (5.27) in succession, we get

$$\left|\frac{\varphi(\zeta)}{\omega(\zeta)}\right| \leq \frac{|a_{|k|}|}{|\psi(\zeta)|} \leq \frac{\exp(\varepsilon_k 2^{n\rho-1})}{|\psi(\zeta)|} \leq \frac{1}{|\psi(\zeta)|^2}, \quad \zeta \in \gamma^1_{kn}; \tag{5.42}$$

therefore, using (5.38), we get

$$(5.43)\qquad |H_m(z) - H_n(z)| \le \frac{1}{2\pi}\int_{\Gamma_{1m}\setminus\Gamma_{1n}} |\zeta^{-2}d\zeta| \le \frac{1}{2\pi}\int_{\Gamma_1\setminus\Gamma_{1n}} |\zeta^{-2}d\zeta|.$$

By (5.24), this implies that the sequence $\{H_n(z)\}$ converges uniformly in each bounded region contained in $\{\operatorname{Re} z > 0\}$.

Now, by the Cauchy formula we have

$$\frac{\varphi(z)}{\omega(z)} = \frac{1}{2\pi i}\int_{\Gamma_{1n}} \frac{\varphi(\zeta)}{\omega(\zeta)}\frac{d\zeta}{\zeta - z}, \qquad z \in E_0, \quad |z| \le 2^n;$$

therefore

$$H_n(z) - \frac{\varphi(z)}{\omega(z)} = \frac{1}{2\pi i}\int_{\Gamma_{1n}} \frac{\varphi(\zeta)}{\omega(\zeta)}\left\{Q_1(\zeta, z) - \frac{1}{\zeta - z}\right\} d\zeta, \qquad z \in E_0, \quad |z| \le 2^n.$$

Using (5.42), (5.38), and (5.24), we get

$$\left|H_n(z) - \frac{\varphi(z)}{\omega(z)}\right| \le \frac{1}{2\pi}\int_{\Gamma_{1n}} |\zeta^{-2}d\zeta| < \frac{1}{4}, \qquad z \in E_0, \quad |z| \le 2^n,$$

whence inequality (5.40) follows.

Let us prove (5.41). Letting $m \to +\infty$ in (5.43), we see that

$$|H(z) - H_n(z)| < \frac{1}{4}, \qquad |z| \le 2^N, \quad n > N + 2,$$

and, in particular,

$$(5.44)\qquad |H(z) - H_{N+3}(z)| < \frac{1}{4}, \qquad |z| \le 2^N.$$

Suppose that N is chosen in such a way that $2^{N-1} \le |z| < 2^N$. Using (5.42) and (5.39) we have

$$\begin{aligned}|H_{N+3}(z)| &= \left|\frac{1}{2\pi i}\int_{\Gamma_{1,N+3}} \frac{\varphi(\zeta)}{\omega(\zeta)} Q_1(\zeta, z)\, d\zeta\right| \\ &\le \frac{1}{2\pi}\int_{\Gamma_{1,N+3}} \left|\frac{1}{\psi(\zeta)}\right|^2 \exp(|\zeta|^\rho)\, |d\zeta| \\ &\le \frac{1}{2\pi}\max_{\zeta\in\Gamma_{1,N+3}} \left\{\left|\frac{\zeta}{\psi(\zeta)}\right|^2 \exp(|\zeta|^\rho)\right\} \int_{\Gamma_{1,N+3}} |\zeta^{-2}d\zeta| \\ &\le \frac{1}{2\pi}\max_{\zeta\in\Gamma_{1,N+3}} \{|\zeta|^2 \exp(3|\zeta|^\rho)\}.\end{aligned}$$

Since for $\zeta \in \Gamma_{1,N+3}$, $2^{N-1} \le |z| < 2^N$ the inequality

$$|\zeta| < \frac{9}{4}2^{N+3} < 36 \cdot 2^{N-1} \le 36|z|$$

holds, we have

$$|H_{N+3}(z)| \le \exp\{D_3(|z| + 1)^\rho\},$$

where $D_3 > 0$ is an absolute constant. Using this and (5.44) we get (5.41).

The function $F(z) = H(z)\omega(z)$ satisfies conditions (5.36) and (5.37). In fact, (5.40) implies

$$|F(z) - \varphi(z)| < \frac{1}{4}|\omega(z)|, \qquad z \in E_0,$$

from which, using (5.28), we get (5.36). The estimate (5.37) follows from (5.41) and (5.29). The lemma is proved. □

Theorem 5.1 follows from Lemmas 5.2 and 5.4. Let us show this. Let $F(z)$ be the function in Lemma 5.4. By Lemma 5.2 there exists an entire function $G(z)$ not exceeding normal type of order ρ, such that

$$G(z) = F(z) + g(z), \tag{5.45}$$

where $g(z)$ admits the estimate (5.20) in the angle $\{|\arg z| \leq \alpha\}$ for sufficiently large $|z|$. We show that each of the numbers $a_k \in \{a_k\}_{k=1}^{\infty}$ is a deficient value for the function $G(z)$.

It is clear that for sufficiently large $|z|$, the inequality $|g(z)| < \frac{1}{2}|\psi(z)|$ (for $|\arg z| < \alpha$) holds; therefore (5.45) and (5.36) imply

$$|G(z) - \varphi(z)| \leq |\psi(z)|, \qquad z \in E_0, \quad |z| > r_0. \tag{5.46}$$

Fix a positive integer k. If $r > 2^{n_k}$, the circle $\{|z| = r\}$ intersects the part of E_0 for which $\varphi(z) = a_k$ in one or two arcs of length $\frac{1}{2}\eta_k r$. For each of these arcs, by (5.46), the inequality (for $|z| > \max(2^{n_k}, r_0)$)

$$|G(z) - a_k| \leq \exp(-\varepsilon_k r^\rho \cos\rho\varphi) \leq \exp\left(-\frac{1}{2}\varepsilon_k r^\rho\right), \qquad z = re^{i\varphi}$$

holds. Therefore

$$\begin{aligned} m(r, a_k, G) &= \frac{1}{2\pi}\int_0^{2\pi} \ln^+ \frac{1}{|G(re^{i\varphi})_- a_k|} d\varphi \\ &> \frac{1}{2\pi}\frac{\eta_k}{2}\frac{1}{2}\varepsilon_k r^\rho > \frac{1}{8\pi}\eta_k\varepsilon_k r^\rho, \qquad r > \max(2^{n_k}, r_0). \end{aligned}$$

Since $T(r, G) \leq \ln^+ M(r, G) = O(r^\rho)$, we have $\delta(a_k, G) > 0$. The theorem is proved. □

6. Stability of deficiencies

Let $f(z)$ be a meromorphic function in the complex plane. All functions of the form $f(az + b)$, $a \neq 0$, map the complex plane onto the same Riemann surface F. Studying the asymptotic behavior of meromorphic functions given by the Riemann surfaces onto which they map their domains, it makes sense to study only those properties which are the same for all functions of the form $f(az + b)$. It is natural to require that the quantities characterizing the asymptotic behavior of $f(z)$ do not change when we pass to $f(az + b)$.

In Theorem 6.7 of Chapter 1 we described the changes in the main quantities of the Nevanlinna theory when we pass from $f(z)$ to $f(az)$, $a \neq 0$. We saw (Theorem 1.6 of Chapter 2) that $f(z)$ and $f(az)$ have the same growth category, and, as was noted at the end of Section 1, the sets E_P, E_B, E_N, E_V for the functions $f(z)$ and $f(az)$ are the same, and the Nevanlinna and the Valiron deficiencies are the same. Therefore it suffices to compare the functions $f(z)$ and $f(z + h)$.

We write $f(z + h) = f_h(z)$ and mark all quantities related to $f_h(z)$ with the subscript h; the absence of the subscript means that the quantity is related to $f(z)$. In Section 1 of Chapter 2 (see the proof of Theorem 1.6) we obtained the following inequalities:

$$n(r-2|h|,a)\le n_h(r-|h|,a)\le n(r,a)\quad (r>2|h|), \tag{6.1}$$

$$(1+o(1))N(r-2|h|,a)\le N_h(r-|h|,a)\le (1+o(1))N(r,a), \tag{6.2}$$

$$(1+o(1))T(r-2|h|,f)\le T_h(r-|h|,f)\le (1+o(1))T(r,f). \tag{6.3}$$

These inequalities imply that $n(r,a)$ and $n_h(r,a)$, $N(r,a)$ and $N_h(r,a)$, $T(r,f)$ and $T_h(r,f)$, respectively, have the same growth categories (Theorem 1.6 of Chapter 2). Therefore the sets of Borel exceptional values for $f(z)$ and $f(z+h)$ are the same. For the Picard exceptional values this is obvious. This makes the following example constructed by Dugué even more unexpected. This is an example of functions $f(z)$ and $f(z+h)$ with different deficiencies.

EXAMPLE 1 (Dugué). Let $\psi(z)=e^{e^z}-1$. The entire function $\psi(z)$ has simple zeros at points $z_{pq}=\ln(2\pi p)+i\left(\frac{\pi}{2}+\pi q\right)$, $p=1,2,\dots$, $q=0,\pm1,\pm2,\dots$. All these zeros are in the half-plane $\{\operatorname{Re} z>0\}$. Let us show that

$$T(r,\psi)=(1+o(1))\frac{e^r}{\sqrt{2\pi^3 r}},\qquad r\to\infty. \tag{6.4}$$

We have

$$\begin{aligned}T(r,\psi)&=T(r,e^{e^z})+O(1)=m(r,e^{e^z})+O(1)\\&=\frac{1}{2\pi}\int_0^{2\pi}\{\operatorname{Re} e^{re^{i\varphi}}\}^+d\varphi+O(1)\\&=\frac{1}{2\pi}\int_0^{2\pi}e^{r\cos\varphi}\{\cos(r\sin\varphi)\}^+d\varphi+O(1)\\&=\frac{1}{\pi}\int_0^{\pi/2}e^{r\cos\varphi}\{\cos(r\sin\varphi)\}^+d\varphi+O(1).\end{aligned}$$

Applying Lemma 4.1 with $\varphi(t)=\{\cos t\}^+$, and taking into account that

$$\lim_{r\to\infty}\frac{1}{r}\int_0^r\{\cos t\}^+dt=\frac{1}{\pi},$$

we get (6.4).

Applying the second main Nevanlinna inequality (2.8) of Chapter 3 to $\psi(z)$, we get the inequality

$$(1+o(1))T(r,\psi)<N(r,0,\psi)+N(r,-1,\psi)+N(r,\infty,\psi)=N(r,0,\psi),$$

which is valid outside some set of intervals of finite total length. Since the function $N(r,0,\psi)$ is increasing, the inequality is valid everywhere on the real line. In fact, if r is in an exceptional interval, then the left end of the interval is $r'=r+o(1)$, and

$$\begin{aligned}N(r,0,\psi)&\ge N(r',0,\psi)\ge(1+o(1))T(r',\psi)\\&=(1+o(1))\frac{e^{r'}}{\sqrt{2\pi^3r'}}=(1+o(1))\frac{e^r}{\sqrt{2\pi^3 r}}=(1+o(1))T(r,\psi).\end{aligned}$$

In addition, we have $N(r,0,\psi)\le T(r,\psi)+O(1)$; therefore

$$N(r,0,\psi)=(1+o(1))T(r,\psi). \tag{6.5}$$

Now consider the function

$$f(z) = \psi(z)/\psi(-z).$$

The function $f(z)$ has simple zeros at the points z_{pq} and simple poles at the points $-z_{pq}$; therefore $N(r,0,f) = N(r,\infty,f) = N(r,0,\psi)$.

Observe that for $x > 0$ the inequality

$$|\psi(-z)| \le 1 + |e^{e^{-z}}| = 1 + e^{e^{-x}\cos y} \le 1 + e, \qquad z = x + iy,$$

holds, and that $|\psi(z)| \le 1 + e$ for $x \le 0$. Therefore

$$\begin{aligned} m(r,f) &\ge \frac{1}{2\pi}\int_{-\pi/2}^{\pi/2} \ln^+ \left| \frac{\psi(re^{i\varphi})}{\psi(-re^{i\varphi})} \right| d\varphi \\ &\ge \frac{1}{2\pi}\int_{-\pi/2}^{\pi/2} \ln^+ \frac{|\psi(re^{i\varphi})|}{1+e}\, d\varphi = \frac{1}{2\pi}\int_{-\pi/2}^{\pi/2} \ln^+ |\psi(re^{i\varphi})|\, d\varphi + O(1) \\ &= m(r,\psi) + O(1) = T(r,\psi) + O(1). \end{aligned} \tag{6.6}$$

Now, by (6.5) and (6.6), we get

$$T(r,f) \le T(r,\psi(z)) + T(r,\psi(-z)) + O(1) = 2T(r,\psi) + O(1)$$

and

$$T(r,f) = m(r,f) + N(r,f) \ge (1 + o(1))2T(r,\psi).$$

Hence

$$T(r,f) = (2 + o(1))T(r,\psi),$$

$$\Delta(0,f) = \delta(0,f) = 1 - \lim_{r\to\infty} \frac{N(r,0,f)}{T(r,f)} = 1 - \lim_{r\to\infty} \frac{N(r,0,\psi)}{2T(r,\psi)} = \frac{1}{2},$$

$$\Delta(\infty,f) = \delta(\infty,f) = 1 - \lim_{r\to\infty} \frac{N(r,\infty,\psi)}{2T(r,\psi)} = \frac{1}{2}.$$

Let h be a non-zero real number. Consider the function $f_h(z) = f(z+h)$. It can be investigated in the same way as before. First of all,

$$\begin{aligned} T(r,\psi_h(z)) &= T(r, e^{e^{z+h}}) + O(1) = \frac{1}{2\pi}\int_{-\pi}^{\pi} \{\text{Re } e^{h+re^{i\varphi}}\}^+ d\varphi + O(1) \\ &= e^h \frac{1}{2\pi}\int_{-\pi}^{\pi} \{\text{Re } e^{re^{i\varphi}}\}^+ d\varphi + O(1) = e^h T(r, e^{e^z}) + O(1). \end{aligned}$$

In the same way as we got (6.5) we obtain

$$N(r,0,\psi_h(z)) = (1 + o(1))T(r,\psi_h(z)), \qquad r \to \infty.$$

Thus

$$\begin{aligned} N(r,0,f_h) &= N(r,0,\psi_h(z)) = (1 + o(1))e^h T(r, e^{e^z}), \\ N(r,\infty,f_h) &= N(r,0,\psi(-z-h)) = N(r,0,\psi_{-h}(-z)) \\ &= N(r,0,\psi_{-h}(z)) = (1 + o(1))e^{-h} T(r, e^{e^z}). \end{aligned}$$

Also,

$$T(r,f_h) \le T(r,\psi_h(z)) + T(r,\psi_{-h}(-z)) + O(1) = (e^h + e^{-h})T(r, e^{e^z}) + O(1).$$

Since for $\text{Re}\, z > 0$ the inequality $|\psi(-z-h)| \le 1 + e^{e^{-h}}$ holds, and for $\text{Re}\, z \le 0$ we have $|\psi(z+h)| \le 1 + e^{e^h}$, in the same way as we proved (6.6) we get $m(r, f_h) \ge T(r, \psi_h) + O(1)$ and $T(r, f_h) = m(r, f_h) + N(r, f_h) \ge (1+o(1))(e^h + e^{-h})T(r, e^{e^z})$. Thus

$$T(r, f_h) = (1+o(1))(e^h + e^{-h})T(r, e^{e^z}), \qquad r \to \infty.$$

Now we easily get

$$\Delta(0, f_h) = \delta(0, f_h) = \frac{e^{-h}}{e^h + e^{-h}},$$

$$\Delta(\infty, f_h) = \delta(\infty, f_h) = \frac{e^h}{e^h + e^{-h}}.$$

If $h = 0$, we have the same formulas as before.

Thus, we see that the deficiencies of the function $f(z+h)$ at 0 and ∞ are different from the deficiencies of the function $f(z)$. This could be shown in a more elementary way, following Dugué, but for us it is important to note that in the presented example the Nevanlinna and the Valiron deficiencies coincide for 0 and for ∞.

In Dugué's example the function $f(z)$ is of infinite order. Besides, although the deficiencies $\delta_h(0)$ and $\delta_h(\infty)$ change as h varies, they remain positive. We now give an example of a meromorphic function $f(z)$ having the maximal type of order one, for which $\delta(\infty) = 0$ but $\delta_{-2}(\infty) = 1$. We start with investigating some auxiliary functions.

Consider the entire function

$$g(z) = g(z; r_k, n_k) = \prod_{k=1}^{\infty} \left\{1 - \left(\frac{z}{r_k}\right)^{n_k}\right\}, \tag{6.7}$$

where the n_k are arbitrary positive integers, and the r_k are arbitrary positive numbers satisfying the conditions $r_{k+1} \ge 2r_k$, $r_1 > 1$. Let us show that $m(r, 0, g) \le 4$ for each such function. In fact, $g(0) = 1$ implies $T(r, g) = m(r, 0, g) + N(r, 0, g)$. On the other hand,

$$\begin{aligned} T(r, g) \le \ln M(r, g) &\le \sum_{k=1}^{\infty} \ln \left\{1 + \left(\frac{r}{r_k}\right)^{n_k}\right\} \\ &= \left(\sum_{r_k > r} + \sum_{r_k \le r}\right) \ln \left\{1 + \left(\frac{r}{r_k}\right)^{n_k}\right\} \\ &\le \sum_{r_k > r} \left(\frac{r}{r_k}\right)^{n_k} + \sum_{r_k \le r} n_k \ln \frac{r}{r_k} + \sum_{r_k \le r} \left(\frac{r_k}{r}\right)^{n_k} \\ &\le \sum_{k=0}^{\infty} 2^{-k} + N(r, 0, g) + \sum_{k=0}^{\infty} 2^{-k} = N(r, 0, g) + 4. \end{aligned}$$

Hence, $m(r, 0, g) \le 4$.

Consider now the meromorphic function

$$\begin{aligned} \Phi(z) = \Phi(z; a_j, b_j, m_j, \sigma_j, \varepsilon) &= \prod_{j=1}^{\infty} \varphi_j(z), \\ \varphi_j(z) &= \left(\frac{z - b_j}{z - a_j}\right)^{m_j}, \end{aligned} \tag{6.8}$$

where $0 < \varepsilon < \frac{1}{2}$; $\{a_j\}$ is an arbitrary sequence of distinct complex numbers approaching ∞; $|a_j| > 1$; the m_j are arbitrary positive integers; σ_j is a sequence of positive numbers such that the discs $\{|z - a_j| < \sigma_j\}$ are pairwise disjoint and

$$\sigma_j < 1, \qquad \sum_{j=1}^{\infty} m_j \sigma_j < \infty; \tag{6.9}$$

complex numbers b_j are close to a_j, so that for $|z - a_j| \geq \sigma_j$ the inequalities $|\varphi_j(z) - 1| < \varepsilon 2^{-j-1}$ and $|1/\varphi_j(z) - 1| < \varepsilon 2^{-j-1}$ hold. It is clear that under these conditions the product in (6.8) is absolutely and uniformly convergent in each disc on the complex plane. Moreover, for $z \notin \bigcup_j \{|z - a_j| < \sigma_j\}$, we have

$$\begin{aligned} |\Phi(z) - 1| &= \left| \prod_{j=1}^{\infty} \{1 + (\varphi_j(z) - 1)\} - 1 \right| \\ &\leq \max \left[\left\{ \prod_{j=1}^{\infty} \left(1 + \frac{\varepsilon}{2^{j+1}}\right) - 1 \right\}, \left\{ 1 - \prod_{j=1}^{\infty} \left(1 - \frac{\varepsilon}{2^{j+1}}\right) \right\} \right] \\ &\leq \max \left[e^{\frac{\varepsilon}{2}} - 1, 1 - e^{-\frac{8}{7}\frac{\varepsilon}{2}} \right] < \varepsilon. \end{aligned} \tag{6.10}$$

We have used the inequality $\ln(1 - x) \geq -\frac{x}{1-x} \geq -\frac{8}{7}x$ valid for $0 < x < \frac{1}{8}$. Inequality (6.10) implies, in particular, that $|\Phi(z)| \leq 2$ for

$$z \notin \bigcup_j \{|z - a_j| < \sigma_j\}.$$

By the maximum modulus principle,

$$|\Phi(z)| \leq 2 \left(\frac{\sigma_j}{|z - a_j|} \right)^{m_j}$$

for each z in the disc $\{|z - a_j| < \sigma_j\}$. Therefore $(a_j = |a_j| e^{i\alpha_j})$,

$$\begin{aligned} m(r, \Phi) &\leq O(1) + \sum_{j=1}^{\infty} \frac{1}{2\pi} \int_{\substack{|z-a_j|<\sigma_j \\ |z|=r}} \ln \left\{ 2 \left(\frac{\sigma_j}{|z - a_j|} \right)^{m_j} \right\} d\varphi \\ &\leq O(1) + \sum_{j=1}^{\infty} \frac{m_j}{2\pi} \int_{-\pi}^{\infty} \ln^+ \frac{\sigma_j}{|re^{i\varphi} - |a_j| e^{i\alpha_j}|} d\varphi. \end{aligned}$$

On the other hand,

$$\begin{aligned} \int_{-\pi}^{\pi} \ln^+ \frac{\sigma_j}{|re^{i\varphi} - |a_j| e^{i\alpha_j}|} d\varphi &= \int_{\alpha_j - \frac{\pi}{2}}^{\alpha_j + \frac{\pi}{2}} \ln^+ \frac{\sigma_j}{|r - |a_j| e^{i(\alpha_j - \varphi)}|} d\varphi \\ &\leq \int_{\alpha_j - \frac{\pi}{2}}^{\alpha_j + \frac{\pi}{2}} \ln^+ \frac{\sigma_j}{|a_j| |\sin(\alpha_j - \varphi)|} d\varphi \leq \int_{-\pi/2}^{\pi/2} \ln^+ \frac{\sigma_j}{|\sin\theta|} d\theta \\ &\leq 2 \int_0^{\pi/2} \ln^+ \frac{2\sigma_j}{\pi\theta} d\theta = 2 \int_0^{2\sigma_j/\pi} \ln \frac{2\sigma_j}{\pi\theta} d\theta = \frac{4\sigma_j}{\pi}. \end{aligned}$$

Therefore, by (6.9),

$$m(r, \Phi) \leq O(1) + \frac{2}{\pi^2} \sum_{j=1}^{\infty} m_j \sigma_j = O(1). \tag{6.11}$$

In the same way we show that

$$m(r,1/\Phi)=O(1). \tag{6.12}$$

Relations (6.11) and (6.12) imply

$$N(r,\infty,\Phi)=N(r,0,\Phi)+O(1). \tag{6.13}$$

We assume that the zeros of $g(z)$ and the poles of $\Phi(z)$ are distinct. Then, by (6.11), (6.12), (6.13), we easily get that for $f(z)=g(z)\Phi(z)$,

$$m(r,0,f)\le m(r,0,g)+m(r,0,\Phi)=O(1), \tag{6.14}$$

$$N(r,\infty,f)=N(r,\infty,\Phi)=\sum_{j=1}^{\infty} m_j\ln^+\frac{r}{|a_j|}, \tag{6.15}$$

$$\begin{aligned} T(r,f)&=N(r,0,f)+m(r,0,f)+O(1)\\ &=N(r,0,g)+N(r,0,\Phi)+O(1)=N(r,0,g)+N(r,\infty,\Phi)+O(1)\\ &=\sum_{k=1}^{\infty} n_k\ln^+\frac{r}{r_k}+\sum_{j=1}^{\infty} m_j\ln^+\frac{r}{|a_j|}+O(1). \end{aligned} \tag{6.16}$$

Since there is much freedom in the choice of n_k, r_k, m_j, a_j from the definition of $f(z)$, this class of functions is suitable for constructing various examples.

EXAMPLE 2. Choose the sequence $\{r_k\}$ to be rapidly increasing, so that

$$\sum_{k=1}^{\kappa-1} r_k(\ln r_k)^4<\ln r_\kappa.$$

Let $a_k=r_k-1$, $n_k=[r_k(\ln r_k)^4]$, $m_k=[r_k(\ln r_k)^3]$. Then, by (6.16),

$$\begin{aligned} T(r_\kappa,f)&=\sum_{k=1}^{\kappa-1} r_k(\ln r_k)^4\ln\frac{r_\kappa}{r_k}+\sum_{k=1}^{\kappa} r_k(\ln r_k)^3\ln\frac{r_\kappa}{r_k-1}+O(\ln r_\kappa)\\ &=O(\ln^2 r_\kappa)+r_\kappa(\ln r_\kappa)^3\ln\frac{r_\kappa}{r_\kappa-1}=(1+o(1))\ln^3 r_\kappa. \end{aligned}$$

By (6.15),

$$N(r_\kappa,\infty,f)=\sum_{k=1}^{\kappa} r_k(\ln r_k)^3\ln\frac{r_\kappa}{r_k-1}+O(\ln r_\kappa)=(1+o(1))\ln^3 r_\kappa.$$

Hence, $N(r_\kappa,\infty,f)\sim T(r_\kappa,f)$ and $\delta(\infty,f)=0$.

Now we consider the function $f_{-2}(z)=f(z-2)$. Let $r_\kappa+1\le r<r_{\kappa+1}+1$. Then

$$N(r,\infty,f_{-2})=\sum_{k=1}^{\kappa}[r_k(\ln r_k)^3]\ln\frac{r}{r_k+1}\le\sum_{k=1}^{\kappa} r_k(\ln r_k)^3\ln\frac{r}{r_k-1}. \tag{6.17}$$

Observe that for sufficiently large r_k, $k\ge\kappa_0$, the measure (in radians) of the arc $\{|z|=r_k\}$ contained in the disc $\{|z-2|<r_k-1\}$ exceeds $\pi/2$ (the measure approaches $2\pi/3$ as $k\to\infty$). Estimating $N(r,0,g_{-2})$ from below we count only those zeros of $g(z)$ which are in the angle $\{|\arg z-\pi|<\pi/4\}$. Then

$$\begin{aligned} T(r,f_{-2})&\ge N(r,0,f_{-2})+O(1)\ge N(r,0,g_{-2})+O(1)\\ &\ge\sum_{k=\kappa_0}^{\kappa}\frac{1}{4}r_k(\ln r_k)^4\ln\frac{r}{r_k-1}+O(\ln r). \end{aligned} \tag{6.17'}$$

Fix some $\kappa_1 > \kappa_0$. Then from (6.17) and (6.17′) we get, for $r_\kappa + 1 \le r < r_{\kappa+1} + 1$,

$$N(r, \infty, f_{-2}) \le \sum_{k=\kappa_1}^{\kappa} r_k (\ln r_k)^3 \ln \frac{r}{r_k - 1} + O(\ln r),$$

$$\begin{aligned} T(r, f_{-2}) &\ge \frac{1}{4} \sum_{k=\kappa_1}^{\kappa} r_k (\ln r_k)^4 \ln \frac{r}{r_k - 1} + O(\ln r) \\ &\ge \frac{1}{4} \ln r_{\kappa_1} \sum_{k=\kappa_1}^{\kappa} r_k (\ln r_k)^3 \ln \frac{r}{r_k - 1} + O(\ln r) \\ &\ge \frac{1}{4} (\ln r_{\kappa_1}) N(r, \infty, f_{-2}) + O(\ln r). \end{aligned}$$

Hence

$$\limsup_{r\to\infty} \frac{N(r, \infty, f_{-2})}{T(r, f_{-2})} \le \frac{4}{\ln r_{\kappa_1}}.$$

Letting $\kappa_1 \to \infty$ in this inequality, we get

$$\limsup_{r\to\infty} \frac{N(r, \infty, f_{-2})}{T(r, f_{-2})} = 0,$$

that is, $\delta(\infty, f_{-2}) = 1$.

It is easy to show that the functions $n(r, 0, g)$ and $n(r, \infty, \Phi)$ have the maximal type of order one, and, by (6.16), the function $f(z)$ also has the maximal type of order one.

EXAMPLE 3. If we let $n_k = [r_k (\ln r_k)^3]$, $m_k = [r_k (\ln r_k)^4]$ in the previous example, leaving all other parameters unchanged, we get an example with varying Valiron deficiency: $\Delta(\infty, f(z)) = 0$, $\Delta(\infty, f(z-2)) = 1$. We leave details of this example to interested readers.

Observe that in Example 2 we have $\Delta(\infty, f) = 1 > \delta(\infty, f) = 0$, but if, before passing to the limit, we excluded from consideration the values of r in the intervals $[r_k - 1, r_k + 1]$, we would have $\delta(\infty, f) = 1$. The situation is similar for $f(z-2)$ in Example 3. This is a fundamental difference between Examples 2 and 3, and Dugué's Example 1. Examples 2 and 3 show how we should "correct" the definition of deficiency to get its shift invariance for meromorphic functions of finite order.

Let E be a measurable set, $E \subset (1, \infty)$, $\chi(r, E)$ the indicator function of this set, and $CE = (1, \infty) \backslash E$. We write $E \in L$ if

$$\limsup_{r\to\infty} \frac{1}{\ln r} \int_1^r \chi(t, E)\, dt < \infty. \tag{6.18}$$

It is easy to show that for $E \in L$,

$$\int_1^\infty \chi(t, E) t^{-\sigma} dt < \infty$$

for each $\sigma > 0$. In fact, if (6.18) is satisfied, then $\int_1^r \chi(t, E)\, dt = O(\ln r)$, and therefore

$$\begin{aligned} \int_1^r \chi(t, E) t^{-\sigma} dt &= r^{-\sigma} \int_1^r \chi(t, E)\, dt + \sigma \int_1^r \frac{dt}{t^{1+\sigma}} \int_1^t \chi(\tau, E)\, d\tau \\ &= r^{-\sigma} O(\ln r) + \sigma \int_1^r \frac{O(\ln t)\, dt}{t^{1+\sigma}} = O(1). \end{aligned}$$

In particular ($\sigma = 1$), all sets of the class L have finite logarithmic measure (see the definition in Exercise 2 of Chapter 3, Section 1). Observe also that if $E' \in L$ and $E'' \in L$, then $E' \cup E'' \in L$, $E' \cap E'' \in L$.

Now we introduce the following definitions (all limits are taken as $r \to \infty$):

$$\delta^*(a) = \sup_{CE \in L} \liminf_{r \in E} \frac{m(r,a)}{T(r,f)} = 1 - \inf_{CE \in L} \limsup_{r \in E} \frac{N(r,a)}{T(r,f)}, \tag{6.19}$$

$$\Delta^*(a) = \inf_{CE \in L} \limsup_{r \in E} \frac{m(r,a)}{T(r,f)} = 1 - \sup_{CE \in L} \liminf_{r \in E} \frac{N(r,a)}{T(r,f)}.$$

It is easy to check that in Example 2 the following relations hold: $\delta(\infty) = 0$, $\Delta(\infty) = 1$, $\delta_{-2}(\infty) = \Delta_{-2}(\infty) = 1$, $\delta^*(\infty) = \Delta^*(\infty) = 1$, $\delta^*_{-2}(\infty) = \Delta^*_{-2}(\infty) = 1$.

THEOREM 6.1. *For meromorphic functions of finite order, the equalities*

$$\delta_h^*(a) = \delta^*(a), \qquad \Delta_h^*(a) = \Delta^*(a)$$

are satisfied for each h.

PROOF. We give the proof for $\delta^*(a)$; the proof for $\Delta^*(a)$ is similar. In this section we use the spherical characteristic $T(r,f)$, without reflecting this in our notation. Denote by M_ε the set of intervals on which $\frac{d}{dr} \ln T(r) < \varepsilon$, $\varepsilon > 0$. Let us show that $CM_\varepsilon \in L$. In fact, let $\Delta_r = CM_\varepsilon \cap (1,r)$. Then

$$\ln T(r) - \ln T(1) \geq \int_{\Delta_r} d\ln T(r) \geq \varepsilon \int_{\Delta_r} dr = \varepsilon \int_1^r \chi(t, CM_\varepsilon)\, dt.$$

Since the function $f(z)$ has finite order ρ, we have

$$\limsup_{r \to \infty} \frac{1}{\ln r} \int_1^r \chi(t, CM_\varepsilon)\, dt \leq \frac{\rho}{\varepsilon} < \infty,$$

that is, $CM_\varepsilon \in L$. By (6.2) and (6.3) we get

$$\frac{N_h(r - |h|, a)}{T_h(r - |h|)} \leq (1 + o(1)) \frac{N(r,a)}{T(r - 2|h|)}. \tag{6.20}$$

Since $T(r)$ is a convex function of $\ln r$, for $r > 4|h|$, $0 < \theta = \theta(r) < 1$, we have

$$\begin{aligned} T(r) &= T(r - 2|h|) + \frac{dT(r - \theta 2|h|)}{d\ln r} \ln \frac{r}{r - 2|h|} \\ &\leq T(r - 2|h|) + \frac{dT(r)}{d\ln r} \frac{2|h|}{r - 2|h|} \\ &= T(r - 2|h|) + T'(r) \frac{2|h| r}{r - 2|h|}. \end{aligned} \tag{6.21}$$

If $r \in M_\varepsilon$, $\varepsilon < (8|h|)^{-1}$, then (6.21) implies

$$\begin{aligned} \frac{T(r)}{T(r - 2|h|)} &\leq 1 + \frac{2|h| r}{r - 2|h|} \frac{T'(r)}{T(r)} \frac{T(r)}{T(r - 2|h|)} \\ &\leq 1 + 4|h| \varepsilon \frac{T(r)}{T(r - 2|h|)}, \end{aligned} \tag{6.22}$$

$$\frac{T(r)}{T(r - 2|h|)} \leq \frac{1}{1 - 4|h|\varepsilon} < 1 + 8|h|\varepsilon.$$

By (6.20) and (6.22) we get, for $r \in M_\varepsilon$,

$$\frac{N_h(r-|h|,a)}{T_h(r-|h|)} \le (1+o(1))(1+8|h|\varepsilon)\frac{N(r,a)}{T(r)}.$$

Let E_ε be an arbitrary set satisfying $E_\varepsilon \subset M_\varepsilon$ and $CE_\varepsilon \in L$. Then

$$\limsup_{r\in E_\varepsilon} \frac{N_h(r-|h|,a)}{T_h(r-|h|)} \le (1+8|h|\varepsilon)\limsup_{r\in E_\varepsilon}\frac{N(r,a)}{T(r)},$$

$$\inf_{\varepsilon,E_\varepsilon}\limsup_{r\in E_\varepsilon} \frac{N_h(r-|h|,a)}{T_h(r-|h|)} \le \inf_{\varepsilon,E_\varepsilon}\limsup_{r\in E_\varepsilon}\frac{N(r,a)}{T(r)} \tag{6.23}$$

(the infimum is taken over all E_ε for all $0<\varepsilon<(8|h|)^{-1}$). But

$$\inf_{\varepsilon,E_\varepsilon}\limsup_{r\in E_\varepsilon}\frac{N}{T} = \inf_{CE\in L}\limsup_{r\in E}\frac{N}{T}. \tag{6.24}$$

In fact, on one hand,

$$\inf_{CE\in L}\limsup_{r\in E}\frac{N}{T} \le \inf_{\substack{E\subset M_\varepsilon\\ CE\in L}}\limsup_{r\in E}\frac{N}{T} = \inf_{\varepsilon,E_\varepsilon}\limsup_{r\in E_\varepsilon}\frac{N}{T}.$$

On the other hand, if $CE\in L$, then $C(E\cap M_\varepsilon) = CE\cup CM_\varepsilon \in L$; therefore

$$\inf_{CE\in L}\limsup_{r\in E}\frac{N}{T} \ge \inf_{CE\in L}\limsup_{r\in E\cap M_\varepsilon}\frac{N}{T} = \inf_{\varepsilon,E_\varepsilon}\limsup_{r\in E_\varepsilon}\frac{N}{T}.$$

By (6.23) and (6.24) we get

$$\inf_{CE\in L}\limsup_{r\in E}\frac{N_h(r-|h|,a)}{T_h(r-|h|)} \le \inf_{CE\in L}\limsup_{r\in E}\frac{N(r,a)}{T(r)},$$

that is, $\delta^*(a,f(z+h)) \ge \delta^*(a,f(z))$.

Applying this relation to $f(z+h)$ and $f((z+h)-h)$, we obtain

$$\delta^*(a,f(z)) \ge \delta^*(a,f(z+h))$$

and hence

$$\delta^*(a,f(z)) = \delta^*(a,f(z+h)). \qquad \square$$

REMARK. We could define the deficiency in such a way that it would be shift invariant for functions (not necessarily of finite order) satisfying

$$\lim_{r\to\infty} r^{-1}\ln T(r,f) = 0. \tag{6.25}$$

It suffices to define the deficiency in the same way as $\delta^*(a)$, but with L consisting of sets E with zero relative measure, that is, satisfying

$$\lim_{r\to\infty}\frac{1}{r}\int_1^r \chi(t,E)\,dt = 0.$$

The Dugué example shows that if we drop the restriction (6.25), the shift invariance of deficiency cannot be achieved no matter what sets (even the whole interval $(1,\infty)$ without a sequence $r_n\to\infty$) we would exclude from consideration.

We preferred to define $\delta^*(a)$ and $\Delta^*(a)$ as above, since in the value distribution theory there are many results which are valid for r outside intervals of finite logarithmic measure, whereas consideration of larger exceptional sets would require significant extensions of the Nevanlinna theory.

THEOREM 6.2. *The deficiencies of meromorphic functions satisfy the relation*

$$\sum_a \delta^*(a) \leq 2.$$

PROOF. Consider arbitrary distinct points a_j, $j = 1, 2, \ldots, q$. By the definition of $\delta^*(a)$, for each $\varepsilon > 0$ there exists a set E_j such that CE_j has finite logarithmic measure and $m(r, a_j)/T(r, f) \geq \delta^*(a_j) - \varepsilon$ for $r \in E_j$, $j = 1, \ldots, q$. The set $E = \bigcap_{j=1}^q E_j$ is non-empty because the set CE has finite logarithmic measure. By Theorem 2.1 of Chapter 3 the inequalities

$$\sum_{j=1}^q \frac{m(r, a_j)}{T(r, f)} \leq 2 + \frac{O(\ln r) + O(\ln T(r, f))}{T(r, f)},$$

$$\sum_{j=1}^q \delta^*(a_j) \leq 2 + \varepsilon q + \frac{O(\ln r) + O(\ln T(r, f))}{T(r, f)}$$

hold on E, since the intervals on which the inequalities are not satisfied have finite measure, and without loss of generality we may assume that they are contained in CE.

Letting $r \to \infty$ and taking into account that ε can be chosen arbitrarily, we get the inequality

$$\sum_{j=1}^q \delta^*(a_j) \leq 2;$$

hence

$$\sum_a \delta^*(a) \leq 2. \qquad \square$$

THEOREM 6.3. *The following relation holds:*

$$\delta(a) \leq \delta^*(a) \leq \Delta^*(a) \leq \Delta(a). \tag{6.26}$$

PROOF. Only the middle inequality must be verified; the other two follow from the definitions of $\delta^*(a)$ and $\Delta^*(a)$. Consider a set E_1 such that $CE_1 \in L$ and

$$\delta^*(a) \geq \liminf_{r \in E_1} \frac{m(r, a)}{T(r, f)} > \delta^*(a) - \varepsilon,$$

and a set E_2 such that $CE_2 \in L$ and

$$\Delta^*(a) \leq \limsup_{r \in E_2} \frac{m(r, a)}{T(r, f)} < \Delta^*(a) + \varepsilon.$$

Then

$$\begin{aligned} \delta^*(a) - \varepsilon < \liminf_{r \in E_1} \frac{m(r, a)}{T(r, f)} &\leq \liminf_{r \in E_1 \cap E_2} \frac{m(r, a)}{T(r, f)} \\ &\leq \limsup_{r \in E_1 \cap E_2} \frac{m(r, a)}{T(r, f)} \leq \limsup_{r \in E_2} \frac{m(r, a)}{T(r, f)} < \Delta^*(a) + \varepsilon. \end{aligned}$$

Since ε can be chosen to be arbitrarily small, we get $\delta^*(a) \leq \Delta^*(a)$. $\square$

The inequality (6.26) implies that $\delta^*(a)$ is a more precise characteristic of the distribution of exceptional a-points than the Nevanlinna deficiency $\delta(a)$. The following result is also a corollary of (6.26).

THEOREM 6.4. *If* $\delta(a, f) = \Delta(a, f)$, *then* $\delta(a, f) = \delta^*(a, f) = \Delta^*(a, f)$.

Together with Theorem 6.1 this result implies that for functions of finite order the equality $\delta(a,f)=\Delta(a,f)$ implies $\delta_h^*(a,f)=\Delta_h^*(a,f)=\delta(a,f)$. For functions of infinite order this implication does not hold, as is shown in Example 1.

Observe also, that it may happen that $\delta^*(a)>\delta_h(a)$ for each h satisfying $0\le|h|<\infty$. To show this it is enough to modify Example 2 as follows. In addition to $a_k=r_k-1$ we introduce

$$a_k'=-r_k+1,\quad a_k''=i(r_k-1),\quad a_k'''=-i(r_k-1),\quad m_k'=m_k''=m_k'''=m_k.$$

The modified function $\Phi(z)$ is

$$\Phi(z)=\prod_{k=1}^{\infty}\left\{\frac{(z-b_k)(z-b_k')(z-b_k'')(z-b_k''')}{(z-a_k)(z-a_k')(z-a_k'')(z-a_k''')}\right\}^{m_k},$$

where b_k,b_k',b_k'',b_k''' are sufficiently close to a_k,a_k',a_k'',a_k'''; the rest of the construction remains unchanged. Repeating the argument from the example with minor modifications, we get that $\delta_h(\infty)=0$ for each h whereas $\delta^*(\infty)=1$.

Our next result distinguishes a rather wide class of functions for which $\delta^*(a)=\delta(a)$.

THEOREM 6.5. *If the difference between the order and the lower order for a meromorphic function $f(z)$ satisfies $\rho-\lambda<1$, then*

$$\delta^*(a)=\delta(a),\qquad \Delta^*(a)=\Delta(a)\tag{6.27}$$

for all a. The equalities (6.27) hold also for all meromorphic functions whose growth does not exceed the normal type of order one.

COROLLARY. *If the conditions of Theorem 6.5 are satisfied, then $\delta_h(a)=\delta(a)$ for each h; that is, the deficiencies are shift invariant.*

PROOF. To prove $\delta^*(a)=\delta(a)$ and $\Delta^*(a)=\Delta(a)$, it suffices to show that for each $0<K<\infty$ and each R, $r<R<r+K\ln r$, the inequality

$$(1+o(1))\frac{N(r,a)}{T(r,f)}\le\frac{N(R,a)}{T(R,f)}\le(1+o(1))\frac{N(r+K\ln r,a)}{T(r+K\ln r,f)}\tag{6.28}$$

holds, since this inequality shows that the exclusion from consideration of sets of class L does not affect the value of $\limsup\frac{N}{T}$ and $\liminf\frac{N}{T}$. On the other hand,

$$\frac{T(r,f)}{T(r+K\ln r,f)}\frac{N(r,a)}{T(r,f)}\le\frac{N(R,a)}{T(R,f)}\le\frac{N(r+K\ln r,a)}{T(r+K\ln r,f)}\frac{T(r+K\ln r,f)}{T(r,f)}.$$

Therefore in order to prove (6.28) it is enough to show that

$$\lim_{r\to\infty}\frac{T(r+K\ln r)}{T(r)}=1.\tag{6.29}$$

We have

$$\frac{T(r+K\ln r)}{T(r)}=1+\frac{T'(r+\theta(r)K\ln r)}{T(r)}K\ln r,$$

where $0\le\theta(r)\le1$. Assume that (6.29) does not hold. Then there exists a sequence $r_n\to\infty$ such that

$$\frac{T'(r_n')}{T(r_n)}>\frac{\eta}{\ln r_n},\qquad \eta>0,\quad r_n'=r_n+\theta(r_n)K\ln r_n.$$

Since

$$\frac{d}{dr}rT'(r) = \overset{\circ}{A}{}'(r) > 0,$$

the function $rT'(r)$ is an increasing function, and therefore

$$T'(r) \geq \frac{r'_n T'(r'_n)}{r}, \qquad r \geq r'_n,$$

whence

$$\int_{r'_n}^{er'_n} T'(r)\,dr \geq \int_{r'_n}^{er'_n} \frac{r'_n T'(r'_n)}{r}\,dr,$$

$$T(er'_n) \geq r'_n T'(r'_n) + T(r'_n) > \frac{r'_n \eta}{\ln r_n} T(r_n). \tag{6.30}$$

For all $n \geq n_0(\varepsilon)$, we have $r_n \leq r'_n < 2r_n$ and $T(r_n) > r_n^{\lambda-\varepsilon}$, $\varepsilon > 0$.

From (6.30) we get

$$T(6r_n) > \frac{r_n^{1+\lambda-\varepsilon}}{\ln r_n}.$$

Hence, $\rho \geq 1 + \lambda - \varepsilon$ and, letting $\varepsilon \to 0$, we get $\rho \geq 1 + \lambda$. Hence, if (6.27) does not hold, then $\rho - \lambda \geq 1$. Since for each transcendental meromorphic function $\lim_{r\to\infty} \frac{T(r)}{\ln r} = \infty$, the inequality (6.30) implies

$$\lim_{n\to\infty} \frac{T(6r_n)}{6r_n} = \infty;$$

that is, if (6.27) does not hold, then $f(z)$ has growth of at least maximal type of order one. □

Examples 2, 3, and other examples based on the same ideas (we leave the constructions of such examples to interested readers) show that the assumptions of Theorem 6.5 cannot be significantly weakened. Nevertheless some additions to it can be made. For example, (6.30) immediately implies

THEOREM 6.6. *Let $f(z)$ be a meromorphic function satisfying $\rho = 1 + \lambda$. If one of the following conditions is satisfied:*

(a) $\lim_{r\to\infty} T(r)/r^\lambda \ln r = \infty$ *and* $\limsup_{r\to\infty} T(r)/r^\rho < \infty$,

(b) $\liminf_{r\to\infty} T(r)/r^\lambda > 0$ *and* $T(r) = o(r^\rho (\ln r)^{-1})$,

then (6.27) *holds.*

Shift invariance of deficiencies holds under weaker assumptions.

THEOREM 6.7. *Let $f(z)$ be a meromorphic function satisfying $\rho = 1 + \lambda$. If one of the following conditions is satisfied:*

(a) $\lim_{r\to\infty} T(r)/r^\lambda = \infty$ *and* $\limsup_{r\to\infty} T(r)/r^\rho < \infty$,

(b) $\liminf_{r\to\infty} T(r)/r^\lambda > 0$ *and* $T(r) = o(r^\rho)$,

then for all h,

$$\delta_h(a) = \delta(a), \qquad \Delta_h(a) = \Delta(a). \tag{6.31}$$

The proof is similar to the proof of Theorem 6.5. For (6.31) to be satisfied it is enough that for each K, $0 < K < \infty$, the equality

$$\lim_{r\to\infty} \frac{T(r+K)}{T(r)} = 1 \tag{6.32}$$

holds. Clearly, it is enough to show that

$$\limsup_{r\to\infty}\frac{T(r+1)}{T(r)}=1. \tag{6.33}$$

But $T(r+1)=T(r)+T'(r+\theta(r))$, $0\le\theta(r)\le 1$. If (6.33) does not hold, then there exists a sequence $r_n\to\infty$ such that

$$\frac{T'(r_n')}{T(r_n)}>\eta,\qquad \eta>0,$$

where $r_n'=r_n+\theta(r_n)$. As before, we get that $T(er_n')\ge r_n'T'(r_n')+T(r_n')$; therefore

$$T(er_n')>\eta r_n' T(r_n). \tag{6.34}$$

Theorem 6.7 immediately follows from inequality (6.34).

Note that the conditions of Theorem 6.7 may be satisfied even when $\delta^*(a)>\delta(a)$ or $\Delta^*(a)<\Delta(a)$. We only give an example for which condition (a) of Theorem 6.7 is satisfied and $\delta^*(\infty)>\delta(\infty)$.

EXAMPLE 4. Let $\lambda>0$, $\rho=\lambda+1$. Consider the equation

$$b^\rho\ln\frac{r}{b}=r^\lambda(\ln r)^{1/8},\qquad b>2^{2\lambda+8}. \tag{6.35}$$

Taking into account the fact that the left-hand side of the equation is a linear function of $\ln r$ and the right-hand side is convex with respect to $\ln r$, it is easy to see that equation (6.35) has two solutions in $[b,\infty)$, and these solutions, denoted by $R_1(b)$ and $R_2(b)$, $R_1(b)<R_2(b)$, satisfy $b<R_1(b)<2b$, $\ln R_2(b)\sim\frac{\rho}{\lambda}\ln b$ as $b\to\infty$. If $R_1(b)<r<R_2(b)$, the left-hand side of (6.35) is greater than the right-hand side.

We take the function $g(z;r_k,n_k)$ (see (6.7)), with parameters r_k and n_k chosen as follows. Let $r_1>1$ be large so that $R_2(b)\ge 2b$ for each $b\ge r_1$. Set $r_{k+1}=R_2(r_k)$, $k=1,2,\dots$, and $n_k=[r_k^\rho]$. Consider the meromorphic function $\Phi(z;a_j,b_j,m_j,\sigma_j,\varepsilon)$ (see (6.8)), where $a_k=r_k-\sqrt{\ln r_k}$, $m_k=\left[r_k^\rho(\ln r_k)^{-\frac14}\right]$. Let $f(z)=g(z)\Phi(z)$. By (6.16) we have $T(r,f)=N(r,0,g)+N(r,\infty,\Phi)+O(1)$. It is easy to see that $n(r,0,g)\le(1-2^{-\rho})^{-1}r^\rho$ and hence $N(r,0,g)=O(r^\rho)$ and $N(r,\infty,\Phi)=o(r^\rho)$. Thus, $T(r,f)=O(r^\rho)$. For $r\in[r_k,r_{k+1})$, taking into account (6.35), we get

$$\begin{aligned}T(r,f)\ge N(r,0,g)+O(1)&\ge r_{k-1}^\rho\ln\frac{r}{r_{k-1}}+r_k^\rho\ln\frac{r}{r_k}+O(\ln r)\\&\ge r_{k-1}^\rho\ln\frac{r_k}{r_{k-1}}+r_k^\rho\ln\frac{r}{r_k}+O(\ln r)\\&=r_k^\lambda(\ln r_k)^{1/8}+r_k^\rho\ln\frac{r}{r_k}+O(\ln r).\end{aligned}$$

For $r_k\le r<2r_k$ the inequality

$$T(r,f)\ge r_k^\lambda(\ln r_k)^{1/8}+O(\ln r)$$

holds, and for $2r_k\le r<r_{k+1}$ we have

$$T(r,f)\ge r_k^\rho\ln\frac{r}{r_k}+O(\ln r)\ge r^\lambda(\ln r)^{1/8}+O(\ln r).$$

Hence

$$\liminf_{r\to\infty}\frac{T(r,f)}{r^\lambda(\ln r)^{1/8}}>0,$$

and the function $f(z)$ satisfies the condition (a) of Theorem 6.7. Using (6.15) we get

$$N(r_k,\infty,f)=N(r_k,\infty,\Phi)\sim\sum_{j=1}^{k}r_j^{\rho}(\ln r_j)^{-\frac14}\ln\frac{r_k}{r_j-\sqrt{\ln r_j}}$$
$$\geq r_k^{\rho}(\ln r_k)^{-\frac14}\ln\frac{r_k}{r_k-\sqrt{\ln r_k}}\sim r_k^{\lambda}(\ln r_k)^{\frac14},$$

$$N(r_k,0,g)\leq\sum_{j=1}^{k}r_j^{\rho}\ln\frac{r_k}{r_j}=\sum_{j=1}^{k-1}r_j^{\rho}\ln\frac{r_k}{r_j}$$
$$=O\left(r_{k-1}^{\rho}\ln\frac{r_k}{r_{k-1}}\right)=O(r_k^{\lambda}(\ln r_k)^{1/8}),$$

$$T(r_k,f)=N(r_k,0,g)+N(r_k,\infty,\Phi)+O(1)\sim N(r_k,\infty,\Phi)=N(r_k,\infty,f).$$

Hence $\delta(\infty,f)=0$ and, by Theorem 6.7, $\delta_h(\infty,f)=0$ for all h.

Now let $r_k+\sqrt{\ln r_k}\leq r\leq r_{k+1}-\sqrt{\ln r_{k+1}}$. Then

$$N(r,\infty,f)\leq\sum_{j=1}^{k}r_j^{\rho}(\ln r_j)^{-\frac14}\ln\frac{r}{r_j-\sqrt{\ln r_j}}=O\left(r_k^{\rho}(\ln r_k)^{-\frac14}\ln\frac{r}{r_k-\sqrt{\ln r_k}}\right),$$

$$N(r,0,g)\geq r_k^{\rho}\ln\frac{r}{r_k}.$$

Since for $r_k+\sqrt{\ln r_k}\leq r\leq r_{k+1}-\sqrt{\ln r_{k+1}}$, the inequality

$$\frac{\ln\frac{r}{r_k-\sqrt{\ln r_k}}}{\ln\frac{r}{r_k}}\leq\frac{\ln\frac{r_k+\sqrt{\ln r_k}}{r_k-\sqrt{\ln r_k}}}{\ln\frac{r_k+\sqrt{\ln r_k}}{r_k}}\to 2,\qquad k\to\infty,$$

holds, we have $N(r,\infty,f)=o(N(r,0,g))$ as $r\to\infty$ if we exclude from $[0,\infty)$ the set $E=\bigcup_{k=1}^{\infty}\{|r-r_k|<\sqrt{\ln r_k}\}$. It is easy to see that $E\in L$. Since for $r\in CE$ the relation $N(r,\infty,f)=o(N(r,0,g))=o(T(r,f))$ holds, we get $\delta^*(\infty,f)=1$, that is, $\delta^*(\infty,f)>\delta(\infty,f)$.

REMARK 1. In many situations the easiest way to verify that $\delta^*(a)=\delta(a)$ is to check that condition (6.33) is satisfied.

REMARK 2. It is clear that for fixed a the statements of Theorems 6.5, 6.6, and 6.7 remain true if the conditions of the theorems are imposed on $N(r,a)$ instead of $T(r,f)$, for example, if in Theorem 6.5 we require

$$\rho[N(r,a)]-\lambda[N(r,a)]<1.$$

In fact, $N(r,a)$ is also an increasing function convex with respect to the logarithm, and our proof is based on these properties of $T(r,f)$ only.

The passage from $f(z)$ to $f(z+h)$ can be regarded as a passage from the exhaustion of the plane by the circles $\{|z|=r\}$ to the exhaustion of the plane by the circles $\{|z+h|=r\}$. We can generalize this problem and consider the exhaustion of the complex z-plane by a family γ_r of curves which are the images of the family of circles $\{|z|=r\}$ under a homeomorphic mapping of the complex plane onto itself. We assume, for simplicity, that for $r\leq r_0$ we have $\gamma_r=\{|z|=r\}$. Let D_r be the closed region bounded by the curve γ_r. Denote by $n_\gamma(r,a)$ the number of a-points of the function $f(z)$ in D_r, and by $\pi\overset{\circ}{A}_\gamma(r,f)$ the (measured in the spherical metric) area of the piece of the Riemann surface onto which the function $w = f(z)$ maps D_r. After this the quantities $N_\gamma(r,a)$, $T_\gamma(r,f)$, $\delta_\gamma(a,f)$, $\Delta_\gamma(a,f)$, $\delta^*_\gamma(a,f)$, and $\Delta^*_\gamma(a,f)$ are introduced in the usual manner. Let $r_1(r) = \min_{z\in\gamma_r}|z|$, $r_2(r) = \max_{z\in\gamma_r}|z|$.

THEOREM 6.8. *If*

$$Ar - B < r_1(r) \le r_2(r) < Ar + B, \qquad 0 < A, B < \infty,$$

then $\rho[T_\gamma(r)] = \rho$, $\lambda[T_\gamma(r)] = \lambda$ *and*

(1) *If* $\rho < \infty$, *then* $\delta^*_\gamma(a) = \delta^*(a)$, $\Delta^*_\gamma(a) = \Delta^*(a)$;
(2) *If* $\rho - \lambda < 1$, *then* $\delta^*_\gamma(a) = \delta_\gamma(a)$, $\Delta^*_\gamma(a) = \Delta_\gamma(a)$.

Without loss of generality we may assume that $A = 1$. Analysis of the proofs of Theorems 6.1 and 6.5, and also of Theorem 1.6 in Chapter 2 shows that we did not use the fact that the exhaustion is done by concentric circles, but only the fact that $\{|z+h| = r\}$ lies between $\{|z| = r + |h|\}$ and $\{|z| = r - |h|\}$. Therefore the same argument works in the present situation.

THEOREM 6.9. *Let*

$$\lim_{r\to\infty} \frac{r_2(r)}{r_1(r)} = 1. \tag{6.36}$$

If there exist constants K_1 *and* K_2 *such that* $(0 < \rho < \infty)$

$$0 < K_1 \le r^{-\rho} T(r,f) \le K_2 < \infty, \qquad r \ge r_0, \tag{6.37}$$

then $\delta_\gamma(a) = \delta(a)$ *and* $\Delta_\gamma(a) = \Delta(a)$ *for all* a.

PROOF. It is easy to see (compare with the proof of Theorem 6.5) that to establish $\delta_\gamma(a) = \delta(a)$ and $\Delta_\gamma(a) = \Delta(a)$ it suffices to show that for each function $\varepsilon(r) > 0$ satisfying $\lim_{r\to\infty}\varepsilon(r) = 0$, we have

$$\lim_{r\to\infty} \frac{T(r(1+\varepsilon(r)))}{T(r)} = 1. \tag{6.38}$$

Furthermore,

$$\frac{T(r(1+\varepsilon(r)))}{T(r)} = 1 + \frac{T'(r(1+\theta(r)\varepsilon(r)))}{T(r)} r\varepsilon(r),$$

where $0 \le \theta(r) \le 1$.

If (6.38) is not satisfied, then there exist a function $\varepsilon_1(r) > 0$, $\varepsilon_1(r) \to 0$ as $r \to \infty$, and a sequence $r_n \uparrow \infty$, $r_n \ge r_0$, such that

$$\frac{T'(r'_n)}{T(r_n)} > \frac{\eta}{r_n \varepsilon_1(r_n)}, \quad \eta > 0; \qquad r'_n = r_n(1 + \theta(r_n)\varepsilon_1(r_n)) < 2r_n.$$

Therefore, arguing in the same way as in the proof of (6.30), we get

$$T(er'_n) > r'_n T'(r'_n) + T(r'_n) > \frac{r'_n \eta}{r_n \varepsilon(r_n)} T(r_n) \ge \frac{\eta}{\varepsilon(r_n)} T(r_n) > \frac{\eta}{\varepsilon(r_n)} K_1 r_n^\rho. \tag{6.39}$$

Since $r'_n < 2r_n$, we get $T(er'_n) < K_2(2e)^\rho r_n^\rho$, a contradiction with (6.39).

REMARK. If condition (6.36) is satisfied and, furthermore, $r_1(r) \sim Kr$, $0 < K < \infty$, then we may use $T_\gamma(r,f)$ instead of $T(r,f)$ in the definition of the order and the type of a function $f(z)$, and also replace $T(r,f)$ by $T_\gamma(r,f)$ in condition

(6.37) of Theorem 6.9. In fact, the inequality $\mathring{A}(r_1(r)) \le \mathring{A}_\gamma(r) \le \mathring{A}(r_2(r))$ implies that for $r > r_0(\varepsilon)$,

$$\mathring{A}((K-\varepsilon)r) \le \mathring{A}_\gamma(r) \le \mathring{A}((K+\varepsilon)r), \qquad K-\varepsilon > 0,$$

from which

$$\int_{r_0(K-\varepsilon)}^{r(K-\varepsilon)} \frac{\mathring{A}(t)}{t}\,dt \le \int_{r_0}^{r} \frac{\mathring{A}_\gamma(t)}{t}\,dt \le \int_{r_0(K+\varepsilon)}^{r(K+\varepsilon)} \frac{\mathring{A}(t)}{t}\,dt,$$

$$T((K-\varepsilon)r) + O(1) \le T_\gamma(r) \le T((K+\varepsilon)r) + O(1). \tag{6.40}$$

Inequality (6.40) immediately implies that $T(r)$ and $T_\gamma(r)$ have the same order and the same type, and also, that the following inequality is a necessary and sufficient condition for (6.37):

$$0 < K_1' \le r^{-\rho} T_\gamma(r) \le K_2' < \infty, \ r \ge r_0'. \tag{6.41}$$

The proof of the theorem is complete. □

Condition (6.37) of Theorem 6.9 in general cannot be weakened. We give only one example of this, showing that if K_2 is a function of r, $K_2 = K_2(r)$, arbitrarily slowly approaching ∞ as $r \to \infty$, then it can happen that $\delta_\gamma(a) \ne \delta(a)$. An analogous example with $K_1 = K_1(r) \to 0$ can be constructed in a similar way.

EXAMPLE 5. Without loss of generality we may assume that $K_2(r) > 1$ is an increasing slowly varying function. Take the function

$$w = F(\zeta) = g(\zeta; r_k, n_k)\Phi(\zeta; a_k, b_k, m_k, \sigma_k, \varepsilon)$$

(see (6.7), (6.8)) with $\varepsilon < 1/2$, $r_k = 2^k$, $a_k = r_k - 2$, and

$$n_k = m_k = \begin{cases} [2^{\rho k}], & k \ne k_i, \quad i = 1, 2, \dots, \\ [2^{\rho k} K_2(2^k)], & k = k_i, \end{cases}$$

where the choice of $k_i\,(\ge 4)$ will be specified later; now we require only that $K_2(r_{k_i}) > 64$.

Independently of the choice of the sequence $\{k_i\}$ the inequalities

$$n(r, 0, g) \ge (2^\rho - 1)^{-1} r^\rho,$$
$$n(r, \infty, \Phi) \ge (2^\rho - 1)^{-1} r^\rho$$

hold. Therefore, taking into account (6.16), we get $T(r, f) \ge \{\rho(2^\rho - 1)\}^{-1} r^\rho + O(1)$. On the other hand,

$$n(r, 0, g) = O(r^\rho K_2(r)),$$
$$n(r, \infty, \Phi) = O(r^\rho K_2(r)),$$
$$T(r, f) = O(r^\rho K_2(r)).$$

It is now clear that the sequence $\{k_i\}$ can be chosen to be rapidly increasing so that

$$N(r_{k_i}, 0, g) \le C r_{k_i}^\rho, \tag{6.42}$$

where C is a constant greater than $2^\rho\{\rho(2^\rho - 1)\}^{-1}$, because if $n_k = [r_k^\rho]$ for all k, then we would have

$$\limsup_{r\to\infty} \frac{n(r, 0, g)}{r^\rho} \le \frac{2^\rho}{2^\rho - 1}.$$

It is easy to check that $N(r,0,g) \sim N(r,\infty,\Phi)$ and, using formulas (6.14), (6.15), (6.16), to get that $\delta(\infty, F(\zeta)) = \Delta(\infty, F(\zeta)) = 1/2$, and $\delta(0, F(\zeta)) = \Delta(0, F(\zeta)) = 0$.

Consider the strips

$$S_i = \left\{ r_{k_i} - 1 > \operatorname{Re}\zeta > r_{k_i}\left(1 - K_2^{-\frac{1}{2}}(r_{k_i})\right) - 1, |\operatorname{Im}\zeta| < \eta_i \right\},$$

where the sequence $\{\eta_i\}$ is chosen to be rapidly converging to 0 so that the area (in the spherical metric) of the regions on the Riemann surface corresponding to $\bigcup_{i=1}^{\infty} S_i$ under the mapping $w = F(\zeta)$ is finite. Now we consider a homeomorphism $z = z(\zeta)$ of the complex plane onto itself satisfying the following conditions: $z(\zeta) = \zeta$ outside $\bigcup_{i=1}^{\infty} S_i$, and on S_i the mapping $z(\zeta)$ leaves the boundary of S_i fixed and maps the point $r_{k_i} - 2$ to the point $r_{k_i}(1 - K_2^{-\frac{1}{2}}(r_{k_i}))$. We denote the inverse of the mapping $z(\zeta)$ by $\zeta(z)$. Let γ_r be the image of $\{|z| = r\}$ under $\zeta(z)$. It is easy to compute that

$$N_\gamma(r_{k_i}, \infty, F) \sim r_{k_i}^{\rho} K_2(r_{k_i}) \ln\left(1 - K_2^{-\frac{1}{2}}(r_{k_i})\right)^{-1} \sim r_{k_i}^{\rho}\sqrt{K_2(r_{k_i})},$$

$$T_\gamma(r_{k_i}, F) \sim N(r_{k_i}, 0, F) \sim O(r_{k_i}^{\rho}) + r_{k_i}^{\rho} K_2(r_{k_i}) \ln\left(1 - K_2^{-\frac{1}{2}}(r_{k_i})\right)^{-1}$$

$$\sim r_{k_i}^{\rho}\sqrt{K_2(r_{k_i})}.$$

Hence $\delta_\gamma(\infty, F) = \delta_\gamma(0, F) = 0$ and $\delta_\gamma(\infty, F) < \delta(\infty, F)$.

Note that it is still unknown whether the deficiencies can be non-shift-invariant for entire functions of finite order. For entire functions of infinite order such an example was constructed by Hayman [**Hay53**]. We reproduce this example.

EXAMPLE 6. Consider the entire function

$$f(z) = \prod_{n=1}^{\infty} \left\{1 + \left(\frac{z}{n}\right)^{3n}\right\}^{2^n}. \tag{6.43}$$

Since for $|z| \le r$ we have

$$\sum_{n=1}^{\infty} 2^n \left(\frac{|z|}{n}\right)^{3n} \le \sum_{n=1}^{\infty} \left(\frac{2r^3}{n^3}\right)^n < \infty,$$

the infinite product in (6.43) is absolutely and uniformly convergent in each disc $\{|z| \le r\}$, and hence represents an entire function.

Since for $n \to \infty$ we have

$$\left(\frac{n + \frac{1}{2}}{n+1}\right)^{3(n+1)} \to e^{-\frac{3}{2}}, \qquad \left(\frac{n}{n + \frac{1}{2}}\right)^{3n} \to e^{-\frac{3}{2}},$$

and

$$\ln\left(1 - e^{-\frac{3}{2}}\right) > -0.3, \qquad e^{-\frac{3}{2}} < 0.3,$$

there exists $N_0 > 1$ such that for $n \ge N_0$,

$$\left|\ln\left\{1 - \left(\frac{n}{n+\frac{1}{2}}\right)^{3n}\right\}\right| < 0.3, \tag{6.44}$$

$$\left(\frac{n+\frac{1}{2}}{n+1}\right)^{3(n+1)} < 0.3, \tag{6.45}$$

$$\left(\frac{1}{n+\frac{1}{2}}\right)^3 < \left(\frac{n}{n+\frac{1}{2}}\right)^{3n}, \tag{6.46}$$

$$\ln\left(\frac{n+\frac{1}{2}}{k}\right) > \frac{8}{9}\frac{n+\frac{1}{2}-k}{k}, \qquad k = n-2,\ n-1,\ n. \tag{6.47}$$

Now we estimate $|f(z)|$ on the circle $\{|z| = n+\frac{1}{2}\}$, $n \geq N_0$. We have

$$\begin{aligned}\ln|f(z)| &= \sum_{k=1}^{\infty} 2^k \ln\left|1+\left(\frac{z}{k}\right)^{3k}\right| \\ &= \sum_{k=1}^{n} 2^k 3k \ln\frac{|z|}{k} + \sum_{k=1}^{n} 2^k \ln\left|1+\left(\frac{k}{z}\right)^{3k}\right| + \sum_{k=n+1}^{\infty} 2^k \ln\left|1+\left(\frac{z}{k}\right)^{3k}\right| \\ &= \sum_{k=1}^{n} 2^k 3k \ln\frac{n+\frac{1}{2}}{k} + \Phi_1(z) + \Phi_2(z) \\ &= N\left(n+\frac{1}{2}, 0, f\right) + \Phi_1(z) + \Phi_2(z).\end{aligned} \tag{6.48}$$

Taking into account the equality $\sum_{k=0}^{\infty} kx^k = x(1-x)^{-2}$, $|x|<1$, we get

$$\begin{aligned}N\left(n+\frac{1}{2}, 0, f\right) &= \sum_{k=1}^{n} 2^k 3k \ln\left(1+\frac{n+\frac{1}{2}-k}{k}\right) \\ &< \sum_{k=1}^{n} 2^k 3k \frac{n+\frac{1}{2}-k}{k} = 3\cdot 2^n \sum_{\nu=0}^{n-1}\left(\nu+\frac{1}{2}\right)2^{-\nu} \\ &< 3\cdot 2^n \sum_{\nu=0}^{\infty}\left(\nu+\frac{1}{2}\right)2^{-\nu} = 9\cdot 2^n.\end{aligned} \tag{6.49}$$

On the other hand, taking into account (6.47) we get

$$\begin{aligned}N\left(n+\frac{1}{2}, 0, f\right) &> \sum_{k=n-2}^{n} 2^k 3k \ln\frac{n+\frac{1}{2}}{k} > \frac{8}{3}\sum_{k=n-2}^{n} 2^k\left(n+\frac{1}{2}-k\right) \\ &= \frac{8}{3}\left(\frac{1}{2}+\frac{1}{2}\cdot\frac{3}{2}+\frac{1}{4}\cdot\frac{5}{2}\right)2^n = 5\cdot 2^n.\end{aligned} \tag{6.50}$$

An elementary investigation of the function $\left(\frac{x}{n+\frac{1}{2}}\right)^{3x}$ shows that it decreases on the interval $\left[1, \left(n+\frac{1}{2}\right)e^{-1}\right]$ and increases on the interval $\left[\left(n+\frac{1}{2}\right)e^{-1}, \infty\right)$. Taking into account inequality (6.46), we see that this function attains its maximum on the interval $[1, n]$ at the point $x = n$, and its minimum value, on the interval $[n+1, \infty)$, at the point $x = n+1$. Therefore

$$\left(\frac{k}{n+\frac{1}{2}}\right)^{3k} \leq \left(\frac{n}{n+\frac{1}{2}}\right)^{3n}, \qquad 1 \leq k \leq n,$$

$$\left(\frac{n+\frac{1}{2}}{k}\right)^{3k} \leq \left(\frac{n+\frac{1}{2}}{n+1}\right)^{3(n+1)} < 0.3, \qquad k \geq n+1, \tag{6.51}$$

and (6.44) implies the following inequality:

$$\left|\ln\left\{1-\left(\frac{k}{n+\frac{1}{2}}\right)^{3k}\right\}\right| < 0.3, \qquad 1 \le k \le n. \tag{6.52}$$

Now we estimate $|\Phi_1(z)|$ for $|z| = n+\frac{1}{2}$. Using (6.52) and the fact that for $0 < r < 1$, $0 \le \varphi < 2\pi$ the inequality

$$|\ln|1+re^{i\varphi}|| \le |\ln(1-r)| \le \frac{r}{1-r} \tag{6.53}$$

holds, we get

$$\begin{aligned}|\Phi_1(z)| &\le \sum_{k=1}^{n} 2^k \left|\ln\left|1+\left(\frac{k}{z}\right)^{3k}\right|\right| \\ &\le \sum_{k=1}^{n} 2^k \left|\ln\left\{1-\left(\frac{k}{n+\frac{1}{2}}\right)^{3k}\right\}\right| < 0.3\sum_{k=1}^{n} 2^k < 2^n \cdot 0.6.\end{aligned} \tag{6.54}$$

Now we turn to an estimate of $|\Phi_2(z)|$ for $|z| = n+1/2$. Using (6.51) and (6.53), we get

$$\begin{aligned}|\Phi_2(z)| &\le \sum_{k=n+1}^{\infty} 2^k \left|\ln\left|1+\left(\frac{z}{k}\right)^{3k}\right|\right| \\ &\le \sum_{k=n+1}^{\infty} 2^k \frac{\left(\frac{n+\frac{1}{2}}{k}\right)^{3k}}{1-\left(\frac{n+\frac{1}{2}}{k}\right)^{3k}} < \frac{1}{0.7}\sum_{k=n+1}^{\infty} 2^k\left(\frac{n+\frac{1}{2}}{k}\right)^{3k} = \frac{1}{0.7}\sum_{k=n+1}^{\infty} u_k.\end{aligned}$$

For $k \ge n+1$ we have

$$\begin{aligned}\frac{u_{k+1}}{u_k} &= 2^{k+1}\left(\frac{n+\frac{1}{2}}{k+1}\right)^{3(k+1)} \frac{1}{2^k}\left(\frac{k}{n+\frac{1}{2}}\right)^{3k} \\ &= 2\left(\frac{n+\frac{1}{2}}{k+1}\right)^3 \left\{\left(1+\frac{1}{k}\right)^k\right\}^{-3} < 2\left\{\left(1+\frac{1}{k}\right)^k\right\}^{-3} < \frac{1}{4},\end{aligned}$$

hence

$$u_k \le \frac{1}{4^{k-n-1}} u_{n+1}.$$

Therefore, by (6.45),

$$\begin{aligned}|\Phi_2(z)| &\le \frac{1}{0.7}\sum_{k=n+1}^{\infty} \frac{1}{4^{k-n-1}} u_{n+1} = \frac{1}{0.7}\cdot\frac{4}{3}u_{n+1} \\ &= \frac{1}{0.7}\cdot\frac{4}{3}2^{n+1}\left(\frac{n+\frac{1}{2}}{n+1}\right)^{3(n+1)} < \frac{1}{0.7}\cdot\frac{4}{3}2^{n+1}\cdot 0.3 = \frac{8}{7}\cdot 2^n < 1.2\cdot 2^n.\end{aligned} \tag{6.55}$$

Relations (6.48)–(6.50), (6.54), and (6.55) imply that on the circle $\{|z| = n+\frac{1}{2}\}$, $n \ge N_0$,

$$3\cdot 2^n < 3.2\cdot 2^n < \ln|f(z)| < 10.8\cdot 2^n < 11\cdot 2^n. \tag{6.56}$$

Inequality (6.56) implies that $|f(z)| > 1$ on the circles $\{|z| = n+\frac{1}{2}\}$; therefore, $m(n+\frac{1}{2}, 0, f) = 0$ and $\delta(0, f) = 0$.

Now we prove that $\delta_1(0, f) > 0$. Let $0 < \eta < \frac{1}{4}e^{-22}$, and let $z_0 = ne^{i\theta}$ be a zero of the function $f(z)$ of order 2^n, $n \geq N_0$. Then the inequality

$$\ln|f(z)| < 2^{n-1}\ln\eta \tag{6.57}$$

holds in the disc $\{|z - z_0| < \eta\}$.

In fact, since $\{|z - z_0| \leq \frac{1}{2}\} \subset \{|z| \leq n + \frac{1}{2}\}$, we conclude using (6.56) that $\ln|f(z)| \leq 11 \cdot 2^n$ in the disc $\{|z - z_0| \leq \frac{1}{2}\}$. On the circle $\{|z - z_0| = \frac{1}{2}\}$ we have

$$\left|\frac{f(z)}{(z - z_0)^{2^n}}\right| \leq \frac{e^{11\cdot 2^n}}{2^{-2^n}} = e^{11\cdot 2^n} 2^{2^n};$$

therefore, applying the maximum modulus principle for the disc $\{|z - z_0| < \eta\}$, we get

$$|f(z)| \leq e^{11\cdot 2^n} 2^{2^n} |z - z_0|^{2^n} < (e^{11}2\eta)^{2^n} \leq \left(\frac{1}{\sqrt{\eta}}\eta\right)^{2^n} = \eta^{2^{n-1}},$$

which is equivalent to (6.57).

Now let r be an arbitrary number exceeding $N_0 + 1$, and let $p = p_r$ be an integer satisfying $r - 1 < p < r + 1$. The circles $\{|z| = p\}$ and $\{|z - 1| = r\}$ intersect; therefore there exists φ_0 such that $|pe^{i\varphi_0} - 1| = r$. We show that for $|\varphi - \varphi_0| \leq \frac{1}{4}\eta$ the inequality

$$||pe^{i\varphi} - 1| - r| < \frac{1}{2}\eta \tag{6.58}$$

holds. In fact, by the Lagrange theorem, we get

$$||pe^{i\varphi} - 1| - r| = \left|\frac{d}{d\varphi}|pe^{i\varphi} - 1|\right|_{\varphi = \varphi'} |\varphi - \varphi_0|,$$

where φ' is a number between φ and φ_0. But

$$\left|\frac{d}{d\varphi}|pe^{i\varphi} - 1|\right| = \left|\frac{d}{d\varphi}\sqrt{p^2 - 2p\sin\varphi + 1}\right| = \left|\frac{-p\cos\varphi}{|pe^{i\varphi} - 1|}\right| \leq \frac{p}{p - 1} < 2;$$

therefore

$$||pe^{i\varphi} - 1| - r| < 2|\varphi - \varphi_0| \leq \frac{1}{2}\eta.$$

Since the circle $\{|z| = p\}$ contains $3p$ zeros of the function $f(z)$ (of order 2^p each), and these zeros divide the circle into arcs of $2\pi/3p$ radians each, the arc $\{|z| = p, |\arg z - \varphi_0| \leq \frac{1}{4}\eta\}$ contains at least

$$\nu = \left[\frac{\eta}{2}\frac{3p}{2\pi} - 1\right] = \left[\frac{3\eta}{4\pi}p - 1\right], \qquad p \geq \frac{8\pi}{3\eta},$$

zeros $z_j = pe^{i\varphi_j}$, $\varphi_1 < \varphi_2 < \cdots < \varphi_\nu$, $|\varphi_j - \varphi_0| \leq \frac{1}{4}\eta$. Since each zero z_j satisfies (6.58), we can find a point $z_j' = 1 + re^{i\theta_j}$ such that $|z_j' - z_j| < \frac{1}{2}\eta$. Denote the set of arcs $C_r^j = \{z - 1 = re^{i\theta}, |\theta - \theta_j| < \eta/(2r)\}$, $1 \leq j \leq \nu$, by C_r. The inequality $|z - z_j'| = |(z - 1) - (z_j' - 1)| = |re^{i\theta} - re^{i\theta_j}| = r|e^{i\theta} - e^{i\theta_j}| < r|\theta - \theta_j| < \frac{1}{2}\eta$ holds on C_r^j; therefore $|z - z_j| \leq |z - z_j'| + |z_j' - z_j| < \eta$. The arcs C_r^j are pairwise disjoint. In fact, if there is a point z in the intersection of C_r^j and C_r^{j+1}, then $|z_j - z_{j+1}| < |z_j - z| + |z - z_{j+1}| < 2\eta$. On the other hand,

$$\begin{aligned}|z_j - z_{j+1}| &= |pe^{i\varphi_j} - pe^{i\varphi_{j+1}}| = p|1 - e^{i(\varphi_{j+1} - \varphi_j)}| \\ &= p|1 - e^{2\pi i/3p}| \geq p\sin\frac{2\pi}{3p} \geq p\frac{2}{\pi}\cdot\frac{2\pi}{3p} = \frac{4}{3},\end{aligned}$$

and we get a contradiction. Thus the total measure of $C_r = \bigcup_{j=1}^{\nu} C_r^j$ in radians is

$$\frac{\eta}{r}\nu = \frac{\eta}{r}\left[\frac{3\eta}{4\pi}p - 1\right].$$

Since $p/r \to 1$ as $r \to \infty$, the measure approaches $3\eta^2/(4\pi)$ as $r \to \infty$, and is greater than $\eta^2/(2\pi)$ if $r > r_0(\eta) = r_0$.

Now we consider the function $f(z+1)$ on the circle $\{|z| = r\}$, $r > r_0$. Let $n = [r + 3/2]$. Then for $|z| = r$ the inequality $|z+1| \leq n + 1/2$ holds. By (6.56), for $\{|z| = r\}$ we have

$$\ln|f(z+1)| < 11 \cdot 2^n$$

and

$$T(r, f(z+1)) \leq \ln^+ M(r, f(z+1)) < 11 \cdot 2^n. \tag{6.59}$$

On the other hand, the circle $\{|z+1| = n-1\}$ intersects the circle $\{|z| = r\}$. It is clear that $r - 1 < n - 1 < r + 1$, and we can use the estimates presented above. For each arc C_r^j there exists a zero $z_j = (n-1)e^{i\varphi_j}$ of the function $f(z)$ such that $|z - z_j| < \eta$ if $z \in C_r^j$. Therefore by (6.57), for $z \in C_r$, the inequality

$$\ln|f(1 + re^{i\theta})| = \ln|f(z)| < 2^{n-2}\ln\eta \tag{6.60}$$

holds. Since the measure of the set K_r of those θ for which (6.60) holds exceeds $\eta^2/(2\pi)$ for $r > r_0$, we have

$$\begin{aligned} m(r, 0, f(z+1)) &\geq \frac{1}{2\pi}\int_{K_r} \ln^+ \frac{1}{|f(1 + re^{i\theta})|}\,d\theta \\ &\geq \frac{1}{2\pi}\int_{K_r} 2^{n-2}\ln\frac{1}{\eta}\,d\theta \\ &> \frac{1}{2\pi}2^{n-2}\left(\ln\frac{1}{\eta}\right)\frac{\eta^2}{2\pi} = 2^n\frac{\eta^2}{16\pi^2}\ln\frac{1}{\eta}. \end{aligned}$$

Taking into account (6.59) we get

$$\begin{aligned} \delta_1(0, f) &= \delta(0, f(z+1)) \\ &= \liminf_{r\to\infty}\frac{m(r, 0, f(z+1))}{T(r, f(z+1))} \\ &\geq \frac{2^n\frac{\eta^2}{16\pi^2}\ln\frac{1}{\eta}}{2^n 11} = \frac{\eta^2}{176\pi^2}\ln\frac{1}{\eta} > 0. \end{aligned}$$

Thus, zero is a deficient value for $f(z+1)$, but is not a deficient value for $f(z)$.

Observe also, that for $f_1(z) = f(z+1)$ we have

$$\lim_{r\to\infty}\frac{\ln\mu(r, f_1)}{\ln M(r, f_1)} = -\infty, \quad \text{where } \mu(r, f_1) = \min_{|z|=r}|f_1(z)|. \tag{6.61}$$

In fact, (6.60) implies that $\ln\mu(r, f_1) \leq 2^{n-2}\ln\eta$, and (6.59) implies that $\ln M(r, f_1) < 11 \cdot 2^n$. Therefore for $r > r_0(\eta)$ the inequality

$$\frac{\ln\mu(r, f_1)}{\ln M(r, f_1)} < \frac{-2^{n-2}\ln\frac{1}{\eta}}{11 \cdot 2^n} = -\frac{1}{44}\ln\frac{1}{\eta}$$

holds. Since η can be chosen arbitrarily small, we get (6.61).

In conclusion we mention that it is possible to do the work similar to that done in this section for the index $\varepsilon(a)$ instead of the defect $\delta(a)$ considered here (see

(6.19)). That is, we can introduce the values $\varepsilon^*(a)$, $\varepsilon_\gamma(a)$, $\varepsilon_\gamma^*(a)$ and prove for them analogs of all theorems of this section except Theorem 6.3. The analog of Theorem 6.4 is stated as follows: If the limit

$$\lim_{r\to\infty} \frac{N_1(r,a)}{T(r,f)} = \varepsilon(a)$$

exists, then $\varepsilon(a) = \varepsilon^*(a)$.

CHAPTER 5

Asymptotic Properties of Meromorphic Functions and Deficiencies

1. Asymptotic values

Let $f(z)$ be a meromorphic function, C a continuous curve in $\{|z| < \infty\}$ given by an equation $z = z(t)$, $0 \le t < \infty$, $|z(t)| < \infty$, and $z(t) \to \infty$ as $t \to \infty$. If there exists a finite or infinite a such that

$$\lim_{\substack{z\to\infty \\ z\in C}} f(z) = \lim_{t\to\infty} f(z(t)) = a,$$

then a is called an *asymptotic value* of the function $f(z)$, and C is called an *asymptotic curve.* The pair $\{a, C\}$ is also called an *asymptotic spot* (sometimes such pair is called an asymptotic value). Two asymptotic spots $\{a_1, C_1\}$ and $\{a_2, C_2\}$ are considered to be equal if:

(1) $a_1 = a_2 = a$.
(2) There exists a sequence of continuous curves γ_k such that one end of γ_k belongs to C_1, the other end belongs to C_2, and

$$\lim_{k\to\infty} \min_{z\in\gamma_k} |z| = \infty, \qquad \lim_{\substack{z\to\infty \\ z\in\bigcup_k \gamma_k}} f(z) = a.$$

It is clear that asymptotic spots $\{a_1, C_1\}$ and $\{a_2, C_2\}$ are equal if the curves C_1 and C_2 have points of intersection in each neighborhood of infinity.

Significant amount of research is devoted to relations between exceptional and asymptotic values of a meromorphic functions; however, many problems are still open. We present some of the known results.

First of all, it is clear that an asymptotic value does not have to be exceptional in any sense. Indeed the value $w = 0$ is an asymptotic value for $f(z) = \sin z/z$ (it corresponds to two different asymptotic spots with asymptotic curves $C_1 : z(t) = t$ and $C_2 : z(t) = -t$) without being a Valiron exceptional value, and hence without being an exceptional value in any other sense. The following theorem due to F. Iversen is a result in the opposite direction.

THEOREM 1.1. *Each Picard exceptional value is an asymptotic value.*

To prove this theorem we need the following lemma.

LEMMA 1.1. *Let G be an unbounded region, and let Γ be the part of the boundary of G in the complex plane (that is, ∞ is not regarded as a point in Γ). Let $f(z)$ be analytic in G and continuous in $G \cup \Gamma$. If $|f(z)|$ is bounded in G and $|f(z)| \le M$ for $z \in \Gamma$, then $|f(z)| \le M$ in G.*

PROOF. If $f(z)$ is constant, the result is obvious. Suppose that $f(z)$ is not constant and $|f(z)| \le K$ for $z \in G$. We may assume that $K > M$ because otherwise there is nothing to prove. Suppose that $z_0 \in G$ is such that $|f(z_0)| > M$ and prove that in such a case for an arbitrary $z' \in G$, $z' \neq z_0$ we have $|f(z')| \leq |f(z_0)|$. It leads to a contradiction with the maximum modulus principle. Let ε be an arbitrarily small positive number. Consider a disc $D = \{|z - z_0| \leq \eta\}$ with a small radius so that it is contained in G, does not contain z', and the inequality $|f(z)| \leq |f(z_0)| + \varepsilon$ holds for z in $\gamma = \{|z - z_0| = \eta\}$. Let $1 > \delta > 0$. We consider the function $\varphi_\delta(z) = f(z)(\eta/(z - z_0))^\delta$ in the region $G \backslash D = G_\eta$. This function is multivalent, but it has a univalent modulus. It is analytic in G_η, continuous in the closure of G_η, including $z = \infty$, and $\varphi_\delta(\infty) = 0$. We have $|\eta/(z - z_0)| = 1$ for $z \in \gamma$ and $|\eta/(z - z_0)| < 1$ for $z \in \Gamma$. Therefore $|\varphi_\delta(z)| \leq |f(z_0)| + \varepsilon$ for $z \in \gamma$ and $|\varphi_\delta(z)| \leq M < |f(z_0)| + \varepsilon$ for $z \in \Gamma$. Applying the maximum modulus principle to $\varphi_\delta(z)$ we get $|\varphi_\delta(z')| \leq |f(z_0)| + \varepsilon$. Letting $\varepsilon \to 0$ and $\delta \to 0$ in this inequality, we get $|f(z')| \leq |f(z_0)|$. □

PROOF. Now we turn to the proof of Theorem 1.1. Without loss of generality we assume that $a = \infty$ is a Picard exceptional value. Otherwise we would consider the function $F(z) = 1/\{f(z) - a\}$. Thus the transcendental function $f(z)$ has at most finitely many poles. First we assume that there are no poles, that is, the function is entire. We need to show the existence of a continuous curve C approaching infinity, such that $f(z)$ approaches infinity along this curve as $z \to \infty$.

Denote by G_n, $n = 1, 2, \ldots$, the open set of those z for which $|f(z)| > n$. The equality $|f(z)| = n$ holds for each point of the boundary of G_n in the complex plane. Each connected component of G_n is an unbounded region, since otherwise, by the maximum modulus principle, we would have $|f(z)| \leq n$ for each $z \in G_n$. By Lemma 1.1 each connected component of G_n contains at least one point of G_{n+1}. Therefore we can find a sequence $\{G'_n\}_{n=1}^\infty$ of connected components of sets G_n such that $G'_1 \supset G'_2 \supset G'_3 \supset \cdots$. Since the function $f(z)$ is bounded in each disc $\{|z| \leq R\}$, it follows that starting with some $n = n(R)$, the regions G'_n do not intersect this set. In each region G'_n consider a point z_n and join it by a Jordan arc $C_n \subset G'_n$ with z_{n+1}. Then $C = C_1 \cup C_2 \cup \cdots$ is the desired curve. In fact, if z successively traverses $C_1, C_2, \ldots$, then $z \to \infty$ since $\bigcup_{j=n(R)}^\infty C_j$ does not intersect the disc $\{|z| \leq R\}$. Besides, $\bigcup_{j=n}^\infty C_j \subset G'_n$ and hence $|f(z)| > n$ for $z \in \bigcup_{j=n}^\infty C_j$; that is, $f(z) \to \infty$ as $z \to \infty$, $z \in C$.

If $f(z)$ has poles at finitely many points $\zeta_1, \ldots, \zeta_p$, we denote by $H_j(z)$ the principal part of the Laurent expansion of $f(z)$ in a neighborhood of ζ_j. The function $H_j(z)$ is rational, has zero at ∞ and the only pole at ζ_j. The function

$$g(z) = f(z) - \sum_{j=1}^{p} H_j(z)$$

is an entire transcendental function; hence, as we have proved above, there exists an asymptotic curve C such that $g(z) \to \infty$ along C. The function $\sum_{j=1}^p H_j(z)$ approaches zero along C, hence $f(z)$ approaches ∞. □

REMARK. The argument used in the proof of Theorem 1.1 can be used to prove the following statement. Let G be an unbounded region, and Γ the part of its boundary in the complex plane. Let $f(z)$ be analytic in G and continuous in $G \cup \Gamma$. If the function $f(z)$ is bounded on Γ but is unbounded on G, then there exists an asymptotic curve in G approaching ∞ and such that $f(z)$ approaches ∞ along C.

EXERCISE. Prove that for each transcendental entire function $f(z)$ and each positive integer n there exists an asymptotic curve C_n such that $f(z)z^{-n} \to \infty$ as $z \to \infty$, $z \in C_n$.

HINT. Apply Theorem 1.1 to $\varphi(z) = z^{-n} f(z)$.

Later we shall see that, in general, Borel or Nevanlinna exceptional values, and hence, Valiron exceptional values are not necessarily asymptotic values.

Now we prove the following important theorem due to Ahlfors.

THEOREM 1.2. *If an entire function $f(z)$ has $p \geq 2$ different asymptotic spots $\{\infty, C_j\}$, $1 \leq j \leq p$, then*

$$\liminf_{r\to\infty} r^{-\frac{p}{2}} T(r, f) > 0. \tag{1.1}$$

The statement remains valid for $p = 1$ if we assume, in addition, that

$$\mu(r, f) = \min_{|z|=r} |f(z)| = O(1), \qquad r \to \infty. \tag{1.2}$$

PROOF. We may assume, without loss of generality, that the curves C_j do not intersect each other. Consider the open set G_n of those points for which $|f(z)| > n$. Denote by $G_n^{(j)}$ the connected component of the set G_j containing $C_j \cap \{|z| > R\}$ for sufficiently large R. It is clear that the region $G_n^{(j)}$ is uniquely determined by this condition. The regions $G_n^{(j)}$, $1 \leq j \leq p$, are pairwise disjoint for sufficiently large n. In fact, if the regions $G_n^{(j_1)}$ and $G_n^{(j_2)}$ intersected for all $n \geq 1$, they would coincide: $G_n^{(j_1)} = G_n^{(j_2)}$, and joining the curves C_{j_1} and C_{j_2} by continuous curves γ_k lying in $G_k^{(j_1)} = G_k^{(j_2)}$, we could easily show that the asymptotic spots $\{\infty, C_{j_1}\}$ and $\{\infty, C_{j_2}\}$ coincide, a contradiction. Increasing n, if necessary, we get into the situation in which none of the regions $G_n^{(j)}$ completely contains any of the circles $\{|z| = r\}$. If $p = 1$, the possibility to choose such n is guaranteed by condition (1.2): it suffices to choose $n > \sup_{0<r<\infty} \mu(r, f)$. For $p \geq 2$ such n can be chosen because the complement of $G_n^{(j)}$ contains at least one unbounded region. Thus, we have found p unbounded pairwise disjoint regions $G_{n_0}^{(j)}$, $1 \leq j \leq p$, in the complex plane having the following properties:

(a) The region $G_{n_0}^{(j)}$ does not completely contain any of the circles $\{|z| = r\}$.
(b) The function $f(z)$ is unbounded in $G_{n_0}^{(j)}$ and satisfies the inequality $|f(z)| > n_0$ in $G_{n_0}^{(j)}$.
(c) $|f(z)| = n_0$ on the part of the boundary of $G_{n_0}^{(j)}$ in the complex plane.

We may assume, without loss of generality, that $n_0 = 1$; otherwise we would consider the function $f(z)/n_0$. In what follows we assume that $n_0 = 1$ and drop the subscript n_0 for $G_{n_0}^{(j)}$.

We need the following *Wirtinger inequality.* Let $y(x)$ be a continuously differentiable function on $[a, b]$, $y(a) = y(b) = 0$. Then

$$\int_a^b [y'(x)]^2 dx \geq \left(\frac{\pi}{b-a}\right)^2 \int_a^b [y(x)]^2 dx. \tag{1.3}$$

We may assume, without loss of generality, that $a = 0$ and $b = \pi$, since the general case can be reduced to this case using the substitution $x = a + \frac{b-a}{\pi} t$. So, we need to prove that

$$\int_0^\pi [y'(x)]^2 dx \geq \int_0^\pi [y(x)]^2 dx. \tag{1.3$'$}$$

Since $y(0) = y(\pi) = 0$, we have $y(x) = O(x)$ as $x \to 0$ and $y(x) = O(\pi - x)$ as $x \to \pi$. Therefore the function $y^2(x)\cot x$ is continuous on the interval $[0, \pi]$ and vanishes at the ends of this interval, and the function $y(x)\cot x$ is continuous on $[0, \pi]$.

From this we get

$$\begin{aligned} 0 = \int_0^\pi d(y^2 \cot x) &= \int_0^\pi \{2yy'\cot x - y^2(1+\cot^2 x)\}dx \\ &= \int_0^\pi \{(y')^2 - y^2 - (y' - y\cot x)^2\}dx \leq \int_0^\pi ((y')^2 - y^2)dx, \end{aligned}$$

that is, inequality (1.3′).

Let r_0 be a large number so that the circle $\{|z| = r_0\}$ intersects all the regions $G^{(j)}$, $1 \leq j \leq p$. We fix a region $G^{(j)}$. Denote by $r_0^{(j)}$ the minimum value of $|z|$ for $z \in \bar{G}^{(j)}$, and by $S^{(j)}(r)$ the intersection $G^{(j)} \cap \{|z| = r\}$ for $r > r_0^{(j)}$. The set $S^{(j)}(r)$ consists of finitely many arcs since the part of the boundary of $G^{(j)}$ contained in an arbitrary disc $\{|z| \leq r\}$ consists of finitely many analytic curves. The last statement can be easily verified using the fact that $|f(z)| = 1$ on the boundary of $G^{(j)}$. Denote by $rl_j(r)$ the length of the longest arc among the arcs forming $S^{(j)}(r)$. It is clear that

$$\sum_{j=1}^p l_j(r) \leq 2\pi. \tag{1.4}$$

Consider the function $\ln|f(re^{i\varphi})| = \ln|f(e^{t+i\varphi})| = u(t,\varphi)$, $t = \ln r$ in the region $G^{(j)}$. It is clear that $u(\ln r, \varphi) > 0$ in the region $G^{(j)}$, and that $u(t,\varphi) = \operatorname{Re}\ln f(e^{t+i\varphi})$ is a harmonic function. Therefore

$$\frac{\partial^2 u}{\partial t^2} + \frac{\partial^2 u}{\partial \varphi^2} = 0. \tag{1.5}$$

We introduce the notation

$$m_j(r) = \int_{S^{(j)}(r)} (\ln|f(re^{i\varphi})|)^2 d\varphi,$$

$$\mu_j(t) = m_j(e^t) = \int_{S^{(j)}(r)} u^2(t,\varphi)\, d\varphi.$$

Taking into account the fact that $u(t,\varphi) = 0$ at the endpoints of the intervals of integration, we get

$$\mu_j'(t) = 2\int_{S^{(j)}(r)} u\frac{\partial u}{\partial t} d\varphi,$$

$$\mu_j''(t) = 2\int_{S^{(j)}(r)} \left\{u\frac{\partial^2 u}{\partial t^2} + \left(\frac{\partial u}{\partial t}\right)^2\right\} d\varphi. \tag{1.6}$$

Integrating by parts and using (1.5), we get

$$\int_{S^{(j)}(r)} u\frac{\partial^2 u}{\partial t^2} d\varphi = -\int_{S^{(j)}(r)} u\frac{\partial^2 u}{\partial \varphi^2} d\varphi = \int_{S^{(j)}(r)} \left(\frac{\partial u}{\partial \varphi}\right)^2 d\varphi.$$

Substituting this in (1.6), we obtain

$$\mu_j''(t) = 2\int_{S^{(j)}(r)} \left\{ \left(\frac{\partial u}{\partial t}\right)^2 + \left(\frac{\partial u}{\partial \varphi}\right)^2 \right\} d\varphi. \tag{1.7}$$

Obviously, for $t > \ln r_0^{(j)}$ we have $\mu_j''(t) \geq 0$; hence $\mu_j'(t)$ is nondecreasing. Since $\mu_j(t) > 0$ for $t > \ln r_0^{(j)}$, and $\mu_j(t) \to 0$ as $t \to \ln r_0^{(j)}$, it is easy to see that $\mu_j'(t) > 0$ for $t > \ln r_0^{(j)}$. When $re^{i\varphi} \in S^{(j)}(r)$, the variable φ ranges over the finite system of intervals $(\varphi_1, \varphi_1'), (\varphi_2, \varphi_2'), \dots, (\varphi_s, \varphi_s')$, where $s, \varphi_1, \varphi_1', \dots, \varphi_s, \varphi_s'$ depend on j and r. For each of the intervals $(\varphi_\sigma, \varphi_\sigma')$ we use the Wirtinger inequality (1.3) and get

$$\int_{\varphi_\sigma}^{\varphi_\sigma'} \left(\frac{\partial u}{\partial \varphi}\right)^2 d\varphi \geq \left(\frac{\pi}{\varphi_\sigma' - \varphi_\sigma}\right)^2 \int_{\varphi_\sigma}^{\varphi_\sigma'} u^2 d\varphi \geq \left(\frac{\pi}{l_j(r)}\right)^2 \int_{\varphi_\sigma}^{\varphi_\sigma'} u^2 d\varphi.$$

Summing these inequalities for all σ from 1 to s, we get

$$\int_{S^{(j)}(r)} \left(\frac{\partial u}{\partial \varphi}\right)^2 d\varphi \geq \left(\frac{\pi}{l_j(r)}\right)^2 \int_{S^{(j)}(r)} u^2 d\varphi = \left(\frac{\pi}{l_j(r)}\right)^2 \mu_j(t). \tag{1.8}$$

Using the Cauchy–Bunyakovskii inequality we get

$$\begin{aligned}(\mu_j'(t))^2 = 4\left\{\int_{S^{(j)}(r)} u\frac{\partial u}{\partial t} d\varphi\right\}^2 &\leq 4\int_{S^{(j)}(r)} u^2 d\varphi \int_{S^{(j)}(r)} \left(\frac{\partial u}{\partial t}\right)^2 d\varphi \\ &= 4\mu_j(t) \int_{S^{(j)}(r)} \left(\frac{\partial u}{\partial t}\right)^2 d\varphi.\end{aligned} \tag{1.9}$$

From (1.7), (1.8), and (1.9) we derive

$$\mu_j''(t) \geq \frac{1}{2}\frac{\{\mu_j'(t)\}^2}{\mu_j(t)} + 2\left(\frac{\pi}{l_j(r)}\right)^2 \mu_j(t),$$

from which, taking into account

$$\left(\frac{\mu_j'}{\mu_j}\right)' = \frac{\mu_j''}{\mu_j} - \left(\frac{\mu_j'}{\mu_j}\right)^2,$$

we get

$$2\omega_j''(t) + {\omega_j'}^2(t) \geq \left(\frac{2\pi}{l_j(r)}\right)^2, \qquad \omega_j(t) = \ln \mu_j(t).$$

This implies

$$\left\{\omega_j'(t) + \frac{\omega_j''(t)}{\omega_j'(t)}\right\}^2 \geq {\omega_j'}^2(t) + 2\omega_j''(t) \geq \left(\frac{2\pi}{l_j(r)}\right)^2.$$

Since $\omega_j'(t) > 0$ for $t > \ln r_0^{(j)}$, the last inequality implies that

$$\omega_j'(t) + \frac{\omega_j''(t)}{\omega_j'(t)} \geq \frac{2\pi}{l_j(r)}. \tag{1.10}$$

It is easy to check that the left-hand side of (1.10) is equal to

$$\frac{d}{dt}\left\{\ln\left(\frac{d}{dt}e^{\omega_j(t)}\right)\right\} = \frac{d}{d\ln r}\left\{\ln\left(\frac{dm_j(r)}{d\ln r}\right)\right\};$$

therefore ($r \geq r_0$)

$$\frac{d}{d\ln r}\left\{\ln\left(\frac{dm_j(r)}{d\ln r}\right)\right\} \geq \frac{2\pi}{l_j(r)}.$$

Integrating this inequality we get

$$\frac{dm_j(r)}{d\ln r} \geq A_j \exp\left\{\int_{r_0}^{r} \frac{2\pi}{tl_j(t)}dt\right\},$$

where

$$A_j = \left.\frac{dm_j(r)}{d\ln r}\right|_{r=r_0} > 0,$$

and

$$m_j(r) \geq m_j(r_0) + A_j \int_{r_0}^{r} \frac{d\tau}{\tau} \exp\left\{\int_{r_0}^{\tau} \frac{2\pi}{l_j(t)t}dt\right\}. \tag{1.11}$$

From (1.11) we get ($A = \min_{1\leq j\leq p} A_j$)

$$\begin{aligned} m_j(er) &\geq A\int_{r}^{er} \frac{d\tau}{\tau} \exp\left\{\int_{r_0}^{\tau} \frac{2\pi}{l_j(t)t}dt\right\} \\ &\geq A\exp\left\{\int_{r_0}^{r} \frac{2\pi}{l_j(t)t}dt\right\}\int_{r}^{er} \frac{d\tau}{\tau} = A\exp\left\{\int_{r_0}^{r} \frac{2\pi}{l_j(t)t}dt\right\}, \end{aligned}$$

from which

$$\ln m_j(er) \geq 2\pi\int_{r_0}^{r} \frac{dt}{l_j(t)t} + \ln A.$$

Since $m_j(er) \leq 2\pi \ln^2 M(er,f)$, we get

$$\ln\ln M(er,f) \geq \pi\int_{r_0}^{r} \frac{dt}{l_j(t)t} + \frac{1}{2}\ln\frac{A}{2\pi}. \tag{1.12}$$

Summing these inequalities we get

$$\ln\ln M(er,f) \geq \frac{\pi}{p}\int_{r_0}^{r} \frac{dt}{t}\sum_{j=1}^{p}\frac{1}{l_j(t)} + \frac{1}{2}\ln\frac{A}{2\pi}. \tag{1.12$'$}$$

Using the the Cauchy–Bunyakovskii inequality and (1.4) we obtain

$$p^2 = \left(\sum_{j=1}^{p}\sqrt{l_j(t)}\frac{1}{\sqrt{l_j(t)}}\right)^2 \leq \sum_{j=1}^{p} l_j(t)\sum_{j=1}^{p}\frac{1}{l_j(t)} \leq 2\pi\sum_{j=1}^{p}\frac{1}{l_j(t)}.$$

Therefore (1.12$'$) implies

$$\ln\ln M(er,f) \geq \frac{p}{2}\int_{r_0}^{r}\frac{dt}{t} + \frac{1}{2}\ln\frac{A}{2\pi} = \ln(Kr^{p/2}),$$

where $K = r_0^{-p/2}\sqrt{A/(2\pi)}$. If $r/2 > r_0$, we get

$$\ln M\left(\frac{r}{2}, f\right) \geq K_1 r^{p/2}, \qquad K_1 = K(2e)^{-p/2}.$$

Hence

$$T(r,f) \geq \frac{1}{3}\ln M\left(\frac{r}{2}, f\right) \geq \frac{1}{3}K_1 r^{p/2},$$

that is, (1.1) is satisfied. □

Our next example shows that the estimate given by Theorem 1.2 is sharp.

Let $f(z) = \cos(z^{p/2})$, $p = 1, 2, \dots$. It is clear that $f(z)$ is an entire function. Using Theorem 6.1 of Chapter 1 or the following asymptotic equality directly ($z = re^{i\varphi}, |\varphi| < \pi$):

$$|\cos(z^{p/2})| = \frac{1}{2}\exp\left(r^{p/2}\left|\cos\frac{p\varphi}{2}\right|\right) + O(1),$$

we get

$$T(r, f) \sim \frac{2}{\pi} r^{p/2}.$$

If p is an even number, $p = 2m$, the function $f(z)$ has p asymptotic spots $\{\infty, C_k\}$, where C_k is the ray $\{\arg z = \varphi_k = \frac{\pi}{2m} + k\frac{\pi}{m}\}$, $k = -m, -m+1, \dots, 0, \dots, m-1$. In fact, if $z = re^{i\varphi_k}$, then $\cos(z^{p/2}) = \cos z^m = \cos\{r^m i(-1)^k\} = \cosh r^m$. Asymptotic spots $\{\infty, C_k\}$ are distinct since $|\cos z^m| \le 1$ on the rays $\{\arg z = \varphi'_k = k\frac{\pi}{m}\}$, $k = -m, -m+1, \dots, m$. If p is odd, $p = 2m-1$, then the function $f(z)$ has p asymptotic spots $\{\infty, C_k\}$, where C_k is the ray

$$\left\{\arg z = \theta_k = \frac{\pi}{2m-1} + k\frac{2\pi}{2m-1}\right\}, \qquad k = -m+1, \dots, m-1.$$

In fact, for $z = re^{i\theta_k}$, the equality $\cos(z^{p/2}) = \cos\{r^{p/2} i(-1)^k\} = \cosh r^{p/2}$ holds.

Since $|\cos(z^{p/2})| \le 1$ on the rays

$$\left\{\arg z = \theta'_k = k\frac{2\pi}{2m-1}\right\}, \qquad k = -m+1, -m+2, \dots, m-1,$$

the asymptotic spots $\{\infty, C_k\}$ are distinct in this case, too.

The following classical theorem is an immediate corollary of Theorem 1.2.

THEOREM 1.3 (Wiman, Heins). *If the lower order λ of an entire function $f(z)$ is less than $1/2$, then*

$$\limsup_{r\to\infty} \mu(r, f) = \infty. \tag{1.13}$$

This conclusion remains true if

$$\liminf_{r\to\infty} r^{-1/2} T(r, f) = 0. \tag{1.14}$$

Relation (1.13) means that there exists a sequence of circles $\{|z| = r_k\}$, $r_k \to \infty$, on which $|f(z)|$ tends to ∞ uniformly with respect to $\arg z$.

To prove Theorem 1.3, we note, that by Theorem 1.1, the function $f(z)$ has at least one asymptotic spot at ∞. If condition (1.13) were not satisfied, then relation (1.2) would hold, and, by (1.1), we would have

$$\liminf_{r\to\infty} r^{-1/2} T(r, f) > 0,$$

a contradiction.

COROLLARY 1. *If the lower order λ of an entire function $f(z)$ is less than $1/2$, then $\delta(a, f) = 0$ for all $a \ne \infty$.*

In fact, suppose that $\mu(r_k, f) \to \infty$ as $r_k \to \infty$. Then $|f(r_k e^{i\varphi}) - a| \ge \mu(r_k, f) - |a| \to \infty$ and $m(r_k, a, f) = 0$ for all $k \ge k_0$.

In Section 3 we show that for $\lambda = 1/2$ an entire function $f(z)$ cannot have finite deficient values either. If $\lambda > 1/2$, then, as we have seen in Section 5 of Chapter 4, the set $E_N(f)$ may contain an arbitrary preassigned countable set.

COROLLARY 2. *Deficiencies of an entire function of order $\rho < 3/2$ are shift invariant.*

In fact, if $\lambda < 1/2$, then, by Corollary 1, there are no finite deficient values, and if $\lambda \geq 1/2$, $\rho < 3/2$, then $\rho - \lambda < 1$, and we can apply the Corollary of Theorem 6.5 in Chapter 4.

COROLLARY 3. *Let $f(z)$ be an entire function of finite lower order λ. Then*

$$\limsup_{r\to\infty} \frac{\ln \mu(r,f)}{\ln M(r,f)} \geq -[2\lambda]. \tag{1.15}$$

If $\lambda < 1/2$, inequality (1.15) follows from (1.13). Therefore we may assume that $\lambda \geq 1/2$ and, hence, $n = [2\lambda] + 1 \geq 2$. We form the function

$$F(z) = f(z)f(\omega z)\cdots f(\omega^{n-1}z), \qquad \omega = e^{2\pi i/n}.$$

It is clear that $F(z) = \Phi(z^n)$, where $\Phi(z)$ is an entire function. Since $M(r,\Phi) = M(r^n, F)$, $M(r,F) \leq \{M(r,f)\}^n$, it is easy to see that the lower order of the function $F(z)$ does not exceed λ, and the lower order of the function $\Phi(z)$ does not exceed $\lambda/n < 1/2$. Equality (1.13) implies that there exists a sequence $r_j \to \infty$ such that $\mu(r_j, \Phi) > 1$. Then $\mu(R_j, F) > 1$, where $R_j = r_j^n$. But

$$1 < \mu(R_j, F) \leq \mu(R_j, f)\{M(R_j, f)\}^{n-1},$$

from which $\ln \mu(R_j, f) > -(n-1)\ln M(R_j, f) = -[2\lambda] \ln M(R_j, f)$, and we get (1.15).

Note also, that as is shown by Example 6 in Section 6 of Chapter 4 (see (6.61)), without the assumption $\lambda < \infty$ we cannot claim that the upper limit in (1.15) is finite.

The estimate (1.15) is not sharp for $\lambda \neq 1/2$. A sharp estimate in the case $\lambda \leq 1$ is obtained in Theorem 3.4 and Corollary 2 from it. No sharp estimate is known for $\lambda > 1$.[1]

Theorem 1.3 can be generalized as follows.

THEOREM 1.3′. *Let $f(z)$ be an entire function of finite lower order λ, and let l be a number satisfying the conditions*

$$0 < l \leq 2\pi, \qquad l < \frac{\pi}{\lambda}. \tag{1.14′}$$

Then there exists a sequence of arcs $\{|z| = r_k,\ \theta_k \leq \arg z \leq \theta_k + l\}$, $r_k \to \infty$, $0 \leq \theta_k < 2\pi$, on which $|f(z)|$ approaches ∞ uniformly with respect to $\arg z$.

PROOF. If $\limsup_{r\to\infty} \mu(r,f) = \infty$, we can take $l = 2\pi$; therefore it suffices to consider the case where $\mu(r,f) = O(1)$. In such a case we are under the assumptions of Theorem 1.2. Using the same notation as in the proof of Theorem 1.2, we can write inequality (1.12). If the inequality $\limsup_{r\to\infty} l_j(r) < \frac{\pi}{\lambda}$ were satisfied, then, taking γ such that $\limsup_{r\to\infty} l_j(r) < \gamma < \frac{\pi}{\lambda}$, by (1.12) we would get

$$\ln\ln M(er, f) \geq \frac{\pi}{\gamma} \ln r + O(1).$$

[1] Sharp estimate is still unknown. See Appendix B at the end of the book for a survey of known results.

Since $\frac{\pi}{\gamma} > \lambda$, we would get a contradiction with the assumption that the lower order of $f(z)$ is equal to λ. Thus $\limsup_{r\to\infty} l_j(r) \geq \frac{\pi}{\lambda}$.

We see from this that if the number l satisfies conditions (1.14′), then there exists a sequence $r_k^{(1)} \to \infty$ such that $l_j(r_k^{(1)}) > l$, $k = 1, 2, \ldots$. By the definition of $l_j(r)$, this indicates that there exists a sequence of arcs $S_k^{(1)} = \{|z| = r_k^{(1)}, \theta_k^{(1)} \leq \arg z \leq \theta_k^{(1)} + l\}$ such that $|f(z)| \geq 1$ for $z \in S_k^{(1)}$, $k = 1, 2, \ldots$. Considering instead of $f(z)$ the function $f(z)/n$, where n is an arbitrary positive integer, we get a sequence of arcs $S_k^{(n)} = \{|z| = r_k^{(n)}, \theta_k^{(n)} \leq \arg z \leq \theta_k^{(n)} + l\}$, $0 \leq \theta_k^{(n)} < 2\pi$, $\lim_{k\to\infty} r_k^{(n)} = \infty$, such that $|f(z)| \geq n$ for $z \in S_k^{(n)}$, $k = 1, 2, \ldots$. The sequence $S_n^{(n)}$, $n = 1, 2, \ldots$, has the desired property. □

THEOREM 1.4 (Denjoy, Carleman, Ahlfors). *If an entire function $f(z)$ has p asymptotic spots $\{a_j, C_j\}$, $1 \leq j \leq p$, $a_j \neq \infty$, then it satisfies* (1.1).

PROOF. Let r_0 be large so that the circle $\{|z| = r_0\}$ intersects each curve C_j, $1 \leq j \leq p$, and the curves are pairwise disjoint for $|z| \geq r_0$. We may assume without loss of generality that the initial points z_j of curves C_j lie on the circle $\{|z| = r_0\}$, and the rest of the curve is situated in $\{|z| > r_0\}$ (if necessary, we delete the part of C_j between the initial point and the last point of intersection with $\{|z| = r_0\}$). We may assume that the curves C_j are numbered in such a way that the subscripts of the points $z_1, \ldots, z_p$ are growing as we follow the circle $\{|z| = r_0\}$ in the positive direction. The curves $C_1, \ldots, C_p$ separate the region $\{r_0 < |z| < \infty\}$ into simply connected regions $D_1, \ldots, D_p$; the region D_j lies between the curves C_j and C_{j+1}, $C_{p+1} = C_1$. Obviously the function $f(z)$ is bounded on the boundary of D_j in the complex plane. Let us show that $f(z)$ is unbounded in the region D_j.

LEMMA 1.2 (Lindelöf). *Let $\Phi(z)$ be analytic and bounded in $\{\operatorname{Re} z > 0\}$ and continuous in $\{\operatorname{Re} z \geq 0\}$. Let $\Phi(iy) \to a_1$ as $y \to +\infty$ and $\Phi(iy) \to a_2$ as $y \to -\infty$. Then $a_1 = a_2 = a$, and the function $\Phi(z)$ approaches a as $z \to \infty$, $\operatorname{Re} z \geq 0$.*

PROOF. First we assume that $a_1 = a_2 = a$. For each $\varepsilon > 0$ there exists $r_1 = r_1(\varepsilon)$ such that $|\Phi(iy) - a| < \varepsilon$ for $|y| \geq r_1(\varepsilon)$. Let $q > r_1$ be a large number so that

$$\left|\frac{z}{z+q}\{\Phi(z) - a\}\right| \leq \frac{r_1}{q - r_1} \max_{z\in\gamma} |\Phi(z) - a| < \varepsilon$$

on the arc $\gamma = \{|z| = r_1,\ \operatorname{Re} z \geq 0\}$. Since

$$\left|\frac{z}{z+q}\right| < 1 \quad \text{for} \quad \operatorname{Re} z > 0,$$

it follows that on the boundary of the region $G = \{|z| > r_1\} \cap \{\operatorname{Re} z > 0\}$ in the complex plane we have

$$\left|\frac{z}{z+q}\{\Phi(z) - a\}\right| \leq \varepsilon, \tag{1.16}$$

and the left-hand side of (1.16) is bounded in the region G. Lemma 1.1 implies that inequality (1.16) holds everywhere in G. Now let $|z| > q$, $\operatorname{Re} z \geq 0$. Inequality (1.16) implies

$$|\Phi(z) - a| \leq \varepsilon \left|\frac{z+q}{z}\right| \leq \varepsilon \left(1 + \frac{q}{|z|}\right) < 2\varepsilon;$$

that is, $\Phi(z)$ approaches a as $z \to \infty$, $\operatorname{Re} z \geq 0$.

Now we assume that $a_1 \neq a_2$. Then the function

$$F(z) = \left\{\Phi(z) - \frac{a_1 + a_2}{2}\right\}^2 \tag{1.17}$$

satisfies the hypothesis of the lemma and $F(iy) \to (\frac{a_1 - a_2}{2})^2$ as $y \to +\infty$ and $y \to -\infty$. As was proved above, the function $F(z)$ approaches $(\frac{a_1-a_2}{2})^2$ as $z \to \infty$, $\operatorname{Re} z \geq 0$.

From (1.17) we see that there is a single-valued branch of $\sqrt{F(z)}$ in $\{\operatorname{Re} z \geq 0\}$ such that

$$\Phi(z) - \frac{a_1 + a_2}{2} = \sqrt{F(z)}.$$

But $\sqrt{F(iy)}$ approaches either $(a_1 - a_2)/2$ or $(a_2 - a_1)/2$ as $|y| \to \infty$; that is, $\Phi(iy)$ approaches a_1 or a_2 as $|y| \to \infty$, which contradicts the assumption that $\Phi(iy)$ has different limits as $y \to +\infty$ and $y \to -\infty$. □

We continue the proof of the theorem. Assume that the function $f(z)$ is bounded in the region D_j. Let function $z = \omega(\zeta)$ be a univalent conformal mapping of the half-plane $\{\operatorname{Re}\zeta > 0\}$ onto D_j, $\omega(\infty) = \infty$. Then the function $\Phi(\zeta) = f(\omega(\zeta))$ satisfies the hypothesis of Lemma 1.2 with the roles of a_1 and a_2 played by a_{j+1} and a_j, respectively. The lemma implies that $a_{j+1} = a_j = a$ and $\Phi(\zeta)$ approaches a as $\zeta \to \infty$, $\operatorname{Re}\zeta \geq 0$ and, hence, as $\zeta \to \infty$, $\operatorname{Re}\zeta \geq 0$, $|\zeta| = k$, $k = 1, 2, \ldots$. Denote by γ_k the image of the semicircle $\{\operatorname{Re}\zeta \geq 0, |\zeta| = k\}$ under the mapping $z = \omega(\zeta)$. It is clear that the distance between γ_k and $z = 0$ tends to ∞ as $k \to \infty$. Then $f(z) \to a$ as $z \to \infty$, $z \in \bigcup_k \gamma_k$. By definition, the asymptotic spots $\{a_j, C_j\}$ and $\{a_{j+1}, C_{j+1}\}$ coincide, and we get a contradiction. Thus the function $f(z)$ is unbounded in the region D_j, $1 \leq j \leq p$. By the remark to Theorem 1.1, there exists an asymptotic curve C'_j in D_j such that $f(z)$ approaches ∞ along this curve. Obviously $\{\infty, C'_j\}$, $1 \leq j \leq p$, are different asymptotic spots, because the function $f(z)$ is bounded on C_j. Hence we are under the assumptions of Theorem 1.2; therefore (1.1) is satisfied. □

COROLLARY. *An entire function of finite lower order λ cannot have more than $[2\lambda]$ distinct finite asymptotic values or, even stronger, $[2\lambda]$ distinct asymptotic spots with finite asymptotic values.*

EXERCISE. Use the example of

$$f(z) = z^{-p/2}\sin(z^{p/2}), \qquad p = 1, 2, \ldots,$$

to show that the estimate in the corollary can be attained.

2. Non-asymptotic deficient values

For a long time the following conjecture of R. Nevanlinna was considered to be plausible: an entire function of a finite order ρ can have at most $[2\rho]$ finite deficient values. The point is that it seemed likely that each deficient value of an entire function is asymptotic. Theorem 5.1 of Chapter 4 not only disproves this conjecture but also shows that an entire function of an arbitrary order ρ, $1/2 < \rho < \infty$, may have an infinite set of deficient non-asymptotic values, because the number of asymptotic values does not exceed $[2\rho]$. The restriction $\rho > 1/2$ is natural for entire functions, since for $\rho \leq 1/2$ an entire function, as will be shown in Section 3 (for $\rho < 1/2$ this was shown in Corollary 1 of Theorem 1.3), does not have finite deficient values, and a Picard exceptional value at ∞ is necessarily an asymptotic value. Now we show that for meromorphic functions a deficient value can be non-asymptotic even for $0 < \rho \le 1/2$. It is still unknown whether this is true for $\rho = 0$.[2]

Since the construction of the example is rather simple, we will not restrict ourselves to the case $0 < \rho \le 1/2$, but will consider the general case $0 < \rho < \infty$.

EXAMPLE 1. Let $a = 4^\rho$, $0 < \rho < \infty$. Consider the function

$$\begin{aligned} f(z) &= \sum_{k=1}^{\infty} \left(\frac{3 \cdot 4^k}{z - 5 \cdot 4^k} \right)^{[a^k]} + \sum_{k=1}^{\infty} \left(\frac{6 \cdot 4^k}{z + 10 \cdot 4^k} \right)^{[a^k]} \\ &= \sum_{k=1}^{\infty} H_k(z) + \sum_{k=1}^{\infty} H_{-k}(z). \end{aligned} \tag{2.1}$$

It is easy to check, using the convergence of the series $\sum_{k=1}^{\infty} (\frac{3}{5})^{[a^k]}$, that both series in (2.1) are absolutely and uniformly convergent in each disc of finite radius, and therefore, the function $f(z)$ determined by (2.1) is meromorphic.

We introduce the following notation ($k = 1, 2, \dots$):

$$C_{+k} = \{|z - 5 \cdot 4^k| < 3 \cdot 4^k\},$$

$$\tilde{C}_{+k} = \{|z - 5 \cdot 4^k| < 4 \cdot 4^k\},$$

$$C_{-k} = \{|z + 10 \cdot 4^k| < 6 \cdot 4^k\},$$

$$\tilde{C}_{-k} = \{|z + 10 \cdot 4^k| < 8 \cdot 4^k\}.$$

It is easy to check that the discs C_j, $j = \pm 1, \pm 2, \dots$, are pairwise disjoint and that no point in the plane is covered by more than two discs $\tilde{C}_j$. Note that for $z \notin \tilde{C}_k$ ($k \ge 1$), the inequality

$$\left| \frac{3 \cdot 4^k}{z - 5 \cdot 4^k} \right| \le \frac{3 \cdot 4^k}{4 \cdot 4^k} = \frac{3}{4}$$

holds, and for $z \notin \tilde{C}_k$ ($k \le -1$), the inequality

$$\left| \frac{6 \cdot 4^{|k|}}{z + 10 \cdot 4^{|k|}} \right| \le \frac{6 \cdot 4^{|k|}}{8 \cdot 4^{|k|}} = \frac{3}{4}$$

holds. Therefore, if $z \notin \bigcup_{j \in A} \tilde{C}_j$, where A is an arbitrary subset of the set B of non-zero integers, then

$$\left| \sum_{j \in A} H_j(z) \right| \le \sum_{j \in A} \left(\frac{3}{4} \right)^{[a^{|j|}]} \le 2 \sum_{j=1}^{\infty} \left(\frac{3}{4} \right)^{a^j} = M < \infty. \tag{2.2}$$

It is easy to see that outside $\bigcup_{j \in B} C_j$ the function $f(z)$ is bounded. In fact, if $z \notin \bigcup_{j \in B} \tilde{C}_j$, then, by (2.2), we have $|f(z)| \le M$. If z is covered by discs $\tilde{C}_j$, then it is covered by at most two such discs, so it is either covered by $\tilde{C}_{j_0}$ only, or by $\tilde{C}_{j_0}$

[2] Ter-Israelian [**151**] constructed a meromorphic function of order 0 with a deficient non-asymptotic value.

and $\tilde{C}_{j_0+1}$. Consider, for example, the second case. Let $z \notin \tilde{C}_j$, $j \neq 0, j_0, j_0+1$; $z \in \tilde{C}_j$, $j = j_0, j_0+1$. By (2.2) we have

$$|f(z)| \leq |H_{j_0}(z)| + |H_{j_0+1}(z)| + \left| \sum_{j \neq 0, j_0, j_0+1} H_j(z) \right| \leq |H_{j_0}(z)| + |H_{j_0+1}(z)| + M.$$

But if $z \notin C_j$, then $|H_j(z)| \leq 1$. Therefore $|f(z)| \leq 2 + M$. If z belongs to only one of the sets $\tilde{C}_{j_0}$, then, using a similar argument we get $|f(z)| \leq 1 + M$. In any case we have

$$|f(z)| \leq 2 + M = M_1, \qquad z \notin \bigcup_{j \in B} C_j. \tag{2.3}$$

Since the discs C_j are pairwise disjoint, it is obvious that for each continuous curve γ tending to ∞ we have

$$\liminf_{\substack{z \to \infty \\ z \in \gamma}} |f(z)| \leq M_1$$

and hence, ∞ cannot be an asymptotic value for $f(z)$.

If $z \in C_j$ then, using (2.2) and the same argument as in the proof of (2.3), we get

$$|H_j(z)| - M_1 \leq |f(z)| \leq |H_j(z)| + M_1, \qquad z \in C_j. \tag{2.4}$$

By (2.3) and (2.4), everywhere in the complex plane the relations

$$\ln^+ |f(z)| = \sum_{j \in B} \ln^+ |H_j(z)| + \Omega(z), \quad |\Omega(z)| \leq \ln(2M_1)$$

hold; hence

$$m(r, f) = \sum_{j \in B} m(r, H_j) + O(1). \tag{2.5}$$

Let $k \geq 1$. Then

$$\begin{aligned} m(r, H_k) &= [a^k] \frac{1}{2\pi} \int_0^{2\pi} \ln^+ \frac{3 \cdot 4^k}{|re^{i\varphi} - 5 \cdot 4^k|} \, d\varphi \\ &= [a^k] \frac{1}{2\pi} \int_0^{2\pi} \ln^+ \frac{3}{|4^{-k} r e^{i\varphi} - 5|} \, d\varphi = [a^k] I_1 \left(\frac{r}{4^k} \right), \end{aligned} \tag{2.6}$$

where the function

$$I_1(t) = \frac{1}{2\pi} \int_0^{2\pi} \ln^+ \frac{3}{|te^{i\varphi} - 5|} d\varphi$$

is continuous for $t \geq 0$, positive for $2 < t < 8$, and is equal to zero outside this interval. Similarly

$$m(r, H_{-k}) = [a^k] I_2 \left(\frac{r}{4^k} \right), \tag{2.7}$$

where the function

$$I_2(t) = \frac{1}{2\pi} \int_0^{2\pi} \ln^+ \frac{6}{|te^{i\varphi} + 10|} d\varphi$$

is continuous for $t \geq 0$, positive for $4 < t < 16$, and is equal to zero outside this interval.

Let $2 \cdot 4^k \le r \le 8 \cdot 4^k$, $k = 2, 3, \dots$. Then

$$\begin{aligned} m(r,f) + O(1) &= \sum_{j=1}^{\infty} [a^j] I_1\left(\frac{r}{4^j}\right) + \sum_{j=1}^{\infty} [a^j] I_2\left(\frac{r}{4^j}\right) \\ &= [a^k] I_1\left(\frac{r}{4^k}\right) + [a^k] I_2\left(\frac{r}{4^k}\right) + [a^{k-1}] I_2\left(\frac{r}{4^{k-1}}\right) \\ &= O(1) + a^k\left\{ I_1\left(\frac{r}{4^k}\right) + I_2\left(\frac{r}{4^k}\right) + \frac{1}{a} I_2\left(\frac{r}{4^{k-1}}\right)\right\}. \end{aligned} \tag{2.8}$$

Let $r4^{-k} = t$. The function

$$I(t) = I_1(t) + I_2(t) + \frac{1}{a} I_2(4t)$$

is continuous and positive on the closed interval $[2, 8]$ because $I_j(t) \ge 0$, $j = 1, 2$, $I_1(t) > 0$ on the interval $(2, 8)$, $I_2(t) > 0$ on the interval $(4, 16)$, and $I_2(4t) > 0$ is positive on the interval $(1, 4)$. Therefore

$$0 < \mu_1 = \min_{[2,8]} I(t) \le \max_{[2,8]} I(t) = \mu_2 < \infty.$$

Relation (2.8) implies that for $2 \cdot 4^k \le r \le 8 \cdot 4^k$ the inequality

$$\mu_1 a^k + O(1) \le m(r,f) \le \mu_2 a^k + O(1)$$

holds. But $a = 4^\rho$; therefore

$$\left(\frac{r}{8}\right)^\rho = a^k = (4^k)^\rho \le \left(\frac{r}{2}\right)^\rho,$$

and

$$8^{-\rho}\mu_1 r^\rho + O(1) \le m(r,f) \le 2^{-\rho}\mu_2 r^\rho + O(1). \tag{2.9}$$

It is easy to see that

$$n(r,f) = \sum_{k=1}^{p(r)} [a^k] + \sum_{k=1}^{q(r)} [a^k], \qquad p(r) = \left[\frac{\ln \frac{r}{5}}{\ln 4}\right], \qquad q(r) = \left[\frac{\ln \frac{r}{10}}{\ln 4}\right].$$

Therefore

$$\begin{aligned} n(r,f) &\le \sum_{k=1}^{p(r)} a^k + \sum_{k=1}^{q(r)} a^k = \frac{a^{p(r)+1} - a}{a-1} + \frac{a^{q(r)+1} - a}{a-1} \le \frac{a}{a-1}\{a^{p(r)} + a^{q(r)}\} \\ &\le \frac{2a}{a-1} a^{p(r)} \le \frac{2a}{a-1} 4^{\rho \frac{\ln \frac{r}{5}}{\ln 4}} = \frac{2a}{a-1}\left(\frac{r}{5}\right)^\rho = K_1 r^\rho. \end{aligned}$$

Hence ($r > 20$)

$$N(r,f) \le \frac{K_1}{\rho} r^\rho = K_2 r^\rho. \tag{2.10}$$

Formulas (2.9) and (2.10) imply that

$$T(r,f) \le (2^{-\rho}\mu_2 + K_2) r^\rho + O(1) = K r^\rho + O(1). \tag{2.11}$$

From (2.9) and (2.11) we get

$$\delta(\infty, f) = \liminf_{r\to\infty} \frac{m(r,f)}{T(r,f)} \ge \frac{8^{-\rho}\mu_1}{K} > 0,$$

that is, ∞ is a deficient value. Formulas (2.9) and (2.11) imply that the order of $f(z)$ is equal to ρ. Thus the function $f(z)$ has the desired properties.

We see that in Iversen's Theorem 1.1 the condition $a \in E_P(f)$ cannot be replaced by a weaker condition $a \in E_N(f)$. Now we show that even the condition $\delta(a, f) = 1$ (or the condition $a \in E_B(f)$) does not imply that a is an asymptotic value. Here is an example.

EXAMPLE 2. Let $1 < \rho < \infty$. Let $l_j(r)$, $j = 1, 2$, be two continuously differentiable functions on $[0, \infty)$ satisfying the following conditions:

A. $\alpha \le l_j(r) \le \rho$, where α is a number such that
$$\frac{\rho}{2\rho - 1} < \alpha < 1.$$

B. $|l_j'(r)| = O\{(r \ln r \ln_2 r \ln_3 r)^{-1}\}$ as $r \to \infty$, where $\ln_j r = \ln \ln_{j-1} r$, $\ln_1 r = \ln r$.

C. There exist sequences of closed intervals $[a_n^{(j)}, b_n^{(j)}]$, $n = 1, 2, \dots$; $j = 1, 2$, such that $a_n^{(j)} \to \infty$, $b_n^{(j)} - a_n^{(j)} \ge n^2$, and $l_j(r) \equiv \alpha$ for $a_n^{(j)} \le r \le b_n^{(j)}$.

D. $l(r) = \max\{l_1(r), l_2(r)\} \equiv \rho$ for $0 \le r < \infty$.

It is easy to see that such functions $l_j(r)$ can be constructed.

In what follows $z = re^{i\varphi}$, $|\varphi| \le \pi$. We consider the functions analytic in the plane with a cut along the negative real semiaxis,
$$\lambda_j(z) = z \int_0^\infty \frac{l_j(t)\, dt}{(t+z)^2}, \qquad j = 1, 2.$$
We show that in each angle $\{|\varphi| \le \pi - \delta\}$, $\delta > 0$, the relation
$$\lambda_j(z) = l_j(r) + o\left(\frac{1}{\ln r}\right) \tag{2.12}$$
holds uniformly with respect to φ. As we already observed in Section 5 of Chapter 2, for $|\varphi| \le \pi - \delta$, $0 \le t < \infty$, the inequality
$$|t + z| \ge (t + r) \sin \frac{\delta}{2}$$
holds. Since
$$z \int_0^\infty \frac{dt}{(t+z)^2} = 1 \quad \text{for} \quad |\varphi| < \pi,$$
we have
$$\begin{aligned} |\lambda_j(z) - l_j(r)| &= \left| z \int_0^\infty \frac{l_j(t) - l_j(r)}{(t+z)^2} dt \right| \le \frac{r}{\sin^2 \frac{\delta}{2}} \int_0^\infty \frac{|l_j(t) - l_j(r)|}{(t+r)^2} dt \\ &= \frac{r}{\sin^2 \frac{\delta}{2}} \left(\int_0^{\frac{r}{(\ln r)^2}} + \int_{\frac{r}{(\ln r)^2}}^{r(\ln r)^2} + \int_{r(\ln r)^2}^\infty \right) \frac{|l_j(t) - l_j(r)|}{(t+r)^2} dt. \end{aligned} \tag{2.13}$$

In the first and the third integrals we use the estimate $|l_j(t) - l_j(r)| \le \rho$. If $r(\ln r)^{-2} \le t \le r(\ln r)^2$, then $|\ln(t/r)| \le 2 \ln_2 r$. In view of the mean value theorem, we get
$$l_j(t) - l_j(r) = \left. \frac{dl_j(t)}{d \ln t} \right|_{t=\xi} \ln \frac{t}{r}, \qquad r(\ln r)^{-2} \le \xi \le r(\ln r)^2,$$
from which, using condition B, we obtain
$$|l_j(t) - l_j(r)| = O\{(\ln \xi \ln_2 \xi \ln_3 \xi)^{-1}\} 2 \ln_2 r = O\{(\ln r \ln_3 r)^{-1}\}, \quad r \to \infty.$$

Substituting these estimates in (2.13) we get

$$|\lambda_j(z) - l_j(r)| \le \frac{r}{\sin^2\frac{\delta}{2}}\left(\rho\int_0^{r(\ln r)^{-2}} \frac{dt}{(t+r)^2}\right.$$
$$\left. + O\{(\ln r\ln_3 r)^{-1}\}\int_{r(\ln r)^{-2}}^{r(\ln r)^2} \frac{dt}{(t+r)^2} + \rho\int_{r(\ln r)^2}^{\infty}\frac{dt}{(t+r)^2}\right)$$
$$\le \frac{r}{\sin^2\frac{\delta}{2}}\left(\frac{\rho}{r^2}\int_0^{r(\ln r)^{-2}} dt\right.$$
$$\left. + O\{(\ln r\ln_3 r)^{-1}\}\int_{r(\ln r)^{-2}}^{r(\ln r)^2} \frac{dt}{(t+r)^2} + \rho\int_{r(\ln r)^2}^{\infty}\frac{dt}{t^2}\right)$$
$$= O\{(\ln r)^{-2}\} + O\{(\ln r\ln_3 r)^{-1}\} = o\{(\ln r)^{-1}\},$$

that is, (2.12) has been proved. Note that (2.12) remains true if condition A is replaced by $l_j(r) = o(\ln r)$. Relation (2.12) implies that

$$z^{\lambda_j(z)} = (1+o(1))z^{l_j(r)}, \quad r\to\infty \tag{2.14}$$

in the angle $\{|\varphi| \le \pi - \delta\}$.

Let θ be a number such that

$$1 < \theta < \frac{2\rho\alpha}{\rho+\alpha} < 2.$$

Denote by $\Gamma_j(t)$ the boundary of the region $D_j(t) = \{r > t, |\varphi| < \pi\theta/2l_j(r)\}, t\ge 1$, oriented in such a way that the region $D_j(t)$ is on the right-hand side of $\Gamma_j(t)$. By (2.14) it is clear that for $z\in\Gamma_j(1)$ we have

$$\operatorname{Re} z^{\lambda_j(z)} = r^{l_j(r)}\left(\cos\frac{\pi}{2}\theta + o(1)\right).$$

In the complement of $\bar{D}(1)$ we define an analytic function $F_j(z)$ by a Cauchy-type integral

$$F_j(z) = \frac{1}{2\pi i}\int_{\Gamma_j(1)} \frac{\exp(\zeta^{\lambda_j(\zeta)})}{\zeta - z}d\zeta, \qquad z\notin \bar{D}_j(1),$$

and consider its analytic continuation to the complex z-plane,

$$F_j(z) = \frac{1}{2\pi i}\int_{\Gamma_j(t)} \frac{\exp(\zeta^{\lambda_j(\zeta)})}{\zeta - z}d\zeta, \qquad t > |z|.$$

Thus, $F_j(z)$ is an entire function.

Let θ' and θ'' be arbitrary numbers satisfying $1 < \theta' < \theta < \theta'' < 2$. If in the definition of $\Gamma_j(t)$ and $D_j(t)$ we replace θ by θ' (or θ''), we obtain the functions denoted by $\Gamma_j'(t)$ and $D_j'(t)$ (or by $\Gamma_j''(t)$ and $D_j''(t)$, respectively). Using (2.14), it is easy to check that

$$\operatorname{Re} z^{\lambda_j(z)} \le r^{l_j(r)}\left(\cos\frac{\pi}{2}\theta' + o(1)\right)$$

on the arcs $\{|z| = r,\ \pi\theta'/2l_j(r) < |\varphi| < \pi\theta''/2l_j(r)\}$. Therefore, applying the integral Cauchy theorem (compare with Example 3 in Section 5, Chapter 2), we get:

1) If $z \notin D_j(1)$, then

$$F_j(z) = \frac{1}{2\pi i}\int_{\Gamma_j'(1)} \frac{\exp(\zeta^{\lambda_j(\zeta)})}{\zeta - z}\,d\zeta.$$

2) If $z \in D_j(1)$, then

$$F_j(z) = \frac{1}{2\pi i}\int_{\Gamma_j''(t)} \frac{\exp(\zeta^{\lambda_j(\zeta)})}{\zeta - z}\,d\zeta, \quad t > |z|.$$

If $|z| > R$, where $R > 1$ is sufficiently large, then $z \notin D_j(1)$ implies that the distance from z to $\Gamma_j'(1)$ is greater than 1, and $z \in D_j(1)$ implies that the distance from z to $\Gamma_j''(1)$ is greater than 1. Assume that $|z| > R$. If $z \notin D_j(1)$, then

$$|F_j(z)| \le \frac{1}{2\pi}\int_{\Gamma_j'(1)} \exp(\operatorname{Re}\zeta^{\lambda_j(\zeta)})|d\zeta| < \infty.$$

If $z \in D_j(1)$, then

$$F_j(z) = \frac{1}{2\pi i}\int_{\Gamma_j''(1)} \frac{\exp(\zeta^{\lambda_j(\zeta)})}{\zeta - z}\,d\zeta + \frac{1}{2\pi i}\int_{\Gamma_j''(t)-\Gamma_j''(1)} \frac{\exp(\zeta^{\lambda_j(\zeta)})}{\zeta - z}\,d\zeta,$$

where $\Gamma_j''(t) - \Gamma_j''(1)$ is the boundary of the region $\{1 < r < t,\ |\varphi| < \pi\theta''/2l_j(r)\}$. Since

$$\left|\frac{1}{2\pi i}\int_{\Gamma_j''(1)} \frac{\exp(\zeta^{\lambda_j(\zeta)})}{\zeta - z}\,d\zeta\right| \le \frac{1}{2\pi}\int_{\Gamma_j''(1)} \exp(\operatorname{Re}\zeta^{\lambda_j(\zeta)})|d\zeta| < \infty$$

and

$$\frac{1}{2\pi i}\int_{\Gamma_j''(t)-\Gamma_j''(1)} \frac{\exp(\zeta^{\lambda_j(\zeta)})}{\zeta - z}\,d\zeta = \exp(z^{\lambda_j(z)}),$$

we have

$$F_j(z) = \exp(z^{\lambda_j(z)}) + O(1).$$

As a result we get the following relations for the entire function $F_j(z)$:

$$F_j(z) = \begin{cases} O(1), & z \notin D_j(1), \\ \exp(z^{\lambda_j(z)}) + O(1), & z \in D_j(1). \end{cases}$$

Denote by Δ_1 the region $D_1(1)$, and by Δ_2 the region symmetric to $D_2(1)$ about the imaginary axis. By condition D,

$$\frac{1}{l_1(r)} + \frac{1}{l_2(r)} \le \frac{1}{\rho} + \frac{1}{\alpha} < \frac{2}{\theta}, \quad r \ge 0;$$

therefore

$$\pi - \frac{\pi\theta}{2l_2(r)} > \frac{\pi\theta}{2l_1(r)},$$

and the closed regions $\bar{\Delta}_1$ and $\bar{\Delta}_2$ do not intersect. Denote by Δ_3 the complement of $\Delta_1 \cup \Delta_2$ in the complex plane. Then the function $\Psi(z) = F_1(z) + F_2(-z)$ satisfies the condition

$$\Psi(z) = \begin{cases} \exp\{z^{\lambda_1(z)}\} + O(1), & z \in \Delta_1, \\ \exp\{(-z)^{\lambda_2(-z)}\} + O(1), & z \in \Delta_2, \\ O(1), & z \in \Delta_3. \end{cases} \tag{2.15}$$

Let $\alpha < \beta < 1$, and

$$V(z) = \prod_{n=1}^{\infty} \left(1 - \frac{z}{n^{1/\beta}}\right), \qquad \Phi(z) = V(z)V(-z).$$

Since all zeros of $V(z)$ are positive and $n(r, 0, V) = [r^{\beta}]$, we can use for $V(z)$ the asymptotic formulas (5.3_2) of Chapter 2. Let η be an arbitrarily small number satisfying $0 < \eta < \pi\theta/2\rho$. Then

$$\ln|V(re^{i\varphi})| = r^{\beta}\frac{\pi}{\sin\pi\beta}\cos\beta(\varphi - \pi) + o(r^{\beta}) \tag{2.16}$$

uniformly in φ satisfying $\eta \le \varphi \le 2\pi - \eta$. If we replace the equality sign in (2.16) by $\le$, we get a relation valid for all φ, $0 \le \varphi < 2\pi$. We also need the following lemma.

LEMMA 2.1. *Let* $\mu(r, V) = \min_{|z|=r} |V(z)|$, $r_n = \left(n + \frac{1}{2}\right)^{1/\beta}$, $n = 1, 2, 3, \dots$. *Then*

$$\ln\mu(r_n, V) = \ln|V(r_n)| = r_n^{\beta}\pi\cot\pi\beta + o(r_n^{\beta}). \tag{2.17}$$

Combining (2.16) and (2.17), it is easy to see that (2.16) holds for all φ, $0 \le \varphi < 2\pi$ as $r = r_n \to \infty$. We do not want to interrupt our presentation of Example 2; for this reason we will prove Lemma 2.1 later.

Since

$$\Phi(z) = \prod_{n=1}^{\infty}\left\{1 - \left(\frac{z}{n^{1/\beta}}\right)^2\right\},$$

we obviously have $\mu(r, \Phi) = |\Phi(r)| = |V(r)||V(-r)| = \mu(r, V)M(r, V)$. Therefore

$$\begin{aligned} \ln\mu(r_n, \Phi) &= \ln\mu(r_n, V) + \ln M(r_n, V) \\ &= r_n^{\beta}\left\{\pi\cot\pi\beta + \frac{\pi}{\sin\pi\beta}\right\} + o(r_n^{\beta}) = r_n^{\beta}\pi\cot\frac{\pi\beta}{2} + o(r_n^{\beta}). \end{aligned} \tag{2.18}$$

Hence $\mu(r_n, \Phi) \to \infty$ as $n \to \infty$. Relation (2.16) implies that for $\eta \le |\varphi| \le \pi - \eta$,

$$\begin{aligned} \ln|\Phi(re^{i\varphi})| &= (1 + o(1))A(\beta)\cos\beta\left(|\varphi| - \frac{\pi}{2}\right)r^{\beta}, \\ A(\beta) &= \pi\csc\frac{\pi\beta}{2}, \end{aligned} \tag{2.19}$$

uniformly in φ. Furthermore, as $r = r_n \to \infty$, relation (2.19) holds uniformly in φ, $|\varphi| \le \pi$.

We show that the function $f(z) = \Psi(z)/\Phi(z)$ has the desired properties. Relation (2.19) implies that for $\eta \le |\varphi| \le \pi - \eta$ the function $|\Phi(re^{i\varphi})|$ uniformly approaches ∞ as $r \to \infty$. By (2.15), the function $f(z)$ uniformly approaches 0 as $z \to \infty$, $z \in \Delta_3$. For $r = r_n$ and $|\varphi| \le \pi$ we get from (2.15) and (2.19) that

$$\begin{aligned} |f(re^{i\varphi})| &= \exp\left\{(1 + o(1))\left[r^{l_j(r)}\cos(l_j(r)\varphi) - A(\beta)\cos\beta\left(|\varphi| - \frac{\pi}{2}\right)r^{\beta}\right]\right\} \\ &\quad + o(1) \quad \text{for} \quad re^{i\varphi} \in \Delta_j,\ j = 1, 2. \end{aligned} \tag{2.20}$$

Taking into account condition D, we get

$$(2.21)\qquad (1+o(1))\frac{1}{\pi\rho}r_n^\rho \le m(r_n, f) \le (1+o(1))\frac{2}{\pi\rho}r_n^\rho.$$

It is easy to see that $n(r,f) = n(r,0,\Phi) + O(1) \sim 2r^\beta$ and

$$(2.22)\qquad N(r,f) \sim \frac{2}{\beta}r^\beta.$$

From (2.21) and (2.22) we get

$$(1+o(1))\frac{1}{\pi\rho}r_n^\rho \le T(r_n, f) \le (1+o(1))\frac{2}{\pi\rho}r_n^\rho.$$

Since the function $T(r,f)$ is increasing and $r_{n+1}/r_n \to 1$, for all $r > 0$ we have

$$(2.23)\qquad (1+o(1))\frac{1}{\pi\rho}r^\rho \le T(r, f) \le (1+o(1))\frac{2}{\pi\rho}r^\rho.$$

Formulas (2.22) and (2.23) imply that $\delta(\infty, f) = 1$ and that ∞ is a Borel exceptional value.

Now we show that ∞ is not an asymptotic value. Since $\beta > 1/2$, we have $r_{n+1} - r_n = O(n) = o(n^2)$. Taking into account condition C, we see that for all sufficiently large k, the interval $[a_k^{(j)}, b_k^{(j)}]$ contains a point $r_{n_k}^{(j)}$. On the arcs $\{|z| = r_{n_k}^{(j)},\ z \in \Delta_j\}$, $j = 1,2$, by condition C and relation (2.20) we have ($r = r_{n_k}^{(j)}$)

$$\begin{aligned}|f(re^{i\varphi})| &= \exp\left\{(1+o(1)\left[O(r^\alpha) - A(\beta)\cos\beta\left(|\varphi| - \frac{\pi}{2}\right)r^\beta\right]\right\} + o(1)\\ &\le \exp\left\{-(1+o(1))\pi\cot\frac{\pi\beta}{2}r^\beta\right\} + o(1) = o(1).\end{aligned}$$

Since, as we have already observed, $|f(z)| = o(1)$ as $z \to \infty$, $z \in \Delta_3$, it is clear that ∞ cannot be an asymptotic value.

A slight modification of the example is necessary in the case $\rho = \infty$. Let $1/2 < \alpha < \beta < 1$, $1 < \theta < 2\alpha$. We choose the functions $l_j(r)$ such that they satisfy conditions B and C and also the following conditions:

A′. $\alpha \le l_j(r) < \infty$.

D′. $l(r) \to \infty$, $l(r) = o(\ln r)$ as $r \to \infty$.

E′. If $l_1(r) < 2\theta\alpha/(2\alpha - \theta)$, then $l_2(r) > \theta\alpha$, and if $l_2(r) < 2\theta\alpha/(2\alpha-\theta)$, then $l_1(r) > \theta\alpha$.

With these modifications the previous argument can be used. We leave a verification of this as an exercise for interested readers. Later in this section we will construct an example of an entire function of infinite order for which $\delta(0, f) = 1$, but 0 is not an asymptotic value. However, in that example $0 \notin E_B(f)$.

Now we prove Lemma 2.1.

PROOF. It is clear that $\mu(r, V) = |V(r)|$. Since (2.16) with $=$ replaced by $\le$ holds also for $\varphi = 0$, it remains to show that

$$(2.24)\qquad \ln|V(r_n)| \ge r_n^\beta \pi\cot\pi\beta + o(r_n^\beta).$$

We have

$$
\begin{aligned}
\ln|V(r_n)| &= \sum_{k=1}^{n} \ln\left(\frac{r_n}{k^{1/\beta}}-1\right) + \sum_{k=n+1}^{\infty} \ln\left(1-\frac{r_n}{k^{1/\beta}}\right) \\
&= \sum_{k=1}^{n} \ln\frac{r_n}{k^{1/\beta}} + \sum_{k=1}^{n} \ln\left[1-\left(\frac{k}{n+\frac{1}{2}}\right)^{1/\beta}\right] \\
&\quad + \sum_{k=n+1}^{\infty} \ln\left[1-\left(\frac{n+\frac{1}{2}}{k}\right)^{1/\beta}\right].
\end{aligned} \tag{2.25}
$$

Since $n(r,0,V)=(1+o(1))r^{\beta}$, we get

$$
\sum_{k=1}^{n} \ln\frac{r_n}{k^{1/\beta}} = N(r_n,0,V) = (1+o(1))\frac{r_n^{\beta}}{\beta}. \tag{2.26}
$$

It is easy to check that the functions $\ln(1-x^{\beta})$ and $\ln(1-x^{-\beta})$ are concave on the intervals $(0,1)$ and $(1,\infty)$, respectively. Therefore for $|t|\le 1/2$ we have

$$
\ln\left[1-\left(\frac{k}{n+\frac{1}{2}}\right)^{1/\beta}\right] \ge \frac{1}{2}\left\{\ln\left[1-\left(\frac{k+t}{n+\frac{1}{2}}\right)^{1/\beta}\right] + \ln\left[1-\left(\frac{k-t}{n+\frac{1}{2}}\right)^{1/\beta}\right]\right\}
$$

if $k\le n$, and

$$
\ln\left[1-\left(\frac{n+\frac{1}{2}}{k}\right)^{1/\beta}\right] \ge \frac{1}{2}\left\{\ln\left[1-\left(\frac{n+\frac{1}{2}}{k+t}\right)^{1/\beta}\right] + \ln\left[1-\left(\frac{n+\frac{1}{2}}{k-t}\right)^{1/\beta}\right]\right\}
$$

if $k\ge n+1$. Integrating the first inequality with respect to t from 0 to 1/2, we get

$$
\begin{aligned}
\ln\left[1-\left(\frac{k}{n+\frac{1}{2}}\right)^{1/\beta}\right] &\ge \int_0^{1/2}\left\{\ln\left[1-\left(\frac{k+t}{n+\frac{1}{2}}\right)^{1/\beta}\right] + \ln\left[1-\left(\frac{k-t}{n+\frac{1}{2}}\right)^{1/\beta}\right]\right\} dt \\
&= \int_{-1/2}^{1/2} \ln\left[1-\left(\frac{k+t}{n+\frac{1}{2}}\right)^{1/\beta}\right] dt \\
&= \int_{k-1/2}^{k+1/2} \ln\left[1-\left(\frac{x}{n+\frac{1}{2}}\right)^{1/\beta}\right] dx, \quad k\le n.
\end{aligned}
$$

Similarly,

$$
\ln\left[1-\left(\frac{n+\frac{1}{2}}{k}\right)^{1/\beta}\right] \ge \int_{k-1/2}^{k+1/2} \ln\left[1-\left(\frac{n+\frac{1}{2}}{x}\right)^{1/\beta}\right] dx, \quad k\ge n+1.
$$

Hence

$$
\begin{aligned}
&\sum_{k=1}^{n} \ln\left[1-\left(\frac{k}{n+\frac{1}{2}}\right)^{1/\beta}\right] + \sum_{k=n+1}^{\infty} \ln\left[1-\left(\frac{n+\frac{1}{2}}{k}\right)^{1/\beta}\right] \\
&\geq \int_{1/2}^{n+1/2} \ln\left[1-\left(\frac{x}{n+\frac{1}{2}}\right)^{1/\beta}\right] dx + \int_{n+1/2}^{\infty} \ln\left[1-\left(\frac{n+\frac{1}{2}}{x}\right)^{1/\beta}\right] dx \\
&\geq \int_{0}^{n+1/2} \ln\left[1-\left(\frac{x}{n+\frac{1}{2}}\right)^{1/\beta}\right] dx + \int_{n+1/2}^{\infty} \ln\left[1-\left(\frac{n+\frac{1}{2}}{x}\right)^{1/\beta}\right] dx \\
&= \beta r_n^\beta \left\{\int_0^1 u^{\beta-1}\ln(1-u)du + \int_1^\infty u^{\beta-1}\ln\frac{u-1}{u}du\right\} = \beta r_n^\beta I(\beta).
\end{aligned} \tag{2.27}
$$

Here we have changed the variable $x = \left(n+\frac{1}{2}\right)u^\beta = r_n^\beta u^\beta$. Now we compute $I(\beta)$, $0 < \beta < 1$. Note that the function

$$
I(z) = \int_0^1 u^{z-1}\ln(1-u)\,du + \int_1^\infty u^{z-1}\ln\frac{u-1}{u}\,du \tag{2.28}
$$

is analytic in the strip $\{|\operatorname{Re} z| < 1\}$. Let us compute $I(z)$ for $z = -\sigma$, $0 < \sigma < 1$. We can rewrite

$$
I(-\sigma) = \int_0^\infty u^{-\sigma-1}\ln|1-u|\,du - \int_1^\infty u^{-\sigma-1}\ln u\,du.
$$

The first of these integrals can be easily computed using the residue calculus:

$$
\begin{aligned}
&\int_0^\infty u^{-\sigma-1}\ln|1-u|\,du \\
&\quad= \lim_{\varepsilon\to+0}\left\{\int_{1+\varepsilon}^\infty u^{-\sigma-1}\ln(u-1)\,du + \int_0^{1-\varepsilon} u^{-\sigma-1}\ln(1-u)\,du\right\} \\
&\quad= \lim_{\varepsilon\to+0}\left\{-\frac{1}{\sigma}u^{-\sigma}\ln(u-1)\big|_{1+\varepsilon}^\infty + \frac{1}{\sigma}\int_{1+\varepsilon}^\infty \frac{u^{-\sigma}}{u-1}du\right. \\
&\qquad\qquad\left. - \frac{1}{\sigma}u^{-\sigma}\ln(1-u)\big|_0^{1-\varepsilon} + \frac{1}{\sigma}\int_0^{1-\varepsilon}\frac{u^{-\sigma}}{u-1}du\right\} \\
&\quad= \lim_{\varepsilon\to+0}\frac{1}{\sigma}\ln\varepsilon\{(1+\varepsilon)^{-\sigma}-(1-\varepsilon)^{-\sigma}\} + \frac{1}{\sigma}\,\text{P.V.}\int_0^\infty \frac{u^{-\sigma}}{u-1}du \\
&\quad= \frac{\pi}{\sigma}\cot\pi\sigma.
\end{aligned} \tag{2.29}
$$

The second integral can be easily computed using integration by parts:

$$
\int_1^\infty u^{-\sigma-1}\ln u\,du = \frac{1}{\sigma^2}. \tag{2.30}
$$

Thus

$$
I(-\sigma) = \frac{\pi}{\sigma}\cot\pi\sigma - \frac{1}{\sigma^2}. \tag{2.31}
$$

The function

$$
\frac{\pi}{z}\cot\pi z - \frac{1}{z^2}
$$

is analytic in the strip $\{|\operatorname{Re} z|<1\}$ and coincides with $I(z)$ on the interval $\{\operatorname{Im} z=0, -1<\operatorname{Re} z<0\}$. Hence

$$I(z)=\frac{\pi}{z}\cot\pi z-\frac{1}{z^2} \tag{2.32}$$

everywhere in the strip $\{|\operatorname{Re} z|<1\}$ and, in particular,

$$I(\beta)=\frac{\pi}{\beta}\cot\pi\beta-\frac{1}{\beta^2}. \tag{2.33}$$

Combining (2.25), (2.26), (2.27), and (2.33), we get (2.24). □

REMARK. If we use the representation

$$\pi\cot\pi z=\frac{1}{z}+\sum_{n=1}^{\infty}\frac{2z}{z^2-n^2},$$

we can come to equality (2.32) in an easier way. Replacing u by u^{-1} in the second integral in (2.28), we get

$$\begin{aligned}I(z)&=\int_0^1 u^{z-1}\ln(1-u)\,du+\int_0^1 u^{-z-1}\ln(1-u)\,du\\&=-\int_0^1\left\{\sum_{n=1}^{\infty}\frac{u^{n+z-1}+u^{n-z-1}}{n}\right\}du=2\sum_{n=1}^{\infty}\frac{1}{z^2-n^2}=\frac{\pi}{z}\cot\pi z-\frac{1}{z^2}.\end{aligned}$$

It is worth mentioning that we will use equalities (2.29) and (2.30) in Section 3 (Lemma 3.3).

Now we construct an example showing that a meromorphic function of order ρ, $0<\rho<\infty$, can have a Borel exceptional value which is not asymptotic. In the same example, a Borel exceptional value is not deficient.

EXAMPLE 3. Let $0<\rho<\infty$, $0<\alpha<\beta<\min(\rho,1/2)$, and let n be an odd integer exceeding 2ρ. Let $V(z)$ be a function from Lemma 2.1. Take a function $l(r)$ continuously differentiable for $r\ge 0$ and such that $l'(r)r\ln r\to 0$ as $r\to\infty$ and $\alpha/n\le l(r)\le\rho/n<1/2$, where $l(r_k^n)=\rho/n$ for a sequence $r_k\to\infty$, and for some increasing sequence $\{n_j\}$ of positive integers, we have

$$l\left(\left(n_j+\frac{1}{2}\right)^{n/\beta}\right)=\frac{\alpha}{n}. \tag{2.34}$$

It is easy to see that such a function $l(r)$ exists. Let $\Phi(z)$ be a canonical product of genus zero with positive zeros a_j such that $n(r,0,\Phi)\sim r^{l(r)}$ as $r\to\infty$.

Using relations $(5.2_2')$ and (5.4) in Chapter 2 we get

$$\ln M(r,\Phi)\sim\pi\csc(\pi l(r))r^{l(r)},$$

$$T(r,\Phi)\sim\frac{1}{l(r)}r^{l(r)}.$$

Let $\Psi(z)=\Phi(z^n)$, $l_1(r)=l(r^n)$. By Theorem 6.6 of Chapter 1 we have

$$\ln M(r,\Psi)\sim\pi\csc(\pi l_1(r))r^{nl_1(r)}, \tag{2.35}$$

$$T(r,\Psi)\sim\frac{1}{l_1(r)}r^{nl_1(r)}. \tag{2.36}$$

Since n is odd, $\Psi(z)$ does not have negative zeros. Then the function $f(z) = \Psi(z)/V(-z)$ has the desired properties. It is obvious that $N(r,\infty,f) = N(r,0,V) \sim \frac{1}{\beta}r^{\beta}$. The function $\Psi(z)$ has order ρ and the function $V(-z)$ has order $\beta < \rho$; therefore the order of the function $f(z)$ is equal to ρ, and $\infty \in E_B(f)$. We show that ∞ is not an asymptotic value. Let $q_j = (n_j + \frac{1}{2})^{1/\beta}$. By (2.34) we have $nl_1(q_j) = \alpha$. Taking into account (2.17) and (2.35) we get

$$\begin{aligned}\ln M(q_j, f) &\leq \ln M(q_j, \Psi) - \ln \mu(q_j, V)\\ &= q_j^{\alpha}\pi \csc\frac{\pi\alpha}{n} + o(q_j^{\alpha}) - q_j^{\beta}\pi\cot\pi\beta + o(q_j^{\beta})\\ &= -q_j^{\beta}\pi\cot\pi\beta + o(q_j^{\beta}) \to -\infty.\end{aligned}$$

Hence $M(q_j, f) = o(1)$ and ∞ cannot be an asymptotic value.

By (2.36) we have

$$\begin{aligned}T(q_j, f) &\geq T(q_j, V) - T(q_j, \Psi) + O(1)\\ &= \frac{1}{\beta}q_j^{\beta} + o(q_j^{\beta}) - \frac{1}{l_1(q_j)}q_j^{\alpha} + o(q_j^{\alpha}) = \frac{1}{\beta}q_j^{\beta} + o(q_j^{\beta}),\\ N(q_j,\infty,f) &= \frac{1}{\beta}q_j^{\beta} + o(q_j^{\beta});\end{aligned}$$

hence $\delta(\infty, f) = 0$.

REMARK 1. The same idea makes it possible to construct a similar example with $\rho = \infty$. It will be necessary to replace the function $\Psi(z)$ by the entire function that was constructed in Section 6 of Chapter 4 (see (6.7) in Chapter 4).

REMARK 2. In Example 2 we not only had the value ∞ as a non-asymptotic Borel exceptional value, but also the order of $N(r,\infty,f)$ was less than the lower order of $T(r,f)$. We do not know[3] whether it is possible to construct such an example with $\rho \leq 1$. In Example 3 we can take $\rho \leq 1$, but there, the order of $N(r,\infty,f)$ exceeds the order of $T(r,f)$.

We have seen that if for a meromorphic function $f(z)$ the order of $N(r,a,f)$ is less than the order of $T(r,f)$, then the value a does not have to be asymptotic. However the following assertion is true.

THEOREM 2.1. *Let $f(z)$ be a meromorphic function of order $\rho < \infty$, and let the orders of $N(r,a)$ and $N(r,b)$, $a \neq b$, be less than ρ. Then a and b are asymptotic values.*

PROOF. It is clear that without loss of generality we may assume that $a = 0$, $b = \infty$. The values 0 and ∞ are Borel exceptional values; hence, as was shown in Section 1 of Chapter 4, the order ρ is a positive integer. Let $\pi_1(z)$ and $\pi_2(z)$ be canonical products built using the zeros and poles of $f(z)$, respectively. The order of $\pi_j(z)$ is equal to $\rho_j < \rho$, and the genus is equal to $p_j \leq \rho - 1$; here ρ_1 is the order of $N(r,0)$, and ρ_2 is the order of $N(r,\infty)$.

The function $f(z)$ can be represented in the form (see Theorem 4.1 of Chapter 2):

$$f(z) = z^m e^{\alpha z^{\rho}} e^{Q(z)} \frac{\pi_1(z)}{\pi_2(z)}, \tag{2.37}$$

[3]This seems to be still unknown.

where m is an integer, $\alpha \neq 0$, and $Q(z)$ is a polynomial of degree at most $\rho - 1$. Let

$$1 > \varepsilon > 0; \quad \rho_j + \varepsilon < \rho,\ j = 1, 2; \quad h = \max(\rho - 1 + \varepsilon, \rho_1 + \varepsilon, \rho_2 + \varepsilon) < \rho.$$

Since for $r \leq r_0$ the inequalities $\ln M(r, e^{-Q}) \leq r^h$, $\ln M(r, \pi_j) \leq r^h$ hold, we have

$$\begin{aligned} \ln|f(z)| &\geq \operatorname{Re}\{\alpha z^\rho\} + \ln|\pi_1(z)| + m \ln|z| - \ln M(r, e^{-Q}) - \ln M(r, \pi_2) \\ &= \operatorname{Re}\{\alpha z^\rho\} + \ln|\pi_1(z)| + O(|z|^h). \end{aligned} \tag{2.38}$$

Let $a_j \neq 0$ be a sequence of zeros of the function $f(z)$. Then for $r \geq r_1$ we have $n(r, 0, f) \leq r^{\rho_1 + \frac{\varepsilon}{2}}$ and

$$\sum_j |a_j|^{-h} < \infty.$$

Let $|z| = r > 1$, $R = 2r$. Denote by $g_R(z)$ the function

$$g_R(z) = \pi_1(z) \prod_{|a_j| \leq R} \frac{R^2 - \bar{a}_j z}{R(z - a_j)} = \pi_1(z)\omega_R(z).$$

Since $|\omega_R(z)| = 1$ for $|z| = R$, the equality

$$T(R, g_R) = m(R, g_R) = m(R, \pi_1) = T(R, \pi_1) \tag{2.39}$$

holds. We have

$$\ln|\pi_1(z)| \geq \ln \mu(|z|, g_R) - \ln|\omega_R(z)|. \tag{2.40}$$

But

$$\begin{aligned} \ln \mu(r, g_R) &= -\ln M\left(r, \frac{1}{g_R}\right) \geq -\frac{R+r}{R-r} T\left(R, \frac{1}{g_R}\right) \\ &= -3T\left(R, \frac{1}{g_R}\right) = -3\{T(R, g_R) - \ln|g_R(0)|\} \\ &= -3\{T(R, \pi_1) - \ln|\omega_R(0)|\} \geq -3T(R, \pi_1) \\ &\geq -3R^h = -3 \cdot 2^h r^h. \end{aligned} \tag{2.41}$$

We exclude from the complex plane the set $M = \bigcup_j \{|z - a_j| \leq |a_j|^{-h}\}$. Outside this set, for $|z| = r \geq r_1$,

$$\begin{aligned} \ln|\omega_R(z)| &= \sum_{|a_j| \leq R} \ln\left|\frac{R^2 - \bar{a}_j z}{R(z - a_j)}\right| \leq \sum_{|a_j| \leq R} \ln \frac{2R}{|z - a_j|} \\ &\leq \sum_{|a_j| \leq R} \ln(2R|a_j|^h) \leq n(R, 0, f) \ln(2R^{h+1}) \\ &\leq R^{\rho_1 + \frac{\varepsilon}{2}} \ln(2R^{h+1}) < R^h = 2^h r^h. \end{aligned} \tag{2.42}$$

Combining inequalities (2.38), (2.40), (2.41), (2.42), we see that outside the set M,

$$\ln|f(z)| \geq \operatorname{Re}\{\alpha z^\rho\} + O(|z|^h). \tag{2.43}$$

Let $0 < \eta < \frac{\pi}{2}$. In the angle $W = \left\{|\arg \alpha + \rho \arg z| < \frac{\pi}{2} - \eta\right\}$ the inequality $\operatorname{Re}\{\alpha z^\rho\} \geq |z|^\rho |\alpha| \sin \eta$ holds. On the other hand, the sum of the radii of the discs forming M is finite. Therefore the open set $W \backslash M$ has one unbounded connected component, in which we can draw a continuous curve C approaching ∞. On the curve C

$$\ln|f(z)| \geq |z|^\rho|\alpha|\sin\eta + O(|z|^h) = (1+o(1))|z|^\rho|\alpha|\sin\eta \to \infty$$

as $z \to \infty$; that is, the curve C is an asymptotic curve with ∞ as an asymptotic value. Applying this statement to $1/f(z)$, we get that 0 is also an asymptotic value. □

It is still unknown whether Theorem 2.1 remains true without the assumption $\rho < \infty$.[4]

The assumption that the orders of $N(r,a)$ and $N(r,b)$ are less than the order of $T(r,f)$ in Theorem 2.1 cannot be replaced by the assumption that a and b are Borel exceptional values. This is shown in our next example.

EXAMPLE 4. We construct an entire function $f(z)$ of the first order such that $0 \in E_B(f)$, but 0 is not an asymptotic value. Let $f(z)$ be a canonical product of genus 1,

$$f(z) = \prod_{m=1}^{\infty}\left(1 - \frac{z}{a_m}\right)e^{\frac{z}{a_m}},$$

such that its roots are real,

$$n(r,0) = O(r), \tag{2.44}$$

and there exists a sequence $4 < r_1 < r_2 < r_3 < \cdots$ such that the following conditions are satisfied:

(a)

$$r_{3k+3} > 2e^{r_{3k+2}}, \quad k = 0,1,\dots. \tag{2.45}$$

(b) The function $f(z)$ does not have zeros in the annuli $Q_k = \{2r_{3k+2} \leq |z| \leq r_{3k+3}\}$, $k = 0,1,\dots$.

(c)

$$\sum_{|a_m|\leq r_{3k+1}} \frac{1}{a_m} > k+2, \quad k = 0,1,\dots. \tag{2.46}$$

(d)

$$\sum_{|a_m|<r} \frac{1}{a_m} = 0, \; r_{3k+2} < r < r_{3k+3}, \quad k = 0,1,\dots. \tag{2.47}$$

Let us assume for a moment that sequences $\{a_m\}$ and $\{r_k\}$ with the desired properties exist and show that $f(z)$ is the desired function.

The order of $n(r,0)$ cannot be less than one because otherwise the series $\sum_m \frac{1}{|a_m|}$ would converge in contradiction with (2.46). Since $f(z)$ is a canonical product of genus one, its order is also equal to one. To determine the type of $T(r,f)$, we use the Lindelöf theorem (Theorem 4.4 of Chapter 2). For $r = r_{3k+1}$ we have, using (2.46),

$$K(r) = \sum_{|a_m|\leq r} \frac{1}{a_m} > k \to \infty;$$

[4]Not anymore. Hayman [**76**] constructed an entire function f of infinite lower order for which $N(r,0,f)$ has order $1/2$, but zero is not an asymptotic value.

therefore $T(r,f)$ has maximal type and, by (2.44), $0 \in E_B(f)$.

Let us show that zero is not an asymptotic value. For this it is enough to establish that the function $f(z)$ approaches ∞ uniformly with respect to $\arg z$ on the sequence $\{|z| = R_k = \exp\left(\frac{1}{2}r_{3k+2}\right)\}$ of circles. It is clear that $2r_{3k+2} < R_k < \frac{1}{2}r_{3k+3}$. Since $f(z)$ does not have zeros in the annuli Q_k, by (2.47), we can write

$$f(z) = g_k(z)h_k(z), \quad k = 0, 1, 2, \dots, \tag{2.48}$$

where

$$g_k(z) = \prod_{|a_m|<2r_{3k+2}} \left(1 - \frac{z}{a_m}\right), \qquad h_k(z) = \prod_{|a_m|>r_{3k+3}} E\left(\frac{z}{a_m}, 1\right).$$

Since for $|u| < 1/2$ we have $|\ln E(u,1)| \le \frac{|u|^2}{2}(1 + |u| + |u|^2 + \cdots) \le |u|^2$, it follows, taking into account (2.44), that

$$\begin{aligned}
|\ln h_k(z)| &\le \sum_{|a_m|>r_{3k+3}} \left|\ln E\left(\frac{z}{a_m}, 1\right)\right| \le \sum_{|a_m|>r_{3k+3}} \left|\frac{z}{a_m}\right|^2 \\
&= R_k^2 \int_{r_{3k+3}}^{\infty} \frac{dn(t,0)}{t^2} \le 2R_k^2 \int_{r_{3k+3}}^{\infty} \frac{n(t,0)}{t^3} dt \\
&= O\left(R_k^2 \int_{r_{3k+3}}^{\infty} t^{-2} dt\right) = O(R_k^2 r_{3k+3}^{-1}) = O(1)
\end{aligned} \tag{2.49}$$

for $|z| = R_k$ as $k \to \infty$. On the other hand, for $|z| = R_k$ we have

$$\ln|g_k(z)| \ge \sum_{|a_m|<2r_{3k+2}} \ln\left(\frac{R_k}{|a_m|} - 1\right) \ge \ln\left(\frac{R_k}{r_{3k+2}} - 1\right) \to \infty.$$

Therefore $\ln|f(z)| \ge \ln|g_k(z)| - |\ln|h_k(z)|| = \ln|g_k(z)| + O(1)$ on the circles $\{|z| = R_k\}$, and the function $f(z)$ uniformly approaches ∞. From this it follows, among other things, that $\delta(0,f) = 0$.

Now we describe the construction of the sequences $\{a_m\}$ and $\{r_k\}$. This construction is based on the divergence of the series $\sum_{n=1}^{\infty} \frac{1}{n}$.

Consider an interval $(0, r_1)$, where $r_1 > 4$ is not an integer. Assume that the interval is long so that

$$\sum_{n\in(0,r_1)} \frac{1}{n} > 2.$$

Now we consider an interval (r_1, r_2), $r_1 < r_2$, where r_2 is not an integer and is such that

$$\sum_{n\in(0,r_1)} \frac{1}{n} - \sum_{n\in(r_1,r_2)} \frac{1}{n} \ge 0, \qquad \sum_{n\in(0,r_1)} \frac{1}{n} - \sum_{n\in(r_1,r_2+1)} \frac{1}{n} < 0. \tag{2.50}$$

We include into the sequence $\{a_m\}$ all integers in $(0, r_1)$ and all integers in $(-r_2, -r_1)$.

Further, we include into $\{a_m\}$ a number b_1 from the interval $(-[r_2] - 1, -[r_2])$, chosen in such a way that

$$\sum_{n\in(0,r_1)} \frac{1}{n} - \sum_{n\in(r_1,r_2)} \frac{1}{n} + \frac{1}{b_1} = 0.$$

Inequalities (2.50) imply that such choice is possible.

Let r_3 be an arbitrary non-integer satisfying $r_3 > 2e^{r_2}$.

Now we choose an interval (r_3, r_4), $r_3 < r_4$, where r_4 is non-integer, in such a way that

$$\sum_{n\in(r_3,r_4)} \frac{1}{n} > 3,$$

and interval (r_4, r_5), $r_4 < r_5$, where r_5 is non-integer, in such a way that

$$\sum_{n\in(r_3,r_4)} \frac{1}{n} - \sum_{n\in(r_4,r_5)} \frac{1}{n} \geq 0, \qquad \sum_{n\in(r_3,r_4)} \frac{1}{n} - \sum_{n\in(r_4,r_5+1)} \frac{1}{n} < 0.$$

We include into $\{a_m\}$ all integers in (r_3, r_4) and all integers in $(-r_5, -r_4)$, and also a number b_2 satisfying $-[r_5] - 1 < b_2 < -[r_5]$ and such that

$$\sum_{n\in(r_3,r_4)} \frac{1}{n} - \sum_{n\in(r_4,r_5)} \frac{1}{n} + \frac{1}{b_2} = 0.$$

Further, we choose a non-integer $r_6 > 2e^{r_5}$ and continue in an obvious way. It is clear that the sequences $\{a_m\}$ and $\{r_k\}$ chosen using this procedure have the desired properties.

REMARK 1. The constructed function $f(z)$ satisfies $m(R_k, 0, f) \to 0$; hence $\delta(0, f) = 0$. Thus $0 \in E_B(f)\backslash E_N(f)$.

REMARK 2. The lower order of the constructed function $f(z)$ is equal to zero. In fact, by (2.48), (2.49), and (2.44), we get

$$\begin{aligned} \ln M(R_k, f) &= \ln M(R_k, g_k) + O(1) \\ &\leq \sum_{|a_m|<2r_{3k+2}} \ln\left(1 + \frac{R_k}{|a_m|}\right) + O(1) \\ &\leq n(2r_{3k+2}, 0)\ln(1 + R_k) + O(1) \\ &= O(r_{3k+2}\ln R_k) = O(\ln^2 R_k). \end{aligned}$$

The order ρ of the function $f(z)$ in our example is equal to 1. If we take $F(z) = f(z^n)$, $n = 2, 3, \ldots$, we get an analogous example with $\rho = n$. It is not known whether such an example exists for functions of infinite order.[5]

We see that in Theorem 2.1 the condition that the orders of $N(r, a)$ and $N(r, b)$ are lower than the order of $T(r, f)$ cannot be replaced by the condition that the categories of $N(r, a)$ and $N(r, b)$ are lower than the category of $T(r, f)$. However, Edrei and Fuchs [**EF59a**] proved the following theorem: *Let $f(z)$ be a meromorphic function of a finite order and $\delta(a) = \delta(b) = 1$, $a \neq b$ (that is, $N(r, a) = o(T(r, f))$ and $N(r, b) = o(T(r, f))$). Then a and b are asymptotic values.*[6] We cannot present a proof of this theorem here; we only give an example showing that this theorem is not valid for functions of infinite order.

EXAMPLE 5. We construct an entire function $f(z)$ of infinite order such that $\delta(0, f) = 1$, but 0 is not an asymptotic value (since now we are dealing with functions of infinite order, this example is stronger than Example 2).

---

[5]Now it is known to exist; see Hayman [**76**].

[6]For meromorphic functions of finite lower order with sum of deficiencies equal to two, all deficient values are asymptotic. See Appendix B at the end of the book.

First of all, we need a function $E(z)$ defined in the following way. Let a be a real number, $A(a)$ a half-strip $\{x > a,\ |y| < \pi\}$, $z = x + iy$, and $L(a)$ the boundary of $A(a)$ oriented in such a way that $A(a)$ is on the right. We define the function $E(z)$ in the complement of $\overline{A(0)}$ by the Cauchy-type integral:

$$E(z) = \frac{1}{2\pi i}\int_{L(0)} \frac{e^{e^t}}{t-z}\,dt. \tag{2.51}$$

Obviously, the integral

$$\frac{1}{2\pi i}\int_{L(a)} \frac{e^{e^t}}{t-z}\,dt$$

for $a > 0$ is an analytic continuation of $E(z)$ onto the complement of $\overline{A(a)}$. Hence the function (2.51) has analytic continuation to the entire complex plane, which means that it is an entire function.

Since by the Cauchy integral theorem, the integral of the function e^{e^t} over the boundary of the rectangle $\{a_1 < x < a_2,\ |y| < \pi\}$ is equal to zero, we have

$$\frac{1}{2\pi i}\int_{L(a_1)} e^{e^t}\,dt = \frac{1}{2\pi i}\int_{L(a_2)} e^{e^t}\,dt,$$

and the integral

$$I = \frac{1}{2\pi i}\int_{L(a)} e^{e^t}\,dt$$

does not depend on a. Since the function e^{e^t} is periodic with period $2\pi i$, the integrals over the rays $\{y = \pm\pi,\ x \geq a\}$ cancel each other, and we get

$$I = \frac{1}{2\pi}\int_{-\pi}^{\pi} e^{e^{a+iy}}\,dy.$$

Letting $a \to -\infty$ in the right-hand side of this equality, we get $I = 1$. Let $z \in \overline{A(0)}$. Then

$$\begin{aligned} E(z) + \frac{1}{z} &= \frac{1}{2\pi i}\int_{L(2)} \frac{e^{e^t}}{t-z}\,dt + \frac{1}{2\pi i}\int_{L(2)} \frac{e^{e^t}}{z}\,dt = \frac{1}{2\pi i}\int_{L(2)} \frac{t}{z(t-z)} e^{e^t}\,dt \\ &= \frac{1}{z^2}\frac{1}{2\pi i}\int_{L(2)} \left(-t + \frac{t^2}{t-z}\right) e^{e^t}\,dt = \frac{1}{z^2}\Omega(z). \end{aligned} \tag{2.52}$$

If the distance from z to $L(2)$ is not less than $\pi/3$, then we have $|-t+t^2(t-z)^{-1}| \leq |t| + |t|^2 \leq 2|t|^2$. Since the equality $|\exp(e^{x+iy})| = \exp(-e^x)$ holds on the rays $\{x > 1,\ y = \pm\pi\}$, the function $\Omega(z)$ is bounded by a constant which does not depend on z. If the distance from z to $L(2)$ is less than $\pi/3$, we replace the part of $L(2)$ inside the disc $\{|t-z| < \pi/3\}$ by an arc of the circle $\{|t-z| = \pi/3\}$ lying in $A(2)$. By the Cauchy integral theorem, the value of the last integral in (2.52) is not affected by this change of the path of integration. As before, we have $|-t+t^2(t-z)^{-1}| \leq 2|t|^2$, and for points z on the arc $\{|t-z| = \pi/3\}$ we have the inequality $|\exp(e^t)| \leq \exp\left(-\frac{1}{2}e^{\operatorname{Re} t}\right)$. Thus in this case the function $\Omega(z)$ is also bounded. If $z \in L(0)$ we can use similar reasoning. We get that

$$E(z) = -\frac{1}{z} + O\left(\frac{1}{|z|^2}\right), \qquad z \notin A(0). \tag{2.53}$$

Now let $z \in A(0)$. Than, taking $a > \operatorname{Re} z > 0$, we have

$$\begin{aligned} E(z) &= \frac{1}{2\pi i}\int_{L(a)} \frac{e^{e^t}}{t-z}\,dt = \frac{1}{2\pi i}\int_{L(-2)} \frac{e^{e^t}}{t-z}\,dt + \frac{1}{2\pi i}\int_{L(a)-L(-2)} \frac{e^{e^t}}{t-z}\,dt \\ &= \frac{1}{2\pi i}\int_{L(-2)} \frac{e^{e^t}}{t-z}\,dt + e^{e^z}. \end{aligned}$$

As before, we get that

$$E(z) - e^{e^z} + \frac{1}{z} = \frac{1}{z^2}\frac{1}{2\pi i}\int_{L(-2)} \left(-t + \frac{t^2}{t-z}\right) e^{e^t}\,dt.$$

The boundedness of the last integral is proved as before; only in this case, when the distance from z to $L(-2)$ is less than $\pi/3$, we replace part of $L(-2)$ by an arc of the circle $\{|t-z| = \pi/3\}$ lying outside $A(-2)$. Thus

$$E(z) = e^{e^z} - \frac{1}{z} + O\left(\frac{1}{|z|^2}\right), \qquad z \in A(0). \tag{2.54}$$

We write A instead of $A(0)$. Relations (2.53) and (2.54) imply that

$$E(z) = \begin{cases} e^{e^z} + \dfrac{\Psi_1(z)}{z}, & z \in A, \\ \dfrac{\Psi_2(z)}{z}, & z \notin A, \end{cases} \tag{2.55}$$

where $|\Psi_j(z)| < 2$ for $|z| > r_0 \geq 1$, $j = 1, 2$.

Let $K > 8$ be a real number such that the open set $G \subset A$ defined as the set of all solutions of the inequality $|e^{e^z} + K| < 7$ lies in the half-plane $\{x > 4r_0\}$ and the diameter of each connected component of G is less than π. Obviously, sufficiently large numbers K have this property. Let

$$q_n = \sqrt{(2\pi 2^n)^2 - 1}, \qquad \theta_n = \arcsin(2\pi 2^n)^{-1},$$

$$\Phi(z) = K + \sum_{n=1}^{\infty} E\{e^{i\theta_n}(z - iq_n)\}. \tag{2.56}$$

Uniform convergence of the series (2.56) in each finite disc can be easily checked; hence $\Phi(z)$ is an entire function.

Denote by A^0 the half-strip $\{x > 0,\ |y| < \pi/2\}$, and by A_n, A_n^0, G_n the images of A, A^0, G, respectively, under the mapping $\lambda_n(z) = e^{-i\theta_n}z + iq_n$. It is easy to see that the half-strips A_n are pairwise disjoint, $G_n \subset A_n$ and $\bigcup_{n=1}^{\infty} G_n \subset H = \{x > 2r_0\}$. Denote by C_n the disc $\{|z - iq_n| < r_0\}$. Let us show that

$$\Omega(z) = \sum_{n=1}^{\infty} |z - iq_n|^{-1} < 3 \tag{2.57}$$

for $z \notin C = \bigcup_{n=1}^{\infty} C_n$. In fact, if $q_m \le y \le q_{m+1}$, $m = 1, 2, \ldots$, then

$$\begin{aligned}
\Omega(z) &\le 2 + \sum_{n=1}^{m-1} |z - iq_n|^{-1} + \sum_{n=m+2}^{\infty} |z - iq_n|^{-1} \\
&\le 2 + \sum_{n=1}^{m-1} |y - q_n|^{-1} + \sum_{n=m+2}^{\infty} |y - q_n|^{-1} \\
&\le 2 + \sum_{n=1}^{m-1} (q_m - q_n)^{-1} + \sum_{n=m+2}^{\infty} (q_n - q_{m+1})^{-1} \\
&\le 2 + \sum_{n=1}^{m-1} \frac{3}{2\pi 2^m} + \sum_{n=m+2}^{\infty} \frac{3}{2\pi 2^n} \\
&\le 2 + \frac{3}{2\pi}\frac{m-1}{2^m} + \frac{3}{8\pi} \le 2 + \frac{3}{8\pi} + \frac{3}{8\pi} < 3.
\end{aligned}$$

If $y \le q_1$, the same estimate works. Relations (2.55), (2.56), (2.57) imply that

$$\Phi(z) = \begin{cases} Q(z), & z \notin \{\bigcup_{n=1}^{\infty} A_n\} \cup C, \\ \exp\exp\{e^{i\theta_n}(z - iq_n)\} + Q(z), & z \in A_n \backslash C, \end{cases} \tag{2.58}$$

where $|Q(z) - K| < 6$. Denote by D the set of those $z \notin C$ for which $|\Phi(z)| < 1$. It is easy to see that $D \subset \bigcup_{n=1}^{\infty} G_n$. In fact, $|Q(z)| \ge K - 6 \ge 2$; therefore $D \subset \bigcup_{n=1}^{\infty} A_n$. Let $z \in D \cap A_n$. Then

$$|\exp\exp\{e^{i\theta_n}(z - iq_n)\} + K| \le |\Phi(z)| + |K - Q(z)| < 1 + 6 = 7,$$

that is, $z \in G_n$. Hence $D \subset H$ and the diameter of each connected component of D is less than π.

Consider the function $g(z) = \Phi(z)\exp\exp z$. Let us show that zero is not an asymptotic value for this function. In the half-plane $H_1 = \{x \le 2r_0\}$ the inequality $|\exp\exp z| \ge \exp(-e^{2r_0}) = \mu$, $0 < \mu < 1$, holds, and for $z \notin C$ we have $|\Phi(z)| \ge 1$ since H_1 does not intersect D. Hence, in $H_1 \backslash C$ we have $|g(z)| \ge \mu$. The sets $B_k = \{x \ge 2r_0,\ |y - 2\pi k| \le \pi/2\} \backslash D$, $k = 0, \pm 1, \pm 2, \ldots$, are unbounded continua, since the diameter of each connected component of D is less than π. On the other hand, on B_k we have $|\Phi(z)| \ge 1$ and $|\exp\exp z| = \exp\{e^x \cos y\} \ge 1$, $|g(z)| \ge 1$. On the rays $S_n = \{\arg(z - iq_n) = -\theta_n,\ x \ge 2r_0\}$, $n \ge 1$, the inequalities

$$\begin{aligned}
|\Phi(z)| &= |\exp\exp|z - iq_n| + Q(z)| \\
&\ge \exp\exp|z - iq_n| + \operatorname{Re} Q(z) \\
&> \exp\exp|z - iq_n| \\
&= \exp\exp\{x \sec\theta_n\}, \\
|g(z)| &= \exp\{e^{x\sec\theta_n} + e^x \cos y\} \ge 1
\end{aligned}$$

hold. Obviously, the open set $H \backslash (B \cup S)$, where $B = \bigcup_{k=-\infty}^{\infty} B_k$, $S = \bigcup_{n=1}^{\infty} S_n$, does not have unbounded connected components.

As the final result we get that the open set of those z for which $|g(z)| < \mu$ does not have unbounded connected components. Hence zero is not an asymptotic value for $g(z)$.

Let us estimate $T(r, \Phi(z))$. It is convenient for us to estimate $T(r, \Phi(z-1))$ first. Observe that by the maximum modulus principle, $|\Phi(z)| = O(1)$ for $z \in C$.

Let $z = re^{i\theta}$. Then, by (2.58), we get

$$\begin{aligned} T(r, \Phi(z-1)) &= \frac{1}{2\pi}\int_{|z|=r} \ln^+ |\Phi(z-1)|\, d\theta \\ &\le O(1) + \sum_n \frac{1}{2\pi}\int_{\substack{|z|=r \\ z-1\in A_n^0}} \ln^+ |\Phi(z-1)|\, d\theta \qquad (2.59) \\ &= O(1) + \sum_n \frac{1}{2\pi}\int_{\substack{|z|=r \\ z-1\in A_n^0}} \operatorname{Re}\exp\{e^{i\theta_n}(z-1-iq_n)\}\, d\theta. \end{aligned}$$

But

$$-1 - iq_n = -2\pi 2^n e^{i(\frac{\pi}{2}-\theta_n)} = -i2\pi 2^n e^{-i\theta_n};$$

therefore $\exp\{e^{i\theta_n}(z-1-iq_n)\} = \exp\{re^{i(\theta+\theta_n)}\}$. Suppose that the intersection of the set of those z for which $z-1 \in A_n^0$ with the circle $\{|z| = r\}$ is non-empty, that is, $2\pi 2^n - \frac{\pi}{2} \le r$. Then

$$\begin{aligned} \int_{\substack{|z|=r \\ z-1\in A_n^0}} \operatorname{Re}\exp\{re^{i(\theta+\theta_n)}\}d\theta &= \int_{\alpha_n-\theta_n}^{\beta_n-\theta_n} e^{r\cos(\theta+\theta_n)}\cos\{r\sin(\theta+\theta_n)\}d\theta \qquad (2.60) \\ &= \int_{\alpha_n}^{\beta_n} e^{r\cos\theta}\cos\{r\sin\theta\}d\theta, \end{aligned}$$

where

$$\alpha_n = \arcsin\frac{2\pi 2^n - \pi/2}{r},$$
$$\beta_n = \arcsin\min\left(1, \frac{2\pi 2^n + \pi/2}{r}\right).$$

Substituting (2.60) into (2.59) we get

$$T(r, \Phi(z-1)) \le O(1) + \sum_{n=1}^{\nu(r)} \frac{1}{2\pi}\int_{\alpha_n}^{\beta_n} e^{r\cos\theta}\cos\{r\sin\theta\}d\theta, \qquad (2.61)$$

where $\nu(r) = \left[(\ln 2)^{-1}\ln\frac{r+\pi/2}{2\pi}\right]$. We define a continuous function $\varphi(y)$ on $[0,\infty)$ in the following way. We let $\varphi(y) = \cos y$ for $|y - 2\pi 2^n| \le \frac{\pi}{2}$, $n = 1, 2, \dots$, and let $\varphi(y) = 0$ for all other $y \ge 0$. Then the inequality (2.61) can be rewritten in the following way:

$$T(r, \Phi(z-1)) \le O(1) + \frac{1}{2\pi}\int_0^{\pi/2} e^{r\cos\theta}\varphi(r\sin\theta)d\theta. \qquad (2.62)$$

It is clear that

$$\frac{1}{r}\int_0^r \varphi(y)dy \to 0 \quad \text{as} \quad r\to\infty.$$

Applying Lemma 4.1 of Chapter 4 to the integral in the right-hand side of (2.62), we get $T(r, \Phi(z-1)) = o(e^r/\sqrt{r})$. Since $T(r, \Phi(z)) \le (1+o(1))T(r+1, \Phi(z-1))$ (see (1.4) in Chapter 2), we have

$$T(r, \Phi(z)) = o(e^r/\sqrt{r}). \qquad (2.63)$$

On the other hand, as was shown in the construction of Example 1 in Section 6 of Chapter 4, we have

$$T(r, e^{e^z}) = (1+o(1))\frac{e^r}{\sqrt{2\pi^3 r}}, \qquad r\to\infty. \qquad (2.64)$$

Since $T(r, e^{e^z}) - T(r, \Phi^{-1}) \le T(r, g) \le T(r, e^{e^z}) + T(r, \Phi)$, relations (2.63) and (2.64) imply that

$$T(r, g) = (1 + o(1))\frac{e^r}{\sqrt{2\pi^3 r}}, \qquad r \to \infty.$$

But $N(r, 0, g) = N(r, 0, \Phi) \le T(r, \Phi) + O(1) = o(T(r, g))$. Hence $\delta(0, g) = 1$ and the function $g(z)$ has the desired properties.

3. Meromorphic functions of order less than 1/2

By the Wiman–Heins Theorem 1.3 for an entire function of order $\lambda < 1/2$, there exists a sequence of circles $\{|z| = r_k\}$, $r_k \to \infty$, on which $f(z)$ approaches ∞ uniformly in $\arg z$. The following theorem is a generalization of this result for meromorphic functions.

THEOREM 3.1. *Let $f(z)$ be a meromorphic function of lower order $\lambda < 1/2$. If*

$$\delta(\infty, f) > 1 - \cos \pi\lambda, \tag{3.1}$$

then there exists a sequence of circles $\{|z| = r_k\}$, $r_k \to \infty$, on which $f(z)$ approaches ∞ uniformly in $\arg z$.

Observe that if condition (3.1) is replaced by the condition $\delta(a, f) > 1 - \cos \pi\lambda$, $a \ne \infty$, we can claim that there exists a sequence of circles $\{|z| = r_k\}$, $r_k \to \infty$, on which $f(z)$ approaches a uniformly in $\arg z$. In fact, for the function $f_1(z) = \frac{1}{f(z) - a}$ we have $\lambda[f_1] = \lambda[f] = \lambda$, $\delta(\infty, f_1) = \delta(a, f)$, so it suffices to apply Theorem 3.1 to $f_1(z)$. By virtue of this remark, Theorem 3.1 immediately implies the following corollary, containing the Corollary of Theorem 1.3 and Valiron's Theorem 4.6 of Chapter 2.

COROLLARY. *Let $f(z)$ be a meromorphic function of lower order $\lambda < 1/2$. If $\delta(a, f) > 1 - \cos \pi\lambda$, then a is the only deficient value of the function $f(z)$. In particular, a meromorphic function of lower order $\lambda = 0$ cannot have more than one deficient value.*

In fact, since the function $f(z)$ tends to a on the sequence of circles $\{|z| = r_k\}$, $r_k \to \infty$, uniformly in $\arg z$, we see that $m(r_k, b) = O(1)$ for $b \ne a$, and hence $\delta(b, f) = 0$.

Theorem 3.1 is a corollary of a subtler theorem that gives a quantitative estimate of the degree of convergence of $f(z)$ to ∞.

THEOREM 3.2. *Let $f(z)$ be a meromorphic function of lower order $\lambda < 1/2$. Then the inequality*

$$\limsup_{r\to\infty} \frac{\ln^+ \mu(r, f)}{T(r, f)} \ge \frac{\pi\lambda}{\sin \pi\lambda}(\delta(\infty, f) - 1 + \cos \pi\lambda) \tag{3.2}$$

holds.

The proof of Theorem 3.2 relies on several lemmas. The first of them will be also used in Section 4. For this reason we prove it in a more general formulation than is needed for the proof of Theorem 3.2.

LEMMA 3.1. *Let $f(z)$, $f(0) = 1$, be a meromorphic function with zeros a_μ and poles b_ν, and let $q \geq 0$ be an integer. The following representation holds for each $R > 0$:*

$$f(z) = \alpha_{R,q}(z)\omega_{R,q}(z), \tag{3.3}$$

where

$$\alpha_{R,q}(z) = \prod_{|a_\mu|<R} E\left(\frac{z}{a_\mu}, q\right)\left\{\prod_{|b_\nu|<R} E\left(\frac{z}{b_\nu}, q\right)\right\}^{-1}, \tag{3.4}$$

and $\ln \omega_{R,q}(z)$ (the branch of $\ln$ is chosen in such a way that $\ln \omega_{R,q}(0) = 0$) satisfies the estimate

$$|\ln \omega_{R,q}(z)| \leq K_1 \left(\frac{r}{R}\right)^{q+1} T(2R, f) + K_2 q(r^q + 1), \tag{3.5}$$

$$|z| = r \leq \frac{1}{2}R,$$

where K_1 and K_2, $0 < K_1, K_2 < \infty$, are constants independent of r and R.

PROOF. By formula (2.5) in Chapter 1 we have

$$\begin{aligned}\{\ln f(\zeta)\}^{(q+1)} = (-1)^q q! \sum_{|a_\mu|<R} \frac{1}{(\zeta - a_\mu)^{q+1}} - (-1)^q q! \sum_{|b_\nu|<R} \frac{1}{(\zeta - b_\nu)^{q+1}} \\ + I_{R,q}(\zeta) + J_{R,q}(\zeta),\end{aligned} \tag{3.6}$$

where

$$I_{R,q}(\zeta) = \frac{(q+1)!}{2\pi} \int_0^{2\pi} \ln|f(Re^{i\theta})| \frac{2Re^{i\theta}}{(Re^{i\theta} - \zeta)^{q+2}}\, d\theta,$$

$$J_{R,q}(\zeta) = q! \sum_{|a_\mu|<R} \left(\frac{a_\mu}{R^2 - \bar{a}_\mu \zeta}\right)^{q+1} - q! \sum_{|b_\nu|<R} \left(\frac{b_\nu}{R^2 - b_\nu \zeta}\right)^{q+1}.$$

A straightforward calculation shows that

$$\{\ln \alpha_{R,q}(\zeta)\}^{(q+1)} = (-1)^q q! \sum_{|a_\mu|<R} \frac{1}{(\zeta - a_\mu)^{q+1}} - (-1)^q q! \sum_{|b_\nu|<R} \frac{1}{(\zeta - b_\nu)^{q+1}}.$$

Subtracting this equality from (3.6) we get

$$\{\ln \omega_{R,q}(\zeta)\}^{(q+1)} = I_{R,q}(\zeta) + J_{R,q}(\zeta).$$

Let us estimate $I_{R,q}(\zeta)$ and $J_{R,q}(\zeta)$ for $|\zeta| \leq \frac{1}{2}R$. We have

$$\begin{aligned}|I_{R,q}(\zeta)| &\leq \frac{(q+1)!2R}{(R - |\zeta|)^{q+2}} \frac{1}{2\pi} \int_0^{2\pi} |\ln|f(Re^{i\theta})||\, d\theta \\ &\leq \frac{(q+1)!2^{q+3}}{R^{q+1}} \left(m(R, f) + m\left(R, \frac{1}{f}\right)\right) \leq \frac{(q+1)!2^{q+4}}{R^{q+1}} T(R, f),\end{aligned}$$

$$\begin{aligned}|J_{R,q}(\zeta)| &\leq q! \left(\frac{R}{R^2 - R|\zeta|}\right)^{q+1} \left(n(R, f) + n\left(R, \frac{1}{f}\right)\right) \\ &\leq \frac{q!2^{q+1}}{R^{q+1}} n(R; 0, \infty) \leq \frac{q!2^{q+2}}{R^{q+1}} N(2R; 0, \infty) \leq \frac{q!2^{q+3}}{R^{q+1}} T(2R, f)\end{aligned}$$

(passing from $n(R; 0, \infty)$ to $N(2R; 0, \infty)$ we used Lemma 7.1 of Chapter 1).

Thus

$$|\{\ln \omega_{R,q}(\zeta)\}^{(q+1)} \leq \frac{(q+1)!2^{q+5}}{R^{q+1}} T(2R,f), \qquad |\zeta| \leq \frac{1}{2}R. \tag{3.7}$$

Later on, we will need the following relation ($|z| < R$):

$$\ln \omega_{R,q}(z) = \frac{1}{q!}\int_0^z (z-\zeta)^q \{\ln \omega_{R,q}(\zeta)\}^{(q+1)} d\zeta + \sum_{k=0}^{q} \frac{z^k}{k!}\{\ln \omega_{R,q}(\zeta)\}^{(k)}_{\zeta=0} \tag{3.8}$$

(the integral is taken over a line segment joining 0 and z). This relation follows from the observation that the derivatives of order $q+1$ of the left and right sides are the same, and the derivatives of orders $k = 0, 1, \dots, q$ coincide at $z = 0$.

Using (3.7) and (3.8) we get the estimate ($z = |z| \leq R/2$)

$$|\ln \omega_{R,q}(z)| \leq (q+1)2^{q+5}\left(\frac{r}{R}\right)^{q+1} T(2R,f) + \left|\sum_{k=0}^{q} \frac{z^k}{k!}\{\ln \omega_{R,q}(\zeta)\}^{(k)}_{\zeta=0}\right|.$$

The values of $\{\ln \omega_{R,q}(\zeta)\}^{(k)}_{\zeta=0}$, $k = 0, 1, \dots, q$, do not depend on R since $\ln \omega_{R,q}(\zeta) = \ln f(\zeta) - \ln \alpha_{R,q}(\zeta)$ and $\{\ln \alpha_{R,q}(\zeta)\}^{(k)}_{\zeta=0} = 0$, $k = 0, 1, \dots, q$. Besides, $\ln \omega_{R,q}(0) = 0$; therefore the estimate

$$\left|\sum_{k=0}^{q} \frac{z^k}{k!}\{\ln \omega_{R,q}(\zeta)\}^{(k)}_{\zeta=0}\right| \leq K_2 q(r^q + 1)$$

holds, where $0 < K_2 < \infty$ does not depend on R and r. The proof of the lemma is complete. □

LEMMA 3.2. *Let $f(z)$, $f(0) = 1$, be a meromorphic function with zeros a_μ and poles b_ν. Set*

$$H_R(r) = \sum_{|a_\mu|<R} \ln\left(1 + \frac{r}{|a_\mu|}\right) + \sum_{|b_\nu|<R} \ln\left(1 + \frac{r}{|b_\nu|}\right). \tag{3.9}$$

The estimate

$$H_R(r) \leq 4T(2R,f), \qquad 0 \leq r \leq R, \tag{3.10}$$

holds.

PROOF. By the inequality $\ln(1+x) \leq \ln^+ x + \ln 2$, $0 \leq x < \infty$, we have

$$\begin{aligned} H_R(r) &\leq \sum_{|a_\mu|<R} \left(\ln^+ \frac{r}{|a_\mu|} + \ln 2\right) + \sum_{|b_\nu|<R} \left(\ln^+ \frac{r}{|b_\nu|} + \ln 2\right) \\ &\leq N(r,0) + n(R,0)\ln 2 + N(r,\infty) + n(R,\infty)\ln 2 \\ &= N(r;0,\infty) + n(R;0,\infty)\ln 2. \end{aligned}$$

Using Lemma 7.1 of Chapter 1 we get

$$H_R(r) \leq N(r;0,\infty) + N(2R;0,\infty) \leq 2T(r,f) + 2T(2R,f) \leq 4T(2R,f). \qquad \square$$

LEMMA 3.3. *For each a and b, $0 < a < b$, and each σ, $0 < \sigma < 1$, the inequalities*

$$\begin{aligned} &\int_a^b \{\ln|1-x| - \pi\sigma \cot \pi\sigma \ln^+ x\} x^{-1-\sigma} dx \\ &\qquad \geq C_1 a^{-\sigma}\ln(1+a) - C_2 b^{-\sigma}\ln(1+b), \end{aligned} \tag{3.11}$$

$$(3.12)\qquad \int_a^b \{\pi\sigma \csc \pi\sigma \ln^+ x - \ln(1+x)\}x^{-1-\sigma}dx \geq C_3 a^{-\sigma}\ln(1+a) - C_4 b^{-\sigma}\ln(1+b)$$

hold, where C_k, $0 < C_k < \infty$, $k = 1,2,3,4$, are constants which do not depend on a and b.

PROOF. We use the equalities

$$\text{(a)}\ \int_0^\infty \frac{\ln^+ x}{x^{1+\sigma}}dx = \frac{1}{\sigma^2}; \qquad \text{(b)}\ \int_0^\infty \frac{\ln(1+x)}{x^{1+\sigma}}dx = \frac{\pi}{\sigma \sin \pi\sigma};$$

$$(3.13)\qquad \text{(c)}\ \int_0^\infty \frac{\ln|1-x|}{x^{1+\sigma}}dx = \frac{\pi}{\sigma \tan \pi\sigma}.$$

To get (b) we need to integrate by parts and use the Residue Calculus:

$$\int_0^\infty \frac{\ln(1+x)}{x^{1+\sigma}}dx = \frac{1}{\sigma}\int_0^\infty \frac{dx}{x^\sigma(1+x)} = \frac{\pi}{\sigma \sin \pi\sigma}.$$

Equalities (a) and (c) were proved in Section 2 (see (2.29) and (2.30)).

Consider the function

$$\varphi(a) = \int_a^\infty \{\ln|1-x| - \pi\sigma \cot \pi\sigma \ln^+ x\}x^{-1-\sigma}dx, \quad 0 \leq a < \infty.$$

By equalities (a) and (c), we have $\varphi(0) = 0$; it is clear that $\lim_{a\to\infty}\varphi(a) = 0$. Since the derivative

$$\varphi'(a) = -\{\ln|1-a| - \pi\sigma \cot \pi\sigma \ln^+ a\}a^{-1-\sigma},$$

as is easy to check, changes sign only once on the half-axis $\{0 < a < \infty\}$, from plus to minus, we have $\varphi(a) > 0$ for $0 < a < \infty$.

Set

$$\psi(a) = \frac{\varphi(a)}{a^{-\sigma}\ln(1+a)}.$$

This function is continuous and positive for $0 < a < \infty$. Using the l'Hôpital rule we find its limits as $a \to +0$ and $a \to +\infty$ and see that both are finite and positive. Therefore there exist constants C_1 and C_2, $0 < C_1 < C_2 < \infty$, such that

$$C_1 \leq \psi(a) \leq C_2, \quad 0 < a < \infty.$$

The last inequality can be rewritten in the form

$$C_1 a^{-\sigma}\ln(1+a) \leq \varphi(a) \leq C_2 a^{-\sigma}\ln(1+a), \quad 0 < a < \infty,$$

from which we get

$$\varphi(a) - \varphi(b) \geq C_1 a^{-\sigma}\ln(1+a) - C_2 b^{-\sigma}\ln(1+b), \quad 0 < a < b < \infty;$$

the last inequality is equivalent to (3.11).

To prove inequality (3.12), we use a similar argument for

$$\varphi(a) = \int_a^\infty \{\pi\sigma \csc \pi\sigma \ln^+ x - \ln(1+x)\}x^{-1-\sigma}dx. \qquad \square$$

LEMMA 3.4. *Let $f(z)$, $f(0) = 1$, be a meromorphic function with lower order $\lambda < 1$. Then for each σ, $\lambda < \sigma < 1$, the inequality*

$$(3.14)\qquad \limsup_{r\to\infty}\left\{\ln \mu(r,f) + \frac{\pi\sigma}{\sin \pi\sigma}[N(r,\infty) - \cos \pi\sigma N(r,0)]\right\} \geq 0$$

holds.

PROOF. We may assume without loss of generality that $f(z)$ is not identically constant. Let a_μ be the zeros of the function $f(z)$, and b_ν its poles. Using the change of variable in the integrals in (3.11) and (3.12), we get the inequalities

$$\tag{3.15} \begin{aligned} &\int_\xi^\eta \left\{\ln\left|1-\frac{r}{|a_\mu|}\right| - \pi\sigma\cot\pi\sigma\ln^+\frac{r}{|a_\mu|}\right\}\frac{dr}{r^{1+\sigma}} \\ &\quad\geq C_1\xi^{-\sigma}\ln\left(1+\frac{\xi}{|a_\mu|}\right) - C_2\eta^{-\sigma}\ln\left(1+\frac{\eta}{|a_\mu|}\right), \end{aligned}$$

$$\tag{3.16} \begin{aligned} &\int_\xi^\eta \left\{\frac{\pi\sigma}{\sin\pi\sigma}\ln^+\frac{r}{|b_\nu|} - \ln\left(1+\frac{r}{|b_\nu|}\right)\right\}\frac{dr}{r^{1+\sigma}} \\ &\quad\geq C_3\xi^{-\sigma}\ln\left(1+\frac{\xi}{|b_\nu|}\right) - C_4\eta^{-\sigma}\ln\left(1+\frac{\eta}{|b_\nu|}\right), \end{aligned}$$

where $0 < \sigma < 1$, and ξ and η are arbitrary numbers satisfying the condition $0 < \xi < \eta < \infty$. We sum inequalities of the form (3.15) over all a_μ satisfying $|a_\mu| < R$, and sum all inequalities of the form (3.16) over all b_ν satisfying $|b_\nu| < R$. We get two inequalities. Summing them, we obtain

$$\tag{3.17} \begin{aligned} &\int_\xi^\eta \left\{\ln|\breve{\alpha}_{R,0}(r)| - \frac{\pi\sigma}{\tan\pi\sigma}N(r,0) + \frac{\pi\sigma}{\sin\pi\sigma}N(r,\infty)\right\}\frac{dr}{r^{1+\sigma}} \\ &\quad\geq C_5\xi^{-\sigma}H_R(\xi) - C_6\eta^{-\sigma}H_R(\eta), \quad 0<\xi<\eta<R, \end{aligned}$$

where $C_5 = C_1 + C_3$, $C_6 = C_2 + C_4$, $H_R(r)$ is determined by (3.9), and

$$\tag{3.18} \breve{\alpha}_{R,0}(z) = \prod_{|a_\mu|<R}\left(1-\frac{z}{|a_\mu|}\right)\left\{\prod_{|b_\nu|<R}\left(1+\frac{z}{|b_\nu|}\right)\right\}^{-1}.$$

By Lemma 3.1 with $q = 0$, we have a representation

$$\tag{3.19} f(z) = \alpha_{R,0}(z)\omega_{R,0}(z),$$

where

$$\alpha_{R,0}(z) = \prod_{|a_\mu|<R}\left(1-\frac{z}{a_\mu}\right)\left\{\prod_{|b_\nu|<R}\left(1-\frac{z}{b_\nu}\right)\right\}^{-1},$$

and $\ln\omega_{R,0}(z)$ admits the estimate

$$|\ln\omega_{R,0}(z)| \leq K_1\frac{r}{R}T(2R,f), \qquad |z| = r \leq \frac{1}{2}R$$

(K_1 does not depend on r and R). Obviously,

$$\mu(r,\alpha_{R,0}) \geq |\breve{\alpha}_{R,0}(r)|;$$

therefore

$$\tag{3.19'} \mu(r,f) \geq \mu(r,\alpha_{R,0})\mu(r,\omega_{R,0}) \geq |\breve{\alpha}_{R,0}(r)|\mu(r,\omega_{R,0}),$$

from which, using estimate (3.5), we get

$$\ln|\breve{\alpha}_{R,0}(r)| \leq \ln\mu(r,f) + |\ln\mu(r,\omega_{R,0}))| \leq \ln\mu(r,f) + K_1\frac{r}{R}T(2R,f),\ 0\leq r\leq\frac{1}{2}R,$$

whence

$$\tag{3.20} \int_\xi^\eta \ln|\breve{\alpha}_{R,0}(r)|\frac{dr}{r^{1+\sigma}} \leq \int_\xi^\eta \ln\mu(r,f)\frac{dr}{r^{1+\sigma}} + \frac{K_1}{1-\sigma}\frac{\eta^{1-\sigma}}{R}T(2R,f),$$

$0<\sigma<1$, $0<\xi<\eta\le\frac{1}{2}R$. The next inequality follows from (3.17) and (3.20):

$$\int_\xi^\eta\left\{\ln\mu(r,f)-\frac{\pi\sigma}{\tan\pi\sigma}N(r,0)+\frac{\pi\sigma}{\sin\pi\sigma}N(r,\infty)\right\}\frac{dr}{r^{1+\sigma}}$$
$$\ge C_5\xi^{-\sigma}H_R(\xi)-C_6\eta^{-\sigma}H_R(\eta)-\frac{K_1}{1-\sigma}\frac{\eta^{1-\sigma}}{R}T(2R,f).$$

Let $\eta=\frac{1}{2}R$ in this inequality. Since, by Lemma 3.2, $H_R(\eta)\le 4T(2R,f)$, we get the inequality

$$\int_\xi^{\frac{1}{2}R}\left\{\ln\mu(r,f)-\frac{\pi\sigma}{\tan\pi\sigma}N(r,0)+\frac{\pi\sigma}{\sin\pi\sigma}N(r,\infty)\right\}\frac{dr}{r^{1+\sigma}}$$
$$\ge C_5\xi^{-\sigma}H_R(\xi)-C_7R^{-\sigma}T(2R,f),\qquad 0<\xi<\frac{1}{2}R, \tag{3.21}$$

where the constant C_7, $0<C_7<\infty$, does not depend on ξ and R.

Let $\xi_0>0$ be an arbitrarily large number. Choose $\xi>\xi_0$ such that $H_R(\xi)>0$ for $R>2\xi$. This choice is possible because otherwise the function $f(z)$ would be an entire function without zeros, and Lemma 6.2 of Chapter 1 would imply that $f(z)\equiv f(0)=1$.

Assume that σ in inequality (3.21) satisfies the condition $\lambda<\sigma<1$. Then $\liminf_{R\to\infty}R^{-\sigma}T(2R,f)=0$; therefore the number $R>2\xi$ can be chosen in such a way that the right-hand side of (3.21) is positive. From this it follows that for some $r>\xi_0$ the integrand in the left-hand side of (3.21) is also positive. Since ξ_0 is arbitrarily large, we get the statement of the lemma. □

PROOF. Now we prove Theorem 3.2. We may assume that the function is transcendental, because for rational functions the theorem is trivial.

By Theorem 2.1 of Chapter 4 we can find a number $a\ne\infty$ such that $N(r,a,f)\sim T(r,f)$. Let

$$f_1(z)=\frac{f(z)-a}{c_kz^k},$$

where c_k is the first non-zero coefficient of the Laurent series of the function $f(z)-a$ at $z=0$. Observe that $f_1(0)=1$,

$$\begin{aligned}&N(r,\infty,f_1)=N(r,\infty,f)+O(\ln r),\\&N(r,0,f_1)=N(r,a,f)+O(\ln r)\sim T(r,f),\\&\ln^+\mu(r,f_1)\le\ln^+\mu(r,f)+O(\ln r).\end{aligned}\tag{3.22}$$

By Lemma 3.4, for each σ, $\lambda<\sigma<1$, and each $\varepsilon>0$ there exists a sequence $r_n\to\infty$, such that

$$\ln^+\mu(r_n,f_1)\ge\frac{\pi\sigma}{\sin\pi\sigma}\{\cos\pi\sigma N(r_n,0,f_1)-N(r_n,\infty,f_1)\}-\varepsilon.$$

Dividing both sides of this relation by $T(r_n,f)$ and using relations (3.22), we get

$$\limsup_{r\to\infty}\frac{\ln^+\mu(r,f)}{T(r,f)}\ge\frac{\pi\sigma}{\sin\pi\sigma}\{\cos\pi\sigma-1+\delta(\infty,f)\}.$$

Letting σ tend to λ, we get the statement of the theorem. Note that we have not used the condition $\lambda<1/2$; however, for $1/2\le\lambda<1$, the right-hand side of (3.2) cannot be positive and the theorem becomes trivial. □

The following corollary of Lemma 3.4, complementing Corollary 1 of Theorem 1.3, is of interest.

COROLLARY. *An entire function $f(z)$ of lower order $\lambda \leq 1/2$ cannot have finite deficient values.*

Obviously, it suffices to prove that $\delta(0, f) = 0$. We may assume without loss of generality that $f(0) = 1$. By Lemma 3.4, for each $\varepsilon > 0$ and each σ, $1/2 < \sigma < 1$, there exists a sequence $r_k \to \infty$ such that

$$\ln \mu(r_k, f) \geq \pi\sigma \cot \pi\sigma N(r_k, 0) - \varepsilon,$$

whence

$$\ln \frac{1}{\mu(r_k, f)} \leq -\pi\sigma \cot \pi\sigma N(r_k, 0) + \varepsilon.$$

Since the right-hand side of this inequality is positive $(1/2 < \sigma < 1)$, we can replace ln by $\ln^+$ in the left-hand side. Taking this into account, we get

$$\begin{aligned}\delta(0, f) &= \liminf_{r\to\infty} \frac{m(r,0)}{T(r,f)} \leq \liminf_{r\to\infty} \frac{\ln^+ \frac{1}{\mu(r,f)}}{T(r,f)} \\ &\leq \liminf_{k\to\infty} \frac{\ln^+ \frac{1}{\mu(r_k,f)}}{T(r_k,f)} \leq -\pi\sigma \cot \pi\sigma \limsup_{k\to\infty} \frac{N(r_k,0)}{T(r_k,f)} \\ &\leq -\pi\sigma \cot \pi\sigma (1 - \delta(0,f)),\end{aligned}$$

whence

$$\delta(0, f) \leq \frac{-\pi\sigma \cot \pi\sigma}{1 - \pi\sigma \cot \pi\sigma}.$$

Letting σ tend to 1/2, we get $\delta(0, f) = 0$.

The method used in the proof of Theorem 3.2 allows one to get various relations connecting the asymptotic behavior of a meromorphic function with the distribution of its zeros and poles. We restrict our attention to the following result.

THEOREM 3.3. *Let $f(z)$ be a meromorphic function of lower order $\lambda < 1/2$ such that*

$$\lim_{r\to\infty} \frac{\ln M(r,f)}{\ln r} = +\infty.$$

If $0 < \lambda < 1/2$, the inequality

$$\limsup_{r\to\infty} \frac{\ln \mu(r,f)}{\ln M(r,f)} + \pi\lambda \sin \pi\lambda \limsup_{r\to\infty} \frac{N(r,f)}{\ln M(r,f)} \geq \cos \pi\lambda \tag{3.23}$$

holds. This inequality remains true for $\lambda = 0$ provided

$$\limsup_{r\to\infty} \frac{N(r,f)}{\ln M(r,f)} < \infty.$$

To prove Theorem 3.3 we need the following lemma, analogous to Lemma 3.3.

LEMMA 3.5. *For all a and b, $0 < a < b < \infty$, and σ, $0 < \sigma < \frac{1}{2}$, we have the inequalities*

$$\begin{aligned}&\int_a^b \{\ln|1 - x| - \cos \pi\sigma \ln(1 + x)\} x^{-1-\sigma} dx \\ &\qquad \geq C_8 a^{-\sigma} \ln(1 + a) - C_9 b^{-\sigma} \ln(1 + b),\end{aligned} \tag{3.24}$$

$$\int_a^b \{\cos\pi\sigma\ln|1-x| - \ln(1+x) + \pi\sigma\sin\pi\sigma\ln^+ x\}x^{-1-\sigma}dx \tag{3.25}$$
$$\geq C_{10}a^{-\sigma}\ln(1+a) - C_{11}b^{-\sigma}\ln(1+b),$$

where C_k, $0 < C_k < \infty$, $k = 8, 9, 10, 11$, *are constants which do not depend on* a *and* b.

This lemma is a corollary of Lemma 3.3. In fact, if we multiply (3.12) by $\cos\pi\sigma$ and add the result to (3.11), we get (3.24). If we multiply (3.11) by $\cos\pi\sigma$ and add the result to (3.12), we get (3.25) (we take into account that $0 < \sigma < 1/2$ and hence $\cos\pi\sigma > 0$).

Next we prove a lemma analogous to Lemma 3.4.

LEMMA 3.6. *Let* $f(z)$, $f(0) = 1$, *be a meromorphic function of lower order* $\lambda < 1/2$. *Then for each* σ, $\lambda < \sigma < 1/2$, *the inequality*

$$\limsup_{r\to\infty}\{\ln\mu(r,f) + \pi\sigma\sin\pi\sigma N(r,\infty) - \cos\pi\sigma\ln M(r,f)\} \geq 0 \tag{3.26}$$

holds.

PROOF. Using the change of variable in (3.24) and (3.25), we get the inequalities

$$\int_\xi^\eta \left\{\ln\left|1-\frac{r}{|a_\mu|}\right| - \cos\pi\sigma\ln\left(1+\frac{r}{|a_\mu|}\right)\right\} r^{-1-\sigma}dr \tag{3.24'}$$
$$\geq C_8\xi^{-\sigma}\ln\left(1+\frac{\xi}{|a_\mu|}\right) - C_9\eta^{-\sigma}\ln\left(1+\frac{\eta}{|a_\mu|}\right),$$

$$\int_\xi^\eta \left\{\cos\pi\sigma\ln\left|1-\frac{r}{|b_\nu|}\right| - \ln\left(1+\frac{r}{|b_\nu|}\right) + \pi\sigma\sin\pi\sigma\ln^+\frac{r}{|b_\nu|}\right\}\frac{dr}{r^{1+\sigma}}$$
$$\geq C_{10}\xi^{-\sigma}\ln\left(1+\frac{\xi}{|b_\nu|}\right) - C_{11}\eta^{-\sigma}\ln\left(1+\frac{\eta}{|b_\nu|}\right),$$

where ξ and η are arbitrary numbers satisfying the condition $0 < \xi < \eta < \infty$, a_μ are the zeros of the function $f(z)$, and b_ν are its poles. From these inequalities and the argument similar to the one used in Lemma 3.4, we get

$$\int_\xi^\eta \{\ln|\breve{\alpha}_{R,0}(r)| - \cos\pi\sigma\ln|\breve{\alpha}_{R,0}(-r)| + \pi\sigma\sin\pi\sigma N(r,\infty)\}\frac{dr}{r^{1+\sigma}} \tag{3.17'}$$
$$\geq C_{12}\xi^{-\sigma}H_R(\xi) - C_{13}\eta^{-\sigma}H_R(\eta),$$

where $0 < \xi < \eta < R$, $C_{12} = C_8 + C_{11}$, $C_{13} = C_9 + C_{11}$, and $\breve{\alpha}_{R,0}(z)$ is defined by formula (3.18).

As in the proof of Lemma 3.4, we represent the function $f(z)$ in the form (3.19). Obviously

$$M(r,\alpha_{R,0}) \leq |\breve{\alpha}_{R,0}(-r)|, \tag{3.27}$$

and therefore

$$M(r,f) \leq M(r,\alpha_{R,0})M(r,\omega_{R,0}) \leq |\breve{\alpha}_{R,0}(-r)|M(r,\omega_{R,0}),$$

from which, using the estimate (3.5), we get

$$-\ln|\breve{\alpha}_{R,0}(-r)| \leq -\ln M(r,f) + \ln M(r,\omega_{R,0})$$
$$\leq -\ln M(r,f) + K_1\frac{r}{R}T(2R,f), \qquad 0 \leq r \leq \frac{1}{2}R,$$

whence

$$\int_\xi^\eta (-\ln|\breve{a}_{R,0}(-r)|)\frac{dr}{r^{1+\sigma}} \le \int_\xi^\eta (-\ln M(r,f))\frac{dr}{r^{1+\sigma}} + \frac{K_1}{1-\sigma}\frac{\eta^{1-\sigma}}{R}T(2R,f), \tag{3.28}$$

$$0<\sigma<\frac{1}{2}, \qquad 0<\xi<\eta\le\frac{1}{2}R.$$

Using (3.17′), (3.20), and (3.28), we get the inequality

$$\int_\xi^\eta \{\ln\mu(r,f) - \cos\pi\sigma\ln M(r,f) + \pi\sigma\sin\pi\sigma N(r,\infty)\}\frac{dr}{r^{1+\sigma}}$$
$$\le C_{12}\xi^{-\sigma}H_R(\xi) - C_{13}\eta^{-\sigma}H_R(\eta) - \frac{2K_1}{1-\sigma}\frac{\eta^{1-\sigma}}{R}T(2R,f).$$

We now repeat the argument used in the proof of Lemma 3.3 to derive (3.14) from (3.21), and get relation (3.26). □

PROOF OF THEOREM 3.3. We apply Lemma 3.6 to the function $f_1(z) = \frac{f(z)}{c_k z^k}$, where c_k is the first non-zero Laurent coefficient of $f(z)$ at $z=0$. We get that for each σ, $\lambda<\sigma<1/2$, and each $\varepsilon>0$ there exists a sequence $r_n\to\infty$ such that

$$\ln\mu(r_n,f_1) + \pi\sigma\sin\pi\sigma N(r_n,\infty,f_1) \ge \cos\pi\sigma\ln M(r_n,f_1) - \varepsilon. \tag{3.29}$$

Taking into account that

$$N(r,\infty,f_1) = N(r,\infty,f) + O(\ln r),$$

$$\ln M(r,f_1) = \ln M(r,f) + O(\ln r), \qquad \ln\mu(r,f_1) = \ln\mu(r,f) + O(\ln r),$$

and dividing both sides of (3.29) by $\ln M(r_n,f)$, we obtain

$$\limsup_{r\to\infty}\frac{\ln\mu(r,f)}{\ln M(r,f)} + \pi\sigma\sin\pi\sigma\limsup_{r\to\infty}\frac{N(r,\infty,f)}{\ln M(r,f)} \ge \cos\pi\sigma.$$

Letting σ tend to λ, we get relation (3.23). □

EXERCISE. Prove that

$$\limsup_{r\to\infty}\frac{T(r,f)}{\ln M(r,f)} \ge \frac{\sin\pi\lambda}{\pi\lambda} \tag{3.30}$$

for each entire function $f(z)$ of lower order $0\le\lambda<1$, with the help of the method used in the proofs of Theorems 3.2 and 3.3.

HINT. It suffices to prove that

$$\limsup_{r\to\infty}\frac{N(r,0)}{\ln M(r,f)} \ge \frac{\sin\pi\lambda}{\pi\lambda}. \tag{3.31}$$

We may assume without loss of generality that $f(0)=1$. Let a_μ be the zeros of the function $f(z)$. Using inequality (3.12), we get

$$\int_\xi^\eta \left\{\frac{\pi\sigma}{\sin\pi\sigma}\ln^+\frac{r}{|a_\mu|} - \ln\left(1+\frac{r}{|a_\mu|}\right)\right\}\frac{dr}{r^{1+\sigma}}$$
$$\ge C_3\xi^{-\sigma}\ln\left(1+\frac{\xi}{|a_\mu|}\right) - C_4\eta^{-\sigma}\ln\left(1+\frac{\eta}{|a_\mu|}\right), \quad 0<\sigma<1,\ 0<\xi<\eta<\infty.$$

Summing over all a_μ, $|a_\mu| < R$, we have

$$\int_\xi^\eta \left\{\frac{\pi\sigma}{\sin\pi\sigma} N(r,0) - \ln|\breve{\alpha}_{R,0}(-r)|\right\} \frac{dr}{r^{1+\sigma}}$$
$$\geq C_3\xi^{-\sigma} H_R(\xi) - C_4\eta^{-\sigma} H_R(\eta), \quad 0 < \sigma < 1,\ 0 < \xi < \eta < R,$$

where $\breve{\alpha}_{R,0}(z)$ are defined by relation (3.18) (no poles b_ν in this case). Next, using representation (3.19) and estimate (3.28), we arrive at the inequality ($0 < \sigma < 1$, $0 < \xi < \eta \leq \frac{1}{2}R$)

$$\int_\xi^\eta \left\{\frac{\pi\sigma}{\sin\pi\sigma} N(r,0) - \ln M(r,f)\right\} \frac{dr}{r^{1+\sigma}}$$
$$\geq C_3\xi^{-\sigma} H_R(\xi) - C_4\eta^{-\sigma} H_R(\eta) - \frac{K_1}{1-\sigma}\frac{\eta^{1-\sigma}}{R} T(2R,f),$$

from which we get (3.31) using the same argument as at the end of the proof of Lemma 3.3.

The relation between $\ln\mu(r,f)$ and $\ln M(r,f)$ is of interest for entire functions.

THEOREM 3.4. *Let $f(z)$ be an entire function of lower order $\lambda < 1$. Then*

$$\limsup_{r\to\infty} \frac{\ln\mu(r,f)}{\ln M(r,f)} \geq \cos\pi\lambda. \tag{3.32}$$

PROOF. In the case $0 \leq \lambda < 1/2$ this theorem is an immediate corollary of Theorem 3.3. Consider the case $1/2 \leq \lambda < 1$.

Observe that inequality (3.24) holds also for $0 < \sigma < 1$. To show this, it is enough to set

$$\varphi(a) = \int_a^\infty \{\ln|1-x| - \cos\pi\sigma \ln(1+x)\} \frac{dx}{x^{1+\sigma}},$$

and, using (3.13) (b) and (c), to argue in the same way as in the proof of (3.11) in Lemma 3.3. Therefore (3.24) holds also for $0 < \sigma < 1$. From this, summing over a_μ, $|a_\mu| < R$, we get

$$\begin{aligned}&\int_\xi^\eta \{\ln|\breve{\alpha}_{R,0}(r)| - \cos\pi\sigma \ln|\breve{\alpha}_{R,0}(-r)|\} \frac{dr}{r^{1+\sigma}}\\ &\quad\geq C_8\xi^{-\sigma} H_R(\xi) - C_9\eta^{-\sigma} H_R(\eta), \quad \lambda < \sigma < 1,\ 0 < \xi < \eta \leq \frac{1}{2}R,\end{aligned} \tag{3.33}$$

where $\breve{\alpha}_{R,0}(z)$ is defined by (3.18) (there are no poles b_μ in this case). Next we use the representation of $f(z)$ in the form (3.19).

Since $\cos\pi\sigma < 0$, we cannot, using (3.20) and (3.28), get from (3.33) an inequality with the left-hand side equal to

$$\int_\xi^\eta \{\ln\mu(r,f) - \cos\pi\sigma \ln M(r,f)\} \frac{dr}{r^{1+\sigma}}.$$

To prove the theorem we argue as follows.

By (3.5) we have $(0 < \xi < \eta \leq \frac{1}{2}R)$

$$\int_\xi^\eta \frac{\ln\mu(r,\omega_{R,0})}{r^{1+\sigma}} dr \geq -\frac{K_1}{1-\sigma}\frac{\eta^{-\sigma}}{R} T(2R,f).$$

Multiplying this inequality by $1 - \cos \pi\sigma$ and adding the resulting inequality to (3.33), we get

$$\int_\xi^\eta \{\ln |\breve{\alpha}_{R,0}(r)\mu(r,\omega_{R,0})| - \cos\pi\sigma \ln |\breve{\alpha}_{R,0}(-r)\mu(r,\omega_{R,0})|\}\frac{dr}{r^{1+\sigma}}$$
$$\geq C_8\xi^{-\sigma}H_R(\xi) - C_9\eta^{-\sigma}H_R(\eta) - \frac{2K_1}{1-\sigma}\frac{\eta^{-\sigma}}{R}T(2R,f),$$
$$0 < \sigma < 1, \quad 0 < \xi < \eta \leq \frac{1}{2}R.$$

From this, arguing in the same way as at the end of the proof of Lemma 3.4, we conclude that for each fixed σ, $\lambda < \sigma < 1$, there exist sequences $\{R_k\}_{k=1}^\infty$ and $\{r_k\}_{k=1}^\infty$, $\lim_{k\to\infty} R_k = \lim_{k\to\infty} r_k = \infty$, $r_k \leq \frac{1}{2}R_k$, such that for $R = R_k$, $r = r_k$, the inequality

$$\ln |\breve{\alpha}_{R,0}(r)\mu(r,\omega_{R,0})| - \cos\pi\sigma \ln |\breve{\alpha}_{R,0}(-r)\mu(r,\omega_{R,0})| \geq 0$$

holds. Adding $\ln |\breve{\alpha}_{R,0}(-r)\mu(r,\omega_{R,0})|$ to both sides and exponentiating, we get

$$|\breve{\alpha}_{R,0}(r)\breve{\alpha}_{R,0}(-r)|\mu^2(r,\omega_{R,0}) \geq |\breve{\alpha}_{R,0}(-r)\mu(r,\omega_{R,0})|^{1+\cos\pi\sigma}. \tag{3.34}$$

If z_0 is a point on the circle $\{|z| = r\}$ for which $|\alpha_{R,0}(z)| = \mu(r,\alpha_{R,0})$, then

$$M(r,\alpha_{R,0})\mu(r,\alpha_{R,0}) \geq |\alpha_{R,0}(-z)\alpha_{R,0}(z_0)|$$
$$= \prod_{|a_\mu|<R}\left|1 - \frac{z_0^2}{a_\mu^2}\right| \geq \prod_{|a_\mu|<R}\left|1 - \frac{r^2}{|a_\mu|^2}\right| = |\breve{\alpha}_{R,0}(r)\breve{\alpha}_{R,0}(-r)|.$$

Therefore, by (3.34) and (3.27), we get

$$M(r,\alpha_{R,0})\mu(r,\alpha_{R,0})\mu^2(r,\omega_{R,0}) \geq \{M(r,\alpha_{R,0})\mu(r,\omega_{R,0})\}^{1+\cos\pi\sigma},$$

whence

$$\mu(r,\alpha_{R,0})\mu(r,\omega_{R,0}) \geq \{M(r,\alpha_{R,0})\mu(r,\omega_{R,0})\}^{\cos\pi\sigma}.$$

Observing that

$$M(r,f) \geq M(r,\alpha_{R,0})\mu(r,\omega_{R,0})$$

and using (3.19′), we get

$$\mu(r,f) \geq \{M(r,f)\}^{\cos\pi\sigma}, \qquad r = r_k,$$

and thus

$$\limsup_{r\to\infty}\frac{\ln\mu(r,f)}{\ln M(r,f)} > \cos\pi\sigma.$$

Letting σ tend to λ, we conclude the proof of the theorem. □

Theorems 3.2 and 3.4 imply the following result.

COROLLARY 1. *For each entire function $f(z)$ of lower order* 0, *we have*

$$\limsup_{r\to\infty}\frac{\ln\mu(r,f)}{T(r,f)} = \limsup_{r\to\infty}\frac{\ln\mu(r,f)}{\ln M(r,f)} = \limsup_{r\to\infty}\frac{T(r,f)}{\ln M(r,f)} = 1.$$

In fact, Theorems 3.2 and 3.4 imply the inequalities

$$\limsup_{r\to\infty}\frac{\ln\mu(r,f)}{T(r,f)} \geq 1, \qquad \limsup_{r\to\infty}\frac{\ln\mu(r,f)}{\ln M(r,f)} \geq 1.$$

Since $\ln \mu(r,f) \le m(r,f) = T(r,f) \le \ln^+ M(r,f)$, the inequalities in the opposite direction are trivial. If the inequality

$$\limsup_{r\to\infty} \frac{T(r,f)}{\ln M(r,f)} < 1$$

were true,[7] we would have

$$\limsup_{r\to\infty} \frac{\ln \mu(r,f)}{T(r,f)} > \limsup_{r\to\infty} \frac{\ln \mu(r,f)}{\ln M(r,f)}.$$

Theorem 3.4 also implies the following corollary.

COROLLARY 2. *Each entire function $f(z)$ of lower order $\lambda \le 1$ satisfies*

$$\limsup_{r\to\infty} \frac{\ln \mu(r,f)}{\ln M(r,f)} \ge -1.$$

In fact, let $F(z) = f(z)f(-z)$. Then $\lambda[F(\sqrt{z})] = \lambda_1 \le 1/2$. Inequality (3.32) implies that

$$\limsup_{r\to\infty} \frac{\ln \mu(r,F(z))}{\ln M(r,F(z))} = \limsup_{r\to\infty} \frac{\ln \mu(r,F(\sqrt{z}))}{\ln M(r,F(\sqrt{z}))} \ge \cos \pi\lambda_1 \ge 0.$$

Hence for each $\varepsilon > 0$ there exists a sequence $r_k \to \infty$ such that

$$\ln \mu(r_k, F(z)) \ge -\varepsilon \ln M(r_k, F(z)).$$

But it is clear that $\mu(r,F(z)) \le \mu(r,f(z))M(r,f(z))$, $M(r,F(z)) \le \{M(r,f(z))\}^2$. Therefore

$$\ln \mu(r_k, f) \ge -(1+2\varepsilon)\ln M(r_k, f),$$

from which the statement of Corollary 2 follows. The sharpness of the obtained estimate is shown by the example of $f(z) = e^z$.

Now we consider examples showing that Theorems 3.1–3.4 are precise.

To verify this for Theorems 3.2 and 3.3 it suffices to show that for every λ, $0 \le \lambda < 1/2$, there exists a meromorphic function $f(z)$ for which (3.2) and (3.23) are equalities. We can ask the following more general question. Let the numbers $0 \le \lambda < 1/2$, $\lambda \le \rho \le \infty$, $1 - \cos \pi\lambda \le \delta_0 \le 1$ be given. Does there exist a meromorphic function $f(z)$ of order ρ, lower order λ, with $\delta(\infty, f) = \delta_0$, and such that (3.2) and (3.23) are equalities for $f(z)$? The answer to this question in full generality is unknown. We now give the affirmative answer under the additional conditions $\lambda > 0$ and $\rho < 1$.

EXAMPLE 1. Let $0 < \lambda \le \rho < 1$, and let $f(z,u)$ be a canonical product of genus 0 with positive zeros such that $n(r,0) \sim ur^{l(r)}$, where $0 < u \le 1$, and $l(r)$ is a differentiable function on $[r_0, \infty)$, $\lim_{r\to\infty} rl'(r) = 0$,

$$\limsup_{r\to\infty} l(r) = \rho, \qquad \liminf_{r\to\infty} l(r) = \lambda. \tag{3.35}$$

By asymptotic formula (5.2_2) in Chapter 2, for each $\eta > 0$ the relation

$$\ln |f(re^{i\varphi},\ u)| = \pi u \frac{\cos l(r)(\varphi - \pi)}{\sin \pi l(r)} r^{l(r)} + o(r^{l(r)}) \tag{3.36}$$

[7] Note that the inequality $\limsup_{r\to\infty} T(r,f)/\ln M(r,f) \ge 1$ for the entire functions of lower order 0 follows also from (3.30).

holds uniformly in φ, $\eta \le \varphi \le 2\pi - \eta$. Defining $f(z,0) \equiv 1$, consider the function

$$f_{uv}(z) = \frac{f(z,u)}{f(-z,v)}, \qquad u \ge v \ge 0,\ u > 0. \tag{3.37}$$

By (3.36), for each $\eta > 0$ the relation

$$\ln|f_{uv}(re^{i\varphi})| = \frac{\pi}{\sin \pi l(r)}\{u\cos l(r)(\varphi - \pi) - v\cos l(r)\varphi\}r^{l(r)} + o(r^{l(r)})$$

holds uniformly in φ, $\eta \le \varphi \le \pi - \eta$. Since $f_{uv}(z) = \overline{f_{uv}(\bar{z})}$, we have

$$m(r, f_{uv}) = \frac{1}{\pi}\int_0^\pi \ln^+|f_{uv}(re^{i\varphi})|\,d\varphi,$$

and using Theorem 7.4 of Chapter 1, we get

$$m(r, f_{uv}) = \frac{1}{l(r)} I(u, v, l(r))r^{l(r)} + o(r^{l(r)}),$$

where

$$I(u,v,\gamma) = \frac{\gamma}{\sin \pi\gamma}\int_0^\pi \{u\cos\gamma(\varphi-\pi) - v\cos\gamma\varphi\}^+ d\varphi, \quad 0 < \gamma < 1.$$

By formula (5.6) of Chapter 2 we have

$$N(r, f_{uv}) = N(r, 0, f(z,v)) = \frac{v}{l(r)}r^{l(r)} + o(r^{l(r)}), \tag{3.38}$$

and hence

$$T(r, f_{uv}) = \frac{1}{l(r)}\{I(u,v,l(r)) + v\}r^{l(r)} + o(^{l(r)}), \tag{3.39}$$

$$\delta(\infty, f_{uv}) = \liminf_{r\to\infty} \frac{I(u,v,l(r))}{I(u,v,l(r)) + v}.$$

By virtue of (3.35) and (3.39) the function $f_{uv}(z)$ has order ρ and lower order λ.

A straightforward computation gives:

$$I(u,v,\gamma) = u - v \qquad \text{if } u\cos\gamma\pi \ge v,$$

$$I(u,v,\gamma) = \left\{u^2 + \left(\frac{v - u\cos\gamma\pi}{\sin\gamma\pi}\right)^2\right\}^{1/2} - v \qquad \text{if } u \ge v \ge u\cos\gamma\pi.$$

It is not difficult to determine that $I(u,v,\gamma)$ is a non-decreasing function of γ. Therefore

$$\delta(\infty, f_{uv}) = \frac{I(u,v,\lambda)}{I(u,v,\lambda) + v}. \tag{3.40}$$

Imposing the conditions $0 < \lambda < 1/2$ and $u\cos\lambda\pi \ge v$ on the parameters λ, u, and v, we get

$$\delta(\infty, f_{uv}) = \frac{u - v}{u}. \tag{3.40$'$}$$

It is easy to see that selecting v, $0 \le v \le u\cos\lambda\pi$, in a suitable way, we can give $\delta(\infty, f_{uv})$ an arbitrary preassigned value δ_0, $1 - \cos\lambda\pi \le \delta_0 \le 1$.

Now we find an upper estimate for

$$\limsup_{r\to\infty} \frac{\ln\mu(r, f_{uv})}{T(r, f_{uv})}.$$

Obviously,

$$\ln \mu(r, f_{uv}) = \ln |f_{uv}(r)| = \ln |f(r, u)| - \ln |f(-r, v)|.$$

Formula (3.36) does not give an asymptotic formula for $\ln |f(r, u)|$. However, since $|f(re^{i\eta}, u)| \geq |f(r, u)|$ for each $\eta > 0$, we have the following upper estimate:

$$\begin{aligned} \ln \mu(r, f_{uv}) &\leq \ln |f(re^{i\eta},\ u)| - \ln |f(-r, v)| \\ &= \frac{\pi}{\sin \pi l(r)} \{u \cos l(r)(\pi - \eta) - v\} r^{l(r)} + o(r^{l(r)}). \end{aligned} \tag{3.41}$$

Using (3.39) we get

$$\limsup_{r\to\infty} \frac{\ln \mu(r, f_{uv})}{T(r, f_{uv})} \leq \limsup_{r\to\infty} \frac{\vartheta(\mu, v, l(r), \eta)}{I(u, v, l(r)) + v},$$

where

$$\vartheta(u, v, \gamma, \eta) = \frac{\pi\gamma}{\sin \pi\gamma} (u \cos \gamma(\pi - \eta) - v), \quad 0 < \gamma < 1.$$

It is easy to see that

$$\frac{1}{\pi} \frac{\partial}{\partial\gamma} \vartheta(u, v, \gamma, 0) = u \frac{\sin 2\pi\gamma - 2\pi\gamma}{2 \sin^2 \pi\gamma} + v \frac{\pi\gamma \cos \pi\gamma - \sin \pi\gamma}{\sin^2 \pi\gamma}.$$

Since for $0 < \gamma < 1$ the inequalities $2\pi\gamma > \sin 2\pi\gamma$ and $\sin \pi\gamma > \pi\gamma \cos \pi\gamma$ hold, it follows that

$$\frac{\partial}{\partial\gamma} \vartheta(u, v, \gamma, 0) < 0, \qquad 0 < \gamma < 1;$$

therefore the number $\eta > 0$ can be chosen to be small so that the conditions

$$\frac{\partial}{\partial\gamma} \vartheta(u, v, \gamma, \eta) < 0, \qquad \frac{\lambda}{2} \leq \gamma < 1$$

be satisfied. Then the function $\vartheta(u, v, \gamma, \eta)$ is non-increasing for $\frac{\lambda}{2} < \gamma < 1$. Taking into account that $I(u, v, \gamma)$ is a non-decreasing function of γ, we get

$$\limsup_{r\to\infty} \frac{\ln \mu(r, f_{uv})}{T(r, f_{uv})} \leq \frac{\vartheta(u, v, \lambda, \eta)}{I(u, v, \lambda) + v}.$$

Since this inequality holds for each sufficiently small $\eta > 0$, we have

$$\limsup_{r\to\infty} \frac{\ln \mu(r, f_{uv})}{T(r, f_{uv})} \leq \frac{\vartheta(u, v, \lambda, 0)}{I(u, v, \lambda) + v} = \frac{\pi\lambda}{\sin \pi\lambda} \frac{u \cos \pi\lambda - v}{u}.$$

Together with (3.40′) this gives

$$\limsup_{r\to\infty} \frac{\ln \mu(r, f_{uv})}{T(r, f_{uv})} \leq \frac{\pi\lambda}{\sin \pi\lambda} \{\cos \pi\lambda - 1 + \delta(\infty, f_{uv})\};$$

moreover, since the right-hand side of this inequality is non-negative, we may replace $\ln \mu(r, f_{uv})$ by $\ln^+ \mu(r, f_{uv})$ in the left-hand side.

Comparing with (3.2), we see that for the function $f_{uv}(z)$ it becomes an equality.

Let us show that for the function $f_{uv}(z)$, inequality (3.23) also becomes an equality. Since for each $\eta > 0$ the inequality $|f(re^{i\eta}, v)| \geq |f(r, v)|$ holds, we have

$$\begin{aligned} \ln M(r, f_{uv}) &= \ln |f(-r, u)| - \ln |f(r, v)| \geq \ln |f(-r, u)| - \ln |f(re^{i\eta}, v)| \\ &= \frac{\pi}{\sin \pi l(r)} \{u - v \cos l(r)(\pi - \eta)\} r^{l(r)} + o(r^{l(r)}), \end{aligned}$$

from which, by (3.41) and (3.38), we get

$$\limsup_{r\to\infty}\frac{\ln\mu(r,f_{uv})}{\ln M(r,f_{uv})}\le\limsup_{r\to\infty}\vartheta_1(u,v,l(r),\eta),\tag{3.42}$$

$$\limsup_{r\to\infty}\frac{N(r,f_{uv})}{\ln M(r,f_{uv})}\le\limsup_{r\to\infty}\vartheta_2(u,v,l(r),\eta),\tag{3.43}$$

where

$$\vartheta_1(u,v,\gamma,\eta)=\frac{u\cos\gamma(\pi-\eta)-v}{u-v\cos\gamma(\pi-\eta)},$$

$$\vartheta_2(u,v,\gamma,\eta)=\frac{\sin\pi\gamma}{\pi\gamma}\,\frac{v}{u-v\cos\gamma(\pi-\eta)}.$$

Since $\vartheta_j(u,v,\gamma,\eta)$, $j=1,2$, are non-increasing functions of γ for sufficiently small η, we may replace $l(r)$ by λ in the right-hand sides of (3.42) and (3.43). Letting, after that, η tend to zero, we get

$$\limsup_{r\to\infty}\frac{\ln\mu(r,f_{uv})}{\ln M(r,f_{uv})}\le\frac{u\cos\lambda\pi-v}{u-v\cos\lambda\pi},\tag{3.44}$$

$$\limsup_{r\to\infty}\frac{N(r,f_{uv})}{\ln M(r,f_{uv})}\le\frac{\sin\pi\lambda}{\pi\lambda}\,\frac{v}{u-v\cos\lambda\pi}.$$

Thus

$$\begin{aligned}\limsup_{r\to\infty}\frac{\ln\mu(r,f_{uv})}{\ln M(r,f_{uv})}&+\pi\lambda\sin\pi\lambda\limsup_{r\to\infty}\frac{N(r,f_{uv})}{\ln M(r,f_{uv})}\\ &\le\frac{u\cos\lambda\pi-v}{u-v\cos\lambda\pi}+\sin^2\pi\lambda\frac{v}{u-v\cos\lambda\pi}=\cos\lambda\pi.\end{aligned}$$

Hence, if $0<\lambda<1/2$, then the sides of (3.23) are equal for the function $f_{uv}(z)$.

Observe that in our proof of inequality (3.44) we did not use the conditions $\lambda<1/2$, $u\cos\pi\lambda\ge v$; hence, (3.44) holds for $0<\lambda<1$ and $u\ge v$. For $v=0$ we get

$$\limsup_{r\to\infty}\frac{\ln\mu(r,f_{u0})}{\ln M(r,f_{u0})}\le\cos\lambda\pi.$$

But $f_{u0}(z)\equiv f(z,u)$ is an entire function and, thus, for each λ, $0<\lambda<1$, and each ρ, $\lambda\le\rho<1$, we have an example of an entire function of order ρ and lower order λ, for which (3.32) is an equality.

EXERCISE. Prove that the function $f_{u0}(z)$ satisfies

$$\limsup_{r\to\infty}\frac{T(r,f_{u0})}{\ln M(r,f_{u0})}=\begin{cases}\dfrac{\sin\pi\lambda}{\pi\lambda}, & 0<\lambda<\frac12,\\[2mm] \dfrac{1}{\pi\lambda}, & \frac12\le\lambda<1.\end{cases}$$

Thus, for each λ and ρ, $0<\lambda\le1/2$, $\lambda\le\rho<1$, there exists an entire function of order ρ and lower order λ, for which (3.30) is an equality.[8]

[8] There exists an entire function $g(z)$ of order ρ and lower order λ, $0\le\lambda\le\rho\le\infty$, for which

$$\limsup_{r\to\infty}\frac{T(r,g)}{\ln M(r,g)}=\begin{cases}\dfrac{\sin\pi\lambda}{\pi\lambda}, & 0\le\lambda\le\frac12,\\[2mm] \dfrac{1}{\pi\lambda}, & \frac12\le\lambda\le\infty\end{cases}$$

(see M. N. Sheremeta [**She69**]).

EXAMPLE 2. We show that for each δ_0, $0 \le \delta_0 \le 1$, there exists a meromorphic function $f(z)$ with $\lambda = \rho = 0$ such that $\delta(\infty, f) = \delta_0$ and

$$\limsup_{r\to\infty} \frac{\ln \mu(r,f)}{T(r,f)} = \delta(\infty, f), \tag{3.45}$$

$$\limsup_{r\to\infty} \frac{\ln \mu(r,f)}{\ln M(r,f)} = 1, \qquad \limsup_{r\to\infty} \frac{N(r,f)}{\ln M(r,f)} < \infty; \tag{3.46}$$

that is, we have an equality in (3.2) and (3.23).

Let $\rho(r) \to 0$ be a proximate order, such that $r^{\rho(r)} \uparrow \infty$ as $r \uparrow \infty$. Denote by $g(z,u)$ the canonical product of genus 0 with positive zeros such that $n(r,0) \sim ur^{\rho(r)}$. By asymptotic formula (5.28_2) in Chapter 2, for each $\eta > 0$ the relation

$$\ln |g(re^{i\varphi}, u)| = (1 + o(1))N(r, 0, g(z,u)) = (u + o(1)) \int_1^r t^{\rho(t)-1} dt \tag{3.47}$$

holds uniformly in φ, $\eta \le \varphi \le 2\pi - \eta$. Defining $g(z, 0) \equiv 1$, consider the function

$$g_{uv}(z) = \frac{g(z,u)}{g(-z,v)}, \qquad 0 \le v \le u < \infty,\ u > 0.$$

Relation (3.47) implies that for each $\eta > 0$ the relation

$$\ln |g_{uv}(re^{i\varphi})| = (u - v + o(1)) \int_1^r t^{\rho(t)-1} dt$$

holds uniformly in φ, $\eta \le \varphi \le 2\pi - \eta$. Whence, with the help of Theorem 7.4 of Chapter 1, we get

$$m(r, g_{uv}) = \frac{1}{\pi} \int_0^\pi \ln^+ |g_{uv}(re^{i\varphi})| d\varphi = (u - v + o(1)) \int_1^r t^{\rho(t)-1} dt.$$

Since

$$N(r, g_{uv}) = N(r, 0, g(z,v)) = (v + o(1)) \int_1^r t^{\rho(t)-1} dt,$$

we have

$$T(r, g_{uv}) = m(r, g_{uv}) + N(r, g_{uv}) = (u + o(1)) \int_1^r t^{\rho(t)-1} dt,$$

and hence,

$$\delta(\infty, g_{uv}) = \liminf_{r\to\infty} \frac{m(r, g_{uv})}{T(r, g_{uv})} = \frac{u - v}{u}.$$

Obviously, fixing $u > 0$ and choosing v, $0 \le v \le u$, we can give $\delta(\infty, g_{uv})$ an arbitrary preassigned value δ_0, $0 \le \delta_0 \le 1$. Since $\ln \mu(r,f) \le m(r,f)$, we have

$$\limsup_{r\to\infty} \frac{\ln \mu(r, g_{uv})}{T(r, g_{uv})} \le \limsup_{r\to\infty} \frac{m(r, g_{uv})}{T(r, g_{uv})} = \frac{u - v}{u} = \delta(\infty, g_{uv}).$$

Comparing with (3.2) for $\lambda = 0$, we see, that (3.45) holds for $f(z) = g_{uv}(z)$.

Since $\ln^+ M(r,f) \ge m(r,f)$, for $v < u$ we have

$$\limsup_{r\to\infty} \frac{N(r, g_{uv})}{\ln^+ M(r, g_{uv})} \le \limsup_{r\to\infty} \frac{N(r, g_{uv})}{m(r, g_{uv})} = \frac{v}{u - v} < \infty;$$

that is, the second relation from (3.46) holds. The first relation in (3.46) follows from the trivial inequality $\mu(r,f) \le M(r,f)$ and inequality (3.23) for $\lambda = 0$.

EXAMPLE 3. We show that Theorem 3.1 ceases to be true if we replace $<$ by $\leq$ in (3.1) (it is clear that this does not follow from the previous examples).

To this end, for arbitrary λ and ρ satisfying $0 < \lambda \leq \rho < 1/2$, we construct a meromorphic function $f(z)$ of order ρ and lower order λ, such that $|f(-r)| \leq 2$, $0 \leq r < \infty$, and $\delta(\infty, f) = 1 - \cos \pi\lambda$.

Let $g(z, u)$ be a function defined in the same way as $f(z, u)$ in Example 1, with the only difference that condition (3.35) is replaced by the conditions

$$\liminf_{r\to\infty} l(r) = 2\lambda, \qquad \limsup_{r\to\infty} l(r) = 2\rho.$$

Let

$$g_1(z) = \frac{g(z,1)}{g(-z,1)}.$$

As in Example 1, we get that for each $\eta > 0$ the relation

$$\begin{aligned} \ln|g_1(re^{i\varphi})| &= \frac{\pi}{\sin \pi l(r)}\{\cos l(r)(\varphi - \pi) - \cos l(r)\varphi\}r^{l(r)} + o(r^{l(r)}) \\ &= \frac{\pi}{\cos\frac{\pi l(r)}{2}}\left\{\sin l(r)\left(\varphi - \frac{\pi}{2}\right)\right\} r^{l(r)} + o(r^{l(r)}) \end{aligned} \tag{3.48}$$

holds uniformly in φ, $\eta \leq \varphi \leq \pi - \eta$.

Consider the function

$$g_2(z) = g_1(z) + g_1(-z).$$

From (3.48) we see that

$$g_1(re^{i\varphi}) \to 0, \quad g_1(-re^{i\varphi}) \to \infty \quad \text{as } r \to +\infty$$

uniformly in φ for $\eta \leq \varphi \leq \frac{\pi}{2} - \eta$, and hence

$$g_2(re^{i\varphi}) = (1 + o(1))g_1(-re^{i\varphi})$$

and

$$g_1(re^{i\varphi}) \to \infty, \quad g_1(-re^{i\varphi}) \to 0 \quad \text{as } r \to +\infty$$

uniformly in φ satisfying $\frac{\pi}{2} + \eta \leq \varphi \leq \pi - \eta$, and hence

$$g_2(re^{i\varphi}) = (1 + o(1))g_1(re^{i\varphi}).$$

Therefore the relation

$$\ln|g_2(re^{i\varphi})| = \frac{\pi}{\cos\frac{\pi l(r)}{2}}\left\{\sin\left(l(r)\left|\varphi - \frac{\pi}{2}\right|\right)\right\} r^{l(r)} + o(r^{l(r)})$$

holds uniformly in φ satisfying $\eta \leq \varphi \leq \frac{\pi}{2} - \eta$ or $\frac{\pi}{2} + \eta \leq \varphi \leq \pi - \eta$. Applying Theorem 7.4 of Chapter 1 we have

$$\begin{aligned} m(r, g_2) &= \frac{1}{\pi}\int_0^{\pi} \ln^+ |g_2(re^{i\varphi})|d\varphi \\ &= r^{l(r)} \sec\frac{\pi l(r)}{2}\int_0^{\pi} \sin\left(l(r)\left|\varphi - \frac{\pi}{2}\right|\right) d\varphi + o(r^{l(r)}) \\ &= r^{l(r)}\frac{2}{l(r)} \sec\frac{\pi l(r)}{2}\left(1 - \cos\frac{\pi l(r)}{2}\right) + o(r^{l(r)}). \end{aligned}$$

Since

$$N(r, \infty, g_2) = 2N(r, \infty, g_1) = \frac{2}{l(r)}r^{l(r)} + o(r^{l(r)}),$$

we get

$$T(r,g_2)=\frac{2}{l(r)}\sec\frac{\pi l(r)}{2}r^{l(r)}+o(r^{l(r)}).$$

Since the function $g_2(z)$ is even, the function $f(z)=g_2(\sqrt{z})$ is univalent and, hence, meromorphic. By the Corollary of Theorem 6.6 in Chapter 1 we have

$$\begin{aligned}T(r,f)&=T(\sqrt{r},g_2)\\&=\frac{2}{l(\sqrt{r})}\sec\frac{\pi l(\sqrt{r})}{2}r^{\frac{1}{2}l(\sqrt{r})}+o\left(r^{\frac{1}{2}l(\sqrt{r})}\right),\end{aligned}\tag{3.49}$$

$$\begin{aligned}\delta(\infty,f)&=\liminf_{r\to\infty}\frac{m(r,f)}{T(r,f)}=\liminf_{r\to\infty}\frac{m(\sqrt{r},g_2)}{T(\sqrt{r},g_2)}\\&=\liminf_{r\to\infty}\left(1-\cos\frac{\pi l(\sqrt{r})}{2}\right)=1-\cos\pi\lambda,\end{aligned}$$

and also, it is clear from (3.49) that $f(z)$ has order ρ and lower order λ.

Since $g(z,1)$ is a canonical product with positive zeros, we have $g(z,1)=\overline{g(\bar{z},1)}$. Therefore $|g(iy,1)|=|g(-iy,1)|$, $-\infty<y<\infty$, and hence $|g_1(iy)|=1$ and $|g_2(iy)|\le 2$, $-\infty<y<\infty$. Thus $|f(-r)|\le 2$, $0\le r<\infty$, what was to be proved.

4. Bounds for the sum of two deficiencies and related problems

In Section 1 of Chapter 4 we saw that a meromorphic function of non-integer order cannot have more than one Borel exceptional value. Examples given in Section 4 of Chapter 4 show that this result does not immediately carry over to deficient values. However, it turns out that a meromorphic function $f(z)$ of a non-integer order ρ cannot have two deficient values, a and b, with the sum $\delta(a,f)+\delta(b,f)$ close to 2.

More precisely, for each non-integer $\rho>0$ we can find a positive number $d(\rho)$ such that for each meromorphic function of order ρ and for each $a\neq b$, the inequality

$$\delta(a,f)+\delta(b,f)\le 2-d(\rho)\tag{4.1}$$

holds.

This result can be easily obtained using the Corollary of Theorem 4.2 in Chapter 2 (p. 60). First we observe that it is enough to prove (4.1) for $a=0$, $b=\infty$. In fact, if $a\neq\infty$, $b\neq\infty$, then for the function $f_1(z)=\frac{f(z)-a}{f(z)-b}$ we have $\rho[f_1]=\rho[f]=\rho$, $\delta(0,f_1)=\delta(a,f)$, $\delta(\infty,f_1)=\delta(b,f)$, and the inequality $\delta(0,f_1)+\delta(\infty,f_1)\le 2-d(\rho)$ implies (4.1). If $a\neq\infty$, $b=\infty$, we let $f_1(z)=f(z)-a$. Further, since

$$\delta(0,f)+\delta(\infty,f)\le 2-\limsup_{r\to\infty}\frac{N(r;0,\infty)}{T(r,f)},\tag{4.2}$$

and, by the Corollary of Theorem 4.2 in Chapter 2,

$$\limsup_{r\to\infty}\frac{N(r;0,\infty)}{T(r,f)}\ge\frac{\sin\pi(\rho-[\rho])}{C_1([\rho])},$$

where $0<C_1([\rho])<\infty$ does not depend on $f(z)$, we get (4.1) with $d(\rho)=|\sin\pi\rho|/C_1([\rho])$.

The problem of the best possible value of $d(\rho)$ arises. By (4.2) this problem is connected with the problem of the best lower estimate in terms of ρ of the value

$$\kappa(f) = \limsup_{r\to\infty} \frac{N(r;0,\infty)}{T(r,f)}$$

for meromorphic functions $f(z)$ of order ρ. More general is a problem of the best lower estimate of $\kappa(f)$ for meromorphic functions $f(z)$ of order ρ and lower order λ.

At present, none of these problems is completely solved.[9] The existing results are summarized in the following theorem.

THEOREM 4.1. *Let*

$$\nu(x) = \begin{cases} 1, & 0 \le x \le 1/2, \\ \sin \pi x, & 1/2 \le x \le 1, \\ |\sin \pi x|\{2.2x + 0.5|\sin \pi x|\}^{-1}, & 1 \le x < \infty, \\ 0, & x = \infty. \end{cases}$$

For each meromorphic function $f(z)$ of order $\rho \le \infty$ and lower order $\lambda < \infty$, the following estimates hold:

$$\kappa(f) \ge \max_{\lambda \le x \le \rho} \nu(x), \tag{4.3}$$

$$\delta(a,f) + \delta(b,f) \le 2 - \max_{\lambda \le x \le \rho} \nu(x), \quad a \ne b. \tag{4.3'}$$

In particular, for each meromorphic function $f(z)$ of order ρ,

$$\kappa(f) \ge \nu(\rho), \tag{4.4}$$

$$\delta(a,f) + \delta(b,f) \le 2 - \nu(\rho), \quad a \ne b. \tag{4.4'}$$

To characterize the accuracy of estimates (4.4) and (4.4′), define a function $\nu_1(x)$, $x \ge 0$, as follows:

$$\nu_1(x) = \begin{cases} |\sin \pi x|\{[x] + |\sin \pi x|\}^{-1}, & [x] \le x \le [x] + \frac{1}{2}, \\ |\sin \pi x|\{[x]+1\}^{-1}, & [x] + \frac{1}{2} \le x \le [x] + 1. \end{cases}$$

Obviously

$$\nu_1(x) = \nu(x), \quad 0 \le x \le 1,$$
$$0.3\nu_1(x) \le \nu(x) \le \nu_1(x), \quad x \ge 1$$

(where the equality in the last inequality holds for integer x only).

Exercise 1 in Section 1 of Chapter 4 contains an example of a function of order ρ, such that

$$1 - \delta(0,f) = \kappa(f) = \nu_1(\rho), \qquad \delta(\infty, f) = 1.$$

Thus, estimates (4.4) and (4.4′) are the best possible for $0 < \rho \le 1$, and, in a certain sense, are close to the best possible for $\rho > 1$.

As for the accuracy of estimates (4.3) and (4.3′), note that Exercise 2 in Section 1 of Chapter 4 contains, for each integer $p \ge 0$ and for numbers λ and ρ satisfying $p < \lambda \le \rho < p+1$, an example of a function of order ρ and lower order λ, such that

$$1 - \delta(0,f) = \kappa(f) = \max_{\lambda \le x \le \rho} \nu_1(x), \qquad \delta(\infty, f) = 1.$$

---

[9]If $\lambda < \infty$ and the sum of deficiencies is 2, then $\rho = \lambda = n/2$, where n is an integer [**44**].

Thus estimates (4.3) and (4.3′) are the best possible for $0 < \lambda \le \rho < 1$ and, in a certain sense, are close to the best possible if $[\rho] < \lambda \le \rho$, $[\rho] \ge 1$.

We can also add that estimates (4.3) and (4.3′) are the best possible for $0 \le \lambda \le 1/2$. In fact, in this case $\max_{\lambda \le x \le \rho} \nu(x) = 1$ for each $\rho \ge \lambda$. On the other hand, for each entire function $f(z)$ of order ρ and lower order λ, the relations

$$1 - \delta(0, f) = \kappa(f) = \limsup_{r \to \infty} \frac{N(r, 0)}{T(r, f)} \le 1, \qquad \delta(\infty, f) = 1$$

hold, and hence we have the equality in (4.3) and (4.3′).

Note that Theorem 4.1 immediately implies the following statement, which has already been proved as a corollary of Lemma 3.4: an entire function of lower order $\lambda \le 1/2$ cannot have finite deficient values. In fact, since $\delta(\infty, f) = 1$ and $\max_{\lambda \le x \le \rho} \nu(x) = 1$, letting $b = \infty$ in (4.3′), we get $\delta(a, f) = 0$ for each $a \ne \infty$.

Theorem 4.1 is a corollary of the following two theorems.

THEOREM 4.2. *Let*

$$\nu_2(x) = |\sin \pi x| \{2.2x + 0.5 |\sin \pi x|\}^{-1}, \quad x \ge 0.$$

For each meromorphic function $f(z)$ of order $\rho \le \infty$ and lower order $\infty > \lambda > 0$ the estimate

$$\kappa(f) \ge \max_{\lambda \le x \le \rho} \nu_2(x) \tag{4.5}$$

holds.

THEOREM 4.3. *For each meromorphic function of lower order $0 \le \lambda < 1$ the estimate*

$$\kappa(f) \ge \nu(\lambda) \tag{4.6}$$

holds.

In order to derive Theorem 4.1 from these theorems, it suffices to observe that for $\lambda \ge 1$, estimate (4.3) coincides with (4.5), and for $\lambda < 1$,

$$\max_{\lambda \le x \le \rho} \nu(x) = \max \left\{ \nu(\lambda), \ \max_{\lambda \le x \le \rho} \nu_2(x) \right\},$$

and estimate (4.3) follows from (4.5) and (4.6). Estimate (4.3′) is an easy consequence of (4.3) and (4.2).

Now we turn to the lemmas used in the proof of Theorem 4.2.

LEMMA 4.1. *Let $f(z)$, $f(0) = 1$, be a meromorphic function, and $q \ge 0$ an integer. If $0 \le r \le \frac{1}{2}R$, then*

$$2T(r, f) - N(r; 0, \infty) \le r^q \int_0^R n(t; 0, \infty) \Phi\left(\frac{t}{r}\right) t^{-q-1} dt$$
$$+ C_1 \left(\frac{r}{R}\right)^{q+1} T(2R, f) + C_2 q (r^q + 1),$$

where $0 < C_1, C_2 < \infty$ are constants that do not depend on r and R, and

$$\Phi(t) = \frac{1}{2\pi} \int_0^{2\pi} \frac{d\theta}{|te^{i\theta} - 1|}.$$

PROOF. We have

$$
\begin{aligned}
2T(r,f) - N(r;0,\infty) &= m(r,f) + m\left(r, \frac{1}{f}\right) \\
&= \frac{1}{2\pi}\int_0^{2\pi} |\ln|f(re^{i\theta})||\, d\theta.
\end{aligned}
\tag{4.7}
$$

Applying Lemma 3.1, we represent the function $f(z)$ in the form (3.3). We get

$$
\begin{aligned}
&\frac{1}{2\pi}\int_0^{2\pi} |\ln|f(re^{i\theta})||\, d\theta \\
&\quad\le \frac{1}{2\pi}\int_0^{2\pi} |\ln|\alpha_{R,q}(re^{i\theta})||\, d\theta + \frac{1}{2\pi}\int_0^{2\pi} |\ln|\omega_{R,q}(re^{i\theta})||\, d\theta,
\end{aligned}
\tag{4.8}
$$

from which, by (3.4) and (3.5), we get the inequality

$$
\begin{aligned}
\frac{1}{2\pi}\int_0^{2\pi} |\ln|f(re^{i\theta})||\, d\theta &\le \sum_{|a_\mu|<R} I(a_\mu) + \sum_{|b_\nu|<R} I(b_\nu) \\
&\quad + K_1\left(\frac{r}{R}\right)^{q+1} T(2R,f) + K_2 q(r^q+1), \\
&\quad 0 \le r \le \frac{1}{2}R,
\end{aligned}
\tag{4.9}
$$

where

$$
I(a) = \frac{1}{2\pi}\int_0^{2\pi} \left|\ln\left|E\left(\frac{re^{i\theta}}{a},\ q\right)\right|\right| d\theta, \qquad 0 < |a| < \infty.
$$

Let us estimate $I(a)$. Since

$$
\frac{\partial}{\partial t}\ln|E(te^{i\theta},\ q)| = \operatorname{Re}\frac{\partial}{\partial t}\ln E(te^{i\theta},\ q) = \operatorname{Re}\frac{t^q e^{i(q+1)\theta}}{te^{i\theta}-1},
$$

for $0 < \theta < 2\pi$, $0 < t < \infty$, we have

$$
|\ln|E(re^{i\theta},\ q)|| \le \int_0^r \left|\frac{\partial}{\partial t}\ln|E(te^{i\theta},\ q)|\right| dt \le \int_0^r \frac{t^q\,dt}{|te^{i\theta}-1|},
$$

and hence,

$$
\begin{aligned}
I(a) = I(|a|) &= \frac{1}{2\pi}\int_0^{2\pi} \left|\ln\left|E\left(\frac{r}{|a|}e^{i\theta}, q\right)\right|\right| d\theta \\
&\le \int_0^{r/|a|} \left\{\frac{1}{2\pi}\int_0^{2\pi} \frac{d\theta}{|te^{i\theta}-1|}\right\} t^q\,dt = \int_0^{r/|a|} t^q\Phi(t)\,dt.
\end{aligned}
$$

It is easy to see that $\Phi(t) = \frac{1}{t}\Phi\left(\frac{1}{t}\right)$; therefore we have the estimate

$$
I(a) \le \int_0^{r/|a|} t^{q-1}\Phi\left(\frac{1}{t}\right) dt = r^q \int_{|a|}^{\infty} u^{-q-1}\Phi\left(\frac{u}{r}\right) du.
$$

Using this estimate we get

$$\begin{aligned}
&\sum_{|a_\mu|<R} I(a_\mu) + \sum_{|b_\nu|<R} I(b_\nu) \\
&\quad \le r^q \sum_{|a_\mu|<R} \int_{|a_\mu|}^{\infty} u^{-q-1}\Phi\left(\frac{u}{r}\right) du + r^q \sum_{|b_\nu|<R} \int_{|b_\nu|}^{\infty} u^{-q-1}\Phi\left(\frac{u}{r}\right) du \\
&\quad \le r^q \int_0^R dn(s;0,\infty) \int_s^\infty u^{-q-1}\Phi\left(\frac{u}{r}\right) du \\
&\quad = r^q \left\{ n(R;0,\infty) \int_R^\infty u^{-q-1}\Phi\left(\frac{u}{r}\right) du + \int_0^R n(s;0,\infty) s^{-q-1}\Phi\left(\frac{s}{r}\right) ds \right\}.
\end{aligned}$$

For $0 \le r \le \frac{1}{2}R$, $u \ge R$, the inequality

$$\Phi\left(\frac{u}{r}\right) = \frac{1}{2\pi}\int_0^{2\pi} \frac{d\theta}{\left|\frac{u}{r}e^{i\theta} - 1\right|} \le \frac{r}{u-r} \le 2\frac{r}{u}$$

holds, and therefore

$$\int_R^\infty u^{-q-1}\Phi\left(\frac{u}{r}\right) du \le 2r\int_R^\infty u^{-q-2} du = \frac{2}{q+1}\frac{r}{R^{q+1}}.$$

Taking into the account the inequality $n(R;0,\infty) \le 4T(2R,f)$ (Lemma 7.1 of Chapter 1), we get the estimate

$$\begin{aligned}
&\sum_{|a_\mu|<R} I(a_\mu) + \sum_{|b_\nu|<R} I(b_\nu) \\
&\quad \le r^q \left\{ \frac{8}{q+1}\frac{r}{R^{q+1}} T(2R,f) + \int_0^R n(s;0,\infty)\Phi\left(\frac{s}{r}\right) s^{-q-1} ds \right\}.
\end{aligned} \tag{4.10}$$

The conclusion of the lemma follows from (4.7), (4.9), and (4.10). □

LEMMA 4.2. *Let*

$$J(\beta) = \int_0^\infty t^{-\beta-1}\Phi\left(\frac{1}{t}\right) dt, \qquad 0 < \beta < 1.$$

The estimate

$$J(\beta) \le 4.4 \csc \pi\beta$$

holds.

PROOF. We have

$$J(\beta) = \frac{1}{\pi}\int_0^\pi d\theta \int_0^\infty \frac{t^{-\beta} dt}{\sqrt{t^2 - 2t\cos\theta + 1}}.$$

Making the change of variable

$$t = \cos\theta + \sin\theta \tan\varphi = \frac{\cos(\theta-\varphi)}{\cos\varphi},$$

we get

$$J(\beta) = \frac{1}{\pi}\int_0^\pi d\theta \int_{\theta-\frac{\pi}{2}}^{\pi/2} \{\cos(\theta-\varphi)\}^{-\beta}\{\cos\varphi\}^{\beta-1} d\varphi.$$

Changing the order of integration, we have

$$J(\beta) = \frac{1}{\pi} \int_{-\pi/2}^{\pi/2} \{\cos\varphi\}^{\beta-1} d\varphi \int_0^{\varphi+\frac{\pi}{2}} \{\cos(\theta-\varphi)\}^{-\beta} d\theta.$$

Since the second integral in the right-hand side of the equality

$$\int_0^{\varphi+\frac{\pi}{2}} \{\cos(\theta-\varphi)\}^{-\beta} d\theta = \int_0^{\pi/2} \{\cos\omega\}^{-\beta} d\omega + \int_{-\varphi}^{0} \{\cos\omega\}^{-\beta} d\omega$$

is an odd function of φ, we get

$$J(\beta) = \frac{1}{\pi} \int_{-\pi/2}^{\pi/2} \{\cos\varphi\}^{\beta-1} d\varphi \int_0^{\pi/2} \{\cos\omega\}^{-\beta} d\omega.$$

Using the well-known formula

$$\int_0^{\pi/2} \{\cos\alpha\}^x d\alpha = \frac{\sqrt{\pi}\Gamma\left(\frac{x+1}{2}\right)}{\Gamma\left(\frac{x}{2}+1\right)}, \qquad -1 < x < \infty,$$

we get

$$J(\beta) = \frac{\Gamma\left(\frac{\beta}{2}\right)\Gamma\left(\frac{1-\beta}{2}\right)}{2\Gamma\left(\frac{\beta+1}{2}\right)\Gamma\left(1-\frac{\beta}{2}\right)},$$

from which, by virtue of $\Gamma(x)\Gamma(1-x) = \pi \csc \pi x$, we obtain

$$J(\beta) = \frac{\pi^2}{\sin \pi\beta \Gamma^2\left(\frac{1+\beta}{2}\right)\Gamma^2\left(1-\frac{\beta}{2}\right)}.$$

Letting $t = \frac{1}{4} - \frac{\beta}{2}$, we see that $-\frac{1}{4} < t < \frac{1}{4}$, and

$$\Gamma\left(\frac{1+\beta}{2}\right)\Gamma\left(1-\frac{\beta}{2}\right) = \Gamma\left(\frac{3}{4}+t\right)\Gamma\left(\frac{3}{4}-t\right).$$

Let us find the minimum of the function $g(t) = \Gamma\left(\frac{3}{4}+t\right)\Gamma\left(\frac{3}{4}-t\right)$ on the interval $\left(-\frac{1}{4}, \frac{1}{4}\right)$. Using the well-known formula (see Whittaker, Watson [**WW96**, 12·16])

$$\frac{\Gamma'(z)}{\Gamma(z)} = -\gamma - \frac{1}{z} - \sum_{n=1}^{\infty} \left\{\frac{1}{z+n} - \frac{1}{n}\right\}$$

(γ is the Euler–Mascheroni constant), we get

$$\frac{g'(t)}{g(t)} = \frac{\Gamma'\left(\frac{3}{4}+t\right)}{\Gamma\left(\frac{3}{4}+t\right)} - \frac{\Gamma'\left(\frac{3}{4}-t\right)}{\Gamma\left(\frac{3}{4}-t\right)} = \sum_{n=0}^{\infty} \frac{2t}{\left(n+\frac{3}{4}\right)^2 - t^2}.$$

Hence we see that

$$\operatorname{sgn} g'(t) = \operatorname{sgn} t$$

on the interval $\left(-\frac{1}{4}, \frac{1}{4}\right)$. This indicates that

$$\min_{-\frac{1}{4}<t<\frac{1}{4}} g(t) = g(0) = \Gamma^2\left(\frac{3}{4}\right).$$

Thus,

$$J(\beta) \le \frac{\pi^2}{\sin \pi\beta \Gamma^4\left(\frac{3}{4}\right)} < \frac{4.4}{\sin \pi\beta}.$$

□

LEMMA 4.3. *Let $f(z)$, $f(0) = 1$, be a meromorphic function, $q \ge 0$ an integer, and $\kappa' > \kappa(f)$, $\varepsilon > 0$. It is possible to find $R_0 \ge 1$ such that for $R \ge 2R_0$, $q + \varepsilon \le x \le q + 1 - \varepsilon$, the inequality*

$$(\nu_2(x) - \kappa') \int_{R_0}^{\frac{1}{2}R} T(r, f) r^{-x-1} dr \le C_3 R^{-x} T(2R, f) + C_4 \tag{4.11}$$

holds, where $0 < C_3, C_4 < \infty$ are constants that do not depend on R and x.

PROOF. Let $R_0 \ge 1$ be such that $N(r; 0, \infty) \le \kappa' T(r, f)$ holds for $r \ge R_0$. Lemma 4.1 implies the inequality

$$\begin{aligned}(2 - \kappa')T(r, f) \le r^q \int_0^R n(t; 0, \infty)\Phi\left(\frac{t}{r}\right) t^{-q-1} dt \\ + C_1 \left(\frac{r}{R}\right)^{q+1} T(2R, f) + 2C_2 q r^q, \qquad R_0 \le r \le \frac{1}{2}R.\end{aligned}$$

We multiply both sides of this inequality by r^{-x-1} and integrate with respect to r from R_0 to $\frac{1}{2}R$. We have

$$\begin{aligned}&(2 - \kappa') \int_{R_0}^{\frac{1}{2}R} T(r, f) r^{-x-1} dr \\ &\quad\le \int_{R_0}^{\frac{1}{2}R} r^{q-x-1} dr \int_0^R n(t; 0, \infty) \Phi\left(\frac{t}{r}\right) t^{-q-1} dt \\ &\quad\quad + \frac{C_1}{q+1-x} R^{-x} T(2R, f) + \frac{2C_2 q}{x - q}.\end{aligned} \tag{4.12}$$

Denote the iterated integral in the right-hand side of (4.12) by I. Changing the order of integration, we get

$$\begin{aligned}I &= \int_0^R n(t; 0, \infty) t^{-q-1} dt \int_{R_0}^{\frac{1}{2}R} r^{q-x-1} \Phi\left(\frac{t}{r}\right) dr \\ &= \int_0^R n(t; 0, \infty) t^{-x-1} dt \int_{R_0/t}^{R/(2t)} u^{q-x-1} \Phi\left(\frac{1}{u}\right) du \\ &\le \int_0^R n(t; 0, \infty) t^{-x-1} dt \int_0^{\infty} u^{q-x-1} \Phi\left(\frac{1}{u}\right) du \\ &= J(x - q) \int_0^R n(t; 0, \infty) t^{-x-1} dt \\ &= J(x - q) \left\{ N(R; 0, \infty) R^{-x} + x \int_0^R N(t; 0, \infty) t^{-x-1} dt \right\} \\ &= J(x - q) \left\{ N(R; 0, \infty) R^{-x} + x \int_{\frac{1}{2}R}^R N(t; 0, \infty) t^{-x-1} dt \right. \\ &\quad \left. + x \int_{R_0}^{\frac{1}{2}R} N(t; 0, \infty) t^{-x-1} dt + x \int_0^{R_0} N(t; 0, \infty) t^{-x-1} dt \right\}.\end{aligned}$$

Since

$$N(R; 0, \infty) \le 2T(R, f) \le 2T(2R, f),$$

$$x \int_{\frac{1}{2}R}^R N(t; 0, \infty) t^{-x-1} dt \le N(R; 0, \infty) \left(\frac{1}{2}R\right)^{-x} \le 2^{q+2} R^{-x} T(2R, f),$$

$$x\int_{R_0}^{\frac{1}{2}R} N(t;0,\infty)t^{-x-1}dt \le x\kappa' \int_{R_0}^{\frac{1}{2}R} T(t,f)t^{-x-1}dt,$$

$$x\int_0^{R_0} N(t;0,\infty)t^{-x-1}dt$$
$$\le (q+1)\left\{\int_0^1 N(t;0,\infty)t^{-q-2}dt + \int_1^{R_0} N(t;0,\infty)t^{-q-1}dt\right\},$$

and, by Lemma 4.2,

$$J(x-q) \le 4.4|\csc \pi x|,$$

we get the estimate

$$I \le \frac{4.4x\kappa'}{|\sin \pi x|}\int_{R_0}^{\frac{1}{2}R} T(r,f)r^{-x-1}dr + C_5R^{-x}T(2R,f) + C_6,$$

where $0 < C_5, C_6 < \infty$ do not depend on R and x, $q+\varepsilon < x < q+1-\varepsilon$. Using this estimate, from (4.12) we get

$$\left(2-\kappa' - \frac{4.4x\kappa'}{|\sin \pi x|}\right)\int_{R_0}^{\frac{1}{2}R} T(r,f)r^{-x-1}dr \le C_7R^{-x}T(2R,f) + C_8, \tag{4.13}$$

where $0 < C_7, C_8 < \infty$ do not depend on R and x, $q+\varepsilon < x < q+1-\varepsilon$. Since

$$2-\kappa' - \frac{4.4x\kappa'}{|\sin \pi x|} = (\nu_2(x) - x')\left(1 + \frac{4.4x}{|\sin \pi x|}\right),$$

multiplying both sides of (4.13) by $(1+\frac{4.4x}{|\sin \pi x|})^{-1}$, we get inequality (4.11). □

PROOF. Now we prove Theorem 4.2. We consider separately the following two cases: (a) $0 < \lambda < \rho \le \infty$, (b) $1 \le [\rho] < \rho = \lambda$. We do not need to consider the case $1 \le [\rho] = \rho = \lambda$, since in this case the right-hand side of inequality (4.5) vanishes, and the conclusion of Theorem 4.2 is trivial.

Case (a). By the continuity of the function $\nu_2(x)$ it suffices to prove that for each non-integer x, $\lambda < x < \rho$, the inequality

$$\kappa(f) \ge \nu_2(x) \tag{4.14}$$

holds. Letting $q = [x]$, $\varepsilon = \min(x-q, q+1-x)$, we apply Lemma 4.3. Note that the quantities C_3 and C_4 in (4.11) could depend on q and ε; therefore now they can depend on x, but this does not interfere with our argument.

Since $x < \rho$, we get

$$\lim_{R\to\infty}\int_{R_0}^{\frac{1}{2}R} T(r,f)r^{-x-1}dr = \infty.$$

Since $x > \lambda$, there exists a sequence $R_k \to \infty$ such that

$$\lim_{k\to\infty} R_k^{-x}T(2R_k,f) = 0.$$

Therefore, letting R tend to ∞ along the sequence R_k in (4.11), we get

$$\nu_2(x) - \kappa' \le 0.$$

Since this inequality holds for each $\kappa' > \kappa(f)$, it holds also for $\kappa' = \kappa(f)$; that is, (4.14) holds.

Case (b). We prove that

$$\kappa(f) \ge \nu_2(\lambda).$$

Let $\lambda(r)$ be the lower proximate order of the function $T(r,f)$, $q = [\lambda]$, $\varepsilon = \frac{1}{2}\min(\lambda - q, q+1-\lambda)$. We choose $R_0' \geq R_0$ large, so that for $r \geq R_0'$ the inequalities $q+\varepsilon < \lambda(r) < q+1-\varepsilon$ and $T(r,f) \geq \frac{1}{2}r^{\lambda(r)}$ hold. Letting $x = \lambda(R)$ in (4.11), we get the inequality

$$(4.15)\qquad (\nu_2(\lambda(R)) - \kappa')\int_{R_0}^{\frac{1}{2}R} T(r,f)r^{-\lambda(R)-1}dr \leq C_3R^{-\lambda(R)}T(2R,f) + C_4.$$

By the result of Exercise 2 in Section 2 of Chapter 2 we have

$$\begin{aligned}\lim_{R\to\infty}&\int_{R_0}^{\frac{1}{2}R} T(r,f)r^{-\lambda(R)-1}dr\\ &\geq \lim_{R\to\infty}\left\{\int_{R_0}^{R_0'} T(r,f)r^{-\lambda(R)-1}dr + \frac{1}{2}\int_{R_0'}^{\frac{1}{2}R} r^{\lambda(r)-\lambda(R)-1}dr\right\} = \infty.\end{aligned}$$

The definition of a lower proximate order implies that there exists a sequence $R_k \to \infty$, such that

$$\lim_{k\to\infty} R_k^{-\lambda(R_k)}T(2R_k,f) < \infty.$$

Letting R tend to ∞ along the sequence R_k, we get the inequality $\nu_2(\lambda) - \kappa' \leq 0$, which, in turn, implies (4.5). The proof of Theorem 4.2 is complete. $\square$

Now we turn to lemmas used in the proof of Theorem 4.3.

LEMMA 4.4. *Let*

$$\alpha(z) = \prod_{\mu}\left(1 - \frac{z}{a_\mu}\right)\left\{\prod_{\nu}\left(1 - \frac{z}{b_\nu}\right)\right\}^{-1}$$

be a meromorphic function of genus zero. Let

$$\breve{\alpha}(z) = \prod_{\mu}\left(1 - \frac{z}{|a_\mu|}\right)\left\{\prod_{\nu}\left(1 + \frac{z}{|b_\nu|}\right)\right\}^{-1}.$$

Then

$$m(r,\breve{\alpha}) \geq m(r,\alpha), \qquad T(r,\breve{\alpha}) \geq T(r,\alpha).$$

PROOF. Let $a_\mu = |a_\mu|e^{i\alpha_\mu}$, $b_\nu = |b_\nu|e^{i\beta_\nu}$. We fix $r > 0$ and denote by D the set of those values of θ, $-\pi \leq \theta < \pi$, for which $|\alpha(re^{i\theta})| \geq 1$. Let $D(a_\mu)$ be the set of those values of θ, $-\pi \leq \theta < \pi$, for which $\theta + \alpha_\mu + \pi \in D$ (mod 2π), and let $D(b_\nu)$ be the set of those values of θ, $-\pi \leq \theta < \pi$, for which $\theta + \beta_\nu \in D$ (mod 2π). It is clear that $d = \operatorname{meas} D = \operatorname{meas} D(a_\mu) = \operatorname{meas} D(b_\nu)$. We have

$$\begin{aligned}m(r,\alpha) &= \frac{1}{2\pi}\int_{-\pi}^{\pi}\ln^+|\alpha(re^{i\theta})|\,d\theta = \frac{1}{2\pi}\int_D \ln|\alpha(re^{i\theta})|\,d\theta\\ &= \sum_\mu \frac{1}{2\pi}\int_D \ln\left|1 - \frac{re^{i\theta}}{a_\mu}\right|d\theta + \sum_\nu \frac{1}{2\pi}\int_D -\ln\left|1 - \frac{re^{i\theta}}{b_\nu}\right|d\theta\\ &= \sum_\mu \frac{1}{2\pi}\int_{D(a_\mu)} \ln\left|1 + \frac{re^{i\theta}}{|a_\mu|}\right|d\theta + \sum_\nu \frac{1}{2\pi}\int_{D(b_\nu)} -\ln\left|1 - \frac{re^{i\theta}}{|b_\nu|}\right|d\theta.\end{aligned}$$

Since the functions $\ln\left|1+\frac{re^{i\theta}}{|a_\mu|}\right|$ and $-\ln\left|1-\frac{re^{i\theta}}{|b_\nu|}\right|$ are decreasing on $[0,\pi]$ and even, by Lemma 7.2 of Chapter 1, we have

$$\int_{D(a_\mu)} \ln\left|1+\frac{re^{i\theta}}{|a_\mu|}\right| d\theta \le \int_{-\frac{1}{2}\operatorname{meas} D(a_\mu)}^{\frac{1}{2}\operatorname{meas} D(a_\mu)} \ln\left|1+\frac{re^{i\theta}}{|a_\mu|}\right| d\theta = \int_{-\frac{1}{2}d}^{\frac{1}{2}d} \ln\left|1+\frac{re^{i\theta}}{|a_\mu|}\right| d\theta,$$

$$\int_{D(b_\nu)} -\ln\left|1-\frac{re^{i\theta}}{|b_\nu|}\right| d\theta \le \int_{-\frac{1}{2}\operatorname{meas} D(b_\nu)}^{\frac{1}{2}\operatorname{meas} D(b_\nu)} -\ln\left|1-\frac{re^{i\theta}}{|b_\nu|}\right| d\theta = \int_{-\frac{1}{2}d}^{\frac{1}{2}d} -\ln\left|1-\frac{re^{i\theta}}{|b_\nu|}\right| d\theta.$$

We get

$$\begin{aligned} m(r,\alpha) &\le \sum_\mu \frac{1}{2\pi}\int_{-\frac{1}{2}d}^{\frac{1}{2}d} \ln\left|1+\frac{re^{i\theta}}{|a_\mu|}\right| d\theta + \sum_\nu \frac{1}{2\pi}\int_{-\frac{1}{2}d}^{\frac{1}{2}d} -\ln\left|1-\frac{re^{i\theta}}{|b_\nu|}\right| d\theta \\ &= \frac{1}{2\pi}\int_{-\frac{1}{2}d}^{\frac{1}{2}d} \ln|\breve{\alpha}(-re^{i\theta})|\, d\theta \le \frac{1}{2\pi}\int_{-\pi}^{\pi} \ln^+|\breve{\alpha}(-re^{i\theta})|\, d\theta = m(r,\breve{\alpha}). \end{aligned}$$

Since $N(r,\alpha)=N(r,\breve{\alpha})$, we conclude that

$$T(r,\alpha)=m(r,\alpha)+N(r,\alpha)\le m(r,\breve{\alpha})+N(r,\breve{\alpha})=T(r,\breve{\alpha}). \qquad \square$$

LEMMA 4.5. *Let $f(z)$, $f(0)=1$, be a meromorphic function of of genus zero with positive zeros and negative poles. Denote by A the (possibly, empty) set of those values of r, $0<r<\infty$, for which at least one of the following equalities is true:*

$$T(r,f)=N(r,f), \qquad T(r,f)=N\left(r,\frac{1}{f}\right).$$

For each $r\notin A$ there exists $\beta=\beta(r)$, $0<\beta(r)<\pi$, such that

$$T(r,f)=\int_0^\infty N\left(t,\frac{1}{f}\right)K(t,r,\beta)\,dt+\int_0^\infty N(t,f)K(t,r,\pi-\beta)\,dt, \tag{4.16}$$

where

$$K(t,r,\varphi)=\frac{1}{\pi}\frac{r\sin\varphi}{t^2+r^2-2rt\cos\varphi}=\frac{1}{\pi}\operatorname{Im}\frac{1}{t-re^{i\varphi}}, \qquad 0<\varphi<\pi.$$

PROOF. We choose a single-valued branch of $\ln f(z)$ satisfying $\ln f(0)=0$ in the region $\{\operatorname{Im} z>0\}\cup\{|z|<q\}$, where q is small so that $f(z)\ne 0,\infty$ for $|z|<q$. We have the representation

$$\ln f(z)=\int_0^\infty \ln\left(1-\frac{z}{t}\right)dn\left(t,\frac{1}{f}\right)-\int_0^\infty \ln\left(1+\frac{z}{t}\right)dn(t,f), \qquad \operatorname{Im} z>0,$$

from which, integrating by parts twice, we get

$$\ln f(z)=-z\int_0^\infty N\left(t,\frac{1}{f}\right)\frac{dt}{(t-z)^2}-z\int_0^\infty N(t,f)\frac{dt}{(t+z)^2}, \qquad \operatorname{Im} z>0. \tag{4.17}$$

Fix $r\notin A$; then $m(r,f)>0$ and $m(r,\frac{1}{f})>0$. Since the function $\ln|f(re^{i\theta})|$ is an increasing function of θ on $[0,\pi]$, we get from this that $\ln|f(r)|<0$ and $\ln|f(-r)|>0$. Denote by $\beta=\beta(r)$, $0<\beta(r)<\pi$, the number determined by the condition $\ln|f(re^{i\beta})|=0$. Observing that $\ln|f(re^{i\theta})|$ is an even function of θ, we get for $r\notin A$,

$$m(r,f)=\frac{1}{\pi}\int_0^\pi \ln^+|f(re^{i\theta})|\,d\theta=\frac{1}{\pi}\int_\beta^\pi \ln|f(re^{i\theta})|\,d\theta=\lim_{\delta\to\pi}\frac{1}{\pi}\operatorname{Im}\int_{re^{i\beta}}^{re^{i\delta}} \ln f(z)\frac{dz}{z}.$$

By virtue of representation (4.17) we get

$$
\begin{aligned}
m(r,f) = -\frac{1}{\pi}\lim_{\delta\to\pi}\Bigg\{ &\operatorname{Im}\int_0^\infty N\left(t,\frac{1}{f}\right)dt\int_{re^{i\beta}}^{re^{i\delta}}\frac{dz}{(z-t)^2}\\
&+\operatorname{Im}\int_0^\infty N(t,f)dt\int_{re^{i\beta}}^{re^{i\delta}}\frac{dz}{(z+t)^2}\Bigg\}\\
= \frac{1}{\pi}\lim_{\delta\to\pi}\Bigg\{&\int_0^\infty N\left(t,\frac{1}{f}\right)\operatorname{Im}\left[\frac{1}{re^{i\delta}-t}-\frac{1}{re^{i\beta}-t}\right]dt\\
&+\int_0^\infty N(t,f)\operatorname{Im}\left[\frac{1}{re^{i\delta}+t}-\frac{1}{re^{i\beta}+t}\right]dt\Bigg\}\\
= \int_0^\infty &N\left(t,\frac{1}{f}\right)K(t,r,\beta)\,dt+\int_0^\infty N(t,f)K(t,r,\pi-\beta)\,dt\\
-\lim_{\delta\to\pi}&\int_0^\infty N\left(t,\frac{1}{f}\right)K(t,r,\delta)\,dt-\lim_{\delta\to\pi}\int_0^\infty N(t,f)K(t,r,\pi-\delta)\,dt.
\end{aligned}
$$

From the estimate

$$K(t,r,\delta)\le\frac{1}{\pi}\frac{r\sin\delta}{r^2+t^2},\qquad \frac{\pi}{2}\le\delta<\pi,$$

and the convergence of the integral

$$\int_0^\infty N\left(t,\frac{1}{f}\right)t^{-2}dt,$$

it easily follows that

$$\lim_{\delta\to\pi}\int_0^\infty N\left(t,\frac{1}{f}\right)K(t,r,\delta)\,dt=0.$$

We show that

$$\lim_{\delta\to\pi}\int_0^\infty N(t,f)K(t,r,\pi-\delta)\,dt=N(r,f). \tag{4.18}$$

Choose an arbitrary $\varepsilon>0$. By the continuity of the function $N(t,f)$, we can choose $\eta>0$ such that for $|t-r|<\eta$ the inequality $|N(t,f)-N(r,f)|<\varepsilon$ holds. We have

$$K(t,r,\pi-\delta)\le\frac{1}{\pi}\frac{r\sin\delta}{(r-t)^2}. \tag{4.19}$$

Since $\int_{-\infty}^\infty K(t,r,\varphi)\,dt=1$, $0<\varphi<\pi$, it follows that

$$
\begin{aligned}
&\int_0^\infty N(t,f)K(t,r,\pi-\delta)\,dt-N(r,f)\\
&\quad=\int_0^\infty N(t,f)K(t,r,\pi-\delta)\,dt-\int_{-\infty}^\infty N(r,f)K(t,r,\pi-\delta)\,dt\\
&\quad=\int_{|t-r|<\eta}\{N(t,f)-N(r,f)\}K(t,r,\pi-\delta)\,dt\\
&\qquad+\int_{\substack{|t-r|\ge\eta\\ t>0}}N(t,f)K(t,r,\pi-\delta)\,dt-\int_{|t-r|\ge\eta}N(r,f)K(t,r,\pi-\delta)\,dt\\
&\quad=I_1+I_2-I_3.
\end{aligned}
$$

Noticing that $K(t,r,\varphi)>0$, $0<\varphi<\pi$, we get

$$|I_1|\le\varepsilon\int_{|t-r|<\eta}K(t,r,\pi-\delta)\,dt<\varepsilon\int_{-\infty}^{\infty}K(t,r,\pi-\delta)\,dt=\varepsilon.$$

Next, by virtue of estimate (4.19) we have

$$|I_2|\le\frac{1}{\pi}\sin\delta\int_{\substack{|t-r|\ge\eta\\ t>0}}N(t,f)\frac{r}{(r-t)^2}dt,$$

$$|I_3|\le\frac{1}{\pi}\sin\delta N(r,f)\int_{|t-r|\ge\eta}\frac{r}{(r-t)^2}dt.$$

Therefore for all δ sufficiently close to π, the inequality $|I_2|+|I_3|<\varepsilon$ holds, and hence, $|I_1|+|I_2|+|I_3|<2\varepsilon$. Thus, we have proved equality (4.18). So we have $(r\notin A)$

$$m(r,f)=\int_0^\infty N\left(t,\frac{1}{f}\right)K(t,r,\beta)\,dt+\int_0^\infty N(t,f)K(t,r,\pi-\beta)\,dt-N(r,f),$$

an equality that is equivalent to (4.16). □

LEMMA 4.6. *Let $V(r)>0$ be a continuous non-decreasing function of r.*

(A) *If for some σ, $0<\sigma<\infty$,*

$$(\alpha)\ \liminf_{r\to\infty}r^{-\sigma}V(r)=0,\qquad(\beta)\ \limsup_{r\to\infty}r^{-\sigma}V(r)=\infty,$$

then for each ε, $0<\varepsilon<1$, there are two sequences, $r_k\to\infty$ and $R_k\to\infty$, such that

$$\text{(a) } V(t)\le\left(\frac{t}{r_k}\right)^{\sigma}V(r_k),\quad 1\le t\le R_k;\qquad\text{(b) } r_kR_k^{-1}<\varepsilon. \tag{4.20}$$

(B) *If*

$$\limsup_{r\to\infty}\frac{\ln V(r)}{\ln r}=\rho<\infty,$$

then for each ε, $0<\varepsilon<1$, there exists a sequence $r_k\to\infty$, such that

$$V(t)\le\left(\frac{t}{r_k}\right)^{\rho-\varepsilon}V(r_k),\qquad 1\le t\le r_k, \tag{4.21}$$

$$V(t)\le\left(\frac{t}{r_k}\right)^{\rho+\varepsilon}V(r_k),\qquad r_k\le t<\infty. \tag{4.22}$$

PROOF. We prove statement (B) first. Since $\limsup_{r\to\infty}r^{-\rho+\varepsilon}V(r)=\infty$, the set of those values of r for which

$$r^{-\rho+\varepsilon}V(r)=\max_{1\le t\le r}\{t^{-\rho+\varepsilon}V(t)\},$$

is not bounded from above. We select a sequence $s_k\uparrow\infty$ in this set. Since $\lim_{t\to\infty}t^{-\rho-\varepsilon}V(t)=0$, for each s_k we can find $r_k\ge s_k$ such that

$$r_k^{-\rho-\varepsilon}V(r_k)=\max_{s_k\le t\le<\infty}t^{-\rho-\varepsilon}V(t). \tag{4.23}$$

The sequence $\{r_k\}$ has the desired properties. In fact, (4.22) follows from (4.23) and the inequality $r_k\ge s_k$.

Next, for $1 \le t \le s_k$ we have

$$\frac{V(t)}{t^{\rho-\varepsilon}} \le \frac{V(s_k)}{s_k^{\rho-\varepsilon}} = \frac{V(s_k)}{s_k^{\rho+\varepsilon}} s_k^{2\varepsilon} \le \frac{V(r_k)}{r_k^{\rho+\varepsilon}} s_k^{2\varepsilon} \le \frac{V(r_k)}{r_k^{\rho+\varepsilon}} r_k^{2\varepsilon} = \frac{V(r_k)}{r_k^{\rho-\varepsilon}},$$

and for $s_k \le t \le r_k$ we have

$$\frac{V(t)}{t^{\rho-\varepsilon}} = \frac{V(t)}{t^{\rho+\varepsilon}} t^{2\varepsilon} \le \frac{V(r_k)}{r_k^{\rho+\varepsilon}} t^{2\varepsilon} \le \frac{V(r_k)}{r_k^{\rho+\varepsilon}} r_k^{2\varepsilon} = \frac{V(r_k)}{r_k^{\rho-\varepsilon}}.$$

Therefore inequality (4.21) is also valid.

Now we prove statement (A). By (α), there exists a sequence $R_k \uparrow \infty$ such that $\lim_{k\to\infty} R_k^{-\sigma} V(R_k) = 0$. Since $V(r)$ is non-decreasing, we have

$$\max_{\varepsilon R_k \le r \le R_k} r^{-\sigma} V(r) \le \varepsilon^{-\sigma} R_k^{-\sigma} V(R_k) \to 0 \quad \text{as} \quad k \to \infty. \tag{4.23$'$}$$

Denote by F the closed set consisting of those $r \ge 1$ for which

$$r^{-\sigma} V(r) = \max_{1 \le t \le r} t^{-\sigma} V(t).$$

By (β) the set F is not bounded from above and $\lim_{\substack{r\to\infty \\ r\in F}} r^{-\sigma} V(r) = \infty$. Therefore (4.23$'$) implies that for $k \ge k_0$ none of the intervals $[\varepsilon R_k, R_k]$ intersects the set F. Denote by r_k the largest number in the set $F \cap [1, \varepsilon R_k]$, $k \ge k_0$. The sequences $\{r_k\}_{k_0}^{\infty}$ and $\{R_k\}_{k_0}^{\infty}$ have the desired properties. In fact, since $r_k \in F \cap [1, \varepsilon R_k]$, inequality (b) in (4.20) holds. Part (a) of (4.20) for $1 \le t \le r_k$ follows from $r_k \in F$. Since the half-open interval $(r_k, R_k]$ does not intersect the set F, part (a) of (4.20) holds also for $r_k < t \le R_k$.

The proof of the lemma is complete. □

PROOF. Now we prove Theorem 4.3. We may assume without loss of generality that $f(0) = 1$. Consider separately two cases: (a) $\lambda < \rho$, (b) $\lambda = \rho$ (by the assumptions of the theorem we have $0 \le \lambda < 1$ in both cases).

CASE (a). Using Lemma 3.1 we represent the function $f(z)$ in the form $f(z) = \alpha_{R,0}(z)\omega_{R,0}(z)$. By virtue of estimate (3.5) and Lemma 4.4 we have ($r \le \frac{1}{2}R$)

$$T(r, f) \le T(r, \alpha_{R,0}) + T(r, \omega_{R,0}) \le T(r, \check{\alpha}_{R,0}) + K_1 \frac{r}{R} T(2R, f), \tag{4.24}$$

where K_1 is a constant that does not depend on r and R.

Let σ be an arbitrary number satisfying the condition $\lambda < \sigma < \min(\rho, 1)$. Then the function $V(r) = T(r, f)$ satisfies conditions (α) and (β) of Lemma 4.6(A), and therefore, for each ε, $0 < \varepsilon < 1/4$, we can find sequences $\{r_k\}$ and $\{R_k\}$ satisfying (4.20)(a), (b). In what follows we assume that

$$r = r_k, \quad 2R = R_k. \tag{4.25}$$

Note that then

$$T(t, f) \le \left(\frac{t}{r}\right)^{\sigma} T(r, f), \quad 1 \le t \le 2R \text{ and } \frac{r}{R} < 2\varepsilon, \tag{4.26}$$

and the inequality (4.24) implies

$$T(r, f) \le T(r, \check{\alpha}_{R,0}) + 2K_1 \varepsilon^{1-\sigma} T(r, f). \tag{4.27}$$

To estimate $T(r, \check{\alpha}_{R,0})$ we use Lemma 4.5.

Suppose that there is an infinite sequence of pairs (r, R) satisfying conditions (4.25) and such that

$$T(r, \breve{\alpha}_{R,0}) = N(r, \breve{\alpha}_{R,0}). \tag{4.28}$$

Since for $r \leq R$ we have

$$N(r, \breve{\alpha}_{R,0}) = N(r, f),$$

by (4.27) we get

$$T(r, f) \leq N(r, f) + 2K_1\varepsilon^{1-\sigma}T(r, f),$$

whence

$$\limsup_{r\to\infty} \frac{N(r, f)}{T(r, f)} \geq 1 - 2K_1\varepsilon^{1-\sigma}; \tag{4.28$'$}$$

hence, a fortiori,

$$\kappa(f) \geq 1 - 2K_1\varepsilon^{1-\sigma}.$$

Letting ε tend to 0, we get $\kappa(f) \geq 1$, and inequality (4.6) is proved in this case. Similarly, inequality (4.6) holds also in the case where there is an infinite sequence of pairs (r, R) satisfying conditions (4.25), such that

$$T(r, \breve{\alpha}_{R,0}) = N(r, 1/\breve{\alpha}_{R,0}). \tag{4.29}$$

If there is neither an infinite sequence of pairs (r, R) satisfying (4.25) and (4.28), nor an infinite sequence of pairs satisfying (4.25) and (4.29), then by Lemma 4.5, for all pairs (r, R) satisfying (4.25) for $k \geq k_0$, there is $\beta = \beta(r)$, $0 < \beta < \pi$, such that the equality

$$\begin{aligned} T(r, \breve{\alpha}_{R,0}) &= \int_0^\infty N(t, 1/\breve{\alpha}_{R,0})K(t, r, \beta)\,dt + \int_0^\infty N(t, \breve{\alpha}_{R,0})K(t, r, \pi - \beta)\,dt \\ &= I_1 + I_2 \end{aligned} \tag{4.30}$$

holds. Since for $t \leq R$ we have $N(t, \breve{\alpha}_{R,0}) = N(t, f)$, and for $t > R$,

$$\begin{aligned} N(t, \breve{\alpha}_{R,0}) &= N(R, f) + n(R, f)\ln\frac{t}{R} \\ &\leq T(R, f) + 2T(2R, f)\ln\frac{t}{R} \leq 2T(2R, f)\ln\frac{et}{R}, \end{aligned}$$

we conclude that

$$I_2 \leq \int_0^R N(t, f)K(t, r, \pi - \beta)\,dt + 2T(2R, f)\int_R^\infty \ln\left(\frac{et}{R}\right)K(t, r, \pi - \beta)\,dt.$$

Observing that for $t \geq R (\geq 2r)$, the inequality

$$K(t, r, \beta) \leq \frac{1}{\pi}\frac{r\sin\beta}{(r-t)^2} \leq 2\frac{r}{t^2}$$

holds, and

$$\int_R^\infty t^{-2}\ln\left(\frac{et}{R}\right)dt = \frac{2}{R},$$

we get

$$I_2 \leq \int_0^R N(t, f)K(t, r, \pi - \beta)\,dt + 8\frac{r}{R}T(2R, f). \tag{4.31}$$

Similarly we get

$$I_1 \leq \int_0^R N\left(t, \frac{1}{f}\right)K(t, r, \beta)\,dt + 8\frac{r}{R}T(2R, f). \tag{4.32}$$

The combination of (4.27), (4.30)–(4.32), and (4.26) implies the inequality

$$T(r,f) \le \int_0^R N\left(t,\frac{1}{f}\right) K(t,r,\beta)\,dt + \int_0^R N(t,f)K(t,r,\pi-\beta)\,dt + (2K_1+32)\varepsilon^{1-\sigma}T(r,f). \tag{4.33}$$

Let $\gamma = \min(\beta, \pi-\beta)$. Then, obviously, $0<\gamma\le\frac{\pi}{2}$ and

$$K(t,r,\gamma) = \max(K(t,r,\beta), K(t,r,\pi-\beta)). \tag{4.34}$$

Therefore (4.33) implies

$$T(r,f) \le \int_0^R N(t;0,\infty)K(t,r,\gamma)\,dt + (2K_1+32)\varepsilon^{1-\sigma}T(r,f). \tag{4.35}$$

Let $\kappa' > \kappa(f)$. Then there is $R_0 \ge 1$ such that for $t \ge R_0$ the inequality $N(t;0,\infty) \le \kappa' T(t,f)$ holds. Taking into account (4.26), we derive from (4.35) the inequality

$$T(r,f) \le \kappa' T(r,f) \int_{R_0}^R \left(\frac{t}{r}\right)^\sigma K(t,r,\gamma)\,dt + \frac{N(R_0;0,\infty)rR_0}{\pi(r-R_0)^2} + (2K_1+32)\varepsilon^{1-\sigma}T(r,f). \tag{4.36}$$

Next we observe that

$$\int_{R_0}^R \left(\frac{t}{r}\right)^\sigma K(t,r,\gamma)\,dt \le \int_0^\infty \left(\frac{t}{r}\right)^\sigma K(t,r,\gamma)\,dt = \int_0^\infty u^\sigma K(u,1,\gamma)\,du, \tag{4.37}$$

and that using the Residue Calculus, it is easy to get

$$\int_0^\infty u^\sigma K(u,1,\varphi)\,du = \frac{\sin(\pi-\varphi)\sigma}{\sin\pi\sigma}, \qquad 0<\varphi<\pi,\ -1<\sigma<1. \tag{4.37$'$}$$

Therefore (4.36) implies the inequality

$$1 \le \kappa' \frac{\sin(\pi-\gamma)\sigma}{\sin\pi\sigma} + (2K_1+32)\varepsilon^{1-\sigma} + \frac{N(R_0;0,\infty)rR_0}{\pi(r-R_0)^2 T(r,f)}.$$

Letting r tend to ∞ in this inequality, and then letting ε tend to 0, we get

$$1 \le \kappa' \max_{0\le\gamma\le\frac{\pi}{2}} \frac{\sin(\pi-\gamma)\sigma}{\sin\pi\sigma}. \tag{4.38}$$

But

$$\max_{0\le\gamma\le\frac{\pi}{2}} \frac{\sin(\pi-\gamma)\sigma}{\sin\pi\sigma} = \left\{\begin{array}{ll} 1, & 0<\sigma\le\frac12 \\ \dfrac{1}{\sin\pi\sigma}, & \frac12<\sigma<1 \end{array}\right\} = \frac{1}{\nu(\sigma)}; \tag{4.39}$$

therefore (4.38) can be rewritten in the form

$$\kappa' \ge \nu(\sigma). \tag{4.40}$$

Here σ is an arbitrary number satisfying the inequality $\lambda<\sigma<\min(\rho,1)$, and κ' is an arbitrary number satisfying the inequality $\kappa'>\kappa(f)$. Therefore (4.40) implies (4.6).

CASE (b). We prove that for each function $f(z)$, $f(0)=1$, of order $\rho<1$, the inequality $\kappa(f)\ge\nu(\rho)$ holds (we do not use the condition $\lambda=\rho$). First we assume in addition that all zeros of the function $f(z)$ are positive, and all of its poles are negative. Then, since $f(z)$ has genus zero, we can apply Lemma 4.5.

If there is an infinite sequence of values of r for which at least one of the equalities

$$T(r,f)=N(r,f), \qquad T(r,f)=N\left(r,\frac{1}{f}\right)$$

is valid, then $\kappa(f)\ge 1$, and hence (4.6) holds. Therefore, by Lemma 4.5, we may assume that for all sufficiently large r, there is $\beta=\beta(r)$, $0<\beta(r)<\pi$, such that inequality (4.16) holds. Taking $\gamma=\min(\beta,\pi-\beta)$, as in Case (a), and using (4.34), we get

$$T(t,f)\le\int_0^\infty N(t;0,\infty)K(t,r,\gamma)\,dt. \tag{4.41}$$

Let ε, $0<\varepsilon<\min(\frac14,1-\rho)$, be an arbitrary prescribed number. Let $V(r)=T(r,f)$ in Lemma 4.6(B) and denote by $\{r_k\}$ the sequence whose existence is established in the lemma. In what follows we assume that r takes only those values that belong to the sequence $\{r_k\}$. Note that by Lemma 4.6(B) we have

$$T(t,f)\le\left(\frac{t}{r}\right)^{\rho-\varepsilon}T(r,f), \qquad 1\le t\le r, \tag{4.42}$$

$$T(t,f)\le\left(\frac{t}{r}\right)^{\rho+\varepsilon}T(r,f), \qquad r\le t<\infty. \tag{4.42'}$$

Let $\kappa'>\kappa(f)$ and let $R_0\ge 1$ be such that for $t\ge R_0$ the inequality $N(t;0,\infty)\le\kappa' T(t,f)$ holds. Taking into account (4.42), (4.42′), we get from (4.41) the inequality

$$\begin{aligned} T(r,f)\le\kappa' T(r,f)\Big\{&\int_{R_0}^r\left(\frac{t}{r}\right)^{\rho-\varepsilon}K(t,r,\gamma)\,dt\\ &+\int_r^\infty\left(\frac{t}{r}\right)^{\rho+\varepsilon}K(t,r,\gamma)\,dt\Big\}\\ &+\frac{N(R_0;0,\infty)rR_0}{\pi(r-R_0)^2}. \end{aligned} \tag{4.43}$$

We estimate the expression in braces. We have

$$\begin{aligned} &\int_{R_0}^r\left(\frac{t}{r}\right)^{\rho-\varepsilon}K(t,r,\gamma)\,dt+\int_r^\infty\left(\frac{t}{r}\right)^{\rho+\varepsilon}K(t,r,\gamma)\,dt\\ &\quad=\int_{R_0/r}^1 u^{\rho-\varepsilon}K(u,1,\gamma)\,du+\int_1^\infty u^{\rho+\varepsilon}K(u,1,\gamma)\,du\\ &\quad\le\int_0^\infty u^{\rho+\varepsilon}K(u,1,\gamma)\,du+\int_0^1(u^{\rho-\varepsilon}-u^{\rho+\varepsilon})K(u,1,\gamma)\,du. \end{aligned}$$

By virtue of (4.37′),

$$\int_0^\infty u^{\rho+\varepsilon}K(u,1,\gamma)\,du=\frac{\sin(\pi-\gamma)(\rho+\varepsilon)}{\sin\pi(\rho+\varepsilon)}.$$

Next, since $1-u^{2\varepsilon}<8\varepsilon u^{-\frac14}$, $u^{\rho-\varepsilon}<u^{-\frac14}$ for $0<u<1$, we get

$$\begin{aligned} \int_0^1(u^{\rho-\varepsilon}-u^{\rho+\varepsilon})K(u,1,\gamma)\,du&\le 8\varepsilon\int_0^1 u^{-\frac12}K(u,1,\gamma)\,du\\ &<8\varepsilon\int_0^\infty u^{-\frac12}K(u,1,\gamma)\,du=8\varepsilon\sin\frac{\pi-\gamma}{2}\le 8\varepsilon. \end{aligned}$$

Therefore (4.43) implies

$$1 \leq \kappa' \left\{ \frac{\sin(\pi - \gamma)(\rho + \varepsilon)}{\sin \pi(\rho + \varepsilon)} + 8\varepsilon \right\} + \frac{N(R_0; 0, \infty) r R_0}{\pi (r - R_0)^2 T(r, f)}.$$

Letting r tend to ∞, we get

$$1 \leq \kappa' \left\{ \max_{0 \leq \gamma \leq \frac{\pi}{2}} \frac{\sin(\pi - \gamma)(\rho + \varepsilon)}{\sin \pi(\rho + \varepsilon)} + 8\varepsilon \right\},$$

whence, by (4.39),

$$1 \leq \kappa' \left\{ \frac{1}{\nu(\rho + \varepsilon)} + 8\varepsilon \right\}.$$

Letting ε tend to 0, and then letting κ' tend to $\kappa(f)$, we get $\kappa(f) \geq \nu(\rho)$, what was to be proved.

Now we get rid of the assumption that all zeros of the function $f(z)$ are positive and all of its poles are negative. Let $f(z)$, $f(0) = 1$, be a meromorphic function of order $\rho < 1$. We form the function $\check{f}(z)$. Obviously,

$$N(r, f) = N(r, \check{f}), \qquad N\left(r, \frac{1}{f}\right) = N\left(r, \frac{1}{\check{f}}\right). \tag{4.44}$$

Since the functions $f(z)$ and $\check{f}(z)$ are quotients of canonical products, we get, using Theorem 3.4 of Chapter 2, that

$$\rho[f] = \max\left\{\rho[N(r, f)], \rho\left[N\left(r, \frac{1}{f}\right)\right]\right\},$$

$$\rho[\check{f}] = \max\left\{\rho[N(r, \check{f}], \rho\left[N\left(r, \frac{1}{\check{f}}\right)\right]\right\},$$

and hence $\rho[f] = \rho[\check{f}] = \rho$. All zeros of the function $\check{f}(z)$ are positive and all of its poles are negative. Therefore, since the result has been already proved in this case, $\kappa(\check{f}) \geq \nu(\rho)$. But $\kappa(f) \geq \kappa(\check{f})$; this follows from (4.44) and the inequality $T(r, f) \leq T(r, \check{f})$, which is a consequence of Lemma 4.4. The proof of Theorem 4.3 is complete. $\square$

The method used in the proof of Theorem 4.3 can be used to prove the following result concerning deficiencies.

THEOREM 4.4. *Let $f(z)$ be a meromorphic function of lower order $\lambda < 1$, and let $a \neq b$ be two numbers in the extended plane. Let*

$$u = 1 - \delta(a, f), \qquad v = 1 - \delta(b, f).$$

Then

$$u^2 + v^2 - 2uv \cos \pi\lambda \geq \sin^2 \pi\lambda, \tag{4.45}$$

where $v = 1$ if $u < \cos \pi\lambda$, and $u = 1$ if $v < \cos \pi\lambda$.

Observe that the equation $u^2 + v^2 - 2uv \cos \pi\lambda = \sin^2 \pi\lambda$ represents an ellipse in the (u, v)-plane, with $\{u = v\}$ and $\{u = -v\}$ being its axes of symmetry, inscribed into the square $\{-1 \leq u \leq 1,\ -1 \leq v \leq 1\}$. This ellipse is tangent to the lines $\{u = 1\}$ and $\{v = 1\}$ at the points $(1, \cos \pi\lambda)$ and $(\cos \pi\lambda, 1)$, respectively. For $1/2 < \lambda < 1$, both points of tangency are outside the square $Q = \{0 \leq u \leq 1,$ $0 \leq v \leq 1\}$, and the statement of Theorem 4.4 means that the point (u, v) lies in the square Q, but not in the interior of the ellipse. For $0 < \lambda \leq 1/2$, the statement

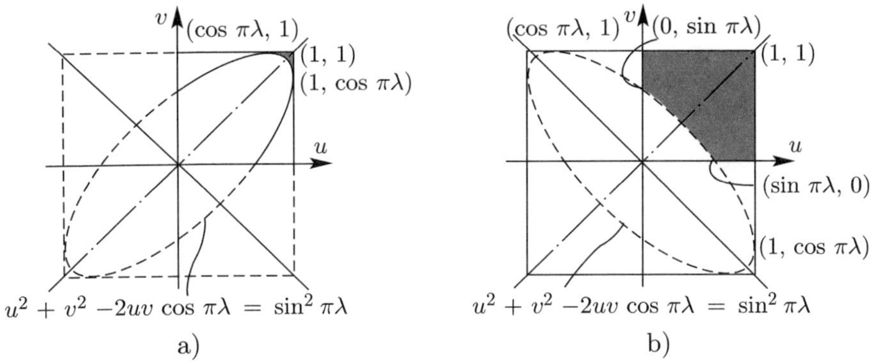

FIGURE 4

of Theorem 4.4 means that the point (u, v) lies either on one of the lines $\{u = 1\}$ and $\{v = 1\}$ or in the piece of the square Q near the vertex $(1, 1)$, which is cut off the square by the arc of the ellipse joining the points $(1, \cos\pi\lambda)$ and $(\cos\pi\lambda, 1)$. (Figure 4, (a) shows the case $0 \le \lambda < 1/2$; Figure 4, (b) shows the case $1/2 \le \lambda < 1$. In both cases the set of possible values of (u, v) is shaded.)

For $\lambda = 0$ the ellipse degenerates into a line segment joining the points $(-1, -1)$ and $(1, 1)$, and the statement of Theorem 4.4 means that the point (u, v) lies on one of the lines $\{u = 1\}$ and $\{v = 1\}$. This last result has been obtained earlier (see the Corollary of Theorem 3.1).

PROOF. Proving Theorem 4.4 we may assume without loss of generality that $a = 0$, $b = \infty$, $f(0) = 1$. As in the proof of Theorem 4.3 we consider two cases: (a) $\lambda < \rho$, and (b) $\lambda = \rho$.

CASE (a). Arguing in the same way as in the proof of Theorem 4.3, we arrive at inequality (4.27). If there is a sequence of pairs (r, R) satisfying conditions (4.25) and (4.28), then (4.28′) implies $\delta(\infty, f) \le 2K_1\varepsilon^{1-\sigma}$, and since ε can be chosen arbitrarily small, $\delta(\infty, f) = 0$, $v = 1 - \delta(\infty, f) = 1$. Similarly, if there is a sequence of pairs (r, R) satisfying conditions (4.25) and (4.29), then $\delta(0, f) = 0$, $u = 1$. It is clear that the points of the form $(u, 1)$ and $(1, v)$ satisfy the conditions of Theorem 4.4.

If there is neither a sequence of pairs (r, R) satisfying conditions (4.25) and (4.28) nor a sequence of pairs satisfying conditions (4.25) and (4.29), then, arguing in the same way as in the proof of Theorem 4.3, we arrive at inequality (4.33).

Let X and Y be an arbitrary pair of numbers satisfying the conditions $X > u$, $Y > v$. Then there is $R_0 \ge 1$ such that for $t \ge R_0$ the inequalities $N(t, \frac{1}{f}) \le XT(t, f)$, $N(t, f) \le YT(t, f)$ hold. Taking into account (4.26) and (4.33), we get

$$\begin{aligned} T(r, f) \le XT(r, f) &\int_{R_0}^{R} \left(\frac{t}{r}\right)^{\sigma} K(t, r, \beta)\, dt \\ &+ YT(r, f) \int_{R_0}^{R} \left(\frac{t}{r}\right)^{\sigma} K(t, r, \pi - \beta)\, dt + \frac{N(R_0; 0, \infty) r R_0}{\pi(r - R_0)^2} \\ &+ (2K_1 + 32)\varepsilon^{1-\sigma} T(r, f), \end{aligned}$$

where r and R satisfy condition (4.25). Hence, using (4.37) and (4.37′), we have

$$(4.46)\qquad 1 \le X\frac{\sin(\pi-\beta)\sigma}{\sin\pi\sigma} + Y\frac{\sin\beta\sigma}{\sin\pi\sigma} + (2K_1+32)\varepsilon^{1-\sigma} + \frac{N(R_0;0,\infty)rR_0}{\pi(r-R_0)^2T(r,f)}.$$

Since

$$\begin{aligned} &X\sin(\pi-\beta)\sigma + Y\sin\beta\sigma \\ &\qquad = (Y - X\cos\pi\sigma)\sin\beta\sigma + (X\sin\pi\sigma)\cos\beta\sigma \\ &\qquad \le \{(Y-X\cos\pi\sigma)^2 + (X\sin\pi\sigma)^2\}^{1/2} = \{X^2+Y^2-2XY\cos\pi\sigma\}^{1/2}, \end{aligned}$$

inequality (4.46) implies

$$1 \le \frac{1}{\sin\pi\sigma}\{X^2+Y^2-2XY\cos\pi\sigma\}^{1/2} + (2K_1+32)\varepsilon^{1-\sigma} + \frac{N(R_0;0,\infty)rR_0}{\pi(r-R_0)^2T(r,f)}.$$

Letting r tend to ∞ in this inequality, and then ε tend to 0 and σ to λ, we get

$$\sin^2\pi\lambda \le X^2+Y^2-2XY\cos\pi\lambda.$$

Since X and Y are arbitrary numbers satisfying the inequalities $X > u$ and $Y > v$, inequality (4.45) is proved.

The fact that $v = 1$ for $u < \cos\pi\lambda$ and that $u = 1$ for $v < \cos\pi\lambda$ follows immediately from the Corollary of Theorem 3.1

Case (b) can be treated in the same way as case (b) in Theorem 4.3; we leave the details to the reader. □

It turns out that Theorem 4.4 cannot be improved; that is, for each pair u, v of numbers, $0 \le u, v \le 1$, satisfying the conditions described in the statement of the theorem, there exists a meromorphic function $f(z)$ of lower order λ such that $\delta(a,f) = 1-u$, $\delta(b,f) = 1-v$. We do not prove this result in the general case, referring to the paper of Edrei and Fuchs [**EF60**]. We only give the following example.

EXAMPLE. Let numbers ρ and λ, $0 < \lambda \le \rho < 1$, and u and v, $0 \le u, v \le 1$, $u^2+v^2-2uv\cos\pi\lambda = \sin^2\pi\lambda$, be given. If $\lambda < 1/2$, we assume in addition that $u, v \ge \cos\pi\lambda$. We find a meromorphic function $f(z)$ of order ρ and lower order λ, such that

$$(4.47)\qquad \delta(0,f) = 1-u, \qquad \delta(\infty,f) = 1-v.$$

In the case $u \ge v$, the function $f_{uv}(z)$ considered in Example 1 of Section 3 has these properties. In fact, by (3.40) we have

$$1-\delta(\infty,f_{uv}) = \frac{v}{I(u,v,\lambda)+v}.$$

But for u and v satisfying the conditions described above, the equality

$$(4.48)\qquad I(u,v,\lambda) = \frac{1}{\sin\pi\lambda}\{u^2+v^2-2uv\cos\pi\lambda\}^{1/2} - v = 1-v$$

holds; therefore $\delta(\infty,f_{uv}) = 1-v$. Next, since

$$N\left(r,\frac{1}{f_{uv}}\right) = N(r,0,f(z,u)) = \frac{u}{l(r)}r^{l(r)} + o(r^{l(r)}),$$

using (3.38) and (4.48) we get

$$1-\delta(0,f_{uv})=\limsup_{r\to\infty}\frac{N\left(r,\frac{1}{f_{uv}}\right)}{T(r,f_{uv})}=\limsup_{r\to\infty}\frac{u}{I(u,v,l(r))+v}=\frac{u}{I(u,v,\lambda)+v}=u,$$

and hence $\delta(0,f_{uv})=1-u$.

If $v\le u$, then $1/f_{uv}(z)$ is a desired function.

If instead of the restriction on u and v just considered, we required that $u=1$, $0\le v<\cos\pi\lambda$, then the function $f_{uv}(z)$ would also satisfy (4.47). This follows from relation (3.40′) and the equality $\delta(0,f_{uv})=0$, which is true by the Corollary of Theorem 3.1. If $0\le u<\cos\pi\lambda$, $v=1$, then $1/f_{uv}(z)$ is a desired function.

Theorem 4.4 also implies the following result containing both the Corollary of Theorem 3.1 and the Corollary of Lemma 3.4.

THEOREM 4.5. *Let $f(z)$ be a meromorphic function of lower order $\lambda\le 1/2$. If $\delta(a,f)\ge 1-\cos\pi\lambda$, $0<\lambda\le 1/2$, or $\delta(a,f)>0$, $\lambda=0$, then a is the only deficient value of the function $f(z)$.*

By the Corollary of Theorem 3.1 it is enough to consider the case $\delta(a,f)=1-\cos\pi\lambda$, $0<\lambda\le 1/2$. Let $b\ne a$ and $u=1-\delta(a,f)$, $v=1-\delta(b,f)$. Then $u=\cos\pi\lambda$, $0\le v\le 1$, and inequality (4.45) implies $v\ge 1$. Therefore $v=1$, $\delta(b,f)=0$, what was to be proved.

5. The growth of a meromorphic function with exceptional values

The presence of one, even Picard's, exceptional value does not impose any restrictions on the order and lower order of a meromorphic function. Let us show that for arbitrary λ and ρ satisfying $0\le\lambda\le\rho\le\infty$, there is an entire function of order ρ and lower order λ.

EXAMPLE 1 (The case $0<\lambda\le\rho<1$). Take a function $l(r)$ that is differentiable on $[1,\infty)$ and satisfies

$$\liminf_{r\to\infty} l(r)=\lambda,\qquad \limsup_{r\to\infty} l(r)=\rho.$$

Let $f(z)$ be a canonical product of genus 0 with positive zeros such that $n(r,0)\sim r^{l(r)}$. By formula (5.4) in Chapter 2, the function $f(z)$ has order ρ and lower order λ.

EXAMPLE 2 (The case $0<\lambda\le\rho<\infty$). Let $n\ge 2\rho$ be a positive integer, $\lambda_1=\lambda/n$, $\rho_1=\rho/n$. Then $0<\lambda_1\le\rho_1\le 1/2$. Let $f(z)$ be an entire function of order ρ_1 and lower order λ_1. The function $f(z^n)$ has order ρ and lower order λ.

EXAMPLE 3 (The case $\lambda=0$, $0<\rho<\infty$). In Section 6 of Chapter 4 we proved (p. 152) that the entire function

$$g(z)=\prod_{k=1}^{\infty}\left\{1-\left(\frac{z}{r_k}\right)^{n_k}\right\},\tag{5.1}$$

where n_k are arbitrary positive integers, and r_k are arbitrary positive integers subject to the condition $r_{k+1}\ge 2r_k$, $r_1>1$, satisfies the relation

$$T(r,g)=N(r,0)+O(1).$$

Let $r_1 = e$, $r_{k+1} = e^{r_k}$, $n_k = [r_k^\rho]$. Then

$$n(r,0) = \sum_{r_k \le r} [r_k^\rho] = O(r^\rho), \qquad n(r_n, 0) \sim r_n^\rho.$$

Since $\rho[N(r,0)] = \rho[n(r,0)]$, the function $g(z)$ has order ρ. Next,

$$\begin{aligned} N(r_n, 0) &= \int_e^{r_n} \frac{n(t,0)}{t}\,dt \le n(r_{n-1}, 0) \ln \frac{r_n}{e} \\ &= O(r_{n-1}^\rho) \ln \frac{r_n}{e} = O\left((\ln r_n)^{\rho+1}\right). \end{aligned}$$

Hence the lower order of the function $g(z)$ is equal to 0.

EXAMPLE 4 (The case $\lambda = 0$, $\rho = \infty$). Consider a somewhat more general case. Let $\varphi(r)$ and $\psi(r)$ be two arbitrary non-decreasing functions such that $\psi(r) \ge \varphi(r) \ge 1$ and

$$\lim_{r\to\infty} \frac{\varphi(r)}{\ln r} = \infty.$$

We construct an entire function $g(z)$ for which

$$\liminf_{r\to\infty} \frac{T(r,g)}{\varphi(r)} \le 1, \qquad \limsup_{r\to\infty} \frac{T(r,g)}{\psi(r)} \ge 1. \tag{5.2}$$

Select a sequence $\{r_k\}$ of positive numbers such that $r_1 > 1$, $r_{k+1} \ge 2r_k$,

$$\sum_{k=1}^{n-1} [\psi(r_k e)] \ln \frac{r_n}{r_k} \le \varphi(r_n). \tag{5.3}$$

This is possible since $\ln r = o(\varphi(r))$. Letting $n_k = [\psi(r_k e)]$, consider the function (5.1). Relation (5.3) is equivalent to

$$N(r_n, 0) \le \varphi(r_n).$$

Besides, it is clear that

$$N(er_n, 0) \ge n(r_n, 0) \ge [\psi(r_n e)].$$

Therefore the function (5.1) satisfies conditions (5.2).

EXAMPLE 5 (The case $0 < \lambda < \infty$, $\rho = \infty$). Let $g(z)$ be an entire function of order ∞ and lower order 0, and let $h_\lambda(z)$ be a canonical product with positive zeros such that $n(r, 0, h_\lambda) \sim r^\lambda$. Then $\ln T(r, h_\lambda) \sim \lambda \ln r$; therefore the lower order of the function $f(z) = g(z) h_\lambda(z)$ is equal to λ, but the order of $f(z)$ is infinite.

In the cases $\lambda = \rho = 0$, $\lambda = \rho = \infty$ the corresponding examples are trivial.

If a function $f(z)$ has two Picard exceptional values, then necessarily $\rho[f] = \lambda[f]$, and it is either a positive integer or ∞. This result follows from the next theorem.

THEOREM 5.1. *If a function $f(z)$ is such that for some a and b, $a \ne b$, the inequality $\rho[N(r; a, b)] < \rho[f]$ holds, then $\rho[f] = \lambda[f]$, and it is either a positive integer or ∞.*

PROOF. We may assume without loss of generality that $a = 0$, $b = \infty$, $f(0) \ne 0, \infty$. Let $\pi_1(z)$ be a canonical product whose zeros and their multiplicities coincide with those of the function $f(z)$, and let $\pi_2(z)$ be a canonical product whose zeros and their multiplicities coincide with poles and their multiplicities of $f(z)$. Then we have a representation

$$f(z) = g(z)\frac{\pi_1(z)}{\pi_2(z)},$$

where $g(z)$ is an entire function without zeros. Since

$$T(r, f) \le T(r, g) + T\left(r, \frac{\pi_1}{\pi_2}\right),$$

$$T(r, g) \le T(r, f) + T\left(r, \frac{\pi_1}{\pi_2}\right),$$

and

$$\rho\left[\frac{\pi_1}{\pi_2}\right] = \rho[N(r; 0, \infty)],$$

we have the inequalities

$$\rho[f] \le \max\{\rho[g],\ \rho[N(r; 0, \infty]\} = \rho[g],$$
$$\lambda[g] \le \max\{\lambda[f],\ \rho[N(r; 0, \infty)]\}. \tag{5.4}$$

But $\rho[g] = \lambda[g]$, and is either an integer or ∞. In fact, if $\lambda[g] < \infty$, then by Lemma 6.2 of Chapter 1 we have $g(z) = e^{P(z)}$, where $P(z)$ is a polynomial; hence $\lambda[g] = \rho[g]$ and is equal to the degree of the polynomial $P(z)$.

Therefore (5.4) implies that $\rho[f] \le \max\{\lambda[f], \rho[N(r; 0, \infty)]\} = \lambda[f]$, whence $\rho[f] = \lambda[f]$, and is either an integer or ∞. It remains to observe that $\rho[f] > 0$, because $\rho[f] > \rho[N(r; 0, \infty)]$. □

However, the presence of two Borel exceptional values does not imply, in general, the equality $\rho[f] = \lambda[f]$, although in this case, by Theorem 1.1 of Chapter 4, the number $\rho[f]$ should be either a positive integer[10] or ∞.

EXAMPLE 6. We show that for each positive integer ρ and each λ, $0 \le \lambda \le \rho$, there exists an entire function $g_{\lambda\rho}(z)$ of order ρ and lower order λ for which zero is a Borel exceptional value.

For $g_{01}(z)$ we take the function $f(z)$ from Example 4 of Section 2 (its lower order is equal to zero by Remark 2 to that example). In the case $0 < \lambda < 1$ we let

$$g_{\lambda 1}(z) = g_{01}(z)\omega_\lambda(z),$$

where

$$\omega_\lambda(z) = \prod_{n=1}^{\infty}\left(1 - \frac{z}{n^{1/\lambda}}\right).$$

Since $g_{01}(z) = f(z)$ is a function of maximal type of order 1, and the order of $\omega_\lambda(z)$ is equal to $\lambda < 1$, it follows that $g_{\lambda 1}(z)$ is also of maximal type of order 1. Taking into account that $n(r, 0, f) = O(r)$, we get $n(r, 0, g_{\lambda 1}) = O(r)$, and hence $0 \in E_B(g_{\lambda 1})$. Since $\ln T(r, \omega_\lambda) \sim \lambda \ln r$, and the lower order of the function $g_{01}(z) = f(z)$ is equal to zero, the lower order of the function $g_{\lambda 1}(z)$ is equal to λ.

Let $g_{11}(z) = e^z$. Thus, the function $g_{\lambda\rho}(z)$ is constructed for $\rho = 1$, $0 \le \lambda \le 1$. Now let ρ be an arbitrary positive integer, $0 \le \lambda \le \rho$. Then, taking

$$g_{\lambda\rho}(z) = g_{\mu 1}(z^\rho), \quad \text{where} \quad \mu = \lambda/\rho,$$

we obviously get a desired function.

---

[10]The possibility $\rho[f] = 0$ is excluded, since in this case $n(r; a, b) = O(1)$ implies that $f(z)$ is a rational function.

If the function $f(z)$ has two deficient values, then by the Corollary of Theorem 3.1, necessarily $\lambda[f] > 0$. Let us show that for arbitrary λ and ρ, $0 < \lambda \leq \rho < \infty$, there exists a meromorphic function $f(z)$ such that $\lambda[f] = \lambda$, $\rho[f] = \rho$, and $\delta(0, f) = \delta(\infty, f) > 0$.

EXAMPLE 7. Suppose first that $0 < \lambda \leq \rho \leq 1/2$. Letting $u = v = \cos\frac{\pi\lambda}{2}$ in Example 1 of Section 4, we get a function $f_{\lambda\rho}(z)$ of order ρ and lower order λ with $\delta(0, f_{\lambda\rho}) = \delta(\infty, f_{\lambda\rho}) = 1 - \cos\frac{\pi\lambda}{2}$. If λ and ρ are arbitrary numbers satisfying $0 < \lambda \leq \rho < \infty$, a desired function can be obtained by selecting a positive integer $n \geq 2\rho$ and taking

$$f_{\lambda\rho}(z) = f_{\lambda_1\rho_1}(z^n), \quad \text{where} \quad \lambda_1 = \lambda/n, \ \rho_1 = \rho/n.$$

It turns out that the existence of two deficient values a and b, $a \neq b$ with deficiencies $\delta(a, f)$ and $\delta(b, f)$ close to one, implies proximity of $\rho[f]$ and $\lambda[f]$ to some positive integer.

We may assume without loss of generality that $a = 0$, $b = \infty$. By (4.2) the inequality

$$\{1 - \delta(0, f)\} + \{1 - \delta(\infty, f)\} \geq \kappa(f)$$

holds. We show that smallness of $\kappa(f)$ implies the proximity of $\rho[f]$ and $\lambda[f]$ to some positive integer.

THEOREM 5.2. *Let $f(z)$ be a meromorphic function of order $\rho \leq \infty$ and lower order $\lambda < \infty$, and p an integer determined by the condition*

$$p - \frac{1}{2} \leq \lambda < p + \frac{1}{2}.$$

If

$$\kappa(f) \leq \frac{\varepsilon}{1.1(p+1)}, \quad \textit{where} \quad 0 \leq \varepsilon < \frac{1}{2},$$

then $p \geq 1$ and

$$p - \varepsilon \leq \lambda \leq \rho \leq p + \varepsilon. \tag{5.5}$$

PROOF. By Theorem 4.1,

$$\kappa(f) \geq \max_{\lambda \leq x \leq \rho} \nu(x);$$

therefore

$$\max_{\lambda \leq x \leq \rho} \nu(x) \leq \frac{\varepsilon}{1.1(p+1)}. \tag{5.6}$$

Since $p \geq 0$, the right-hand side of this inequality is less than one. But $\nu(x) = 1$ for $0 \leq x \leq 1/2$; therefore (5.6) implies that $\lambda > 1/2$ and $p \geq 1$.

Next, since

$$\nu(x) \geq \frac{|\sin \pi x|}{2.2x + 0.5|\sin \pi x|},$$

we obtain from (5.6) that

$$\frac{|\sin \pi x|}{2.2x + 0.5|\sin \pi x|} \leq \frac{\varepsilon}{1.1(p+1)}$$

for all x, $\lambda \leq x \leq \rho$. But this does not hold for $x = p + 1/2$, and therefore $\rho < p + 1/2$. Since for $|x - p| < 1/2$ we have $|\sin \pi x| \geq 2|x - p|$, we come to the conclusion that for all x, $\lambda \leq x \leq \rho$, the inequality

$$\frac{2|x-p|}{2.2x + |x-p|} \leq \frac{\varepsilon}{1.1(p+1)}$$

and a fortiori the inequality

$$\frac{2|x-p|}{2.2\left(p+\frac{1}{2}\right)+\frac{1}{2}} \leq \frac{\varepsilon}{1.1(p+1)}$$

must hold. Therefore for all x, $\lambda \leq x \leq \rho$, we have $|x - p| \leq \varepsilon$. From this we get (5.5). □

Theorem 5.2 and the Corollary of Theorem 6.5 in Chapter 4 immediately imply the following.

COROLLARY. *Let $f(z)$ be a meromorphic function of finite lower order λ. If*

$$\kappa(f) < \frac{1}{2.2(\lambda + \frac{3}{2})},$$

then the deficiencies of the function $f(z)$ are shift-invariant; that is, the deficiencies of $f(z+h)$ are independent of h.

We mention also the following theorem, which can be proved with the help of Theorem 5.2.

THEOREM 5.3. *Let $f(z)$ be a meromorphic function of order $\rho < \infty$ and lower order λ, and p an integer determined by the condition*

$$p - \frac{1}{2} \leq \lambda < p + \frac{1}{2}.$$

If

$$\sum_{a \neq \infty} \delta(a, f) \geq 1 - \gamma, \qquad \delta(\infty, f) \geq 1 - \gamma, \tag{5.7}$$

where

$$\gamma = \frac{\varepsilon}{5(p+1)}, \qquad 0 \leq \varepsilon < \frac{1}{2}, \tag{5.8}$$

then $p \geq 1$ and

$$p - \varepsilon \leq \lambda \leq \rho \leq p + \varepsilon.$$

COROLLARY 1. *Under the assumptions of Theorem 5.3 all deficiencies are shift-invariant.*

COROLLARY 2. *If $f(z)$ is an entire function of finite order and*

$$\sum_{a \neq \infty} \delta(a, f) = 1,$$

then $\rho[f] = \lambda[f]$ is a positive integer and all deficiencies are shift-invariant.

To prove Theorem 5.3 we need the following lemma.

LEMMA 5.1. *Let $f(z)$ be a meromorphic function of finite order and let conditions (5.7), where $0 \le \gamma < 1$, be satisfied. Then*

$$\liminf_{r\to\infty} \frac{T(r,f')}{T(r,f)} \ge 1-\gamma, \qquad \limsup_{r\to\infty} \frac{T(r,f')}{T(r,f)} \le 1+\gamma, \tag{5.9}$$

and

$$\kappa(f') \le \frac{4\gamma}{1-\gamma^2}. \tag{5.10}$$

PROOF. We have

$$\begin{aligned} T(r,f') = m(r,f') + N(r,f') &\le m(r,f) + m\left(r,\frac{f'}{f}\right) + 2N(r,f) \\ &= T(r,f) + N(r,f) + m\left(r,\frac{f'}{f}\right) \\ &\le T(r,f) + (1-\delta(\infty,f)+o(1))T(r,f) + O(\ln r) \end{aligned}$$

(we used Theorem 1.3 of Chapter 3). The function $f(z)$ is transcendental since a rational function cannot have more that one deficient value. Therefore $\ln r = o(T(r,f))$, and we get

$$\limsup_{r\to\infty} \frac{T(r,f')}{T(r,f)} \le 2-\delta(\infty,f) \le 1+\gamma;$$

that is, the second inequality in (5.9) is proved.

To prove the first inequality in (5.9), we use inequality (2.5) in Chapter 3. Adding $N(r,\frac{1}{f'})$ to both sides of the latter, we get

$$T(r,f') \ge N\left(r,\frac{1}{f'}\right) + \sum_{\nu=1}^{q} m(r,a_\nu,f) + O(\ln r), \tag{5.11}$$

where $\{a_1,\dots,a_q\}$ is an arbitrary finite collection of (finite) complex numbers. We take for $a_1,\dots,a_q$ finite deficient values of the function $f(z)$. From (5.11) we get

$$\liminf_{r\to\infty} \frac{T(r,f')}{T(r,f)} \ge \liminf_{r\to\infty} \frac{N\left(r,\frac{1}{f'}\right)}{T(r,f)} + \sum_{\nu=1}^{q} \liminf_{r\to\infty} \frac{m(r,a_\nu,f)}{T(r,f)} \ge \sum_{\nu=1}^{q} \delta(a_\nu,f).$$

Since q is an arbitrary positive integer, we get from this the first inequality in (5.9).

Now we prove inequality (5.10). We have

$$\begin{aligned} \limsup_{r\to\infty} \frac{N(r,f')}{T(r,f')} &\le \limsup_{r\to\infty} \frac{2N(r,f)}{T(r,f')} \\ &\le \limsup_{r\to\infty} \frac{2N(r,f)}{T(r,f)} \limsup_{r\to\infty} \frac{T(r,f)}{T(r,f')} \le 2(1-\delta(\infty,f))\frac{1}{1-\gamma} \le \frac{2}{1-\gamma}. \end{aligned}$$

Next, inequality (5.11) implies

$$\limsup_{r\to\infty}\frac{N\left(r,\frac{1}{f'}\right)}{T(r,f')}\le 1-\liminf_{r\to\infty}\frac{1}{T(r,f')}\sum_{\nu=1}^{q}m(r,a_\nu,f)$$
$$\le 1-\left\{\liminf_{r\to\infty}\frac{T(r,f)}{T(r,f')}\right\}\liminf_{r\to\infty}\sum_{\nu=1}^{q}\frac{m(r,a_\nu,f)}{T(r,f)}$$
$$\le 1-\left\{\liminf_{r\to\infty}\frac{T(r,f)}{T(r,f')}\right\}\sum_{\nu=1}^{q}\delta(a_\nu,f).$$

Since q can be chosen arbitrarily, we get

$$\limsup_{r\to\infty}\frac{N\left(r,\frac{1}{f'}\right)}{T(r,f')}\le 1-\left\{\liminf_{r\to\infty}\frac{T(r,f)}{T(r,f')}\right\}\sum_{a\ne\infty}\delta(a,f)$$
$$\le 1-\frac{1}{1+\gamma}(1-\gamma)=\frac{2\gamma}{1+\gamma}.$$

Thus

$$\kappa(f')=\limsup_{r\to\infty}\frac{N\left(r,\frac{1}{f'}\right)+N(r,f')}{T(r,f')}\le\frac{2\gamma}{1+\gamma}+\frac{2\gamma}{1-\gamma}=\frac{4\gamma}{1-\gamma^2}.\qquad\square$$

PROOF OF THEOREM 5.3. Now let $f(z)$ be a function satisfying the hypothesis of Theorem 5.3. Then by (5.9) we get $\ln T(r,f')\sim\ln T(r,f)$ and hence, $\lambda[f']=\lambda[f]=\lambda$, $\rho[f']=\rho[f]=\rho$. By (5.10) and (5.8) we have $\kappa(f')\le\frac{\varepsilon}{1.1(p+1)}$, therefore Theorem 5.2 is applicable to $f'(z)$, and we conclude that $p\ge 1$, $p-\varepsilon\le\lambda\le\rho\le p+\varepsilon$. $\square$

REMARK 1. N. Toda [**153**] proved that the assumption $\rho<\infty$ in Theorem 5.3 can be dropped.

REMARK 2. Performing a linear-fractional transformation of the function $f(z)$, we see that condition (5.7) in Theorem 5.3 can be replaced by the condition

$$\sum_{a\ne b}\delta(a,f)\ge 1-\gamma,\qquad \delta(b,f)\ge 1-\gamma,$$

where b is an arbitrary number in the extended complex plane.

In connection with Theorem 5.3 the question arises: whether the proximity of the sum of all deficiencies $\sum_a\delta(a,f)$ to 2 implies the proximity of $\rho[f]$ and $\lambda[f]$ to a positive integer. It turns out that this question has a negative answer.

EXAMPLE 8. For each integer $q\ge 3$ there exists a meromorphic function $f(z)$ such that[11]

$$\sum_a\delta(a,f)=2,\qquad \rho[f]=\frac{q}{2}.$$

Let

$$I_n(z)=\sum_{k=0}^{\infty}\frac{\left(\frac{1}{2}z\right)^{n+2k}}{k!\Gamma(n+k+1)},\qquad |\arg z|<\pi,$$

[11]Other examples of functions with such properties will be given in Section 5 of Chapter 7.

be the Bessel function with purely imaginary argument. We assume that the parameter n is real and $|n| < 1/2$. Then for each ε, $0 < \varepsilon < \pi/2$, the following asymptotic formulas hold uniformly with respect to $\arg z$ in the region $\left\{|\arg z| < \frac{\pi}{2} - \varepsilon\right\}$ (see Whittaker, Watson [**WW96**, 17·7]):

$$I_n(z) \sim \frac{e^z}{(2\pi z)^{1/2}}, \qquad z \to \infty, \tag{5.12}$$

$$I_{-n}(z) - I_n(z) \sim 2\sin n\pi \frac{e^{-z}}{2\pi z)^{1/2}}, \qquad z \to \infty. \tag{5.13}$$

For a given integer $q \geq 3$ let

$$w_1(z) = z^{1/2} I_{1/q}(2z^{q/2}), \qquad w_2(z) = z^{1/2} I_{-1/q}(2z^{q/2}).$$

Since

$$w_1(z) = z\sum_{k=0}^{\infty} \frac{z^{kq}}{k!\Gamma\left(\frac{1}{q} + k + 1\right)}, \qquad w_2(z) = \sum_{k=0}^{\infty} \frac{z^{kq}}{k!\Gamma\left(-\frac{1}{q} + k + 1\right)}, \tag{5.14}$$

$w_1(z)$ and $w_2(z)$ are entire functions. By (5.12) and (5.13), for each ε, $0 < \varepsilon < \pi/q$, the asymptotic formulas

$$w_1(z) \sim z^{1/2} \frac{e^{2z^{q/2}}}{(4\pi z^{q/2})^{1/2}}, \qquad w_2(z) \sim z^{1/2} \frac{e^{2z^{q/2}}}{(4\pi z^{q/2})^{1/2}}, \tag{5.15}$$

$$w_2(z) - w_1(z) \sim z^{1/2} 2\sin\frac{\pi}{q} \frac{e^{-2z^{q/2}}}{(4\pi z^{q/2})^{1/2}}, \qquad z \to \infty, \tag{5.15$'$}$$

hold uniformly in $\arg z$ in the region $\left\{|\arg z| < \frac{\pi}{q} - \varepsilon\right\}$. We need the relations

$$T(r, w_l) = \frac{4r^{q/2}}{\pi} + o(r^{q/2}), \quad l = 1, 2, \tag{5.16}$$

$$m\left(r, \frac{w_1}{w_2}\right) = o(r^{q/2}),$$

$$m\left(r, \frac{w_2}{w_1 - w_2}\right) \geq \frac{8r^{q/2}}{q\pi} + o(r^{q/2}).$$

To prove these relations, observe that (5.15) and (5.15′) imply that for each ε, $0 < \varepsilon < \frac{\pi}{q}$, uniformly in φ, $|\varphi| < \frac{\pi}{q} - \varepsilon$,

$$\begin{aligned} \ln|w_l(re^{i\varphi})| &= 2r^{q/2}\cos\frac{q}{2}\varphi + O(\ln r), \qquad l = 1, 2, \\ \ln\left|\frac{w_2(re^{i\varphi})}{w_1(re^{i\varphi}) - w_2(re^{i\varphi})}\right| &= 4r^{q/2}\cos\frac{q}{2}\varphi + O(\ln r). \end{aligned} \tag{5.17}$$

From (5.14) we see that $M(r, w_l) = w_l(r)$, $l = 1, 2$, $0 \leq r < \infty$, and hence $T(r, w_l) \leq \ln^+ M(r, w_l) = 2r^{q/2} + O(\ln r)$. By (5.14) we also have

$$w_1(\omega z) = \omega w_1(z), \qquad w_2(\omega z) = w_2(z), \tag{5.18}$$

where $\omega = \exp\{2\pi i/q\}$; therefore

$$T(r, w_l) = m(r, w_l) = \frac{q}{2\pi}\int_{-\pi/q}^{\pi/q} \ln^+|w_l(re^{i\varphi})|\, d\varphi,$$

and

$$m\left(r, \frac{w_1}{w_2}\right) = \frac{q}{2\pi}\int_{-\pi/q}^{\pi/q} \ln^+ \left|\frac{w_1(re^{i\varphi})}{w_2(re^{i\varphi})}\right| d\varphi.$$

Using (5.17) and Theorem 7.4 of Chapter 1, we get

$$T(r, w_l) = 2r^{q/2}\frac{q}{2\pi}\int_{-\pi/q}^{\pi/q} \cos\frac{q}{2}\varphi\, d\varphi + o(r^{q/2}) = \frac{4r^{q/2}}{\pi} + o(r^{q/2}),$$

$$m\left(r, \frac{w_1}{w_2}\right) = o(r^{q/2}),$$

$$\begin{aligned} m\left(r, \frac{w_2}{w_1 - w_2}\right) &\geq \frac{1}{2\pi}\int_{-\pi/q}^{\pi/q} \ln^+ \left|\frac{w_2(re^{i\varphi})}{w_1(re^{i\varphi}) - w_2(re^{i\varphi})}\right| d\varphi \\ &= 4r^{q/2}\frac{1}{2\pi}\int_{-\pi/q}^{\pi/q} \cos\frac{q}{2}\varphi\, d\varphi + o(r^{q/2}) = \frac{8r^{q/2}}{q\pi} + o(r^{q/2}). \end{aligned}$$

Consider the meromorphic function

$$f(z) = \frac{w_1(z)}{w_2(z)}.$$

We show that $\rho[f] = q/2$, $\delta(\omega^k, f) = 2/q$ $(k = 0, 1, \dots, q-1)$. This indicates that $f(z)$ is a desired example.

We have

$$\begin{aligned} T(r, f) = N(r, f) + m(r, f) &\leq N\left(r, \frac{1}{w_2}\right) + m\left(r, \frac{w_1}{w_2}\right) \\ &\leq T(r, w_2) + o(r^{q/2}) = \frac{4r^{q/2}}{\pi} + o(r^{r/2}). \end{aligned} \tag{5.19}$$

Next,

$$m\left(r, \frac{1}{f-1}\right) = m\left(r, \frac{w_2}{w_1 - w_2}\right) \geq \frac{8r^{q/2}}{q\pi} + o(r^{q/2}). \tag{5.20}$$

Inequalities (5.19) and (5.20) imply that $\rho[f] = q/2$, $\delta(1, f) \geq 2/q$.

By (5.18) we have $f(\omega^{-k}z) = \omega^{-k}f(z)$, $k = 0, 1, \dots, q-1$. As was observed at the end of Section 1 of Chapter 4 (p. 115), the deficiencies of the functions $f(z)$ and $f(az)$, $a \neq 0$, coincide. Therefore $\delta(1, f(z)) = \delta(1, f(\omega^{-k}z)) = \delta(1, \omega^{-k}f(z))$. But, obviously, $\delta(1, af(z)) = \delta(a^{-1}, f(z))$, $a \neq 0$, and hence

$$\delta(\omega^k, f) = \delta(1, f) \geq 2/q, \quad k = 0, 1, \dots, q-1. \tag{5.21}$$

Since $\sum_{k=0}^{q-1} \delta(\omega^k, f) = q\delta(1, f) \leq 2$, (5.21) is an equality, what was to be proved.

Another question arising in connection with Theorem 5.3 is the the following. What can be said about the relation between $\rho[f]$ and $\lambda[f]$ in the case where $f(z)$ has at least two deficient values and one of them is close to 1? The answer to this question is unknown. It is possible that such a relation does not exist at all, and all the values of $\rho[f]$ and $\lambda[f]$ satisfying the condition $1/2 \leq \lambda[f] \leq \rho[f] \leq \infty$ are admissible. The last assumption is partially supported by the following example.

EXAMPLE 9. There exists an entire function $f(z)$ of order ρ and lower order λ, $1/2 < \lambda \leq \rho < \infty$, such that $\delta(0, f) > 0$.

Note that the restriction $\lambda > 1/2$ cannot be dropped (see Corollary 1 from Theorem 1.3 and Theorem 4.5). It is an open problem whether the condition $\rho < \infty$ can be dropped.[12]

Let α be a given number satisfying $1/2 < \alpha < \min(1, \lambda)$, and n a positive integer satisfying $n > \rho$. Suppose that there exists an entire function $g(z)$ having the following properties:

(a) The function $g(z)$ can be represented by a power series of the form

$$g(z) = 1 + c_n z^n + c_{n+1} z^{n+1} + \cdots \tag{5.22}$$

with real coefficients and $c_n < 0$;

(b) $|g(x)| < 1$ for $0 < x < \infty$;

(c) There exists η, $0 < \eta < \frac{\pi}{2} |\cot \pi\alpha|$, such that for $r > R_0(\eta)$, $0 \le \varphi \le 2\pi$, the inequality

$$\ln |g(re^{i\varphi}) < r^\alpha \left\{ \eta + \frac{\pi \cos \alpha(\varphi - \pi)}{\sin \pi\alpha} \right\} \tag{5.23}$$

holds.

We postpone the proof of the existence of such a function $g(z)$ because we do not want to interrupt the principal argument.

Let r_k be a sequence such that $1 < r_1 \le r_2 \le \cdots$, $r_k \to \infty$ as $k \to \infty$, and

$$\sum_{k=1}^\infty r_k^{-n} < \infty. \tag{5.24}$$

By Theorem 3.2 of Chapter 2, the absolutely convergent infinite product

$$f(z) = \prod_{k=1}^\infty g\left(\frac{z}{r_k}\right)$$

represents an entire function. Denote by $\nu(r)$ the number of points r_k on the interval $[0, r]$. Since (5.23) implies that $\ln M(r, g) = O(r^\alpha)$, by Theorem 3.3 of Chapter 2 we get

$$\ln M(r, f) < C \int_0^\infty \frac{\nu(r\tau)}{\tau^{1+\alpha}} \frac{d\tau}{1 + \tau^{n-\alpha}}, \qquad 0 < r < \infty, \tag{5.25}$$

where C is a positive constant depending on $g(z)$ only.

Representation (5.22) implies that

$$\ln |g(re^{i\varphi})| = c_n r^n \cos n\varphi + O(r^{n+1})$$

as $r \to \infty$, whence there exist R_1, $0 < R_1 < R_0$, and φ_1, $0 < \varphi_1 < \frac{\pi}{2n}$, such that for all $r \in [0, R_1]$ and $|\varphi| \le \varphi_1$ the inequality

$$\ln |g(re^{i\varphi})| \le -C_1 r^n \tag{5.26}$$

holds (we use the notation C_j for positive constants depending on $g(z)$ only). Inequality (5.23) implies the existence of $R_2 > R_0$ and φ_2, $0 < \varphi_2 < \pi(1 - \frac{1}{2\alpha})$ such that for all $r > R_2$ and $|\varphi| \le \varphi_2$ the inequality

$$\ln |g(re^{i\varphi})| \le -C_2 r^\alpha \tag{5.27}$$

---

[12] There exist similar examples with $\rho = \infty$; L. J. Kotman [**100**].

holds. If $x \in [R_1, R_2]$, by condition (b) we have $\ln|g(x)| < 0$. Since the function $|g(re^{i\varphi})|$ is uniformly continuous for $r \in [R_1, R_2]$, $\varphi \in [-\pi, \pi]$, there exists φ_3, $0 < \varphi_3 < \pi$, such that for all $r \in [R_1, R_2]$ and $|\varphi| \leq \varphi_3$ the estimate

$$\ln|g(re^{i\varphi})| \leq -C_3 \tag{5.28}$$

holds. Combining (5.26), (5.27), and (5.28) we get that for all r, $0 \leq r < \infty$, and φ, $|\varphi| \leq \varphi_0 = \min(\varphi_1, \varphi_2, \varphi_3)$, the inequality

$$\ln|g(re^{i\varphi})| \leq -C_4\Psi(r), \qquad \Psi(r) = \frac{r^n}{1 + r^{n-\alpha}},$$

holds. Hence, for $|\varphi| \leq \varphi_0$, we have

$$\ln|f(re^{i\varphi})| \leq -C_4 \sum_{k=1}^{\infty} \Psi\left(\frac{r}{r_k}\right) = -C_4\Phi(r)$$

and

$$m(r, 0, f) \geq \frac{1}{2\pi} \int_{-\varphi_0}^{\varphi_0} \ln \frac{1}{|f(re^{i\varphi})|} d\varphi \geq \frac{\varphi_0}{\pi} C_4\Phi(r).$$

Taking into account the following lower estimate of $\Phi(r)$:

$$\Phi(r) \geq C_5 \int_0^{\infty} \frac{\nu(r\tau)}{\tau^{1+\alpha}} \frac{d\tau}{1 + \tau^{n-\alpha}}, \qquad 0 < r < \infty$$

(this estimate was obtained in the proof of Theorem 3.3 of Chapter 2), we get

$$m(r, 0, f) \geq C_6 \int_0^{\infty} \frac{\nu(r\tau)}{\tau^{1+\alpha}} \frac{d\tau}{1 + \tau^{n-\alpha}}, \qquad 0 < r < \infty. \tag{5.29}$$

Comparing (5.25) and (5.29) we find that

$$\delta(0, f) \geq \liminf_{r\to\infty} \frac{m(r, 0, f)}{\ln M(r, f)} \geq \frac{C_6}{C} > 0.$$

Thus, for each sequence $\{r_k\}$ satisfying condition (5.24) we have $0 \in E_N(f)$.

Now we show that the sequence $\{r_k\}$ can be chosen in such a way that the function $f(z)$ has a prescribed order ρ and lower order λ. Observe that (5.25) and (5.29) imply

$$CQ(r) \geq T(r, f) \geq C_6Q(r), \qquad Q(r) = \int_0^{\infty} \frac{\nu(r\tau)}{\tau^{1+\alpha}} \frac{d\tau}{1 + \tau^{n-\alpha}};$$

therefore the order and the lower order of the functions $T(r, f)$ and $Q(r)$ coincide.

If $\rho = \lambda$, we choose the sequence $\{r_k\}$ in such a way that $\nu(r) = [r^\rho]$ for $r \geq 2^{1/\rho} = r_0$ and $\nu(r) = 0$ for $0 \leq r < r_0$. Then for all $r \geq r_0$ we have $\frac{1}{2}r^\rho \leq \nu(r) \leq r^\rho$ and

$$\frac{1}{2}r^\rho I(\rho) - \frac{1}{2}r^\rho \int_0^{r_0/r} \frac{\tau^{\rho-\alpha-1}}{1 + \tau^{n-\alpha}} d\tau \leq Q(r) \leq r^\rho I(\rho),$$

where

$$I(\rho) = \int_0^{\infty} \frac{\tau^{\rho-\alpha-1}}{1 + \tau^{n-\alpha}} < \infty,$$

and hence the order and lower order of the function $f(z)$ are equal to ρ. In the case $\rho = \lambda$ there are easier ways to construct an example (see Exercise 1 in Section 1, Chapter 4).

Now let $\lambda < \rho$. We choose a number $q_1 > 1$ such that $r^\rho \geq \frac{1}{2}r^\rho + 1 \geq r^\lambda \geq \frac{1}{2}r^\lambda + 1$ for $r \geq q_1$. We define sequences q_k, p_k, p'_k, s_k, t_k, $k = 1, 2, \ldots$, by the following recursions: $p_k = \exp\exp q_k$, $p'_k = p_k^{\rho/\lambda}$, $q_{k+1} = \exp\exp p'_k$, $s_k = \exp q_k$, $t_k = \exp p'_k$, $k = 1, 2, \ldots$. It is clear that these sequences tend to ∞. We select a sequence r_k in such a way that

$$\nu(r) = \begin{cases} 0 & \text{for } 0 \leq r < q_1, \\ [r^\rho] & \text{for } q_k \leq r \leq p_k, \\ [r^\lambda] & \text{for } p'_k \leq r < q_{k+1}, \\ [p_k^\rho] = [p_k'^\lambda] & \text{for } p_k \leq r \leq p'_k. \end{cases}$$

Then for all $r \geq q_1$ we have $\frac{1}{2}r^\lambda \leq \nu(r) \leq r^\rho$. Since $\rho < n$, condition (5.24) is satisfied. It is easy to see that

$$\frac{1}{2}I(\lambda)r^\lambda + o(r^\lambda) = \frac{1}{2}I(\lambda)r^\lambda - \frac{1}{2}r^\lambda \int_0^{q_1/r} \frac{\tau^{\lambda-\alpha-1}}{\tau^{n-\alpha}+1}\,d\tau \leq Q(r) \leq I(\rho)r^\rho.$$

Hence

$$\lambda \leq \liminf_{r\to\infty} \frac{\ln Q(r)}{\ln r} \leq \limsup_{r\to\infty} \frac{\ln Q(r)}{\ln r} \leq \rho. \tag{5.30}$$

On the other hand

$$\begin{aligned} Q(s_k) &= \int_0^\infty \frac{\nu(\tau s_k)\,d\tau}{\tau^{1+\alpha}(\tau^{n-\alpha}+1)} \geq \int_{q_k/s_k}^{p_k/s_k} \frac{\nu(\tau s_k)\,d\tau}{\tau^{1+\alpha}(\tau^{n-\alpha}+1)} \\ &\geq \frac{1}{2}s_k^\rho \int_{q_k/\exp q_k}^{p_k/\ln p_k} \frac{\tau^{\rho-\alpha-1}\,d\tau}{\tau^{n-\alpha}+1} = \frac{1}{2}s_k^\rho\{I(\rho) + o(1)\}. \end{aligned}$$

Therefore

$$\limsup_{r\to\infty} \frac{\ln Q(r)}{\ln r} \geq \limsup_{k\to\infty} \frac{\ln Q(s_k)}{\ln s_k} \geq \rho,$$

and together with (5.30) this shows that the order of the function $Q(r)$ is equal to ρ. Next,

$$Q(t_k) = \left\{\int_{p'_k/t_k}^{q_{k+1}/t_k} + \int_{q_{k+1}/t_k}^\infty + \int_0^{p'_k/t_k}\right\} \frac{\nu(\tau t_k)\,d\tau}{\tau^{1+\alpha}(\tau^{n-\alpha}+1)}.$$

We estimate each of the three integrals from above:

$$\int_{p'_k/t_k}^{q_{k+1}/t_k} \frac{\nu(\tau t_k)\,d\tau}{\tau^{1+\alpha}(\tau^{n-\alpha}+1)} \leq t_k^\lambda \int_{p'_k/t_k}^{q_{k+1}/t_k} \frac{\tau^{\lambda-\alpha-1}\,d\tau}{\tau^{n-\alpha}+1} \leq t_k^\lambda I(\lambda),$$

$$\begin{aligned} \int_{q_{k+1}/t_k}^\infty \frac{\nu(\tau t_k)\,d\tau}{\tau^{1+\alpha}(\tau^{n-\alpha}+1)} &\leq t_k^\rho \int_{q_{k+1}/t_k}^\infty \frac{\tau^{\rho-\alpha-1}\,d\tau}{\tau^{n-\alpha}+1} \leq t_k^\rho \int_{q_{k+1}/t_k}^\infty \tau^{\rho-n-1}\,d\tau \\ &= \frac{t_k^\rho}{n-\rho}\frac{t_k^{n-\rho}}{q_{k+1}^{n-\rho}} = \frac{1}{n-\rho}\frac{t_k^n}{(\exp t_k)^{n-\rho}} = o(1), \end{aligned}$$

$$\begin{aligned} \int_0^{p'_k/t_k} \frac{\nu(\tau t_k)\,d\tau}{\tau^{1+\alpha}(\tau^{n-\alpha}+1)} &\leq t_k^\rho \int_0^{p'_k/t_k} \frac{\tau^{\rho-\alpha-1}\,d\tau}{\tau^{n-\alpha}+1} \\ &\leq t_k^\rho \int_0^{\ln t_k/t_k} \tau^{\rho-\alpha-1}\,d\tau = \frac{t_k^\rho}{\rho-\alpha}\left(\frac{\ln t_k}{t_k}\right)^{\rho-\alpha} = \frac{t_k^\alpha}{\rho-\alpha}(\ln t_k)^{\rho-\alpha} = o(t_k^\lambda). \end{aligned}$$

Joining these estimates, we get

$$Q(t_k) \le t_k^\lambda I(\lambda) + o(t_k^\lambda);$$

therefore

$$\liminf_{r\to\infty} \frac{\ln Q(r)}{\ln r} \le \liminf_{k\to\infty} \frac{\ln Q(t_k)}{\ln t_k} \le \lambda.$$

Together with (5.30) this shows that the lower order of $Q(r)$ is equal to λ. As was mentioned above, this proves that the order and the lower order of the function $f(z)$ are equal to ρ and λ, respectively.

It remains to show that there exists a function $g(z)$ having properties (a), (b), and (c). Take

$$V(z) = \prod_{k=1}^{\infty} \left(1 - \frac{z}{k^{1/\alpha}}\right).$$

Formula (5.3_2) in Chapter 2 implies that the entire function

$$g(z) = R(z)V(z), \tag{5.31}$$

where $R(z)$ is an arbitrary rational function, all poles of which are zeros of $V(z)$, satisfies condition (c). We show that the function $R(z)$ can be selected in such a way that the function (5.31) also satisfies conditions (a) and (b).

Let p be a positive integer, $p > 2n$; the choice of p will be specified later. Let

$$V_p(z) = \prod_{k=p+1}^{\infty} \left(1 - \frac{z}{k^{1/\alpha}}\right), \qquad v_p(z) = \prod_{k=1}^{p} \left(1 - \frac{z}{k^{1/\alpha}}\right).$$

We have a power series representation

$$\frac{1}{V_p(z)} = \sum_{k=0}^{\infty} a_k z^k, \qquad a_k = a_k^{(p)}, \quad a_0 = 1, \tag{5.32}$$

in the circle $\{|z| < (p+1)^{1/\alpha}\}$. Since

$$\frac{1}{V_p(z)} = \prod_{k=p+1}^{\infty} \sum_{j=0}^{\infty} \left(\frac{z}{k^{1/\alpha}}\right)^j,$$

obviously $a_k > 0$, $k = 0, 1, 2, \ldots$. The function $V_p(z)$ tends to 1 as $p \to \infty$ uniformly over the circle $\{|z| = 1\}$. Therefore $|V_p(z)| \ge 1/2$ for $\{|z| = 1\}$ if $p \ge p_0$, and

$$0 < a_k = \left| \frac{1}{2\pi i} \int_{|z|-1} \frac{dz}{z^{k+1} V_p(z)} \right| \le 2, \qquad k = 0, 1, \ldots. \tag{5.33}$$

Take $R(z) = \sum_{k=0}^{n-1} a_k z^k / v_p(z)$ in (5.31); then

$$g(z) = \frac{\sum_{k=0}^{n-1} a_k z^k}{v_p(z)} V(z) = V_p(z) \sum_{k=0}^{n-1} a_k z^k = 1 + c_n z^n + c_{n+1} z^{n+1} + \cdots$$

(see Example 2 in Section 3 of Chapter 2). Since the function $V_p(z)$ takes real values on the real axis, it is easy to see that all coefficients c_k, $k \ge n$, are real numbers. Since

$$V_p(z) \sum_{k=0}^{n-1} a_k z^k = 1 - V_p(z) \left(\frac{1}{V_p(z)} - \sum_{k=0}^{n-1} a_k z^k \right) = 1 - V_p(z) \sum_{k=n}^{\infty} a_k z^k$$

for $|z| < (p+1)^{1/\alpha}$, it is clear that $c_n = -a_n < 0$. Thus the function $g(z)$ satisfies condition (a).

We show that $|V(x)| < 1$ for $0 < x < \infty$. For $x > 0$, $x \neq k^{1/\alpha}$, $k = 1, 2, \dots$, the inequality

$$0 < \frac{1 - \frac{x}{k^{1/\alpha}}}{1 - \frac{x^{2\alpha}}{k^2}} = \frac{1 - \frac{x}{k^{1/\alpha}}}{1 - \left(\frac{x}{k^{1/\alpha}}\right)^{2\alpha}} < 1$$

holds; therefore

$$|V(x)| = \prod_{k=1}^{\infty} \frac{1 - \frac{x}{k^{1/\alpha}}}{1 - \frac{x^{2\alpha}}{k^2}} \left|1 - \frac{x^{2\alpha}}{k^2}\right| \leq \prod_{k=1}^{\infty} \left|1 - \frac{x^{2\alpha}}{k^2}\right| = \frac{|\sin \pi x^\alpha|}{\pi x^\alpha} < 1.$$

For $0 < x < (p+1)^{1/\alpha}$ we have

$$(5.34) \qquad |g(x)| = g(x) = V_p(x) \sum_{k=0}^{n-1} a_k x^k < V_p(x) \sum_{k=0}^{\infty} a_k x^k = 1.$$

If $x \geq (p+1)^{1/\alpha}$, then

$$(5.35) \qquad |g(x)| = |V(x)| \frac{\sum_{k=0}^{n-1} a_k x^k}{|v_p(x)|} < \frac{\sum_{k=0}^{n-1} a_k x^k}{|v_p(x)|}.$$

Letting $x^\alpha = \xi$, $\xi \geq p+1$, and using the inequality $a^{1/\alpha} - 1 \geq \frac{1}{\alpha}(a-1)$, $a \geq 1$, we get

$$\begin{aligned}
|v_p(x)| &= \prod_{k=1}^{p} \left(\frac{x}{k^{1/\alpha}} - 1\right) = \prod_{k=1}^{p} \left\{\left(\frac{\xi}{k}\right)^{1/\alpha} - 1\right\} \\
&\geq \frac{1}{\alpha^p} \prod_{k=1}^{p} \left(\frac{\xi}{k} - 1\right) = \frac{1}{\alpha^p p!} \prod_{k=1}^{2n} (\xi - k) \prod_{k=2n+1}^{p} (\xi - k) \\
&\geq \frac{1}{a^p p!} (\xi - 2n)^{2n} (p-2n)! \geq \frac{1}{\alpha^p p^{2n}} (\xi - 2n)^{2n}.
\end{aligned}$$

Taking into account (5.33), (5.35), and $1/2 < \alpha < 1$, for $p \geq p_0$ we get

$$|g(x)| < \frac{2n \xi^{n/\alpha} \alpha^p p^{2n}}{(\xi - 2n)^{2n}} < 2n \left(\frac{\xi}{\xi - 2n}\right)^{2n} p^{2n} \alpha^p \leq 2n \left(\frac{p+1}{p+1-2n}\right)^{2n} p^{2n} \alpha^p.$$

Since the right-hand side of the last inequality tends to 0 as $p \to \infty$, the number p can be chosen so large that the right-hand side is less than 1. Taking into account (5.34), we find that $|g(x)| < 1$ for $0 < x < \infty$, that is, condition (b) is satisfied. We have proved the existence of a desired function $g(z)$.

REMARK. If in Example 9 we required the inequality $1 < \lambda \leq \rho < \infty$ only, it would be possible, using the same idea, to construct an example much more easily. Let $g(z) = e^{-z} \sum_{k=0}^{n-1} z^k / k!$ and $\alpha = 1$. We can repeat the previous argument (inequality (5.23) will be replaced by

$$\ln |g(re^{i\varphi})| < r(\eta - \cos \varphi)$$

for $r > R_0(\eta)$), but the investigation of properties of the function $g(z)$ will be significantly simplified.

6. Meromorphic functions representable by series of simple fractions

In this section we consider meromorphic functions of the form

$$f(z)=\sum_{k=1}^{\infty}\frac{A_k}{z-h_k},\qquad h_k\to\infty, \tag{6.1}$$

where the series is assumed to be absolutely convergent, which is equivalent to the condition

$$\sum_{|h_k|>1}|A_k/h_k|<\infty. \tag{6.2}$$

Obviously, the rational functions that admit representations of the form

$$R(z)=\sum_{k=1}^{n}\frac{A_k}{z-h_k}\qquad (1\le n<\infty)$$

satisfy the following conditions:

(a) $R(z)\to 0$ as $z\to\infty$.

(b) For each $a\neq 0$ the number of a-points of $R(z)$ is equal to n, and the number of zeros is equal to $n-1$ if $\sum_{k=1}^{n}A_k\neq 0$.

The main results of this section are the following two theorems. They show that the transcendental meromorphic functions of the form (6.1) also have properties that are in a certain sense close to (a) and (b).

THEOREM 6.1. *Let $f(z)$ be a meromorphic function of the form* (6.1). *Then*

$$m(r,f)=o(1),\qquad r\to\infty. \tag{6.3}$$

Moreover, for each $0<p<1$,

$$\int_0^{2\pi}|f(re^{i\varphi})|^p d\varphi=o(1),\qquad r\to\infty. \tag{6.4}$$

THEOREM 6.2. *Let $f(z)$ be a meromorphic function of the form* (6.1) *for which $\lambda[f]<\infty$. Then $\delta(a,f)=0$ for all $a\neq 0$, and the order of $N(r,a)$ is equal to the order of $T(r,f)$. This statement remains true also for $a=0$ if we assume, in addition, that*

$$\sum_{k=1}^{\infty}|A_k|<\infty \tag{6.5}$$

and

$$\sum_{k=1}^{\infty}A_k\neq 0. \tag{6.6}$$

We assume, without loss of generality, that $h_k\neq 0$, $k=1,2,\dots$.

To prove Theorem 6.1 we need the following lemma.

LEMMA 6.1. *If a function $F(z)$ is analytic in the disc $\{|z|<R\}$ and is such that the functions $\operatorname{Re}F(z)$ and $\operatorname{Im}F(z)$ have constant sign, then the inequality*

$$\int_0^{2\pi}u^p(\varphi)\,d\varphi\le\frac{2\pi}{\cos\frac{\pi p}{2}}|F(0)|^p$$

holds for each $0 < p < 1$, *where*

$$u(\varphi) = \liminf_{\substack{z \to Re^{i\varphi} \\ |z| < R}} |F(z)|.$$

PROOF. We may assume without loss of generality that $\operatorname{Re} f(z) > 0$ for $|z| < R$. The function $F(z)$ has no zeros in the disc $\{|z| < R\}$, and we can fix a branch of $\arg F(z)$ such that $|\arg F(z)| < \pi/2$ for $|z| < R$. The function

$$F^p(z) = |F(z)|^p \exp\{ip \arg F(z)\}$$

is analytic in the disc $\{|z| < R\}$, and hence $\operatorname{Re}\{F(z)\}^p$ is harmonic there. Since

$$\operatorname{Re}\{F(z)\}^p = |F(z)|^p \cos(p \arg F(z)) \ge |F(z)|^p \cos \frac{p\pi}{2},$$

we have

$$\int_0^{2\pi} |F(re^{i\varphi})|^p d\varphi \le \frac{1}{\cos \frac{p\pi}{2}} \int_0^{2\pi} \operatorname{Re}\{F(re^{i\varphi})\}^p d\varphi = \frac{2\pi}{\cos \frac{p\pi}{2}} \operatorname{Re}\{F(0)\}^p;$$

thus

$$\int_0^{2\pi} |F(re^{i\varphi})|^p d\varphi \le \frac{2\pi}{\cos \frac{p\pi}{2}} |F(0)|^p, \qquad r < R.$$

Taking limits as $r \to R$ and using the Fatou lemma, we complete the proof of the lemma. □

PROOF. We prove Theorem 6.1. Relation (6.3) follows from (6.4) and the inequality $\ln^+ a \le \frac{1}{p} a^p$, $a > 0$; therefore it suffices to prove (6.4).

Let $\theta_k = \arg h_k$, $A_k = \alpha_k + i\beta_k$. Fixing $R > 0$, we write the function $f(z)$ in the form

$$\begin{aligned} f(z) = & \sum_{|h_k|>R}{}' \frac{e^{i\theta_k} \operatorname{Re}(A_k e^{-i\theta_k})}{z - h_k} + \sum_{|h_k|>R}{}'' + i \sum_{|h_k|>R}{}' \frac{e^{i\theta_k} \operatorname{Im}(A_k e^{-i\theta_k})}{z - h_k} \\ & + i \sum_{|h_k|>R}{}'' + \sum_{|h_k|\le R}{}' \frac{\alpha_k}{z - h_k} + \sum_{|h_k|\le R}{}'' + i \sum_{|h_k|\le R}{}' \frac{\beta_k}{z - h_k} + i \sum_{|h_k|\le R}{}'' \\ = & F_1(z) + \cdots + F_8(z), \end{aligned}$$

where the sums $\sum'$ contain those terms for which $\operatorname{Re}(A_k e^{-i\theta_k})$, $\operatorname{Im}(A_k e^{-i\theta_k})$, α_k, β_k, respectively, are positive, and the sums $\sum''$ contain all other terms.

Since for $|h_k| > R$, $|z| < R$, we have

$$\operatorname{Re} \frac{e^{i\theta_k}}{z - h_k} < 0,$$

the functions $F_j(z)$, $j = 1, 2, 3, 4$, satisfy the hypothesis of Lemma 6.1. Hence $(j = 1, 2, 3, 4)$

$$\int_0^{2\pi} |F_j(Re^{i\varphi})|^p d\varphi \le \frac{2\pi}{\cos \frac{p\pi}{2}} |F_j(0)|^p \le \frac{2\pi}{\cos \frac{p\pi}{2}} \left(\sum_{|h_k|>R} \left| \frac{A_k}{h_k} \right| \right)^p.$$

Since for $|z| = R$ we have

$$\left| \sum_{|h_k|\le R}{}' \frac{\alpha_k}{z - h_k} \right| = \left| \sum_{|h_k|\le R}{}' \frac{R\alpha_k}{R^2 - h_k z} \right|,$$

and for $|h_k| \le R$, $|z| < R$, the inequality

$$\operatorname{Re} \frac{R}{R^2 - h_k z} > 0$$

holds, we see that the function

$$F_5^{(1)}(z) = \sum_{|h_k| \le R}' \frac{R\alpha_k}{R^2 - h_k z}$$

satisfies the hypothesis of Lemma 6.1, and moreover, $|F_5^{(1)}(Re^{i\varphi})| = |F_5(Re^{i\varphi})|$, $0 \le \varphi < 2\pi$. Hence

$$\int_0^{2\pi} |F_5(Re^{i\varphi})|^p d\varphi \le \frac{2\pi}{\cos \frac{p\pi}{2}} |F_5^{(1)}(0)|^p \le \frac{2\pi}{\cos \frac{p\pi}{2}} \left(\frac{1}{R} \sum_{|h_k| \le R} |A_k| \right)^p .$$

In a similar way we get

$$\int_0^{2\pi} |F_j(Re^{i\varphi})|^p d\varphi \le \frac{2\pi}{\cos \frac{p\pi}{2}} \left(\frac{1}{R} \sum_{|h_k| \le R} |A_k| \right)^p , \qquad j = 6, 7, 8.$$

Using the obtained estimates, we come to the following conclusion:

$$\begin{aligned}
\int_0^{2\pi} |f(Re^{i\varphi})|^p d\varphi &\le \sum_{j=1}^{8} \int_0^{2\pi} |F_j(Re^{i\varphi})|^p d\varphi \\
&\le \frac{8\pi}{\cos \frac{p\pi}{2}} \left\{ \left(\sum_{|h_k| > R} \left| \frac{A_k}{h_k} \right| \right)^p + \left(\frac{1}{R} \sum_{|h_k| < R} |A_k| \right)^p \right\} \\
&\le \frac{8\pi}{\cos \frac{p\pi}{2}} \left\{ \left(\sum_{|h_k| > R} \left| \frac{A_k}{h_k} \right| \right)^p + \left(\frac{1}{\sqrt{R}} \sum_{|h_k| < \sqrt{R}} \left| \frac{A_k}{h_k} \right| + \sum_{\sqrt{R} \le |h_k| \le R} \left| \frac{A_k}{h_k} \right| \right)^p \right\} \\
&= o(1), \qquad R \to \infty.
\end{aligned}$$

Thus, relation (6.4) has been proved. □

Theorem 6.1 immediately implies the following.

COROLLARY. *Functions of the form* (6.1) *satisfy the relations*

$$T(r, f) = N(r, f) + o(1), \qquad \delta(\infty, f) = 0.$$

Let $\varphi(r)$ be an arbitrary function such that $\frac{d\varphi(r)}{d \ln r}$ monotonically tends to $+\infty$. If we choose the poles of the function (6.1) in such a way that

$$n(r, f) = \left[\frac{d\varphi(r)}{d \ln r} \right] = \frac{d\varphi(r)}{d \ln r} + O(1),$$

then

$$T(r, f) = N(r, f) + o(1) = \varphi(r) + O(\ln r);$$

that is, the characteristic $T(r, f)$ has the prescribed asymptotic behavior.

Upon a modification of the method used in the proof of Theorem 6.1 we can get information on the behavior of a function of the form (6.1) along the rays $\{\arg z = \text{const}\}$.

THEOREM 6.3. *Let $f(z)$ be a meromorphic function of the form (6.1). Then, for arbitrary α and p, $0 \le \alpha < 2\pi$, $0 < p < 1$, we have*

$$\int_1^\infty \frac{\ln^+ |f(te^{i\alpha})|}{t^{1+p}}\, dt < \infty, \tag{6.7}$$

$$\int_1^\infty \frac{|f(te^{i\alpha})|^p}{t^2}\, dt < \infty. \tag{6.8}$$

We need the following lemma, similar to Lemma 6.1.

LEMMA 6.2. *If the function $g(z)$ is analytic in the half-plane $\{\operatorname{Im} z > 0\}$ and is such that the functions $\operatorname{Re} g(z)$ and $\operatorname{Im} g(z)$ have a constant sign, then for each p, $0 < p < 1$, we have*

$$\int_{-\infty}^\infty \frac{v^p(t)}{1+t^2}\, dt \le \frac{\pi}{\cos \frac{p\pi}{2}} |g(i)|^p,$$

where

$$v(t) = \liminf_{\substack{z \to t \\ \operatorname{Im} z > 0}} |g(z)|.$$

PROOF. The function $F(z) = g\left(\frac{1}{i}\frac{z-1}{z+1}\right)$ satisfies the hypothesis of Lemma 6.1 ($R = 1$). Observing that for this function we have $u(\varphi) = v\left(\frac{1}{i}\frac{e^{i\varphi}-1}{e^{i\varphi}+1}\right) = v\left(\tan\frac{\varphi}{2}\right)$ and $F(0) = g(i)$, we get the inequality

$$\int_{-\pi}^{\pi} v^p\left(\tan\frac{\varphi}{2}\right) d\varphi \le \frac{2\pi}{\cos\frac{p\pi}{2}} |g(i)|^p.$$

Making the change of variables $t = \tan\frac{\varphi}{2}$, we get the desired inequality. □

PROOF OF THEOREM 6.3. Obviously, it is enough to prove Theorem 6.3 for $\alpha = 0$. Observe that relation (6.7) follows from (6.8). In fact, using the inequality $\ln^+ a < \frac{1}{p} a^p$, $p > 0$, we get

$$\begin{aligned}\int_1^\infty \frac{\ln^+|f(t)|}{t^{1+p}}\, dt &\le \int_1^\infty \frac{\ln^+\left|f(t)t^{\frac{p-1}{p}}\right| + \frac{1-p}{p}\ln t}{t^{1+p}}\, dt \\ &\le \frac{1}{p}\int_1^\infty \frac{|f(t)|^p}{t^2}\, dt + \frac{1-p}{p}\int_1^\infty \frac{\ln t}{t^{1+p}}\, dt.\end{aligned}$$

We let $A_k = \alpha_k + i\beta_k$ and represent the function $f(z)$ in the form

$$\begin{aligned} f(z) = &\sum_{\operatorname{Im} h_k \le 0}{}' \frac{\alpha_k}{z-h_k} + \sum_{\operatorname{Im} h_k \le 0}{}'' + i \sum_{\operatorname{Im} h_k \le 0}{}' \frac{\beta_k}{z-h_k} + i \sum_{\operatorname{Im} h_k \le 0}{}'' \\ &+ \sum_{\operatorname{Im} h_k > 0}{}' \frac{\alpha_k}{z-h_k} + \sum_{\operatorname{Im} h_k > 0}{}'' + i \sum_{\operatorname{Im} h_k > 0}{}' \frac{\beta_k}{z-h_k} + i \sum_{\operatorname{Im} h_k > 0}{}'' \\ = &\, g_1(z) + \cdots + g_8(z), \end{aligned}$$

where the sums $\sum'$ contain those terms for which α_k and β_k are positive, and the sums $\sum''$ contain all other terms.

Since for $\operatorname{Im} h_k \le 0$, $\operatorname{Im} z > 0$ the inequality

$$\operatorname{Im} \frac{1}{z - h_k} < 0$$

holds, Lemma 6.2 can be applied to the functions $g_j(z)$, $j = 1,2,3,4$, whence

$$\int_{-\infty}^{\infty} \frac{|g_j(t)|^p}{1+t^2} dt \le \frac{\pi}{\cos \frac{\pi p}{2}} |g_j(i)|^p \le \frac{\pi}{\cos \frac{\pi p}{2}} \left(\sum_{\operatorname{Im} h_k \le 0} \left| \frac{A_k}{h_k} \right| \right)^p, \quad j = 1,2,3,4.$$

Observe that Lemma 6.2 can also be applied to the functions $g_j(-z)$, $j = 5,6,7,8$, whence

$$\int_{-\infty}^{\infty} \frac{|g_j(t)|^p}{1+t^2} dt \le \frac{\pi}{\cos \frac{\pi p}{2}} |g_j(-i)|^p \le \frac{\pi}{\cos \frac{\pi p}{2}} \left(\sum_{\operatorname{Im} h_k > 0} \left| \frac{A_k}{h_k} \right| \right)^p, \quad j = 5,6,7,8.$$

Thus

$$\begin{aligned} \int_{-\infty}^{\infty} \frac{|f(t)|^p}{1+t^2} dt &\le \sum_{j=1}^{8} \int_{-\infty}^{\infty} \frac{|g_j(t)|^p}{1+t^2} dt \\ &\le \frac{4\pi}{\cos \frac{\pi p}{2}} \left\{ \left(\sum_{\operatorname{Im} h_k \le 0} \left| \frac{A_k}{h_k} \right| \right)^p + \left(\sum_{\operatorname{Im} h_k > 0} \left| \frac{A_k}{h_k} \right| \right)^p \right\} < \infty, \end{aligned}$$

whence $\int_1^\infty |f(t)|^p t^{-2} dt < \infty$. □

Now we turn to the proof of Theorem 6.2. If $a = \infty$, the statement of Theorem 6.2 is contained in the Corollary of Theorem 6.1. The statement of Theorem 6.2 for $a \ne 0, \infty$ follows from its statement for $a = 0$. In fact, for $a \ne 0, \infty$ we have

$$\frac{f(z) - a}{z} = -\frac{a}{z} + \sum_{k=1}^{\infty} \frac{A_k}{z(z - h_k)} = \frac{-a - \sum_{k=1}^\infty A_k/h_k}{z} + \sum_{k=1}^{\infty} \frac{A_k/h_k}{z - h_k}.$$

Letting

$$f_1(z) = \frac{f(z) - a}{z}, \quad h_0 = 0, \quad B_0 = -a - \sum_{k=1}^{\infty} A_k/h_k, \quad B_k = A_k/h_k \ (k \ge 1),$$

we obtain the representation

$$f_1(z) = \sum_{k=0}^{\infty} \frac{B_k}{z - h_k},$$

and, obviously,

$$T(r, f_1) = T(r, f) + O(\ln r), \qquad N(r, 0, f_1) = N(r, a, f) + O(\ln r),$$

$$\delta(0, f_1) = \delta(a, f), \quad \sum_{k=0}^{\infty} |B_k| < \infty, \quad \sum_{k=0}^{\infty} B_k = -a \ne 0.$$

Now we prove Theorem 6.2 in the case $a = 0$.

By virtue of (6.5) and (6.6), we may assume without loss of generality that $\sum_{k=1}^\infty A_k = 1$. Let

$$\psi(z) = z f(z) - 1 = \sum_{k=1}^{\infty} A_k \left(\frac{z}{z - h_k} - 1 \right) = \sum_{k=1}^{\infty} \frac{A_k h_k}{z - h_k}.$$

The function $\psi(z)$ satisfies the hypothesis of Theorem 6.1; therefore for each p, $0 < p < 1$, the relation

$$\int_0^{2\pi} |\psi(re^{i\varphi})|^p d\varphi = o(1), \qquad r \to \infty,$$

holds. Denote by E_r the set of those values of $\varphi \in [0, 2\pi)$ for which $|\psi(re^{i\varphi})| \geq 1/2$, and let $\varepsilon_r = \operatorname{meas} E_r$. Observe that

$$\varepsilon_r \leq 2^p \int_{E_r} |\psi(re^{i\varphi})|^p d\varphi \leq 2^p \int_0^{2\pi} |\psi(re^{i\varphi})|^p d\varphi = o(1), \qquad r \to \infty.$$

For $\varphi \in CE_r = [0, 2\pi)\backslash E_r$ we have $|f(re^{i\varphi})| \geq \frac{1}{2r}$; hence,

$$m\left(r, \frac{1}{f}\right) \leq \frac{1}{2\pi} \int_{E_r} \ln^+ \frac{1}{|f(re^{i\varphi})|} d\varphi + \ln(2r).$$

By Theorem 7.3 of Chapter 1 (p. 39), for each $k > 1$, the estimate

$$\frac{1}{2\pi} \int_{E_r} \ln^+ \frac{1}{|f(re^{i\varphi})|} d\varphi \leq \left(\frac{6r}{k-1} \varepsilon_r \ln \frac{2\pi e}{\varepsilon_r}\right) T\left(kr, \frac{1}{f}\right)$$

holds. Since $\ln(2r) = o(T(r, f))$ (we assume that the function $f(z)$ is transcendental), and $T(r, \frac{1}{f}) = T(r, f) + O(1)$, we obtain the relation

$$m\left(r, \frac{1}{f}\right) \leq \theta(k, r) T(kr, f), \tag{6.9}$$

where $\theta(k, r) = o(1)$ as $r \to \infty$ and k is fixed.

Suppose that $\delta(0, f) > 0$. Then for all sufficiently large r the inequality

$$T(r, f) \leq \frac{2}{\delta(0, f)} m\left(r, \frac{1}{f}\right) \leq \frac{2\theta(k, r)}{\delta(0, f)} T(kr, f)$$

holds. Hence for each $\beta > 1$ we can find $r_0 = r_0(k, \beta)$ such that for $r \geq r_0$ the inequality

$$T(kr, f) \geq \beta T(r, f)$$

holds. From this we get

$$T(k^n r_0, f) \geq \beta^n T(r_0, f), \qquad n = 1, 2, \ldots.$$

Letting

$$n = \left[\frac{\ln r - \ln r_0}{\ln k}\right], \qquad r \geq r_0,$$

we get

$$r \geq k^n r_0, \qquad \beta^n \geq \beta^{-\frac{\ln r_0}{\ln k} - 1} r^{\frac{\ln \beta}{\ln k}},$$

and hence the inequality

$$T(r, f) \geq r^{\frac{\ln \beta}{\ln k}} \beta^{-\frac{\ln r_0}{\ln k} - 1} T(r_0, f).$$

Thus $\lambda[f] \geq \frac{\ln \beta}{\ln k}$. Since k can be chosen to be arbitrarily close to 1, we get $\lambda[f] = \infty$ contrary to the hypothesis of the theorem.

The statement that the orders of $N(r, 0)$ and $T(r, f)$ coincide requires a separate proof because (see Example 3 in Section 2) a Borel exceptional value can have zero deficiency.

Adding $N(r, \frac{1}{f})$ to both sides of (6.9), we get the relation

$$T(r,f) \le \vartheta(k,r)T(kr,f) + N\left(r, \frac{1}{f}\right), \tag{6.10}$$

where $\vartheta(k,r) = o(1)$ as $r \to \infty$ and k is fixed.

Suppose that the order of $N(r, \frac{1}{f})$ is less than the order of $T(r,f)$ and choose a number μ situated between these orders. Then $N(r, \frac{1}{f}) < r^\mu$ for $r > r_1$. Fixing $k > 1$, we choose $r_2 > r_1$ such that $T(r_2, f) \ge 2r_2^\mu$ and $4k^\mu \vartheta(k,r) < 1$ for $r \ge r_2$. Then by (6.10) we have

$$T(kr_2, f) \ge \frac{1}{\vartheta(k,r_2)} \left\{ T(r_2,f) - N\left(r_2, \frac{1}{f}\right)\right\} \ge \frac{1}{2\vartheta(k,r_2)} T(r_2,f) \ge 2(kr_2)^\mu.$$

Using (6.10) repeatedly, we get

$$T(k^n r_2, f) \ge \frac{1}{2^n \vartheta(k,r_2)\vartheta(k,kr_2)\cdots\vartheta(k,k^{n-1}r_2)} T(r_2,f) \ge (2k^\mu)^n T(r_2,f).$$

From this, selecting $n = [\frac{\ln r - \ln r_2}{\ln k}]$, $r \ge r_2$, we get the inequality

$$T(r,f) > (2k^\mu)^{\frac{\ln r - \ln r_2}{\ln k} - 1} T(r_2, f), \qquad r \ge r_2,$$

and conclude that $\lambda[f] \ge \frac{\mu \ln k + \ln 2}{\ln k}$. Because k can be chosen to be arbitrarily close to 1, we again get a contradiction.

REMARK. The statement of Theorem 6.2 is no longer true if we omit condition (6.6).

In fact, suppose $\delta(0,f) = 0$ for each function $f(z)$ of the form (6.1), satisfying condition (6.5). From this assumption we deduce the following corollary: Meromorphic functions of finite order cannot have deficient values different from 0 and ∞. This is clearly a nonsense.

Let $g(z)$ be a meromorphic function of finite order. Denoting the genus of this function by p, we can, by the Remark to Theorem 4.1 of Chapter 2 (p. 57), represent the function $g(z)$ in the form

$$g(z) = z^s e^{Q(z)} \frac{\prod_m E\left(\frac{z}{a_m}, p\right)}{\prod_n E\left(\frac{z}{b_n}, p\right)},$$

where s is an integer, $Q(z)$ is a polynomial of degree at most p, a_m are the zeros of the function $g(z)$, and b_n are its poles. From this it follows that

$$\frac{g'(z)}{g(z)} = \frac{s}{z} + Q'(z) + z^p \sum_m \frac{1}{a_m^p (z - a_m)} - z^p \sum_n \frac{1}{b_n^p (z - b_n)}.$$

Let $h(z) = (z - c_1)\cdots(z - c_p)$, where $c_1, \dots, c_p$ are pairwise distinct and different from a_m, b_n, and 0. It is easy to see that the function

$$f(z) = \frac{g'(z)}{h(z)g(z)}$$

is representable in the form (6.1), and moreover, condition (6.5) holds.

By virtue of the above assumption, the condition $\delta(0,f) = 0$ must hold. Hence, there exists a sequence $\{r_n\}_{n=1}^\infty$, $r_n \uparrow \infty$, such that

$$m(r_n, 0, f) = o(T(r_n, f)).$$

Since

$$m(r,0,f) = m\left(r, \frac{hg}{g'}\right) \geq m\left(r, \frac{g}{g'}\right) + O(\ln r),$$

$$T(r,f) = T\left(r, \frac{g'}{hg}\right) \leq T\left(r, \frac{g'}{g}\right) + O(\ln r),$$

$$\begin{aligned} T\left(r, \frac{g'}{g}\right) &= m\left(r, \frac{g'}{g}\right) + N\left(r, \frac{g'}{g}\right) \\ &= O(\ln r) + \bar{N}(r,g) + \bar{N}\left(r, \frac{1}{g}\right) \leq 2T(r,g) + O(\ln r), \end{aligned}$$

we have

$$m\left(r_n, \frac{g}{g'}\right) = o(T(r_n, g)).$$

Using Lemma 2.1 of Chapter 3 (p. 98), we get that the condition

$$m(r_n, a, g) = o(T(r_n, g))$$

holds for each $a \neq 0, \infty$; hence $\delta(a,g) = 0$.

CHAPTER 6

Value Distribution with Respect to the Arguments

As was already mentioned in Section 4 of Chapter 1, the function $N(r, a)$ characterizes only the moduli of a-points. Therefore Chapters 4 and 5 do not contain any information on the role of the arguments of a-points.

In the present chapter we show that the restrictions on the arguments of a-points can have an important impact on the asymptotic properties of a meromorphic function. In particular, restrictions of this kind can imply: (a) growth estimates; (b) regularity of growth; (c) presence of deficient values.

Note that among the theorems presented in the previous chapters it is easy to point out those theorems in which (a) or (b) follows from restrictions imposed on the moduli of a-points. For example, by Theorem 2.2 of Chapter 3, if the order of $N(r, a)$ does not exceed ρ for three different values of a, then the order of $f(z)$ does not exceed ρ. By Theorem 5.1 of Chapter 5, if the order of $N(r, a)$ is less than the order of $f(z)$ for two different values of a, then $f(z)$ has regular growth in the sense that $\rho[f] = \lambda[f]$. Finally, we note that from the very definition of the quantity $\delta(a, f)$,

$$\delta(a, f) = 1 - \limsup_{r\to\infty} \frac{N(r, a)}{T(r, f)},$$

we see that the condition $\delta(a, f) > 0$ can be regarded as a condition imposed on $N(r, a)$, that is, on the moduli of a-points.

Results related to (a) are considered in Sections 2, 4, and 5; results related to (b) are considered in Sections 1 and 2; results related to (c) are considered in Section 1. Section 3 contains proofs of results on asymptotic properties of functions meromorphic in an angle, which are used in Section 2.

The methods of Sections 2–5 make substantial use of the apparatus of angular Nevanlinna characteristics; the Tsuji characteristics play essential role in Section 5.

1. Meromorphic functions with separated poles and zeros

Let $f(z)$ be a meromorphic function with zeros a_m and poles b_n, let p be a positive integer, and let η be a number satisfying the condition $0 \le \eta < \frac{\pi}{2p}$. We set

$$D_1^p(\eta) = \bigcup_{j=0}^{p-1} \left\{ \left| \arg z - \pi\frac{2j}{p} \right| \le \eta \right\},$$

$$D_2^p(\eta) = \bigcup_{j=0}^{p-1} \left\{ \left| \arg z - \pi\frac{2j+1}{p} \right| \le \eta \right\}.$$

DEFINITION. If the condition

$$\sum_{a_m \notin D_1^p(\eta)} \frac{1}{|a_m|^p} + \sum_{b_n \notin D_2^p(\eta)} \frac{1}{|b_n|^p} < \infty \tag{1.1}$$

is satisfied, then we say that the zeros and the poles of $f(z)$ are *(p,η)-separated.*

In particular, the zeros and the poles of a function $f(z)$ are (p,η)-separated if all of its zeros are in $D_1^p(\eta)$, and all of its poles are in $D_2^p(\eta)$.

In what follows we assume, without loss of generality, that $f(0) = 1$.

THEOREM 1.1. *If the poles and the zeros of $f(z)$ are (p,η)-separated, then the limit $\lim_{r\to\infty} r^{-p}T(r,f)$, finite or infinite, exists.*

This theorem shows that the growth of $f(z)$ cannot be arbitrarily irregular. In particular, it immediately implies that if $\lambda[f] < p$, then $\rho[f] \le p$.

To prove Theorem 1.1 we need the following lemma, which will be used repeatedly in this section.

LEMMA 1.1. *If the zeros and the poles of a function $f(z)$ are (p,η)-separated, then*

$$\begin{aligned} &\frac{1}{2\pi}\int_0^{2\pi} \ln|f(re^{i\varphi})| \cos p\varphi \, d\varphi \\ &\ge \cos p\eta \int_0^r n(t;0,\infty)\cosh\left(p\ln\frac{r}{t}\right)\frac{dt}{t} + O(r^p) \\ &= \cos p\eta \left\{ N(r;0,\infty) + p\int_0^r N(t;0,\infty)\sinh\left(p\ln\frac{r}{t}\right)\frac{dt}{t}\right\} + O(r^p). \end{aligned}$$

PROOF. We start from formula (2.6) of Chapter 1. Separating the real part, we get the relation

$$\begin{aligned} &\frac{1}{2\pi}\int_0^{2\pi} \ln|f(re^{i\varphi})| \cos p\varphi \, d\varphi \\ &\quad = \frac{1}{2p}\sum_{|a_m|<r}\left(\frac{r^p}{|a_m|^p} - \frac{|a_m|^p}{r^p}\right)\cos p\alpha_m \\ &\quad - \frac{1}{2p}\sum_{|b_n|<r}\left(\frac{r^p}{|b_n|^p} - \frac{|b_n|^p}{r^p}\right)\cos p\beta_n + r^p \operatorname{Re}\frac{d^p}{dz^p}\ln f(z)|_{z=0}, \end{aligned}$$

where $\alpha_m = \arg a_m$, $\beta_n = \arg b_n$. This relation can be written in the form

$$\begin{aligned} \frac{1}{2\pi}\int_0^{2\pi} \ln|f(re^{i\varphi})| \cos p\varphi \, d\varphi &= \frac{1}{p}\sum_{|a_m|<r}\sinh\left(p\ln\frac{r}{|a_m|}\right)\cos p\alpha_m \\ &- \frac{1}{p}\sum_{|b_n|<r}\sinh\left(p\ln\frac{r}{|b_n|}\right)\cos p\beta_n + r^p\operatorname{Re}\frac{d^p}{dz^p}\ln f(z)|_{z=0}. \end{aligned} \tag{1.2}$$

Since for $a_m \in D_1^p(\eta)$ the inequality $\cos p\alpha_m \geq \cos p\eta > 0$ holds, we have

$$\begin{aligned}
&\sum_{|a_m|<r} \sinh\left(p\ln\frac{r}{|a_m|}\right)\cos p\alpha_m \\
&\quad = \sum_{\substack{|a_m|<r\\ a_m\in D_1^p(\eta)}} \sinh\left(p\ln\frac{r}{|a_m|}\right)\cos p\alpha_m + \sum_{\substack{|a_m|<r\\ a_m\notin D_1^p(\eta)}} \sinh\left(p\ln\frac{r}{|a_m|}\right)\cos p\alpha_m \\
&\quad \geq \cos p\eta \sum_{\substack{|a_m|<r\\ a_m\in D_1^p(\eta)}} \sinh\left(p\ln\frac{r}{|a_m|}\right) - \sum_{\substack{|a_m|<r\\ a_m\notin D_1^p(\eta)}} \sinh\left(p\ln\frac{r}{|a_m|}\right) \\
&\quad = \cos p\eta \sum_{|a_m|<r} \sinh\left(p\ln\frac{r}{|a_m|}\right) - (1+\cos p\eta) \sum_{\substack{|a_m|<r\\ a_m\notin D_1^p(\eta)}} \sinh\left(p\ln\frac{r}{|a_m|}\right).
\end{aligned}$$

By the inequality $\sinh x \leq \frac{1}{2}e^x$ and condition (1.1) we get

$$\sum_{\substack{|a_m|<r\\ a_m\notin D_1^p(\eta)}} \sinh\left(p\ln\frac{r}{|a_m|}\right) \leq \frac{1}{2} \sum_{\substack{|a_m|<r\\ a_m\notin D_1^p(\eta)}} \left(\frac{r}{|a_m|}\right)^p \leq \frac{1}{2}r^p \sum_{a_m\notin D_1^p(\eta)} \frac{1}{|a_m|^p} = O(r^p);$$

therefore

$$\begin{aligned}
&\sum_{|a_m|<r} \sinh\left(p\ln\frac{r}{|a_m|}\right)\cos p\alpha_m \geq \cos p\eta \sum_{|a_m|<r} \sinh\left(p\ln\frac{r}{|a_m|}\right) + O(r^p) \\
&\qquad = \cos p\eta \int_0^r \sinh\left(p\ln\frac{r}{t}\right) dn(t,0) + O(r^p) \\
&\qquad = p\cos p\eta \int_0^r n(t,0)\cosh\left(p\ln\frac{r}{t}\right)\frac{dt}{t} + O(r^p) \qquad (1.3)\\
&\qquad = p\cos p\eta \left\{N(r,0) + p\int_0^r N(t,0)\sinh\left(p\ln\frac{r}{t}\right)\frac{dt}{t}\right\} + O(r^p).
\end{aligned}$$

Using a similar argument, we conclude that

$$\begin{aligned}
&-\sum_{|b_n|<r} \sinh\left(p\ln\frac{r}{|b_n|}\right)\cos p\beta_n \\
&\qquad \geq p\cos p\eta \int_0^r n(t,\infty)\cosh\left(p\ln\frac{r}{t}\right)\frac{dt}{t} + O(r^p) \qquad (1.4)\\
&\qquad = p\cos p\eta \left\{N(r,\infty) + p\int_0^r N(t,\infty)\sinh\left(p\ln\frac{r}{t}\right)\frac{dt}{t}\right\} + O(r^p).
\end{aligned}$$

Relations (1.2), (1.3), and (1.4) imply the conclusion of the lemma. □

PROOF OF THEOREM 1.1. Obviously it is sufficient to consider the case where $\liminf_{r\to\infty} r^{-p}T(r,f) < \infty$. Since

$$\frac{1}{2\pi}\int_0^{2\pi} \ln|f(re^{i\varphi})|\cos p\varphi\, d\varphi \leq m(r,f) + m\left(r,\frac{1}{f}\right) \leq 2T(r,f), \qquad (1.5)$$

by Lemma 1.1 we get

$$2T(r,f) \geq \cos p\eta \int_0^r n(t;0,\infty)\cosh\left(p\ln\frac{r}{t}\right)\frac{dt}{t} + O(r^p).$$

Since $\cosh x \geq \frac{1}{2}e^x$, it follows that

$$2T(r,f) \geq \frac{1}{2}\cos p\eta \cdot r^p \int_0^r n(t;0,\infty)t^{-p-1}dt + O(r^p).$$

Dividing both sides of this inequality by r^p and letting r tend to ∞ along a sequence $r_k \uparrow \infty$ such that $T(r_k, f) = O(r_k^p)$, we come to the conclusion that the integral

$$\int_0^\infty n(t;0,\infty)t^{-p-1}dt$$

converges.

Convergence of this integral implies, by Theorem 1.8 of Chapter 2, convergence of the series

$$\sum_m \frac{1}{|a_m|^p} \quad \text{and} \quad \sum_n \frac{1}{|b_n|^p};$$

therefore (see Theorem 3.2 of Chapter 2 and Example 1 on p. 55) the canonical products

$$\pi_1(z) = \prod_m E\left(\frac{z}{a_m}, p-1\right), \qquad \pi_2(z) = \prod_n E\left(\frac{z}{b_n}, p-1\right)$$

converge. The function

$$F(z) = f(z)\frac{\pi_2(z)}{\pi_1(z)}$$

is an entire function without zeros. Since

$$T(r,F) \leq T(r,f) + T(r,\pi_1) + T(r,\pi_2)$$

and, by Remark 2 to Theorem 3.3 of Chapter 2, $T(r,\pi_j) = o(r^p)$, $j = 1,2$, we have $\liminf_{r\to\infty} r^{-p}T(r,F) < \infty$. By Lemma 6.2 of Chapter 1 we get a representation $F(z) = e^{P(z)}$, where $P(z) = c_p z^p + \cdots + c_0$ is a polynomial of degree at most p. Thus

$$f(z) = e^{P(z)}\frac{\pi_1(z)}{\pi_2(z)},$$

from which

$$T(r,f) = T(r,e^{P(z)}) + o(r^p) = \frac{|c_p|}{\pi}r^p + o(r^p),$$

and hence the limit $\lim_{r\to\infty} r^{-p}T(r,f)$ exists. The theorem has been proved. □

Let us mention the following fact, established in our proof of Theorem 1.1.

REMARK. If the zeros and the poles of a function $f(z)$ are (p,η)-separated and

$$\liminf_{r\to\infty} r^{-p}T(r,f) < \infty,$$

then

$$\int_0^\infty \frac{n(t;0,\infty)}{t^{p+1}}dt < \infty, \qquad \sum_m \frac{1}{|a_m|^p} + \sum_n \frac{1}{|b_n|^p} < \infty.$$

Now we turn to the corollaries of Theorem 1.1.

COROLLARY 1. *Let $f(z)$ be an entire function of a finite lower order λ with positive zeros. Then the growth of $f(z)$ does not exceed the minimal type of order $[\lambda]+1$. If λ is a positive integer, then there is a finite or infinite limit $\lim_{r\to\infty} r^{-\lambda}T(r,f)$. In either case, the numbers $\rho[f]$ and $\lambda[f]$ belong to the same interval $[k-1,k]$, where k is a positive integer.*

In fact, the zeros and poles (the latter are absent) of $f(z)$ are $(p,0)$-separated for each positive integer p. For $p=[\lambda]+1$ we have $\liminf_{r\to\infty} r^{-p}T(r,f)=0$. By Theorem 1.1 we then also have $\lim_{r\to\infty} r^{-p}T(r,f)=0$; that is, the growth of $f(z)$ does not exceed the minimal type of order $p=[\lambda]+1$. If λ is a positive integer, we can apply Theorem 1.1 with $p=\lambda$.

The statement of Corollary 1 is sharp in the sense that for each ρ and λ ($\lambda\le\rho$) in the same interval $[k-1,k]$, where k is a positive integer, there exist entire functions with positive zeros of order ρ and lower order λ. For $k-1<\lambda\le\rho<k$, the canonical products considered in Section 5 of Chapter 2, p. 67 (we let $\eta_1=\lambda$, $\eta_2=\rho$ in their construction), are such functions. Now we construct an example of an entire function with positive zeros and $\rho=k$, $\lambda=k-1$, $k=1,2,\dots$.

EXAMPLE 1. Let $a_1=e$, $a_{n+1}=\exp\exp a_n$ $(n=1,2,\dots)$, $q_n=a_n^k(\ln a_n)^{-2}$. Since the series

$$\sum_{n=1}^{\infty} q_n a_n^{-k}$$

converges, the canonical product

$$f_k(z)=\prod_{n=1}^{\infty}\left\{E\left(\frac{z}{a_n},\ k-1\right)\right\}^{q_n}$$

converges by Theorem 3.2 of Chapter 2. Since

$$\sum_{n=1}^{m}\frac{a_n^k}{(\ln a_n)^2}=\frac{a_m^k}{(\ln a_m)^2}(1+o(1)),\qquad m\to\infty,$$

we have

$$n(r,0,f_k)=\frac{r^k}{(\ln r)^2}(1+o(1)),\qquad r\to\infty,$$

for $r=a_m$ and

$$n(r,0,f_k)\le\frac{r^k}{(\ln r)^2}(1+o(1)),\qquad r\to\infty,$$

for all $r>0$. Thus, the order of $n(r,0,f_k)$ is equal to k. By Theorem 3.4 of Chapter 2, the order of $f_k(z)$ is also equal to k; moreover, in view of Remark 2 to Theorem 3.3 of Chapter 2,

$$\ln M(r,f_k)=o(r^k). \tag{1.6}$$

Next, by estimate (3.6) in Chapter 2 (p. 56) we have

$$\ln M(r,f_k)\le C(k-1)\left\{r^{k-1}\int_0^r\frac{n(t,0,f_k)}{t^k}dt+r^k\int_r^{\infty}\frac{n(t,0,f_k)}{t^{k+1}}dt\right\}.$$

For $r = e^{a_m} = \ln a_{m+1}$ we have

$$\int_0^r \frac{n(t,0,f_k)}{t^k}\,dt \le n(a_m,0,f_k)\int_1^r \frac{dt}{t^k} \le \frac{n(a_m,0,f_k)}{k-1}$$
$$= \frac{1}{k-1}\frac{a_m^k}{(\ln a_m)^2}(1+o(1)) = O(\ln^k r) \quad \text{if } k>1,$$

$$\int_0^r \frac{n(t,0,f_1)}{t}\,dt \le n(a_m,0,f_1)\int_1^r \frac{dt}{t} = n(a_m,0,f_1)\ln r$$
$$= \frac{a_m}{(\ln a_m)^2}(1+o(1))\ln r = O(\ln^2 r),$$

$$\int_r^\infty \frac{n(t,0,f_k)}{t^{k+1}}\,dt = n(a_m,0,f_k)\int_r^{a_{m+1}} \frac{dt}{t^{k+1}} + \int_{a_{m+1}}^\infty \frac{n(t,0,f_k)}{t^{k+1}}\,dt$$
$$\le \frac{n(a_m,0,f_k)}{kr^k} + 2\int_{a_{m+1}}^\infty \frac{dt}{t(\ln t)^2}$$
$$= \frac{1}{kr^k}\frac{a_m^k}{(\ln a_m)^2}(1+o(1)) + \frac{2}{\ln a_{m+1}} = O\left(\frac{\ln^k r}{r^k}\right) + O\left(\frac{1}{r}\right).$$

Thus, for $r = e^{a_m} = \ln a_{m+1}$ we have

$$\ln M(r,f_k) = O(r^{k-1}\ln^k r), \quad k>1; \qquad \ln M(r,f_1) = O(\ln^2 r). \tag{1.7}$$

Hence, the lower order of the function $f_k(z)$ does not exceed $k-1$. Since the order of $f_k(z)$ is equal to k, it follows from Theorem 1.1 that the lower order of $f_k(z)$ cannot be less than $k-1$.

We leave the construction of an example of an entire function with positive zeros and $k-1=\lambda\le\rho<k$ or $k-1<\lambda\le\rho=k$ $(k=1,2,\dots)$ as an exercise.

Corollary 1 and Theorems 6.5 and 6.7 of Chapter 4 immediately imply the following result.

COROLLARY 2. *Deficiencies of entire functions of finite lower order with positive zeros are shift-invariant.*

Using the observation that poles and zeros of a function $f(z)$ with positive zeros and negative poles are $(p,0)$-separated for each odd p, we get from Theorem 1.1 the following corollary.

COROLLARY 3. *Let $f(z)$ be a meromorphic function of finite lower order λ, with positive zeros and negative poles. Then the growth of the function $f(z)$ does not exceed the minimal type of order $2[\frac{\lambda+1}{2}]+1$. If $\lambda\ge 1$ is an odd number, then the limit $\lim_{r\to\infty} r^{-\lambda}T(r,f)$ exists. In any case, the numbers $\rho[f]$ and $\lambda[f]$ belong to the same interval of the form $[2k-3, 2k-1]$, where k is a positive integer.*

The following corollary of Theorem 1.1 is also of interest.

COROLLARY 4. *Let $f(z)$ be an entire function of finite lower order λ, and its zeros a_m satisfy the condition*

$$\sum_m \left|\operatorname{Im}\frac{1}{a_m}\right| < \infty. \tag{1.8}$$

Then the growth of $f(z)$ does not exceed the minimal type of order $2[\frac{\lambda}{2}]+2$. If $\lambda \geq 2$ is an even number, then the limit $\lim_{r\to\infty} r^{-\lambda}T(r,f)$ exists. In any case, the numbers $\rho[f]$ and $\lambda[f]$ belong to the same interval of the form $[2k-2, 2k]$, where k is a positive integer.

To prove this corollary it suffices to observe that, by condition (1.8), the zeros and poles of the function $f(z)$ are $(p, \frac{\pi}{4p})$-separated for each even $p \geq 2$.

Now we show that the functions with sufficiently large growth that have separated zeros and poles, have "few" zeros and poles in the sense that the number

$$\kappa(f) = \limsup_{r\to\infty} \frac{N(r;0,\infty)}{T(r,f)}$$

is strictly less than 2.

THEOREM 1.2. *Let the zeros and poles of a function $f(z)$ be (p,η)-separated.*

(a) *If $0 < \limsup_{r\to\infty} r^{-p}T(r,f) < \infty$, then $\kappa(f) = 0$.*

(b) *If $\limsup_{r\to\infty} r^{-p}T(r,f) = \infty$, then*

$$\kappa(f) \leq \frac{2}{1+\cos p\eta}. \tag{1.9}$$

PROOF. In case (a), by Theorem 1.1 we have $\lim_{r\to\infty} r^{-p}T(r,f) > 0$, that is, $r^p = O(T(r,f))$. By the Remark to Theorem 1.1 on p. 262, the relation $n(r;0,\infty) = o(r^p)$ holds, and hence $N(r;0,\infty) = o(r^p)$. Therefore $\kappa(f) = 0$.

In case (b), by Theorem 1.1, we have $\lim_{r\to\infty} r^{-p}T(r,f) = \infty$, that is, $r^p = o(T(r,f))$. Therefore, from Lemma 1.1 and relation (1.5) it follows that

$$m(r,f) + m\left(r, \frac{1}{f}\right) \geq \cos p\eta N(r;0,\infty) + o(T(r,f)).$$

Adding $N(r;0,\infty)$ to both sides, we get

$$2T(r,f) \geq (1+\cos p\eta)N(r;0,\infty) + o(T(r,f)).$$

Dividing both sides by $T(r,f)$ and letting r tend to ∞, we get the inequality (1.9). □

Let us show that the parameter $\kappa(f)$ can be equal to 2 for a function with $(p,0)$-separated zeros and poles, and such that $\lim_{r\to\infty} r^{-p}T(r,f) = 0$.

EXAMPLE 2. For $p = 1$ the function

$$f(z) = f_1(z)/f_1(-z),$$

where $f_k(z)$ is the function in Example 1, is a desired example. In fact, the poles and zeros of the function $f(z)$ are $(1,0)$-separated. Since

$$T(r,f) \leq 2T(r,f_1) \leq 2\ln^+ M(r,f_1), \tag{1.10}$$

relation (1.6) implies that $\lim_{r\to\infty} r^{-1}T(r,f) = 0$. Now we show that $\kappa(f) = 2$. From (1.7) and (1.10) we get that

$$T(r,f) = O(\ln^2 r)$$

for $r = e^{a_m}$, and thus the function $f(z)$ has lower order zero. By the Corollary of Theorem 3.1 of Chapter 5 (p. 201), the function $f(z)$ cannot have more than one deficient value. Since, obviously, $N(r,f) = N(r,\frac{1}{f})$, we have $\delta(0,f) = \delta(\infty,f)$ and hence $\delta(0,f) = \delta(\infty,f) = 0$. From this we get

$$\kappa(f) = \limsup_{r\to\infty} \frac{N(r,f) + N(r,\frac{1}{f})}{T(r,f)} = 2\limsup_{r\to\infty} \frac{N(r,f)}{T(r,f)} = 2(1-\delta(\infty,f)) = 2.$$

To get an example for $p > 1$ we consider $f(z^p)$.

As we already mentioned in Chapter 5, the inequality $\kappa(f) \le 2 - \delta(0,f) - \delta(\infty,f)$ holds. The question arises: is it possible to improve Theorem 1.2 supplementing it with the statement that $2 - \delta(0,f) - \delta(\infty,f) < 2$, that is, $\delta(0,f) + \delta(\infty,f) > 0$. In case (a) such an improvement is possible: the relation $\kappa(f) = 0$ and the obvious inequality

$$\min\{\delta(0,f), \delta(\infty,f)\} \ge 1 - \kappa(f) \tag{1.11}$$

imply that $\delta(0,f) = \delta(\infty,f) = 1$. We show that in case (b) it is possible that $\delta(0,f) = \delta(\infty,f) = 0$ for each value of $\eta > 0$.

EXAMPLE 3. Let a positive integer p and a number η, $0 < \eta < \frac{\pi}{2p}$, be given. Let ρ be a non-integer satisfying $\rho \ge \frac{\pi}{2\eta}$. We will construct a meromorphic function $f(z)$ of order ρ with zeros on the rays $\{\arg z = 0\}$ and $\{\arg z = \frac{\pi}{\rho}\}$, and poles on the rays $\{\arg z = \frac{\pi}{p}\}$ and $\{\arg z = \frac{\pi}{p} + \frac{\pi}{\rho}\}$, and such that $\delta(0,f) = \delta(\infty,f) = 0$. Since $\frac{\pi}{\rho} \le 2\eta$, the zeros and poles of the function $f(z)$ are (p,η)-separated, and since $\rho \ge \frac{\pi}{2\eta} > \rho$, we have $\limsup_{r\to\infty} r^{-p}T(r,f) = \infty$. Thus, the desired example will be constructed.

Let $r_1 = 1$, $r_{m+1} = \exp\exp r_m$ $(m = 1,2,\dots)$. Denote by $g_1(z)$ $(g_2(z))$ the canonical product of genus $q = [\rho]$ with zeros at the points $k^{1/\rho}$, $k = 1,2,\dots$, in the set $\bigcup_{j=1}^{\infty}[r_{2j-1}, r_{2j})$ $\left(\bigcup_{j=1}^{\infty}[r_{2j}, r_{2j+1})\right)$. Let us show that the function

$$f(z) = \frac{g_1(z)g_1(ze^{-i\alpha})}{g_2(ze^{-i\beta})g_2(ze^{-i\gamma})}, \qquad \alpha = \frac{\pi}{\rho}, \quad \beta = \frac{\pi}{p}, \quad \gamma = \frac{\pi}{\rho} + \frac{\pi}{p},$$

has all the desired properties.

It is clear that its poles and zeros belong to the indicated rays and that the order of $f(z)$ does not exceed ρ.

Since

$$n(r,0,g_1) + n(r,0,g_2) = [r^\rho],$$

we have

$$n(r;0,\infty) = n(r,f) + n\left(r,\frac{1}{f}\right) = 2[r^\rho]$$

and

$$N(r;0,\infty) \sim \frac{2}{\rho} r^\rho.$$

Therefore

$$T(r,f) \ge \left(\frac{1}{\rho} + o(1)\right) r^\rho, \tag{1.12}$$

and the order of the function $f(z)$ is equal to ρ.

Let us prove that $\delta(0,f) = \delta(\infty,f) = 0$.

Using the estimate

$$\ln M(r,g_1) \le C(q)\left\{r^q \int_0^r \frac{n(t,0,g_1)}{t^{q+1}}dt + r^{q+1}\int_r^\infty \frac{n(t,0,g_1)}{t^{q+2}}dt\right\},$$

which is valid by Remark 1 to Theorem 3.3 of Chapter 2, and taking into account the fact that for $r = e^{r_{2j}} = \ln r_{2j+1}$ the inequalities

$$\int_0^r \frac{n(t,0,g_1)}{t^{q+1}}dt \le \frac{n(r_{2j},0,g_1)}{q} = O(r_{2j}^\rho) = O(\ln^\rho r),$$

$$\begin{aligned}\int_r^\infty \frac{n(t,0,g_1)}{t^{q+2}}dt &\le \frac{n(r_{2j},0,g_1)}{(q+1)r^{q+1}} + \int_{r_{2j+1}}^\infty \frac{n(t,0,g_1)}{t^{q+2}}dt \\ &= O\left(\frac{\ln^\rho r}{r^{q+1}}\right) + \int_{r_{2j+1}}^\infty \frac{O(t^\rho)}{t^{q+2}}dt = O\left(\frac{\ln^\rho r}{r^{q+1}}\right) + O(r_{2j+1}^{\rho-q-1}) \\ &= O\left(\frac{\ln^\rho r}{r^{q+1}}\right) + O(e^{r(\rho-q-1)}) = O\left(\frac{\ln^\rho r}{r^{q+1}}\right)\end{aligned}$$

hold, we get

$$\ln M(r,g_1) = O(r^q \ln^\rho r) + O(\ln^\rho r) = o(r^\rho), \qquad r = e^{r_{2j}}. \tag{1.13}$$

In a similar manner, we get

$$\ln M(r,g_2) = o(r^\rho), \qquad r = e^{r_{2j+1}}. \tag{1.14}$$

The function

$$g(z) = g_1(z)g_2(z)$$

is a canonical product with zeros at the points $k^{1/\rho}$, $k = 1,2,\dots$, and by the asymptotic formula (5.3_2) in Chapter 2, for each $\delta > 0$ the relation

$$\ln|g(re^{i\varphi})| = \frac{\pi}{\sin \pi\rho}\cos\rho(\varphi-\pi)r^\rho + o(r^\rho)$$

holds uniformly in φ, $\delta \le \varphi \le 2\pi - \delta$. Therefore the function

$$G(z) = g(z)g(zc^{-i\alpha}), \qquad \alpha = \frac{\pi}{\rho},$$

for each δ, $0 < \delta < \frac{\pi}{2\rho}$, uniformly in φ, satisfies the asymptotic equalities

$$\ln|G(re^{i\varphi})| = 2\pi \sin\rho\varphi r^\rho + o(r^\rho), \qquad \delta \le \varphi \le \alpha - \delta,$$

$$\ln|G(re^{i\varphi})| = o(r^\rho), \qquad \alpha + \delta \le \varphi \le 2\pi - \delta.$$

Applying Theorem 7.4 of Chapter 1, we get

$$m\left(r, \frac{1}{G}\right) = \left\{\frac{1}{2\pi}\int_0^\alpha (-2\pi\sin\rho\varphi)^+ d\varphi\right\} r^\rho + o(r^\rho) = o(r^\rho). \tag{1.15}$$

The function $f(z)$ can be represented in the form

$$f(z) = \frac{G(z)}{g_2(ze^{-i\beta})g_2(ze^{-i\gamma})g_2(z)g_2(ze^{-i\alpha})},$$

from which we get

$$m\left(r,\frac{1}{f}\right) \le m\left(r,\frac{1}{G}\right) + 4m(r,g_2) \le m\left(r,\frac{1}{G}\right) + 4\ln^+ M(r,g_2).$$

Taking into account (1.15) and (1.14) we obtain

$$m\left(r,\frac{1}{f}\right) = o(r^\rho), \qquad r = e^{r_{2j+1}},$$

and applying (1.12), we conclude that $\delta(0,f) = 0$.

Similarly, representing the function $f(z)$ in the form

$$f(z) = \frac{g_1(z)g_1(ze^{-i\alpha})g_1(ze^{-i\beta})g_1(ze^{-i\gamma})}{G(z)}$$

and using (1.15), (1.13), and (1.12), we conclude that $\delta(\infty, f) = 0$.

The question whether it is possible that $\delta(0, f) = \delta(\infty, f) = 0$ holds under assumption (b) of Theorem 1.2 with $\eta = 0$, is still unanswered in full generality. Under some additional conditions, the answer to this question turns out to be negative.

The following two theorems are fairly easy to prove.

THEOREM 1.3. *Suppose that assumption* (b) *of Theorem* 1.2 *is satisfied with* $\eta = 0$, *and suppose, in addition, that the limits*

$$q_1 = \liminf_{r\to\infty} \frac{N(r, f)}{N\left(r, \frac{1}{f}\right)} \quad \text{and} \quad q_2 = \limsup_{r\to\infty} \frac{N(r, f)}{N\left(r, \frac{1}{f}\right)}$$

are finite and positive. Then

$$\delta(0, f) \geq \frac{q_1}{1 + q_1} > 0, \qquad \delta(\infty, f) \geq \frac{1}{1 + q_2} > 0. \tag{1.16}$$

THEOREM 1.4. *Suppose that assumption* (b) *of Theorem* 1.2 *is satisfied with* $\eta = 0$, *and suppose that, in addition, there exist numbers* $\mu > 1$ *and* $1 > \nu > 0$ *such that*

$$\frac{N(r; 0, \infty)}{N(\mu r; 0, \infty)} \geq \nu, \qquad r \geq r_0. \tag{1.17}$$

Then

$$\min\{\delta(0, f), \delta(\infty, f)\} \geq \frac{\nu(\cosh(p \ln \mu) - 1)}{2 + \nu(\cosh(p \ln \mu) - 1)} > 0. \tag{1.18}$$

PROOF OF THEOREM 1.3. We have

$$\begin{aligned}\kappa(f) &= \limsup_{r\to\infty} \frac{N(r; 0, \infty)}{T(r, f)} \geq \limsup_{r\to\infty} \frac{N\left(r, \frac{1}{f}\right)}{T(r, f)} \liminf_{r\to\infty} \left(1 + \frac{N(r, f)}{N\left(r, \frac{1}{f}\right)}\right) \\ &= (1 - \delta(0, f))(1 + q_1),\end{aligned}$$

$$\kappa(f) \geq \limsup_{r\to\infty} \frac{N(r, f)}{T(r, f)} \liminf_{r\to\infty} \left(1 + \frac{N\left(r, \frac{1}{f}\right)}{N(r, f)}\right) = (1 - \delta(\infty, f)) \left(1 + \frac{1}{q_2}\right).$$

Since inequality (1.9) with $\eta = 0$ gives us $\kappa(f) \leq 1$, we get inequalities (1.16). □

PROOF OF THEOREM 1.4. First we establish that for functions satisfying assumption (b) of Theorem 1.2 with $\eta \geq 0$ and condition (1.17), the following estimate holds:

$$\kappa(f) \leq \frac{2}{1 + \cos p\eta\{1 - \nu + \nu \cosh(p \ln \mu)\}}, \tag{1.19}$$

which is somewhat sharper than (1.9). Using Lemma 1.1 and inequality (1.5), we get

$$2T(r,f) \geq (1+\cos p\eta)N(r;0,\infty) + p\cos p\eta \int_0^r N(t;0,\infty)\sinh\left(p\ln\frac{r}{t}\right)\frac{dt}{t} + o(T(r,f)).$$

By condition (1.17) we have

$$p\int_0^r N(t;0,\infty)\sinh\left(p\ln\frac{r}{t}\right)\frac{dt}{t} \geq p\int_{r/\mu}^r N(t;0,\infty)\sinh\left(p\ln\frac{r}{t}\right)\frac{dt}{t}$$
$$\geq N\left(\frac{r}{\mu};0,\infty\right)p\int_{r/\mu}^r \sinh\left(p\ln\frac{r}{t}\right)\frac{dt}{t} = N\left(\frac{r}{\mu};0,\infty\right)\{\cosh(p\ln\mu)-1\}$$
$$\geq N(r;0,\infty)\nu\{\cosh(p\ln\mu)-1\}.$$

We come to the relation

$$2T(r,f) \geq \{1+\cos p\eta(1-\nu+\nu\cosh(p\ln\mu))\}N(r;0,\infty) + o(T(r,f)),$$

whence (1.19) follows.

From (1.19) with $\eta = 0$ we get

$$\kappa(f) \leq \frac{2}{2+\nu(\cosh(p\ln\mu)-1)} < 1$$

and, using (1.11), obtain (1.18). □

The proof of the following theorem is much more difficult.

THEOREM 1.5. *If assumption* (b) *of Theorem* 1.2 *is satisfied and, in addition, there exists a number* $\xi < \infty$ *such that*

$$\sum_m \frac{1}{|a_m|^\xi} + \sum_n \frac{1}{|b_n|^\xi} < \infty, \tag{1.20}$$
$$\eta \leq \frac{\pi}{60\xi},$$

then

$$\min\{\delta(0,f),\delta(\infty,f)\} \geq A > 0, \tag{1.21}$$

where A *is an absolute constant.*

Let us mention some of the corollaries of Theorem 1.5.

COROLLARY 1. *If assumption* (b) *of Theorem* 1.2 *is satisfied with* $\eta = 0$ *and, in addition, there exists a number* $\xi < \infty$ *such that* (1.20) *is true, then* (1.21) *holds.*

Observe that (1.20) is always true for $\xi > \rho[f]$. Therefore Corollary 1 immediately implies the following.

COROLLARY 2. *If assumption* (b) *of Theorem* 1.2 *is satisfied with* $\eta = 0$ *and, in addition, the function* $f(z)$ *has finite order, then* (1.21) *holds.*

Let us mention, also, the following statements.

COROLLARY 3. *Let* $f(z)$ *be a function with positive zeros and negative poles. If* $f(z)$ *has finite lower order and* $\limsup_{r\to\infty} r^{-1}T(r,f) = \infty$, *then* (1.21) *holds.*

COROLLARY 4. *Let $f(z)$ be an entire function with real zeros. If $f(z)$ has a finite lower order and* $\limsup_{r\to\infty} r^{-2}T(r,f)=\infty$, *then* (1.21) *holds.*

Corollary 3 follows from Corollary 2, since, by Corollary 3 of Theorem 1.1, the function $f(z)$ has finite order. Corollary 4 can be derived from Corollary 2 using Corollary 4 of Theorem 1.1.

Now we turn to auxiliary results needed to prove Theorem 1.5.

LEMMA 1.2. *Let $\{c_k\}_{k=1}^{\infty}$ be a sequence of points in the complex plane that lie in the angle $\{|\arg z|\le \frac{\pi}{6}-\varepsilon\}$, where $0<\varepsilon\le\frac{\pi}{6}$, and satisfy the condition*

$$\sum_{k=1}^{\infty}\frac{1}{|c_k|^3}<\infty.$$

Let[1]

$$g(z)=\prod_{k=1}^{\infty}E\left(\frac{z}{c_k},2\right).$$

Then

$$\int_{\frac{\pi}{3}-\frac{\varepsilon}{2}}^{\frac{\pi}{3}+\frac{\varepsilon}{2}}\ln^+\left|\frac{g(re^{i\varphi})}{g(-re^{i\varphi})}\right|d\varphi\ge\varepsilon\sin\frac{\varepsilon}{2}N(r,0,g).$$

PROOF. It is easy to see that for $\operatorname{Im}\zeta>0$ the equality

$$\ln\frac{E(\zeta,2)}{E(-\zeta,2)}=\ln\left(\frac{1-\zeta}{1+\zeta}\right)+2\zeta=2\int_0^{\zeta}\frac{t^2\,dt}{t^2-1}$$

holds, where the branches on the logarithm in both cases are chosen in such a way that they have zero value at $\zeta=0$, and the integral is taken over the segment joining the points 0 and ζ. Separating the real part, we get

$$\ln\left|\frac{E(\rho e^{i\theta},2)}{E(-\rho e^{i\theta},2)}\right|=2\operatorname{Re}\left\{\int_0^{\rho}\frac{\tau^2e^{2i\theta}}{\tau^2e^{2i\theta}-1}e^{i\theta}d\tau\right\}=2\int_0^{\rho}\frac{\tau^4\cos\theta-\tau^2\cos3\theta}{\tau^4-2\tau^2\cos2\theta+1}d\tau,$$

whence

$$\ln\left|\frac{E\left(\frac{re^{i\varphi}}{c_k},2\right)}{E\left(-\frac{re^{i\varphi}}{c_k},2\right)}\right|=2\int_0^{r/|c_k|}\frac{\tau^4\cos(\varphi-\varphi_k)-\tau^2\cos3(\varphi-\varphi_k)}{\tau^4-2\tau^2\cos2(\varphi-\varphi_k)+1}d\tau,\tag{1.22}$$

where $\varphi_k=\arg c_k$.

Since $|\varphi_k|\le\frac{\pi}{6}-\varepsilon$, for $|\varphi-\frac{\pi}{3}|\le\frac{\varepsilon}{2}$ we have

$$\frac{\pi}{6}+\frac{\varepsilon}{2}\le\varphi-\varphi_k\le\frac{\pi}{2}-\frac{\varepsilon}{2},$$

hence

$$\cos(\varphi-\varphi_k)\ge\cos\left(\frac{\pi}{2}-\frac{\varepsilon}{2}\right)=\sin\frac{\varepsilon}{2},$$
$$\cos3(\varphi-\varphi_k)\le\cos3\left(\frac{\pi}{6}+\frac{\varepsilon}{2}\right)=-\sin\frac{3\varepsilon}{2}.$$

Therefore (1.22) implies that for $|\varphi-\frac{\pi}{3}|\le\frac{\varepsilon}{2}$ the inequality

$$\ln\left|\frac{E\left(\frac{re^{i\varphi}}{c_k},2\right)}{E\left(-\frac{re^{i\varphi}}{c_k},2\right)}\right|\ge2\int_0^{r/|c_k|}\frac{\tau^4\sin\frac{\varepsilon}{2}+\tau^2\sin\frac{3\varepsilon}{2}}{\tau^4+2\tau^2+1}d\tau\ge2\sin\frac{\varepsilon}{2}\int_0^{r/|c_k|}\frac{\tau^2\,d\tau}{\tau^2+1}$$

[1] By Theorem 3.2 of Chapter 2, the product converges absolutely and uniformly in each finite disc.

holds. Summing these inequalities for different values of k, we get ($|\varphi - \frac{\pi}{3}| \le \frac{\varepsilon}{2}$)

$$\begin{aligned}\ln\left|\frac{g(re^{i\varphi})}{g(-re^{i\varphi})}\right| &\ge 2\sin\frac{\varepsilon}{2}\sum_{k=1}^{\infty}\int_0^{r/|c_k|}\frac{\tau^2 d\tau}{\tau^2+1}\\ &= 2\sin\frac{\varepsilon}{2}\int_0^{\infty}dn(t,0,g)\int_0^{r/t}\frac{\tau^2 d\tau}{\tau^2+1} = 2r^3\sin\frac{\varepsilon}{2}\int_0^{\infty}\frac{n(t,0,g)}{t^2(r^2+t^2)}dt\\ &\ge 2r^3\sin\frac{\varepsilon}{2}\int_0^{r}\frac{n(t,0,g)}{t^2(r^2+t^2)}dt \ge \sin\frac{\varepsilon}{2}\int_0^{r}\frac{n(t,0,g)}{t}dt = \sin\frac{\varepsilon}{2}N(r,0,g).\end{aligned}$$

Integrating with respect φ from $\frac{\pi}{3}-\frac{\varepsilon}{2}$ to $\frac{\pi}{3}+\frac{\varepsilon}{2}$, we get the desired inequality. □

LEMMA 1.3. *Let $G(z)$, $G(0)=1$, be a meromorphic function with all zeros in the angle $\{|\arg z| \le \frac{\pi}{60}\}$ and all poles in the angle $\{|\pi - \arg z| \le \frac{\pi}{60}\}$, and let*

$$\lim_{r\to\infty} r^{-1}T(r,G) = \infty. \tag{1.23}$$

Suppose, in addition, that $G(z)$ is of the form

$$G(z) = e^{S(z)}G_1(z), \tag{1.24}$$

where $G_1(z)$ is a meromorphic function of genus at most 2, and $S(z)$ is an entire function.

Then

$$\min\{\delta(0,G),\delta(\infty,G)\} \ge A > 0, \tag{1.25}$$

where A is an absolute constant.

PROOF. First we consider the case where the genus of the function $G(z)$ itself does not exceed 2. Let

$$g(z) = \prod_m E\left(\frac{z}{a_m},2\right)\prod_n E\left(-\frac{z}{b_n},2\right).$$

Then

$$\frac{G(z)}{G(-z)} = e^{Cz}\frac{g(z)}{g(-z)},$$

where C is some constant. Applying Lemma 1.2 with $\varepsilon = \frac{9}{10}\cdot\frac{\pi}{6}$, we get

$$\begin{aligned}\int_{\frac{\pi}{3}-\frac{\varepsilon}{2}}^{\frac{\pi}{3}+\frac{\varepsilon}{2}}\ln^+\left|\frac{G(re^{i\varphi})}{G(-re^{i\varphi})}\right|d\varphi &\ge \int_{\frac{\pi}{3}-\frac{\varepsilon}{2}}^{\frac{\pi}{3}+\frac{\varepsilon}{2}}\ln^+\left|\frac{g(re^{i\varphi})}{g(-re^{i\varphi})}\right|d\varphi - |C|\varepsilon r\\ &\ge \varepsilon\sin\frac{\varepsilon}{2}N(r,0,g) + O(r)\\ &= \left(\frac{3\pi}{20}\sin\frac{3\pi}{40}\right)\{N(r,0,G)+N(r,\infty,G)\} + O(r).\end{aligned} \tag{1.26}$$

Obviously, the zeros and poles of the function $G(z)/G(-z)$ are $(1,\frac{\pi}{60})$-separated. By Lemma 1.1 we have the inequality

$$\begin{aligned}\frac{1}{2\pi}\int_0^{2\pi}\ln\left|\frac{G(re^{i\varphi})}{G(-re^{i\varphi})}\right|\cos\varphi\,d\varphi &\ge \cos\frac{\pi}{60}\left\{N\left(r,\frac{G(z)}{G(-z)}\right)+N\left(r,\frac{G(-z)}{G(z)}\right)\right\}+O(r)\\ &= 2\cos\frac{\pi}{60}\{N(r,0,G)+N(r,\infty,G)\}+O(r).\end{aligned}$$

By 2π-periodicity of the function $G(re^{i\varphi})$ in φ, we have the equality

$$\frac{1}{2\pi}\int_0^{2\pi} \ln\left|\frac{G(-re^{i\varphi})}{G(re^{i\varphi})}\right| d\varphi = 0.$$

Adding it to the preceding inequality, we get

$$\frac{1}{2\pi}\int_0^{2\pi} \ln\left|\frac{G(-re^{i\varphi})}{G(re^{i\varphi})}\right| (1-\cos\varphi)\,d\varphi \geq 2\cos\frac{\pi}{60}\{N(r,0,G)+N(r,\infty,G)\}+O(r).$$

Since

$$\begin{aligned}2m\left(r,\frac{G(-z)}{G(z)}\right) &\geq \frac{1}{2\pi}\int_0^{2\pi}\ln^+\left|\frac{G(-re^{i\varphi})}{G(re^{i\varphi})}\right|(1-\cos\varphi)\,d\varphi \\ &= \frac{1}{2\pi}\int_0^{2\pi}\ln^+\left|\frac{G(re^{i\varphi})}{G(-re^{i\varphi})}\right|(1-\cos\varphi)\,d\varphi + \frac{1}{2\pi}\int_0^{2\pi}\ln\left|\frac{G(-re^{i\varphi})}{G(re^{i\varphi})}\right|(1-\cos\varphi)\,d\varphi,\end{aligned}$$

it follows from this that ($\varepsilon = \frac{9}{10}\cdot\frac{\pi}{6}$)

$$\begin{aligned}2m\left(r,\frac{G(-z)}{G(z)}\right) &\geq \frac{1}{2\pi}\int_0^{2\pi}\ln^+\left|\frac{G(re^{i\varphi})}{G(-re^{i\varphi})}\right|(1-\cos\varphi)\,d\varphi \\ &\quad + 2\cos\frac{\pi}{60}\{N(r,0,G)+N(r,\infty,G)\}+O(r) \\ &\geq \frac{1}{2\pi}\left(1-\cos\frac{\pi}{4}\right)\int_{\frac{\pi}{3}-\frac{\varepsilon}{2}}^{\frac{\pi}{3}+\frac{\varepsilon}{2}}\ln^+\left|\frac{G(re^{i\varphi})}{G(-re^{i\varphi})}\right|d\varphi \\ &\quad + 2\cos\frac{\pi}{60}\{N(r,0,G)+N(r,\infty,G)\}+O(r).\end{aligned}$$

Using (1.26) we get

$$\begin{aligned}&2m\left(r,\frac{G(-z)}{G(z)}\right) \\ &\quad\geq \left\{\frac{3}{40}\left(1-\cos\frac{\pi}{4}\right)\sin\frac{3\pi}{40}+2\cos\frac{\pi}{60}\right\}\{N(r,0,G)+N(r,\infty,G)\}+O(r).\end{aligned}$$

Since

$$\begin{aligned}m\left(r,\frac{G(-z)}{G(z)}\right) &\leq m(r,G)+m\left(r,\frac{1}{G}\right) \\ &= 2T(r,G)-\{N(r,0,G)+N(r,\infty,G)\} = 2T(r,G)-N(r;0,\infty),\end{aligned}$$

we have

$$T(r,G) \geq \frac{1}{2}\left\{1+\frac{3}{80}\left(1-\cos\frac{\pi}{4}\right)\sin\frac{3\pi}{40}+\cos\frac{\pi}{60}\right\}N(r;0,\infty)+O(r).$$

Straightforward computation shows that the number in braces exceeds 2.0004. Denoting this number by $2+2B$, where $B>0.0002$, we get

$$T(r,G) \geq (1+B)N(r;0,\infty)+O(r),$$

and by (1.23), this inequality implies

$$\kappa(G) \leq 1/(1+B).$$

Using inequality (1.11) and letting $A = B/(1+B)$, we get (1.25).

In the case where the genus of $G(z)$ exceeds 2, the function $S(z)$ in the representation (1.24) is either a polynomial of degree at least 3, or an entire transcendent function. In both cases, by Lemma 6.2, we have

$$r^3 = O(T(r, e^{S(z)})).$$

Since the genus of $G_1(z)$ does not exceed 2, we have

$$T(r, G_1) = o(r^3),$$

and hence

$$N(r,0,G) = N(r,0,G_1) = o(r^3), \quad N(r,\infty,G) = N(r,\infty,G_1) = o(r^3),$$

$$r^3 = O(T(r, e^{S(z)}) + T(r, G_1)) = O(T(r, G)).$$

Therefore $\delta(0,G) = 1$, $\delta(\infty, G) = 1$. The proof of the lemma is complete. □

Now we turn to the proof of Theorem 1.5.

Without loss of generality we may assume that all zeros of the function $f(z)$ are in the region $D_1^p(\eta)$ and all of its poles are in the region $D_2^p(\eta)$. In fact, by (1.1) the products

$$\pi_1(z) = \prod_{a_m \notin D_1^p(\eta)} E\left(\frac{z}{a_m}, p-1\right), \qquad \pi_2(z) = \prod_{b_n \notin D_2^p(\eta)} E\left(\frac{z}{b_n}, p-1\right)$$

are absolutely convergent. Let

$$f_1(z) = f(z)\frac{\pi_2(z)}{\pi_1(z)}.$$

Then all zeros of the function $f_1(z)$ are in $D_1^p(\eta)$, and all of its poles are in $D_2^p(\eta)$, the function $f_1(z)$ also satisfies (1.20), and since $T(r, \pi_j) = o(r^p)$, $j = 1, 2$, we have

$$T(r, f_1) = T(r, f) + o(r^p),$$

$$N(r,\infty,f_1) = N(r,\infty,f) + o(r^p), \qquad N(r,0,f_1) = N(r,0,f) + o(r^p),$$

and because (see the proof of Theorem 1.2 (b)) $r^p = o(T(r,f))$, we get

$$\lim_{r\to\infty} r^{-p}T(r,f_1) = \infty, \qquad \delta(0,f_1) = \delta(0,f), \qquad \delta(\infty,f_1) = \delta(\infty,f).$$

We may assume also that

$$\sum_m \frac{1}{|a_m|^p} + \sum_n \frac{1}{|b_n|^p} = \infty, \tag{1.27}$$

since otherwise we have $N(r,\infty,f) = o(r^p)$, $N(r,0,f) = o(r^p)$, and hence $\delta(0,f) = \delta(\infty,f) = 1$.

Thus, we assume that all zeros of the function $f(z)$ are in the region $D_1^p(\eta)$, and all of its poles are in the region $D_2^p(\eta)$, and condition (1.27) is satisfied. Note that then $\xi > p$, where ξ is the number appearing in the statement of the theorem; we denote by q a positive integer satisfying

$$\sum_m \frac{1}{|a_m|^q} + \sum_n \frac{1}{|b_n|^q} = \infty, \qquad \sum_m \frac{1}{|a_m|^{q+1}} + \sum_n \frac{1}{|b_n|^{q+1}} < \infty. \tag{1.28}$$

It is clear that

$$p \le q \le [\xi].$$

By (1.28) the function $f(z)$ can be represented in the form

$$f(z) = e^{H(z)} \frac{\prod_m E(\frac{z}{a_m}, q)}{\prod_n E(\frac{z}{b_n}, q)},$$

where $H(z)$ is an entire function. We choose an odd number s such that

$$ps \le q < p(s+2), \tag{1.29}$$

and let

$$l = ps, \quad \omega = \exp\left(\frac{2\pi i}{l}\right).$$

Consider the function

$$F(z) = \prod_{k=0}^{l-1} f(\omega^k z). \tag{1.30}$$

Using the equality ($\nu = 0, 1, 2, \dots$)

$$\sum_{k=0}^{l-1} \omega^{k\nu} = \begin{cases} l & \text{if } \nu \text{ is divisible by } l, \\ 0 & \text{if } \nu \text{ is not divisible by } l, \end{cases}$$

we easily get

$$\prod_{k=0}^{l-1} E(\omega^k z, q) = E\left(z^l, \left[\frac{q}{l}\right]\right), \qquad \sum_{k=0}^{l-1} H(\omega^k z) = S(z^l),$$

where $S(z)$ is an entire function. Therefore we have

$$F(z) = e^{S(z^l)} \frac{\prod_m E\left(\frac{z^l}{a_m^l}, \left[\frac{q}{l}\right]\right)}{\prod_n E\left(\frac{z^l}{b_n^l}, \left[\frac{q}{l}\right]\right)}.$$

Let us show that the meromorphic function

$$G(z) = F(z^{1/l}) = e^{S(z)} \frac{\prod_m E\left(\frac{z}{a_m^l}, \left[\frac{q}{l}\right]\right)}{\prod_n E\left(\frac{z}{b_n^l}, \left[\frac{q}{l}\right]\right)}$$

satisfies the assumptions of Lemma 1.3 and the inequalities

$$\delta(0, G) \le \delta(0, f), \qquad \delta(\infty, G) \le \delta(\infty, f). \tag{1.31}$$

Since (1.29) implies $q < l + 2p \le 3l$, we have $\left[\frac{q}{l}\right] \le 2$. The conditions

$$\left|\arg a_m - \pi\frac{2j}{p}\right| \le \eta, \qquad \left|\arg b_n - \pi\frac{2j+1}{p}\right| \le \eta$$

($j \ge 0$ is an integer) imply that

$$|\arg(a_m^l) - s\pi 2j| \le l\eta, \qquad |\arg(b_n^l) - s\pi(2j+1)| \le l\eta,$$

where s is defined by (1.29). Taking into account that s is odd, and $l\eta \le q\eta \le [\xi]\eta \le \frac{\pi}{60}$, we observe that all zeros of the function $G(z)$ are in the angle $\{|\arg z| \le \frac{\pi}{60}\}$, and all of its poles are in the angle $\{|\pi - \arg z| \le \frac{\pi}{60}\}$.

Now we show that the zeros and poles of different factors in (1.30) cannot cancel each other; this fact will be used later on. The zeros of all functions $f(\omega^k z)$, $k = 0, 1, \dots, l-1$, belong to the set

$$\bigcup_{k=0}^{l-1}\bigcup_{j=0}^{p}\left\{\left|\arg z - k\frac{2\pi}{l} - \pi\frac{2j}{p}\right| \le \eta,\ 0 < |z| < \infty\right\},$$

and their poles belong to the set

$$\bigcup_{k=0}^{l-1}\bigcup_{j=0}^{p}\left\{\left|\arg z - k\frac{2\pi}{l} - \pi\frac{2j+1}{p}\right| \le \eta,\ 0 < |z| < \infty\right\}.$$

Let us show that these sets are disjoint.

Assume the contrary, and let z be a point belonging to both sets. Then there exist integers k_1, j_1, k_2, j_2, such that

$$\left|\arg z - \pi\frac{2k_1}{l} - \pi\frac{2j_1}{p}\right| \le \eta, \qquad \left|\arg z - \pi\frac{2k_2}{l} - \pi\frac{2j_2+1}{p}\right| \le \eta.$$

Hence

$$\left|\pi\frac{2k_1}{l} + \pi\frac{2j_1}{p} - \pi\frac{2k_2}{l} - \pi\frac{2j_2+1}{p}\right| \le 2\eta,$$

and since $l = ps$, we have

$$\frac{\pi}{p}\left|\frac{2(k_1-k_2)}{s} - 2(j_2 - j_1) - 1\right| \le 2\eta.$$

Because the distance from a fraction with an even numerator and odd denominator s to an arbitrary odd number is not less than $1/s$, we get

$$\frac{\pi}{ps} \le 2\eta,$$

and since $ps \le q < \xi$, $\eta \le \frac{\pi}{60\xi}$, we get a contradiction.

It is clear that all points a_m^l are zeros of the function $G(z)$, and all points b_n^l are its poles. If we assumed that $\liminf_{r\to\infty} r^{-1}T(r, G) < \infty$, we would get

$$\sum_m \frac{1}{|a_m^l|} + \sum_n \frac{1}{|b_n^l|} < \infty.$$

But this contradicts (1.28) since $l \le q$. Therefore (1.23) holds, and hence the function $G(z)$ satisfies the assumptions of Lemma 1.3.

To prove (1.31), we observe that (1.30) implies

$$T(r, G) = T(r^{1/l}, F) \le lT(r^{1/l}, f). \tag{1.32}$$

Next, since, as we have already shown, the zeros and poles of different factors in (1.30) cannot cancel each other, the equalities

$$\begin{aligned} N(r, \infty, G) &= N(r^{1/l}, \infty, F) = lN(r^{1/l}, \infty, f),\\ N(r, 0, G) &= N(r^{1/l}, 0, F) = lN(r^{1/l}, 0, f) \end{aligned} \tag{1.33}$$

hold. Inequalities (1.31) immediately follow from (1.32) and (1.33).

Applying Lemma 1.3 to the function $G(z)$ and using (1.31), we get the conclusion of Theorem 1.5.

2. Meromorphic functions with poles and zeros located close to a system of rays

In this section we systematically use the angular Nevanlinna characteristics and the results of Sections 5 and 6 in Chapter 1, and Section 3 in Chapter 3.

Denote by $D = D(\alpha_1, \dots, \alpha_n)$ the system of rays

$$\bigcup_{j=1}^{n} \{\arg z = \alpha_j\}, \qquad \alpha_1 < \alpha_2 < \dots < \alpha_n < \alpha_{n+1} = \alpha_1 + 2\pi.$$

Let

$$\omega_j(D) = \frac{\pi}{\alpha_{j+1} - \alpha_j} \quad (j = 1, \dots, n), \qquad \omega(D) = \max_{1 \le j \le n} \omega_j(D).$$

Let $f(z)$ be a meromorphic function, and a a number from the extended complex plane. We introduce the function[2]

$$U_D(r, a, f) = \max_{1 \le t \le r} \sum_{j=1}^{n} t^{\omega_j(D) - \omega(D)} \bar{C}_{\alpha_j \alpha_{j+1}}(t, a, f).$$

The growth of the function $U_D(r, a, f)$ as $r \to \infty$ is one of the natural ways to measure how distant are the a-points of the function $f(z)$ from the system $D(\alpha_1, \dots, \alpha_n)$ of rays. Note that the relation $U_D(r, a, f) \equiv 0$ means that all a-points of the function $f(z)$ lying outside the disc $\{|z| \le 1\}$ belong to the system $D(\alpha_1, \dots, \alpha_n)$ of rays. The relation $U_D(r, a, f) = O(1)$ holds if the number of a-points not lying on the rays of the system $D(\alpha_1, \dots, \alpha_n)$ is finite, but it can also hold in cases where the number of such points is infinite.

EXAMPLE 1. The condition $U_D(r, a, f) = O(1)$ for the system $D(0, \pi)$ is equivalent to the condition

$$\sum_{r_m > 1} \frac{|\sin \varphi_m|}{r_m} < \infty, \tag{2.1}$$

where the $r_m e^{i\varphi_m}$ are a-points of the function $f(z)$, listed without taking into account their multiplicities. In fact, we have

$$\begin{aligned} \bar{C}_{0\pi}(r, a, f) &= 2 \sum_{\substack{1 < r_m < r \\ 0 < \varphi_m < \pi}} \left(\frac{1}{r_m} - \frac{r_m}{r^2} \right) \sin \varphi_m \\ &\le 2 \sum_{\substack{1 < r_m < r \\ 0 < \varphi_m < \pi}} \frac{\sin \varphi_m}{r_m} \le 2 \sum_{\substack{r_m > 1 \\ 0 < \varphi_m < \pi}} \frac{\sin \varphi_m}{r_m}; \end{aligned}$$

$$\begin{aligned} \bar{C}_{0\pi}(r, a, f) &\ge 2 \sum_{\substack{1 < r_m < \frac{r}{2} \\ 0 < \varphi_m < \pi}} \left(\frac{1}{r_m} - \frac{r_m}{r^2} \right) \sin \varphi_m \\ &= 2 \sum_{\substack{1 < r_m < \frac{r}{2} \\ 0 < \varphi_m < \pi}} \left(1 - \frac{r_m^2}{r^2} \right) \frac{\sin \varphi_m}{r_m} \ge 2 \cdot \frac{3}{4} \sum_{\substack{1 < r_m < \frac{r}{2} \\ 0 < \varphi_m < \pi}} \frac{\sin \varphi_m}{r_m}. \end{aligned}$$

---

[2]The parameters $\bar{C}_{\alpha\beta}(r, a, f)$ were defined on p. 111.

Estimating in a similar way the parameter $\bar{C}_{\pi,2\pi}(r,a,f)$ and observing that for the system $D(0,\pi)$ we have $\omega_1(D)=\omega_2(D)=\omega(D)=1$ and

$$U_D(r,a,f)=\max_{1\le t\le r}\{\bar{C}_{0\pi}(t,a,f)+\bar{C}_{\pi,2\pi}(t,a,f)\}=\bar{C}_{0\pi}(r,a,f)+\bar{C}_{\pi,2\pi}(r,a,f),$$

we verify the claim stated above. It is clear that (2.1) holds, for example, when $r_m=m$, $\varphi_m=1/m$.

Observe that letting $z_m=r_me^{i\varphi_m}$, we can rewrite condition (2.1) in the form

$$\sum_{|z_m|>1}\left|\operatorname{Im}\frac{1}{z_m}\right|<\infty.$$

For $a\ne b$ we let

$$U_D(r,a,b;f)=U_D(r,a,f)+U_D(r,b,f).$$

If f is clear from the context, we write $U_D(r,a,b)$ and $U_D(r,a)$, respectively, instead of $U_D(r,a,b;f)$ and $U_D(r,a,f)$.

The quantity

$$\mu(E)=\limsup_{r\to\infty}r^{-1}\operatorname{meas}(E\cap(0,r)),$$

defined for a measurable subset $E\subset[0,\infty)$, will be called the *upper relative measure* of E.

It is easy to see that the following statements are true.

(a) $0\le\mu(E)\le 1$ for each E.

(b) If $E\subset\bigcup_{k=1}^m E_k$, $m<\infty$, then $\mu(E)\le\sum_{k=1}^m\mu(E_k)$.

(c) If the logarithmic measure of E is finite (that is, $\int_{E\cap[1,\infty)}r^{-1}dr<\infty$), then[3] $\mu(E)=0$.

Denote by L' the class of all measurable subsets $E\subset[0,\infty)$ satisfying $\mu(E)<1$. We shall repeatedly use the property of the class L' described in the following lemma.

LEMMA 2.1. *Let $u(r)$ and $v(r)$ be two non-decreasing functions such that $u(r)\le v(r)$ for all $r\ge 0$, excluding, possibly, a set $E\in L'$. Then there exists a constant K, $1\le K<\infty$, such that for all sufficiently large r the inequality $u(r)\le v(Kr)$ holds.*

PROOF. Suppose that there exists a sequence $r_n\uparrow\infty$ such that

$$u(r_n)>v\left(\frac{2}{1-\mu(E)}r_n\right).$$

Consider the set $E_1=\bigcup_{n=1}^{\infty}(r_n,r_{n1})$, where $r_{n1}=\frac{2}{1-\mu(E)}r_n$. It is clear that

$$\begin{aligned}\mu(E_1)&\ge\limsup_{n\to\infty}r_{n1}^{-1}\operatorname{meas}(E_1\cap(0,r_{n1}))\\&\ge\limsup_{n\to\infty}r_{n1}^{-1}(r_{n1}-r_n)=\frac{1+\mu(E)}{2}>\mu(E).\end{aligned}$$

---

[3]In fact,

$$\operatorname{meas}(E\cap(0,r))=\int_{E\cap(0,\sqrt{r})}dt+\int_{E\cap(\sqrt{r},r)}dt\le\sqrt{r}+r\int_{E\cap(\sqrt{r},r)}t^{-1}dt=o(r).$$

Therefore there exists a point $r \in E_1 \backslash E$. Let $r \in (r_k, r_{k1})$; then we have $u(r_k) \leq u(r) \leq v(r) \leq v(r_{k1}) = v(\frac{2}{1-\mu(E)} r_k)$, and we get a contradiction. $\square$

We introduce the parameter

$$\delta'(a,f) = \sup_{CE \in L'} \liminf_{\substack{r \to \infty \\ r \in E}} \frac{m(r,a)}{T(r,f)} = 1 - \inf_{CE \in L'} \limsup_{\substack{r \to \infty \\ r \in E}} \frac{N(r,a)}{T(r,f)}.$$

Comparing this with the parameter $\delta(a,f)$ and the parameter $\delta^*(a,f)$ introduced in Section 6 of Chapter 4 (p. 156), we see that

$$\delta(a,f) \leq \delta^*(a,f) \leq \delta'(a,f),$$

thus, for each deficient value a the inequality $\delta'(a,f) > 0$ holds.

The following theorem is the main result of this section.

THEOREM 2.1. *Let $D = D(\alpha_1, \ldots, \alpha_n)$ be a system of rays, and $f(z)$ a meromorphic function satisfying $\delta'(a,f) > 0$ for some $a \neq 0, \infty$. Then the growth category of $f(z)$ does not exceed the growth category of*

$$r^{\omega(D)}\{U_D(r,0,\infty) + 1\}.$$

Let us agree for the remainder of this section that K will denote different finite positive constants independent of r.

The condition $\delta'(a,f) > 0$ implies that

$$T(r,f) \leq \frac{2}{\delta'(a,f)} m(r,a), \qquad r \notin B_1, \tag{2.2}$$

where $B_1 \subset [0,\infty)$ is a set satisfying $\mu(B_1) < 1$. By Lemma 2.1 of Chapter 3 (p. 98) we have

$$m(r,a) \leq m\left(r, \frac{f}{f'}\right) + Q(r,f). \tag{2.3}$$

We assume that $f(z)$ is transcendental, since otherwise Theorem 2.1 is trivial. Then

$$Q(r,f) = O(\ln r) + O(\ln T(r,f)) \leq \frac{1}{2} T(r,f)$$

outside some set $B_2 \subset [0,\infty)$ of finite measure, and (2.2) and (2.3) imply that for $r \notin B_3 = B_1 \cup B_2$ the inequality

$$T(r,f) \leq K m\left(r, \frac{f}{f'}\right) \tag{2.4}$$

holds. Observe that $\mu(B_3) \leq \mu(B_1) + \mu(B_2) = \mu(B_1) < 1$.

Next, we need to estimate $m(r, \frac{f}{f'})$ from above in terms of $U_D(r,0,\infty)$. Since such estimates are of independent interest, we state them as a separate theorem.

THEOREM 2.2. *Let $D = D(\alpha_1, \ldots, \alpha_n)$ be a system of rays, $f(z)$ a meromorphic function, and $\varepsilon > 0$, $d > 1$ preassigned numbers.*

(A) *For all $r \geq 0$ except, possibly, a subset $E \subset [0,\infty)$ satisfying $\mu(E) < \varepsilon$, we have the inequality*

$$m\left(r, \frac{f}{f'}\right) \leq K r^{d\omega(D)} \{U_D(r,0,\infty) + \ln r + 1\}^d \{\ln^+ T(r,f) + 1\}^d. \tag{2.5}$$

(B) *If the function $f(z)$ satisfies the condition*

$$\int_1^\infty r^{-q(D)-1} \ln^+ T(r,f)\, dr < \infty, \qquad q(D) = \min_{1 \leq j \leq n} \omega_j(D), \tag{2.6}$$

then for all $r \geq 0$ *except, possibly, a subset* $E \subset [0,\infty)$ *satisfying* $\mu(E) < \varepsilon$, *the inequality*

$$m\left(r, \frac{f}{f'}\right) \leq K r^{\omega(D)}\{U_D(dr, 0, \infty) + 1\} \tag{2.7}$$

holds.

To prove Theorem 2.2 we need upper estimates for

$$m_{\alpha\beta}(r, f) = \frac{1}{2\pi}\int_{\alpha}^{\beta} \ln^+ |f(re^{i\varphi})|\, d\varphi,$$

where $f(z)$ is an arbitrary meromorphic function and $0 < \beta - \alpha \leq 2\pi$, in terms of the Nevanlinna characteristic $S_{\alpha\beta}(r, f)$. Such estimates are given by the following theorem. Its proof is rather tedious, and will be postponed till Section 3 of the present chapter.

THEOREM 2.3. *Let* $f(z)$ *be a meromorphic function in the angle* $\{\alpha \leq \arg z \leq \beta\}$, $0 < \beta - \alpha \leq 2\pi$; *let* $\varepsilon > 0$ *and* $d > 1$ *be preassigned numbers; and let* $\omega = \pi/(\beta - \alpha)$. *For all* $r \geq 0$ *except, possibly, a subset* $E \subset [0,\infty)$ *satisfying* $\mu(E) < \varepsilon$, *we have the inequalities*

$$m_{\alpha\beta}(r, f) \leq K r^{\omega}\{S_{\alpha\beta}(r, f) + 1\}^d,$$

$$m_{\alpha\beta}(r, f) \leq K r^{\omega}\{S_{\alpha\beta}(dr, f) + 1\}.$$

PROOF OF THEOREM 2.2. Now we prove statement (A) of Theorem 2.2.

By Theorem 2.3, for all $r \geq 0$ except, possibly, a subset $E_j \subset [0,\infty)$ satisfying $\mu(E_j) < \frac{\varepsilon}{n}$, the inequality

$$m_{\alpha_j\alpha_{j+1}}\left(r, \frac{f}{f'}\right) \leq K r^{\omega_j(D)} \left\{S_{\alpha_j\alpha_{j+1}}\left(r, \frac{f}{f'}\right) + 1\right\}^d$$

holds ($j = 1, \ldots, n$). Applying Theorem 5.2 of Chapter 1 (p. 26) and Theorem 3.1 of Chapter 3 (p. 105), we get

$$\begin{aligned} &S_{\alpha_j\alpha_{j+1}}\left(r, \frac{f}{f'}\right) \\ &\quad = A_{\alpha_j\alpha_{j+1}}\left(r, \frac{f'}{f}\right) + B_{\alpha_j\alpha_{j+1}}\left(r, \frac{f'}{f}\right) + C_{\alpha_j\alpha_{j+1}}\left(r, \frac{f'}{f}\right) + O(1) \\ &\quad = C_{\alpha_j\alpha_{j+1}}\left(r, \frac{f'}{f}\right) + Q_{\alpha_j\alpha_{j+1}}(r, f). \end{aligned} \tag{2.8}$$

We recall that, by the definition of $Q_{\alpha_j\alpha_{j+1}}(r, f)$ (see p. 105),

$$Q_{\alpha_j\alpha_{j+1}}(r, f) = O(\ln r) + O(\ln T(r, f))$$

outside some set $E_j^{(1)} \subset [0,\infty)$ of finite measure. From this, using the obvious equality

$$C_{\alpha_j\alpha_{j+1}}\left(r, \frac{f'}{f}\right) = \bar{C}_{\alpha_j\alpha_{j+1}}(r, f) + \bar{C}_{\alpha_j\alpha_{j+1}}\left(r, \frac{1}{f}\right), \tag{2.9}$$

we get, for $r \notin E = \bigcup_{j=1}^{n}(E_j \cup E_j^{(1)})$, the inequality

$$\begin{aligned} m\left(r, \frac{f}{f'}\right) &= \sum_{j=1}^{n} m_{\alpha_j\alpha_{j+1}}\left(r, \frac{f}{f'}\right) \\ &\le K\sum_{j=1}^{n} r^{\omega_j(D)}\left\{\bar{C}_{\alpha_j\alpha_{j+1}}(r,f) + \bar{C}_{\alpha_j\alpha_{j+1}}\left(r, \frac{1}{f}\right) + \ln r + \ln^+ T(r,f)\right\}^d \\ &\le Kr^{d\omega(D)}\sum_{j=1}^{n}\left\{r^{\omega_j(D)-\omega(D)}\left(\bar{C}_{\alpha_j\alpha_{j+1}}(r,f)\right.\right. \\ &\qquad\qquad \left.\left. + \bar{C}_{\alpha_j\alpha_{j+1}}\left(r, \frac{1}{f}\right) + \ln r + \ln^+ T(r,f)\right)\right\}^d. \end{aligned}$$

From this, using the inequality $\sum_j a_j^d \le (\sum_j a_j)^d$, $a_j \ge 0$, $d > 1$, we get, for $r \notin E$, the inequality

$$\begin{aligned} m\left(r, \frac{f}{f'}\right) &\le Kr^{d\omega(D)}\left\{\sum_{j=1}^{n} r^{\omega_j(D)-\omega(D)}\left(\bar{C}_{\alpha_j\alpha_{j+1}}(r,f)\right.\right. \\ &\qquad\qquad \left.\left. + \bar{C}_{\alpha_j\alpha_{j+1}}\left(r, \frac{1}{f}\right) + \ln r + \ln^+ T(r,f)\right)\right\}^d \\ &\le Kr^{d\omega(D)}\{U_D(r,0,\infty) + \ln r + \ln^+ T(r,f)\}^d \\ &\le Kr^{d\omega(D)}\{U_D(r,0,\infty) + \ln r + 1\}^d\{\ln^+ T(r,f) + 1\}^d. \end{aligned}$$

Since

$$\mu(E) \le \sum_{j=1}^{n}\mu(E_j \cup E_j^{(1)}) \le \sum_{j=1}^{n}\{\mu(E_j) + \mu(E_j^{(1)})\} < \varepsilon,$$

statement (A) of Theorem 2.2 has been proved.

Next, by Theorem 2.3, for all $r \ge 0$ except, possibly, a subset $E_j' \subset [0,\infty)$ satisfying $\mu(E_j') < \frac{\varepsilon}{n}$, the inequality

$$m_{\alpha_j\alpha_{j+1}}\left(r, \frac{f}{f'}\right) \le Kr^{\omega_j(D)}\left\{S_{\alpha_j\alpha_{j+1}}\left(dr, \frac{f}{f'}\right) + 1\right\}$$

holds ($j = 1, \ldots, n$). Using (2.8), (2.9) and observing that, by condition (2.6) and the definition of $Q_{\alpha_j\alpha_{j+1}}(r,f)$ (see p. 105), the relation $Q_{\alpha_j\alpha_{j+1}}(r,f) = O(1)$ holds, we conclude that for $r \notin E' = \bigcup_{j=1}^{n} E_j'$ the inequality

$$\begin{aligned} m\left(r, \frac{f}{f'}\right) &= \sum_{j=1}^{n} m_{\alpha_j\alpha_{j+1}}\left(r, \frac{f}{f'}\right) \\ &\le K\sum_{j=1}^{n} r^{\omega_j(D)}\left\{\bar{C}_{\alpha_j\alpha_{j+1}}(dr,f) + \bar{C}_{\alpha_j\alpha_{j+1}}\left(dr, \frac{1}{f}\right) + K\right\} \\ &\le Kr^{\omega(D)}\{U_D(dr,0,\infty) + 1\} \end{aligned}$$

holds. Since $\mu(E') \le \sum_{j=1}^{n}\mu(E_j') < \varepsilon$, we have proved statement (B) of Theorem 2.2. □

Now we turn to the proof of Theorem 2.1.

We may assume that the order of the function $U_D(r,0,\infty)$ is finite; otherwise Theorem 2.1 is trivial.

First we show that this assumption implies that the order of the function $f(z)$ is also finite.

Suppose inequality (2.4) holds outside a set B_3. We choose $\varepsilon > 0$ to be so small that $\mu(B_3)+\varepsilon < 1$. To estimate $m(r, \frac{f}{f'})$ we use statement (A) of Theorem 2.2. Let E be a set such that $\mu(E) < \varepsilon$ and for $r \notin E$ inequality (2.5) holds. Inequalities (2.4) and (2.5) imply that for $r \notin B_4 = B_3 \cup E$ the inequality

$$T(r,f) \le Kr^{d\omega(D)}\{U_D(r,0,\infty) + \ln r + 1\}^d\{\ln^+ T(r,f) + 1\}^d \tag{2.10}$$

holds. For each ε_1, $0 < \varepsilon_1 < 1$, we can find $r_1 > 0$ such that

$$\{\ln^+ T(r,f) + 1\}^d \le T^{\varepsilon_1}(r,f), \qquad r \ge r_1. \tag{2.11}$$

Therefore (2.10) implies that for $r \notin B_5 = B_4 \cup (0, r_1)$ we have

$$T(r,f) \le Kr^{\frac{d\omega(D)}{1-\varepsilon_1}}\{U_D(r,0,\infty)+1\}^{\frac{d}{1-\varepsilon_1}}. \tag{2.12}$$

We denote by $v(r)$ the function on the right-hand side of this inequality. It is a non-decreasing function, and for all $r \notin B_5$ the inequality $T(r,f) \le v(r)$ holds. Since

$$\mu(B_5) \le \mu(B_3) + \mu(E) + \mu((0,r_1)) < \mu(B_3) + \varepsilon < 1,$$

by Lemma 2.1, $T(r,f) \le v(Kr)$ for all sufficiently large r. Since $U_D(r,0,\infty)$, and hence $v(r)$ has finite order, $f(z)$ also has finite order.

To prove Theorem 2.1 it suffices to show that

$$T(r,f) \le Kr^{\omega(D)}\{U_D(Kr,0,\infty)+1\}, \qquad r \ge K. \tag{2.13}$$

This inequality can be obtained using (2.4) and statement (B) of Theorem 2.2 (condition (2.6) is satisfied since, as we have proved, the order of $f(z)$ is finite). In fact, choose $\varepsilon > 0$ so small that $\mu(B_3) + \varepsilon < 1$, and denote by E a set satisfying $\mu(E) < \varepsilon$ and such that for $r \notin E$ inequality (2.7) holds. Inequalities (2.4) and (2.7) imply that for $r \notin B_4 = B_3 \cup E$,

$$T(r,f) \le Kr^{\omega(D)}\{U_D(dr,0,\infty)+1\}. \tag{2.14}$$

Since $\mu(B_4) \le \mu(B_3) + \varepsilon < 1$, applying Lemma 2.1 we get (2.13). Our proof of Theorem 2.1 is complete. □

The statement of Theorem 2.1 is no longer true if we drop the condition that $\delta'(a,f) > 0$ for some $a \ne 0, \infty$. In fact, for each entire function $f(z)$ without zeros, we have $\delta(0,f) = \delta(\infty,f) = 1$, and for each system D of rays we have $U_D(r,0,\infty) \equiv 0$. However, the growth of such an entire function $f(z)$ can be arbitrarily rapid.

We consider also the following example.

EXAMPLE 2. Let $\{h_k\}_{k=1}^{\infty}$ and $\{A_k\}_{k=1}^{\infty}$ be sequences of positive numbers such that $\lim_{k\to\infty} h_k = \infty$, $\sum_{k=1}^{\infty} A_k < \infty$. Consider the meromorphic function

$$f(z) = \sum_{k=1}^{\infty} \frac{A_k}{z - h_k}.$$

By the Corollary of Theorem 6.1 in Chapter 5 (p. 253) we have

$$T(r,f) = N(r,f) + o(1).$$

The growth of $N(r,f)$ depends on the choice of the sequence h_k only. Therefore, choosing an appropriate sequence, we can assure that the function $f(z)$ has a preassigned growth category. Next, since for $\operatorname{Im} z \neq 0$ we have

$$\operatorname{Im} f(z) = -\operatorname{Im} z \cdot \sum_{k=1}^{\infty} \frac{A_k}{|z-h_k|^2} \neq 0,$$

and for negative real z the function $f(z)$ is negative, it follows that the function $f(z)$ takes non-negative real values only on the ray $\{\arg z = 0\}$. This ray can be regarded as the system $D = D(0)$ for which $\omega(D) = 1/2$.

Thus, there exist functions of an arbitrary preassigned growth, such that $U_{D(0)}(r,b) \equiv 0$ for infinitely many values of b. This shows that if we do not assume that $\delta'(a,f) > 0$ for some $a \neq 0, \infty$, then we cannot estimate the growth category of $f(z)$ only by means of the growth category of the function $r^{\omega(D)}\{U_D(r,0,\infty)+1\}$, even by means of the growth categories of the functions of the form

$$r^{\omega(D)}\left\{\sum_{\nu=1}^{q} U_D(r,b_\nu)+1\right\},$$

for an arbitrarily large q.

Let us state some corollaries of Theorem 2.1.

COROLLARY 1. *Let a, b, and c be three different numbers on the extended complex plane, and let $D = D(\alpha_1,\dots,\alpha_n)$ be a system of rays. If a meromorphic function $f(z)$ satisfies the condition $\delta'(a,f) > 0$, then its category of growth does not exceed the category of the function $r^{\omega(D)}\{U_D(r,b,c)+1\}$.*

To prove this corollary it suffices to apply Theorem 2.1 to the function $L(f(z))$, where $L(w)$ is a linear-fractional transformation mapping the points b and c onto the points 0 and ∞, respectively.

COROLLARY 2. *Let a, b, and c be three different numbers on the extended complex plane, and let $D = D(\alpha_1,\dots,\alpha_n)$ be a system of rays. If a meromorphic function $f(z)$ satisfies the conditions $\delta'(a,f) > 0$, $U_D(r,b,c) = O(1)$, then the growth of the function $f(z)$ does not exceed normal type of order $\omega(D)$.*

Since for an entire function $f(z)$ the condition $\delta(\infty,f) = 1$ holds, we have

COROLLARY 3. *Let a and b, $a \neq b$, be complex numbers, $D = D(\alpha_1,\dots,\alpha_n)$ a system of rays, and $f(z)$ an entire function. If $U_D(r,a,b) = O(1)$, then the growth of $f(z)$ does not exceed normal type of order $\omega(D)$.*

This growth estimate cannot be improved.

EXAMPLE 3. Let us show that for each system $D = D(\alpha_1,\dots,\alpha_n)$ of rays it is possible to construct an entire function $\Phi_D(z)$ of normal type of order $\omega(D)$ such that for each a the relation $U_D(r,a) = O(1)$ holds.

Let $\{\alpha_{j_0} \leq \arg z \leq \alpha_{j_0+1}\}$ be an angle formed by the rays of the system D such that $\omega_{j_0}(D) = \omega(D)$. As in Section 5 of Chapter 2, denote by $E_\rho(z)$ the

Mittag-Leffler function and let

$$\Phi_D(z) = E_{\omega(D)}\left(z \exp\left(-i\frac{\alpha_{j_0} + \alpha_{j_0+1}}{2}\right)\right).$$

We show that $\Phi_D(z)$ is a desired function.

By the result proved in Section 5 of Chapter 2 (see (5.40)), the function $\Phi_D(z)$ is a function of normal type of order $\omega(D)$, and its modulus is bounded outside the angle $\{\alpha_{j_0} < \arg z < \alpha_{j_0+1}\}$. Therefore we have $(j = 1, \dots, n)$

$$A_{\alpha_j\alpha_{j+1}}(r, \Phi_D) = O(1), \qquad B_{\alpha_j\alpha_{j+1}}(r, \Phi_D) = O(1),$$

and hence $(j = 1, \dots, n)$

$$S_{\alpha_j\alpha_{j+1}}(r, \Phi_D) = O(1).$$

But then, by Theorem 5.1 of Chapter 1, for each a we have

$$C_{\alpha_j\alpha_{j+1}}(r, a) \le S_{\alpha_j\alpha_{j+1}}(r, \Phi_D) + O(1) = O(1),$$

from which

$$U_D(r, a) = \max_{1\le t\le r} \sum_{j=1}^{n} t^{\omega_j(D)-\omega(D)} \bar{C}_{\alpha_j\alpha_{j+1}}(t, a) \le \sum_{j=1}^{n} C_{\alpha_j\alpha_{j+1}}(r, a) = O(1).$$

Observe that the condition $U_D(r, a) = O(1)$ will also be satisfied for the entire function

$$\sum_{j=1}^{n} E_{\omega_j(D)}\left(z \exp\left(-i\frac{\alpha_j + \alpha_{j+1}}{2}\right)\right).$$

Applying Corollary 3 to the system $D(0, \pi)$ and using the comments made at the beginning of this section (see Example 1) we get

COROLLARY 4. *Let $f(z)$ be an entire function, $a \ne b$ complex numbers, $\{z_m\}$ the set of all solutions of the equations $f(z) = a$ and $f(z) = b$ in which each solution is listed only once, regardless of its multiplicity. If the condition*

$$\sum_{|z_m|>1} \left|\operatorname{Im}\frac{1}{z_m}\right| < \infty$$

is satisfied, then the growth of the function $f(z)$ does not exceed normal type of order 1.

Note that the function $f(z) = \sin z$ takes all real values a, $|a| \le 1$, on the real line only.

We cannot deduce the following result from Theorem 2.1.

THEOREM 2.4. *Let $D = D(\alpha_1, \dots, \alpha_n)$ be a system of rays, and $f(z)$ a meromorphic function having a Borel exceptional value $a \ne 0, \infty$. Then the category of growth of $f(z)$ does not exceed that of the function $r^{\omega(D)}\{U_D(r, 0, \infty)+1\}$.*

PROOF. By the first fundamental theorem we have

$$T(r, f) = N(r, a) + m(r, a) + O(1).$$

Using (2.3) and assuming that the function $f(z)$ is transcendental, we get that for all $r \ge 0$ except, possibly, a set of finite length, the inequality

$$T(r, f) \le K\left\{N(r, a) + m\left(r, \frac{f}{f'}\right)\right\}$$

holds. To estimate $m(r, \frac{f}{f'})$ we use statement (A) of Theorem 2.2. For all $r \geq 0$ except, possibly, a set B_1 such that $\mu(B_1) < \varepsilon < 1$ we have

$$T(r,f) \leq KN(r,a) + Kr^{d\omega(D)}\{U_D(r,0,\infty) + \ln r + 1\}^d\{\ln^+ T(r,f) + 1\}^d$$
$$\leq K\{N(r,a) + r^{d\omega(D)}(U_D(r,0,\infty) + \ln r + 1)^d\}\{\ln^+ T(r,f) + 1\}^d.$$

Taking into account (2.11) we get

$$T(r,f) \leq K\{N(r,a) + r^{d\omega(D)}(U_D(r,0,\infty) + \ln r + 1)^d\}^{\frac{1}{1-\varepsilon_1}}$$
$$(r \notin B_1,\ 0 < \varepsilon_1 < 1).$$

Denote by $v(r)$ the function in the right-hand side. By Lemma 2.1, for sufficiently large r the inequality $T(r,f) \leq v(Kr)$ holds. By the definition of a Borel exceptional value, the order of $N(r,a)$ is finite. We may assume that the order of the function $U_D(r,0,\infty)$ is also finite; otherwise Theorem 2.4 is trivial. Therefore the order of $v(r)$, and hence of the function $f(z)$, is finite.

To estimate $m(r, \frac{f}{f'})$ we use statement (B) of Theorem 2.2. We get that for all $r \geq 0$ except, possibly, a set B_2 satisfying $\mu(B_2) < \varepsilon < 1$, the inequality

$$T(r,f) \leq K\{N(r,a) + r^{\omega(D)}(U_D(dr,0,\infty) + 1)\}$$

holds. Using Lemma 2.1 we get, for sufficiently large r, the estimate

$$T(r,f) \leq K\{N(Kr,a) + r^{\omega(D)}(U_D(Kr,0,\infty) + 1)\}.$$

Since the category of the characteristic $N(r,a)$ is lower than that of the characteristic $T(r,f)$, the conclusion of the theorem follows. □

The condition of positivity of $\delta'(a,f)$ for some $a \neq 0,\infty$ in Theorem 2.1 can be replaced by a condition of a more general character which can be stated as follows: there exists a finite set $\{a_1,\dots,a_q\}$, $a_\nu \neq 0,\infty$, $\nu = 1,\dots,q$, of numbers such that the quantity

$$\delta'(a_1,\dots,a_q;f) = \sup_{CE\in L'} \liminf_{\substack{r\to\infty \\ r\in E}} \frac{\sum_{\nu=1}^q m(r,a_\nu)}{T(r,f)}$$

is positive. In fact, this condition implies that outside a set B_1, $\mu(B_1) < 1$, the inequality

$$T(r,f) \leq \frac{2}{\delta'(a_1,\dots,a_q;f)} \sum_{\nu=1}^q m(r,a_\nu)$$

holds. By Lemma 2.1 of Chapter 3 we have

$$\sum_{\nu=1}^q m(r,a_\nu) \leq m\left(r, \frac{f}{f'}\right) + Q(r,f).$$

Therefore we again get inequality (2.4) outside a set of relative upper measure < 1. From this, as has already been shown, the conclusion of Theorem 2.1 follows.

Theorem 2.1 admits the following generalization.

THEOREM 2.5. *Let $D = D(\alpha_1,\dots,\alpha_n)$ be a system of rays, and $f(z)$ a meromorphic function such that for some $a \neq 0,\infty$ and some integer $l \geq 0$ the inequality $\delta'(a, f^{(l)}) > 0$ holds. Then the category of the function $f(z)$ does not exceed the category of the function*

$$r^{\omega(D)}\{U_D(r,0,\infty;f) + 1\}.$$

PROOF. Now the role of inequality (2.4) will be played by the inequality

$$T(r,f) \le K \sum_{\nu=0}^{l} m\left(r, \frac{f^{(\nu)}}{f^{(\nu+1)}}\right). \tag{2.15}$$

Let us show that this inequality holds for all $r \ge 0$ except, possibly, a set $B_1 \subset [0,\infty)$ such that $\mu(B_1) < 1$.

The condition $\delta'(a, f^{(l)}) > 0$ together with Lemma 2.1 of Chapter 3, applied to $f^{(l)}(z)$, implies that for $r \notin B_2$, where $B_2 \subset [0,\infty)$ is a set satisfying $\mu(B_2) < 1$, the inequality

$$T(r, f^{(l)}) \le \frac{2}{\delta'(a, f^{(l)})} m(r, a, f^{(l)}) \le \frac{2}{\delta'(a, f^{(l)})} m\left(r, \frac{f^{(l)}}{f^{(l+1)}}\right) + Q(r, f^{(l)})$$

holds. Since

$$\begin{aligned} T(r,f) = N(r,f) + m(r,f) &\le N(r, f^{(l)}) + m\left(r, \frac{f}{f'}\cdot\frac{f'}{f''}\cdot \dots \cdot \frac{f^{(l-1)}}{f^{(l)}} f^{(l)}\right) \\ &\le N(r, f^{(l)}) + \sum_{\nu=0}^{l-1} m\left(r, \frac{f^{(\nu)}}{f^{(\nu+1)}}\right) + m(r, f^{(l)}) \\ &= T(r, f^{(l)}) + \sum_{\nu=0}^{l-1} m\left(r, \frac{f^{(\nu)}}{f^{(\nu+1)}}\right), \end{aligned}$$

we get

$$T(r,f) \le \frac{2}{\delta'(a, f^{(l)})} \sum_{\nu=0}^{l} m\left(r, \frac{f^{(\nu)}}{f^{(\nu+1)}}\right) + Q(r, f^{(l)}).$$

By Theorem 2.3 of Chapter 3, for all $r \ge 0$ except, possibly, a set of finite measure, we have the inequality

$$Q(r, f^{(l)}) = Q(r,f) \le \frac{1}{2} T(r,f)$$

(we assume that the function $f(z)$ is transcendental). From this it follows that (2.15) holds for all $r \ge 0$ except, possibly, a set $B_1 \subset [0,\infty)$ satisfying $\mu(B_1) < 1$.

Next, we need to estimate $m\left(r, \frac{f^{(\nu)}}{f^{(\nu+1)}}\right)$, $\nu = 0, \dots, l$, from above in terms of $U_D(r, 0, \infty; f)$. For this, besides Theorem 2.2, we need the following inequality:

$$U_D(r, 0, \infty; f^{(\nu)}) \le K U_D(r, 0, \infty; f) + \sum_{j=1}^{n} Q_{\alpha_j \alpha_{j+1}}(r, f). \tag{2.16}$$

To prove this inequality we use induction. For $\nu = 0$ the inequality is trivial. Suppose that it is true for some $\nu \ge 0$. We have

$$\bar{C}_{\alpha_j \alpha_{j+1}}(r, f^{(\nu+1)}) = \bar{C}_{\alpha_j \alpha_{j+1}}(r, f^{(\nu)}),$$

$$\bar{C}_{\alpha_j \alpha_{j+1}}\left(r, \frac{1}{f^{(\nu+1)}}\right) \le \bar{C}_{\alpha_j \alpha_{j+1}}\left(r, \frac{1}{f^{(\nu)}}\right) + \bar{C}_{\alpha_j \alpha_{j+1}}\left(r, \frac{f^{(\nu)}}{f^{(\nu+1)}}\right). \tag{2.17}$$

By Theorem 5.2 of Chapter 1,

$$\bar{C}_{\alpha_j\alpha_{j+1}}\left(r, \frac{f^{(\nu)}}{f^{(\nu+1)}}\right) \leq S_{\alpha_j\alpha_{j+1}}\left(r, \frac{f^{(\nu+1)}}{f^{(\nu)}}\right) + O(1)$$
$$= A_{\alpha_j\alpha_{j+1}}\left(r, \frac{f^{(\nu+1)}}{f^{(\nu)}}\right) + B_{\alpha_j\alpha_{j+1}}\left(r, \frac{f^{(\nu+1)}}{f^{(\nu)}}\right) + C_{\alpha_j\alpha_{j+1}}\left(r, \frac{f^{(\nu+1)}}{f^{(\nu)}}\right) + O(1).$$

Using Theorem 3.1 of Chapter 3 and the fact that relation (2.14) in Chapter 3 (p. 99) remains true if we replace $Q(r, f)$ by $Q_{\alpha_j\alpha_{j+1}}(r, f)$ (this was established in Theorem 3.3 of Chapter 3), we get

$$A_{\alpha_j\alpha_{j+1}}\left(r, \frac{f^{(\nu+1)}}{f^{(\nu)}}\right) + B_{\alpha_j\alpha_{j+1}}\left(r, \frac{f^{(\nu+1)}}{f^{(\nu)}}\right) = Q_{\alpha_j\alpha_{j+1}}(r, f^{(\nu)}) = Q_{\alpha_j\alpha_{j+1}}(r, f);$$

hence

$$\bar{C}_{\alpha_j\alpha_{j+1}}\left(r, \frac{f^{(\nu)}}{f^{(\nu+1)}}\right) \leq Q_{\alpha_j\alpha_{j+1}}(r, f) + C_{\alpha_j\alpha_{j+1}}\left(r, \frac{f^{(\nu+1)}}{f^{(\nu)}}\right)$$
$$= \bar{C}_{\alpha_j\alpha_{j+1}}(r, f^{(\nu)}) + \bar{C}_{\alpha_j\alpha_{j+1}}\left(r, \frac{1}{f^{(\nu)}}\right) + Q_{\alpha_j\alpha_{j+1}}(r, f).$$

From this and from (2.17) we get the inequality

$$\bar{C}_{\alpha_j\alpha_{j+1}}(r, f^{(\nu+1)}) + \bar{C}_{\alpha_j\alpha_{j+1}}\left(r, \frac{1}{f^{(\nu+1)}}\right)$$
$$\leq 2\left\{\bar{C}_{\alpha_j\alpha_{j+1}}(r, f^{(\nu)}) + \bar{C}_{\alpha_j\alpha_{j+1}}\left(r, \frac{1}{f^{(\nu)}}\right)\right\} + Q_{\alpha_j\alpha_{j+1}}(r, f)$$

and conclude that

$$U_D(r, 0, \infty; f^{(\nu+1)}) \leq 2U_D(r, 0, \infty; f^{(\nu)}) + \sum_{j=1}^{n} Q_{\alpha_j\alpha_{j+1}}(r, f).$$

By the induction hypothesis, we get (2.16).

Let $\varepsilon > 0$ be so small that $\mu(B_1)+l\varepsilon < 1$, where B_1 is a set such that inequality (2.15) holds outside of B_1. By statement (A) of Theorem 2.2, for all $r \geq 0$ except, possibly, a subset $E_\nu \subset [0, \infty)$, $\mu(E_\nu) < \varepsilon$, the following inequality holds ($d > 1$):

$$m\left(r, \frac{f^{(\nu)}}{f^{(\nu+1)}}\right) \leq Kr^{d\omega(D)}\{U_D(r, 0, \infty; f^{(\nu)}) + \ln r + 1\}^d\{\ln^+ T(r, f^{(\nu)}) + 1\}^d.$$

In view of inequality (2.16) and Theorem 2.3 of Chapter 3 we get that for all $r \geq 0$ except, possibly, a subset $E_\nu^{(1)} \subset [0, \infty)$ satisfying $\mu(E_\nu^{(1)}) < \varepsilon$, the inequality

$$m\left(r, \frac{f^{(\nu)}}{f^{(\nu+1)}}\right) \leq Kr^{d\omega(D)}\{U_D(r, 0, \infty; f) + \ln r + 1\}^d\{\ln^+ T(r, f) + 1\}^d$$

holds. From this, using (2.15) and (2.11), we conclude that inequality (2.12) holds outside a set whose upper relative measure does not exceed $\mu(B_1) + l\varepsilon < 1$, and, as in the proof of Theorem 2.1, we derive the conclusion that the function $f(z)$ has finite order.

Next, using statement (B) of Theorem 2.2 and inequality (2.16), we get that for all $r \geq 0$ except, possibly a set $E_\nu^{(2)}$ satisfying $\mu(E_\nu^{(2)}) < \varepsilon$,

$$m\left(r, \frac{f^{(\nu)}}{f^{(\nu+1)}}\right) \leq Kr^{\omega(D)}\{U_D(dr, 0, \infty; f^{(\nu)}) + 1\} \leq Kr^{\omega(D)}\{U_D(dr, 0, \infty; f) + 1\}.$$

From this and from (2.15) we conclude that outside a set whose upper relative measure does not exceed $\mu(B_1) + l\varepsilon < 1$, inequality (2.14) holds. □

COROLLARY. *Let $D = D(\alpha_1, \dots, \alpha_n)$ be a system of rays, and $f(z)$ a meromorphic function such that for some $a \neq 0, \infty$ and some integer $l \geq 0$, the inequality $\delta'(a, f^{(l)}) > 0$ holds. If $U_D(r, 0, \infty; f) = O(1)$, then the growth of $f(z)$ does not exceed normal type of order $\omega(D)$.*

Under a more restrictive assumption than $U_D(r, 0, \infty; f) = O(1)$, we can get an additional information about the asymptotic behavior of the function $f(z)$.

THEOREM 2.6. *Let $D = D(\alpha_1, \dots, \alpha_n)$ be a system of rays, and $f(z)$ a meromorphic function such that for some $a \neq 0, \infty$ and some integer $l \geq 0$ the inequality $\delta'(a, f^{(l)}) > 0$ holds. If for some $b \neq 0, \infty$ and some integer $m \geq 0$ we have*

$$\sum_{j=1}^{n} \{C_{\alpha_j \alpha_{j+1}}(r, 0, f) + C_{\alpha_j \alpha_{j+1}}(r, \infty, f) + C_{\alpha_j \alpha_{j+1}}(r, b, f^{(m)})\} = O(1), \tag{2.18}$$

then the growth of the function $f(z)$ does not exceed normal type of order $\omega(D)$; besides, as $r \to \infty$ missing, possibly, a set of finite logarithmic measure, we have, uniformly in φ, $\alpha_j \leq \varphi \leq \alpha_{j+1}$, $j = 1, \dots, n$, the relation

$$\ln|f(re^{i\varphi})| = r^{\omega_j(D)} c_j \sin\{\omega_j(D)(\varphi - \alpha_j)\} + o(r^{\omega_j(D)}), \tag{2.19}$$

where c_j, $j = 1, \dots, n$, are real constants.

The proof of Theorem 2.6 is based on the following theorem about the behavior of functions with bounded characteristic $S_{\alpha\beta}(r, f)$ in the angle $\{\alpha \leq \arg z \leq \beta\}$.

THEOREM 2.7. *Let $f(z)$ be a meromorphic function in the angle $\{\alpha \leq \arg z \leq \beta\}$, $0 < \beta - \alpha \leq 2\pi$, and such that $S_{\alpha\beta}(r, f) = O(1)$. As $r \to \infty$ missing, possibly, a set of finite logarithmic measure, the relation*

$$\ln|f(re^{i\varphi})| = r^\omega c \sin\{\omega(\varphi - \alpha)\} + o(r^\omega)$$

holds uniformly in φ, $\alpha \leq \varphi \leq \beta$, where c is a constant and $\omega = \pi/(\beta - \alpha)$.

Theorem 2.7 will be proved in Section 3.
We need also the following lemma.

LEMMA 2.2. *Let $f(z)$ be a meromorphic function in the angle $\{\alpha \leq \arg z \leq \beta\}$. For each $b \neq 0, \infty$ and for each integer $m \geq 0$ we have the inequality*

$$S_{\alpha\beta}(r, f) \leq C_{\alpha\beta}(r, \infty, f) + C_{\alpha\beta}(r, 0, f) + C_{\alpha\beta}(r, b, f^{(m)}) + Q_{\alpha\beta}(r, f).$$

This lemma immediately follows from an analog of relation (2.17) in Chapter 3 for the Nevanlinna characteristics for half-plane. The validity of this analog was established in Theorem 3.3 of Chapter 3.

Now we prove Theorem 2.6. Since condition (2.18) implies $U_D(r, 0, \infty; f) = O(1)$, the growth of the function $f(z)$ does not exceed normal type of order $\omega(D)$ by Theorem 2.5. Next, we apply Lemma 2.2 to the function $f(z)$ in each of the angles $\{\alpha_j \le \arg z \le \alpha_{j+1}\}$, $j = 1, 2, \dots, n$. Since the function $f(z)$ has finite order, $Q_{\alpha_j\alpha_{j+1}}(r, f) = O(1)$, $j = 1, \dots, n$. Therefore, using condition (2.18) and Lemma 2.2, we get $S_{\alpha_j\alpha_{j+1}}(r, f) = O(1)$, $j = 1, \dots, n$. Applying Theorem 2.7 in each of the angles $\{\alpha_j \le \arg z \le \alpha_{j+1}\}$, we prove the statement of Theorem 2.6.

REMARK 1. If a meromorphic function $f(z)$ and a system $D = D(\alpha_1, \dots, \alpha_n)$ of rays satisfy condition (2.18) for some $b \ne 0, \infty$ and some integer $m \ge 0$, and besides

$$\int_1^\infty \frac{\ln^+ T(r, f)}{r^{q(D)+1}}\,dr < \infty, \qquad q(D) = \min_{1\le j\le n} \omega_j(D), \tag{2.20}$$

then, as $r \to \infty$ missing, possibly, a set of finite logarithmic measure, the relation

$$\ln|f(re^{i\varphi})| = r^{\omega_j(D)} c_j \sin\{\omega_j(D)(\varphi - \alpha_j)\} + o(r^{\omega_j(D)})$$

holds uniformly in φ, $\alpha_j \le \varphi \le \alpha_{j+1}$, $j = 1, \dots, n$.

In fact, condition (2.20) implies that $Q_{\alpha_j\alpha_{j+1}}(r, f) = O(1)$. Therefore, using condition (2.18) and Lemma 2.2 we again get $S_{\alpha_j\alpha_{j+1}}(r, f) = O(1)$, and after that we use Theorem 2.7.

REMARK 2. Under the assumptions of Theorem 2.6, as $r \to \infty$, we have

$$\ln|f(re^{i\alpha_j})| = o(r^{q_j(D)}), \qquad q_j(D) = \min(\omega_{j-1}(D), \omega_j(D))$$

$(j = 1, \dots, n)$ outside a set of finite logarithmic measure. These relations can be supplemented by

$$\int_1^\infty \frac{|\ln|f(te^{i\alpha_j})||}{t^{q_j(D)+1}}\,dt < \infty, \qquad j = 1, \dots, n. \tag{2.21}$$

In fact, in the proof of Theorem 2.6 we established that $S_{\alpha_j\alpha_{j+1}}(r, f) = O(1)$, $j = 1, \dots, n$. Therefore, by Theorem 5.2 of Chapter 1, $A_{\alpha_j\alpha_{j+1}}(r, f) + A_{\alpha_j\alpha_{j+1}}(r, \frac{1}{f})$ $= O(1)$, $j = 1, \dots, n$. Since $(\omega_j = \omega_j(D),\ j = 1, \dots, n)$

$$\begin{aligned} A_{\alpha_j\alpha_{j+1}}&(r, f) + A_{\alpha_j\alpha_{j+1}}\left(r, \frac{1}{f}\right) \\ &= \frac{\omega_j}{\pi}\int_1^r \left(\frac{1}{t^{\omega_j}} - \frac{t^{\omega_j}}{r^{2\omega_j}}\right)\{|\ln|f(te^{i\alpha_j})|| + |\ln|f(te^{i\alpha_{j+1}})||\}\frac{dt}{t} \\ &\ge \frac{\omega_j}{\pi}\int_1^{r/2} \left(\frac{1}{t^{\omega_j}} - \frac{t^{\omega_j}}{r^{2\omega_j}}\right)\{|\ln|f(te^{i\alpha_j})|| + |\ln|f(te^{i\alpha_{j+1}})||\}\frac{dt}{t} \\ &\ge \frac{\omega_j}{\pi}\left(1 - \left(\frac{1}{2}\right)^{\omega_j}\right)\int_1^{r/2} t^{-\omega_j}\{|\ln|f(te^{i\alpha_j})|| + |\ln|f(te^{i\alpha_{j+1}})||\}\frac{dt}{t}, \end{aligned}$$

we easily get (2.21).

REMARK 3. The statement of Theorem 2.6 remains true if we replace $f(z)$ in (2.19) by $\frac{\alpha f^{(l)}(z)+\beta}{\gamma f^{(l)}(z)+\delta}$, where $\alpha\delta - \beta\gamma \ne 0$, and $l \ge 0$ is an integer. This follows from the relations

$$S_{\alpha_j\alpha_{j+1}}\left(r, \frac{\alpha f + \beta}{\gamma f + \delta}\right) = S_{\alpha_j\alpha_{j+1}}(r, f) + O(1), \qquad \alpha\delta - \beta\gamma \ne 0,$$

$$S_{\alpha_j\alpha_{j+1}}(r, f^{(l)}) \le (l + 1)S_{\alpha_j\alpha_{j+1}}(r, f) + Q_{\alpha_j\alpha_{j+1}}(r, f).$$

The first relation holds by (6.11) of Chapter 1 (p. 29), and the second is an analog of (2.13) in Chapter 3 (p. 99), which holds by Theorem 3.3 of Chapter 3.

The following corollary of Theorem 2.6 can be obtained using Remark 2.

COROLLARY 1. *Let $f(z)$ be an entire function, $a \neq b$ complex numbers, and $\{z_m\}$ the set of all roots of the equations $f(z) = a$ and $f(z) = b$, where each root is listed according to its multiplicity. If the condition*

$$\sum_m \left| \operatorname{Im} \frac{1}{z_m} \right| < \infty$$

is satisfied, then the function $f(z)$ has growth of at most normal type of order 1. Moreover, as $r \to \infty$ outside a set of finite logarithmic measure, the relation

$$\ln |f(re^{i\varphi})| = \begin{cases} rc_1 \sin \varphi + o(r), & 0 \leq \varphi \leq \pi, \\ rc_2 \sin \varphi + o(r), & \pi \leq \varphi \leq 2\pi \end{cases}$$

holds uniformly in φ, where c_1 and c_2 are constants, and

$$\int_{-\infty}^{\infty} \frac{|\ln |f(t)||}{1+t^2} dt < \infty.$$

The assumptions of this corollary are close to, but are somewhat stronger (we take into account the multiplicity of a- and b-points) than the assumptions of Corollary 3 of Theorem 2.1 for the system $D(0, \pi)$.

Observe that the statement of Corollary 1 remains true for each entire function $f(z)$ having at most normal type of order 1, and satisfying the condition

$$\int_{-\infty}^{\infty} \frac{\ln^+ |f(t)|}{1+t^2} dt < \infty.$$

In fact, for such a function we have

$$\begin{aligned} S_{0\pi}(r, f) &= \frac{1}{\pi} \int_1^r \left(\frac{1}{t^2} - \frac{1}{r^2} \right) \{\ln^+ |f(t)| + \ln^+ |f(-t)|\} \, dt \\ &\quad + \frac{2}{\pi r} \int_0^{\pi} \ln^+ |f(re^{i\varphi})| \sin \varphi \, d\varphi \\ &= O(1), \end{aligned}$$

and, similarly, $S_{\pi,2\pi}(r, f) = O(1)$. Therefore the desired conclusion follows from Theorem 2.7 and Remark 2.

Theorem 2.6 and Remark 3 immediately imply

COROLLARY 2. *Under the assumptions of Theorem 2.6 for each number d on the extended complex plane and each integer $p \geq 0$ the relation*

$$m\left(r, d, f^{(p)}\right) = cr^{\omega(D)} + o(r^{\omega(D)})$$

holds as $r \to \infty$ missing, possibly, a set of finite logarithmic measure ($c \geq 0$ is a constant which, in general, depends on d and p).

Now we prove the following theorem.

THEOREM 2.8. *Let $D = D(\alpha_1, \dots, \alpha_n)$ be a system of rays, and $f(z)$ a meromorphic function such that for some $a \neq 0, \infty$ the relation $\delta'(a, f) > 0$ holds. If for some $b \neq 0, \infty$ and some integer $m \geq 0$ relation* (2.18) *holds, and in addition*

$$\limsup_{r\to\infty} r^{-\omega(D)}T(r, f) > 0, \tag{2.22}$$

then

$$\liminf_{r\to\infty} r^{-\omega(D)}T(r, f) > 0 \tag{2.23}$$

and the function $f(z)$ has at most s deficient values, where s is the number of angles $\{\alpha_j \leq \arg z \leq \alpha_{j+1}\}$ such that $\omega_j(D) = \omega(D)$.

PROOF. For all $r \geq 0$ except, possibly, a subset $B_1 \subset [0, \infty)$ such that $\mu(B_1) < 1$, the inequality

$$T(r, f) \leq \frac{2}{\delta'(a, f)} m(r, a, f)$$

holds. By Corollary 2 of Theorem 2.6, the relation

$$m(r, a, f) = cr^{\omega(D)} + o(r^{\omega(D)}), \qquad c \geq 0, \tag{2.24}$$

holds as $r \to \infty$ missing, possibly, a set B_2 of finite logarithmic measure. If the constant c in (2.24) were 0, then outside a set B_3 satisfying $\mu(B_3) < 1$, the condition

$$T(r, f) = o(r^{\omega(D)})$$

would hold. By Lemma 2.1 we would derive from this that the function $f(z)$ has growth of at most minimal type of order $\omega(D)$, which would contradict (2.22). Thus we have $c > 0$ in (2.24), and since $T(r, f) \geq m(r, a, f) + O(1)$, we have

$$T(r, f) \geq cr^{\omega(D)} + o(r^{\omega(D)}), \qquad r \notin B_2. \tag{2.25}$$

Take an arbitrary sequence $\{r_k\}_{k=1}^{\infty}$, $r_k \uparrow \infty$. If the condition $T(r_k, f) = o(r_k^{\omega(D)})$ were satisfied, we would have

$$\max_{r_k/2 \leq r \leq r_k} T(r, f) = o(r_k^{\omega(D)}), \tag{2.26}$$

since the function $T(r, f)$ is non-decreasing. The set $\bigcup_{k=1}^{\infty}(\frac{1}{2}r_k, r_k)$ has an infinite logarithmic measure. Thus, we can find in it a sequence $\{\rho_\nu\}_{\nu=1}^{\infty}$, $\rho_\nu \uparrow \infty$, such that $\rho_\nu \notin B_2$. By (2.25) we would get $T(\rho_\nu, f) \geq c\rho_\nu^{\omega(D)} + o(\rho_\nu^{\omega(D)})$, which contradicts (2.26). Thus we have established (2.23).

By Remark 3 to Theorem 2.6, for each $d \neq \infty$ the relation

$$\ln|f(re^{i\varphi}) - d| = r^{\omega_j(D)}c_j(d)\sin\{\omega_j(D)(\varphi - \alpha_j)\} + o(r^{\omega_j(D)}) \tag{2.27}$$

holds uniformly in φ, $\alpha_j \leq \varphi \leq \alpha_{j+1}$, as $r \to \infty$ missing, possibly, a set B_4 of finite logarithmic measure. If all the $c_j(d)$ with $\omega_j(D) = \omega(D)$ are non-negative, then (2.27) implies that for $r \notin B_4$ the relation $m(r, d, f) = o(r^{\omega(D)})$ holds; taking into account (2.23) we get $\delta(d, f) = 0$. Thus, if $\delta(d, f) > 0$, then for some j the conditions $c_j(d) < 0$ and $\omega_j(D) = \omega(D)$ must be satisfied simultaneously. But it is obvious that if $c_j(d) < 0$, then $c_j(d') = 0$ for each $d' \neq d$. Therefore the number of deficient values cannot exceed the number of values of j for which $\omega_j(D) = \omega(D)$. □

The estimate for the number of deficient values in Theorem 2.8 cannot be improved in general.

EXAMPLE 4. Let D be a system of rays

$$\bigcup_{j=1}^{q}\left\{\arg z = \frac{2j-1}{q}\pi\right\}.$$

For this system we have $\omega(D) = q/2$, and the number s determined in the statement of Theorem 2.8 is equal to q. Let $f(z)$ be the meromorphic function in Example 8 of Section 5 of Chapter 5. The number of deficient values of this function is equal to q. Let us show that the function $f(z)$ satisfies the conditions of Theorem 2.8 with the above-mentioned system D of rays. Here we shall use the notation of Example 8 in Section 5 of Chapter 5.

Condition (2.22) holds, since by (5.20) (p. 245) we have

$$T(r, f) \geq \frac{8}{q\pi}r^{q/2} + o(r^{q/2}).$$

To verify (2.18), we show that for an arbitrary number b on the extended complex plane,

$$\sum_{j=1}^{q} C_{\alpha_j\alpha_{j+1}}(r, b, f) = O(1), \qquad \alpha_j = \frac{2j-1}{q}\pi, \quad j = 1, \dots, q. \tag{2.28}$$

By the formula (see Whittaker, Watson [**WW96**, 17·7])

$$I_n(z) = \frac{z^n}{2^n\Gamma\left(\frac{1}{2}\right)\Gamma\left(n+\frac{1}{2}\right)}\int_0^{\pi}\cosh(z\cos\varphi)\sin^{2n}\varphi\,d\varphi, \qquad n > -\frac{1}{2},$$

we get that for $|\arg z| = \frac{\pi}{2}$ the relation

$$I_n(z) = O(|z|^n), \qquad |z| \to \infty,$$

holds. Therefore for $|\arg z| = \frac{\pi}{q}$ we have

$$w_l(z) = O(|z|), \qquad |z| \to \infty, \quad l = 1, 2.$$

This relation, by (5.18), p. 244, holds for all rays of the system D. Therefore for all $j = 1, 2, \dots, q$,

$$\begin{aligned} &A_{\alpha_j\alpha_{j+1}}(r, w_l) \\ &= \frac{q}{2\pi}\int_1^r\left(\frac{1}{t^{q/2}} - \frac{t^{q/2}}{r^q}\right)\{\ln^+|w_l(te^{i\alpha_j})| + \ln^+|w_l(te^{i\alpha_{j+1}})|\}\frac{dt}{t} = O(1). \end{aligned}$$

Relation (5.16), p. 244, implies that

$$\begin{aligned} B_{\alpha_j\alpha_{j+1}}(r, w_l) &= \frac{q}{\pi r^{q/2}}\int_{\alpha_j}^{\alpha_{j+1}}\ln^+|w_l(re^{i\varphi})|\sin\frac{q}{2}(\varphi - \alpha_j)d\varphi \\ &\leq \frac{2q}{r^{q/2}}T(r, w_l) = O(1). \end{aligned}$$

Thus,

$$S_{\alpha_j\alpha_{j+1}}(r, w_l) = A_{\alpha_j\alpha_{j+1}}(r, w_l) + B_{\alpha_j\alpha_{j+1}}(r, w_l) = O(1),$$

and using relations (6.11) and (6.12) of Chapter 1, we get

$$\begin{aligned}S_{\alpha_j\alpha_{j+1}}(r,f) &\le S_{\alpha_j\alpha_{j+1}}(r,w_1)+S_{\alpha_j\alpha_{j+1}}\left(r,\frac{1}{w_2}\right)\\ &= S_{\alpha_j\alpha_{j+1}}(r,w_1)+S_{\alpha_j\alpha_{j+1}}(r,w_2)+O(1)=O(1).\end{aligned}$$

By Theorem 5.2 of Chapter 1, for each b we have

$$C_{\alpha_j\alpha_{j+1}}(r,b,f)\le S_{\alpha_j\alpha_{j+1}}(r,f)+O(1)=O(1),\qquad j=1,2,\dots,q,$$

and relation (2.28) has been proved.

The theorems proved in this section have applications to the study of functions of the form

$$f(z)=\sum_{k=1}^{\infty}\frac{A_k}{z-h_k},\qquad h_k\to\infty \tag{2.29}$$

considered in Section 6 of Chapter 5 (the series is assumed to be absolutely convergent). These applications are based on the following lemma.

LEMMA 2.3. *If a function $f(z)$ is representable in the form (2.29), then for each complex number b and every α and β satisfying $0<\beta-\alpha\le 2\pi$,*

$$C_{\alpha\beta}(r,b,f)\le C_{\alpha\beta}(r,\infty,f)+O(1).$$

PROOF. By Theorem 5.2 of Chapter 1 we have

$$C_{\alpha\beta}(r,b,f)\le S_{\alpha\beta}(r,f)+O(1)=A_{\alpha\beta}(r,f)+B_{\alpha\beta}(r,f)+C_{\alpha\beta}(r,f)+O(1).$$

Since (for $w=\pi/(\beta-\alpha)$)

$$\begin{aligned}A_{\alpha\beta}(r,f)&=\frac{\omega}{\pi}\int_1^r\left(\frac{1}{t^\omega}-\frac{t^\omega}{r^{2\omega}}\right)\{\ln^+|f(te^{i\alpha})|+\ln^+|f(te^{i\beta})|\}\frac{dt}{t}\\ &\le\frac{\omega}{\pi}\int_1^\infty\frac{\ln^+|f(te^{i\alpha})|}{t^{1+\omega}}dt+\frac{\omega}{\pi}\int_1^\infty\frac{\ln^+|f(te^{i\beta})|}{t^{1+\omega}}dt,\end{aligned}$$

using Theorem 6.3 of Chapter 5 we get

$$A_{\alpha\beta}(r,f)=O(1).$$

Since

$$B_{\alpha\beta}(r,f)=\frac{2\omega}{\pi r^\omega}\int_\alpha^\beta\ln^+|f(re^{i\varphi})|\sin\omega(\varphi-\alpha)\,d\varphi\le\frac{4\omega}{r^\omega}m(r,f),$$

by Theorem 6.1 of Chapter 5,

$$B_{\alpha\beta}(r,f)=o(1).$$ □

Let $f(z)$ be a function of the form (2.29), and $D=D(\alpha_1,\dots,\alpha_n)$ a system of rays. Lemma 2.3 implies that for each complex number b the relation

$$U_D(r,b,f)\le U_D(r,\infty,f)+O(1)$$

holds. Therefore Corollary 1 of Theorem 2.1 implies the following result.

THEOREM 2.9. *Let $D=D(\alpha_1,\dots,\alpha_n)$ be a system of rays, and $f(z)$ a meromorphic function of the form (2.29) such that $\delta'(a,f)>0$ for some a. Then the category of growth of the function $f(z)$ does not exceed the category of growth of the function*

$$r^{\omega(D)}\{U_D(r,\infty,f)+1\}.$$

Note that, by Theorem 6.1 of Chapter 5, the functions of the form (2.29) satisfy $\delta(\infty, f) = \delta'(\infty, f) = 0$. By Theorem 6.2 of Chapter 5, for $a \neq 0, \infty$ we have $\delta(a, f) = 0$, but we do not know whether $\delta'(a, f) = 0$ always holds. The cases where $\delta(0, f) > 0$ are possible; we discussed this matter at the end of Section 6 of Chapter 5.

We would like to mention the following corollary of Theorem 2.9.

COROLLARY. *Let $D = D(\alpha_1, \ldots, \alpha_n)$ be a system of rays, and $f(z)$ a meromorphic function of the form* (2.29) *such that $\delta'(a, f) > 0$ for some a. If $U_D(r, \infty, f) = O(1)$, then the growth of the function $f(z)$ does not exceed normal type of order $\omega(D)$.*

This Corollary shows that Theorem 6.2 of Chapter 5 can be complemented with the following statement. If a function $f(z)$ of the form (2.29) has growth of at least maximal type of order $\omega(D)$ and $U_D(r, \infty, f) = O(1)$, then (regardless of whether the conditions $\sum_{k=1}^{\infty} |A_k| < \infty$ and $\sum_{k=1}^{\infty} A_k \neq 0$ are satisfied) the equality $\delta'(0, f) = 0$ holds.

By Theorem 2.4, the statements of Theorem 2.9 and its Corollary remain true if the condition of positivity of $\delta'(a, f)$ for some a is replaced by the condition that a is a Borel exceptional value.

Using Theorem 2.6, Remark 2 to it, and Lemma 2.3, we get the following theorem.

THEOREM 2.10. *Let $D = D(\alpha_1, \ldots, \alpha_n)$ be a system of rays, and $f(z)$ a meromorphic function of the form* (2.29) *such that $\delta'(a, f) > 0$ for some a. If*

$$\sum_{j=1}^{n} C_{\alpha_j \alpha_{j+1}}(r, \infty, f) = O(1),$$

then the growth of the function $f(z)$ does not exceed normal type of order $\omega(D)$, and the integrals

$$\int_1^{\infty} \frac{|\ln|f(te^{i\alpha_j})||}{t^{1+q_j(D)}}\,dt, \qquad q_j(D) = \min\{\omega_{j-1}(D),\ \omega_j(D)\}, \quad j = 2, \ldots, n+1,$$

converge. Besides, as $r \to \infty$ missing possibly a set of finite logarithmic measure, the relation

$$\ln|f(re^{i\varphi})| = r^{\omega_j(D)} c_j \sin\{\omega_j(D)(\varphi - \alpha_j)\} + o(r^{\omega_j(D)})$$

holds uniformly in φ, $\alpha_j \leq \varphi \leq \alpha_{j+1}$, $j = 1, \ldots, n$, where c_j, $j = 1, \ldots, n$, are some constants.

COROLLARY 1. *Let $f(z)$ be a meromorphic function of the form* (2.29) *such that $\delta'(a, f) > 0$ for some a. If the condition*

$$\sum_k \left| \operatorname{Im} \frac{1}{h_k} \right| < \infty$$

is satisfied, then the growth of the function $f(z)$ does not exceed normal type of order 1 and, in addition, as $r \to \infty$ outside a set of finite logarithmic measure, the relation

$$\ln|f(re^{i\varphi})| = \begin{cases} rc_1 \sin\varphi + o(r), & 0 \leq \varphi \leq \pi, \\ rc_2 \sin\varphi + o(r), & \pi \leq \varphi \leq 2\pi, \end{cases}$$

holds uniformly in φ, *where* c_1 *and* c_2 *are constants, and*

$$\int_{-\infty}^{\infty} \frac{|\ln|f(t)||}{1+t^2}\,dt < \infty.$$

COROLLARY 2. *If the zeros* $\{a_k\}$ *of an entire function* $f(z)$ *satisfy the condition*

$$\sum_k \left|\operatorname{Im}\frac{1}{a_k}\right| < \infty,$$

and, in addition, the function $1/f(z)$ *is representable in the form* (2.29), *then the statement of Corollary* 1 *remains true for the function* $f(z)$.

The growth estimate in Corollary 2 is sharp, since the function $(z^2-1)\sin z$ is exactly of normal type of order 1, and $\frac{1}{(z^2-1)\sin z}$ is representable in the form (2.29), namely,

$$\frac{1}{(z^2-1)\sin z} = \frac{1}{2\sin 1}\cdot\frac{1}{z-1} + \frac{1}{2\sin 1}\cdot\frac{1}{z+1} + \sum_{k=-\infty}^{\infty} \frac{(-1)^k}{\{(k\pi)^2-1\}(z-k\pi)}.$$

3. Proofs of Theorems 2.3 and 2.7

We prove Theorems 2.3 and 2.7 under the assumption that the angle $\{\alpha \le \arg z \le \beta\}$ is a half-plane

$$\Pi = \{\operatorname{Im} z \ge 0\}.$$

The general case can be easily reduced to this case using the transformation $w = z^{1/\omega}e^{i\alpha}$, $\omega = \pi/(\beta-\alpha)$, which maps the half-plane Π onto the angle $\{\alpha \le \arg w \le \beta\}$. We assume, without mentioning this explicitly, that all functions $f(z)$ appearing in the statements of theorems of this section are of the form $f(z) = g(z^\omega)$, where $\omega > 0$, and $g(\zeta)$ is a function meromorphic in the angle $\{0 \le \arg\zeta \le \pi/\omega\}$.

The proof of Theorem 2.7 is based on the following representation of the functions with bounded characteristic $S(r,f) = S_{0\pi}(r,f)$.

THEOREM 3.1. *If*

$$S(r,f) = O(1),$$

then $(z = re^{i\varphi},\ 0 < \varphi < \pi)$

$$\ln|f(z)| = \frac{r\sin\varphi}{\pi}\int_{-\infty}^{\infty} \frac{\ln|f(t)|}{r^2+t^2-2rt\cos\varphi}\,dt - \sum_m \ln\left|\frac{z-\bar a_m}{z-a_m}\right| + \sum_n \ln\left|\frac{z-\bar b_n}{z-b_n}\right| + \eta r\sin\varphi, \tag{3.1}$$

where a_m *are the zeros of the function* $f(z)$, *and* b_n *are its poles in* Π, η *being a constant which does not depend on* z. *The integral and the series in the right-hand side of* (3.1) *are absolutely convergent.*

PROOF. Observe that the condition $S(r,f) = O(1)$ implies convergence of the integral

$$\int_{-\infty}^{\infty} \frac{|\ln|f(t)||}{1+t^2}\,dt \tag{3.2}$$

and of the series

$$\sum_m \frac{\sin\varphi_m}{r_m}, \quad \sum_n \frac{\sin\psi_n}{\rho_n}, \tag{3.3}$$

where $a_m = r_m e^{i\varphi_m}$, $b_n = \rho_n e^{i\psi_n}$. This can be shown using the argument used in the proof of Remark 2 to Theorem 2.6 and in Example 1 in Section 2.

By Theorem 2.3 of Chapter 1 (p. 6) and the Remark after it, the formula $(z = re^{i\varphi})$

$$\begin{aligned} \ln|f(z)| &= \frac{1}{\pi}\int_{-R}^{R} \ln|f(t)| \left\{\frac{r\sin\varphi}{|z-t|^2} - \frac{R^2 r\sin\varphi}{|R^2-zt|^2}\right\} dt \\ &+ \frac{1}{2\pi}\int_0^{\pi} \ln|(Re^{i\theta})| \left\{\frac{R^2-r^2}{|Re^{i\theta}-z|^2} - \frac{R^2-r^2}{|Re^{-i\theta}-z|^2}\right\} d\theta \\ &- \sum_{|a_m|<R} \ln\left|\frac{R^2-\bar a_m z}{R(z-a_m)} \cdot \frac{R(z-\bar a_m)}{R^2-a_m z}\right| \\ &+ \sum_{|b_n|<R} \ln\left|\frac{R^2-\bar b_n z}{R(z-b_n)} \cdot \frac{R(z-\bar b_n)}{R^2-b_n z}\right| \\ &= I_1(R) + I_2(R) - I_3(R) + I_4(R) \end{aligned} \tag{3.4}$$

holds in the half-disc $\{|z| < R,\ \operatorname{Im} z > 0\}$. To get formula (3.1) we consider the limits as $R \to \infty$.

Convergence of the integral (3.2) clearly implies the absolute convergence of the integral (3.1). Therefore the limit

$$\lim_{R\to\infty} \frac{1}{\pi}\int_{-R}^{R} \ln|f(t)| \frac{r\sin\varphi}{|z-t|^2} dt = \frac{1}{\pi}\int_{-\infty}^{\infty} \ln|f(t)| \frac{r\sin\varphi}{|z-t|^2} dt$$

exists.

Next we note that

$$\begin{aligned} &\left|\int_{-R}^{R} \ln|f(t)| \frac{R^2 r\sin\varphi}{|R^2-zt|^2} dt\right| \\ &\quad \le \int_{-R}^{R} |\ln|f(t)|| \frac{R^2 r\sin\varphi}{(R^2-rt)^2} dt \le \frac{r}{(R-r)^2}\int_{-R}^{R} |\ln|f(t)||\, dt \\ &\quad \le \frac{r}{(R-r)^2}\left\{(1+R)\int_{-\sqrt{R}}^{\sqrt{R}} \frac{|\ln|f(t)||}{1+t^2} dt \right. \\ &\qquad \left. + (1+R^2)\int_{\sqrt{R}}^{R} \frac{|\ln|f(t)||}{1+t^2} dt + (1+R^2)\int_{-R}^{-\sqrt{R}} \frac{|\ln|f(t)||}{1+t^2} dt\right\} \to 0 \end{aligned}$$

as $R \to \infty$. Thus, the limit

$$\lim_{R\to\infty} I_1(R) = \frac{r\sin\varphi}{\pi}\int_{-\infty}^{\infty} \frac{\ln|f(t)|}{r^2+t^2-2rt\cos\varphi} dt \quad (z = re^{i\varphi})$$

exists.

The convergence of the series $\sum_m r_m^{-1}\sin\varphi_m$ implies the convergence of the series

$$\sum_m \ln\left|\frac{z-\bar a_m}{z-a_m}\right|,$$

since for $|a_m| > 2|z|$ we have

$$\ln\left|\frac{z-\bar{a}_m}{z-a_m}\right| = \ln\left|1+\frac{z2i\operatorname{Im}\frac{1}{a_m}}{1-\frac{z}{a_m}}\right| \le \frac{2|z|\left|\operatorname{Im}\frac{1}{a_m}\right|}{\left|1-\frac{z}{a_m}\right|} \le 4|z|\frac{\sin\varphi_m}{r_m}.$$

Therefore the following limit exists:

$$\lim_{R\to\infty}\sum_{|a_m|<R}\ln\left|\frac{z-\bar{a}_m}{z-a_m}\right| = \sum_m \ln\left|\frac{z-\bar{a}_m}{z-a_m}\right|.$$

Next we observe that

$$\begin{aligned}\left|\sum_{|a_m|<R}\ln\left|\frac{R^2-\bar{a}_m z}{R^2-a_m z}\right|\right| &= \sum_{|a_m|<R}\ln\left|\frac{R^2-a_m z}{R^2-\bar{a}_m z}\right| \\ &= \sum_{|a_m|<R}\ln\left|1-\frac{z2i\operatorname{Im}a_m}{R^2-\bar{a}_m z}\right| \le \sum_{|a_m|<R}\frac{2|z||\operatorname{Im}a_m|}{|R^2-\bar{a}_m z|} \\ &\le \frac{2r}{R(R-r)}\sum_{|a_m|<R} r_m\sin\varphi_m \\ &\le \frac{2r}{R(R-r)}\left\{R\sum_{|a_m|<\sqrt{R}}\frac{\sin\varphi_m}{r_m} + R^2\sum_{\sqrt{R}\le|a_m|\le R}\frac{\sin\varphi_m}{r_m}\right\}\to 0\end{aligned}$$

as $R\to\infty$. Thus, the following limit exists:

$$\lim_{R\to\infty} I_3(R) = \sum_m \ln\left|\frac{z-\bar{a}_m}{z-a_m}\right|.$$

In a similar way we show the existence of the limit

$$\lim_{R\to\infty} I_4(R) = \sum_n \ln\left|\frac{z-\bar{b}_n}{z-b_n}\right|.$$

Since the left-hand side of the equality $\ln|f(z)| = I_1(R)+I_2(R)-I_3(R)+I_4(R)$ does not depend on R, the limit $\lim_{R\to\infty} I_2(R)$ also exists. We need to show that

$$\lim_{R\to\infty} I_2(R) = \eta r\sin\varphi,$$

where η does not depend on r and φ. It suffices to show that

$$\lim_{R\to\infty} J(R) = 0,$$

where

$$J(R) = I_2(R) - r\sin\varphi\frac{2}{\pi R}\int_0^\pi \ln|f(Re^{i\theta})|\sin\theta\,d\theta.$$

We have

$$\begin{aligned}J(R) &= \frac{1}{2\pi}\int_0^\pi \ln|f(Re^{i\theta})|\frac{(R^2-r^2)4Rr\sin\varphi\sin\theta\,d\theta}{(R^2+r^2-2Rr\cos(\varphi-\theta))(R^2+r^2-2Rr\cos(\varphi+\theta))} \\ &\quad - r\sin\varphi\frac{2}{\pi R}\int_0^\pi \ln|f(Re^{i\theta})|\sin\theta\,d\theta \\ &= r\sin\varphi\frac{2}{\pi R}\int_0^\pi \ln|f(Re^{i\theta})|\sin\theta\cdot K(R,r,\varphi,\theta)\,d\theta,\end{aligned}$$

where $K(R,r,\varphi,\theta)=O(R^{-1})$ uniformly in θ, $0\le\theta\le\pi$ as $R\to\infty$, and r and φ are fixed. Therefore

$$\begin{aligned}|J(R)| &\le r\frac{2}{\pi R}\int_0^\pi |\ln|f(Re^{i\theta})||\sin\theta\,d\theta\cdot O(R^{-1})\\ &= r\{B(R,f)+B(R,1/f)\}\cdot O(R^{-1})\\ &\le r\{2S(R,f)+O(1)\}\cdot O(R^{-1})\to 0 \quad\text{as } R\to\infty.\end{aligned}$$

□

REMARK. In passing we have proved that

$$\eta=\lim_{R\to\infty}\{B(R,f)-B(R,1/f)\}.$$

Let $(z=re^{i\varphi},\ 0<\varphi<\pi)$

$$u(z)=\frac{r\sin\varphi}{\pi}\int_{-\infty}^{\infty}\frac{|\ln|f(t)||}{r^2+t^2-2rt\cos\varphi}dt + \sum_m \ln\left|\frac{z-\bar a_m}{z-a_m}\right| + \sum_n \ln\left|\frac{z-\bar b_n}{z-b_n}\right|. \tag{3.5}$$

Theorem 3.1 implies that

$$|\ln|f(re^{i\varphi})|-\eta r\sin\varphi|\le u(re^{i\varphi}).$$

Whence to prove Theorem 2.7 it suffices to prove the following lemma.

LEMMA 3.1. *For the function $u(z)$ given by (3.5), the relation*

$$u(re^{i\varphi})=o(r)$$

holds uniformly in φ, $0<\varphi<\pi$, as $r\to\infty$ missing a set $A\subset[1,\infty)$ of finite logarithmic measure.

A proof of the lemma could be based on estimates of the integral and the series on the right-hand side of (3.5). In doing so, we would repeat twice essentially the same argument. This can be avoided if we write the right-hand side of (3.5) as a Lebesgue–Stieltjes integral over the half-plane Π.

We proceed as follows. We define a Lebesgue–Stieltjes measure $\nu(E)$ on Borel subsets $E\subset\Pi$ by the formula

$$\nu(E)=\frac{1}{\pi}\int_{E\cap(-\infty,\infty)}\frac{|\ln|f(t)||}{1+t^2}dt+\sum_{a_m\in E}\frac{2r_m\sin\varphi_m}{1+r_m^2}+\sum_{b_n\in E}\frac{2\rho_n\sin\psi_n}{1+\rho_n^2}. \tag{3.6}$$

This measure takes only finite values, since by the convergence of the integral (3.2) and the series (3.3) we have

$$\nu(\Pi)=\frac{1}{\pi}\int_{-\infty}^{\infty}\frac{|\ln|f(t)||}{1+t^2}dt+\sum_m\frac{2r_m\sin\varphi_m}{1+r_m^2}+\sum_n\frac{2\rho_n\sin\psi_n}{1+\rho_n^2}<\infty.$$

Let $(\zeta=\xi+i\eta,\ \eta\ge 0,\ z=re^{i\varphi})$

$$K(z,\zeta)=\begin{cases}\frac{1+|\zeta|^2}{2\eta}\ln\left|\frac{z-\bar\zeta}{z-\zeta}\right|, & \eta>0,\\ \frac{(1+\xi^2)r\sin\varphi}{r^2+\xi^2-2r\xi\cos\varphi}, & \eta=0.\end{cases} \tag{3.7}$$

Then we have

$$u(z)=\iint_\Pi K(z,\zeta)\,d\nu(\zeta). \tag{3.8}$$

Having in mind also the proof of Theorem 2.3, we obtain the following very general theorem concerning estimates of integrals of the form (3.8).

THEOREM 3.2. *Let $\nu(E)$ be a Lebesgue–Stieltjes measure defined on Borel subsets $E \subset \Pi$ and such that $\nu(\Pi) < \infty$. Define the function $u(z;\nu)$, $z \in \Pi$, by the equality*

$$u(z;\nu) = \iint_{\Pi} K(z,\zeta)\, d\nu(\zeta),$$

where $K(z,\zeta)$ is given by (3.7)

For each $q > 0$, we have the inequality

$$\begin{aligned} u(re^{i\varphi};\nu) \le \frac{D}{q}\nu(\Pi)r &+ \frac{4}{r}\iint_{|\zeta|<\frac{1}{2}r} (1+|\zeta|^2)\, d\nu(\zeta) \\ &+ 18r\iint_{|\zeta|>\frac{3}{2}r} d\nu(\zeta), \qquad 0<\varphi<\pi, \end{aligned} \tag{3.9}$$

where D is an absolute constant, which holds for all $r \in [1,\infty)$ except, possibly, a subset $A_q \subset [1,\infty)$ satisfying

$$\int_{A_q} r^{-1} dr \le q.$$

Before turning to the proof of Theorem 3.2 we derive from it a corollary containing Lemma 3.1.

COROLLARY. *Let $u(z;\nu)$ be the function defined in the statement of Theorem 3.2. Then the relation*

$$u(re^{i\varphi};\nu) = o(r)$$

holds uniformly in φ, $0<\varphi<\pi$, as $r\to\infty$ missing a subset $A \subset [1,\infty)$ of finite logarithmic measure.

Let q be a positive integer, and A_q the corresponding set from the statement of Theorem 3.2. Choose a large number R_q so that $R_q > q$ and

$$\int_{A_q\cap(R_q,\infty)} r^{-1} dr < \frac{1}{2^q}.$$

Let

$$A = \bigcup_{q=1}^{\infty} \{A_q \cap (R_q,\infty)\}.$$

Obviously,

$$\int_A r^{-1} dr \le \sum_{q=1}^{\infty} 2^{-q} = 1.$$

Let r be sufficiently large, $r \notin A$, and let $q(r)$ be the largest integer q satisfying $R_q < r$. Then inequality (3.9) holds with $q = q(r)$. Obviously $q(r)\to\infty$ as $r\to\infty$; therefore

$$\frac{D}{q(r)}\nu(\Pi)r = o(r).$$

Since

$$\begin{aligned}\frac{4}{r}&\iint_{|\zeta|<\frac{1}{2}r}(1+|\zeta|^2)\,d\nu(\zeta)\\&=\frac{4}{r}\left\{\iint_{|\zeta|<\sqrt{r}}(1+|\zeta|^2)\,d\nu(\zeta)+\iint_{\sqrt{r}\le|\zeta|<\frac{1}{2}r}(1+|\zeta|^2)\,d\nu(\zeta)\right\}\\&\le\frac{4}{r}\left\{(1+r)\nu(\Pi)+\left(1+\frac{1}{4}r^2\right)\iint_{|\zeta|\ge\sqrt{r}}d\nu(\zeta)\right\}=o(r),\end{aligned}$$

$$18r\iint_{|\zeta|>\frac{3}{2}r}d\nu(\zeta)=o(r),$$

the Corollary and, thus, Lemma 3.1 have been proved.

Now we prove Theorem 3.2.

Denote by B the set of those values of $r\ge 1$ for which the following is true: for each h, $0<h\le\frac{1}{2}r$, the inequality

$$\nu(\{r-h\le|z|\le r+h\}\cap\Pi)<\frac{8\nu(\Pi)h}{qr}\tag{3.10}$$

holds. The set B is Borel since it can be represented in the form

$$B=\left\{r:\ \sup_{0\le h\le\frac{1}{2}r}\frac{F(r+h)-F(r-h-0)}{h}\le\frac{8\nu(\Pi)}{r}\right\},$$

where $F(r)=\nu(\{|z|\le r\})$ is a non-decreasing function. Therefore the set $A_q=[1,\infty)\backslash B$ is also Borel. Let us show that

$$\int_{A_q}r^{-1}dr\le\frac{3}{2}q.\tag{3.11}$$

Let A_{q1} be an arbitrary compact subset of the set A_q. The definition of the set B implies that for each $x\in A_{q1}$ there exists an interval $I_x=(x-h_x,x+h_x)$, $0<h_x\le\frac{1}{2}x$, such that

$$\nu(\{|z|\in\bar{I}_x\}\cap\Pi)\ge\frac{8\nu(\Pi)h_x}{qx},\tag{3.12}$$

where $\bar{I}_x=[x-h_x,x+h_x]$. By the Heine–Borel lemma, there is a finite system $\{I_{x_k}\}$ of intervals covering A_{q1}. Discarding the intervals contained in unions of two other intervals, we may assume that each point of the set A_{q1} is covered by at most two intervals of the system $\{I_{x_k}\}$. It follows that each point of the set A_{q1} belongs to at most three closed intervals $\bar{I}_{x_k}$. Therefore we have

$$\sum_k\nu\left(\{|z|\in\bar{I}_{x_k}\}\cap\Pi\right)\le 3\nu(\Pi).$$

Hence, using (3.12), we get

$$\begin{aligned}\int_{A_{q1}}r^{-1}dr&\le\sum_k\int_{\bar{I}_{x_k}}r^{-1}dr=\sum_k\ln\left(1+\frac{2h_{x_k}}{x_k-h_{x_k}}\right)\le 4\sum_k\frac{h_{x_k}}{x_k}\\&\le 4\frac{q}{8\nu(\Pi)}\sum_k\nu\left(\{|z|\in\bar{I}_{x_k}\}\cap\Pi\right)\le\frac{3}{2}q.\end{aligned}$$

Let $A_{q1}^{(1)} \subset A_{q1}^{(2)} \subset \cdots$ be a sequence of compact subsets of the set A_q, such that

$$\operatorname{meas}\left(A_q \setminus \bigcup_{n=1}^{\infty} A_{q1}^{(n)}\right) = 0.$$

Then we have

$$\int_{A_q} r^{-1} dr = \int_{\bigcup_{n=1}^{\infty} A_{q1}^{(n)}} r^{-1} dr = \lim_{n\to\infty} \int_{A_{q1}^{(n)}} r^{-1} dr \le \frac{3}{2} q,$$

and inequality (3.11) has been proved.

Now we show that (3.9) is satisfied for $r \in B$, thus completing our proof of Theorem 3.2.

We need the following estimate of the function $K(z,\zeta)$:

$$K(z,\zeta) \le \frac{(1+|\zeta|^2)y}{|z-\zeta|^2} \qquad (z = x+iy,\ y>0). \tag{3.13}$$

For $\eta = 0$ ($\zeta = \xi + i\eta$) this estimate is trivial. For $\eta > 0$ we have

$$\begin{aligned} K(z,\zeta) &= \frac{1+|\zeta|^2}{4\eta} \ln \frac{(x-\xi)^2+(y+\eta)^2}{(x-\xi)^2+(y-\eta)^2} \\ &= \frac{1+|\zeta|^2}{4\eta} \ln\left(1 + \frac{4y\eta}{(x-\xi)^2+(y-\eta)^2}\right) \le \frac{(1+|\zeta|^2)y}{(x-\xi)^2+(y-\eta)^2} = \frac{(1+|\zeta|^2)y}{|z-\zeta|^2}. \end{aligned}$$

We write the equality ($z = re^{i\varphi}$, $0<\varphi<\pi$, $r>1$)

$$\begin{aligned} u(z;\nu) &= \iint_{|\zeta|<\frac{1}{2}r} K(z,\zeta)\,d\nu(\zeta) + \iint_{\frac{1}{2}r\le|\zeta|\le\frac{3}{2}r} K(z,\zeta)\,d\nu(\zeta) \\ &\quad + \iint_{|\zeta|>\frac{3}{2}r} K(z,\zeta)\,d\nu(\zeta) = I_1 + I_2 + I_3. \end{aligned} \tag{3.14}$$

The integrals I_1 and I_3 can be easily estimated, using (3.13), as follows:

$$\begin{aligned} I_1 &\le \iint_{|\zeta|<\frac{1}{2}r} \frac{(1+|\zeta|^2)y}{|z-\zeta|^2}\,d\nu(\zeta) \le \frac{4y}{|z|^2} \iint_{|\zeta|<\frac{1}{2}r} (1+|\zeta|^2)\,d\nu(\zeta) \\ &\le \frac{4}{r} \iint_{|\zeta|<\frac{1}{2}r} (1+|\zeta|^2)\,d\nu(\zeta); \\ I_3 &\le \iint_{|\zeta|>\frac{3}{2}r} \frac{(1+|\zeta|^2)y}{|z-\zeta|^2}\,d\nu(\zeta) \le 9y \iint_{|\zeta|>\frac{3}{2}r} \frac{1+|\zeta|^2}{|\zeta|^2}\,d\nu(\zeta) \\ &\le 18y \iint_{|\zeta|>\frac{3}{2}r} d\nu(\zeta). \end{aligned} \tag{3.15}$$

Let us estimate the integral I_2. Fixing a point $z = re^{i\varphi}$, $0<\varphi<\pi$, $r>1$, denote by C_n, $n = 0, \pm1, \pm2, \ldots$, the (possibly, empty) set of all points $\zeta \in \Pi$ satisfying the following two conditions:

$$2^{n-1}y \le |z-\zeta| < 2^n y, \qquad \frac{1}{2}r \le |\zeta| \le \frac{3}{2}r \qquad (y = \operatorname{Im} z).$$

Obviously the sets C_n are pairwise disjoint and

$$\bigcup_{n=-\infty}^{\infty} C_n = \left\{\frac{1}{2}r \le |\zeta| \le \frac{3}{2}r\right\} \setminus \{z\}.$$

Since $r \in B$, by (3.10) we have

$$\nu(\{z\}) \le \nu(\{|z| = r\}) = 0,$$

hence

$$I_2 = \sum_{n=-\infty}^{\infty} \iint_{C_n} K(z,\zeta)\, d\nu(\zeta).$$

Note that the condition $r \in B$, by (3.10), implies

$$\nu(C_n) \le \nu(\{r-h \le |\zeta| \le r+h\}) < \frac{8\nu(\Pi)}{qr} 2^n y, \qquad h = \min\left(2^n y, \frac{r}{2}\right).$$

For $\zeta \in C_n$ inequality (3.13) implies

$$K(z,\zeta) \le \frac{(1+|\zeta|^2)y}{(2^{n-1}y)^2} \le \frac{1+\frac{9}{4}r^2}{4^{n-1}y} \le \frac{16r^2}{4^n y};$$

therefore

$$\sum_{n=0}^{\infty} \iint_{C_n} K(z,\zeta)\, d\nu(\zeta) \le \sum_{n=0}^{\infty} \frac{16r^2}{4^n y} \nu(C_n) \le \frac{16r^2}{y} \cdot \frac{8\nu(\Pi)y}{qr} \sum_{n=0}^{\infty} \frac{1}{2^n} \le \frac{256\nu(\Pi)r}{q}.$$

For $\zeta \in C_n$, $n < 0$, the condition $\eta \ge \frac{1}{2}y > 0$ holds; therefore we have

$$\begin{aligned} K(z,\zeta) = \frac{1+|\zeta|^2}{2\eta} \ln\left|\frac{z-\bar{\zeta}}{z-\zeta}\right| &\le \frac{1+\frac{9}{4}r^2}{y} \ln\left(1+\frac{2y}{|z-\zeta|}\right) \\ &\le \frac{1+\frac{9}{4}r^2}{y} \ln\left(1+\frac{2y}{2^{n-1}y}\right) \le \frac{4r^2}{y} \ln(6\cdot 2^{|n|}) \end{aligned}$$

and

$$\begin{aligned} \sum_{n=-\infty}^{-1} \iint_{C_n} K(z,\zeta)\, d\nu(\zeta) &\le \sum_{n=-\infty}^{-1} \frac{4r^2}{y} \ln(6\cdot 2^{|n|})\nu(C_n) \\ &\le \frac{4r^2}{y} \cdot \frac{8\nu(\Pi)y}{qr} \sum_{n=-\infty}^{-1} 2^{-|n|} \ln(6\cdot 2^{|n|}) \le \frac{128\nu(\Pi)r}{q}. \end{aligned}$$

Thus,

$$I_2 \le \frac{384\nu(\Pi)r}{q}. \tag{3.16}$$

Formulas (3.14)–(3.16) imply (3.9). As we have already mentioned, this completes the proof of Theorem 3.2. □

We obtain Theorem 2.3 as a corollary of the following theorem.

THEOREM 3.3. *Let $k(r) \ge 1$ be a continuous non-decreasing function, $U(r) = r(1+\frac{1}{k(r)})$. Then for each ε, $0 < \varepsilon < 1$, and for all $r \ge 1$ except, possibly, a set $A_\varepsilon \subset [1,\infty)$ satisfying $\mu(A_\varepsilon) < \varepsilon$, the inequality*

$$\ln^+ |f(re^{i\varphi})| \le Ck^3(r)r\{S(U(r),f)+1\}, \qquad 0 < \varphi < \pi, \tag{3.17}$$

holds, where C is a positive constant which does not depend on r and φ.

To get the second inequality of Theorem 2.3 for $r \notin A_\varepsilon$, we choose $k(r) \equiv \frac{1}{d-1}$. Then $U(r) \equiv dr$ and, integrating (3.17) with respect to φ from 0 to π, we get the desired inequality:

$$m_{0\pi}(r, f) \leq C\pi(d-1)^{-3} r\{S(dr, f) + 1\}.$$

The remaining assertions of Theorem 2.3 can be obtained in the following way. By Theorem 5.4 of Chapter 1 (p. 27) we have $S(r, f) = \overset{\circ}{S}(r, f) + O(1)$, where $\overset{\circ}{S}(r, f)$ is a non-decreasing function. Increasing, if necessary, the constant C, we can replace $S(r, f)$ by $\overset{\circ}{S}(r, f)$ in (3.17); after that we let $k(r) = \ln\{\overset{\circ}{S}(r, f) + 1\}$. Applying the corollary of the Borel–Nevanlinna theorem stated in Exercise 2 of Section 1 in Chapter 3 (p. 91) to the function $u(r) = \overset{\circ}{S}(r, f) + 1$, we get

$$\overset{\circ}{S}(U(r), f) + 1 \leq \{\overset{\circ}{S}(r, f) + 1\}^{\frac{1+d}{2}}$$

for $r \notin A$, where A is a set of finite logarithmic measure. For $r \notin A \cup A_\varepsilon$ we have

$$\begin{aligned} \ln^+ |f(re^{i\varphi})| &\leq C \ln^3\{\overset{\circ}{S}(r, f) + 1\} r \{\overset{\circ}{S}(r, f) + 1\}^{\frac{1+d}{2}} \\ &\leq C_1 r\{\overset{\circ}{S}(r, f) + 1\}^d \leq C_2 r\{S(r, f) + 1\}^d, \qquad 0 < \varphi < \pi, \end{aligned}$$

where $0 < C_1 < C_2 < \infty$ are constants which do not depend on r and φ. Integrating with respect to φ from 0 to π, we get ($r \notin A \cup A_3$)

$$m_{0\pi}(r, f) \leq C_2 \pi r\{S(r, f) + 1\}^d.$$

It remains to observe that $\mu(A \cup A_\varepsilon) \leq \mu(A) + \mu(A_\varepsilon) < \varepsilon$.

Now we turn to the proof of Theorem 3.3.

First we prove the inequality

$$\begin{aligned} \ln^+ |f(re^{i\varphi})| \leq{}& \frac{r \sin\varphi}{\pi} \int_{-R}^{R} \frac{\ln^+ |f(t)|}{r^2 + t^2 - 2rt\cos\varphi} dt \\ &+ \sum_{|b_n| < R} \ln \left| \frac{z - \bar{b}_n}{z - b_n} \right| + 2\left(\frac{R}{R - r}\right)^3 rB(R, f) \end{aligned} \tag{3.18}$$

$$(z = re^{i\varphi},\ 0 < r < R,\ 0 < \varphi < \pi).$$

Now we use formula (3.4). Observe that the expressions in braces in the formulas for $I_1(R)$ and $I_2(R)$ are non-negative by (1.10) and (1.11) in Chapter 1. Therefore

$$\begin{aligned} I_1(R) &\leq \frac{1}{\pi} \int_{-R}^{R} \ln^+ |f(t)| \left\{ \frac{r\sin\varphi}{|z - t|^2} - \frac{R^2 r \sin\varphi}{|R^2 - zt|^2} \right\} dt \\ &\leq \frac{r \sin\varphi}{\pi} \int_{-R}^{R} \frac{\ln^+ |f(t)|}{r^2 + t^2 - 2rt\cos\varphi} dt; \end{aligned}$$

$$\begin{aligned} I_2(R) &\leq \frac{1}{2\pi} \int_0^{\pi} \ln^+ |f(Re^{i\theta})| \left\{ \frac{R^2 - r^2}{|Re^{i\theta} - z|^2} - \frac{R^2 - r^2}{|Re^{-i\theta} - z|^2} \right\} d\theta \\ &= \frac{1}{2\pi} \int_0^{\pi} \ln^+ |f(Re^{i\theta})| \frac{(R^2 - r^2) 4Rr \sin\varphi \sin\theta \, d\theta}{(R^2 + r^2 - 2Rr\cos(\varphi - \theta))(R^2 + r^2 - 2Rr\cos(\varphi + \theta))} \\ &\leq \frac{(R^2 - r^2)4Rr}{(R - r)^4} \cdot \frac{1}{2\pi} \int_0^{\pi} \ln^+ |f(Re^{i\theta})| \sin\theta \, d\theta \leq 2\left(\frac{R}{R - r}\right)^3 rB(R, f). \end{aligned}$$

The summands in $I_3(R)$ and $I_4(R)$ represent, by formula (1.5) in Chapter 1, the values of the Green function for the half-disc $\{|z| < R, \operatorname{Im} z > 0\}$, and hence, are non-negative. Discarding $I_3(R)$ in (3.4) and observing that

$$\left|\frac{R^2 - \bar{b}_n z}{R(z - b_n)} \cdot \frac{R(z - \bar{b}_n)}{R^2 - b_n z}\right| \le \left|\frac{z - \bar{b}_n}{z - b_n}\right|,$$

we get inequality (3.18).

Let us estimate the function

$$u(re^{i\varphi}) = \frac{r \sin\varphi}{\pi} \int_{-R}^{R} \frac{\ln^+ |f(t)|\, dt}{r^2 + t^2 - 2rt\cos\varphi} + \sum_{|b_n|<R} \ln\left|\frac{z - \bar{b}_n}{z - b_n}\right| \tag{3.19}$$

using Theorem 3.2. This function can be represented in the form (3.8) by letting

$$\nu(E) = \frac{1}{\pi} \int_{(-R,R)\cap E} \frac{\ln^+ |f(t)|}{1 + t^2} dt + \sum_{\substack{b_n \in E \\ |b_n| < R}} \frac{2\rho_n \sin\psi_n}{1 + \rho_n^2} \qquad (b_n = \rho_n e^{i\psi_n}).$$

By Theorem 3.2 the inequality

$$\begin{aligned} u(re^{i\varphi}) &\le \frac{D}{q}\nu(\Pi) r + \frac{4}{r} \iint_{|\zeta|<\frac{1}{2}r} (1 + |\zeta|^2) d\nu(\zeta) + 18r \iint_{|\zeta|>\frac{3}{2}r} d\nu(\zeta) \\ &\le \frac{D}{q}\nu(\Pi) r + \frac{4}{r}\left(1 + \frac{1}{4}r^2\right) \iint_{|\zeta|<\frac{1}{2}r} d\nu(\zeta) + 18r \iint_{|\zeta|>\frac{3}{2}r} d\nu(\zeta) \\ &\le \frac{D}{q}\nu(\Pi) r + 18r\nu(\Pi) = \left(\frac{D}{q} + 18\right)\nu(\Pi) r, \qquad 0 < \varphi < \pi, \end{aligned} \tag{3.20}$$

holds for all $r \ge 1$ except, possibly, a subset $A_q \subset [1, \infty)$ satisfying

$$\int_{A_q} r^{-1} dr \le q. \tag{3.21}$$

Let us estimate

$$\nu(\Pi) = \frac{1}{\pi} \int_{-R}^{R} \frac{\ln^+ |f(t)|}{1 + t^2} dt + \sum_{|b_n|<R} \frac{2\rho_n \sin\psi_n}{1 + \rho_n^2}.$$

Let $U > R$. We have

$$\begin{aligned} A(U, f) &= \frac{1}{\pi} \int_1^U \left(\frac{1}{t^2} - \frac{1}{U^2}\right) \{\ln^+ |f(t)| + \ln^+ |f(-t)|\}\, dt \\ &\ge \frac{1}{\pi} \int_1^R \left(\frac{1}{t^2} - \frac{1}{U^2}\right) \{\ln^+ |f(t)| + \ln^+ |f(-t)|\}\, dt \\ &\ge \left(1 - \frac{R^2}{U^2}\right) \frac{1}{\pi} \int_1^R t^{-2} \{\ln^+ |f(t)| + \ln^+ |f(-t)|\}\, dt \\ &\ge \frac{U - R}{U\pi} \left\{\int_{-R}^{R} \frac{\ln^+ |f(t)|}{1 + t^2} dt - \int_{-1}^{1} \frac{\ln^+ |f(t)|}{1 + t^2} dt\right\}, \end{aligned}$$

$$
\begin{aligned}
C(U,f) &= 2\sum_{1<|b_n|<U}\left(\frac{1}{\rho_n}-\frac{\rho_n}{U^2}\right)\sin\psi_n \\
&\geq 2\sum_{1<|b_n|<R}\left(\frac{1}{\rho_n}-\frac{\rho_n}{U^2}\right)\sin\psi_n \geq \left(1-\frac{R^2}{U^2}\right)2\sum_{1<|b_n|<R}\frac{\sin\psi_n}{\rho_n} \\
&\geq \frac{U-R}{U}\left\{\sum_{|b_n|<R}\frac{2\rho_n\sin\psi_n}{1+\rho_n^2}-\sum_{|b_n|\leq 1}\frac{2\rho_n\sin\psi_n}{1+\rho_n^2}\right\}.
\end{aligned}
$$

From this we get

$$
\begin{aligned}
\nu(\Pi) &\leq \frac{U}{U-R}\{A(U,f)+C(U,f)\} \\
&\quad+\frac{1}{\pi}\int_{-1}^{1}\frac{\ln^+|f(t)|}{1+t^2}dt+\sum_{|b_n|\leq 1}\frac{2\rho_n\sin\psi_n}{1+\rho_n^2}.
\end{aligned}
\tag{3.22}
$$

In what follows the letter K will be used to denote finite positive constants which do not depend on variables denoted by r, R, φ, t, q, U, and ε.

By (3.18) and (3.19), for $0<r<R$, $0<\varphi<\pi$, the inequality

$$\ln^+|f(re^{i\varphi})| \leq u(re^{i\varphi})+2\left(\frac{R}{R-r}\right)^3 rB(R,f)$$

holds. From this, using (3.20) and (3.22), we get, for $r\in[1,R)\backslash A_q$, $0<\varphi<\pi$, the inequality

$$
\begin{aligned}
\ln^+|f(re^{i\varphi})| &\leq K\left(\frac{1}{q}+1\right)\frac{Ur}{U-R}\{A(U,f)+C(U,f)+1\} \\
&\quad+2\left(\frac{R}{R-r}\right)^3 rB(R,f) \qquad (U>R).
\end{aligned}
\tag{3.23}
$$

Since, by Theorem 5.4 of Chapter 1, the function $S(r,f)$ is, up to a bounded summand, non-decreasing, we have

$$B(R,f)\leq S(R,f)\leq S(U,f)+K.$$

Taking into account $A(U,f)+C(U,f)\leq S(U,f)$, we get the inequality

$$\ln^+|f(re^{i\varphi})| \leq Kr\left\{\left(\frac{1}{q}+1\right)\frac{U}{U-R}+\left(\frac{R}{R-r}\right)^3\right\}\{S(U,f)+1\}, \tag{3.24}$$

where $r\in[1,R)\backslash A_q$, $0<\varphi<\pi$, $U>R$.

Let (for $k(r)$ and ε from the statement of Theorem 3.3)

$$U(t)=t\left(1+\frac{1}{k(t)}\right),\qquad U_1(t)=\max_{0\leq\tau\leq t}U(\tau),$$

$$R(t)=t\left(1+\frac{1}{2k(t)}\right),\qquad t'=t\left(1+\frac{1}{4k(t)}\right),\qquad q(t)=\frac{\varepsilon}{16k(t)},$$

and denote by $j(t)$ the interval $[t,t']$. We have

$$\frac{U(t)}{U(t)-R(t)}\leq 4k(t),\qquad \frac{1}{q(t)}+1\leq\frac{17k(t)}{\varepsilon},$$

and the inequality

$$\frac{R(t)}{R(t)-r}\leq 6k(t)$$

for $r \in j(t)$. Hence inequality (3.24) implies that for $r \in j(t) \backslash A_{q(t)}$,

$$\ln^+ |f(re^{i\varphi})| \le Krk^3(t)\{S(U(t), f) + 1\}, \qquad 0 < \varphi < \pi.$$

Since, obviously, $t \le r$ for $r \in j(t) \backslash A_{q(t)}$, we also have $k(t) \le k(r)$, $U(t) \le U_1(t) \le U_1(r)$, and, by Theorem 5.4 of Chapter 1,

$$S(U(t), f) \le S(U_1(r), f) + K.$$

Therefore for $r \in j(t) \backslash A_{q(t)}$ we also have the inequality

$$\ln^+ |f(re^{i\varphi})| \le Krk^3(r)\{S(U_1(r), f) + 1\}, \qquad 0 < \varphi < \pi. \tag{3.25}$$

We show that (3.25) holds for all $r \ge 1$ except, possibly, a subset $E_\varepsilon \subset [1, \infty)$ satisfying $\mu(E_\varepsilon) \le \varepsilon$.

Denote by $E(t)$ the set $j(t) \cap A_{q(t)}$. Using (3.21) and the choice of $q(t)$, we get

$$\operatorname{meas} E(t) \le t' \int_{E(t)} \frac{d\tau}{\tau} \le t' \int_{A_{q(t)}} \frac{d\tau}{\tau} \le t' q(t) \le 2t \frac{\varepsilon}{16k(t)} = \frac{t\varepsilon}{8k(t)},$$

whence

$$\operatorname{meas} E(t) \le \frac{\varepsilon}{2} \operatorname{meas} j(t). \tag{3.26}$$

Let

$$t_0 = 1, \quad t_k = t'_{k-1}, \quad k = 1, 2, \dots,$$

and consider the sequence of intervals $j(t_k)$, $k = 0, 1, \dots$. These intervals cover the whole semiaxis $[1, \infty)$. Otherwise we would have $\lim_{k\to\infty} t_k = \alpha < \infty$, and choosing $\delta > 0$ and n so that $4k(\alpha) < \delta^{-1}$ and $t_n > \alpha(1+\delta)^{-1}$, we would arrive at a contradiction:

$$t_{n+1} = t'_n = t_n \left(1 + \frac{1}{4k(t_n)}\right) \ge t_n \left(1 + \frac{1}{4k(\alpha)}\right) > \frac{\alpha}{1+\delta}(1+\delta) = \alpha.$$

Take

$$E_\varepsilon = \bigcup_{k=0}^{\infty} E(t_k).$$

Obviously, for $r \in [1, \infty) \backslash E_\varepsilon$ inequality (3.25) holds. Moreover, by (3.26), we have

$$\begin{aligned}
\mu(E_\varepsilon) &= \limsup_{r\to\infty} r^{-1} \operatorname{meas}\{E_\varepsilon \cap (0, r)\} \\
&\le \limsup_{r\to\infty} r^{-1} \sum_{t_k < r} \operatorname{meas} E(t_k) \le \frac{\varepsilon}{2} \limsup_{r\to\infty} r^{-1} \sum_{t_k < r} \operatorname{meas} j(t_k) \\
&\le \frac{\varepsilon}{2} \lim_{r\to\infty} r^{-1} \operatorname{meas} \left\{ \left[1, r\left(1 + \frac{1}{4k(r)}\right)\right] \right\} \le \frac{\varepsilon}{2} \cdot \frac{5}{4} < \varepsilon.
\end{aligned}$$

The theorem will be proved is we establish that for all $r \ge 1$ except, possibly, a set of finite logarithmic measure, the equality $U_1(r) = U(r)$ holds. Denote by E the set on which $U_1(r) > U(r)$. Since the functions $U_1(r)$ and $U(r)$ are continuous, the set E is open and hence is a union of a finite or a countable set of pairwise disjoint intervals (a_α, b_α).

Since $\lim_{r\to\infty} U(r) = \infty$, none of the intervals can be infinite.

Obviously, $U(a_\alpha) = U(b_\alpha)$; that is,

$$a_\alpha \left(1 + \frac{1}{k(a_\alpha)}\right) = b_\alpha \left(1 + \frac{1}{k(b_\alpha)}\right).$$

This implies

$$b_\alpha - a_\alpha = \frac{a_\alpha}{k(a_\alpha)} - \frac{b_\alpha}{k(b_\alpha)} \le a_\alpha \left(\frac{1}{k(a_\alpha)} - \frac{1}{k(b_\alpha)} \right),$$

and we get

$$\int_E \frac{dt}{t} = \sum_\alpha \int_{a_\alpha}^{b_\alpha} \frac{dt}{t} \le \sum_\alpha \frac{b_\alpha - a_\alpha}{a_\alpha} \le \sum_\alpha \left(\frac{1}{k(a_\alpha)} - \frac{1}{k(b_\alpha)} \right).$$

Using the Leibniz theorem concerning alternating series, we come to the desired conclusion. □

4. Meromorphic functions with poles and zeros in small angles

Consider a system of rays

$$D' = D(\alpha_1, \beta_1, \alpha_2, \beta_2, \ldots, \alpha_n, \beta_n), \tag{4.1}$$

$$\alpha_1 < \beta_1 < \alpha_2 < \beta_2 < \cdots < \alpha_n < \beta_n < \alpha_{n+1} = \alpha_1 + 2\pi, \tag{4.2}$$

such that

$$\max_{1 \le j \le n} (\beta_j - \alpha_j) < \min_{1 \le j \le n} (\alpha_{j+1} - \beta_j). \tag{4.3}$$

The angles $\{\alpha_j \le \arg z \le \beta_j\}$ will be called *small angles* of the system D'. Let

$$\omega_j'(D') = \pi/(\beta_j - \alpha_j), \qquad \omega_j''(D') = \pi/(\alpha_{j+1} - \beta_j), \qquad j = 1, \ldots, n,$$

$$\omega'(D') = \min_{1 \le j \le n} \omega_j'(D'),\ \omega''(D') = \max_{1 \le j \le n} \omega_j''(D').$$

Note that by condition (4.3),

$$\omega'(D') > \omega''(D').$$

DEFINITION. If the condition

$$\sum_{j=1}^{n} r^{\omega_j''(D') - \omega''(D')} \bar{C}_{\beta_j \alpha_{j+1}}(r, a, f) = O(1), \qquad r \to \infty, \tag{4.4}$$

is satisfied, we say that almost all a-points of the function $f(z)$ *are in small angles of the system* D'.

This condition is obviously satisfied if all a-points of the function $f(z)$ except, possibly, finitely many are in small angles of the system D'.

The following theorem is the main result of this section.

THEOREM 4.1. *Let $f(z)$ be a meromorphic function of order $\rho < \infty$, such that almost all its zeros and poles are in small angles of the system D'. If*

$$\delta(a, f) > 1 - \cos \frac{\pi\rho}{2\omega'(D')} \tag{4.5}$$

for some $a \ne 0, \infty$, then the relation

$$\omega''(D') < \rho < \omega'(D')$$

cannot be true.

In particular, if the system D' consists of two rays, $D' = D(\alpha, \beta)$, $0 < \beta - \alpha < \pi$, then $\omega'(D') = \pi/(\beta - \alpha)$, $\omega''(D') = \pi/(2\pi - \beta + \alpha)$, and Theorem 4.1 immediately implies the following.

COROLLARY 1. *Let $f(z)$ be a meromorphic function of order $\rho < \infty$, with almost all poles and zeros in the small angle of the system $D' = D(\alpha, \beta)$, $0 < \beta - \alpha < \pi$. If*

$$\delta(a, f) > 1 - \cos\frac{\rho(\beta - \alpha)}{2} \tag{4.5'}$$

for some $a \neq 0, \infty$, then the relation

$$\frac{\pi}{2\pi - \beta + \alpha} < \rho < \frac{\pi}{\beta - \alpha}$$

cannot be true.

From this we get the following.

COROLLARY 2. *Let $f(z)$ be an entire function of order ρ, with all its b- and c-points ($b \neq c$; b, $c \neq \infty$) except, possibly, finitely many, in the angle $\{\alpha \leq \arg z \leq \beta\}$, $0 < \beta - \alpha < \pi$. Then either $\rho \geq \pi/(\beta - \alpha)$, or $\rho \leq \pi/(2\pi - \beta + \alpha)$.*

In fact, consider the system $D' = D(\alpha, \beta)$ and the meromorphic function $f_1(z) = \frac{f(z)-b}{f(z)-c}$. Almost all zeros and poles of this function are in the small angle of the system $D' = D(\alpha, \beta)$, and $\delta(1, f_1) = 1$. If $\rho < \pi/(\beta - \alpha)$, then the function $f_1(z)$ satisfies condition (4.5′) with $a = 1$, and by Corollary 1 the inequality $\rho > \pi/(2\pi - \beta + \alpha)$ cannot be true.

We show that for each $\rho \geq \pi/(\beta - \alpha)$ and each $\rho \leq \pi/(2\pi - \beta + \alpha)$ it is possible to construct an entire function of order ρ, such that for some b and c ($b \neq c; b, c \neq \infty$), all b- and c-points of $f(z)$, except, possibly, finitely many are in the angle $\{\alpha \leq \arg z \leq \beta\}$.

EXAMPLE 1. As was proved in 3° in Section 5 of Chapter 2, the Mittag-Leffler function $E_\rho(z)$ has order ρ. We assume that $\rho > 1/2$. By formula (5.40) in Chapter 2, the value of

$$M = \sup_{\frac{\pi}{2\rho} \leq \arg z \leq 2\pi - \frac{\pi}{2\rho}} |E_\rho(z)|$$

is finite. Therefore, if the numbers b and c satisfy the condition $\min(|b|, |c|) > M$, then all b- and c-points of the function $E_\rho(z)$ are in the angle $\{-\frac{\pi}{2\rho} \leq \arg z \leq \frac{\pi}{2\rho}\}$. Thus, for each $\rho \geq \pi/(\beta - \alpha)$, all b- and c-points of the function $E_\rho(ze^{-i(\alpha + \frac{\pi}{2\rho})})$ are in the angle $\{\alpha \leq \arg z \leq \alpha + \frac{\pi}{\rho}\} \subset \{\alpha \leq \arg z \leq \beta\}$

Consider now the case $\rho \leq \pi/(2\pi - \beta + \alpha)$. Formula (5.40) in Chapter 2 implies that the number

$$m = \inf_{\substack{-\frac{\pi}{2\rho} \leq \arg z \leq \frac{\pi}{2\rho} \\ |z| \geq R}} |E_\rho(z)|$$

is positive if $R > 0$ is sufficiently large. Therefore, if the numbers b and c satisfy the condition $\max(|b|, |c|) < m$, then all b- and c-points of the function $E_\rho(z)$ except, possibly, finitely many are in the angle $\{\frac{\pi}{2\rho} \leq \arg z \leq 2\pi - \frac{\pi}{2\rho}\}$. Then all b- and

c-points of the function $E_\rho(ze^{-i(\alpha-\frac{\pi}{2\rho})})$ except, possibly, finitely many are in the angle $\{\alpha \le \arg z \le \alpha + 2\pi - \frac{\pi}{\rho}\} \subset \{\alpha \le \arg z \le \beta\}$.

Later on we show that Corollary 1 is no longer true if the sign $>$ in (4.5′) is replaced by $\ge$. Now we observe that condition (4.5′) cannot be dropped. In fact, Example 2 in Section 2 shows that even for a function whose zeros and poles lie on a ray, the growth can be absolutely arbitrary.

Now we point to a connection between Theorems 4.1 and 2.1.

Let all zeros and poles except, possibly, finitely many of a meromorphic function $f(z)$ of order ρ lie on rays of the system $D = D(\alpha_1, \alpha_2, \dots, \alpha_n)$, and $\delta(a, f) > 0$ for some $a \ne 0, \infty$. Then, by Theorem 2.1, the inequality $\rho \le \omega(D)$ holds. Under the additional assumption that $\rho < \infty$, this estimate can also be derived from Theorem 4.1. In fact, we choose numbers β_j in such a way that inequalities (4.2) and (4.3) hold. Then almost all zeros and poles of the function $f(z)$ are in small angles of the system $D' = D(\alpha_1, \beta_1, \dots, \alpha_n, \beta_n)$. Choosing β_j sufficiently close to α_j, we can make $\omega'(D')$ arbitrarily large; in particular, so large that (4.5) holds and $\rho < \omega'(D')$. By Theorem 4.1 we get $\rho \le \omega''(D')$ and, since $\omega''(D') \to \omega(D)$ as $\beta_j \to \alpha_j$, we conclude that $\rho \le \omega(D)$.

Note also that if we formally pass to the limit as $\beta_j \to \alpha_j$ $(j = 1, \dots, n)$ in Theorem 4.1, its assumptions will take the form

$$U_D(r, 0, \infty) = O(1), \qquad \delta(a, f) > 0,$$

where $D = D(\alpha_1, \alpha_2, \dots, \alpha_n)$, and the statement will be reduced to the inequality $\rho \le \omega(D)$; that is, we get a special case of Theorem 2.1.

Now we turn to auxiliary results used in the proof of Theorem 4.1.

LEMMA 4.1. *Let*

$$\chi(z) = \frac{\prod_{\nu=1}^{\infty}\left(1 - \frac{z}{\bar{p}_\nu}\right)}{\prod_{\nu=1}^{\infty}\left(1 - \frac{z}{p_\nu}\right)} \tag{4.6}$$

be a meromorphic function of genus zero, with all its poles p_ν in the half-plane $\{\operatorname{Im} z > 0\}$. Suppose that the order of the function $\chi(z)$ is equal to $\eta < 1$. Then for each σ, $\eta < \sigma < 1$,

$$\int_b^\infty \frac{m(t,\chi)}{t^{\sigma+1}}\,dt \le \left(\sec\frac{\pi\sigma}{2} - 1\right)\int_b^\infty \frac{N(t,\chi)}{t^{\sigma+1}}\,dt, \qquad 0 < b < \infty. \tag{4.7}$$

PROOF. First we establish the inequality

$$\int_b^\infty \frac{q(t)}{t^{1+\sigma}}\,dt \le \left(\sec\frac{\pi\sigma}{2} - 1\right)\int_b^\infty \frac{\ln^+ t}{t^{1+\sigma}}\,dt, \quad 0 < b < \infty, \quad 0 < \sigma < 1, \tag{4.8}$$

where

$$q(t) = \frac{1}{2\pi}\int_{-\pi/2}^{\pi/2} \ln\left|\frac{1 + te^{i\theta}}{1 - te^{i\theta}}\right| d\theta.$$

Choose an arbitrary branch of the function $\arg(\frac{1+z}{1-z})$ on the semicircle

$$C_t = \left\{|z| = t,\ -\frac{\pi}{2} \le \arg z \le \frac{\pi}{2}\right\}$$

oriented counterclockwise. By the Cauchy–Riemann equations we have ($t \neq 1$)

$$\begin{aligned} q'(t) &= \frac{1}{2\pi}\int_{-\pi/2}^{\pi/2} \frac{\partial}{\partial t}\ln\left|\frac{1+te^{i\theta}}{1-te^{i\theta}}\right| dt \\ &= \frac{1}{2\pi t}\int_{-\pi/2}^{\pi/2} \frac{\partial}{\partial\theta}\arg\left(\frac{1+te^{i\theta}}{1-te^{i\theta}}\right) d\theta = \frac{1}{2\pi t}\Delta_{C_t}\arg\left(\frac{1+te^{i\theta}}{1-te^{i\theta}}\right). \end{aligned}$$

Therefore

$$\begin{aligned} q'(t) &= \frac{2}{\pi t}\arctan t, \qquad 0 < t < 1, \\ q'(t) &= \frac{2}{\pi t}\arctan t - \frac{1}{t}, \qquad t > 1; \end{aligned}$$

hence,

$$q(t) = \frac{2}{\pi}\int_0^t \arctan s \frac{ds}{s} - \ln^+ t.$$

Below we use the argument employed in the proof of Lemma 3.3 of Chapter 5. Consider the function

$$\varphi(b) = \int_b^\infty \left\{\left(\sec\frac{\pi\sigma}{2} - 1\right)\ln^+ t - q(t)\right\}\frac{dt}{t^{1+\sigma}}. \tag{4.9}$$

Since

$$\begin{aligned} \int_0^\infty \frac{q(t)}{t^{1+\sigma}}dt &= \frac{1}{\sigma}\int_0^\infty \frac{q'(t)}{t^\sigma}dt = \frac{2}{\pi\sigma}\int_0^\infty \frac{\arctan t}{t^{1+\sigma}}dt - \frac{1}{\sigma}\int_1^\infty \frac{dt}{t^{1+\sigma}} \\ &= \frac{2}{\pi\sigma^2}\int_0^\infty \frac{dt}{t^\sigma(1+t^2)} - \frac{1}{\sigma^2} = \frac{1}{\sigma^2}\left(\sec\frac{\pi\sigma}{2} - 1\right), \end{aligned}$$

$$\int_0^\infty \frac{\ln^+ t}{t^{1+\sigma}}dt = \frac{1}{\sigma^2},$$

we have $\varphi(0) = 0$. Obviously $\lim_{b\to\infty}\varphi(b) = 0$.

The integrand in (4.9) is negative for $0 < t < 1$, and for $t > 1$ we have

$$\left\{\left(\sec\frac{\pi\sigma}{2} - 1\right)\ln^+ t - q(t)\right\}' = \frac{1}{t}\left\{\sec\frac{\pi\sigma}{2} - \frac{2}{\pi}\arctan t\right\} > 0.$$

Therefore the function $\varphi'(b)$ changes its sign only once on the semiaxis $\{0 < b < \infty\}$, from plus to minus. Since $\lim_{b\to\infty}\varphi(b) = 0$, we have $\varphi(b) > 0$ for all b, $0 < b < \infty$; thus, the inequality (4.8) has been proved.

Using Lemma 4.4 of Chapter 5 and the notation in its statement, we have

$$m(r,\chi) \leq m(r,\check{\chi}). \tag{4.10}$$

But, obviously,

$$\begin{aligned} m(r,\check{\chi}) &= \frac{1}{2\pi}\int_0^{2\pi} \ln^+\left|\prod_{\nu=1}^\infty \frac{1-\frac{re^{i\theta}}{|p_\nu|}}{1+\frac{re^{i\theta}}{|p_\nu|}}\right| d\theta \\ &= \frac{1}{2\pi}\int_{\pi/2}^{3\pi/2} \ln\left|\prod_{\nu=1}^\infty \frac{1-\frac{re^{i\theta}}{|p_\nu|}}{1+\frac{re^{i\theta}}{|p_\nu|}}\right| d\theta \\ &= \frac{1}{2\pi}\int_{-\pi/2}^{\pi/2} \ln\left|\prod_{\nu=1}^\infty \frac{1+\frac{re^{i\theta}}{|p_\nu|}}{1-\frac{re^{i\theta}}{|p_\nu|}}\right| d\theta = \sum_{\nu=1}^\infty q\left(\frac{r}{|p_\nu|}\right). \end{aligned}$$

By (4.8), for each $b > 0$, we have

$$\int_b^\infty \frac{q\left(\frac{r}{|p_\nu|}\right)}{r^{1+\sigma}} dr \le \left(\sec\frac{\pi\sigma}{2} - 1\right) \int_b^\infty \frac{\ln^+ \frac{r}{|p_\nu|}}{r^{1+\sigma}} dr.$$

Assuming that $\eta < \sigma < 1$ and summing over ν, we get

$$\int_b^\infty \frac{m(r,\check{\chi})}{r^{1+\sigma}} dr \le \left(\sec\frac{\pi\sigma}{2} - 1\right) \int_b^\infty \frac{N(r,\chi)}{r^{1+\sigma}} dr,$$

from which, by (4.10), the inequality (4.7) follows. □

LEMMA 4.2. *Let $g(t) \ge 0$, $-\infty < t < \infty$, be a function such that for some λ, $0 \le \lambda \le 2$, the inequality*

$$\int_{-\infty}^\infty \frac{g(t)}{1+|t|^\lambda} < \infty$$

holds. Let

$$u(z) = \frac{y}{\pi} \int_{-\infty}^\infty \frac{g(t)dt}{(x-t)^2 + y^2}, \qquad z = x + iy, \quad y > 0.$$

Then

$$u(z) \le K|z|^\lambda y^{-1}, \qquad |z| > 1,$$

where K is a positive constant which does not depend on z.

The statement of the lemma follows from the estimate

$$\begin{aligned}\frac{\pi u(z)}{y} &= \int_{-\infty}^\infty \frac{g(t)}{|z-t|^2} dt = \int_{|t|<2|z|} \frac{g(t)}{1+|t|^\lambda} \cdot \frac{1+|t|^\lambda}{|z-t|^2} dt \\ &\quad + \int_{|t|\ge 2|z|} \frac{g(t)}{1+|t|^\lambda} \cdot \frac{1+|t|^\lambda}{t^2} \cdot \frac{dt}{|zt^{-1}-1|^2} \\ &\le \frac{1+2^\lambda|z|^\lambda}{y^2} \int_{|t|<2|z|} \frac{g(t)\,dt}{1+|t|^\lambda} + \left(\frac{1}{|z|^2} + \frac{2^\lambda}{|z|^{2-\lambda}}\right) \int_{|t|\ge 2|z|} \frac{g(t)\,dt}{1+|t|^\lambda}.\end{aligned}$$

LEMMA 4.3. *Let $f(z)$ be a meromorphic function of finite order, let $a \ne 0, \infty$ be a complex number, and $0 < \beta - \alpha \le 2\pi$, $\omega = \pi/(\beta-\alpha)$. Suppose that*

$$\bar{C}_{\alpha\beta}(r,f) + \bar{C}_{\alpha\beta}\left(r, \frac{1}{f}\right) = O(r^p), \qquad p > 0, \quad r \to \infty.$$

For an arbitrary ε satisfying $0 < \varepsilon < \frac{1}{2}(\beta-\alpha)$, the estimate

$$m_{\alpha+\varepsilon,\ \beta-\varepsilon}\left(r, \frac{1}{f-a}\right) = O(r^{\omega+p}), \qquad r \to \infty, \tag{4.11}$$

holds, and for each $\varepsilon_1 > 0$ the integral

$$I(\varphi) = \int_1^\infty \ln^+ \frac{1}{|f(re^{i\varphi}) - a|} \frac{dr}{r^{1+p+\omega+\varepsilon_1}}$$

converges for all $\varphi \in (\alpha+\varepsilon,\ \beta-\varepsilon)$ except, possibly, a set E satisfying meas $E = 0$.

PROOF. It is easy to see that

$$m_{\alpha+\varepsilon,\ \beta-\varepsilon}\left(r, \frac{1}{f-a}\right) \le \frac{r^\omega}{4\omega \sin\omega\varepsilon} B_{\alpha\beta}\left(r, \frac{1}{f-a}\right).$$

Using the analog of Lemma 2.1 of Chapter 3 for the angular Nevanlinna characteristics, that is valid by Theorem 3.3 of Chapter 3, we have

$$B_{\alpha\beta}\left(r, \frac{1}{f-a}\right) \leq A_{\alpha\beta}\left(r, \frac{f}{f'}\right) + B_{\alpha\beta}\left(r, \frac{f}{f'}\right) + Q_{\alpha\beta}(r, f) \leq S_{\alpha\beta}\left(r, \frac{f}{f'}\right) + O(1).$$

Now we use Theorems 5.2 of Chapter 1 and 3.1 of Chapter 3, and get

$$\begin{aligned} S_{\alpha\beta}\left(r, \frac{f}{f'}\right) &= A_{\alpha\beta}\left(r, \frac{f'}{f}\right) + B_{\alpha\beta}\left(r, \frac{f'}{f}\right) + C_{\alpha\beta}\left(r, \frac{f'}{f}\right) + O(1) \\ &= \bar{C}_{\alpha\beta}(r, f) + \bar{C}_{\alpha\beta}\left(r, \frac{1}{f}\right) + O(1). \end{aligned}$$

Thus,

$$m_{\alpha+\varepsilon,\ \beta-\varepsilon}\left(r, \frac{1}{f-a}\right) \leq \frac{r^{\omega}}{4\omega \sin \omega\varepsilon}\left\{\bar{C}_{\alpha\beta}(r, f) + \bar{C}_{\alpha\beta}\left(r, \frac{1}{f}\right) + O(1)\right\} = O(r^{\omega+p}).$$

Dividing relation (4.11) by $r^{1+p+\omega+\varepsilon_1}$ and integrating with respect to r from 1 to ∞, we get

$$\begin{aligned} \infty &> \int_1^{\infty} m_{\alpha+\varepsilon,\ \beta-\varepsilon}\left(r, \frac{1}{f-a}\right) \frac{dr}{r^{1+p+\omega+\varepsilon_1}} \\ &= \frac{1}{2\pi} \int_1^{\infty} \frac{dr}{r^{1+p+\omega+\varepsilon_1}} \int_{\alpha+\varepsilon}^{\beta-\varepsilon} \ln^+ \frac{1}{|f(re^{i\varphi}) - a|} d\varphi \\ &= \frac{1}{2\pi} \int_{\alpha+\varepsilon}^{\beta-\varepsilon} d\varphi \int_1^{\infty} \ln^+ \frac{1}{|f(re^{i\varphi}) - a|} \cdot \frac{dr}{r^{1+p+\omega+\varepsilon_1}} = \frac{1}{2\pi} \int_{\alpha+\varepsilon}^{\beta-\varepsilon} I(\varphi)\, d\varphi. \end{aligned}$$

From this it follows that $I(\varphi) < \infty$ for all $\varphi \in (\alpha+\varepsilon,\ \beta-\varepsilon)$, except, possibly, a set E, $\operatorname{meas} E = 0$. □

Now we turn to a proof of Theorem 4.1. Suppose that the order ρ of the function $f(z)$ satisfies the inequality $\omega''(D') < \rho < \omega'(D')$, and prove that for each $a \neq 0, \infty$ the inequality

$$\delta(a, f) \leq 1 - \cos \frac{\pi\rho}{2\omega'(D')} \tag{4.12}$$

holds.

Denote by D'_ε the system $D'_\varepsilon = D(\alpha_1-\varepsilon,\ \beta_1+\varepsilon,\ \alpha_2-\varepsilon,\ \beta_2+\varepsilon, \dots, \alpha_n-\varepsilon,\ \beta_n+\varepsilon)$ of rays, where $\varepsilon > 0$ is so small that

$$\max_{1\leq j\leq n} (\beta_j - \alpha_j) + 4\varepsilon < \min_{1\leq j\leq n} (\alpha_{j+1} - \beta_j),$$

$$\omega''(D') < \omega''(D'_\varepsilon) < \omega''(D'_{2\varepsilon}) < \rho < \omega'(D'_{2\varepsilon}) < \omega'(D'_\varepsilon) < \omega'(D'). \tag{4.13}$$

We have

$$\begin{aligned} m(r, a) &= \sum_{j=1}^{n} m_{\alpha_j-\varepsilon,\ \beta_j+\varepsilon}\left(r, \frac{1}{f-a}\right) + \sum_{j=1}^{n} m_{\beta_j+\varepsilon,\ \alpha_{j+1}-\varepsilon}\left(r, \frac{1}{f-a}\right) \\ &= \Phi_1(r) + \Phi_2(r). \end{aligned}$$

The estimate for the sum $\Phi_2(r)$ is obtained using Lemma 4.3. Since almost all of zeros and poles of the function $f(z)$ are in small angles of the system D', we have

$$\bar{C}_{\beta_j\alpha_{j+1}}(r, f) + \bar{C}_{\beta_j\alpha_{j+1}}\left(r, \frac{1}{f}\right) = O\left(r^{\omega''(D') - \omega''_j(D')}\right), \qquad j = 1, \dots, n.$$

Applying Lemma 4.3, we get the estimate

$$m_{\beta_j+\varepsilon,\ \alpha_{j+1}-\varepsilon}\left(r, \frac{1}{f-a}\right) = O(r^{\omega''(D')}), \qquad j = 1, \dots, n;$$

hence,

$$\Phi_2(r) = O(r^{\omega''(D')}). \tag{4.14}$$

The estimate of $\Phi_1(r)$ is significantly more complicated.

By Lemma 4.3, there are numbers φ_j and ψ_j, $\alpha_j - 2\varepsilon < \varphi_j < \alpha_j - \varepsilon$, $\beta_j + \varepsilon < \psi_j < \beta_j + 2\varepsilon$, such that for $\theta = \varphi_j,\ \psi_j\ (j = 1, \dots, n)$ we have

$$\int_1^\infty \ln^+ \frac{1}{|f(re^{i\theta}) - a|} \frac{dr}{r^{1+\omega''(D'_\varepsilon)}} < \infty. \tag{4.15}$$

We map the angle $\{\varphi_j \le \arg z \le \psi_j\}$ onto the half-plane $\{\operatorname{Im} \zeta \ge 0\}$ using the mapping $\zeta = (ze^{-i\varphi_j})^{\pi/(\psi_j - \varphi_j)}$. Under this mapping the angle $\{\alpha_j - \varepsilon \le \arg z \le \beta_j + \varepsilon\}$ is mapped onto an angle of the form $\{\xi_j \le \arg \xi \le \eta_j\}$, $0 < \xi_j < \eta_j < \pi$. Let

$$F_j(\zeta) = \{f(e^{i\varphi_j}\zeta^{(\psi_j - \varphi_j)/\pi}) - a\}^{-1}. \tag{4.16}$$

It is easy to see that

$$m_{\alpha_j-\varepsilon,\ \beta_j+\varepsilon}\left(r, \frac{1}{f-a}\right) = \frac{\psi_j - \varphi_j}{\pi} m_{\xi_j\eta_j}\left(r^{\frac{\pi}{\psi_j - \varphi_j}}, F_j\right); \tag{4.17}$$

therefore to estimate the sum Φ_1 it suffices to estimate the quantities $m_{\xi_j\eta_j}(r, F_j)$, $j = 1, \dots, n$.

Let us show that the function $F_j(z)$ satisfies

$$S(r, F_j) = O(1), \tag{4.18}$$

and hence, Theorem 3.1 is applicable to it.

Note that

$$\int_{-\infty}^\infty \frac{\ln^+ |F_j(t)|}{1 + |t|^{1+\mu_j}} dt < \infty, \tag{4.19}$$

where $\mu_j = (\psi_j - \varphi_j)\omega''(D'_\varepsilon)/\pi$. In fact, by (4.16) and (4.15), we have

$$\begin{aligned}\int_{-\infty}^\infty \frac{\ln^+ |F_j(t)|}{1 + |t|^{1+\mu_j}} dt &= \int_{-1}^1 \frac{\ln^+ |F_j(t)|}{1 + |t|^{1+\mu_j}} dt + \int_1^\infty \frac{\ln^+ |F_j(t)|}{t^{1+\mu_j}} dt + \int_{-\infty}^{-1} \frac{\ln^+ |F_j(t)|}{|t|^{1+\mu_j}} dt \\ &= \int_{-1}^1 \frac{\ln^+ |F_j(t)|}{1 + |t|^{1+\mu_j}} dt + \frac{\pi}{\psi_j - \varphi_j} \int_1^\infty \frac{\ln^+ \left|\{f(te^{i\varphi_j}) - a\}^{-1}\right|}{t^{1+\omega''(D'_\varepsilon)}} dt \\ &\quad + \frac{\pi}{\psi_j - \varphi_j} \int_1^\infty \frac{\ln^+ \left|\{f(te^{i\psi_j}) - a\}^{-1}\right|}{t^{1+\omega''(D'_\varepsilon)}} dt \\ &< \infty.\end{aligned}$$

Next, since

$$\frac{\psi_j - \varphi_j}{\pi} < \{\omega'(D'_{2\varepsilon})\}^{-1}, \tag{4.19'}$$

by (4.13) we have $\mu_j < \{\omega'(D'_{2\varepsilon})\}^{-1}\omega''(D'_\varepsilon) < 1$. Therefore (4.19) implies

$$A(r, F_j) \le \frac{1}{\pi} \int_1^r \{\ln^+ |F_j(t)| + \ln^+ |F_j(-t)|\} \frac{dt}{t^2} = O(1).$$

The quantity $B(r, F_j)$ can be estimated as follows:

$$B(r,F_j) \le \frac{4}{r} m_{0\pi}(r,F_j) = \frac{4\pi}{(\psi_j-\varphi_j)r} m_{\varphi_j\psi_j}\left(r^{\frac{\psi_j-\varphi_j}{\pi}}, \frac{1}{f-a}\right) \le \frac{4\pi}{(\psi_j-\varphi_j)r}\left\{T\left(r^{\frac{\psi_j-\varphi_j}{\pi}}, f\right) + O(1)\right\}.$$

By (4.13) and (4.19′) we have $\frac{\psi_j-\varphi_j}{\pi}\rho < 1$, and we get

$$B(r, F_j) = o(1). \tag{4.20}$$

In a similar way we get

$$B(r, 1/F_j) = o(1). \tag{4.21}$$

Denote by $r_\nu e^{i\theta_\nu}$ the a-points of the function $f(z)$; we have

$$C(r,F_j) = C_{\varphi_j\psi_j}\left(r^{\frac{\psi_j-\varphi_j}{\pi}}, \frac{1}{f-a}\right) \le 2 \sum_{\substack{r_\nu>1\\ \varphi_j\le\theta_\nu\le\psi_j}} r_\nu^{-\frac{\pi}{\psi_j-\varphi_j}} < \infty,$$

since $\frac{\pi}{\psi_j-\varphi_j} > \rho$.

Thus (4.18) has been proved, and, by Theorem 3.1, the function $F_j(\zeta)$ admits a representation

$$\begin{aligned}\ln|F_j(\zeta)| = {} & \frac{r\sin\varphi}{\pi}\int_{-\infty}^{\infty}\frac{\ln|F_j(t)|\,dt}{r^2+t^2-2rt\cos\varphi} \\ & - \sum_\kappa \ln\left|\frac{\zeta-\bar q_\kappa}{\zeta-q_\kappa}\right| + \sum_\nu\left|\frac{\zeta-\bar p_\nu}{\zeta-p_\nu}\right| + \eta r\sin\varphi, \qquad \zeta = re^{i\varphi},\end{aligned} \tag{4.22}$$

where the q_κ are the zeros of the function $F_j(\zeta)$, the p_ν are its poles (lying in $\{\operatorname{Im}\zeta > 0\}$), and η is a constant which does not depend on ζ. Estimates (4.20) and (4.21) and the remark to Theorem 3.1 imply that $\eta = 0$. Therefore (4.22) implies that

$$\ln^+|F_j(\zeta)| \le \frac{y}{\pi}\int_{-\infty}^{\infty}\frac{\ln^+|F_j(t)|\,dt}{(x-t)^2+y^2} + \sum_\nu \ln\left|\frac{\zeta-\bar p_\nu}{\zeta-p_\nu}\right|, \qquad \zeta = x+iy.$$

Relation (4.19) and Lemma 4.2 imply that the estimate

$$\frac{y}{\pi}\int_{-\infty}^{\infty}\frac{\ln^+|F_j(t)|}{(x-t)^2+y^2}dt = O(|\zeta|^{\mu_j}), \qquad \zeta = x+iy,$$

holds in the angle $\{\xi_j \le \arg\zeta \le \eta_j\}$. Defining the function $\chi(z)$ by equality (4.6), we get that

$$\ln^+|F_j(re^{i\varphi})| \le O(r^{\mu_j}) + \ln|\chi(re^{i\varphi})|, \qquad \xi_j \le \arg\zeta \le \eta_j,$$

from which, integrating with respect to φ from ξ_j to η_j, we conclude that

$$m_{\xi_j\eta_j}(r,F_j) \le K_j r^{\mu_j} + m_{\xi_j\eta_j}(r,\chi) \le K_j r^{\mu_j} + m(r,\chi), \qquad r \ge 1, \tag{4.23}$$

where K_j is a constant which does not depend on r.

Let us agree to use the notation $N_{\alpha\beta}(r,f)$, $0 < \beta-\alpha \le 2\pi$, $r \ge 0$, for the sum

$$N_{\alpha\beta}(r,f) = \sum_{\alpha\le\arg a_n\le\beta} \ln^+\frac{r}{|a_n|},$$

where a_n are the poles of the function $f(z)$ counted according to their multiplicity. Obviously,

$$N(r,\chi) = \frac{\pi}{\psi_j - \varphi_j} N_{\varphi_j\psi_j}\left(r^{\frac{\psi_j-\varphi_j}{\pi}}, \frac{1}{f-a}\right). \tag{4.24}$$

Therefore

$$N(r,\chi) \le \frac{\pi}{\psi_j - \varphi_j} N\left(r^{\frac{\psi_j-\varphi_j}{\pi}}, \frac{1}{f-a}\right) \le \frac{\pi}{\psi_j - \varphi_j} T\left(r^{\frac{\psi_j-\varphi_j}{\pi}}, f\right) + O(1).$$

From this it follows that the order of the function $N(r,\chi)$ does not exceed the number $\frac{\psi_j-\varphi_j}{\pi}\rho < \{\omega'(D'_{2\varepsilon})\}^{-1}\rho < 1$. Since $N(r,\chi) = N(r,1/\chi)$, and the function $\chi(z)$ is a quotient of two canonical products of genus zero, it follows (Theorem 3.4 of Chapter 2) that the order of this function also does not exceed $\frac{\psi_j-\varphi_j}{\pi}\rho$.

Using Lemma 4.1, we write inequality (4.7) for our function $\chi(z)$. It holds for all $b > 0$ and σ satisfying the condition $\frac{\psi_j-\varphi_j}{\pi}\rho < \sigma < 1$.

We divide (4.23) by $r^{1+\sigma}$ and integrate it with respect to r from $b > 1$ to ∞. Taking into account (4.7) we get

$$\int_b^\infty \frac{m_{\xi_j\eta_j}(r,F_j)}{r^{1+\sigma}} dr \le \left(\sec\frac{\pi\sigma}{2} - 1\right)\int_b^\infty \frac{K'_j r^{\mu_j} + N(r,\chi)}{r^{1+\sigma}} dr,$$

where K'_j is a constant which does not depend on r and b. Using equalities (4.17) and (4.24) and making the change of variable $r = t^{\pi/(\psi_j-\varphi_j)}$ in the integrals, we see that for $\frac{\psi_j-\varphi_j}{\pi}\rho < \sigma < 1$ and for all $b > 1$,

$$\int_b^\infty \frac{m_{\alpha_j-\varepsilon,\ \beta_j+\varepsilon}(t, \frac{1}{f-a})}{t^{1+\frac{\sigma\pi}{\psi_j-\varphi_j}}} dt$$
$$\le \left(\sec\frac{\pi\sigma}{2} - 1\right)\int_b^\infty \frac{K''_j t^{\omega''(D'_\varepsilon)} + N_{\varphi_j\psi_j}(t, \frac{1}{f-a})}{t^{1+\frac{\sigma\pi}{\psi_j-\varphi_j}}} dt,$$

where K''_j is a constant which does not depend on t and b. Let $\eta = \frac{\sigma\pi}{\psi_j-\varphi_j}$ and note that as σ ranges over the interval $(\rho\frac{\psi_j-\varphi_j}{\pi}, 1)$, the parameter η ranges over the interval $(\rho, \frac{\pi}{\psi_j-\varphi_j})$, which contains the interval $(\rho, \omega'(D'_{2\varepsilon}))$. Note also that $\sigma < \eta\{\omega'(D'_{2\varepsilon})\}^{-1}$. Then we can claim that for $\rho < \eta < \omega'(D'_{2\varepsilon})$ and $b > 1$ the inequality

$$\int_b^\infty \frac{m_{\alpha_j-\varepsilon,\ \beta_j+\varepsilon}(t, \frac{1}{f-a})}{t^{1+\eta}} dt$$
$$\le \left(\sec\frac{\pi\eta}{2\omega'(D'_{2\varepsilon})} - 1\right)\int_b^\infty \frac{K''_j t^{\omega''(D'_\varepsilon)} + N_{\varphi_j\psi_j}(t, \frac{1}{f-a})}{t^{1+\eta}} dt$$

holds. Summing these inequalities over $j = 1, \ldots, n$ and taking into account that

$$\sum_{j+1}^{n} N_{\varphi_j\psi_j}\left(t, \frac{1}{f-a}\right) \le N\left(t, \frac{1}{f-a}\right), \qquad t \ge 1,$$

since the intervals (φ_j, ψ_j) do not intersect, we get

$$\int_b^\infty \frac{\Phi_1(t)}{t^{1+\eta}} dt \le \left(\sec\frac{\pi\eta}{2\omega'(D'_{2\varepsilon})} - 1\right)\int_b^\infty \frac{K t^{\omega''(D'_\varepsilon)} + N(t,a)}{t^{1+\eta}} dt, \tag{4.25}$$

where K is a constant which does not depend on b and t, $\rho < \eta < \omega'(D'_{2\varepsilon})$, $b > 1$.

Relation (4.14) implies that for $\eta > \rho\,(> \omega''(D'_\varepsilon) > \omega''(D'))$ the inequality

$$\int_b^\infty \frac{\Phi_2(t)}{t^{1+\eta}}dt \le \int_b^\infty \frac{K' t^{\omega''(D'_\varepsilon)}}{t^{1+\eta}}dt$$

holds, where K' is a constant which does not depend on b and t. Adding this to (4.25) we get the inequality

$$\int_b^\infty \frac{m(t,a)}{t^{1+\eta}}dt \le \left(\sec\frac{\pi\eta}{2\omega'(D'_{2\varepsilon})} - 1\right)\int_b^\infty \frac{K'' t^{\omega''(D'_\varepsilon)} + N(t,a)}{t^{1+\eta}}dt,$$

where K'' does not depend on b and t, $\rho < \eta < \omega'(D'_{2\varepsilon})$, $b > 1$.

Since $\rho > \omega''(D'_\varepsilon)$, for each $\varepsilon_1 > 0$ there is a sequence $\{b_q\}$, $b_q \uparrow \infty$, such that

$$\int_{b_q}^\infty \frac{K'' t^{\omega''(D_\varepsilon)}}{t^{1+\eta}}dt \le \varepsilon_1 \int_{b_q}^\infty \frac{T(t,f)}{t^{1+\eta}}dt.$$

Therefore we have the inequality

$$\int_{b_q}^\infty \frac{m(t,a)}{t^{1+\eta}}dt \le \left(\sec\frac{\pi\eta}{2\omega'(D'_{2\varepsilon})} - 1\right)\int_{b_q}^\infty \frac{N(t,a) + \varepsilon_1 T(t,f)}{t^{1+\eta}}dt.$$

This inequality implies that there exists a number $t_q > b_q$ such that

$$m(t_q,a) \le \left(\sec\frac{\pi\eta}{2\omega'(D'_{2\varepsilon})} - 1\right)\{N(t_q,a) + \varepsilon_1 T(t_q,f)\}. \tag{4.26}$$

Dividing both sides of (4.26) by $T(t_q,f)$ and taking the limit as $t_q \to \infty$, we get

$$\delta(a,f) \le \left(\sec\frac{\pi\eta}{2\omega'(D'_{2\varepsilon})} - 1\right)\{1 - \delta(a,f) + \varepsilon_1\}.$$

Letting ε and ε_1 tend to 0, and η tend to ρ, we get the inequality

$$\delta(a,f) \le \left(\sec\frac{\pi\rho}{2\omega'(D')} - 1\right)\{1 - \delta(a,f)\},$$

which is equivalent to (4.12). This completes the proof of the theorem. □

Now we construct an example showing that the sign $>$ in condition (4.5′) in Corollary 1 cannot be replaced by $\geq$.

EXAMPLE 2. For arbitrary preassigned γ and ρ, $0 < \gamma < \pi$, $0 < \rho < \pi\gamma^{-1}$, we construct a meromorphic function $f(z)$ of order ρ with all its zeros and poles in the angle $\{|\arg z| \le \frac{\gamma}{2}\}$, and $\delta(1,f) = 1 - \cos\frac{\rho\gamma}{2}$.

Let

$$\Phi(z;\eta) = \prod_{\nu=1}^\infty \frac{1 + z\nu^{-\frac{1}{\eta}}}{1 - z\nu^{-\frac{1}{\eta}}}, \qquad 0 < \eta < 1.$$

The function $\Phi(z;\eta)$ is a special case of the function $f_{uv}(z)$ considered in Example 1 of Section 3 in Chapter 5 ($u = v = 1$, $l(r) \equiv \eta$). Therefore, by formulas (3.38) and (3.39) of Chapter 5 we have

$$N(r,\Phi(z;\eta)) = \frac{1}{\eta}r^\eta + o(r^\eta), \tag{4.27}$$

$$T(r,\Phi(z;\eta)) = \frac{1}{\eta}\{I(1,1,\eta) + 1\}r^\eta + o(r^\eta) = \frac{1}{\eta}\sec\frac{\pi\eta}{2}r^\eta + o(r^\eta). \tag{4.28}$$

Note that

$$|\Phi(z;\eta)| \le 1 \quad \text{for} \quad \operatorname{Re} z \le 0. \tag{4.29}$$

Denote by $L(\theta, a)$ the boundary of the region $D(\theta, a) = \{|\arg z| < \theta,\ |z| > a\}$, oriented clockwise, and by $D_1(\theta, a)$, the region on the left of $L(\theta, a)$. For $z \in D_1(\frac{\gamma}{2}, \frac{1}{2})$ we let

$$\Psi(z) = \frac{1}{2\pi i}\int_{L(\frac{\gamma}{2},\frac{1}{2})} \frac{\Phi\left(\zeta^{\pi/\gamma}; \frac{\rho\gamma}{\pi}\right)}{(\zeta+1)^2(\zeta - z)}\, d\zeta. \tag{4.30}$$

Inequality (4.29) implies that

$$\left|\Phi\left(\zeta^{\pi/\gamma}; \frac{\rho\gamma}{\pi}\right)\right| \le 1 \quad \text{for} \quad \frac{\gamma}{2} \le |\arg\zeta| \le \gamma; \tag{4.31}$$

therefore the integral in (4.30) is absolutely convergent, and the function $\Psi(z)$ is analytic in $D_1(\frac{\gamma}{2}, \frac{1}{2})$.

Using the residue theorem we get the following relation for $R \ge \frac{1}{2}$, $R \ne \nu^{1/\rho}$, $\nu = 1, 2, \dots$ $(z \in D_1(\frac{\gamma}{2}, \frac{1}{2}))$:

$$\begin{aligned} &\frac{1}{2\pi i}\int_{L(\frac{\gamma}{2},\frac{1}{2})} \frac{\Phi\left(\zeta^{\pi/\gamma}; \frac{\rho\gamma}{\pi}\right)}{(\zeta+1)^2(\zeta - z)}\, d\zeta \\ &\quad = \frac{1}{2\pi i}\int_{L(\frac{\gamma}{2},R)} \frac{\Phi\left(\zeta^{\pi/\gamma}; \frac{\rho\gamma}{\pi}\right)}{(\zeta+1)^2(\zeta - z)}\, d\zeta - \sum_{\nu=1}^{[R^\rho]} \operatorname{res}_{\zeta=\nu^{1/\rho}} \frac{\Phi\left(\zeta^{\pi/\gamma}; \frac{\rho\gamma}{\pi}\right)}{(\zeta+1)^2(\zeta - z)}. \end{aligned} \tag{4.32}$$

The right-hand side of this relation represents a function meromorphic in the region $D_1\left(\frac{\gamma}{2}, R\right)$. Since R can be chosen to be arbitrarily large, the function $\Psi(z)$ can be extended to a meromorphic function in the whole complex plane. We shall keep the notation $\Psi(z)$ for this extension.

We show that for $z \in D\left(\gamma, \frac{1}{2}\right)$ the formula

$$\Psi(z) = \frac{1}{2\pi i}\int_{L(\nu,\frac{1}{2})} \frac{\Phi\left(\zeta^{\pi/\gamma}; \frac{\rho\gamma}{\pi}\right)}{(\zeta+1)^2(\zeta - z)}\, d\zeta + \frac{\Phi\left(z^{\pi/\gamma}; \frac{\rho\gamma}{\pi}\right)}{(z+1)^2} \tag{4.33}$$

holds. For $z \in D(\gamma, \frac{1}{2})$ the function $\Psi(z)$ can be represented by the right-hand side of relation (4.32) with $R > |z|$. Using the Cauchy theorem and (4.31), it is easy to get that the integral in the right-hand side of (4.32) does not change if the path of integration is changed from $L(\frac{\gamma}{2}, R)$ to $L(\gamma, R)$. On the other hand, by the residue theorem we have $(z \in D(\gamma, \frac{1}{2}),\ R > |z|)$

$$\begin{aligned} \frac{1}{2\pi i}\int_{L(\gamma,R)} \frac{\Phi\left(\zeta^{\pi/\gamma}; \frac{\rho\gamma}{\pi}\right)}{(\zeta+1)^2(\zeta - z)}\, d\zeta &= \frac{1}{2\pi i}\int_{L(\gamma,\frac{1}{2})} \frac{\Phi\left(\zeta^{\pi/\gamma}; \frac{\rho\gamma}{\pi}\right)}{(\zeta+1)^2(\zeta - z)}\, d\zeta \\ &\quad + \sum_{\nu=1}^{[R^\rho]} \operatorname{res}_{\zeta=\nu^{1/\rho}} \frac{\Phi\left(\zeta^{\pi/\gamma}; \frac{\rho\gamma}{\pi}\right)}{(\zeta+1)^2(\zeta - z)} + \frac{\Phi\left(z^{\pi/\gamma}; \frac{\rho\gamma}{\pi}\right)}{(z+1)^2}. \end{aligned}$$

Thus formula (4.33) has been proved.

Using (4.31), we get from (4.30) and (4.33) the following relations:

$$\begin{aligned} &|\Psi(z)| \le c/(|z|+1), && \frac{\gamma}{2} \le \arg z \le 2\pi - \frac{\gamma}{2}, \\ &\left|\Psi(z) - \frac{\Phi\left(z^{\pi/\gamma}; \frac{\rho\gamma}{\pi}\right)}{(z+1)^2}\right| \le c/(|z|+1), && |\arg z| \le \frac{\gamma}{2}, \end{aligned} \tag{4.34}$$

where c is a constant which does not depend on z. These relations imply the following equalities:

$$N(r,\Psi) = N\left(r^{\pi/\gamma},\ \Phi\left(z;\frac{\rho\gamma}{\pi}\right)\right),$$

$$\begin{aligned} m(r,\Psi) &= m_{-\frac{\gamma}{2},\frac{\gamma}{2}}(r,\Psi) + O(1) = m\left(r^{\pi/\gamma},\ \Phi\left(z;\frac{\rho\gamma}{\pi}\right)\right) + O(\ln r) \\ &= T\left(r^{\pi/\gamma},\ \Phi\left(z;\frac{\rho\gamma}{\pi}\right)\right) - N\left(r^{\pi/\gamma}, \Phi\left(z;\frac{\rho\gamma}{\pi}\right)\right) + O(\ln r). \end{aligned}$$

Taking into account (4.27) and (4.28), we get

$$N(r,\Psi) = \frac{\pi}{\rho\gamma} r^\rho + o(r^\rho),$$

$$T(r,\Psi) = \frac{\pi}{\rho\gamma} \sec\frac{\rho\gamma}{2} r^\rho + o(r^\rho),$$

whence $\delta(\infty,\Psi) = 1 - \cos\frac{\rho\gamma}{2}$.

By (4.34) the quantity

$$Q = \sup_{\frac{\gamma}{2}\le \arg z \le 2\pi - \frac{\gamma}{2}} |\Psi(z)|$$

is finite. Therefore for each b, $|b| > Q$, all b-points of the function $\Psi(z)$ are in the angle $\{|\arg z| \le \gamma/2\}$. Choose arbitrarily two complex numbers b and c, $\min(|b|,|c|) > Q$, and consider the function

$$f(z) = \frac{\Psi(z) - b}{\Psi(z) - c}.$$

It is clear that it is a desired function.

5. Entire functions with derivatives vanishing only near the real axis

As is known, the zeros of the derivatives of a polynomial with real zeros are real. The direct analog of this result for entire transcendental functions does not hold. For example, all zeros of the function $f(z) = (z^2 - 1)e^{z^2/2}$ are real, while $f'(z) = (z^3 + z)e^{z^2/2}$ has also non-real zeros. Sufficient conditions under which the zeros of all derivatives of an entire function with real zeros are real, are given by the following classical theorem.

THEOREM 5.1 (Laguerre). *If an entire function $f(z)$ is representable in the form*

$$f(z) = e^{-\gamma z^2} h(z), \tag{5.1}$$

where $\gamma \ge 0$ and $h(z)$ is an entire function of genus at most 1 that is real on the real line and has only real zeros, then the zeros of all derivatives of the function $f(z)$ are real.

PROOF. Formula (5.1) implies that

$$f'(z) = e^{-\gamma z^2} h_1(z),$$

where

$$h_1(z) = h'(z) - 2\gamma z h(z).$$

Theorem 5.1 will be proved if we establish that $h_1(z)$ is an entire function of genus at most 1 that is real on the real line and has only real zeros.

The function $h(z)$ is representable in the form

$$h(z) = z^m e^{az+b} \prod_k \left(1 - \frac{z}{a_k}\right) e^{\frac{z}{a_k}},$$

where $m \geq 0$ is an integer, a, b, and a_k are real numbers, and

$$\sum_k a_k^{-2} < \infty. \tag{5.2}$$

Therefore we have

$$\frac{f'(z)}{f(z)} = -2\gamma z + \frac{m}{z} + a + \sum_k \left(\frac{1}{z - a_k} + \frac{1}{a_k}\right), \tag{5.3}$$

whence

$$\operatorname{Im} \frac{f'(z)}{f(z)} = -\left\{2\gamma + \frac{m}{|z|^2} + \sum_k \frac{1}{|z - a_k|^2}\right\} \operatorname{Im} z. \tag{5.4}$$

We may assume that $f(z)$ is not of the form e^{az+b}, because otherwise the theorem is trivial. Then the expression in braces in (5.4) is different from zero and hence

$$\operatorname{Im} \frac{f'(z)}{f(z)} \neq 0 \quad \text{for} \quad \operatorname{Im} z \neq 0.$$

This means that all zeros of $f'(z)$, and hence, of $h_1(z)$ are real.

It remains to prove that the genus of the function $h_1(z)$ does not exceed 1. Since the genus of the function $h(z)$ does not exceed 1, it follows from Remark 2 to Theorem 3.3 of Chapter 2 that $T(r, h) = o(r^2)$. By Theorem 2.3 of Chapter 3 we also have $T(r, h') = o(r^2)$, and hence, $T(r, h_1) = T(r, h' - 2\gamma z h) = o(r^2)$. Therefore, in order to show that the genus of the function $h_1(z)$ does not exceed 1, it suffices to establish convergence of the series

$$\sum_k a_{k1}^{-2}, \tag{5.5}$$

where the a_{k1} are the zeros of the function $h_1(z)$ different from $z = 0$ (multiple zeros are repeated according to their multiplicity).

Equality (5.3) implies

$$\left(\frac{f'(x)}{f(x)}\right)' = -2\gamma - \sum_k \frac{1}{(x - a_k)^2}.$$

For real x the expression in the right-hand side is negative. Therefore, in the interval between two consecutive zeros of the function $f(x)$, the function $\frac{f'(x)}{f(x)}$ is decreasing from ∞ to $-\infty$. Thus, between two consecutive zeros of the function $f(x)$ there is exactly one simple zero of the function $f'(x)$. Therefore the convergence of the series (5.2) implies the convergence of the series (5.5). □

We say that an entire function $f(z)$ belongs to *the class* A (notation: $f(z) \in A$) if its zeros $\{z_n\}$ satisfy the condition

$$\sum_{|z_n|>1} \left|\operatorname{Im} \frac{1}{z_n}\right| < \infty$$

(multiple zeros are repeated according to their multiplicity). As we mentioned in Section 2 (see Example 1), a function $f(z)$ belongs to the class A if and only if

$$C(r,0,f(z)) + C(r,0,f(-z)) = O(1).$$

If is clear that all functions with real zeros belong to the class A.

We find conditions under which all derivatives of a function in the class A belong to the class A.

THEOREM 5.2. *Let $f(z)$ be an entire function in the class A that is representable in the form*

$$f(z) = e^{g(z)} h(z), \tag{5.6}$$

where $g(z)$ is an entire function which does not exceed the normal type of order 1 and satisfies the condition

$$\int_{-\infty}^{\infty} \frac{\ln^+ |g'(t)|}{1+t^2}\, dt < \infty, \tag{5.7}$$

and $h(z)$ is an entire function satisfying the condition

$$\int_1^{\infty} t^{-2} \ln^+ T(t,h)\, dt < \infty. \tag{5.8}$$

Then all derivatives of the function $f(z)$ belong to the class A.

Theorem 5.2 can be regarded as an analog of Laguerre's Theorem 5.1 in the case where the condition that all zeros of $f(z)$ are real is replaced by a weaker condition: the function $f(z)$ belongs to the class A.

To prove the theorem it suffices to establish that the function $f'(z)$ belongs to the class A and is representable in the form

$$f'(z) = e^{g(z)} h_1(z), \tag{5.9}$$

where $h_1(z)$ is an entire function satisfying the condition

$$\int_1^{\infty} t^{-2} \ln^+ T(t,h_1)\, dt < \infty. \tag{5.10}$$

We need the following lemma.

LEMMA 5.1. *If a function $f(z)$ satisfies the assumptions of Theorem 5.2, then*

$$S\left(r, \frac{f'}{f}\right) = O(1).$$

PROOF. Since $f(z)$ belongs to the class A, it is clear that

$$C\left(r, \frac{f'}{f}\right) = \bar{C}(r,0,f) \le C(r,0,f) = O(1).$$

Furthermore, we have

$$\begin{aligned} A\left(r, \frac{f'}{f}\right) + B\left(r, \frac{f'}{f}\right) &= A\left(r, g' + \frac{h'}{h}\right) + B\left(r, g' + \frac{h'}{h}\right) \\ &\le A(r,g') + B(r,g') + A\left(r, \frac{h'}{h}\right) + B\left(r, \frac{h'}{h}\right) + O(1). \end{aligned}$$

Condition (5.7) implies that $A(r, g') = O(1)$. Since $T(r, g) = O(r)$, by inequality (2.13) in Chapter 3, we have $T(r, g') = O(r)$ and hence

$$B(r, g') \leq \frac{4}{r} m(r, g') = O(1).$$

By Theorem 1.3 of Chapter 3,

$$A\left(r, \frac{h'}{h}\right) + B\left(r, \frac{h'}{h}\right) = Q_{0\pi}(r, h).$$

Recalling the definition of $Q_{0\pi}(r, h)$ (page 105), from condition (5.8) we get

$$A\left(r, \frac{h'}{h}\right) + B\left(r, \frac{h'}{h}\right) = O(1).$$

Thus,

$$S\left(r, \frac{f'}{f}\right) = A\left(r, \frac{f'}{f}\right) + B\left(r, \frac{f'}{f}\right) + C\left(r, \frac{f'}{f}\right) = O(1).$$

The lemma is proved. □

Since

$$C(r, 0, f') \leq C\left(r, \frac{f}{f'}\right) + C(r, 0, f) \leq S\left(r, \frac{f'}{f}\right) + C(r, 0, f) + O(1)$$

(we used Theorem 5.1 of Chapter 1), we have $C(r, 0, f') = O(1)$ if the function $f(z)$ satisfies the assumptions of Theorem 5.2. Considering the function $f_1(z) = f(-z)$, in an analogous way we get $C(r, 0, f_1') = O(1)$. Thus we have proved that $f'(z)$ belongs to the class A.

It is clear that the function $f'(z)$ can be represented in the form (5.9) with $h_1(z) = g'(z)h(z) + h'(z)$. We have

$$T(r, h_1) \leq T(r, g') + T(r, h) + T(r, h') + \ln 2 \leq T(r, h) + T(r, h') + O(r).$$

Since

$$T(r, h') = m(r, h') \leq m(r, h) + m\left(r, \frac{h'}{h}\right) = T(r, h) + m\left(r, \frac{h'}{h}\right),$$

using estimate (1.13) in Chapter 3, we get

$$T(r, h_1) \leq 2T(r, h) + O(\ln r + \ln T(2r, h)) + O(r) \leq 3T(2r, h) + O(r),$$

$$\ln^+ T(r, h_1) \leq \ln^+ T(2r, h) + O(\ln r).$$

Therefore (5.8) implies (5.10). □

Observe that the assumptions of Theorem 5.2 are satisfied for each entire function $h(z)$ belonging to the class A and satisfying condition (5.8). In particular, Theorem 5.2 has the following corollary.

COROLLARY. *If an entire function of finite order belongs to the class A, then all of its derivatives belong to the class A.*

If we weaken somewhat the assumptions of Theorem 5.2 and suppose only that the function $f(z)$ belongs to the class A and satisfies the condition $\ln^+ T(r, f) = O(r)$, the statement of Theorem 5.2 is no longer true. It suffices to consider the example $f(z) = e^{\cosh z}$. It is clear that $\ln^+ T(r, f) = O(r)$, $f(z) \in A$, but $f'(z) = \sinh z e^{\cosh z}$ does not belong to A.

In 1914 G. Pólya and A. Wiman posed the following problem: Are the conditions indicated in Theorem 5.1 not only sufficient but also necessary for an entire function with real zeros that is real-valued on the real line to have all the derivatives with only real zeros?

A complete answer to this problem is still unknown.

An analogous problem can be posed in connection with Theorem 5.2: Are its conditions not only sufficient but also necessary for an entire function of the class A to have all of its derivatives in the class A? Clearly, any results on the latter problem can be regarded as progress towards a solution of the problem of Pólya and Wiman.

It seems likely that the following conjecture is true. If an entire function $f(z)$ is such that $f(z) \in A$ and $f''(z) \in A$, then $f(z)$ is representable in the form (5.6), where $g(z)$ is an entire function that does not exceed normal type of order 1 and satisfies (5.7), and $h(z)$ satisfies condition (5.8). We can prove the following theorem.

THEOREM 5.3. *Let $f(z)$ be an entire function such that $f(z) \in A$, $f''(z) \in A$. Suppose that the function $f(z)$ is representable in the form*

$$f(z) = e^{g(z)} h(z),$$

where $g(z)$ is an entire function, and $h(z)$ satisfies condition (5.8). *Then the function $g(z)$ does not exceed normal type of order* 1 *and satisfies condition* (5.7).

Note that the condition $f''(z) \in A$ cannot be replaced by the condition $f'(z) \in A$. This can be seen from the example

$$f(z) = \exp\left\{\int_0^z G(\zeta)\, d\zeta\right\},$$

where $G(\zeta)$ is an arbitrary entire function without zeros.

One can observe some analogy between Theorem 5.3 and Theorem 2.7 of Chapter 3. One of the assumptions in the latter was that $f(z)f''(z)$ does not have zeros, and the condition $f(z) \in A$, $f''(z) \in A$ means, roughly speaking, that $f(z)f''(z)$ is "rarely" equal to zero in each of the half-planes $\{\operatorname{Im} z > 0\}$ and $\{\operatorname{Im} z < 0\}$.

In the proof of Theorem 2.7 of Chapter 3 we considered the function

$$F(z) = \frac{f(z)}{f'(z)}$$

and studied it using inequality (2.23) in Chapter 3. Proving Theorem 5.3, we also consider this function, but now we use analogs of inequality (2.23) in Chapter 3 for the Tsuji characteristics and angular Nevanlinna characteristics. As opposed to the proof of Theorem 2.7 in Chapter 3, where we succeeded in establishing that $T(r, F) = O(\ln r)$ and $\frac{f'(z)}{f(z)} \equiv \text{const}$, now we will be able to get only some estimates for $m(r, F)$ and $m(r, 1/F)$. We need an estimate for $m(r, 1/F) = m(r, f'/f)$ only. It is given by the following theorem, interesting by itself.

THEOREM 5.4. *If $f(z)$ is an entire function and $f(z) \in A$, $f''(z) \in A$, then*

$$\int_R^\infty \frac{m(r, \frac{f'}{f})}{r^3} dr = O\left(\frac{\ln R}{R}\right), \qquad R \to \infty.$$

To prove this theorem we need the following lemma, connecting the quantity $m_{0\pi}(r, f)$ with the Tsuji characteristic $\mathfrak{m}(r, f)$.

LEMMA 5.2. *If a function $f(z)$ is meromorphic in the half-plane $\{\operatorname{Im} z > 0\}$, then*

$$\int_R^\infty \frac{m_{0\pi}(r,f)}{r^3}\,dr \le \int_R^\infty \frac{\mathfrak{m}(r,f)}{r^2}\,dr, \qquad R \ge 1. \tag{5.11}$$

Observe that the statement of the lemma is non-trivial only in the case where the integral in the right-hand side of (5.11) is convergent.

Consider the integral

$$I(R) = \frac{1}{2\pi}\iint_{\Omega_R} \frac{\ln^+|f(re^{i\varphi})|}{r^3}\,dr\,d\varphi,$$

where $\Omega_R = \{r > R\sin\varphi,\ r > 1,\ 0 < \varphi < \pi\}$. Making the change of variables $\rho = r/\sin\varphi$, $\theta = \varphi$, we get

$$I(R) = \int_R^\infty \frac{d\rho}{\rho^2}\left\{\frac{1}{2\pi}\int_{\arcsin\rho^{-1}}^{\pi-\arcsin\rho^{-1}} \ln^+|f(\rho\sin\theta e^{i\theta})|\frac{d\theta}{\rho\sin^2\theta}\right\} = \int_R^\infty \frac{\mathfrak{m}(\rho,f)}{\rho^2}\,d\rho.$$

Next, we consider the integral

$$J(R) = \frac{1}{2\pi}\iint_{K_R} \frac{\ln^+|f(re^{i\varphi})|}{r^3}\,dr\,d\varphi,$$

where $K_R = \{r > R, 0 < \varphi < \pi\}$. Because of $K_R \subset \Omega_R$ we have $J(R) \le I(R)$. Since

$$J(R) = \int_R^\infty \frac{dr}{r^3}\left\{\frac{1}{2\pi}\int_0^\pi \ln^+|f(re^{i\varphi})|\,d\varphi\right\} = \int_R^\infty \frac{m_{0\pi}(r,f)}{r^3}\,dr,$$

inequality (5.11) has been proved. □

Now we prove Theorem 5.4. Let $F(z) = f(z)/f'(z)$. Since $f(z) \in A$, all zeros $\{\rho_n e^{i\psi_n}\}$ of the function $F(z)$ lying in the half-plane $\{\operatorname{Im} z > 0\}$ (all of them are simple and are zeros of $f(z)$) satisfy the condition

$$\sum_{\rho_n > 1} \frac{\sin\psi_n}{\rho_n} < \infty.$$

The relation

$$\mathfrak{N}(r,0,F) = \sum_{1<\rho_n\le r\sin\psi_n}\left(\frac{\sin\psi_n}{\rho_n} - \frac{1}{r}\right)$$

implies (cf. Section 5 of Chapter 1, p. 26)

$$\mathfrak{N}(r,0,F) = O(1). \tag{5.12}$$

Since

$$F'(z) = 1 - \frac{f(z)f''(z)}{\{f'(z)\}^2} \tag{5.13}$$

and $f(z) \in A$, $f''(z) \in A$, the zeros $\{\rho_n' e^{i\psi_n'}\}$ of the function $F'(z) - 1$ lying in the half-plane $\{\operatorname{Im} z > 0\}$ also satisfy the condition

$$\sum_{\rho_n' > 1} \frac{\sin\psi_n'}{\rho_n'} < \infty;$$

hence

$$\mathfrak{N}(r,1,F') = O(1). \tag{5.14}$$

By Theorem 3.3 of Chapter 3, the analog of inequality (2.23) of Chapter 3 holds for the Tsuji characteristics. In particular, the inequality

$$\mathfrak{T}(r,F) \le 3\mathfrak{N}(r,0,F) + 4\mathfrak{N}(r,1,F') + \mathfrak{Q}(r,F)$$

holds. Taking into account (5.12) and (5.14), we get

$$\mathfrak{T}(r,F) = \mathfrak{Q}(r,F).$$

By the definition of $\mathfrak{Q}(r,F)$ this means that outside a set E of finite measure,

$$\mathfrak{T}(r,F) = O(\ln r) + O(\ln \mathfrak{T}(r,F)), \qquad r \to \infty;$$

hence

$$\mathfrak{T}(r,F) = O(\ln r), \qquad r \to \infty, \tag{5.15}$$

for $r \notin E$.

By Theorem 5.4 of Chapter 1, the function $\mathfrak{T}(r,F)$ is non-decreasing within a bounded summand. Therefore, applying Lemma 2.1, we conclude that (5.15) holds for $r \in E$.

By Theorem 5.3 of Chapter 1, relation (5.15) implies

$$\mathfrak{m}\left(r, \frac{1}{F}\right) = O(\ln r).$$

Applying Lemma 5.2, we get the estimate

$$\int_R^\infty \frac{m_{0\pi}(r,1/F)}{r^3}\,dr \le \int_R^\infty \frac{\mathfrak{m}(r,1/F)}{r^2}\,dr = O\left(\frac{\ln R}{R}\right). \tag{5.16}$$

Clearly, the above argument remains true for $F_1(z) = F(-z)$; therefore we can replace F by F_1 in (5.16).

Since

$$m\left(r, \frac{f'}{f}\right) = m_{0\pi}\left(r, \frac{1}{F}\right) + m_{0\pi}\left(r, \frac{1}{F_1}\right),$$

Theorem 5.4 has been proved. □

Now we turn to the proof of Theorem 5.3.

First we show that

$$\int_1^\infty t^{-2} \ln^+ T\left(t, \frac{f'}{f}\right) dt < \infty. \tag{5.17}$$

Obviously

$$N\left(r, \frac{f'}{f}\right) \le N\left(r, \frac{1}{h}\right) \le T(r,h) + O(1);$$

hence

$$\begin{aligned}
\ln^+ T\left(r, \frac{f'}{f}\right) &\le \ln^+ T(r,h) + \ln^+ m\left(r, \frac{f'}{f}\right) + O(1) \\
&\le \ln^+ T(r,h) + \ln^+ \left\{r^{-1} m\left(r, \frac{f'}{f}\right)\right\} + O(\ln r) \\
&\le \ln^+ T(r,h) + r^{-1} m\left(r, \frac{f'}{f}\right) + O(\ln r).
\end{aligned}$$

Dividing by r^2 and integrating with respect to r from 1 to ∞, by condition (5.8) and Theorem 5.4, we get (5.17).

Now we establish that

$$S\left(r, \frac{f'}{f}\right) = O(1). \tag{5.18}$$

We argue in the same way as in the proof of Theorem 5.4, but instead of Tsuji characteristics we use the Nevanlinna characteristics for the half-plane. Let $F(z) = f(z)/f'(z)$ as before. Since $f(z) \in A$, $f''(z) \in A$, we get (taking into account (5.13)) the relations $C(r, 0, F) = O(1)$, $C(r, 1, F') = O(1)$. By the analog of inequality (2.23) in Chapter 3, for the Nevanlinna characteristics for the half-plane we have

$$S(r, F) \leq 3C(r, 0, F) + 4C(r, 1, F') + Q_{0\pi}(r, F);$$

hence,

$$S(r, F) = Q_{0\pi}(r, F).$$

Since (5.17) is equivalent to the condition $\int_1^\infty t^{-2} \ln^+ T(t, F)\, dt < \infty$, the definition of $Q_{0\pi}(r, F)$ implies that $Q_{0\pi}(r, F) = O(1)$. Thus, we have proved that

$$S\left(r, \frac{f'}{f}\right) = S\left(r, \frac{1}{F}\right) = O(1).$$

Since

$$\frac{f'(z)}{f(z)} = g'(z) + \frac{h'(z)}{h(z)},$$

we have

$$S(r, g') \leq S\left(r, \frac{f'}{f}\right) + S\left(r, \frac{h'}{h}\right) + O(1).$$

Applying Lemma 5.1 to the function $h(z)$, we see that $S(r, \frac{h'}{h}) = O(1)$. Taking into account (5.18) we get

$$S(r, g') = O(1). \tag{5.19}$$

The above argument can be applied also to the function $f_1(z) = f(-z)$. Therefore, we also have

$$S(r, g_1') = O(1),$$

where $g_1(z) = g(-z)$.

Applying Theorem 2.7 to the functions $g(z)$ and $g_1(z)$, we see that the relation

$$\ln |g'(re^{i\varphi})| = O(r)$$

holds uniformly in φ, $0 \leq \varphi < 2\pi$, as $r \to \infty$ possibly omitting a set E of finite logarithmic measure. Hence for $r \notin E$ we have

$$\ln M(r, g') = O(r). \tag{5.20}$$

Since $g'(z)$ is an entire function, the function $\ln M(r, g')$ is monotonically increasing with r and, by Lemma 2.1, the relation (5.20) holds also for $r \in E$. The equality

$$g(z) = g(0) + \int_0^z g'(\zeta)\, d\zeta$$

implies

$$M(r, g) \leq |g(0)| + rM(r, g'),$$

and by (5.20) we get

$$\ln M(r, g) = O(r),$$

whence

$$T(r, g) = O(r).$$

The validity of (5.7) follows from the relation $A(r, g') = O(1)$, which immediately follows from (5.19). The proof of the theorem is complete. □

COROLLARY. *Let $G(z)$ be an entire function such that the function $G^2(z) + G'(z)$ belongs to the class A. Then $G(z)$ does not exceed normal type of order 1, and*

$$\int_{-\infty}^{\infty} \frac{\ln^+ |G(t)|}{1+t^2}\, dt < \infty.$$

To prove the corollary we consider the function

$$f(z) = \exp\left\{\int_0^z G(\zeta)\, d\zeta\right\}.$$

This function does not have zeros and

$$f''(z) = \{G^2(z) + G'(z)\} f(z) \in A.$$

By Theorem 5.3, the growth of the function $g(z) = \int_0^z G(\zeta)\, d\zeta$ does not exceed the normal type of order 1, and relation (5.7) holds for this function. Since $g'(z) = G(z)$ and, by Theorem 2.3 of Chapter 3, $T(r, G) \le 2T(r, g) + O(\ln r)$, we get the desired result. □

Using Theorem 5.3, we can give an affirmative solution of the Pólya–Wiman problem under the additional assumption that the function considered has finitely many zeros. We are not going to present this result here; we refer the reader to the paper by Levin and Ostrovskiĭ [**LO60**].

The condition of representability of the function $f(z)$ in the form (5.6), where $h(z)$ satisfies (5.8), imposes a priori restrictions on the moduli of the zeros of the function $f(z)$, since

$$N\left(r, \frac{1}{f}\right) = N\left(r, \frac{1}{h}\right) \le T(r, h) + O(1).$$

In the next theorem there are no such restrictions; however, restrictions of a different character are needed.

THEOREM 5.5. *If an entire function $f(z)$ is real on the real line, has only real zeros, and $f''(z) \in A$, then*

$$\ln T(r, f) = O(r \ln r). \tag{5.21}$$

Note that under the assumptions of Theorem 5.3 we have a stronger estimate $\ln T(r, f) = O(r)$. This estimate cannot be improved. In fact, among the functions satisfying the assumptions of Theorem 5.2 may obviously be such that

$$\liminf_{r\to\infty} r^{-1} \ln T(r, f) > 0$$

(for example, $e^{e^{iz}}$, $e^{\cos z}$, etc.). All derivatives of these functions belong to the class A.

Let us prove necessary auxiliary results.

LEMMA 5.3. *Let $g(z)$ be analytic in the half-plane $\{\operatorname{Im} z > 0\}$ and satisfy there the condition $\operatorname{Im} g(z) > 0$. Then the following estimate holds:*

$$\frac{1}{4}|g(i)|\frac{\sin\varphi}{r} \le |g(re^{i\varphi})| \le 4|g(i)|\frac{r}{\sin\varphi}, \qquad 0 < \varphi < \pi. \tag{5.22}$$

PROOF. First we show that if a function $G(\zeta)$ is analytic in the disc $\{|\zeta| < 1\}$ and satisfies the condition $\operatorname{Re} G(\zeta) > 0$, then the estimate

$$|G(\zeta)| \le \frac{2|G(0)|}{1-|\zeta|} \tag{5.23}$$

holds. In fact, by the Schwarz formula, we have

$$G(\zeta) = \frac{1}{2\pi}\int_0^{2\pi} \frac{Re^{i\theta}+\zeta}{Re^{i\theta}-\zeta} \operatorname{Re}\{G(Re^{i\theta})\}\, d\theta + i \operatorname{Im}\{G(0)\}, \qquad |\zeta| < R < 1.$$

Hence

$$\begin{aligned} |G(\zeta)| &\le \frac{R+|\zeta|}{R-|\zeta|} \cdot \frac{1}{2\pi}\int_0^{2\pi} \operatorname{Re}\{G(Re^{i\theta})\}\, d\theta + |\operatorname{Im}\{G(0)\}| \\ &\le \frac{R+|\zeta|}{R-|\zeta|}|G(0)| + |G(0)| = \frac{2R}{R-|\zeta|}|G(0)|. \end{aligned}$$

Letting R tend to 1, we get (5.23).

Now let $g(z)$ be a function satisfying the assumptions of the lemma. Then the function

$$G(\zeta) = \frac{1}{i} g\left(\frac{1}{i}\frac{\zeta-1}{\zeta+1}\right)$$

is analytic for $|\zeta| < 1$ and $\operatorname{Re} G(\zeta) > 0$. Applying inequality (5.23) to it and then letting $\zeta = -\frac{z-i}{z+i}$, we get

$$|g(z)| \le \frac{2|z+i|}{|z+i| - |z-i|}|g(i)| = \frac{2|z+i|(|z+i|+|z-i|)}{|z+i|^2 - |z-i|^2}|g(i)|.$$

Noting that $|z+i|^2 - |z-i|^2 = 4r\sin\varphi$ ($z = re^{i\varphi}$), $|z \pm i| \le 2r$ ($r \ge 1$), we get the right inequality in (5.22). To get the left inequality, it suffices to apply the right part to $g_1(z) = -1/g(z)$. □

LEMMA 5.4. *Let $\{a_k\}_{k=N}^{\infty}$ and $\{b_k\}_{k=N}^{\infty}$, $-\infty \le N < 0$, be two sequences of real numbers such that*

$$a_N < b_N < a_{N+1} < b_{N+1} < \cdots < a_0 < b_0 < 0 < a_1 < b_1 < \cdots, \qquad \lim_{k\to+\infty} a_k = +\infty,$$

if $N > -\infty$, and

$$\cdots < a_0 < b_0 < 0 < a_1 < b_1 < \cdots, \qquad \lim_{k\to-\infty} a_k = -\infty, \qquad \lim_{k\to+\infty} a_k = +\infty,$$

if $N = -\infty$. Let

$$\Pi(z) = \prod_{k=N}^{\infty} \frac{1-\frac{z}{b_k}}{1-\frac{z}{a_k}}. \tag{5.24}$$

Then the infinite product $\Pi(z)$ converges absolutely and uniformly on each bounded subset of the complex plane, and hence $\Pi(z)$ is a meromorphic function with simple poles at the points a_k and simple zeros at the points b_k. This function satisfies the following conditions:

(a) $\operatorname{Im}\Pi(z)\cdot\operatorname{Im} z > 0$ *for* $\operatorname{Im} z \ne 0$.

(b) *For each $\eta > 0$ the relation*

$$\Pi(re^{i\varphi}) = O(r)$$

holds uniformly in φ outside the angles $\{|\arg z| < \eta\}$, $\{|\pi - \arg z| < \eta\}$.

(c) $m(r,\Pi) + m(r, 1/\Pi) = O(\ln r)$.

(d) *The relation*

$$\ln|\Pi(re^{i\varphi})| = O(r)$$

holds uniformly in φ, $0 \le \varphi < 2\pi$, *as* $r \to \infty$ *missing, possibly, a set* E *of finite logarithmic measure.*

PROOF. Note that by the Leibniz theorem concerning alternating series, we have

$$\sum_{k=N}^{\infty}\left(\frac{1}{a_k}-\frac{1}{b_k}\right) < \infty.$$

Since for $|a_k| > 2|z|$ the relation

$$\left|\frac{1-\frac{z}{b_k}}{1-\frac{z}{a_k}} - 1\right| = |z|\frac{\left(\frac{1}{a_k}-\frac{1}{b_k}\right)}{\left|1-\frac{z}{a_k}\right|} \le 2|z|\left(\frac{1}{a_k}-\frac{1}{b_k}\right)$$

holds, the absolute and uniform convergence of the infinite product $\Pi(z)$ on each bounded set follows from Theorem 3.1 of Chapter 2.

Let $\arg(z-\alpha)$, where α is a real number and $\operatorname{Im} z > 0$, be the principal value of the argument, and

$$\arg\frac{1-\frac{z}{b_k}}{1-\frac{z}{a_k}} = \arg(z-b_k) - \arg(z-a_k).$$

The difference $\arg(z-b_k) - \arg(z-a_k)$ represents the size of the angle which the segment $[a_k, b_k]$ subtends at the point z ($\operatorname{Im} z > 0$). Therefore the number

$$\arg\Pi(z) = \sum_{k=N}^{\infty}\arg\frac{1-\frac{z}{b_k}}{1-\frac{z}{a_k}} = \sum_{k=N}^{\infty}\{\arg(z-b_k) - \arg(z-a_k)\}$$

satisfies the inequality

$$0 < \arg\Pi(z) < \pi, \qquad \operatorname{Im} z > 0.$$

Hence

$$\operatorname{Im}\Pi(z) > 0 \quad \text{for} \quad \operatorname{Im} z > 0.$$

Similarly we prove that $\operatorname{Im}\Pi(z) < 0$ for $\operatorname{Im} z < 0$. Thus we have proved that the function $\Pi(z)$ satisfies condition (a).

Applying Lemma 5.3, we get the inequality

$$\frac{1}{4}|\Pi(i)|\frac{|\sin\varphi|}{r} \le |\Pi(re^{i\varphi})| \le 4|\Pi(i)|\frac{r}{|\sin\varphi|}, \qquad 0 \le \varphi \le 2\pi, \quad r \ge 1.$$

From this it easily follows that the function $\Pi(z)$ satisfies conditions (b) and (c).

To prove the last assertion of the lemma, we establish that

$$S(r, \Pi(z)) + S(r, \Pi(-z)) = O(1);$$

then the desired conclusion follows from Theorem 2.7.

Let

$$q(z) = \frac{\Pi(z) - i}{\Pi(z) + i}.$$

Since $\operatorname{Im}\Pi(z) > 0$ for $\operatorname{Im} z > 0$, the function $q(z)$ is analytic in the half-plane $\{\operatorname{Im} z > 0\}$ and admits there the estimate $|q(z)| < 1$. Therefore $A(r,q) = B(r,q) = C(r,q) \equiv 0$, and hence, $S(r,q) \equiv 0$. Using relation (6.11) in Chapter 1, we get $S(r,\Pi(z)) = S(r,q) + O(1) = O(1)$. Similarly we get $S(r,\Pi(-z)) = O(1)$. This completes the proof of the lemma. □

Now we prove Theorem 5.5.

We may restrict ourselves to the case where the set of zeros of the function $f(z)$ is not bounded above. In fact, if the set of zeros of this function is finite, then $f(z) = e^{g(z)}h(z)$, where $g(z)$ is an entire function and $h(z)$ is a polynomial, and using Theorem 5.3 we get the estimate $\ln T(r, f) = O(r)$, which is more precise than (5.21). If the set of zeros is infinite, but is bounded from the right, we consider the function $f(-z)$ instead of $f(z)$.

We enumerate the zeros of the function $f(z)$ into a sequence $\{a_k\}_{k=N}^{\infty}$, $-\infty \leq N < \infty$, in such a way that the inequalities

$$\cdots < a_{k-2} < a_{k-1} < a_k < a_{k+1} < \cdots$$

are satisfied (multiple zeros are counted only once). By the Rolle theorem, each of the intervals (a_k, a_{k+1}) contains zeros of the function $f'(z)$. We choose one zero of $f'(z)$ in every interval, and denote it by b_k. Considering, if necessary, the function $f(z+\alpha)$ instead of $f(z)$, where α is a suitably chosen real number, we may assume that the sequences $\{a_k\}_{k=N}^{\infty}$ and $\{b_k\}_{k=N}^{\infty}$ satisfy the assumptions of Lemma 5.4.

Let $\Pi(z)$ be a function determined by relation (5.24). Since the function $\Pi(z)$ has simple poles at the points a_k and simple zeros at the points b_k, the function $\frac{f'(z)}{f(z)\Pi(z)}$ is an entire function. We denote it by $\psi(z)$; then we have the equality

$$\frac{f'(z)}{f(z)} = \psi(z)\Pi(z). \tag{5.25}$$

Let us estimate the growth of the function $\psi(z)$. Equality (5.25) implies

$$m(r, \psi) \leq m\left(r, \frac{f'}{f}\right) + m\left(r, \frac{1}{\Pi}\right).$$

We divide both sides of this inequality by r^3 and integrate them with respect to r from $R \geq 1$ to ∞. We get

$$\int_R^{\infty} \frac{m(r, \psi)}{r^3} dr \leq \int_R^{\infty} \frac{m(r, \frac{f'}{f})}{r^3} dr + \int_R^{\infty} \frac{m(r, \frac{1}{\Pi})}{r^3} dr.$$

Using Theorem 5.4 and taking into account that, by Lemma 5.4,

$$m\left(r, \frac{1}{\Pi}\right) = O(\ln r),$$

we arrive at the estimate

$$\int_R^{\infty} \frac{m(r, \psi)}{r^3} dr = O\left(\frac{\ln R}{R}\right).$$

Since $\psi(z)$ is an entire function, we have $m(r, \psi) = T(r, \psi)$ and, by monotonicity of the characteristic $T(r, \psi)$, we get

$$\frac{T(R, \psi)}{2R^2} \leq \int_R^{\infty} \frac{T(r, \psi)}{r^3} dr = O\left(\frac{\ln R}{R}\right),$$

whence

$$T(R, \psi) = O(R \ln R).$$

Letting $R = 2r$ in inequality (7.1) of Chapter 1, we get

$$\ln^+ M(r, \psi) \leq 3T(2r, \psi) = O(r \ln r). \tag{5.26}$$

Since the function $\Pi(z)$ satisfies conditions (b) and (d) of Lemma 5.4, we take into account relation (5.25) and estimate (5.26) and come to the conclusion that for each η, $0 < \eta < \frac{\pi}{2}$, and all $z = re^{i\varphi}$ except, possibly, those for which

$$r \in E,\ |\varphi| < \eta \quad \text{or} \quad r \in E,\ |\pi - \varphi| < \eta,$$

where $E \subset (0, \infty)$ is a set of finite logarithmic measure, the inequality

$$\left| \frac{f'(z)}{f(z)} \right| \leq \exp\{Kr \ln r\} \tag{5.27}$$

holds, where $K > 0$ is a constant.

Now we estimate the growth of the function $f(z)$. Let $z = re^{i\varphi}$, $r \notin E$. We integrate both sides of (5.27) along a path L which starts at 0, goes to the point ir along the ray $\{\arg z = \frac{\pi}{2}\}$, and then continues along the arc of the circle $\{|z| = r\}$ counterclockwise from the point ir to the point $re^{i\varphi}$. We get

$$\left| \ln \left| \frac{f(z)}{f(0)} \right| \right| = \left| \operatorname{Re} \int_L \frac{f'(\zeta)}{f(\zeta)} d\zeta \right| \leq \int_L \left| \frac{f'(\zeta)}{f(\zeta)} \right| |d\zeta| \leq (2\pi + 1) r \exp\{Kr \ln r\}.$$

Hence, for $r \notin E$,

$$\ln^+ \ln^+ M(r, f) = O(r \ln r). \tag{5.28}$$

Since the function $M(r, f)$ is non-decreasing, it follows from Lemma 2.1 that (5.28) holds also for $r \in E$. Since $T(r, f) \leq \ln^+ M(r, f)$, we come to the conclusion of the theorem.

Theorem 5.5 shows that, working on the problem of Pólya and Wiman, it suffices to consider only the entire functions $f(z)$ satisfying $\ln T(r, f) = O(r \ln r)$.

CHAPTER 7

Applications of Riemann Surfaces to Value Distribution

In this chapter we will use some facts about Riemann surfaces. Familiarity with the corresponding chapter of the well-known course of S. Stoïlow [**Sto62**, Vol. II, Chapter VII] is more than sufficient for understanding this chapter.

1. Geometric meaning of deficient and index values

Let $w = f(z)$ be a meromorphic function in the complex z-plane. It is a one-to-one mapping of the complex plane $\{|z| < \infty\}$ onto a simply connected Riemann surface F, that is, onto a Riemann surface which is homeomorphic to the plane.[1] If the function $w = f(z)$ is meromorphic in the extended complex plane, that is, if it is rational, then it maps the extended z-plane onto a simply connected closed Riemann surface (of genus zero), that is, onto a surface which is homeomorphic to the closed sphere. In this case the surface F is finite-sheeted, and the number n of sheets is equal to the degree of the rational function. If the function $f(z)$ is transcendental, then the Riemann surface F is open and infinite-sheeted.

On the other hand, it is known (one of the simplest proofs can be found in the book by G. M. Goluzin [**Gol69**, Chapter XI, §2]) that for each open simply connected Riemann surface F there exists a function $w = f(z)$ that is meromorphic in the disc $\{|z| < R\}$, $0 < R \leq \infty$, and maps $\{|z| < R\}$ onto F. If F is a simply connected closed Riemann surface without one point, then the function $f(z)$ is rational, and the surface F is called a surface of elliptic type. Otherwise the surface F is of parabolic or hyperbolic type depending on whether $R = \infty$ or $R < \infty$. In all cases the function $w = f(z)$ will be called a *mapping function* of the surface F, and the Riemann surface F will be called the surface corresponding to $w = f(z)$.

We will be mainly interested in functions which are meromorphic and transcendental in $\{|z| < \infty\}$ and map $\{|z| < \infty\}$ onto surfaces of parabolic type. For this reason, unless stated otherwise, by a Riemann surface we mean an open simply connected Riemann surface of parabolic type.

A mapping function of a given Riemann surface is determined up to a univalent conformal mapping of the complex plane onto itself; that is, if $f(z)$ is a mapping function of F, then all functions of the form $f(az + b)$, $a \neq 0$, and only such functions, are mapping functions of F. If we normalize the mapping functions in

[1]Strictly speaking, the one-to-one mapping is realized by a function $p = \tilde{f}(z)$, $p \in F$, such that $T(\tilde{f}(z)) = f(z)$, where $T(p)$ is the projection of a point p onto the w-plane. We follow the tradition not to distinguish between $\tilde{f}(z)$ and $f(z)$ both in our notation and in our terminology; the meaning will always be clear from the context. We also agree to say that the point p *lies over* $T(p)$.

an appropriate way,[2] we get a one-to-one correspondence between Riemann surfaces and mapping functions. Thus, meromorphic functions can be given by means of the corresponding Riemann surfaces. Therefore it is not surprising that, as was shown by L. Ahlfors [**Ahl35**] (see also S. Stoïlow [**Sto62**, Vol. II, Chapter X, §§1, 2]) that the main results of the value distribution theory and their generalizations can be obtained in a purely geometric way, by studying the corresponding Riemann surfaces. We will not present the Ahlfors theory here. We will be concerned with the determination of the geometric meaning of deficient and index values only.

Universal necessary and sufficient conditions under which a certain value is deficient for a function $f(z)$, which is a mapping function of a surface F, in terms of the structure of the surface F only, cannot be given in principle. In fact, as we saw in Section 6 of Chapter 4, the function $f(z+h)$, which is also a mapping function of F, can have different deficient values for different h. Of course, if we normalize the mapping function, the above-mentioned reason will disappear, but this shows that the problem is very complicated. Up to now necessary and sufficient conditions for the existence of deficient values are known for very restricted classes of Riemann surfaces only. The same can be said about index values.

Here we only present conditions which are necessary for a to be a deficient or index value. It is convenient to state them as conditions which are sufficient for absence of deficient or index values.

THEOREM 1.1. *If the number of algebraic ramification points of a Riemann surface F lying over the point a is finite, then the corresponding mapping functions satisfy $\varepsilon(a) = 0$.*

Suppose that the sum of orders of the algebraic ramification points lying over a is equal to q. Algebraic ramification points correspond to multiple a-points of the mapping function $f(z)$; moreover, if an algebraic ramification point has order m, then the corresponding multiple a-point has order $m+1$. All other points of F lying over a correspond to simple a-points of the function $f(z)$. Therefore $n_1(r,a) = q$ for $r > r_0$, from which we get $N_1(r,a) \sim q \ln r$. Since the function $f(z)$ is transcendental, we have $\ln r = o(T(r,f))$ and $\varepsilon(a) = 0$.

THEOREM 1.2. *Suppose that the points of a Riemann surface F whose projections belong to the η-neighborhood $U(a,\eta)$ of a point a form a set of regions $F_\nu \subset F$.*

Suppose that each point of $U(a,\eta)$ is a projection of exactly λ_ν points of F_ν (each algebraic ramification point of order m is counted $m+1$ times), where $1 \le \lambda_\nu \le \Lambda < \infty$, that is, the Riemann surface F_ν has λ_ν sheets and completely covers $U(a,\eta)$.

Then the value a is not deficient for the mapping function $f(z)$.

We may assume without loss of generality that $a = 0$ and $1 < \eta < \infty$. In fact, if this condition is not satisfied, we consider the meromorphic function $w = f_1(z) = L(f(z))$, where $L(w)$ is a linear-fractional function that maps $U(a,\eta)$ onto $\{|w| < \eta_1\}$, $L(a) = 0$, $1 < \eta_1 < \infty$. It is clear that $\delta(a,f) = \delta(0,f_1)$ and that the Riemann surface corresponding to the function $f_1(z)$ satisfies the assumptions of Theorem 1.2 with $a = 0$, $\eta = \eta_1$. Denote by F'_ν the closed set of points in F_ν lying

[2]For example, we can require that the point $z = 0$ be mapped to a preassigned *simple point* of the Riemann surface F (that is, a point which is not an algebraic ramification point), and $f'(0) = 1$.

over $\{|w| \le 1\}$. Obviously F'_ν consists of at most Λ closed regions. Let D_ν and D'_ν be the inverse images of the sets F_ν and F'_ν, respectively, under the mapping $w = f(z)$. Since F_ν completely covers $\{|w| < \eta\}$, it follows that D_ν, as well as D'_ν, contains as many $e^{i\theta}$-points of the function $f(z)$ as the number of zeros, namely λ_ν. Let Γ_r be a closed curve on F which is the image of the circle $\{|z| = r\}$, $r > 1$, under the mapping $w = f(z)$, and

$$L(r) = \int_0^{2\pi} \overset{\circ}{f}(re^{i\varphi})r\, d\varphi$$

is the length of Γ_r with respect to the spherical metric. Denote by l the distance on the Riemann sphere between the circles whose stereographic projections are the circles $\{|w| = 1\}$ and $\{|w| = \eta\}$.

All $e^{i\theta}$-points and zeros of the function $f(z)$ are contained in $\bigcup_\nu D'_\nu \subset \bigcup_\nu D_\nu$. If $D'_\nu \subset \{|z| \le r\}$, it contributes λ_ν to the functions $n(r, e^{i\theta})$ and $n(r, 0)$. Therefore, if we denote by $k(r)$ the number of sets D'_ν that have nonempty intersections with both $\{|z| \le r\}$ and $\{|z| > r\}$, then

$$n(r, e^{i\theta}) - n(r, 0) \le \Lambda k(r), \tag{1.1}$$

since the number of $e^{i\theta}$-points in D'_ν does not exceed Λ. If $k(r) \ge 2$, then the circle $\{|z| = r\}$ cannot be entirely contained in one of the domains D_ν; that is, the curve Γ_r is not contained in one of the regions F_ν. The curve Γ_r intersects $k(r) \ge 2$ different sets F'_ν; hence, its projection intersects the annulus $\{1 < |w| < \eta\}$ at least $2k(r)$ times, joining points on the circles $\{|w| = 1\}$ and $\{|w| = \eta\}$. Hence $L(r) \ge 2k(r)l$. From (1.1) we get the inequality

$$\begin{aligned} n(r, e^{i\theta}) - n(r, 0) &\le \Lambda \max(1, k(r)) \\ &\le \Lambda \max\left(1, \frac{1}{2l} L(r)\right) \le \Lambda\left(1 + \frac{1}{2l} L(r)\right). \end{aligned} \tag{1.2}$$

Since

$$N(r, a) - N(1, a) = \int_1^r \frac{n(t, a)}{t}\, dt,$$

integrating inequality (1.2) with respect to $\ln r$ we get

$$N(r, e^{i\theta}) - N(1, e^{i\theta}) - N(r, 0) + N(1, 0) \le \Lambda\left(\ln r + \frac{1}{2l}\int_1^r \frac{L(t)}{t}\, dt\right). \tag{1.3}$$

Integrating this inequality with respect to θ and using Cartan's formula (4.13) of Chapter 1, we get

$$T(r, f) - T(1, f) - N(r, 0) + N(1, 0) \le \Lambda\left(\ln r + \frac{1}{2l}\int_1^r \frac{L(t)}{t}\, dt\right),$$

from which, by Jensen's formula (4.1) of Chapter 1,

$$m(r, 0) - m(1, 0) \le \Lambda\left(\ln r + \frac{1}{2l}\int_1^r \frac{L(t)}{t}\, dt\right). \tag{1.4}$$

Now we estimate the integral in the right-hand side of (1.4) using the Cauchy–Bunyakovskii inequality. We have

$$\begin{aligned}\int_1^r \frac{L(t)}{t}dt &= \int_1^r \int_0^{2\pi} \overset{\circ}{f}(te^{i\theta})\,d\theta\,dt \\ &\le \left\{\int_1^r \int_0^{2\pi} \{\overset{\circ}{f}(te^{i\theta})\}^2 t\,dt\,d\theta\right\}^{\frac{1}{2}} \left\{\int_1^r \int_0^{2\pi} \frac{dt\,d\theta}{t}\right\}^{\frac{1}{2}} \\ &\le \sqrt{2\pi \ln r}\left\{\int_0^r \int_0^{2\pi} \{\overset{\circ}{f}(te^{i\theta})\}^2 t\,dt\,d\theta\right\}^{\frac{1}{2}} = \sqrt{2\pi\ln r}\sqrt{\pi \overset{\circ}{A}(r,f)},\end{aligned}$$

by the definition of $\overset{\circ}{A}(r,f)$ in Section 4 of Chapter 1. Thus (1.4) implies

$$m(r,0) \le m(1,0) + \Lambda\left(\ln r + \frac{\pi}{\sqrt{2}l}\sqrt{\ln r}\sqrt{\overset{\circ}{A}(r,f)}\right). \tag{1.5}$$

Now we use the following lemma.

LEMMA 1.1. *Let $\varphi(t)$ be a continuously differentiable, positive function, non-decreasing for $t \ge t_0$. Then $\varphi'(t) \le \{\varphi(t)\}^{1+\varepsilon}$, $\varepsilon > 0$, for all $t \ge t_0$ except, possibly, a set Δ of intervals of finite total length.*

In fact, $\varphi'(t) > \{\varphi(t)\}^{1+\varepsilon}$ on Δ; therefore

$$\int_\Delta dt \le \int_\Delta \frac{\varphi'(t)}{\{\varphi(t)\}^{1+\varepsilon}}dt \le \int_{t_0}^\infty \frac{\varphi'(t)\,dt}{\{\varphi(t)\}^{1+\varepsilon}} \le \frac{1}{\varepsilon}\frac{1}{\{\varphi(t_0)\}^\varepsilon}.$$

By the definition of the Shimizu–Ahlfors spherical characteristic $\overset{\circ}{T}(r,f)$ (see Section 4 of Chapter 1), we have

$$\overset{\circ}{A}(r,f)\ln r = \frac{d\overset{\circ}{T}(r,f)}{d\ln r}\ln r = \frac{d\overset{\circ}{T}(r,f)}{d\ln\ln r}.$$

Therefore, applying Lemma 1.1 to the function $\overset{\circ}{T}(r,f)$ with $t = \ln\ln r$, $\varepsilon = 1/2$, we see that $\overset{\circ}{A}(r,f)\ln r \le \{\overset{\circ}{T}(r,f)\}^{3/2}$ outside a set Δ of intervals for which

$$\int_\Delta d\ln\ln r < \infty.$$

Since $\ln r = o(T(r,f))$, we get from (1.5) that outside Δ, that is, on a set of values of r unbounded from above, the relation

$$m(r,0) \le O(1) + \Lambda\left\{\ln r + \frac{\pi}{\sqrt{2}l}(\overset{\circ}{T}(r,f))^{3/4}\right\} = o(\overset{\circ}{T}(r,f)), \qquad r\to\infty,$$

holds, that is, $\delta(0,f) = 0$.

It is obvious that the assumptions of Theorem 1.2 are not satisfied if a is an asymptotic value for $f(z)$. In fact, if the assumptions of the theorem are satisfied even with the weaker condition $1 \le \lambda_\nu < \infty$, then the regions in which $|f(z) - a| < \eta' < \eta$ (or $|f(z)| > \frac{1}{\eta'} > \frac{1}{\eta}$ if $a = \infty$) are all bounded, and a cannot be an asymptotic value.

Theorem 1.2 does not remain true if we omit the condition $\lambda_\nu \le \Lambda$.

Consider, for example, the meromorphic function $f(z)$ from Example 1 of Section 2 in Chapter 5. The set D of those z for which $|f(z)| > K$, where $K > M$ and

M is the constant defined in the analysis of the example, consists of bounded connected components D_ν since $D \subset \bigcup_{j\in B} C_j$. Obviously, D_ν contains finitely many poles (in our example it is one multiple pole, since each disc C_j contains one multiple pole). By the known theorems (see Stoïlow [**Sto62**, Vol. II, Chapter V, §2]), the image of D_ν under the mapping $f(z)$ onto the Riemann surface F is a finite-sheeted region $F_\nu \subset F$ entirely covering the disc $\{|w| > K\}$. Hence all the conditions of Theorem 1.2 except $\lambda_\nu \leq \Lambda$ are satisfied. In fact, in our example the number λ_ν of sheets of F_ν is equal to the order of the pole of $f(z)$ lying in D_ν, and the orders of poles of $f(z)$ are not bounded. On the other hand, as was shown in the analysis of Example 1, ∞ is a deficient value for $f(z)$.

2. Quasiconformal mappings

Let $w = f(z)$ be a meromorphic function which is a one-to-one mapping of the complex plane onto a Riemann surface F. This mapping is conformal and univalent in sufficiently small neighborhoods of all points z except multiple a-points of the function $f(z)$, which correspond to algebraic ramification points on F. For brevity we call such mapping conformal, without mentioning the exceptional, at most countable, set. Thus, the problem of finding a mapping function for a given Riemann surface F can be described in other words as a problem of finding a conformal mapping of F onto $\{|z| < \infty\}$. With very few exceptions it is not known how to construct such a mapping effectively. For this reason, the so-called quasiconformal mappings of the Riemann surface F onto the complex plane are constructed. The definition of a quasiconformal mapping we give is not the most general, but it is sufficient for our purposes.

We say that a function $w = \varphi(z) = u(z) + iv(z)$ is continuously differentiable in the closed region $\bar{D}$ if there exist functions $a(z)$, $b(z)$, $c(z)$, and $d(z)$, continuous in $\bar{D}$ and such that for each point z ($z = x + iy$) the equalities

$$\begin{aligned} u(z+\Delta z) - u(z) &= a(z)\Delta x + b(z)\Delta y + o(|\Delta z|), \\ v(z+\Delta z) - v(z) &= c(z)\Delta x + d(z)\Delta y + o(|z|) \end{aligned} \tag{2.1}$$

hold as $\Delta z \to 0$, $z + \Delta z \in \bar{D}$. The number $J = J[\varphi] = a(z)d(z) - c(z)b(z)$ is called the *Jacobian* of the function $\varphi(z)$ at the point z. It is clear that

$$a(z) = \frac{\partial u(z)}{\partial x}, \qquad b(z) = \frac{\partial u(z)}{\partial y}, \qquad c(z) = \frac{\partial v(z)}{\partial x}, \qquad d(z) = \frac{\partial v(z)}{\partial y} \tag{2.2}$$

at all interior points of $\bar{D}$.

The same equalities hold for those points on the boundary of $\bar{D}$ where the corresponding partial derivatives exist. However, the partial derivatives of u and v can be undefined at some boundary points. We regard equalities (2.2) as the definitions of the partial derivatives on ∂D. Equalities (2.1) imply that the image of a smooth curve $C \subset \bar{D}$ under the mapping $w = \varphi(z)$ is a smooth curve if $J[\varphi] \neq 0$ for $z \in C$.

Let D_1 and D_2 be two closed regions bounded by a finite number of piecewise smooth closed Jordan curves. A mapping $w = \varphi(z)$ of the region D_1 onto the region D_2 is called *quasiconformal*[3] if: (1) the mappings $\varphi(z)$ and $\varphi^{-1}(w)$ are one-to-one and continuous on D_1 and D_2, respectively; (2) the region D_1 can be divided by a finite set of piecewise smooth curves into closed regions where the function $\varphi(z)$ is

[3] This is different from the modern definition.

continuously differentiable and has a positive Jacobian. In this case it is clear that the inverse mapping $\varphi^{-1}(w)$ is also quasiconformal.

Let F be a simply connected Riemann surface covering the w-plane (the case of a plane region is not excluded). Let p_0 be a point on F whose projection is w_0; we let $m = 1$ if p_0 is a simple point, and $m = k+1$ if p_0 is a ramification point of order k. For sufficiently small $\eta = \eta(p_0) > 0$ the function $w = w_0 + \zeta^m$ maps the disc $\{|\zeta| \leq \eta\}$ onto a closed neighborhood of the point p_0. Let $q = \psi(p)$ be a one-to-one mapping of the Riemann surface F onto the Riemann surface F_1 which covers the z-plane, $p \in F$, $q \in F_1$, and let $T(q)$ be the projection of the point q in the z-plane. The mapping $q = \psi(p)$ is called quasiconformal if the mapping $z = T\{\psi(w_0+\zeta^m)\}$ of the disc $\{|\zeta| \leq \eta\}$ is quasiconformal for each point $p_0 \in F$. In the case where F_1 is a region in the z-plane we have $T\{\psi(w_0+\zeta^m)\} \equiv \psi(w_0+\zeta^m)$.

Thus, the definition of a quasiconformal mapping of a Riemann surface (including the case of open plane regions) is reduced to the definition of a quasiconformal mapping of a closed plane region.

Suppose that a Riemann surface F is mapped conformally onto $\{|z| < \infty\}$ and quasiconformally onto $\{|\zeta| < \infty\}$. Then the mapping $\{|z| < \infty\} \to F \to \{|\zeta| < \infty\}$ is a quasiconformal mapping $\zeta = \zeta(z)$ of the z-plane onto the ζ-plane. The inverse mapping $z = z(\zeta)$ is also quasiconformal. We may assume without loss of generality that $z(0) = 0$. As we have already mentioned, usually it is not known how to construct a conformal mapping $w = f(z)$ of the z-plane onto the Riemann surface F, but a quasiconformal mapping $w = \varphi(\zeta)$ of the ζ-plane onto F can sometimes be efficiently constructed.

Suppose that such a mapping has been constructed. A root ζ_0 of the equation $\varphi(\zeta) = a$ ($1/\varphi(\zeta) = 0$ for $a = \infty$) will be called an *a-point* of the function $\varphi(\zeta)$. Moreover, if ζ_0 is mapped to a simple point of F, then ζ_0 is considered to be an a-point of the first order, and if ζ_0 is mapped to an algebraic ramification point of order m, then ζ_0 is considered to be an a-point of *order* $m+1$. If $\varphi(\zeta)$ is a meromorphic function, this definition is equivalent to the usual definition of an a-point. We denote by $n(r, a, \varphi)$ the number of a-points of the function $\varphi(\zeta)$ in the disc $\{|\zeta| \leq r\}$, and by $\pi \mathring{A}(r, \varphi)$ the spherical area of the Riemann surface F_r which is the image of $\{|\zeta| < r\}$ under $w = \varphi(\zeta)$. The curves γ_r, which are the images of the circles $\{|\zeta| = r\}$ under the mapping $z = z(\zeta)$, exhaust the z-plane. It is clear that

$$n_\gamma(r, a, f) \equiv n(r, a, \varphi), \qquad \mathring{A}_\gamma(r, f) \equiv \mathring{A}(r, f),$$

where the quantities $n_\gamma(r, a, f)$ and $\mathring{A}_\gamma(r, f)$ have the same meaning as in Section 6 of Chapter 4. Thus, if we know the quasiconformal mapping $w = \varphi(\zeta)$, we can study the value distribution of the meromorphic function $w = f(z)$ in the case where the complex plane $\{|z| < \infty\}$ is exhausted by the family of the curves γ_r; moreover, we can find $T_\gamma(r, f)$, $N_\gamma(r, a, f)$, $\delta_\gamma(a, f)$, $\varepsilon_\gamma(a, f)$, etc. Since we are interested in the usual deficiencies and indices, $\delta(a, f)$ and $\varepsilon(a, f)$, rather than in the quantities $\delta_\gamma(a, f)$ and $\varepsilon_\gamma(a, f)$, we need conditions on the family of curves γ_r and on $T(r, f)$ under which $\delta_\gamma(a, f) = \delta(a, f)$ and $\varepsilon_\gamma(a, f) = \varepsilon(a, f)$. Such conditions were found in Section 6 of Chapter 4. In Theorem 6.9 of Chapter 4, the following condition was imposed on the family of curves γ_r:

$$r_2(r) \sim r_1(r), \quad r \to \infty, \tag{2.3}$$

where $r_2(r) = \max_{z\in\gamma_r} |z|$, $r_1(r) = \min_{z\in\gamma_r} |z|$. Since in our case

$$r_2(r) = \max_{|\zeta|=r} |z(\zeta)|, \qquad r_1(r) = \min_{|\zeta|=r} |z(\zeta)|,$$

it is clear that condition (2.3) holds if

$$\lim_{\zeta\to\infty} \frac{|z(\zeta)|}{|\zeta|} = \lim_{z\to\infty} \frac{|z|}{|\zeta(z)|} = A, \qquad 0 < A < \infty. \tag{2.4}$$

In the next section we find a rather large class of quasiconformal mappings $z(\zeta)$ for which relation (2.4) holds.

Now we consider some properties of quasiconformal mappings. We restrict our attention to quasiconformal mappings of plane regions (as we have seen, quasiconformal mappings of Riemann surfaces can be reduced to this case).

First we observe that a composition of two quasiconformal mappings is also a quasiconformal mapping. Let $w = \psi(z) = u(z) + iv(z)$ be a function which quasiconformally maps a region G onto a region D. Let z_0 be a point in G such that the function $\psi(z)$ is continuously differentiable in a sufficiently small neighborhood of z_0. Let $dz = |dz|e^{i\varphi}$; then

$$|dw|^2 = (du)^2 + (dv)^2 = (E\cos^2\varphi + 2F\sin\varphi\cos\varphi + G\sin^2\varphi)|dz|^2,$$

where

$$E = \left(\frac{\partial u}{\partial x}\right)^2 + \left(\frac{\partial v}{\partial x}\right)^2, \qquad F = \frac{\partial u}{\partial x}\frac{\partial u}{\partial y} + \frac{\partial v}{\partial x}\frac{\partial v}{\partial y}, \qquad G = \left(\frac{\partial u}{\partial y}\right)^2 + \left(\frac{\partial v}{\partial y}\right)^2.$$

The modulus of the directional derivative in the direction of $\arg(z - z_0)$ is equal to

$$\left|\frac{dw}{dz}\right|_\varphi = \sqrt{E\cos^2\varphi + 2F\sin\varphi\cos\varphi + G\sin^2\varphi} = \sqrt{h(\varphi)}.$$

It is easy to compute that

$$\max_{0\le\varphi\le 2\pi} h(\varphi) = \frac{1}{2}\{E + G + \sqrt{(E+G)^2 - 4J^2}\}, \tag{2.5$'$}$$

$$\min_{0\le\varphi\le 2\pi} h(\varphi) = \frac{1}{2}\{E + G - \sqrt{(E+G)^2 - 4J^2}\}. \tag{2.5$''$}$$

The quantity

$$p(z_0, w_0) = p_\psi(z_0, w_0) = \left(\max_{0\le\varphi\le 2\pi}\left|\frac{dw}{dz}\right|_\varphi\right) \Big/ \left(\min_{0\le\varphi\le 2\pi}\left|\frac{dw}{dz}\right|_\varphi\right), \quad w_0 = \psi(z_0),$$

is called the *characteristic* of the quasiconformal mapping $w = \psi(z)$. At those points of G that have no neighborhood in which $\psi(z)$ is continuously differentiable, the characteristic $p(z_0, w_0)$ is not defined. However, as is easy to see, even at these points there exists the directional derivative $|dw/dz|_\varphi$ in an arbitrary direction, continuously depending on φ. By the Heine–Borel lemma, in each closed subregion of the region G, the characteristic $p(z, w)$ is defined and is continuous everywhere except, possibly, a finite set of piecewise smooth curves. If the mapping $w = \psi(z)$ is conformal at z_0, then $|dw/dz|$ does not depend on the direction φ, and $p(z_0, w_0) = 1$.

Conversely, if $p(z_0, w_0) = 1$, then the mapping is conformal at z_0. In fact, if $p(z_0, w_0) = 1$, then $(E+G)^2 - 4J^2 = 0$, that is,

$$(E+G-2J)(E+G+2J)$$
$$= \left\{ \left(\frac{\partial u}{\partial x} - \frac{\partial v}{\partial y} \right)^2 + \left(\frac{\partial u}{\partial y} + \frac{\partial v}{\partial x} \right)^2 \right\} \left\{ \left(\frac{\partial u}{\partial x} + \frac{\partial v}{\partial y} \right)^2 + \left(\frac{\partial u}{\partial y} - \frac{\partial v}{\partial x} \right)^2 \right\} = 0.$$

The expression in the second braces cannot be equal to zero, since otherwise we would have

$$\frac{\partial u}{\partial x} = -\frac{\partial v}{\partial y}, \qquad \frac{\partial u}{\partial y} = \frac{\partial v}{\partial x}, \qquad J = -\left(\frac{\partial u}{\partial x} \right)^2 - \left(\frac{\partial u}{\partial y} \right)^2 \le 0,$$

contrary to the assumption $J > 0$. Hence the expression in the first braces is equal to zero; that is, $w = \psi(z)$ satisfies the Cauchy–Riemann equations at the point z_0, which, together with the condition $J > 0$, implies that the mapping is conformal.

If $p(z, w) = 1$ at all points of G where the characteristic is defined, then the function $w = \psi(z)$ is analytic in G. In fact, the function $\psi(z)$ is continuous in G and is analytic in G everywhere except possibly a set whose intersection with each closed subregion of G consists of finitely many piecewise smooth curves. By the well-known result from the theory of analytic functions, the function $w = \psi(z)$ is analytic everywhere in the region G. Since $J[\psi] > 0$, we have $\psi'(z) \neq 0$, and the mapping $w = \psi(z)$ is conformal in G.

It is easy to verify that at the point z_0 where the characteristic $p(z_0, w_0)$ exists, the mapping $w = \psi(z)$ sends the element $dz = |dz|e^{i\varphi}$ into a certain element $dw = |dw|e^{i\theta}$, where θ gets an increment 2π if φ gets an increment 2π. Therefore

$$\max_{\varphi} |dw/dz|_{\varphi} = \frac{1}{\min_{\theta} |dz/dw|_{\theta}},$$

$$\min_{\varphi} |dw/dz|_{\varphi} = \frac{1}{\max_{\theta} |dz/dw|_{\theta}},$$

$$p(z_0, w_0) = \frac{\max_{\varphi} |dw/dz|_{\varphi}}{\min_{\varphi} |dw/dz|_{\varphi}} = \frac{\max_{\theta} |dz/dw|_{\theta}}{\min_{\theta} |dz/dw|_{\theta}} = p(w_0, z_0).$$

In what follows, when writing relations containing the characteristics, we assume that the characteristics are defined at the corresponding points. First of all, note the obvious inequality: $1 \le p(z, w) < \infty$.

Next, if a quasiconformal mapping $w = \psi(z)$ is obtained as a composition of two quasiconformal mappings $w = w(\zeta)$ and $\zeta = \zeta(z)$, then

$$p(z, w) \le p(z, \zeta) p(\zeta, w). \tag{2.6}$$

In fact,

$$p(z, w) = \frac{\max |dw/dz|}{\min |dw/dz|} = \frac{\max\{|dw/d\zeta||d\zeta/dz|\}}{\min\{|dw/d\zeta||d\zeta/dz|\}}$$
$$\le \frac{\max |dw/d\zeta| \max |d\zeta/dz|}{\min |dw/d\zeta| \min |d\zeta/dz|} = p(\zeta, w) p(z, \zeta).$$

If one of the mappings $w = w(\zeta)$ and $\zeta = \zeta(z)$ is conformal, then the equality is attained in (2.6). In fact, let, for example, $\zeta = \zeta(z)$ be a conformal mapping. Then $|d\zeta/dz|_{\varphi} = |d\zeta/dz|$ does not depend on φ and

$$p(z, w) = \frac{\max\{|dw/d\zeta||d\zeta/dz|\}}{\min\{|dw/d\zeta||d\zeta/dz|\}} = \frac{\max |dw/d\zeta|}{\min |dw/d\zeta|} = p(\zeta, w). \tag{2.7}$$

Similarly, if the mapping $w = w(\zeta)$ is conformal, then

$$p(z, w) = p(z, \zeta). \tag{2.8}$$

It is also easy to see that if, along with the quasiconformal mapping $w = \psi(z)$, we consider the "symmetric" mapping $w = \psi_1(z) = \overline{\psi(\bar{z})}$, then $p_\psi(z, w) = p_{\psi_1}(\bar{z}, \bar{w})$. Equalities (2.7) and (2.8) show that the characteristic of a quasiconformal mapping is a conformal invariant.

Equalities (2.5′) and (2.5″) imply that

$$\begin{aligned} p(z, w) &= \frac{E+G}{2J} + \sqrt{\left(\frac{E+G}{2J}\right)^2 - 1} \\ &= \frac{\max_\varphi |dw/dz|_\varphi^2}{J} = \frac{J}{\min_\varphi |dw/dz|_\varphi^2} = K + \sqrt{K^2 - 1}, \end{aligned} \tag{2.9}$$

where $K = (E+G)/2J$. From this we get the estimate

$$p(z, w) \leq (E+G)/J = 2K. \tag{2.10}$$

Now we give several examples of quasiconformal mappings, to which we refer in what follows.

EXAMPLE 1. The function $w = \psi(z) = xl_1/l + iy$ quasiconformally maps the rectangle $\{0 \leq x \leq l,\ 0 \leq y \leq 1\}$ onto the rectangle $\{0 \leq u \leq l_1,\ 0 \leq v \leq 1\}$. In this case $J = l_1/l$, $E = (l_1/l)^2$, $G = 1$. Using formula (2.9) we get

$$p(z, w) = \frac{1 + (l_1/l)^2 + |1 - (l_1/l)^2|}{2l_1/l} = \max\left(\frac{l_1}{l}, \frac{l}{l_1}\right). \tag{2.11}$$

EXAMPLE 2. The function

$$w = \psi(z) = x + i\left\{\left(\frac{q_2 - q_1}{l}x + q_1\right)\frac{y}{m} + \frac{a}{l}x\right\},$$

where $-\infty < a < \infty$, $0 < q_1, q_2 < \infty$, $0 < l < \infty$, quasiconformally maps the rectangle $\{0 \leq x \leq l,\ 0 \leq y \leq m\}$ onto the trapezoid with vertices at the points $w = 0$, $q_1 i$, $l + (a + q_2)i$, $l + ai$. For this mapping we have

$$\frac{\max(q_1, q_2)}{m} \geq J = \frac{\partial v}{\partial y} = \left(\frac{q_2 - q_1}{l}x + q_1\right)\frac{1}{m} \geq \frac{\min(q_1, q_2)}{m},$$

$$E + G = 1 + \left(\frac{q_2 - q_1}{l} \cdot \frac{y}{m} + \frac{a}{l}\right)^2 + \left(\frac{q_2 - q_1}{l}x + q_1\right)^2 \frac{1}{m^2},$$

$$2K = J + \frac{1}{J} + \frac{1}{J}\left(\frac{q_2 - q_1}{l} \cdot \frac{y}{m} + \frac{a}{l}\right)^2.$$

Taking into account (2.10), we get the estimate

$$p(z, w) \leq \frac{m}{\min(q_1, q_2)} + \frac{\max(q_1, q_2)}{m} + m\frac{\max\{a^2, (q_2 - q_1 + a)^2\}}{l^2 \min(q_1, q_2)}. \tag{2.12}$$

In some special cases the estimate can be simplified. If $q_1 = q_2 = m$, then $J \equiv 1$, $E + G = 2 + (a/l)^2$, $K = 1 + \frac{1}{2}(a/l)^2$, and

$$p(z, w) \equiv 1 + \frac{1}{2}\left(\frac{a}{l}\right)^2 + \frac{|a|}{l}\sqrt{\frac{1}{4}\left(\frac{a}{l}\right)^2 + 1} \leq 2 + \left(\frac{a}{l}\right)^2. \tag{2.13}$$

If $a = 0$, $m = q_1 = 1$, and $q_2 = q$, then (2.12) implies

$$p(z, w) \le 1 + \max\left(q, \frac{1}{q}\right) + \frac{(q-1)^2}{l^2 \min(1, q)}.$$

Suppose that a closed region G is mapped quasiconformally onto a closed region D by a function $w = \psi(z)$. Let C be a smooth curve, $C \subset G$. If $|dw/dz|_\varphi = \text{const}$ on C, where the directional derivative is taken in the direction of the tangent to C, we say that the mapping has constant distortion on C. If the curve C is piecewise smooth, and the distortion is equal to the same constant on each of the smooth pieces forming C, we say that the mapping has a constant distortion on C. It is easy to verify that in both preceding examples the distortion is constant on each of the edges of the mapped quadrilaterals.

REMARK. We will have to use quasiconformal mappings that can be obtained from those considered above by additional entire linear mappings or by a passage to "symmetric" mappings. It is clear that the estimates for the characteristic obtained above will remain true. Therefore, referring to one of the examples, we assume that the readers will make these additional transformations themselves if needed. This remark will be applied, also, to the examples that follow.

EXAMPLE 3. Let $u = \omega(x)$ be a function, continuously differentiable on $\{0 \le x \le l\}$ and such that $\omega(0) = 0$, $\omega(l) = l$, $0 < 1/M \le \omega'(x) \le M < \infty$. Our purpose is to construct a quasiconformal mapping of the rectangle $\{0 \le x \le l,\ 0 \le y \le 1\}$ onto the same rectangle $\{0 \le u \le l,\ 0 \le v \le 1\}$, such that the points $z = x + i$ are mapped to the points $w = \omega(x) + i$, and all points on the other three edges of the rectangle are fixed. The mapping

$$u = (\omega(x) - x)y + x,$$
$$v = y$$

is a desired mapping. For this mapping we have

$$J = (\omega'(x) - 1)y + 1, \qquad \min(1, \omega') \le J \le \max(1, \omega'),$$
$$E + G = 1 + \{(\omega'(x) - 1)y + 1\}^2 + (\omega(x) - x)^2.$$

Then, by (2.10),

$$p(z, w) \le \frac{1}{J} + J + \frac{(\omega(x) - x)^2}{J} \tag{2.14}$$
$$\le 1 + \max\left(\omega', \frac{1}{\omega'}\right) + \frac{(\omega(x) - x)^2}{\min(1, \omega')} \le 1 + M + Ml^2.$$

EXAMPLE 4. Let $\omega_j(x)$ $(j = 1, 2)$ be functions, continuously differentiable on $\{0 \le x \le l\}$, $\omega_j(0) = 0$, $\omega_j(l) = l$, $0 < 1/M \le \omega_j'(x) \le M < \infty$, $j = 1, 2$, and let $\omega_j(y)$ $(j = 3, 4)$ be functions, continuously differentiable on $\{0 \le y \le 1\}$, $\omega_j(0) = 0$, $\omega_j(1) = 1$, $0 < 1/M \le \omega_j'(y) \le M < \infty$, $j = 3, 4$. Our purpose is to map the rectangle $R = \{0 \le x \le l,\ 0 \le y \le 1\}$ quasiconformally onto the same rectangle $\{0 \le u \le l,\ 0 \le v \le 1\}$ in such a way that the points $z = x$ are mapped to the points $w = \omega_1(x)$, the points $z = x + i$ are mapped to the points $w = \omega_2(x) + i$, the points $z = iy$ are mapped to the points $w = i\omega_3(y)$, and the points $z = l + iy$ are mapped to the points $w = l + i\omega_4(y)$.

Thus we are given a certain continuous one-to-one mapping of the boundary of the rectangle $\{0 \leq x \leq l,\ 0 \leq y \leq 1\}$ onto the boundary of the same rectangle $\{0 \leq u \leq l,\ 0 \leq v \leq 1\}$, which preserves vertices and satisfies additional smoothness conditions. We need to extend this mapping to a quasiconformal mapping of $\{0 \leq x \leq l,\ 0 \leq y \leq 1\}$ onto $\{0 \leq u \leq l,\ 0 \leq v \leq 1\}$.

The desired mapping can be constructed as follows. First we map the rectangle R onto itself in such a way that the points on the vertical edges are fixed, and the points on the horizontal edges are mapped in the described way. In order to get such a mapping, we apply the mappings of Example 3 to the rectangles $\{0 \leq x \leq l,\ 0 \leq y \leq 1/2\}$ and $\{0 \leq x \leq l,\ 1/2 \leq y \leq 1\}$, the points of the intervals $\{x = 0,\ 0 \leq y \leq 1\}$, $\{x = l,\ 0 \leq y \leq 1\}$, and $\{y = 1/2,\ 0 \leq x \leq l\}$ being fixed points of these mappings. After this, in a similar way, we construct a quasiconformal mapping of R onto R, such that the points on the horizontal edges are fixed, and the points on the vertical edges are mapped in the described way (to do this, we divide R into two rectangles: $\{0 \leq x \leq l/2,\ 0 \leq y \leq 1\}$ and $\{l/2 \leq x \leq l,\ 0 \leq y \leq 1\}$). The composition of these two quasiconformal mappings is a desired mapping. Using (2.6) and (2.14), it is not difficult to get an estimate for its characteristic. We mention only that $p(z, w)$ is bounded by a constant depending on M and l only.

EXAMPLE 5. Let F be a closed Riemann surface of genus zero that covers the w-plane, and D a closed simply connected region on F bounded by a closed Jordan curve C having the following properties:

(1) C does not pass through the algebraic ramification points of the surface F.
(2) There are four points A_1, A_2, A_3, A_4 on C, listed according to their order on C, such that the arcs A_1A_2, A_2A_3, A_3A_4, A_4A_1, which partition C, have continuous curvature.
(3) In a sufficiently small neighborhood of the point A_j, the arcs $A_{j-1}A_j$ and A_jA_{j+1}, $j = 1, 2, 3, 4$, $A_5 = A_1$, $A_0 = A_4$, either consist of two orthogonal line segments or of one line segment and one circular arc, also orthogonal at A_j; moreover, the region D has an angle of magnitude $\pi/2$ at A_j.

Our purpose is to construct a quasiconformal mapping $z = \psi(w)$ which maps D onto some rectangle $R = \{0 \leq x \leq l,\ 0 \leq y \leq 1\}$ in the $(z = x + iy)$-plane, such that the points A_1, A_2, A_3, A_4 are mapped to the vertices of the rectangle and the mapping has a constant distortion on each of the arcs A_jA_{j+1}. To solve this problem, we need some results from the theory of conformal mappings, all of which can be found, for example, in the book by Goluzin [**Gol69**].

We may assume without loss of generality that the projection of C is contained in the w-plane. As is known (see [**Gol69**, Chapter IV, §6]), for a closed region D there exists a one-to-one conformal mapping of D onto a rectangle $R' = \{0 \leq \xi \leq l,\ 0 \leq \eta \leq 1\}$ in the plane $\zeta = \xi + i\eta$, where the constant l depends on D and on the choice of the points A_j and the points A_1, A_2, A_3, and A_4 are mapped to the points $\zeta = 0, l, l+i$, and i, respectively. Denote such a mapping by $\zeta = \zeta(w)$. Since the curves A_jA_{j+1} have continuous curvature, by Kellogg's theorem (see [**Gol69**, Chapter X, §1, Theorem 6]), it is easy to conclude that the meromorphic function $w = w(\zeta)$, inverse to $\zeta(w)$, has a continuous derivative $w'(\zeta) \neq 0$ everywhere on the boundary of R', except the vertices of R'. But, by property (3) and the symmetry principle, the function $w(\zeta)$ can be analytically continued to certain sufficiently small neighborhoods of the vertices; moreover, it establishes a univalent conformal mapping of these neighborhoods. Hence the derivative $w'(\zeta)$ exists also in neighborhoods of vertices of R', where it is also continuous and non-zero. Hence the derivative $w'(\zeta) \neq 0$ exists and is continuous everywhere on the boundary $\partial R'$. Therefore $0 < 1/M_1 \leq |w'(\zeta)| \leq M_1 < \infty$, where M_1 is a constant and $\zeta \in \partial R'$.

Let $z = T(w)$ be a function that establishes a continuous one-to-one correspondence between the curve C and the boundary ∂R of the rectangle, such that the points A_1, A_2, A_3, A_4 are mapped to the vertices $z = 0,\ l,\ l+i,\ i$, respectively, and the distortion is constant on each of the arcs A_jA_{j+1}, that is,

$$\left|\frac{dT}{dw}\right| = \left|\frac{dz}{dw}\right| = \begin{cases} l/\text{length}(A_1A_2) & \text{for } w \in A_1A_2, \\ 1/\text{length}(A_2A_3) & \text{for } w \in A_2A_3, \\ l/\text{length}(A_3A_4) & \text{for } w \in A_3A_4, \\ 1/\text{length}(A_1A_4) & \text{for } w \in A_1A_4. \end{cases} \tag{2.15}$$

It is clear that $0 < 1/M_2 \leq |dz/dw| \leq M_2 < \infty$. On the boundary $\partial R'$ we have defined a function $z = \omega(\zeta) = T(w(\zeta))$ which maps $\partial R'$ onto ∂R in such a way that the vertices are mapped to vertices. The function $z = \omega(\zeta)$ has a continuous derivative on each of the sides of $\partial R'$; moreover

$$1/(M_1M_2) \leq |\omega'(\zeta)| = |T'(w)||w'(\zeta)) \leq M_1M_2.$$

Using the result of Example 4, we can extend this mapping $z = \omega(\zeta)$ to a quasiconformal mapping $z = \Omega(\zeta)$ of the rectangle R' onto R; moreover, the characteristic $p(\zeta, z)$ does not exceed a certain constant depending on l and M_1M_2 only. The quasiconformal mapping $z = \Omega(\zeta(w))$ has the desired properties. In fact, by (2.7), we have $p(w, z) \equiv p(\zeta(w), z)$, and the equality $z = \Omega(\zeta(w)) = \omega(\zeta(w)) = T(w(\zeta)) = T(w)$ holds on C, and, by (2.15), the distortion is constant on each arc A_jA_{j+1}.

EXAMPLE 6. Let $w = \psi(z)$ be a quasiconformal mapping of the disc $\{|z| \leq R\}$ onto the disc $\{|w| \leq R\}$, such that the point $z = re^{i\varphi}$, $0 \leq r \leq R$, $0 \leq \varphi \leq 2\pi$, is mapped to the point $w = re^{i\theta(\varphi)}$, where $\theta(\varphi)$ is a continuously differentiable function on $[0, 2\pi]$, $\theta'(\varphi) > 0$, $\theta(0) = 0$, $\theta(2\pi) = 2\pi$. Let us estimate the characteristic of this mapping. To this end it is convenient to consider an auxiliary mapping of the half-strip $\{\sigma < \ln R,\ 0 < \tau < 2\pi\}$, $\zeta = \sigma + i\tau$ onto the half-strip $\{\sigma_1 < \ln R,\ 0 < \tau_1 < 2\pi\}$, $\zeta_1 = \sigma_1 + i\tau_1$ by means of the function $\zeta_1 = \ln\psi(e^\zeta)$. It is clear that $p(\zeta, \zeta_1) = p(z, w)$. Therefore it suffices to estimate $p(\zeta, \zeta_1)$. Since $\sigma_1 = \sigma$, $\tau_1 = \theta(\tau)$, we have $J = \theta'$, $E + G = 1 + \theta'^2$, and, using formulas (2.9), we get

$$\begin{aligned} p(z, w) = p(\zeta, \zeta_1) &= \frac{1}{2}\left(\frac{1}{\theta'} + \theta'\right) + \sqrt{\frac{1}{4}\left(\frac{1}{\theta'} + \theta'\right)^2 - 1} \\ &= \frac{1}{2}\left(\frac{1}{\theta'} + \theta'\right) + \frac{1}{2}\left|\frac{1}{\theta'} - \theta'\right| = \max\left(\theta', \frac{1}{\theta'}\right). \end{aligned} \tag{2.16}$$

Thus

$$p(z, w) \leq \max\left\{\max_{[0,\ 2\pi]} \theta'(\varphi), \frac{1}{\min_{[0,\ 2\pi]} \theta'(\varphi)}\right\}.$$

EXAMPLE 7. Our purpose is to find a quasiconformal mapping of the disc $\{|z| \le R\}$ onto the disc $\{|w| \le R\}$ such that the point $z = x_0$, $0 < x_0 < R$, is mapped to $w = 0$, and the points $\{|z| = R\}$ are fixed.

First we carry out a linear-fractional transformation of the disc $\{|z| \le R\}$ onto the disc $\{|\zeta| \le R\}$, mapping the point $z = x_0$ to $\zeta = 0$, and $z = R$ to $\zeta = R$:

$$\zeta = \zeta(z) = \frac{R^2(z - x_0)}{R^2 - zx_0}.$$

Under this mapping each point $\zeta = Re^{i\theta}$ has a unique inverse image $z = Re^{i\varphi(\theta)}$, where $\varphi(\theta)$ is an odd continuously differentiable strictly increasing function on $[-\pi, \pi]$, $\varphi(-\pi) = -\pi$, $\varphi(0) = 0$, $\varphi(\pi) = \pi$. Now we construct a quasiconformal mapping which maps the disc $\{|\zeta| \le R\}$ onto the disc $\{|w| \le R\}$ in such a way that the point $\zeta = |\zeta|e^{i\theta}$, $|\theta| \le \pi$, $0 < |\zeta| \le R$, is mapped to the point $w = |\zeta|e^{i\varphi(\theta)}$, $w(0) = 0$. It is easy to see that the mapping $w = w(z) = w[\zeta(z)]$ has the desired properties.

Let us estimate the characteristic $p(z, w)$ of the mapping. Since the mapping $\zeta(z)$ is conformal, we have $p(z, w) = p(\zeta, w)$. Equality (2.16) implies that

$$p(z, w) = \max\left(\varphi', \frac{1}{\varphi'}\right) = \max\left(\frac{d\theta}{d\varphi}, \frac{d\varphi}{d\theta}\right)$$

if $\arg z = \varphi$. But

$$\frac{d\theta}{d\varphi} = \left|\frac{d\zeta}{dz}\right|_{z=Re^{i\varphi}} = \frac{R^2 - x_0^2}{|R - e^{i\varphi}x_0|^2},$$

$$\frac{R - x_0}{R + x_0} \le \frac{d\theta}{d\varphi} \le \frac{R + x_0}{R - x_0}.$$

Hence

$$p(z, w) \le \frac{R + x_0}{R - x_0} = \frac{1 + \frac{x_0}{R}}{1 - \frac{x_0}{R}}. \tag{2.17}$$

This estimate is sharp because

$$p(z, w) = \left.\frac{d\theta}{d\varphi}\right|_{\varphi=0} = \frac{1 + \frac{x_0}{R}}{1 - \frac{x_0}{R}}$$

for $z = x$, $0 < x < R$.

EXAMPLE 8. Let F_1 be the Riemann surface of the function $\sqrt[n]{z}$, $n = 2, 3, \ldots$, over the disc $\{|z| < R\}$, and F_2 the Riemann surface of the function $\sqrt[n]{w - a}$, $0 < |a| < R$, over the disc $\{|w| < R\}$. Our purpose is to find a quasiconformal mapping of F_1 onto F_2 such that the algebraic ramification point (of order $n - 1$) of F_1 over the point $z = 0$ is mapped to the algebraic ramification point of F_2 over $w = a$, and the boundary points remain fixed. Such a mapping can be obtained as $w = w(z) = w\{z_2(z_1(z))\}$, where $z_1(z) = ze^{-i\alpha}$, $\alpha = \arg a$, $z_2 = z_2(z_1)$ is the inverse mapping to the mapping considered in Example 7 (the role of x_0 is played by $|a|$), and $w(z_2) = z_2 e^{i\alpha}$. Here

$$p(z, w) \le \frac{1 + \frac{|a|}{R}}{1 - \frac{|a|}{R}}.$$

EXAMPLE 9. Let F be a closed Riemann surface of genus zero, covering the w-plane, and D a closed simply connected region on F, bounded by a closed Jordan curve C, which does not pass through algebraic ramification points of F and has a continuous curvature. There exists a quasiconformal mapping $z = \psi(w)$ with a bounded characteristic of the region D onto the disc $\{|z| \leq R\}$, which has a constant distortion on C.

Such a quasiconformal mapping can be constructed similarly to the construction of Example 5, but instead of Example 4 we use Example 6, and some parts of the argument are simpler. Note that the estimate of the characteristic $p(w, z)$ does not depend on the radius of the disc $\{|z| \leq R\}$, since we can pass from this disc to any other disc using an additional entire linear transformation.

EXAMPLE 10. Let W be a Riemann surface constructed in the following way. Take two copies, G_1 and G_2, of the disc $\{|w| < 2\}$ with a cut along the segment joining the points $w = \pm 1$. We paste together the upper boundary of the cut in G_1 with the lower boundary of the cut in G_2, and vice versa. We get a two-sheeted Riemann surface W with two algebraic ramification points of the first order over $w = 1$ and $w = -1$. It is clear that the surface W is doubly connected and its boundary consists of two circles, C_1 and C_2, lying over $\{|w| = 2\}$, $C_1 = \partial G_1$, $C_2 = \partial G_2$. Our purpose is to construct a quasiconformal mapping with bounded characteristic of the Riemann surface W onto the one-sheeted annulus $\{1/2 < |z| < 2\}$, such that the point $2e^{i\theta} \in C_1$ is mapped to the point $2e^{i\theta}$, and the point $2e^{-i\theta} \in C_2$ is mapped to the point $e^{i\theta}/2$. We are not interested in precise numerical bounds for the characteristic.

We map the Riemann surface W conformally[4] onto the annulus $\{1/R < |\zeta| < R\}$, where R is some constant, $1 < R < \infty$, in such a way that G_1 is mapped onto $\{1 < |\zeta| < R\}$, G_2 is mapped onto $\{1/R < |\zeta| < 1\}$, the algebraic ramification points over $w = \pm 1$ are mapped to $\zeta = \pm 1$, respectively, the points $\pm 2 \in C_1$ are mapped to the points $\zeta = \pm R$, and the points $\pm 2 \in C_2$ are mapped to $\zeta = \pm 1/R$. To construct such a mapping it suffices to map $\{|w| < 2,\ \operatorname{Im} w > 0\}$ onto $\{1 < |\zeta| < R,\ \operatorname{Im} \zeta > 0\}$ conformally in such a way that the points $-2, -1, 1, 2$ are mapped to the points $-R, -1, 1, R$, respectively, and to continue the obtained function using the symmetry principle. The symmetry principle implies that there exists a continuous, odd, function $\varphi = \varphi(\theta)$, increasing on $[-\pi, \pi]$, such that $\varphi(0) = 0$, $\varphi(\pm\pi) = \pm\pi$, the point $2e^{i\theta} \in C_1$ is mapped to the point $Re^{i\varphi(\theta)}$, and the point $2e^{-i\theta} \in C_2$ is mapped to the point $\frac{1}{R}e^{i\varphi(\theta)}$. Since the mapping is conformal at the points lying on C_1 and C_2, it follows that φ has a continuous derivative $\varphi'(\theta) > 0$ on $[-\pi, \pi]$. Using the function $\zeta_1 = \ln \zeta$, we map $\{1/R < |\zeta| < R,\ |\arg \zeta| < \pi\}$ onto the rectangle $R_1 = \{|\operatorname{Re} \zeta_1| < \ln R,\ |\operatorname{Im} \zeta_1| < \pi\}$. Now we map R_1 quasiconformally onto the rectangle $R_2 = \{|\operatorname{Re} \zeta_2| < \ln 2,\ |\operatorname{Im} \zeta_2| < \pi\}$ in such a way that the point $\xi_1 \pm i\pi$, $|\xi_1| < \ln R$, is mapped to the point $\frac{\ln 2}{\ln R}\xi_1 \pm i\pi = \xi_2 \pm i\pi$, and the point $\pm \ln R + i\varphi(\theta)$ is mapped to the point $\pm \ln 2 + i\theta$.

Examples 1 and 4 imply that the mapping $\zeta_2 = \psi(\zeta_1)$ can be constructed in such a way that its characteristic satisfies $p(\zeta_1, \zeta_2) \leq Q$, where Q is some constant. Now, in order to get the desired mapping, it remains to map R_2 onto $\{1/2 < |z| < 2,\ |\arg z| < \pi\}$ by means of the function $z = e^{\zeta_2}$, and to observe that

[4]The mapping function can be written explicitly using elliptic functions (see Volkovyskiĭ, Lunts, and Aramanovich [**VLA91**, Problem 1370, 7], or Volkovyskiĭ [**Vol54**, p. 146]).

the function $z = z(\zeta) = \exp\{\psi(\ln \zeta)\}$ remains univalent and continuous on the negative radius of the annulus $\{1/R < |\zeta| < R\}$. Our mapping of the Riemann surface W onto $\{1/2 < |z| < 2\}$ is obtained as a composition of mappings all of which, except the mapping $\zeta_2 = \psi(\zeta_1)$, were conformal. Hence, $p(w, z) = p(\zeta_1, \zeta_2) \le Q$.

3. The Teichmüller theorem

In this section we prove the following theorem.

THEOREM 3.1 (O. Teichmüller). *Let $w = w(z) = u(z) + iv(z)$ be a homeomorphic mapping of the complex z-plane onto $\{|w| < R \le \infty\}$, which is quasiconformal for $|z| \ge r_0$, and*

$$\iint_{|z|\ge r_0} \{p(re^{i\varphi},\ w) - 1\}\frac{dr\,d\varphi}{r} < \infty. \tag{3.1}$$

Then R is necessarily equal to ∞ and the limit

$$\lim_{z\to\infty} \frac{|w(z)|}{|z|} = A, \qquad 0 < A < \infty, \tag{3.2}$$

exists as a finite positive limit.

We introduce some notation. Let

$$h(t) = \iint_{|z|\ge t} \{p(re^{i\varphi}, w) - 1\}\frac{dr\,d\varphi}{r}, \qquad t \ge r_0.$$

Inequality (3.1) implies that $h(t)$ has finite values and monotonically tends to 0 as $t \to \infty$. Let $z = re^{i\varphi}$, $w = \rho e^{i\theta}$, and let $d\sigma(z)$ be the area element at z. We can rewrite condition (3.1) in the following way:

$$\iint_{|z|\ge r_0} \{p(z, w) - 1\}\, d\sigma(\ln z) = \iint_{|z|\ge r_0} \{p(z, w) - 1\}\frac{d\sigma(z)}{|z|^2} < \infty. \tag{3.1$'$}$$

The circle $\{|w| = \rho\}$ corresponds to a piecewise smooth closed Jordan curve Γ_ρ in the z-plane. It is clear that if ρ_0 is sufficiently large, then for $\rho \ge \rho_0$ the curve Γ_ρ lies in $\{r_0 < |z| < \infty\}$, contains the point $z = 0$ in its interior, and the increment of $\arg z$ over Γ_ρ is equal to 2π. Let

$$r_1(\rho) = \min_{z\in\Gamma_\rho} |z|, \qquad r_2(\rho) = \max_{z\in\Gamma_\rho} |z|, \qquad \omega(\rho) = \ln \frac{r_2(\rho)}{r_1(\rho)}.$$

It is easy to see that $r_1(\rho)$ and $r_2(\rho)$ are continuous functions, which tend monotonically to ∞ as $\rho \to \infty$. In fact, the regions G_ρ, $\rho > \rho_0$, which are the images of the discs $\{|w| < \rho\}$ under the mapping $z = z(w)$, contain $z = 0$ and exhaust $\{|z| < \infty\}$. Therefore, for each disc $\{|z| \le r\}$, if ρ is sufficiently large, the region G_ρ contains this disc; hence $r_2(\rho) \ge r_1(\rho) \to \infty$. The continuity of the functions $r_1(\rho)$ and $r_2(\rho)$ follows immediately from the continuity of the mapping $z = z(w)$.

The proof of the theorem is rather long. Therefore we start with an outline of its plan, and divide the proof into steps. First we show that $R = \infty$ (Step I). Then (3.2) is equivalent to the relation

$$\lim_{w\to\infty} \frac{|z(w)|}{|w|} = B, \qquad 0 < B = 1/A < \infty. \tag{3.2$'$}$$

To prove (3.2′) it suffices to establish that

$$\lim_{\rho\to\infty}\frac{r_1(\rho)}{\rho}=\lim_{\rho\to\infty}\frac{r_2(\rho)}{\rho}=B.$$

We prove the existence of a finite limit $\lim_{\rho\to\infty}\ln\frac{r_1(\rho)}{\rho}$ using the Bolzano–Cauchy criterion. Let $\rho_2>\rho_1\geq\rho$. We need to estimate the difference

$$Q(\rho_1,\rho_2)=\ln\frac{r_1(\rho_2)}{\rho_2}-\ln\frac{r_1(\rho_1)}{\rho_1}=\ln\frac{r_1(\rho_2)}{r_1(\rho_1)}-\ln\frac{\rho_2}{\rho_1}$$

both from below and from above. Such estimates will be found in Step II of the proof. In order to show that $Q(\rho_1,\rho_2)$ tends to 0 as $\rho_1\to\infty$, we need to show that $\omega(\rho)\to 0$ as $\rho\to\infty$. The proof of this fact is the content of Step III, the most complicated step in the proof. In the final Step IV, using the fact that $\omega(\rho)\to 0$ as $\rho\to\infty$ and the estimates for $Q(\rho_1,\rho_2)$ found in Step II, we prove the existence of the limit $\lim_{\rho\to\infty}r_1(\rho)/\rho=B$, and also the existence of the limit

$$\lim_{\rho\to\infty}\frac{r_2(\rho)}{\rho}=\lim_{\rho\to\infty}e^{\omega(\rho)}\frac{r_1(\rho)}{\rho}=B.$$

Now we turn to the proof.

STEP I. We show that $R=\infty$. Let r' be so large that $r'\geq r_0$, and the image of the disc $\{|z|<r'\}$ under the mapping $w=w(z)$ contains the point $w=0$. Denote by D the image of the region $\{r'<|z|<\infty\}$ under this mapping. If we assume that $R<\infty$, we get that

$$K=\iint_D\frac{d\rho\,d\theta}{\rho}=\iint_D d\sigma(\ln w)=\iint_{|z|>r'}\frac{d\sigma(\ln w)}{d\sigma(\ln z)}d\sigma(\ln z)<\infty.$$

Since the increment of $\arg w(z)$ as z runs over the circle $\{|z|=r\}$, $r>r'$, is equal to 2π, we have

$$2\pi\leq\int_{|z|=r}|d\ln w(z)|=\int_{|z|=r}\left|\frac{d\ln w}{d\ln z}\right|d\varphi=\int_{|z|=r}\left|\frac{z}{w}\right|\left|\frac{dw}{dz}\right|d\varphi.$$

We divide both sides of this inequality by r and integrate with respect to r from r' to $r''>r'$. We get the inequality

$$2\pi\ln\frac{r''}{r'}\leq\iint_{r'<|z|<r''}\left|\frac{z}{w}\right|\left|\frac{dw}{dz}\right|d\sigma(\ln z). \tag{3.3}$$

From (2.9) we obtain that

$$\left|\frac{z}{w}\right|\left|\frac{dw}{dz}\right|\leq\left|\frac{z}{w}\right|\max\left|\frac{dw}{dz}\right|=\left|\frac{z}{w}\right|\sqrt{p(z,w)J[w(z)]}=\sqrt{p(z,w)\frac{d\sigma(\ln w)}{d\sigma(\ln z)}}.$$

Inequality (3.3) implies

$$2\pi\ln\frac{r''}{r'}\leq\iint_{r'<|z|<r''}\sqrt{p(z,w)\frac{d\sigma(\ln w)}{d\sigma(\ln z)}}\,d\sigma(\ln z).$$

Applying the Cauchy–Bunyakovskii inequality, we get

$$\begin{aligned}\left(2\pi \ln \frac{r''}{r'}\right)^2 &\le \iint_{r'<|z|<r''} \frac{d\sigma(\ln w)}{d\sigma(\ln z)} d\sigma(\ln z) \iint_{r'<|z|<r''} p(z,w) d\sigma(\ln z) \\ &\le \iint_{r'<|z|<r''} \frac{d\sigma(\ln w)}{d\sigma(\ln z)} d\sigma(\ln z) \left\{ \iint_{|z|>r'} \{p(z,w)-1\} d\sigma(\ln z) \right. \\ &\qquad \left. + \iint_{r'<|z|<r''} d\sigma(\ln z) \right\}.\end{aligned} \tag{3.4}$$

Hence

$$\left(2\pi \ln \frac{r''}{r'}\right)^2 \le K \left\{ h(r') + 2\pi \ln \frac{r''}{r'} \right\}$$

and

$$2\pi \ln \frac{r''}{r'} \le K \left\{ 1 + \frac{h(r')}{2\pi \ln \frac{r''}{r'}} \right\}.$$

Letting $r'' \to \infty$, we come to a contradiction, since the left-hand side of this inequality tends to ∞, but the right-hand side remains bounded. Thus $R = \infty$.

STEP II. The purpose of this step is to get estimates of the quantity $Q(\rho_1, \rho_2)$ in terms of $\omega(\rho)$ and $h(\rho)$. We fix the value of $\ln z$ at some point $A \in \Gamma_\rho$, and map Γ_ρ onto some curve γ_ρ in the ζ-plane by the branch of $\ln z$ which is continuous on Γ_ρ (that is, such that the function $\ln z(\rho e^{i\theta})$ is continuous as a function of θ on some interval of length 2π). The point A will correspond to the ends of the curve γ_ρ: $\zeta = a$ and $\zeta = b$, $b - a = 2\pi i$. If $\zeta = \xi + i\eta$, then

$$\max_{\zeta\in\gamma_\rho} \xi = \ln r_1(\rho), \qquad \max_{\zeta\in\gamma_\rho} \xi = \ln r_2(\rho).$$

Denote by c and d some points on γ_ρ for which

$$\operatorname{Re} c = \ln r_1(\rho), \qquad \operatorname{Re} d = \ln r_2(\rho).$$

Suppose that the points c and d are located on the curve γ_ρ in the following order: a, c, d, b (we do not exclude the cases $a = c$ and $d = b$). We map the part of the curve γ_ρ located between the points c and d symmetrically with respect to the line $\operatorname{Re}\zeta = \ln r_1(\rho)$, and shift to the left onto $2(\ln r_2(\rho) - \ln r_1(\rho)) = 2\omega(\rho)$ the part of the curve γ_ρ located between the points d and b. We get a continuous curve γ'_ρ with ends at the points $\zeta = a$ and $\zeta = b - 2\omega(\rho) = a - 2\omega(\rho) + 2\pi i$. Hence the length of the curve γ'_ρ is at least $\sqrt{4\pi^2 + 4\omega^2(\rho)}$. But the length of γ_ρ is the same as the length of γ'_ρ; hence it also is at least $\sqrt{4\pi^2 + 4\omega^2(\rho)}$. If the order of the points on the curve γ_ρ is a, d, c, b, we use a similar argument. Thus

$$2\sqrt{\pi^2 + \omega^2(\rho)} \le \int_{\gamma_\rho} |d\zeta| = \int_{\Gamma_\rho} |d\ln z| = \int_{|w|=\rho} \left|\frac{d\ln z}{d\ln w}\right| |d\ln w| = \int_0^{2\pi} \left|\frac{d\ln z}{d\ln w}\right| d\theta.$$

Applying the Cauchy–Bunyakovskii inequality we get

$$2\pi + \frac{2\omega^2(\rho)}{\pi} \le \int_0^{2\pi} \left|\frac{d\ln z}{d\ln w}\right|^2 d\theta,$$

from which, dividing by ρ and integrating with respect to ρ from ρ_1 to ρ_2, $\rho_0 \le \rho_1 < \rho_2 < \infty$, we get

$$2\pi \ln \frac{\rho_2}{\rho_1} + \frac{2}{\pi} \int_{\rho_1}^{\rho_2} \frac{\omega^2(\rho)}{\rho} d\rho \le \int_{\rho_1}^{\rho_2} \int_0^{2\pi} \left|\frac{d\ln z}{d\ln w}\right|^2 \frac{d\rho\, d\theta}{\rho}.$$

By (2.9) we get

$$\left|\frac{d\ln z}{d\ln w}\right|^2 \leq \left|\frac{w}{z}\right|^2 \max\left|\frac{dz}{dw}\right|^2 = \left|\frac{w}{z}\right|^2 p(w,z)J[z(w)] = p(z,w)\frac{d\sigma(\ln z)}{d\sigma(\ln w)}.$$

Thus, we have

$$\int_{\rho_1}^{\rho_2}\int_0^{2\pi} p(z,w)\frac{d\sigma(\ln z)}{d\sigma(\ln w)}\frac{d\rho\,d\theta}{\rho} \geq 2\pi\ln\frac{\rho_2}{\rho_1} + \frac{2}{\pi}\int_{\rho_1}^{\rho_2}\frac{\omega^2(\rho)}{\rho}d\rho. \tag{3.5}$$

But

$$\begin{aligned}\int_{\rho_1}^{\rho_2}\int_0^{2\pi} p(z,w)\frac{d\sigma(\ln z)}{d\sigma(\ln w)}d\sigma(\ln w) &\leq \iint_{r_1(\rho_1)\leq|z|\leq r_2(\rho_2)} p(z,w)\,d\sigma(\ln z)\\ &= \iint_{r_1(\rho_1)\leq|z|\leq r_2(\rho_2)} \{p(z,w)-1\}\,d\sigma(\ln z) + \iint_{r_1(\rho_1)\leq|z|\leq r_2(\rho_2)} d\sigma(\ln z)\\ &\leq \iint_{r_1(\rho_1)\leq|z|<\infty} \{p(z,w)-1\}\,d\sigma(\ln z) + 2\pi\ln\frac{r_2(\rho_2)}{r_1(\rho_1)}\\ &= h(r_1(\rho_1)) + 2\pi\left(\ln\frac{r_1(\rho_2)}{r_1(\rho_1)} + \omega(\rho_2)\right).\end{aligned}$$

Substituting this inequality into (3.5) we obtain

$$\ln\frac{r_1(\rho_2)}{r_1(\rho_1)} + \omega(\rho_2) + \frac{1}{2\pi}h(r_1(\rho_1)) \geq \ln\frac{\rho_2}{\rho_1} + \frac{1}{\pi^2}\int_{\rho_1}^{\rho_2}\frac{\omega^2(\rho)}{\rho}d\rho, \tag{3.6}$$

that is,

$$\begin{aligned}Q(\rho_1,\rho_2) &\geq \frac{1}{\pi^2}\int_{\rho_1}^{\rho_2}\frac{\omega^2(\rho)}{\rho}d\rho - \omega(\rho_2) - \frac{1}{2\pi}h(r_1(\rho_1))\\ &\geq -\omega(\rho_2) - \frac{1}{2\pi}h(r_1(\rho_1)).\end{aligned} \tag{3.7}$$

Denote the function $h(r_1(\rho))$, $\rho_0 \leq \rho < \infty$, by $h_1(\rho)$. It is clear that the function $h_1(\rho)$ monotonically tends to zero as $\rho \to \infty$.

Assume that $r_1(\rho_2) > r_2(\rho_1)$. Letting $r'' = r_1(\rho_2)$ and $r' = r_2(\rho_1)$ in (3.4) and taking into account

$$\iint_{r_2(\rho_1)<|z|<r_1(\rho_2)} \frac{d\sigma(\ln w)}{d\sigma(\ln z)}d\sigma(\ln z) \leq \iint_{\rho_1<|w|<\rho_2} d\sigma(\ln w) = 2\pi\ln\frac{\rho_2}{\rho_1},$$

we find that

$$\begin{aligned}\left(2\pi\ln\frac{r_1(\rho_2)}{r_2(\rho_1)}\right)^2 &\leq 2\pi\ln\frac{\rho_2}{\rho_1}\left\{h(r_2(\rho_1)) + 2\pi\ln\frac{r_1(\rho_2)}{r_2(\rho_1)}\right\}\\ &\leq 2\pi\ln\frac{\rho_2}{\rho_1}\left\{h_1(\rho_1) + 2\pi\ln\frac{r_1(\rho_2)}{r_2(\rho_1)}\right\},\end{aligned}$$

$$2\pi\ln\frac{\rho_2}{\rho_1} \geq \frac{\left(2\pi\ln\frac{r_1(\rho_2)}{r_2(\rho_1)}\right)^2}{2\pi\ln\frac{r_1(\rho_2)}{r_2(\rho_1)} + h_1(\rho_1)} \geq 2\pi\ln\frac{r_1(\rho_2)}{r_2(\rho_1)} - h_1(\rho_1),$$

$$\ln\frac{\rho_2}{\rho_1} \geq \ln\frac{r_1(\rho_2)}{r_2(\rho_1)} - \frac{1}{2\pi}h_1(\rho_1) = \ln\frac{r_1(\rho_2)}{r_1(\rho_1)} - \omega(\rho_1) - \frac{1}{2\pi}h_1(\rho_1), \tag{3.8}$$

that is,

$$Q(\rho_1,\rho_2)\le \omega(\rho_1)+\frac{1}{2\pi}h_1(\rho_1). \tag{3.9}$$

It is clear that inequality (3.9) remains true even for $r_1(\rho_2)\le r_2(\rho_1)$. Inequalities (3.7) and (3.9) imply that for arbitrary ρ_1 and ρ_2 satisfying $\rho_0\le\rho_1\le\rho_2<\infty$, the inequality

$$|Q(\rho_1,\rho_2)|\le \max\{\omega(\rho_1),\ \omega(\rho_2)\}+\frac{1}{2\pi}h_1(\rho_1) \tag{3.10}$$

holds. Thus we have achieved the main purpose of Step II.

STEP III. Now we show that $\omega(\rho)\to 0$ as $\rho\to\infty$. First we observe that (3.6) and (3.8) imply the inequality

$$\omega(\rho_2)+\omega(\rho_1)\ge\frac{1}{\pi^2}\int_{\rho_1}^{\rho_2}\frac{\omega^2(\rho)}{\rho}d\rho-\frac{1}{\pi}h_1(\rho_1), \tag{3.11}$$

for all ρ_1, ρ_2, satisfying $\rho_0\le\rho_1\le\rho_2<\infty$. Let $\rho\ge\max(\rho_0^2,2)$. We introduce the following notation:

$$\begin{aligned}\tau_1(\rho)&=\min_{\sqrt{\rho}\le x\le\rho}\left\{2\omega(x)-\frac{1}{\pi^2}\int_x^\rho\frac{\omega^2(t)}{t}dt\right\}\\ &=2\omega(\rho_1)-\frac{1}{\pi^2}\int_{\rho_1}^{\rho}\frac{\omega^2(t)}{t}dt,\qquad \sqrt{\rho}\le\rho_1\le\rho,\end{aligned} \tag{3.12}$$

$$\begin{aligned}\tau_2(\rho)&=\min_{\rho\le x\le\rho\sqrt{\rho}}\left\{2\omega(x)-\frac{1}{\pi^2}\int_\rho^x\frac{\omega^2(t)}{t}dt\right\}\\ &=2\omega(\rho_2)-\frac{1}{\pi^2}\int_{\rho}^{\rho_2}\frac{\omega^2(t)}{t}dt,\qquad \rho\le\rho_2\le\rho\sqrt{\rho}.\end{aligned} \tag{3.13}$$

We rewrite inequality (3.11), where ρ_1 and ρ_2 are the numbers defined in (3.12) and (3.13), as follows:

$$\begin{aligned}\omega(\rho_1)+\omega(\rho_2)&\le\left\{2\omega(\rho_2)-\frac{1}{\pi^2}\int_\rho^{\rho_2}\frac{\omega^2(t)}{t}dt\right\}\\ &\quad+\left\{2\omega(\rho_1)-\frac{1}{\pi^2}\int_{\rho_1}^{\rho}\frac{\omega^2(t)}{t}dt\right\}+\frac{1}{\pi}h_1(\rho_1)\\ &=\tau_1(\rho)+\tau_2(\rho)+\frac{1}{\pi}h_1(\rho_1).\end{aligned} \tag{3.14}$$

Now we estimate $\tau_1(\rho)$ and $\tau_2(\rho)$. Let

$$\alpha(x,\rho)=\operatorname{sgn}(x-\rho)\frac{1}{\pi^2}\int_\rho^x\frac{\omega^2(t)}{t}dt.$$

Assume that $\rho\sqrt{\rho}\ge x\ge\rho$ and $\tau_2(\rho)>0$. Then (3.13) implies that

$$2\omega(x)\ge\alpha(x,\rho)+\tau_2(\rho),$$

$$\{\alpha(x,\rho)+\tau_2(\rho)\}^2\le 4\omega^2(x)=4\pi^2x\frac{d\alpha(x,\rho)}{dx},$$

whence

$$\frac{1}{2}\ln\rho = \int_\rho^{\rho\sqrt{\rho}} \frac{dx}{x} \le 4\pi^2 \int_\rho^{\rho\sqrt{\rho}} \frac{d\alpha(x,\rho)}{\{\alpha(x,\rho)+\tau_2(\rho)\}^2}$$
$$\le 4\pi^2 \int_{\alpha=0}^{\infty} \frac{d\alpha}{\{\alpha+\tau_2(\rho)\}^2} = \frac{4\pi^2}{\tau_2(\rho)},$$

$$\tau_2(\rho) \le \frac{8\pi^2}{\ln\rho}. \tag{3.15}$$

Inequality (3.15) remains true for $\tau_2(\rho) \le 0$. In a similar way we treat the case where $\sqrt{\rho} \le x \le \rho$ and $\tau_1(\rho) > 0$. In this case (3.12) implies that

$$2\omega(x) \ge \tau_1(\rho) + \alpha(x,\rho),$$

$$\{\tau_1(\rho)+\alpha(x,\rho)\}^2 \le 4\omega^2(x) = 4\pi^2 x \frac{d\{-\alpha(x,\rho)\}}{dx},$$

whence

$$\frac{1}{2}\ln\rho = \int_{\sqrt{\rho}}^{\rho} \frac{dx}{x} \le 4\pi^2 \int_{\sqrt{\rho}}^{\rho} \frac{d\{-\alpha(x,\rho)\}}{\{\tau_1(\rho)+\alpha(x,\rho)\}^2}$$
$$\le 4\pi^2 \int_{\beta=-\infty}^{0} \frac{d\beta}{\{\tau_1(\rho)-\beta\}^2} = \frac{4\pi^2}{\tau_1(\rho)},$$

$$\tau_1(\rho) \le \frac{8\pi^2}{\ln\rho}. \tag{3.16}$$

The last inequality, obviously, holds also when $\tau_1(\rho) \le 0$. Substituting (3.15) and (3.16) in (3.14) we get

$$\omega(\rho_1)+\omega(\rho_2) \le \frac{16\pi^2}{\ln\rho} + \frac{1}{\pi}h_1(\rho_1). \tag{3.17}$$

Let $\eta > \max\{\omega(\rho_1),\ \omega(\rho_2)\}$; later we will make the choice of η more precise. Denote by E the set of t, $\rho_1 \le t \le \rho_2$, for which $\omega(t) > \eta$, and by Δ the logarithmic measure of this set. Inequalities (3.11) and (3.17) imply that

$$\frac{16\pi^2}{\ln\rho} + \frac{1}{\pi}h_1(\rho_1) \ge \frac{1}{\pi^2}\int_{\rho_1}^{\rho_2} \frac{\omega^2(\rho)}{\rho}d\rho - \frac{1}{\pi}h_1(\rho_1)$$
$$\ge \frac{\eta^2}{\pi^2}\int_E \frac{d\rho}{\rho} - \frac{1}{\pi}h_1(\rho_1) = \frac{\eta^2}{\pi^2}\Delta - \frac{1}{\pi}h_1(\rho_1),$$

$$\frac{\eta^2}{\pi^2}\Delta \le \frac{16\pi^2}{\ln\rho} + \frac{2}{\pi}h_1(\rho_1) \le \frac{16\pi^2}{\ln\rho} + \frac{2}{\pi}h_1(\sqrt{\rho}),$$

$$\Delta \le \frac{16\pi^4}{\eta^2\ln\rho} + \frac{2\pi}{\eta^2}h_1(\sqrt{\rho}). \tag{3.18}$$

Assume that $\omega(\rho) > \eta$ and that (ρ', ρ'') is the component of the set E containing ρ. Then $\rho_1 < \rho' < \rho < \rho'' < \rho_2$, $\omega(\rho') = \omega(\rho'') = \eta$. Inequality (3.18) implies that

$$\ln\frac{\rho''}{\rho'} \le \Delta \le \frac{16\pi^4}{\eta^2\ln\rho} + \frac{2\pi}{\eta^2}h_1(\sqrt{\rho}). \tag{3.19}$$

It is clear that

$$\omega(\rho) \le \ln \frac{r_2(\rho'')}{r_1(\rho')} = \ln \frac{r_1(\rho'')}{r_1(\rho')} + \omega(\rho'') = \ln \frac{r_1(\rho'')}{r_1(\rho')} + \eta. \tag{3.20}$$

Replacing ρ_1 and ρ_2 in (3.8) by ρ' and ρ'', respectively, we get

$$\ln \frac{r_1(\rho'')}{r_1(\rho')} \le \omega(\rho') + \ln \frac{\rho''}{\rho'} + \frac{1}{2\pi} h_1(\rho') \le \eta + \ln \frac{\rho''}{\rho'} + \frac{1}{2\pi} h_1(\sqrt{\rho}). \tag{3.21}$$

Inequalities (3.19), (3.20), and (3.21) imply that

$$\omega(\rho) \le 2\eta + \frac{1}{\eta^2}\left\{\frac{16\pi^4}{\ln \rho} + 2\pi h_1(\sqrt{\rho})\right\} + \frac{1}{2\pi} h_1(\sqrt{\rho}). \tag{3.22}$$

It is clear that inequality (3.22) holds also when $\omega(\rho) \le \eta$. Now we choose η such that the right-hand side of (3.22) takes the least possible value. If $A > 0$, then the function $2\eta + A\eta^{-2}$, for positive η takes the least value $3\sqrt[3]{A}$, achieved for $\eta = \sqrt[3]{A}$. We must let

$$\eta = \sqrt[3]{\frac{16\pi^4}{\ln \rho} + 2\pi h_1(\sqrt{\rho})}, \tag{3.23}$$

and we have the right to do so if $\eta > \max\{\omega(\rho_1),\ \omega(\rho_2)\}$. But (3.17) shows that it is sufficient to have

$$\frac{16\pi^2}{\ln \rho} + \frac{1}{\pi} h_1(\sqrt{\rho}) < \sqrt[3]{\frac{16\pi^4}{\ln \rho} + 2\pi h_1(\sqrt{\rho})},$$

which is obviously the case if the expression under the radical sign is less than 1, that is, if $\rho > \bar{\rho}$, where $\bar{\rho} \ge \max(\rho_0^2, 2)$ is sufficiently large. Thus for $\rho > \bar{\rho}$ we choose η according to (3.23), substitute it in (3.22), and get

$$\omega(\rho) \le \sqrt[3]{\frac{16\pi^4}{\ln \rho} + 2\pi h_1(\sqrt{\rho})} + \frac{1}{2\pi} h_1(\sqrt{\rho}),$$

whence $\omega(\rho) \to 0$ as $\rho \to \infty$.

Now we can complete the proof.

STEP IV. Again let ρ_1 and ρ_2 be arbitrary numbers satisfying $\rho_0 \le \rho_1 \le \rho_2 < \infty$. We see that the right-hand side of (3.10) tends to 0 as $\rho_1 \to \infty$, $\rho_2 \ge \rho_1$. By the Bolzano–Cauchy criterion, there exists a finite limit

$$\lim_{\rho \to \infty} \ln \frac{r_1(\rho)}{\rho} = B = e^{\alpha},$$

that is, there exists the limit

$$\lim_{\rho \to \infty} \frac{r_1(\rho)}{\rho} = B = e^{\alpha}, \qquad 0 < B < \infty.$$

Since

$$\frac{r_2(\rho)}{\rho} = \frac{r_1(\rho)}{\rho} e^{\omega(\rho)} \to B, \qquad \rho \to \infty,$$

we have

$$\lim_{w \to \infty} \frac{|z(w)|}{|w|} = B,$$

from which equality (3.2) with $A = 1/B$ follows.

4. Riemann surfaces of the class F_q and line complexes

Surfaces of the so-called class $F_q = F_q(a_1, a_2, \dots, a_q)$ have a particularly simple structure among all Riemann surfaces. We describe this class. Let $a_1, \dots, a_q$ be distinct $q \geq 2$ points in the extended complex w-plane, and let L be a closed Jordan curve passing through the points $a_1, \dots, a_q$. Without loss of generality we assume, to simplify the description, that $a_j \neq \infty$, $1 \leq j \leq q$, and L does not pass through $w = \infty$. We assume that the points a_j are labelled in such a way that they are arranged on L according to the positive orientation, $a_{q+1} = a_1$, $a_q = a_0$. The curve L divides the extended w-plane into two simply connected regions: the interior H_i and the exterior H_e, which will be called half-sheets. The part of the closed Jordan curve L lying between the points a_j and a_{j+1}, excluding these points, will be denoted by L_j. The regions obtained from the closed regions $\bar{H}_i$ and $\bar{H}_e$ after removal of the points a_j will be denoted by H_i° and H_e°, respectively.

Consider a Riemann surface F_0 consisting of a finite or countable set of half-sheets, which are "pasted" in such a way that for each half-sheet H_i (H_e) and each arc L_j, $1 \leq j \leq q$, there is a half-sheet H_e (H_i) which is pasted to a given half-sheet along the arc L_j (that is, the points of L_j from the half-sheets H_i° and H_e° are identified).

We fix a point $a \in H_i$ and a point $b \in H_e$. Denote by C_j a Jordan arc joining the points a and b in the extended complex w-plane and intersecting the curve L only at one point lying on L_j. Besides, we require that the curves C_j, $1 \leq j \leq q$, intersect each other at the endpoints a and b only. Let $\tilde{a}$ and $\tilde{b}$ be points of F_0 that project to a and b, respectively, and let $\tilde{C}_j$ be curves whose projections are the C_j. It is clear that the disjoint curves $\tilde{C}_j$ establish a one-to-one correspondence between their endpoints $\tilde{a}$ and $\tilde{b}$. As a result, for fixed j, we get a one-to-one correspondence between the points $\tilde{a}$ and the points $\tilde{b}$. It follows that the (finite or countable) numbers of half-sheets H_i and H_e in F_0 are the same. For uniformity of notation, we denote the half-sheets contained in F_0, whose projections are H_i and H_e, by $\tilde{H}_i$ and $\tilde{H}_e$, respectively.

Consider the open set in F_0 whose projection is the region D_j bounded by the curves C_{j-1} and C_j ($1 \leq j \leq q$, $C_0 = C_q$, $C_{q+1} = C_1$) and containing the point a_j. This set is decomposed into connected components $\tilde{D}_j^{(k)}$ which can be finite-sheeted or infinite-sheeted. If a component $\tilde{D}_j^{(k)}$ is n-sheeted, $1 \leq n < \infty$, then the part of its boundary whose projection is $C_{j-1} \cup C_j$ is a closed Jordan curve consisting of n arcs $\tilde{C}_j$ and n arcs $\tilde{C}_{j-1}$, which are alternating as we go around the Jordan curve. It is clear that the function

$$z = \left(\frac{w - a_j}{w - b_j}\right)^{1/n}, \qquad b_j \notin \bar{D}_j,$$

maps $\tilde{D}_j^{(k)}$ bijectively onto a doubly connected region in the z-plane, whose boundary consists of a closed Jordan curve and the point $z = 0$. If we add to $\tilde{D}_j^{(k)}$ (and F_0) a point with projection a_j, then $\tilde{D}_j^{(k)}$ becomes a simply connected one-sheeted region if $n = 1$, or a simply connected neighborhood of an algebraic ramification point of order $n - 1$ whose projection is a_j, if $n > 1$. We denote by F_0' the surface F_0 complemented by algebraic ramification points in the way described. If the component $\tilde{D}_j^{(k)}$ is infinite-sheeted, then the part of its boundary whose projection coincides with $C_{j-1} \cup C_j$ is an open Jordan curve consisting of infinitely many arcs $\tilde{C}_j$ and $\tilde{C}_{j-1}$, which are alternating as we go around the Jordan curve. It is easy to see that the function $z = \ln\{(w-a_j)/(w-b_j)\}$, $b_j \notin \bar{D}_j$, maps $\tilde{D}_j^{(k)}$ bijectively onto a simply connected region in the z-plane, which is lying to the left of some open Jordan arc located in a vertical strip and such that $\operatorname{Im} z \to +\infty$ and $\operatorname{Im} z \to -\infty$ as z moves along the curve in one or the other direction. It is clear that $\tilde{D}_j^{(k)}$ is a neighborhood of a logarithmic ramification point of the surface F_0 (and F_0') whose projection is a_j.

If the surface F_0' is simply connected, we say that F_0' belongs to the class F_q.

It is clear that the condition $\infty \notin L$ was imposed only to simplify the verbal description of the construction of surfaces of the class F_q. If $\infty \in L$, then, to define H_i and H_e, we need only to orient the curve L beforehand, and take into account the possibility of $a_j = \infty$ when writing the functions that map $\tilde{D}_j^{(k)}$.

The curve L will be called a *base* curve, and the points $a_1, \ldots, a_q$ will be called *base* points.

Finite-sheeted surfaces F_0' are obviously closed, and if they belong to the class F_q, they have genus zero. It is easy to see that, conversely, each closed Riemann surface of genus zero belongs to the class F_q. In fact, a closed Riemann surface F of genus zero is topologically equivalent to the extended complex plane and hence is simply connected. We draw a closed Jordan curve L through the projections $a_1, \ldots, a_q$ of the algebraic ramification points, and then make cuts of F along all curves whose projections are in L. The surface F decomposes into a finite number of simply connected regions which play the role of half-sheets $\tilde{H}_i$ and $\tilde{H}_e$. It is easy to see that in order to reconstruct the surface F from these half-sheets, we must paste the half-sheets $\tilde{H}_i$ and $\tilde{H}_e$ in the same way as we did in the description of surfaces from F_q.

Open simply connected Riemann surfaces do not necessarily belong to a class F_q. Take, for example, the Riemann surface corresponding to the function $w = e^z - z$. The derivative of this entire function has zeros of order one at the points $z_k = 2k\pi i$, $k = 0, \pm 1, \pm 2, \ldots$; hence the Riemann surface has algebraic ramification points of the first order, whose projections are the points $w_k = 1 - 2k\pi i$. Since the set of all such points w_k is infinite, the surface cannot belong to the class F_q. It is easy to see that if an open surface F of parabolic type belongs to the class F_q, then the mapping meromorphic function $w = f(z)$ may have multiple a-points or have a as an asymptotic value only if a is one of the base points. In this case, asymptotic spots of $f(z)$ correspond bijectively to the logarithmic branch points of F.

Let us show that, conversely, if a meromorphic function $w = f(z)$ in $\{|z| < \infty\}$ has multiple a-points or asymptotic values a only for a finite set $a = a_1, \ldots, a_q$, then the Riemann surface F corresponding to the function $f(z)$ belongs to the class $F_q(a_1, \ldots, a_q)$.

Let Δ be an arbitrary simply connected region in the extended w-plane which does not contain the points $a_1, \ldots, a_q$. Let $w_0 \in \Delta$, and let K be an arbitrary continuous curve lying in Δ and having w_0 as its starting point. We show that an arbitrary analytic element $z = P_j(w, w_0)$ of the function $z = f^{-1}(w)$ centered at the point w_0 admits an analytic continuation along the curve K. Suppose that $P_j(w, w_0)$ has an analytic continuation up to the point $w_1 \in K$, excluding the point

w_1. Let K_1 be the part of the curve K between w_0 and w_1, $K_1 : w = w(t), 0 \le t \le 1$, $w(0) = w_0$, $w(1) = w_1$. We denote by K_1' the curve K_1 without the point w_1, $K_1' : w = w(t)$, $0 \le t < 1$. Let $Q : z = z(t)$, $0 \le t < 1$ be the continuous curve in $\{|z| < \infty\}$ corresponding to K_1' under the mapping based on the analytic continuation of $z = P_j(w, w_0)$. It is obvious that $f(z(t)) \to w_1 \in \Delta$ as $t \to 1$.

Suppose that the function $z = z(t)$ has at least two limit points in the extended z-plane as $t \to 1$. Denote one of them by z_0, $z_0 \ne \infty$. It is clear that $f(z_0) = w_1$. Then $f(z) \ne w_1$ if z is in a sufficiently small neighborhood of z_0 and $z \ne z_0$. Let $\eta > 0$ be so small that the circle $\{|z - z_0| = \eta\}$ is contained in this neighborhood and at least one of the limit points of $z(t)$ is outside the circle. The curve Q intersects the circle $\{|z - z_0| = \eta\}$ infinitely many times. Hence there exists a sequence $t_n \to 1$ such that $z(t_n) \to z_1$, $|z(t_n) - z_0| = \eta$. Then $f(z_1) = w_1$, and we get a contradiction with the condition that $f(z) \ne w_1$ on the circle $\{|z - z_0| = \eta\}$. Hence the function $z(t)$ has a (finite or infinite) limit as $t \to 1$.

Let $z(t) \to z_0 \ne \infty$ as $t \to 1$. Then $f(z_0) = w_1$; moreover, z_0 is not a multiple w_1-point of the function $f(z)$, since otherwise the point z_0 would be mapped onto an algebraic ramification point on F, whose projection is $w_1 \ne a_j$, $1 \le j \le q$. Since the function $f(z)$ is univalent in a sufficiently small neighborhood of the point z_0, continuing the element $P_j(w, w_0)$ along K_1' we can continue it to the point $w_1 \in K_1$, getting an analytic element $P_j(w, w_1)$, $P_j(w_1, w_1) = z_0$. This contradicts the assumption.

If $z(t) \to \infty$ as $t \to 1$, then the curve Q is an asymptotic curve for the function $f(z)$, and w_1 is an asymptotic value, but this again is impossible because $w_1 \ne a_j$, $1 \le j \le q$.

Thus, we have shown that the analytic element $P_j(w, w_0)$ admits an analytic continuation along an arbitrary continuous curve in Δ; hence, by the monodromy theorem (Markushevich [**Mar77**, Vol. 3, Chapter VIII, §40], Stoïlow [**Sto62**, Vol. 1, Chapter IV, §1.6],[5] the function obtained as a result of this analytic continuation is defined in the region Δ and is single-valued in it. In other words, the set of points of the Riemann surface F which are projected into Δ decomposes into infinitely many one-sheeted regions, forming a complete covering of the region Δ.

We draw a closed Jordan curve L through the points $a_1, \dots, a_q$. The curve L divides the extended w-plane into two simply connected regions H_i and H_e. One-sheeted regions on F forming a complete covering of H_i and H_e make up half-sheets $\tilde{H}_i$ and $\tilde{H}_e$. Since for fixed j, $1 \le j \le q$, the region $H_i \cup L_j \cup H_e$ is also simply connected and the parts of the surface F forming a complete covering of this region are one-sheeted, each half-sheet $\tilde{H}_i$ ($\tilde{H}_e$) is pasted along each the arc $\tilde{C}_j$ to one half-sheet $\tilde{H}_e$ ($\tilde{H}_i$). Hence the surface F belongs to the class F_q.

The above discussion shows, in particular, that if F is a closed Riemann surface of genus zero or an open Riemann surface of parabolic type, then its belonging to the class F_q does not depend on the choice of the base curve L, since we have given a complete characterization of mapping meromorphic functions which does not involve base curves. A similar conclusion can be derived for surfaces of hyperbolic type.

The Riemann surfaces corresponding to the functions (1) $w = e^z$ (base points $a_1 = 0$, $a_2 = \infty$); (2) $w = e^{e^z}$ (base points $a_1 = 0$, $a_2 = 1$, $a_3 = \infty$); (3) $w = \sin z$

[5] Or L. Ahlfors, *Complex Analysis*, Chapter 8, 1.6.

(base points $a_1 = -1,\ a_2 = 1,\ a_3 = \infty$) are examples of open Riemann surfaces of the class F_q.

Riemann surfaces of the class F_q can be conveniently presented using so-called line complexes. By a *line complex* B of order q, $2 \leq q < \infty$, we mean a connected set in the extended z-plane having the following properties:

(1) B is a finite or countable set of Jordan arcs, which are called *edges*, and their endpoints are called *vertices*. Two different edges can intersect at vertices only.

(2) All vertices are divided into two classes: internal and external. Each edge has, as its endpoints, one external vertex and one internal vertex.

(3) The edges are labelled using numbers from 1 to q. Each vertex is an endpoint of q edges, which are located around the vertex according to their labels; the order is positive (counterclockwise) if the vertex is interior and is negative if the vertex is exterior.

(4) If B contains infinitely many vertices, then B does not contain ∞, and $z = \infty$ is the only limit point of vertices.

Line complexes B_1 and B_2 are regarded as the same if one of them can be mapped onto the other by a homeomorphism of the extended z-plane, which maps edges to edges with the same label, internal vertices to internal vertices, and external vertices to external vertices.

The complement of B in $\{|z| \leq \infty\}$ if B has a finite number of vertices, and the complement of B in $\{|z| < \infty\}$ if B has infinitely many vertices, is an open set. Each component of this set is a region bounded by either a finite set of edges (algebraic elementary region), or by an infinite set of edges (logarithmic elementary region). By condition (4) logarithmic elementary regions are unbounded. It is easy to see that when we go around an elementary region, the edges with labels $j-1$ and j ($1 \leq j \leq q$, the label 0 is equivalent to the label q, the label $q+1$ is equivalent to the label 1) alternate. Such an elementary region is called a *j-region.* It is clear that the boundary of an algebraic elementary region consists of an even number of edges. Conversely, it is easy to see that each Jordan curve on B that is closed or approaches $z = \infty$ and consists of alternating edges with labels $j-1$ and j, is a boundary of a j-region.

Several edges with consecutive labels, having common endpoints in one internal and one external vertices, form a so-called *edge bundle.* Clearly, two edges belonging to the same edge bundle and having labels $j-1$ and j form a boundary of a digon.

Consider the class of Riemann surfaces $F_q(a_1, \dots, a_q)$ and fix a base curve L. Now we describe a one-to-one correspondence between the Riemann surfaces $F \in F_q(a_1, \dots, a_q)$ and the line complexes B.

1°. Let a Riemann surface $F \in F_q$ be given. If F is open, we map it homeomorphically onto the complex z-plane; if F is closed, we map it homeomorphically onto the extended complex z-plane. Then the set of points on F with projections in $\bigcup_{j=1}^q C_j$ is mapped onto a line complex B; moreover, the $\tilde{C}_j$ are mapped onto the edges with label j, the points $\tilde{a}$ are mapped to internal vertices, and the points $\tilde{b}$ are mapped to external vertices. It is easy to check that conditions (1)–(4) are satisfied. Hence we get a line complex. This complex is uniquely determined by F

and by the base curve. In fact, if we chose points $a' \in H_i$, $b' \in H_e$, and curves C'_j, $1 \le j \le p$, differently, then, as is easy to see, it is possible to map the extended complex w-plane homeomorphically onto itself[6] in such a way that the points a_j, $1 \le j \le p$, are fixed, L is mapped onto L, a' to a, b' to b, and C'_j onto C_j. If we map all half-sheets $\tilde{H}_i$ and $\tilde{H}_e$, and all curves $\tilde{L}_j$ using the same homeomorphism, the Riemann surface F will be mapped homeomorphically onto itself; moreover, the points $\tilde{a}'$ will be mapped to $\tilde{a}$, the point $\tilde{b}'$ will be mapped to $\tilde{b}$, and the curves $\tilde{C}'_j$ will be mapped onto the curves C_j. Hence the line complex B corresponding to the surface F does not depend on the choice of curves C_j, and hence is uniquely determined by the surface F. We would like to mention that there exist examples (see V. G. Tairova [**Tai62**], [**Tai64**]) showing that with different choices of the base curve, the line complexes corresponding to the same Riemann surface $F \in F_q$ may be different.

2°. Now suppose that we are given a line complex B. Since we are given a base curve L, the regions H_i and H_e are determined. We consider a one-to-one correspondence which assigns to each internal vertex of B a copy of the region H_i and to each external vertex of B , a copy of the region H_e. We paste copies of H_i and H_e along the arc L_j, $1 \le j \le q$, if an only if the corresponding internal and external vertices are the endpoints of the same edge of B with label j. By property (3) of line complexes, we get a Riemann surface F_0 (see above). As before, we can choose points $a \in H_i$, $b \in H_e$ and curves C_j, define regions $\tilde{D}_j^{(k)}$, and complement the surface F_0 to the surface F'_0. Now we need to show that the surface F'_0 is simply connected.

In each algebraic j-region of the line complex B, mark a point a'_j; if the j-region is logarithmic, we place the corresponding point a'_j into the boundary point $z = \infty$. Each edge with label j is contained in the boundary of one $(j-1)$-region and one j-region. We join the corresponding points a'_{j-1} and a'_j by a Jordan arc l_j which is contained in the union of the $(j-1)$-region and j-region considered, except for one point on the edge with label j and, possibly, the points a'_{j-1} and a'_j if one or both of them coincide with ∞. We draw these Jordan arcs in such a way that they are pairwise disjoint; it is easy to see that this is always possible. Then the z-plane (the extended z-plane if the number of vertices is finite) is divided into regions S, each of which contains one vertex, and its boundary consists of q Jordan curves l_j with endpoints at a'_j and a'_{j+1}, where, as the index increases, the points $a'_1, \dots, a'_q$ follow along the boundary in the positive direction if the vertex is internal, and in the negative direction if the vertex is external. We established a one-to-one correspondence between the half-sheets $\tilde{H}_i$ $(\tilde{H}_e)$ of our surface F'_0 and internal (external) vertices. We map each half-sheet $\tilde{H}_i$ or $\tilde{H}_e$ homeomorphically onto the region S containing the corresponding vertex, in such a way that the mapping is continuous on the closure of these half-sheets, and in addition the boundary arcs $\tilde{L}_j$ are mapped onto the arcs l_j, and the points $\tilde{a}_j$ on the boundary $\tilde{H}_i$ or $\tilde{H}_e$ are mapped to the points a'_j. Observe that a point a'_j corresponding to a point $\tilde{a}_j \in F'_0$

[6]Here and later on, we use the following fact, without explicitly mentioning it. A homeomorphic mapping between the boundaries of two regions bounded by closed Jordan curves can be extended to a homeomorphic mapping between the regions. In the case where the regions are unit discs, the statement is obvious. The general case can be reduced to this special case by mapping the regions conformally onto the unit disc.

lies inside an algebraic j-region. The constructed mapping F_0' onto the (possibly extended) z-plane in general is not single-valued on the arcs $\tilde{L}_j$, since when we map the half-sheets $\tilde{H}_i$ and $\tilde{H}_e$, pasted along $\tilde{L}_j$, we can map the same point from $\tilde{L}_j$ to two different points in the z-plane. However, using some additional topological transformations of the closed regions $\bar{S}$, we can eliminate this multivalence. In fact, the regions S can be arranged into a (finite or infinite) sequence $S_1, S_2, \dots$ such that $\bar{S}_k$ has at least one common arc l_j with $\bigcup_{j=1}^{k-1} \bar{S}_j$. This is equivalent to writing the line complex as a sequence such that each vertex with number k is joined by an edge with at least one of the vertices having smaller number. Since the line complex is connected, we can always choose such a sequence of vertices. Now we subject $\bar{S}_2$ to an additional topological transformation, mapping $\bar{S}_2$ onto $\bar{S}_2$ in such a way that all points $a_j' \in \partial S_2$ remain fixed, and the arcs l_j belonging to $\partial S_1 \cap \partial S_2$ are transformed so that the points on the corresponding arcs $\tilde{L}_j \subset F_0'$ will have the same image in the z-plane both under the mapping onto ∂S_1 and under the mapping onto ∂S_2. Then we transform $\bar{S}_3$ topologically into $\bar{S}_3$ so that all points $a_j' \in \partial S_3$ remain fixed, and the arcs l_j belonging to $\partial(\bar{S}_1 \cup \bar{S}_2) \cap \partial S_3$ are transformed as above. Continuing this process, we get a topological transformation of the whole Riemann surface F_0' onto $\{|z| < \infty\}$ or $\{|z| \leq \infty\}$ depending on whether F_0' is a closed or an open surface. Thus, the surface is simply connected and belongs to the class F_q.

It is easy to see that if we used the rule described in 1° to get a line complex B_1 corresponding to the obtained Riemann surface F_0', then B_1 would coincide with the line complex B which started our construction of the surface F_0'.

Thus, we established a one-to-one (for a fixed base curve L) correspondence between Riemann surfaces of the class F_q and line complexes. The line complex corresponding to a Riemann surface will be called the *line complex of this surface.* In our diagrams, the internal vertices of a complex will be marked by circles, and the external vertices by crosses. The label of an elementary region (the point a_j) will be written inside the region. We will not write labels inside digons. It is clear that if the labels of elementary regions are marked on a line complex, it is not necessary to label edges, since the edge between the j-region and the $(j+1)$-region has label j. If some j-region is a digon, then, as is easy to see, the corresponding point $\tilde{a}_j \in F_0'$ is not a ramification point. If a j-region is a $2m$-gon, $2 \leq m < \infty$, then the corresponding point $\tilde{a}_j$ is an algebraic ramification point of order $m-1$.

We suggest that the reader analyzes the following examples of line complexes of Riemann surfaces for which the mapping functions are indicated. In all our examples, the base curve is the real line, $H_i = \{\operatorname{Im} w < 0\}$, and $H_e = \{\operatorname{Im} w > 0\}$.

(1) $w = z^2$ (Figure 5);

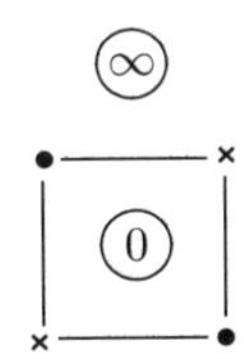

FIGURE 5

(2) $w = z^3$ (Figure 6);

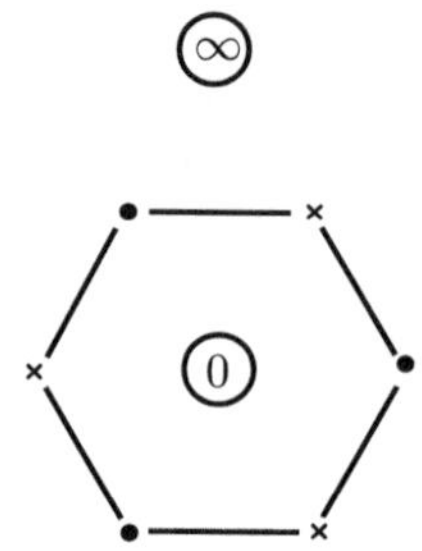

FIGURE 6

(3) $w = \frac{1}{2}\left(z + \frac{1}{z}\right)$ (Figure 7);

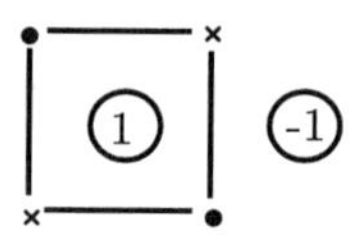

FIGURE 7

(4)$w = \frac{1}{2}\left(z^3 + \frac{1}{z^3}\right)$ (Figure 8);

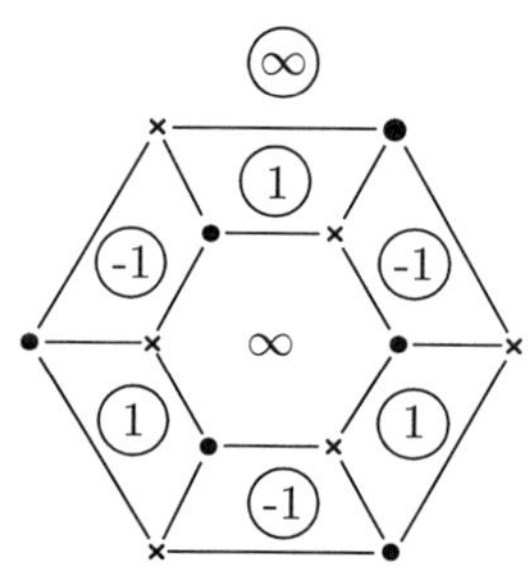

FIGURE 8

(5) $w = e^z$ (Figure 9);

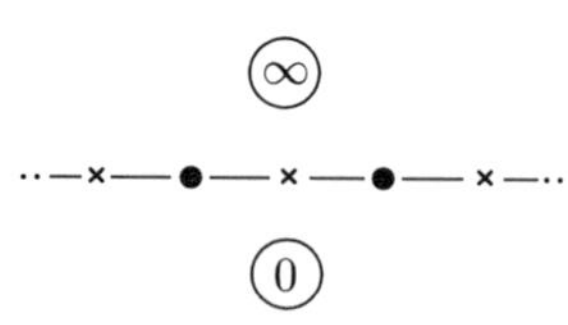

FIGURE 9

(6) $w = e^{z^2}$ (Figure 10);

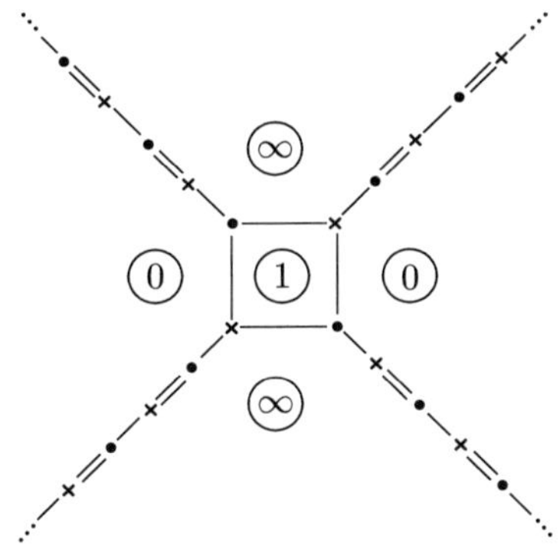

FIGURE 10

(7) $w = e^{z^3}$ (Figure 11);

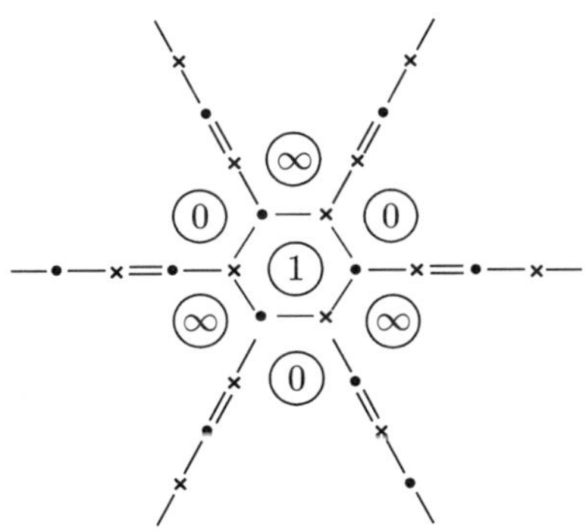

FIGURE 11

(8) $w = e^{e^z}$ (Figure 12);

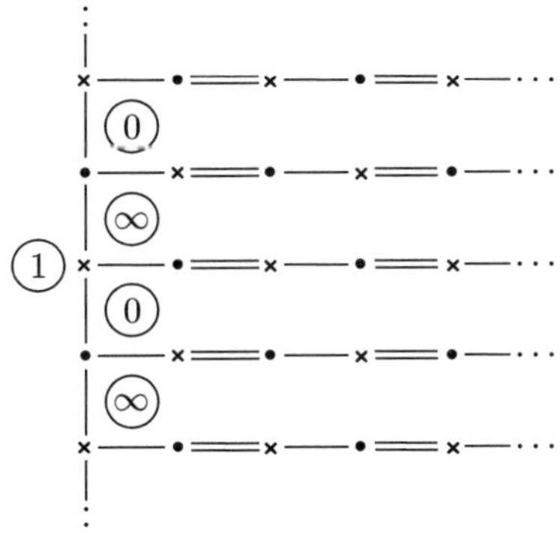

FIGURE 12

(9) $w = \sin z$ (Figure 13);

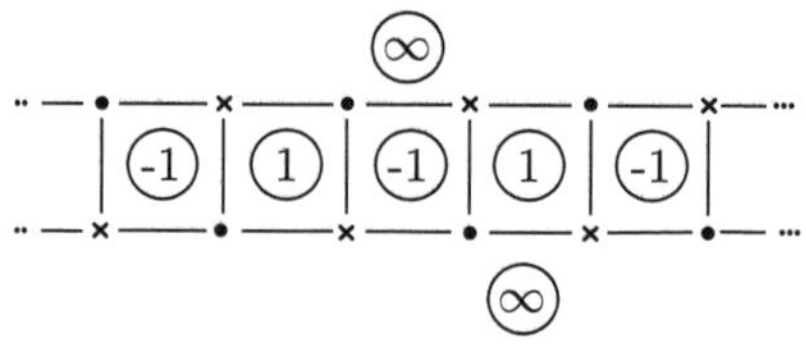

FIGURE 13

(10) $w = \sin z^2$ (Figure 14);

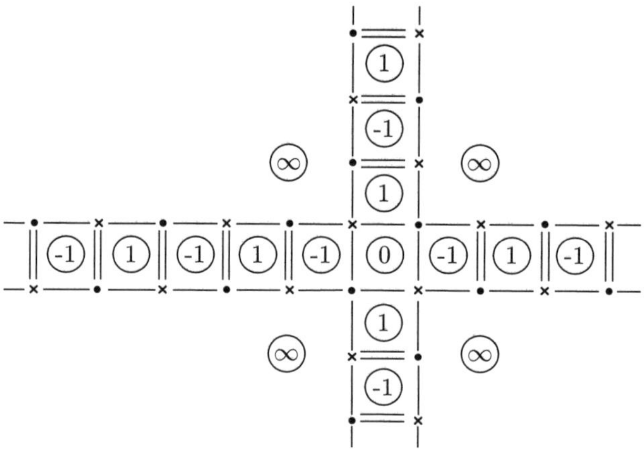

FIGURE 14

(11) $w = \cos\sqrt{z}$ (Figure 15).

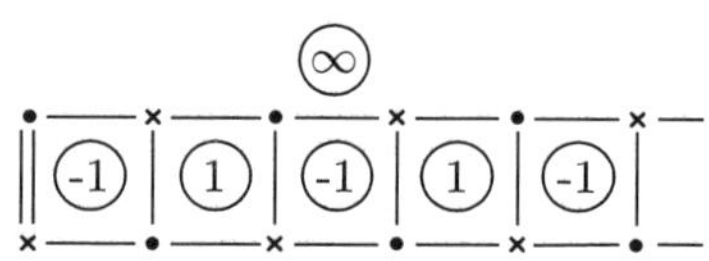

FIGURE 15

We also suggest sketching the location of regions S. As an example, we sketch the corresponding diagrams for the functions (4), (5), and (6) (Figures 16, 17, and 18).

Now we turn our attention to the closed Riemann surfaces F of genus zero. Suppose that $a_1, \dots, a_q$ are the projections of their ramification points (we do not exclude the case of no ramification points over some point a_j). Suppose that a_j is the projection of ν_j algebraic ramification points of F of orders $\lambda_1(a_j), \dots, \lambda_{\nu_j}(a_j)$. Let n be the number of sheets of F, and $\bar{n}(a)$ the number of points of the Riemann surface whose projection is a; here each ramification point is counted once. It is clear that $\bar{n}(a) \le n$ for all a; moreover

$$\bar{n}(a) = \begin{cases} n & \text{for } a \neq a_j,\ j = 1, \dots, q, \\ n - \lambda_1(a_j) - \dots - \lambda_{\nu_j}(a_j) & \text{for } a = a_j. \end{cases}$$

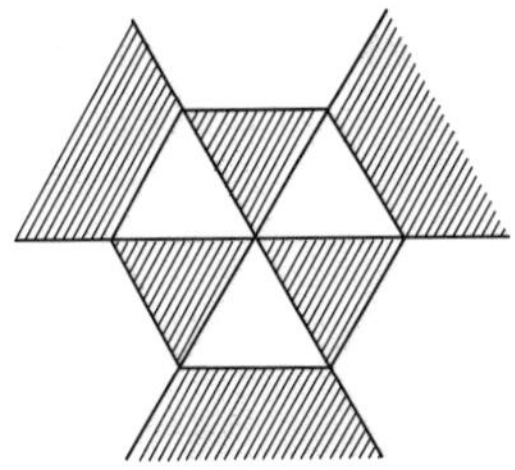

FIGURE 16

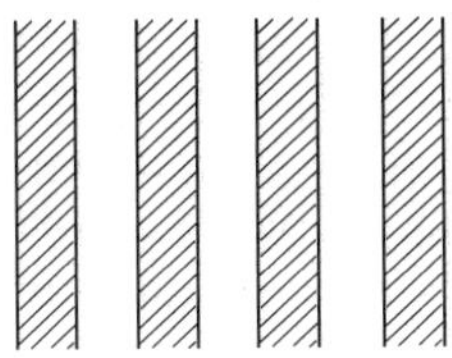

FIGURE 17

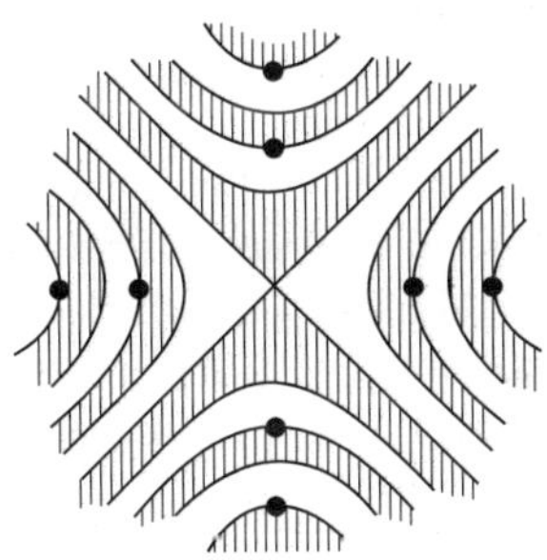

FIGURE 18

Let B be the line complex of a Riemann surface F. Elementary regions, edges, and vertices of B form a topological polyhedron on the Riemann sphere. Using the well-known Euler formula, we get the equality $K - S + E = 2$, where K is the number of elementary regions, S is the number of edges, and E is the number of vertices in the line complex B. The number K is equal to the number of all points of the surface F lying over the base points $a_1, \ldots, a_q$, that is, $\sum_{j=1}^{q} \bar{n}(a_j)$. The number of vertices of E is equal to $2n$, with n internal and n external vertices. Since exactly q edges leave each internal vertex, and, moreover, each edge has one and only one internal vertex as one of its endpoints, we have $S = nq$. Hence

$$\sum_{j=1}^{n} \bar{n}(a_j) - nq + 2n = 2,$$

$$2(n-1) = \sum_{j=1}^{q} \{n - \bar{n}(a_j)\} = \sum_{j=1}^{q} \sum_{k=1}^{\nu_j} \lambda_k(a_j). \tag{4.1}$$

Equality (4.1) represents the well-known Riemann–Hurwitz formula.

The following question arises in a natural way. Assume that we are given complex numbers $a_1, \dots, a_q$ and natural numbers n, ν_j, $\lambda_k(a_j)$, satisfying (4.1) and the inequalities

$$\Lambda(a_j) = \sum_{k=1}^{\nu_j} \lambda_k(a_j) \le n - \nu_j, \qquad j = 1, \dots, q. \tag{4.2}$$

Is it possible to construct a closed Riemann surface F of genus zero, which is n-sheeted, with ν_j ramification points whose projection is a_j, with orders $\lambda_k(a_j)$, $1 \le k \le \nu_j$, $1 \le j \le q$?

It turns out that it is not always possible.

EXAMPLE 1. Let $n = 4$, $q = 3$, $\nu_1 = \nu_2 = 2$, $\nu_3 = 1$, $\lambda_1(a_1) = \lambda_2(a_1) = \lambda_1(a_2) = \lambda_2(a_2) = 1$, $\lambda_1(a_3) = 2$. It is easy to see that these numbers satisfy relations (4.1) and (4.2), but it is impossible to construct the corresponding Riemann surface. In fact, if there were the corresponding surface, its line complex would have eight vertices, its elementary 1- and 2-regions would be quadrangles, one 3-region G_1 would be a hexagon, and the other would be a digon G_2. The boundary of the 3-region G_1 would contain six vertices, each of which is an endpoint of some edge lying outside G_1. There are only two vertices outside $\bar{G}_1$. Suppose that two edges leaving from two vertices on the boundary of G_1 end at one vertex outside $\bar{G}_1$. The exterior of G_1 is divided into two Jordan polygons, a quadrangle and a hexagon. The second vertex lying outside G_1 is either inside the quadrangle or inside the hexagon. Hence one of these Jordan polygons is an elementary region, adjoining to G_1 along two or four consecutive sides of G_1, which is impossible. If both endpoints of some edge lying outside G_1 are on the boundary of G_1, then this edge divides the exterior of G_1 into two quadrangles and, arguing as above, we get a contradiction. Hence, for each vertex on the boundary of G_1 there is an edge lying outside G_1 which ends at a vertex lying outside $\bar{G}_1$, and different edges correspond to different vertices. Thus, there are at least six vertices outside $\bar{G}_1$, and we again come to a contradiction.

We denote by $\mathfrak{F} = \mathfrak{F}(n; a_1 : \lambda_1(a_1), \dots, \lambda_{\nu_1}(a_1); \dots; a_q : \lambda_1(a_q), \dots, \lambda_{\nu_q}(a_q))$, where $a_1, \dots, a_q$ are distinct complex numbers, and n, ν_j, $\lambda_k(a_j)$ are positive integers satisfying (4.1) and (4.2), the class of closed Riemann surfaces of genus zero, which are n-sheeted with ν_j ramification points of orders $\lambda_k(a_j)$, whose projections are a_j. Sometimes it will be convenient for us to allow $\lambda_k(a_j)$ to take zero values also (we will always mention this explicitly). This will mean that we require the Riemann surfaces of the class $\mathfrak{F}$ to have over a_j at least as many unramified sheets as the number of zeros among the numbers $\lambda_1(a_j), \dots, \lambda_{\nu_j}(a_j)$. If there is $a_j : \Lambda(a_j)$ instead of $a_j : \lambda_1(a_j), \dots, \lambda_{\nu_j}(a_j)$ in the notation of the class $\mathfrak{F}$, this means that we require only that the sum of orders of the ramification points projected into a_j be equal to $\Lambda(a_j)$. It is clear that the numbers $\Lambda(a_j)$ satisfy the relations

$$\sum_{j=1}^{q} \Lambda(a_j) = 2(n-1), \tag{4.1$'$}$$

$$\Lambda(a_j) \le n - 1, \qquad j = 1, \dots, q. \tag{4.2$'$}$$

Obviously, for $\Lambda(a_j) = \lambda_1(a_j) + \dots + \lambda_{\nu_j}(a_j)$ we have

$$\mathfrak{F}(n; \dots; a_j : \lambda_1(a_j), \dots, \lambda_{\nu_j}(a_j); \dots) \subset \mathfrak{F}(n; \dots; a_j : \Lambda(a_j); \dots).$$

Example 1 shows that the class $\mathfrak{F}(4; a_1 : 1, 1; a_2 : 1, 1; a_3 : 2)$ is empty.

THEOREM 4.1. *The classes* $\mathfrak{F}(n; a_1 : \lambda_1(a_1); \dots; a_q : \lambda_1(a_q))$, $\lambda_1(a_j) \geq 0$, *are non-empty.*

Theorem 4.1 immediately implies the following.

COROLLARY. *The classes* $\mathfrak{F}(n; a_1 : \Lambda(a_1); \dots; a_q : \Lambda(a_q))$, $\Lambda(a_j) \geq 0$, *are non-empty.*

Before proving Theorem 4.1, we state two more theorems which will be needed in Section 6.

First we introduce some definitions. A Jordan arc C on a Riemann surface F (the case of one-sheeted region F is not excluded) will be called a *J-curve* if it has the following properties: (1) different points of C are projected to different points; only the endpoints of C can be projected to the same point; (2) C does not contain algebraic ramification points of F; the only possible exception is the endpoints of C; (3) there exists a Jordan arc $C_1 \subset C$ with interior points of C as its endpoints, such that $C \backslash C_1$ consists of two line segments; (4) C has a continuous curvature. In some cases, by a J-curve we mean a Jordan arc without one endpoint or without both endpoints, satisfying (1)–(4). Whether or not a J-curve includes its endpoints will be clear from the context.

A closed Jordan curve C in the plane is called a *J'-curve* with endpoints at $a \in C$ if it has a continuous curvature at all points except, possibly, the point a, and in a sufficiently small neighborhood of a, the curve C has the form of a linear angle with a vertex at the point a.

It is clear that the projection of an arbitrary J-curve on a Riemann surface is either a J- or a J'-curve in the plane.

THEOREM 4.2. *Let $q \geq 2$, let $a_1, \dots, a_q$ be distinct complex numbers, and let γ be a J-curve joining the points a_1 and a_2 that does not pass through the points a_j, $3 \leq j \leq q$. Then each of the classes listed below contains a Riemann surface F on which one can draw a J-curve Γ with endpoints $\mathfrak{a}_1$ and $\mathfrak{a}_2 \in F$, such that Γ is projected into γ in a one-to-one way, and $\mathfrak{a}_j$ is a ramification point of order $\lambda_1(a_j) \geq 0$, lying above a_j, $j = 1, 2$. The classes of Riemann surfaces mentioned above are the following:*

(a) $\mathfrak{F}(n; a_1 : \lambda_1(a_1), \lambda_2(a_1); a_2 : \lambda_1(a_2), \lambda_2(a_2); a_3 : \Lambda(a_3); a_4 : \lambda_1(a_4); \dots;$
$a_q : \lambda_1(a_q))$;

(b) $\mathfrak{F}(n; a_1 : \lambda_1(a_1), \lambda_2(a_1); a_2 : \lambda_1(a_2); a_3 : \lambda_1(a_3); \dots; a_q : \lambda_1(a_q))$,
where $\lambda_1(a_1) \geq 0$;

(c) $\mathfrak{F}(n; a_1 : \lambda_1(a_1); a_2 : \lambda_1(a_2), \lambda(a_2); a_3 : \lambda_1(a_3); \dots; a_q : \lambda_1(a_q))$,
where $\lambda_1(a_2) \geq 0$;

(d) $\mathfrak{F}(n; a_1 : \lambda_1(a_1), \lambda_2(a_1); a_2 : 0, \lambda_2(a_2); a_3 : \lambda_1(a_3); \dots; a_q : \lambda_1(a_q))$;

(e) $\mathfrak{F}(n; a_1 : 0, \lambda_2(a_1); a_2 : \lambda_1(a_2), \lambda_2(a_2); a_3 : \lambda_1(a_3); \dots; a_q : \lambda_1(a_q))$;

(f) $\mathfrak{F}(n; a_1 : \lambda_1(a_1); a_2 : \lambda_1(a_2); a_3 : \lambda_1(a_3); \dots; a_q : \lambda_1(a_q))$,
where $\lambda_1(a_1) \geq 0$, $\lambda_1(a_2) \geq 0$;

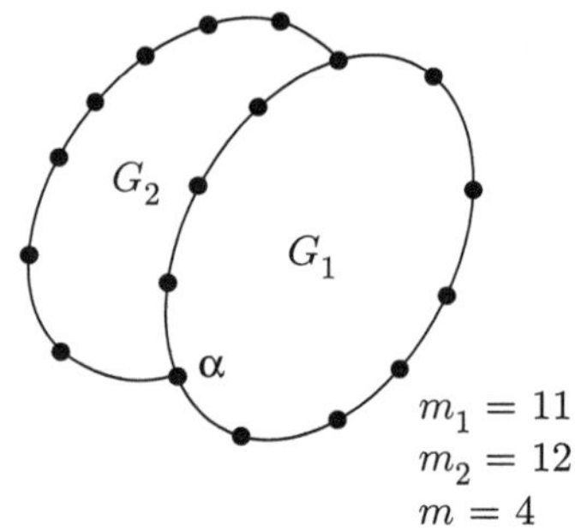

FIGURE 19

(g) $\mathfrak{F}(n; a_1 : \lambda_1(a_1), \lambda_2(a_1); a_2 : 0; a_3 : \lambda_1(a_3) \dots;\ a_q : \lambda_1(a_q))$, *where* $\lambda_1(a_1) \geq 0$;

(h) $\mathfrak{F}(n; a_1 : 0; a_2 : \lambda_1(a_2), \lambda_2(a_2); a_3 : \lambda_1(a_3); \dots; a_q : \lambda_1(a_q))$, *where* $\lambda_1(a_2) \geq 0$;

(i) $\mathfrak{F}(n; a_1 : 0, \lambda_2(a_1); a_2 : 0, \lambda_2(a_2); a_3 : \lambda_1(a_3); \dots; a_q : \lambda_1(a_q))$, *where* $q \geq 4$.

THEOREM 4.3. *Let $q \geq 3$, let $a_1, \dots, a_q$ be distinct complex numbers, and let γ be a J'-curve with endpoints at a_1. Moreover, let the point a_2 be inside the curve γ, and the points $a_3, \dots, a_q$ outside γ. Then each of the following two classes:*

(j) $\mathfrak{F}(n; a_1 : \lambda_1(a_1), \lambda_2(a_1), \lambda_3(a_1); a_2 : \lambda_1(a_2); a_3 : \lambda_1(a_3); \dots; a_q : \lambda_1(a_q))$, *where* $\lambda_k(a_1) \geq 0,\ k = 1, 2, 3;\ \lambda_1(a_2) \geq 2$;

(k) $\mathfrak{F}(n; a_1 : \lambda_1(a_1), \lambda_2(a_1); a_2 : \lambda_1(a_2); \dots; a_q; \lambda_1(a_q))$, *where* $\lambda_k(a_1) \geq 0, k = 1, 2$;

contains a Riemann surface F on which one can draw a J-curve Γ with endpoints $\mathfrak{a}_1$ and $\mathfrak{a}_2 \in F$, such that the curve Γ without endpoints is projected in a one-to-one way into $\gamma \backslash \{a_1\}$, the point $\mathfrak{a}_k$ being a ramification point of order $\lambda_k(a_1)$ over a_1, $k = 1, 2$.

Proofs of Theorems 4.1, 4.2, and 4.3 are not difficult, but they are rather cumbersome and use topological constructions which are far from the main topics of this book. The readers who do not wish to follow these proofs and are ready to accept the results, can omit the proofs without harm to understanding the subsequent material.

Before starting the proofs we fix some notation. Let G_1 be a Jordan m_1-gon. We join it with a Jordan m_2-gon, lying outside G_1 in such a way that $\partial G_2 \cap \partial G_1$ is a Jordan arc that consists of m Jordan arcs which are the sides of G_1 and G_2, $m \leq \min(m_1 - 1, m_2 - 1)$, following one after another, starting with a vertex α, in the negative direction of the boundary of G_1. This operation will be denoted by $U(G_1, G_2, \alpha, m)$ (Figure 19).

PROOF OF THEOREM 4.1. The theorem is trivial if $n = 1$. We assume that $n \geq 2$ and construct the corresponding line complexes. If $q = 2$, then necessarily $\lambda_1(a_1) = \lambda_1(a_2) = n - 1$, and the line complex is a closed Jordan curve consisting of

$2n$ edges (see, for example, Figures 5 and 6). Now let $q \geq 3$. For an elementary 1-region we take the region G_1 which is a Jordan $(2\lambda_1(a_1)+2)$-gon. On the boundary of G_1 we fix an internal vertex α. Let p be the largest integer satisfying

$$2 + 2\sum_{j=1}^{p} \lambda_1(a_j) \leq 2n. \tag{4.3}$$

Since, by (4.1),

$$2\sum_{j=1}^{q} \lambda_1(a_j) = 4(n-1),$$

we have $1 < p < q$. Suppose that there is an equality in (4.3). We join to G_1 an elementary 2-region G_2, which is a Jordan $(2\lambda_1(a_2)+2)$-gon, using the operation $U(G_1, G_2, \alpha, 1)$. Denote by Δ_2 the region $(\bar{G}_1 \cup \bar{G}_2 \backslash \partial(\bar{G}_1 \cup \bar{G}_2)$, which is a Jordan $(2\lambda_1(a_1) + 2\lambda_1(a_2) + 2)$-gon. We join to Δ_2 an elementary 3-region G_3, which is a Jordan $(2\lambda_1(a_3)+2)$-gon, using $U(\Delta_2,\ G_3,\ \alpha,\ 1)$. Suppose that we have already joined an elementary s-region G_s, $1 < s \leq p-1$. Let

$$\Delta_s = \left(\bigcup_1^s \bar{G}_j\right) \backslash \partial \left(\bigcup_1^s \bar{G}_j\right).$$

We join to Δ_s and elementary $(s+1)$-region G_{s+1}, which is a Jordan $(2\lambda_1(a_{s+1})+2)$-gon, using $U(\Delta_s,\ G_{s+1},\ \alpha,\ 1)$. The boundary of Δ_p contains $2\sum_{j=1}^{p}\lambda_1(a_j)+2 = 2n$ vertices. We join to Δ_p an elementary $(p+1)$-region G_{p+1}, which is a Jordan $(2\lambda_1(a_{p+1})+2)$-gon, using $U(\Delta_p,\ G_{p+1},\ \alpha,\ 2\lambda_1(a_{p+1})+1)$. The boundary of Δ_{p+1} contains $2n-2\lambda_1(a_{p+1})$ vertices. Joining to Δ_s, $p \leq s \leq q-2$, an elementary $(s+1)$-region G_{s+1}, which is a $(2\lambda_1(a_{s+1})+2)$-gon, using $U(\Delta_s,\ G_{s+1},\ \alpha,\ 2\lambda_1(a_{s+1})+1)$, we get a Jordan region Δ_{s+1}, the boundary of which contains $2n - 2\sum_{p+1}^{s+1}\lambda_1(a_j)$ vertices. The boundary of the region Δ_{q-1} contains

$$2n - 2\sum_{j=p+1}^{q-1} \lambda_1(a_1) = 2(n-1) - 2\sum_{j=p+1}^{q-1} \lambda_1(a_j) + 2$$

vertices. By (4.1) and the assumption that there is an equality in (4.3), we have

$$2\sum_{j=1}^{p} \lambda_1(a_j) = 2(n-1) = \sum_{j=1}^{q} \lambda_1(a_j),$$

$$\sum_{j=1}^{p} \lambda_1(a_j) = \sum_{j=p+1}^{q} \lambda_1(a_j) = n-1.$$

Therefore the boundary of Δ_{q-1} contains

$$2\sum_{j=p+1}^{q} \lambda_1(a_j) - 2\sum_{j=p+1}^{q-1} \lambda_1(a_j) + 2 = 2\lambda_1(a_q) + 2$$

vertices. For an elementary q-region G_q we take the complement of $\bar{\Delta}_{q-1}$ in the extended complex plane.

Thus, we have constructed all needed elementary regions different from digons. Our system of edges does not yet satisfy condition (3) of the definition of line complexes. However, as is easy to see, it has the property that if some internal (external) vertex lies on the boundary of more than two elementary regions, then the labels of these elementary regions increase (decrease) as we go around the vertex in the positive direction. Clearly, some edges can be replaced by bundles so that we get a line complex having all required properties, that is, representing a Riemann surface of the class $\mathfrak{F}(n; a_1 : \lambda_1(a_1); \ldots; a_q : \lambda_1(a_q))$, $\lambda_1(a_j) \geq 0$.

If there is a strict inequality in (4.3), then the construction is changed at one point only. We construct the elementary 1-, 2-, ..., p-regions $G_1, G_2, \ldots, G_p$, and the regions $\Delta_2, \Delta_3, \ldots, \Delta_p$ the same way as before, but now there are m_1 vertices on the boundary of Δ_p, where

$$m_1 = 2 + 2\sum_{j=1}^{p} \lambda_1(a_j) < 2n. \tag{4.4}$$

On the other hand,

$$2\lambda_1(a_{p+1}) + m_1 = 2 + 2\sum_{j=1}^{p+1} \lambda_1(a_j) > 2n.$$

Inequality (4.4) implies that

$$\lambda_1(a_{p+1}) + \sum_{j=1}^{p} \lambda_1(a_j) < (n-1) + (n-1) = 2(n-1);$$

therefore, by (4.1), we have $p + 1 < q$. Now we join to Δ_p an elementary $(p+1)$-region G_{p+1}, which is a Jordan $(2\lambda_1(a_{p+1})+2)$-gon, using $U(\Delta_p,\ G_{p+1},\ \alpha, m_2+1)$, where $m_2 = 2\lambda_1(a_{p+1}) + m_1 - 2n$. We can use this operation since $m_2 + 1 < m_1$ and $m_2 + 1 < 2\lambda(a_{p+1}) + 2$. Since the number $m_2 + 1$ is odd, the endpoint of the Jordan arc $\partial\Delta_p \cap \partial G_{p+1}$, different from α, is an external vertex. It is easy to count that there are $m_1 + (2\lambda_1(a_{p+1}) + 2) - (m_2 + 2) = 2n$ vertices in $\bar{\Delta}_{p+1}$; moreover, there are $2n - m_2$ vertices on the boundary of Δ_{p+1}.

After having constructed the region Δ_{p+1}, we construct the elementary regions G_s and the regions Δ_s, $p + 2 \leq s \leq q - 1$, the same way as before. The boundary of Δ_{q-1} contains

$$\begin{aligned} 2n - m_2 - 2 \sum_{j=p+1}^{q-1} \lambda_1(a_j) &= 4n - m_1 - 2\lambda_1(a_{p+1}) - 2 \sum_{j=p+2}^{q-1} \lambda_1(a_j) \\ &= 2\{2(n-1)\} + 4 - 2\sum_{j=1}^{p} \lambda_1(a_j) - 2 - 2 \sum_{j=p+1}^{q-1} \lambda_1(a_j) \\ &= 2\sum_{j=1}^{q} \lambda_1(a_j) + 2 - 2\sum_{j=1}^{q-1} \lambda_1(a_j) = 2\lambda_1(a_q) + 2 \end{aligned}$$

vertices. After that the construction of the line complex is completed in the same way as in the previous case. Figures 20 and 21 show the examples of line complexes corresponding to the cases of equality or strict inequality in (4.3). □

REMARK. It is important to note that in the construction used in the proof of Theorem 4.1, there is an internal vertex α that lies simultaneously on the boundaries of the elementary j-regions $G_1, G_2, \ldots, G_q$, which are Jordan $(2\lambda_1(a_j) + 2)$-gons; that is, all ramification points of the Riemann surface lie on the boundary of one internal half-sheet.

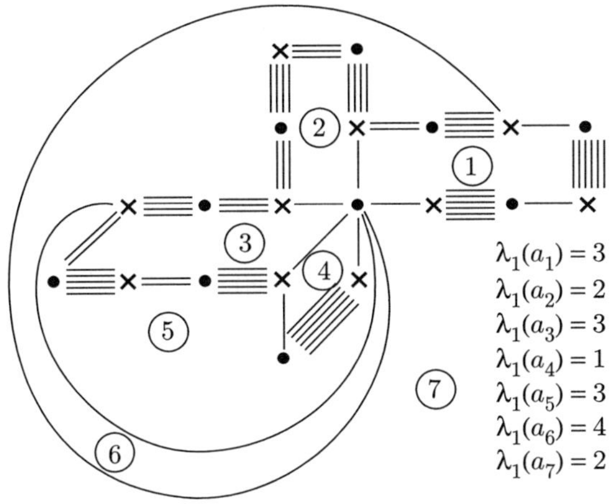

FIGURE 20

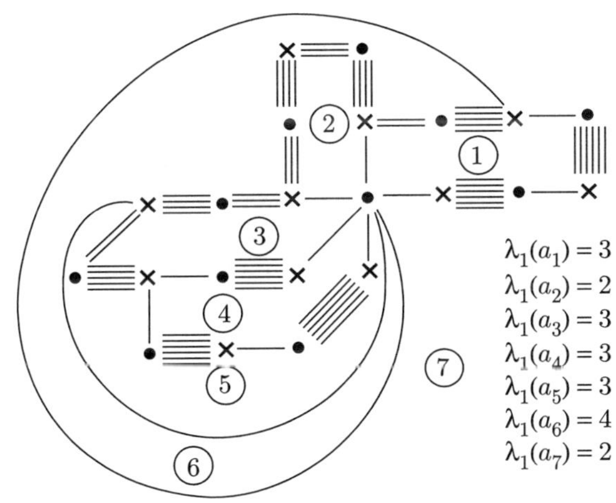

FIGURE 21

PROOF OF THEOREM 4.2. We draw a base curve L through the points $a_1, \dots,$ a_q in such a way that the curve γ, except for its endpoints, is inside L. Now, obviously, it suffices to construct a line complex with prescribed ramification and satisfying the additional condition that the elementary regions corresponding to the points $\mathfrak{a}_1$ and $\mathfrak{a}_2$ have at least one common internal vertex on their boundary.

CASE (a). Since $\tau = \lambda_1(a_1)+\lambda_2(a_1)+\lambda_1(a_2)+\lambda_2(a_2) < 2(n-2) < 2(n-1)$, we have $n \geq 3$, $q \geq 3$, and $\sum_{j=3}^{q} \Lambda(a_j) \geq 2$. Assume first that $\tau \leq n-1$. In this case we consider surfaces of the class $\mathfrak{F}(n;\ a_1:\ \lambda_1(a_1),\ \lambda_2(a_1);\ a_2:\ \lambda_1(a_2), \lambda_2(a_2);\ a_3:$ $\lambda_1(a_3);\ \dots; a_q:\ \lambda_1(a_q))$, where $\lambda_1(a_3) = \Lambda(a_3)$. For an elementary 1-region G_1' take a Jordan $(2\lambda_2(a_1)+2)$-gon and mark some internal vertex α' on its boundary. We join to G_1' an elementary 2-region G_2', which is a Jordan $(2\lambda_2(a_2)+2)$-gon, using $U(G_1',\ G_2',\ \alpha',\ 1)$. Let β be the next external vertex after α' as we go around the boundary of G_2' in the negative direction. We join to $\Delta' = (\bar{G}_1' \cup \bar{G}_2') \backslash \partial(\bar{G}_1' \cup \bar{G}_2')$ an elementary 1-region G_1'', which is a Jordan $(2\lambda_1(a_1)+2)$-gon, using $U(\Delta',\ G_1'',\ \beta,\ 1)$.

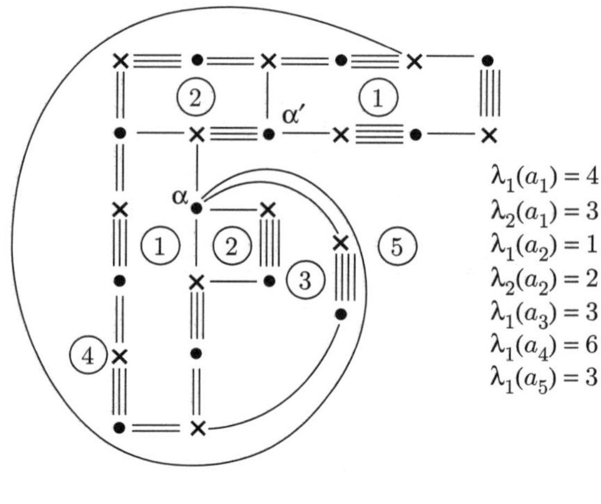

FIGURE 22

Let α be the next internal vertex after β as we go around the boundary of G_1'' in the negative direction. We join to $\Delta'' = (\bar{G}_1'' \cup \bar{\Delta}')\backslash\partial(\bar{G}_1'' \cup \bar{\Delta}')$ an elementary 2-region G_2'', which is a Jordan $(2\lambda_1(a_2)+2)$-gon, using the operation $U(\Delta'', G_2'', \alpha, 1)$. Let $\Delta_2 = (\bar{\Delta}'' \cup \bar{G}_2'')\backslash\partial(\bar{\Delta}'' \cup \bar{G}_2'')$. It is easy to count that the boundary of Δ_2 contains $2 + 2\lambda_1(a_1) + 2\lambda_1(a_2) + 2\lambda_2(a_1) + 2\lambda_2(a_2) = 2 + 2\tau \leq 2n$ vertices, and there are no vertices inside Δ_2. After this we successively join to Δ_2 elementary 3-, ..., q-regions and construct a line complex in exactly the same way as in the proof of Theorem 4.1. In addition, the internal vertex α lies on the boundary of the elementary regions $G_1'', G_2'', G_3, \dots, G_q$ (only the fact that α belongs to $\partial G_1'' \cap \partial G_2''$ is important for us).

A line complex corresponding to the case of $\tau \leq n-1$ is shown in Figure 22.

Now we consider the case where $\tau \geq n$. If $\Lambda(a_3) > 1$, we construct a surface of the class $\mathfrak{F}(n;\ a_1 :\ \lambda_1(a_1),\ \lambda_2(a_1);\ a_2 :\ \lambda_1(a_2),\ \lambda_2(a_2);\ a_3 :\ \lambda_1(a_3),\ \lambda_2(a_3);\ a_4 : \lambda_1(a_4); \dots; a_q :\ \lambda_1(a_q))$, where $\lambda_2(a_3) = 1$, $\lambda_1(a_3) = \Lambda(a_3) - 1$; if $\Lambda(a_3) = 1$, then necessarily $q \geq 4$, and we construct a surface of the class $\mathfrak{F}(n;\ a_1 :\ \lambda_1(a_1),\ \lambda_2(a_1);$ $a_2 :\ \lambda_1(a_2),\ \lambda_2(a_2);\ a_3 : \lambda_1(a_3); \dots;\ a_q :\ \lambda_1(a_q))$, where $\lambda_1(a_3) = 1$. At first we assume that $\tau = n$. The corresponding line complex can be constructed as follows. Let Q_1 be the square with vertices $A = (-1,-1)$, $B = (-1,1)$, $C = (1,1)$, $D = (1,-1)$, and Q_2 the square with vertices $A' = (-2,-2)$, $B' = (-2,2)$, $C' = (2,2)$, $D' = (2,-2)$. The square Q_1 will play the role of an elementary 3-region. Using segments of bisectors of the coordinate angles, we join the corresponding vertices of the squares Q_1 and Q_2, dividing the region $Q_2\backslash\bar{Q}_1$ into four trapezoids $T_1, \dots, T_4$, where T_1 is the lower, T_2 is the left, T_3 is the upper, and T_4 is the right trapezoid. We place internal vertices at the points $\alpha = A', C', B, D$, and external vertices at the remaining vertices of Q_1 and Q_2. Now we place $(2\lambda_1(a_1) - 2)$ vertices in the interval $A'D'$, $(2\lambda_1(a_2) - 2)$ vertices in the interval $A'B'$, $(2\lambda_2(a_1) - 2)$ in the interval $B'C'$, and $(2\lambda_2(a_2) - 2)$ in the interval $C'D'$. We regard T_1 and T_2 as elementary 1-regions and T_2 and T_4 as elementary 2-regions. The closed square $\bar{Q}_2$ will contain $2(\lambda_1(a_1) + \lambda_1(a_2) + \lambda_2(a_1) + \lambda_2(a_2) - 4) + 8 = 2\tau = 2n$ vertices, $2n - 4$ of them lying on the boundary of Q_2. If $\Lambda(a_3) > 1$, we successively join to Q_2 elementary 3-, 4-, ..., $(q-1)$-regions, which are Jordan $(2\lambda_1(a_j)+2)$-gons, using the operations $U(\Delta_{j-1}, G_j, \alpha, 2\lambda_1(a_j)+1)$, $3 \leq j \leq q-1$, where $\Delta_2 = Q_2$,

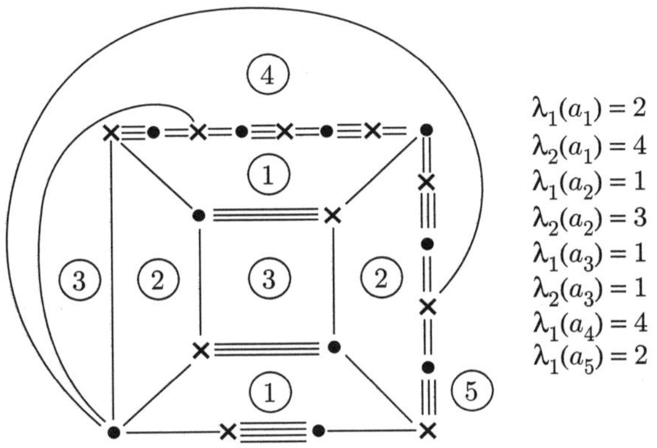

FIGURE 23

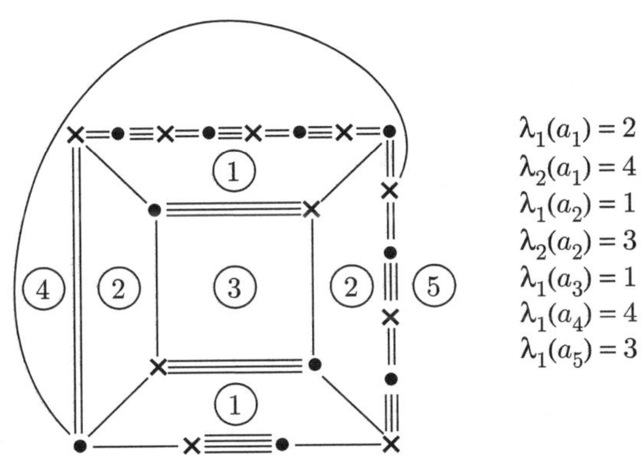

FIGURE 24

$\Delta_j = (\bar{\Delta}_{j-1} \cup \bar{G}_j) \backslash \partial(\bar{\Delta}_{j-1} \cup \bar{G}_j)$. It is easy to count that the boundary of Δ_{q-1} contains $2n - 4 - 2\sum_{j=3}^{q-1} \lambda_1(a_j)$ vertices. But (4.1) implies that

$$\sum_{j=3}^{q} \lambda_1(a_j) = 2(n-1) - \tau - \lambda_2(a_3) = 2(n-1) - n - 1 = n - 3;$$

therefore $2n - 4 = 2\sum_{j=3}^{q} \lambda_1(a_j) + 2$, and the boundary of the region Δ_{q-1} contains $2\lambda(a_q)+2$ vertices. We take the complement of Δ_{q-1} in the extended complex plane for an elementary q-region. Replacing some of the edges by bundles, we get, as before, a required line complex. If $\Lambda(a_3) = 1$, we carry out the construction in the same way, the only difference being that we do not need to take an elementary 3-region in the complement of Q_2. Both cases are illustrated in Figures 23 and 24.

Now we assume that $p = \tau - n > 0$. Then $\lambda_1(a_1) + \lambda_2(a_1) = \tau - \lambda_1(a_2) - \lambda_2(a_2) \geq n + p - (n-2) = p + 2$ and, similarly, $\lambda_1(a_2) + \lambda_2(a_2) \geq p + 2$. We subtract 1 from the largest of the numbers λ_1 and λ_2 and from the largest of the numbers λ_3 and λ_4. We get numbers, which we denote $\lambda_1^{(1)}, \lambda_2^{(1)}, \lambda_3^{(1)}, \lambda_4^{(1)}$. We repeat the same operation with the new numbers, and get $\lambda_1^{(2)}, \dots, \lambda_4^{(2)}$. Repeating the same operation p times, we get the numbers $\lambda_1^{(p)}, \ldots, \lambda_4^{(p)}$. It is easy to see that $\lambda_1^{(p)}, \ldots, \lambda_4^{(p)} \geq 1$. Let p_{13} be the number of ones which were subtracted simultaneously from λ_1 and λ_3 in the process of all p operations. The numbers $p_{14}, p_{23}, p_{24}, p_{31}, p_{32}, p_{41}, p_{42}$ are defined similarly. It is easy to see that $p_{jk} = p_{kj}$, $\lambda_1^{(p)} = \lambda_1 - p_{13} - p_{14}$, $\lambda_2^{(p)} = \lambda_2 - p_{23} - p_{24}$, $\lambda_3^{(p)} = \lambda_3 - p_{31} - p_{32}$, $\lambda_4^{(p)} = \lambda_4 - p_{41} - p_{42}$, and $p_{13} + p_{14} + p_{23} + p_{24} = p_{31} + p_{32} + p_{41} + p_{42} = p$.

Now we construct the squares Q_1 and Q_2 and trapezoids $T_1, \ldots, T_4$, and place vertices of the line complex at the vertices of these squares and trapezoids in the same way as we did for $\tau = n$. In the intervals that join the points shown in the table below, we place the number of vertices shown in the right column of the table:

$A'D'$	$2\lambda_1^{(p)} - 2$	$A'A$	$2p_{13}$
$A'B'$	$2\lambda_3^{(p)} - 2$	$B'B$	$2p_{23}$
$B'C'$	$2\lambda_2^{(p)} - 2$	$C'C$	$2p_{24}$
$C'D'$	$2\lambda_4^{(p)} - 2$	$D'D$	$2p_{14}$

The boundary of T_1 contains $(2\lambda_1^{(p)} - 2) + 2p_{13} + 2p_{14} + 4 = 2\lambda_1 + 2 = 2\lambda_1(a_1) + 2$ vertices, the boundary of T_2 contains $(2\lambda_1(a_2) + 2)$ vertices, the boundary of T_3 contains $(2\lambda_2(a_1) + 2)$ vertices, and the boundary of T_4 contains $(2\lambda_2(a_2) + 2)$ vertices. We take Q_1 for an elementary 3-region, T_1 and T_3 for elementary 1-regions, and T_2 and T_4 for elementary 2-regions. The boundary of Q_2 contains $2(\lambda_1^{(p)} - 1) + \cdots + 2(\lambda_4^{(p)} - 1) + 4 = 2(\lambda_1 - 1 - p_{13} - p_{14}) + \cdots + 2(\lambda_4 - 1 - p_{41} - p_{42}) + 4 = 2\tau - 4p - 4 = 2n - 2p - 4$ vertices, and the interior of Q_2 contains $(2p + 4)$ vertices. Now we successively join to Q_2 elementary 3-, 4-, ..., $(q-1)$-regions. If $\Lambda(a_3) > 1$, or 4-, ..., $(q-1)$-regions if $\Lambda(a_3) = 1$, the same way we did in the case $\tau = n$. Now the boundary of Δ_{q-1} contains $\sigma = 2n - 2p - 4 - 2(\Lambda(a_3) - 1) - 2\sum_{j=4}^{q-1} \lambda_1(a_j)$ vertices. Since $2n - 2 = \tau + \Lambda(a_3) + \sum_{j=4}^{q} \lambda_1(a_j)$ and $\tau = n + p$, we have

$$\tau = 2p + 2 + \Lambda(a_3) + \sum_{j=4}^{q} \lambda_1(a_j)$$

and

$$\sigma = 2\tau - 4p - 2 - 2\Lambda(a_3) - 2\sum_{j=4}^{q-1} \lambda_1(a_j) = 2\lambda_1(a_q) + 2.$$

We take for an elementary q-region the complement of $\bar{\Delta}_{q-1}$ in the extended complex plane, and then complete the proof as in the previous case. A line complex with $\tau > n$ is shown in Figure 25.

Case (b). Denote $\tau = \lambda_1(a_1) + \lambda_2(a_1) + \lambda_1(a_2)$. It is clear that $\tau \leq n - 2 + n - 1 = 2n - 3$; therefore $q \geq 3$. At first we assume that $\tau \leq n - 1$. Take a Jordan $(2\lambda(a_1) + 2)$-gon for a 1-region G_1' and mark an interior vertex α on its boundary. Then we join to G_1' a Jordan $(2\lambda_1(a_2) + 2)$-gon as a 2-region G_2, using the operation $U(G_1', G_2, \alpha, 1)$. We join to the region $\Delta_1 = (\bar{G}_1' \cup \bar{G}_2) \backslash \partial(\bar{G}_1' \cup \bar{G}_2)$ a 1-region G_1'', which is a Jordan $(2\lambda_2(a_1) + 2)$-gon, using the operation $U(\Delta_1, G_1'', \beta, 1)$, where β is an exterior vertex on the boundary of G_2, next after α as we go around the boundary in the negative direction. The boundary of the region $\Delta_2 = (\bar{\Delta}_1 \cup \bar{G}_1'') \backslash \partial(\bar{\Delta}_1 \cup \bar{G}_1'')$ contains $2\lambda_1(a_1) + 2 + 2\lambda_1(a_2) + 2\lambda_2(a_1) = 2\tau + 2 \leq 2n$ vertices. After that we complete the construction of a line complex in the same way as in the proof of Theorem 4.1. If $\tau \geq n$, then $\tau = n - 1 + p$, $p \geq 1$. It is easy to see that

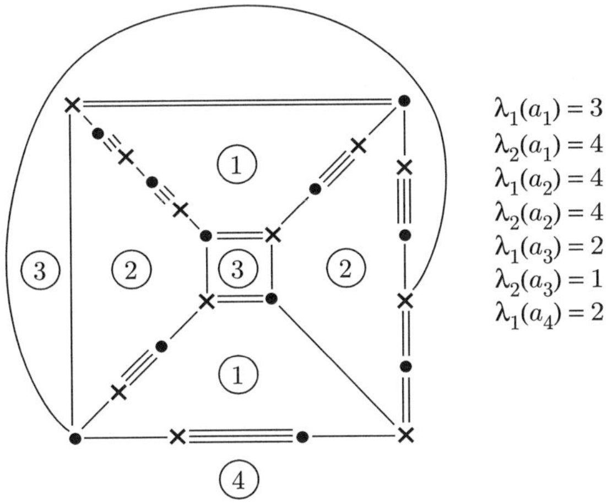

FIGURE 25

$p = \lambda_1(a_1) + \lambda_2(a_1) + \lambda_1(a_2) - (n-1) \leq \lambda_1(a_1) + \lambda_2(a_1)$. Therefore we can find numbers p_1, p_2, such that $0 \leq p_1 \leq \lambda_1(a_1)$, $0 \leq p_2 \leq \lambda_2(a_1)$, $p = p_1 + p_2$. Now we carry out the construction of a line complex the same way as in the case $\tau \leq n-1$, with the only difference that we join G_2 to G_1' using the operation $U(G_1', G_2, \alpha, 2p_1+1)$, and join G_1'' to Δ_1 using the operation $U(\Delta_1, G_1'', \beta, 2p_2+1)$. Now the closed region $\bar{\Delta}_2$ contains $2\lambda_1(a_1)+2+2\lambda_2(a_1)+2+2\lambda_1(a_2)-(2p_1+1)-(2p_2+1) = 2\tau+2-2p = 2n$ vertices, and the boundary of Δ_2 contains $2n-2p_1-2p_2 = 2(n-p)$ vertices. We again successively join 3-, ..., $(q-1)$-regions as in the proof of Theorem 4.1, and get a region Δ_{q-1} whose boundary contains

$$\begin{aligned}
2(n-p) - 2\sum_{j=3}^{q-1}\lambda_1(a_j) &= 2n + 2n - 2 - 2\tau - 2\sum_{j=3}^{q-1}\lambda_1(a_j) \\
&= 2\{2(n-1)\} + 2 - 2\tau - 2\sum_{j=3}^{q-1}\lambda_1(a_j) \\
&= 2\left(\tau + \sum_{j=3}^{q}\lambda_1(a_j)\right) + 2 - 2\tau - 2\sum_{j=3}^{q-1}\lambda_1(a_j) = 2\lambda_1(a_q) + 2
\end{aligned}$$

vertices. After that the construction of a line complex is completed in the usual way. The line complexes corresponding to Case (b) for $\tau \leq n-1$ and $\tau \geq n$ are shown in Figures 26 and 27.

CASE (c). First we construct a line complex for a Riemann surface of the class $\mathfrak{F}(n; a_1 : \lambda_1(a_2), \lambda_2(a_2); a_2 : \lambda_1(a_1); a_3 : \lambda_1(a_q); a_4 : \lambda_1(a_{q-1}); \ldots; a_q : \lambda_1(a_3))$ the same way we did in Case (b). Then we take a mirror image of this line complex and the images of 2-regions for 1-regions, the images of 1 regions for 2-regions, the images of q-regions for 3-regions, the images of $(q-1)$-regions for 4-regions, ..., and the images of 3 regions for q-regions. It is easy to verify that as a result of this transformation (we denote it by S) we get a line complex satisfying all the conditions imposed in Case (c). The line complex obtained as the result of the transformation S for the line complex in Figure 27 is shown in Figure 28.

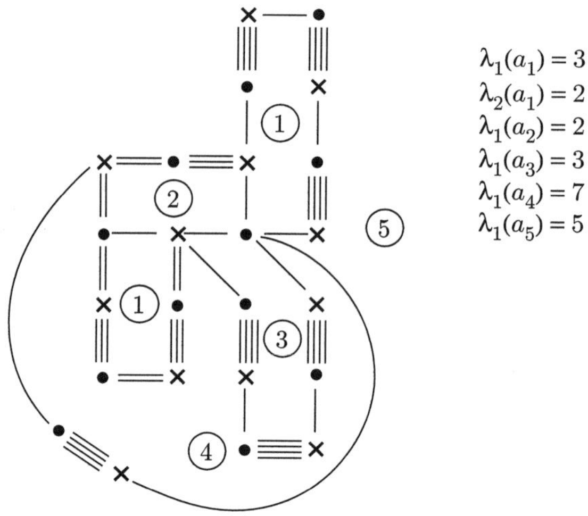

FIGURE 26

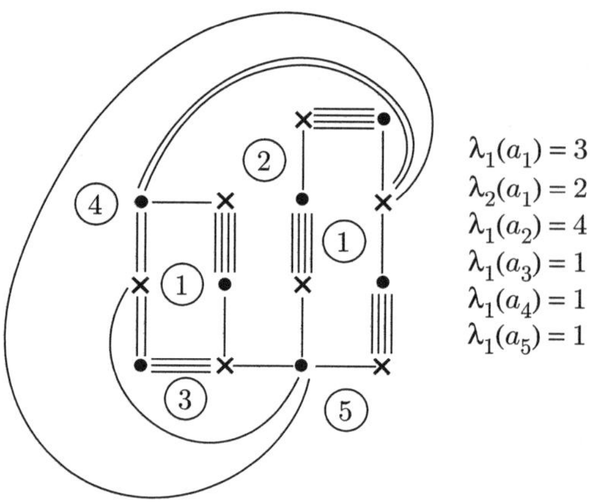

FIGURE 27

CASE (d). Consider the line complex constructed in our study of Case (b), where $\lambda_1(a_2)$ is equal to the number $\lambda_2(a_2)$ prescribed in Case (d), and $\lambda_1(a_1) \geq 1$. Let α' be the exterior vertex on the boundary of the elementary 1-region G_1', and β' the interior vertex following after α' as we go around the boundary of G_1' in the positive direction. Some elementary 2-region G_2' has the edge joining α' and β' for its common boundary with G_1'. If $\tau \leq n-1$, then clearly $G_2' \neq G_2$. If $\tau \geq n$, we observe that $\lambda_1(a_2) \leq n-2$ in our case; therefore $p = \lambda_1(a_1) + \lambda_2(a_1) + \lambda_1(a_2) - (n-1) \leq \lambda_1(a_1)+\lambda_2(a_2)-1$, and we can take $0 \leq p_1 \leq \lambda_1(a_1)-1$, $0 \leq p_2 \leq \lambda_2(a_1)$. With this choice of p_1 we get that $G_2' \neq G_2$ for $\tau \geq n$ also. Thus G_2' is a digon. The line complex satisfies all the conditions, and there is an interior vertex α' in $\partial G_1 \cap \partial G_2'$.

CASE (e). We get the solution by using the solution of Case (d) the same way we used the solution in Case (b) to solve Case (c).

CASE (f). The solution is contained in Theorem 4.1 and the Remark after it.

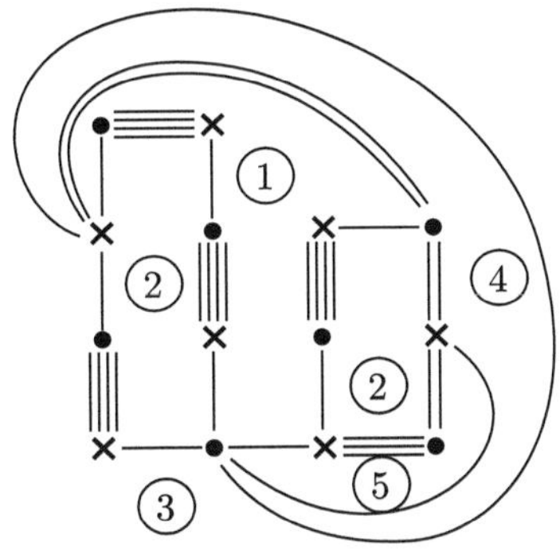

FIGURE 28

CASE (g). First we construct a line complex disregarding the existence of the base point a_2, that is, assuming that there are $q-1$ base points: $a_1, a_3, \dots, a_q$. Then the construction is reduced to Case (b). If we take into account the base point a_2, we need only to replace each edge separating 1- and 3-regions by a digon. Since in our construction in Case (b) the regions G_1' and G_2 in the complex have at least one common edge in their boundaries, in our current case, the boundary of the 1-region G_1', which is a Jordan $(2\lambda_1(a_1)+2)$-gon, and the boundary of some 2-region, which is a digon, have one common internal vertex.

CASE (h). The argument is similar to Case (g).

CASE (i). We construct a line complex the same way as in the proof of Theorem 4.1, for a Riemann surface of the class

$$\mathfrak{F}(n;\ a_2 : \lambda_2(a_2);\ a_3 : \lambda_1(a_3);\dots;\ a_q : \lambda_1(a_q);\ a_1 : \lambda_2(a_1));$$

that is, the region G_j is an elementary $(j+1)$-region, $1 \le j \le q-1$, and the region G_q is an elementary 1-region. We denote by β and α', respectively, the first and the second vertices on the boundary of G_2, counting from α in the negative direction. The vertex β is external, and the vertex α' is internal. These vertices do not lie on the boundary of G_1 since it is a Jordan $(2\lambda_2(a_2)+2)$-gon, and $2\lambda_2(a_2)+2 \le 2n-2$ since $\lambda_2(a_2) \le n-2$. The vertices β and α' lie on the common part of the boundary of a 3-region and a 4-region; therefore they also belong to the common part of the boundary of two digons, which are elementary 1- and 2-regions. Since α' is an internal vertex, the constructed line complex has all required properties.

Note that in Case (i) the requirement $q \ge 4$ is essential. For $q = 3$ the class $\mathfrak{F}(n;\ a_1 :\ 0,\ \lambda_2(a_1);\ a_2 :\ 0,\ \lambda_2(a_2);\ a_3 : \lambda_1(a_3))$ is nonempty, as is easy to show referring to Theorem 4.1. However, it is easy to verify that for $q = 3$ the digons which are elementary 1- and 2-regions have disjoint boundaries. □

REMARK. In Theorem 4.2 we assume that $\lambda_1(a_j) \ge 1$, $3 \le j \le q$ in all cases. This is done for simplicity of the proof only. It is easy to see that it would suffice to require $\lambda_1(a_j) \ge 0$, $3 \le j \le q$ only; in addition, in Case (a), for one of the points a_j, $3 \le j \le q$, we require that $\Lambda(a_{j_0}) \ge 1$ if $\lambda_1(a_j) = 0$ for $3 \le j < j_0$; and in Case (i) we require that $\lambda_1(a_j) \ge 1$ for at least two values of j, $3 \le j \le q$. In fact, in our construction of the line complex, we can first disregard those base points for which $\Lambda(a_j) = 0$, and then complete the constructed line complex with the corresponding digons.

PROOF OF THEOREM 4.3. First we note that since

$$\lambda_1(a_1) + \lambda_2(a_1) + \lambda_3(a_1) \leq n - 3 \tag{4.5}$$

and

$$\lambda_1(a_2) \leq n - 1, \tag{4.6}$$

condition (4.1) implies that the requirement $q \geq 3$ in the statement of the theorem is necessary.

We can draw the base curve L in such a way that it intersects the curve γ in two points only, one of which is a_1. We select interior and exterior vertices on the curve γ, and take the arc of γ joining these vertices and not passing through a_1 for the curve C_2. We select the other curves, $C_1, C_3, \ldots, C_q$, used in the construction of a line complex, without any additional requirements. Now it is easy to verify that it suffices to construct a line complex with the prescribed ramification and the additional restriction that on the boundary of an elementary 1-region corresponding to the point $\mathfrak{a}_1$ (point $\mathfrak{a}_2$) there is an internal (external) vertex with the property that these vertices are the endpoints of the same edge with label 2.

CASE (j). We shall distinguish two subcases: (j_1) where $2\Lambda(a_1)+2\lambda_1(a_2)+2 \leq 2n$; ($j_2$) where $2\Lambda(a_1) + 2\lambda_1(a_2) + 2 = 2n + 2p$, $p > 0$.

SUBCASE (j_1). Take a Jordan $(2\lambda_1(a_1) + 2)$-gon for an elementary 1-region. Mark an internal vertex α on its boundary and join a Jordan $(2\lambda_1(a_2) + 2)$-gon as an elementary 2-region G_2, using the operation $U(G_1,\ G_2,\ \alpha,\ 1)$. We mark on the boundary of G_2 an external vertex β, which is the first after α if we go along the boundary of G_2 in the negative direction, and an internal vertex α' separated from α by four edges. To the region $\Delta = (\bar{G}_1 \cup \bar{G}_2) \backslash \partial(\bar{G}_1 \cup \bar{G}_2)$ we join an elementary 1-region G_3 which is a Jordan $(2\lambda_2(a_1)+2)$-gon, using the operation $U(\Delta_2, G_3, \beta, 1)$. Then we join to $\Delta_3 = (\bar{\Delta}_2 \cup \bar{G}_3) \backslash \partial(\bar{\Delta}_2 \cup \bar{G}_3)$ an elementary 1-region G_4, which is a Jordan $(2\lambda_3(a_1) + 2)$-gon, using the operation $U(\Delta_3, G_4, \alpha', 1)$. It is easy to see that the boundary of the region $\Delta_4 = (\bar{\Delta}_3 \cup \bar{G}_4) \backslash \partial(\bar{\Delta}_3 \cup \bar{G}_4)$ contains

$$2(\lambda_1(a_1) + \lambda_2(a_1) + \lambda_3(a_1) + \lambda_1(a_2)) + 2 = 2\Lambda(a_1) + 2\lambda_1(a_2) + 2 < 2n$$

vertices. After that we successively join to Δ_4 elementary 3-, 4-, $\ldots$, q-regions in the same way as in the proof of Theorem 4.1, but now the role of the region G_1 is played by the region Δ_4, and the number p, $2 \leq p < q$ is defined as the largest integer satisfying $2 + 2\Lambda(a_1) + 2\sum_{j=2}^{p} \lambda_1(a_j) \leq 2n$ (cf. (4.3)). We leave the details of the construction to interested readers. Examples of the line complexes corresponding to the cases of $p = 2$ and $p = 3$ are shown in Figures 29 and 30.

SUBCASE (j_2). By (4.6) we have $p \leq \Lambda(a_1)$. Therefore we can write $p = p_1 + p_2 + p_3$, where $0 \leq p_k \leq \lambda_k(a_1)$, $k = 1, 2, 3$. Take a Jordan $(2\lambda_1(a_1) + 2)$-gon for an elementary 1-region G_1. We mark on its boundary an internal vertex α and join a Jordan $(2\lambda_1(a_2) + 2)$-gon as an elementary 2-region G_2, using the operation $U(G_1, G_2, \alpha, 2p_1 + 1)$. Inequality (4.5) implies that $2p \leq 2\lambda_1(a_2) - 4 = (2\lambda_1(a_2) + 2) - 6$. Therefore the indicated operation of joining, as well as the next two operations, is admissible. We mark on the boundary of G_2 an external vertex β which is the first after α as we go around the boundary of G_2 in the negative direction. We join to the region $\Delta_2 = (\bar{G}_1 \cup \bar{G}_2) \backslash \partial(\bar{G}_1 \cup \bar{G}_2)$ an elementary 1-region

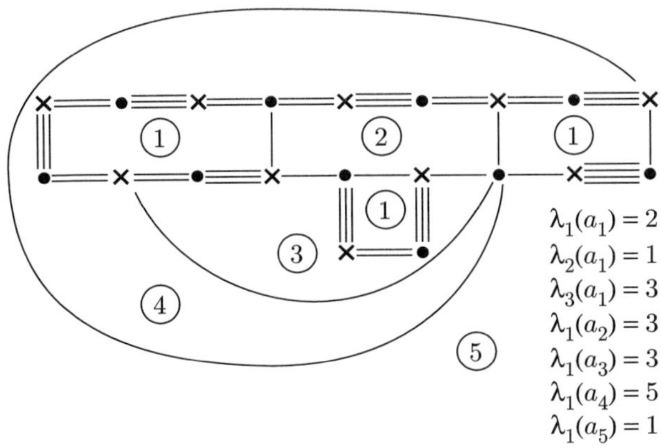

FIGURE 29

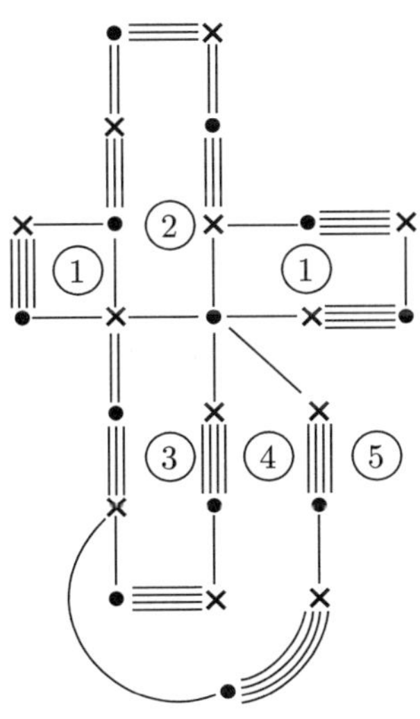

FIGURE 30

G_3, which is a Jordan $(2\lambda_2(a_1)+2)$-gon, using the operation $U(\Delta_2, G_3, \beta, 2p_2+1)$. The Jordan arc $\partial G_3 \cap \partial\Delta_2$ has two endpoints: one is the vertex β, the other is an internal vertex, which we denote by β'. We mark on the boundary of G_2 an external vertex α' which is the first external vertex after β' if we go around the boundary of G_2 in the negative direction. We join to the region $\Delta_3 = (\bar{\Delta}_2 \cup \bar{G}_3)\backslash\partial(\bar{\Delta}_2 \cup \bar{G}_3)$ an elementary 1-region G_4, which is a Jordan $(2\lambda_3(a_1)+2)$-gon, using the operation $U(\Delta_3, G_4, \alpha', 2p_3+1)$. It is easy to count that the region $\Delta_4 = (\bar{\Delta}_3 \cup \bar{G}_4)\backslash\partial(\bar{\Delta}_3 \cup \bar{G}_4)$ contains $2(p_1+p_2+p_3) = 2p$ vertices, and the boundary of Δ_4 contains $2n-2p$

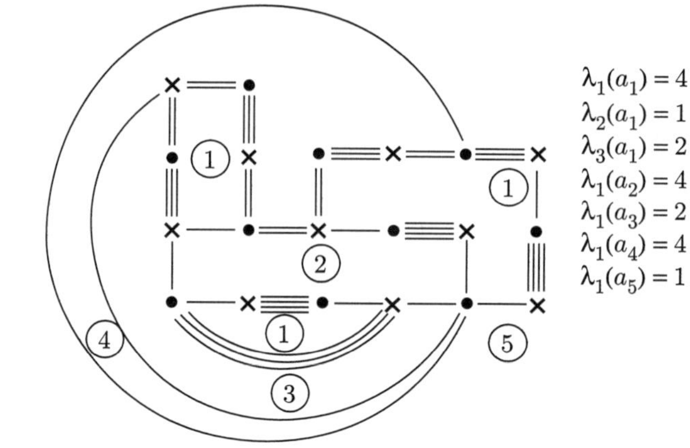

FIGURE 31

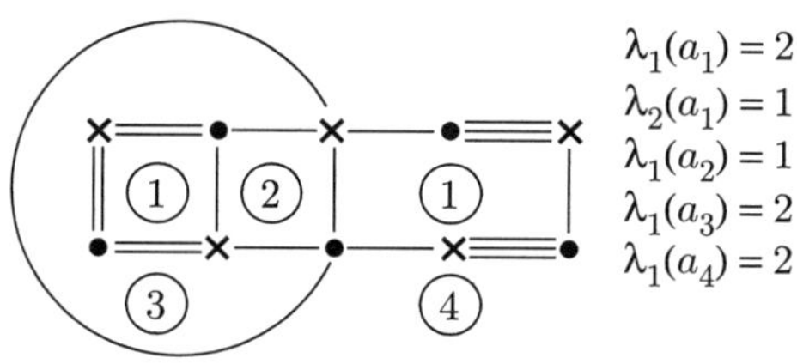

FIGURE 32

vertices. Let

$$\Delta_s = (\Delta_{s-1} \cup \bar{G}_s) \backslash \partial(\bar{\Delta}_{s-1} \cup \bar{G}_s),$$

take $5 \le s \le q+1$, and let G_s be an elementary $(s-2)$-region, which is a Jordan $(2\lambda_1(a_{s-2})+2)$-gon joined to the regon Δ_{s-1} by means of the operation $U(\Delta_{s-1},\ G_s,\ \alpha,\ 2\lambda_1(a_{s-2})+1)$. It is easy to count that the boundary of Δ_{q+1} contains

$$\begin{aligned} 2n - 2p - 2\sum_{j=3}^{q-1} \lambda_1(a_j) &= 2n - (2\Lambda(a_1) + 2\lambda_1(a_2) + 2 - 2n) - 2\sum_{j=3}^{q-1} \lambda_1(a_j) \\ &= 4(n-1) + 2 - 2\Lambda(a_1) - 2\sum_{j=2}^{q-1} \lambda_1(a_j) = 2\lambda_1(a_q) + 2 \end{aligned}$$

vertices. We take the complement of $\bar{\Delta}_{q+1}$ in the extended complex plane for an elementary q-region. We have constructed all elementary regions, and, replacing some of the edges by edge bundles, we get the desired line complex. An example of a line complex corresponding to Subcase (j_2) is shown in Figure 31.

CASE (k). The argument is completely analogous to the argument in Case (j), the only difference being that we do not need to join the third elementary 1-region to Δ_3. We leave the detailed argument in this case to interested readers. An example of a line complex corresponding to Case (k) is shown in Figure 32. Note also that if relation (4.5) is satisfied with $\lambda_1(a_2) \ge 2$, then Case (k) is reduced to Case (j) with $\lambda_3(a_1) = 0$. □

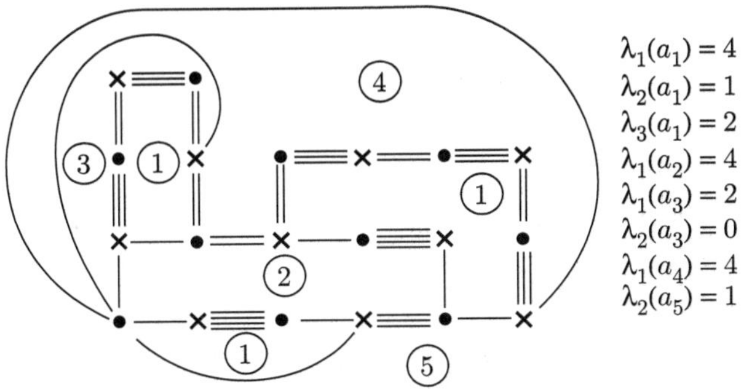

FIGURE 33

REMARK. It is clear that in the hypothesis of Theorem 4.3 we may assume that the point a_2 is outside γ, and the points $a_3, \dots, a_q$ are inside this curve.

EXERCISE. Prove the following statement. Let $q \geq 4$ distinct complex numbers $a_1, \dots, a_q$ and a J'-curve γ with endpoints at a_1 be given; moreover, let the points $a_2, \dots, a_r$, $2 < r < q$, be inside the curve γ, and the points $a_{r+1}, \dots, a_q$ outside the curve γ. Then each of the classes

(l) $\mathfrak{F}(n; a_1 : \lambda_1(a_1), \lambda_2(a_1), \lambda_3(a_1), a_2 : \lambda_1(a_2); a_3 : \lambda_1(a_3), 0; \dots;$
$a_{q-1} : \lambda_1(a_{q-1}), 0; a_q : \lambda_1(a_q)),$ where $\lambda_k(a_1) \geq 0,\ k = 1, 2, 3;\ \lambda_1(a_2) \geq 2;$

(m) $\mathfrak{F}(n; a_1 : \lambda_1(a_1), \lambda_2(a_1); a_2 : \lambda_1(a_2); a_3 : \lambda_1(a_3), 0; \dots; a_{q-1} : \lambda_1(a_{q-1});$
$a_q : \lambda_1(a_q)),$ where $\lambda_k(a_1) \geq 0,\ k = 1, 2,$

contains a Riemann surface F on which we can draw a J-curve Γ with endpoints $\mathfrak{a}_1$ and $\mathfrak{a}_2 \in F$, such that the curve Γ without endpoints is projected in a one-to-one way onto $\gamma \backslash \{a_1\}$, and the point $\mathfrak{a}_k$ is a ramification point of order $\lambda_k(a_1)$ lying over a_1, $k = 1, 2$.

Figure 33, containing a "typical" line complex corresponding to Case (l), can be a hint for the changes to be made in the proof of Theorem 4.3 (cf. Figure 31) in order to get the desired statement.

5. Statement of the inverse problem of the value distribution theory. Riemann surfaces with finitely many logarithmic ends

The direct problem of the value distribution theory is finding the distribution of deficiencies and indices of a given meromorphic function. The inverse problem consists in finding a meromorphic function having a prescribed distribution of deficiencies and indices. In the general form the inverse problem of the value distribution theory for meromorphic function (later on, we call it briefly "the inverse problem") is stated as follows.

Let $a_1, a_2, \dots,$ be a (finite or infinite) sequence of complex numbers. Suppose that for each member of this sequence, numbers $\delta(a_k)$ and $\varepsilon(a_k)$ are given, and these *numbers satisfy the conditions*

$$0 < \delta(a_k) + \varepsilon(a_k) \le 1, \qquad k = 1, 2, \ldots, \tag{5.1}$$

$$\sum_k \{\delta(a_k) + \varepsilon(a_k)\} \le 2. \tag{5.2}$$

The problem is to find a meromorphic function that has deficiencies $\delta(a_k)$ *and indices* $\varepsilon(a_k)$ *at the points* a_k, $k = 1, 2, \ldots,$ *and has no other deficient or index values in the complex plane.*

The most interesting is the study of the distribution of deficiencies. For this reason, the inverse problem is often stated in a narrow form, with no mention of indices. The "restricted" inverse problem, as we call it, is the following.

Let $\{a_k\}$ *be a (finite or infinite) sequence, and* $\delta(a_k)$ *numbers corresponding to the elements of the sequence and satisfying the conditions* $0 < \delta(a_k) \le 1$ *and*

$$\sum_k \delta(a_k) \le 2. \tag{5.3}$$

The problem is to find a meromorphic function $f(z)$ *such that* $\delta(a_k, f) = \delta(a_k)$, $k = 1, 2, \ldots,$ *and* $\delta(a, f) = 0$ *for all* $a \ne a_k$, $k = 1, 2, \ldots,$ *or to prove that such functions do not exist.*

Solutions of the inverse problem can be sought among meromorphic functions belonging to some class C, for example, to the class of entire functions (in which case we, of course, require that $a_1 = \infty$ and $\delta(\infty) = 1$) or to the class of meromorphic functions of finite order. In such a case we say that the inverse problem is solved in the class C.

The complete solution of the inverse problem is still unknown.[7] Fuchs and Hayman [**FH62**] (see also [**Hay64**, §4.1]) found a complete solution of the "restricted" inverse problem in the class of entire functions of infinite order. For this result we refer the reader to the book of Hayman. All other cases in which the solutions of the inverse problem are known today will be presented in this chapter.

We consider the inverse problem in the class of meromorphic functions of finite order without explicitly mentioning this. Further, we restrict our attention to the case when the number of points $\{a_k\}$ is finite: $a_1, a_2, \ldots, a_q$. Although, as we saw in Chapter 4, the set of deficient values of a meromorphic function of a finite positive order can be an arbitrarily prescribed countable set, the deficiencies at the points of this set cannot be prescribed arbitrarily, since they satisfy some additional relations besides inequality (5.3) (see the notes to Chapter 7). The fact that the deficiencies satisfy additional conditions, which are almost unexplored yet, makes the solution of the inverse problem for countable sets $\{a_k\}$ extremely complicated. As for the set of index values, we can easily remove the requirement of its finiteness; we introduce this requirement only to make our presentation more accessible. In fact, as we shall see, in the case where the set $\{a_k\}$ is finite, it suffices to consider rather simple class F_q of Riemann surfaces. However, in the appropriate place we provide hints, which can be used to generalize the obtained results to the case where the set of index values is infinite.

---

[7]Now it is solved. About this and other problems mentioned on this page, see the survey in Appendix B.

Thus, we will work on the inverse problem assuming that the set $\{a_k\} = \{a_1, a_2, \ldots, a_q\}$ is finite. Besides, we assume that all numbers a_j, $1 \le j \le q$, are finite. Obviously the latter assumption does not reduce the generality because we can pass to the general case using a linear-fractional transformation.

Now we prove a theorem due to R. Nevanlinna [**NevR32b**] (see also [**NevR74**, no. 257]). Although this theorem will be considerably generalized in subsequent sections, we present it here not so much because it is historically the first example of a rather general solution of the inverse problem, but because the main idea of the proof (it is different from the original proof due to R. Nevanlinna) will be used in what follows, and here it is not concealed by technicalities.

THEOREM 5.1. *Let $q \ge 2$, let $a_1, \ldots, a_q$ be complex numbers, and let $\delta(a_j)$, $0 < \delta(a_j) \le 1$, be rational numbers satisfying $\sum_{j=1}^q \delta(a_j) = 2$. Then there exists a meromorphic function $f(z)$ whose deficiency at the point a_j is equal to $\delta(a_j)$, $1 \le j \le q$.*

First we introduce a class of Riemann surfaces, the so-called surfaces with finitely many logarithmic ends, and study the value distribution of the meromorphic mapping functions.

Let $a_1, \ldots, a_q$ be a set of distinct complex numbers ($q \ge 2$); for simplicity we assume that they are finite.

Take a finite sequence of numbers $b_1, b_2, \ldots, b_p$ such that all b_j are elements of the set $\{a_1, \ldots, a_q\}$, the number a_j ($1 \le j \le q$) is repeated p_j times in the sequence $\{b_k\}$, and $b_k \ne b_{k-1}, b_{k+1}$ ($b_0 = b_p$, $b_{p+1} = b_1$). It is clear that $\sum_{j=1}^q p_j = p$ and $p_j \le p/2$ for $1 \le j \le q$. Take a simply connected finite-sheeted Riemann surface K with boundary,[8] bounded by a closed Jordan curve C that consists of p J-curves $\sigma_k = B_k B_{k+1}$, where the point B_k is projected to the point b_k. Here and later on, all indices k will be taken modulo p; that is, $k = 0$ should be replaced by $k = p$, and $k = p + 1$ should be replaced by $k = 1$. The surface K can be constructed, for example, in the following way. We draw a closed Jordan curve c through the points $a_1, a_2, \ldots, a_q$ in the w-plane, which has a continuous curvature and consists of line segments in sufficiently small neighborhoods of the points a_j. Let w_0 be a point inside c. Consider a path on the curve c which passes through points $b_1, b_2, \ldots, b_p, b_1$ moving along c in the positive direction only (of course, when we move along c from b_j to b_{j+1}, we may pass through some other points b_k). Then $\arg(w - w_0)$ will get an increment $2\pi\nu$, where ν is a positive integer. We can take for K a part of the Riemann surface of the function $(w - w_0)^{1/\nu}$ (that is, a part of the Riemann surface corresponding to the function $w = w_0 + z^\nu$), covering the interior of the curve c. Then the boundary of K will play the role of C, and, obviously, the points B_k can be placed on the curve C in the desired way. If $p_j = 1$ for all j, we can take for K the interior of the curve c.

Let Λ be a Riemann surface corresponding to the function $w = (be^z - a)/(e^z - 1)$ (that is, the Riemann surface of the (multivalued) function $z = \ln\{(w-a)/(w-b)\}$), with logarithmic ramification points over a and b, $a \ne b$. If we join these logarithmic ramification points by a J-curve Γ on Λ (this curve does not have endpoints since the surface Λ does not cover the points a and b, but as a point moves along Γ in each of the two possible directions, it tends to the corresponding logarithmic ramification

[8] Stoïlow [**Sto62**, Vol. 2, Chapter VII, §5] uses the term polyhedral regions for Riemann surfaces with boundary.

points), and make a cut along this curve, the surface Λ will be decomposed into two parts Λ_1 and Λ_2, each of them called a *logarithmic end.* The curve Γ oriented in such a way that the region Λ_j $(j = 1, 2)$ is on the left, is called the boundary of the logarithmic end Λ_j. For each of the logarithmic ends, we say that it has logarithmic ramification points over a and b.

Now we take p logarithmic ends Λ_k, $1 \leq k \leq p$, where the end Λ_k has logarithmic ramification points over b_k and b_{k+1}, and its boundary is the J-curve $\sigma_k = B_{k+1}B_k$. To the Riemann surface K with the points B_k, $1 \leq k \leq p$, excluded, we paste the logarithmic ends Λ_k along the curves σ_k for all k, $1 \leq k \leq p$. We get a Riemann surface F, which is called a *Riemann surface with p logarithmic ends.* The part K of the Riemann surface F is called a *pseudokernel* (clearly, it is not uniquely determined). If we exclude from K sufficiently small one-sheeted neighborhoods of the boundary points B_k, $1 \leq k \leq p$, we get a Riemann surface with boundary K_0 which is called the *kernel* of the Riemann surface F. The Riemann surface F has at most finitely many algebraic ramification points, all of which belong to K, and p logarithmic ramification points $\tilde{B}_1, \dots, \tilde{B}_p$, which are projected to the points $a_1, \dots, a_q$; moreover, there are p_j logarithmic ramification points over a_j.

A sufficiently small ε-neighborhood of the logarithmic ramification point $\tilde{B}_k$ contains the ε-neighborhood of the boundary point B_k on K and intersects the logarithmic ends Λ_k and Λ_{k-1}. It is clear that the Riemann surface F belongs to the class F_{q_1}, $q_1 \geq q$.

The line complex of a Riemann surface with five logarithmic ends (the sequence b_k is of the form a_1, a_2, a_1, a_3, a_2, $\nu = 3$) is shown in Figure 34. See also Figures 10 and 11.

The line complex of a surface with p logarithmic ends has p logarithmic elementary regions, which are separated from each other by parts of the complex which have the form of a sequence of vertices, where each vertex is the end of exactly two bundles, so that bundles and vertices alternate as follows: bundle, vertex, bundle, vertex, $\dots$. These parts of the line complex also are often called *logarithmic ends* because they, in some sense, correspond to logarithmic ends of the Riemann surface. More precisely, if a point moves along a logarithmic end of the line complex visiting successively all vertices contained in it, then the corresponding point on the Riemann surface moves on a certain logarithmic end Λ, from sheet to sheet, and away from the boundary of Λ.

Now we construct a homeomorphic mapping which is quasiconformal on $F \backslash K_0$, of the Riemann surface F onto the complex ζ-plane. Let us agree that to all regions on F that we consider, we join their boundary points belonging to F.

Take disjoint 2ε-neighborhoods $U(2\varepsilon, \tilde{B}_k)$ of the logarithmic ramification points $\tilde{B}_k$, where ε is so small that $U(2\varepsilon, \tilde{B}_k) \cap K$ is the circular sector with vertex at B_k that does not contain algebraic ramification points and is bounded by two radii and an arc which is projected into the circle $\{|w - b_k| = 2\varepsilon\}$. The radian measure of the arc can exceed 2π, so the sector is not necessarily one-sheeted. We make a cut in $U(\varepsilon, \tilde{B}_k) \cap K$ along a radius

$$S_k = \{|w - b_k| \leq \varepsilon, \ \arg w = \alpha_k\},$$

going out of B_k. Then $U(\varepsilon, \tilde{B}_k)$ decomposes into two parts, $U^+(\varepsilon, \tilde{B}_k)$ and $U^-(\varepsilon, \tilde{B}_k)$, where $U^+(\varepsilon, \tilde{B}_k)$ has a nonempty intersection with Λ_k and $U^-(\varepsilon, \tilde{B}_k)$ has a nonempty intersection with Λ_{k-1}. Denote by K_0 the kernel $K \backslash \bigcup_{k=1}^{p} U(\varepsilon, \tilde{B}_k)$.

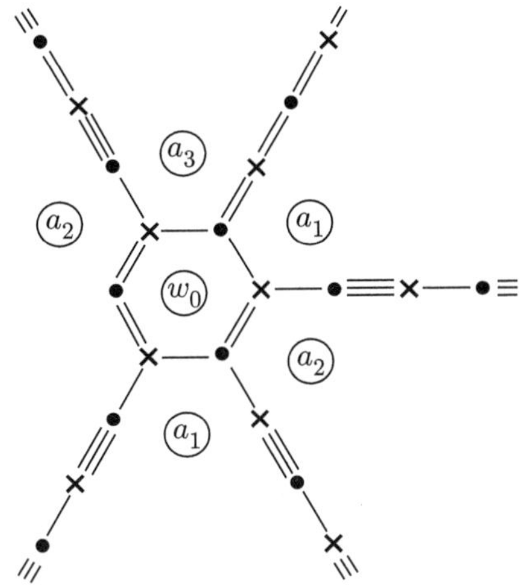

FIGURE 34

Let

$$\lambda_k = \Lambda_k \backslash \{U(\varepsilon, \tilde{B}_k) \cup U(\varepsilon, \tilde{B}_{k+1})\},$$

$$\Lambda'_k = \Lambda_k \cup U^+(\varepsilon, \tilde{B}_k) \cup U^-(\varepsilon, \tilde{B}_{k+1}) = \lambda_k \cup U^+(\varepsilon, \tilde{B}_k) \cup U^-(\varepsilon, \tilde{B}_{k+1}).$$

It is easy to see that

$$F = K_0 \cup \left(\bigcup_{k=1}^{q} \Lambda'_k \right).$$

We map Λ'_k quasiconformally onto the upper half-plane

$$H_k = \{\operatorname{Im} \zeta_k > 0\}$$

with some trapezoid removed. To construct such mapping we do the following. We map $U^-(\varepsilon, \tilde{B}_{k+1})$ conformally onto

$$H_k^- = \{\xi_k < -1,\ \eta_k > 0\}, \qquad \zeta_k = \xi_k + i\eta_k,$$

using the function

$$\zeta_k = \ln \frac{w - b_{k+1}}{\varepsilon} - 1 - i\alpha_{k+1}, \tag{5.4}$$

where the branch of the logarithm is chosen in such a way that the point $w \in S_{k+1}$ is mapped to $\zeta_k = \xi_k = \ln \frac{|w-b_{k+1}|}{\varepsilon} - 1$. Similarly, we map $U^+(\varepsilon, \tilde{B}_k)$ conformally onto

$$H_k^+ = \{\xi_k > 1,\ \eta_k > 0\}$$

using the function

$$\zeta_k = \ln \frac{\varepsilon}{w - b_k} + 1 + i\alpha_k, \tag{5.5}$$

where the branch of the logarithm is chosen in such a way that the point $w \in S_k$ is mapped to $\zeta_k = \xi_k = \ln \frac{\varepsilon}{|w-b_k|} + 1$.

We cut Λ_k along all J-curves lying over σ_k. Then the Riemann surface λ_k decomposes into one-sheeted regions $\Phi_j^{(k)}$, $j = 1, 2, \ldots$, each of which has the form of the extended complex w-plane with the discs $\{|w - b_k| < \varepsilon\}$ and $\{|w - b_{k+1}| < \varepsilon\}$ removed, and a cut σ_k^0 along those part of the curve σ_k which lies outside these discs. We denote the boundaries of this cut by σ_k^+ and σ_k^- depending on whether the positive direction on this boundary (that is, the direction for which the region is on the left) coincides on σ_k with the direction from b_k to b_{k+1} or with the direction from b_{k+1} to b_k. We map $\Phi_j^{(k)}$ quasiconformally onto the rectangle

$$Q = \{-1 < \xi_k < 1,\ 0 < \eta_k < 2\pi\}$$

in such a way that the circle $\{|w-b_{k+1}| = \varepsilon\}$ is mapped onto the segment $\{\xi_k = -1, 0 \leq \eta_k \leq 2\pi\}$, the circle $\{|w - b_k| = \varepsilon\}$ is mapped onto the segment $\{\xi_k = 1, 0 \leq \eta_k \leq 2\pi\}$, σ_k^+ is mapped onto the segment $\{\eta_k = 2\pi,\ -1 \leq \xi_k \leq 1\}$, and σ_k^- is mapped onto the segment $\{\eta_k = 0,\ -1 \leq \xi_k \leq 1\}$, and moreover, the distortion is constant on each of the sides of the rectangle. The existence of such a quasiconformal mapping with bounded characteristic follows from Example 5 of Section 2. Each point of λ_k projected to σ_k lies simultaneously on the boundary σ_k^+ of some $\Phi_j^{(k)}$ and on the boundary σ_k^- of the region $\Phi_{j+1}^{(k)}$. Under the mapping onto Q this point is mapped either to the point $(\xi, 0)$ or to the point $(\xi, 2\pi)$, depending on whether we mapped $\Phi_{j+1}^{(k)}$ or $\Phi_j^{(k)}$, but the x-coordinate in both cases will be the same, since the distortion $|dw/d\zeta_k|$ on the horizontal sides of the rectangle is constant and equal to $\frac{1}{2}\text{length}(\sigma_k^0)$. Therefore, making additional vertical shifts of the rectangles Q corresponding to $\Phi_j^{(k)}$, $j = 1, 2, \ldots$, we can paste them and get a half-strip $L_k = \{-1 < \xi_k < 1,\ \eta_k > 0\}$. This way, we get a quasiconformal mapping $\zeta_k = \omega_k(w)$ of the surface λ_k onto L_k; moreover, $\omega_k(w)$ maps $\Phi_j^{(k)}$ onto

$$Q_j^{(k)} = \{-1 < \xi_k < 1,\ 2\pi(j-1) < \eta_k < 2\pi j\}.$$

The mapping $\omega_k(w)$ has a bounded characteristic, and the distortion

$$|d\zeta_k/dw| = 1/\varepsilon \tag{5.6}$$

is constant on the part of the boundary of λ_k which is projected into $\{|w - b_k| = \varepsilon\}$ or $\{|w - b_{k+1}| = \varepsilon\}$.

The function $\zeta_k = \omega_k(w)$ maps the part of the boundary of λ_k, which is projected into $\{|w - b_{k+1}| = \varepsilon\}$, into $\{\xi_k = -1,\ \eta_k \geq 0\}$. But the same set of points considered as a part of the boundary of $U^-(\varepsilon, \tilde{B}_{k+1})$ is mapped by the function (5.4) into $\{\xi_k = -1,\ \eta_k \geq \theta_{k+1}^-\}$, where $\theta_{k+1}^- > 0$ is the angle between S_{k+1} and σ_k at the point B_{k+1}; moreover, the distortion is constant and satisfies relation (5.6). Similarly, the part of the boundary of λ_k which is projected into $\{|w - b_k| = \varepsilon\}$, on one hand, is mapped by the function $\omega_k(w)$ into $\{\xi_k = 1, \eta_k \geq 0\}$, and on the other hand, is mapped by the function (5.5) into $\{\xi_k = 1,\ \eta_k \geq \theta_k^+\}$, where $\theta_k^+ > 0$ is the angle between S_k and σ_k at the point B_k.

Denote by T_k the trapezoid with vertices at the points $(-1, 0)$, $(1, 0)$, $(1, \theta_k^+)$, $(-1, \theta_{k+1}^-)$. We map quasiconformally L_k onto $L_k \backslash T_k$ using the mapping

$$\zeta_k = \tilde{\omega}_k(\zeta_k) = \xi_k + i\left\{\eta_k + \frac{\theta_k^+ - \theta_{k+1}^-}{2}\xi_k + \frac{\theta_k^+ + \theta_{k+1}^-}{2}\right\}.$$

This quasiconformal mapping (cf. Example 2 in Section 2) is affine, has a bounded characteristic, and reduces to vertical shifts on the vertical sides of L_k. Each rectangle $Q_j^{(k)}$ is mapped onto some parallelogram by the mapping $\tilde{\omega}_k(\zeta_k)$. The function $\zeta_k = \Omega_k(w) = \tilde{\omega}_k(\omega_k(w))$ maps λ_k onto $L_k \backslash T_k$ quasiconformally and with the bounded characteristic $p(w, \zeta_k) \leq M$; moreover, now points of the part of the boundary of λ_k which are projected into $\{|w - b_{k+1}| = \varepsilon\}$ (into $\{|w - b_k| = \varepsilon\}$) are mapped to the same points on $\{\xi_k = -1,\ \eta_k \geq \theta^-_{k+1}\}$ (on $\{\xi_k = 1,\ \eta_k \geq \theta^+_k\}$) both under the mapping $\zeta_k = \Omega_k(w)$ and under the mapping (5.4) (mapping (5.5)).

We have got a quasiconformal mapping $\zeta_k = \Omega^0_k(w)$ of the surface Λ'_k onto $H_k \backslash T_k$, which is represented by the function (5.4) on $U^-(\varepsilon, \tilde{B}_{k+1})$, by the function (5.5) on $U^+(\varepsilon,\ \tilde{B}_k)$, and by the function $\Omega_k(w)$ on λ_k, where the first two functions are conformal mappings, and the last function is a quasiconformal mapping with a bounded characteristic. Therefore

$$
\begin{aligned}
&\iint_{H_k\backslash T_k} \{p(\zeta_k, w) - 1\}\, d\sigma(\ln \zeta_k) \\
&\qquad = \iint_{L_k\backslash T_k} \{p(\zeta_k, w) - 1\}\, d\sigma(\ln \zeta_k) \leq M \iint_{L_k\backslash T_k} d\sigma(\ln \zeta_k) \qquad (5.7)\\
&\qquad = M \iint_{L_k\backslash T_k} \frac{d\sigma(\zeta_k)}{|\zeta_k|^2} \leq M' < \infty.
\end{aligned}
$$

Let $w = \psi_k(\zeta_k)$ be the inverse function of $\zeta_k = \Omega^0_k(w)$, which maps $H_k\backslash T_k$ onto Λ'_k. Denote by $\nu_k(r, a)$ the number of a-points of the function $\psi_k(\zeta_k)$ in $(H_k\backslash T_k) \cap \{|\zeta_k| \leq r\}$. It is easy to see that $\psi_k(\zeta_k) \neq b_k, b_{k+1}$ and $\nu(r, b_k) \equiv 0$, $\nu(r, b_{k+1}) \equiv 0$. If $a \neq b_k, b_{k+1}$, then the a-points $\psi_k(\zeta_k)$ form a sequence $\zeta^{(0)}_k(a) + 2\pi i m$, $m = 0, 1, 2, \ldots$. Therefore $\nu_k(r, a) \sim r/(2\pi)$ as $r \to \infty$ for all $a \neq b_k, b_{k+1}$.

Now we observe that the point $w \in S_k$, as we have already mentioned, is mapped to the point $\xi_k = \ln \frac{\varepsilon}{|w - b_k|} + 1$ on the boundary of $H_k\backslash T_k$, and to the point $\xi_{k-1} = \ln \frac{|w - b_k|}{\varepsilon} - 1 = -\xi_k$ on the boundary of $H_{k-1}\backslash T_{k-1}$. Therefore, if we map $H_k\backslash T_k$ using the function

$$
\zeta = \zeta_k^{\frac{2}{p}} e^{i\frac{2\pi}{p}(k-1)}, \qquad k = 1, 2, \ldots, p, \tag{5.8}
$$

onto a region Ξ_k which is the angle $\{\frac{2\pi}{p}(k-1) < \arg \zeta < \frac{2\pi}{p} k\}$ with some bounded region deleted, then the points on the surface F that correspond to a point ζ in

$$
\partial \Xi_{k-1} \cap \partial \Xi_k = \{\arg \zeta = \frac{2\pi}{p}(k-1),\ |\zeta| \geq 1\}
$$

under the mappings $w = \psi_k((-1)^{k-1}\zeta^{p/2})$ and $w = \psi_{k-1}((-1)^k \zeta^{p/2})$ are the same. Therefore the function

$$
\zeta = \zeta(w) = \{\Omega^0_k(w)\}^{\frac{2}{p}} e^{i\frac{2\pi}{p}(k-1)}, \qquad k = 1, 2, \ldots, p, \tag{5.9}
$$

realizes a quasiconformal mapping of the part of the simply connected Riemann surface $F\backslash K_0 = \bigcup_{k=1}^p \Lambda'_k$ onto $\bigcup_{k=1}^p \bar{\Xi}_k$. We map the Riemann surface K_0 homeomorphically onto the region $\{|\zeta| < \infty\}\backslash \bigcup_{k=1}^p \bar{\Xi}_k$ so that the points $w \in \partial K_0$ be mapped in the same way as by the function $\zeta = \zeta(w)$ defined by formulas (5.9). This will give us a homeomorphic mapping $\zeta = \zeta(w)$ of the Riemann surface F onto $\{|\zeta| < \infty\}$.[9] Thus we have proved that the Riemann surface F is simply connected. Let $z = z(w)$ be a conformal mapping of F onto $\{|z| < R\}$, $R \leq \infty$. We define a homeomorphic mapping $z = z(\zeta)$ of the complex ζ-plane onto $\{|z| < R\}$ as a composition of two mappings $\{|\zeta| < \infty\} \to F \to \{|z| < R\}$; moreover on $\bigcup_{k=1}^p \bar{\Xi}_k$

[9]It is also possible to map K_0 quasiconformally; then the mapping $\zeta = \zeta(w)$ would be quasiconformal everywhere on F.

this mapping is quasiconformal with the characteristic $p(\zeta, z) = p(\zeta, w) = p(\zeta_k, w)$, where ζ and ζ_k are connected by formulas (5.8).

Taking into account (5.7), we get

$$\begin{aligned}\iint_{\bigcup_{k=1}^{p} \bar{\Xi}_k} \{p(\zeta, w) - 1\}\, d\sigma(\ln \zeta) &= \sum_{k=1}^{p} \iint_{\bar{\Xi}_k} \{p(\zeta, w) - 1\}\, d\sigma(\ln \zeta) \\ &= \sum_{k=1}^{p} \iint_{H_k \setminus T_k} \{p(\zeta_k, w) - 1\} \frac{d\sigma(\ln \zeta)}{d\sigma(\ln \zeta_k)}\, d\sigma(\ln \zeta_k) \\ &= \sum_{k=1}^{p} \frac{1}{p^2} \iint_{H_k \setminus T_k} \{p(\zeta_k, w) - 1\}\, d\sigma(\ln \zeta_k) \leq \frac{1}{p} M' < \infty.\end{aligned}$$

Thus, the mapping $z = z(\zeta)$ satisfies the hypothesis of Theorem 3.1 (Teichmüller's theorem). Hence $R = \infty$ and the surface F is of parabolic type. The meromorphic function $w = f(z)$ mapping $\{|z| < \infty\}$ onto F can be normalized in such a way that the function $\zeta = \zeta(z) = \zeta[f(z)]$, mapping the complex z-plane onto the complex ζ-plane, satisfies the condition

$$\lim_{z \to \infty} \left| \frac{\zeta(z)}{z} \right| = \lim_{\zeta \to \infty} \left| \frac{z(\zeta)}{\zeta} \right| = 1. \tag{5.10}$$

Let $w = \varphi(\zeta)$ be the inverse function to $\zeta = \zeta(w)$; then φ maps $\{|\zeta| < \infty\}$ onto the surface F. Outside $\bigcup_{k=1}^{p} \bar{\Xi}_k$, the function $\varphi(\zeta)$ has at most finitely many a-points for each a. It is not difficult to count, using (5.8), that

$$n(r, a, \varphi) = O(1) + \sum_{k=1}^{p} \nu_k(r^{p/2}, a) \sim \frac{p}{2\pi} r^{\frac{p}{2}} \quad \text{for } a \neq a_1, \dots, a_q, \tag{5.11}$$

and

$$n(r, a, \varphi) = O(1) + \sum_{k=1}^{p} \nu_k(r^{p/2}, a) \sim \frac{p - 2p_j}{2\pi} r^{\frac{p}{2}} + O(1) \quad \text{for } a = a_j,\ 1 \leq j \leq q, \tag{5.12}$$

since for $a = a_j$, exactly $2p_j$ of the functions $\nu_k(r, a)$ are identically zero. The summand $O(1)$ in (5.12) is essential only for those a_j for which $2p_j = p$, if there are any. Note that the pseudokernel K can always be chosen in such a way that it does not cover the points a_j, $1 \leq j \leq q$ (this is what happens in the construction of a pseudokernel which was given as an example). Then the summand $O(1)$ in (5.12) is absent (it appears because of the points on K_0 that are projected onto a_j), and if $2p_j = p$, the function $f(z)$ does not take the value a_j.

Equality (5.10) implies

$$n(r, a, f) \sim \begin{cases} \frac{p}{2\pi} r^{\frac{p}{2}} & \text{for } a \neq a_1, \dots, a_q, \\ \frac{p - 2p_j}{2\pi} r^{\frac{p}{2}} + O(1) & \text{for } a = a_j,\ 1 \leq j \leq q; \end{cases}$$

so

$$N(r, a, f) \sim \frac{1}{\pi} r^{\frac{p}{2}} \quad \text{for } a \neq a_1, \dots, a_q, \tag{5.13}$$

and

$$N(r, a, f) \sim \frac{1}{\pi} \left(1 - \frac{2p_j}{p}\right) r^{\frac{p}{2}} + O(\ln r) \quad \text{for } a = a_j,\ 1 \leq j \leq q. \tag{5.14}$$

Take $a \neq a_1, \ldots, a_q$ such that the Valiron deficiency $\Delta(a, f)$ is zero. Then (5.13) implies that

$$T(r, f) \sim \frac{1}{\pi} r^{\frac{p}{2}}. \tag{5.14'}$$

On one hand, relation (5.14′) implies that the function $f(z)$ has a normal type of order $\rho = p/2$; on the other hand, using (5.13) and (5.14) we find that

$$\delta(a, f) = \Delta(a, f) = \begin{cases} 0 & \text{for } a \neq a_1, \ldots, a_q, \\ \frac{2p_j}{p} & \text{for } a = a_j, \ 1 \leq j \leq q; \end{cases} \tag{5.15}$$

moreover, by the Corollary of Theorem 6.5 of Chapter 4, (5.15) remains true for each normalization of the function $f(z)$ that maps $\{|z| < \infty\}$ onto F, that is, for all $f(Az + B)$, $A \neq 0$. We have

$$\sum_{j=1}^{p} \delta(a_j, f) = \frac{2}{p} \sum_{j=1}^{p} p_j = 2. \tag{5.16}$$

If we let p be an odd number ≥ 3, we get examples of meromorphic functions with properties similar to the properties of functions in Example 8 in Section 5 of Chapter 5.

Now it is not difficult to prove Theorem 5.1. We write the rational numbers $\delta(a_j)/2$ in the form $\delta(a_j)/2 = p_j/p$, where p_j and p are positive integers. Since $\sum_{j=1}^{p} \delta(a_j) = 2$, we have $\sum_{j=1}^{q} p_j = p$. Besides, it is obvious that $p_j \leq p/2$, $1 \leq j \leq q$. It is clear that we can find a sequence $b_1, \ldots, b_p$ whose elements belong to the set $\{a_1, \ldots, a_q\}$, and the value a_j $(1 \leq j \leq q)$ occurs p_j times in the sequence $b_1, \ldots, b_p$, and $b_k \neq b_{k-1}, b_{k+1}$. Constructing, for such a sequence $b_1, \ldots, b_p$, a surface F with p logarithmic ends in the way described above, we get that formulas (5.1) hold for the mapping function $f(z)$; that is, the deficiency at the point a_j is equal to $\delta(a_j)$, $1 \leq j \leq q$, and there are no other deficient values.

If $\delta(a_j) = 1$, then the surface F can be chosen in such a way that it does not cover the point a_j; that is, $f(z) \neq a_j$ for all z.

6. Almost periodic ends

In the construction of surfaces with logarithmic ends, presented in the previous section, the main role was played by the notion of a logarithmic end, a sort of a "half" of the Riemann surface of a logarithmic function. In order to solve the inverse problem for a wider class of cases than given by Theorem 5.1, we generalize the class of Riemann surfaces with finitely many logarithmic ends, and to this end we first generalize the notion of a logarithmic end.

Consider a sequence $\{\Phi_k\}$, $k = \pm 1, \pm 2, \ldots$, of closed Riemann surfaces of genus zero, all ramification points of which are projected to the points $a_1, a_2, \ldots, a_q$ only, where the points $a_1, \ldots, a_q$ are distinct. For simplicity we assume that the complex numbers $a_1, \ldots, a_q$ are finite. Let b_1 and b_2 be complex numbers such that either $b_1 = a_1$, $b_2 = a_2$ or $b_1 = b_2 = a_1$. We assume that $\Phi_k = \Phi_{-k}$, $k \geq 1$. For each Riemann surface Φ_k we make a cut Γ_k along some J-curve with endpoints $B_1^k, B_2^k \in \Phi_k$, that is projected to a fixed curve γ with endpoints b_1 and b_2, which does not pass through the points $a_j \neq b_1$, b_2. The case $b_1 = b_2$, which corresponds to a closed Jordan curve γ, is not excluded, but, obviously, in this case the requirement of possibility of making the cut Γ_k imposes a certain restriction on the surface Φ_k; for example, Φ_k cannot be a one-sheeted surface. The Riemann surface obtained from Φ_k after making the cut Γ_k will be called a brick $\tilde{\Phi}_k$, $\tilde{\Phi}_k = \tilde{\Phi}_{-k}$.

Considering γ to be a cut, denote its boundaries by γ^+ and γ^-, so that in a sufficiently small neighborhood of b_2, the angle between γ^- and γ^+ with vertex at b_2 is equal to 2π. Transferring this notation to each of the cuts Γ_k, we get the boundaries Γ_k^+ and Γ_k^-. Now we paste Γ_k^+ with Γ_{k+1}^-, $k \neq -1, 0$, and Γ_{-1}^+ with Γ_1^-. We get an open simply connected Riemann surface Φ which has two logarithmic ramification points $\tilde{B}_1$ and $\tilde{B}_2$ which are projected to b_1 and b_2, respectively (if $b_1 = b_2$, then the correspondence between logarithmic ramification points and b_j, $j = 1, 2$, is fixed in an arbitrary way). The cut on the surface Φ along the curve $\Gamma = \Gamma_1^- = \Gamma_{-1}^+$ divides Φ into two parts, Λ_1 and Λ_2, each of which is called an *end*. The boundary of the end Λ_j is defined in the same way as for logarithmic ends.

We distinguish between the surfaces Φ with $b_1 \neq b_2$ and with $b_1 = b_2$. Following Volkovyskiĭ [**Vol50**], we call the ends arising from Φ the ends of the *first* and the *second kind*, respectively.

If all Φ_k are one-sheeted, then, as is easy to see, the surface Φ is the Riemann surface of the function $\ln \frac{w-a_2}{w-a_1}$, and the ends Λ_j are logarithmic. In the case where all bricks $\tilde{\Phi}_k$ are the same, the ends Λ_j are called *periodic*.[10] Ends satisfying somewhat less restrictive assumptions are called almost periodic. Before listing the assumption, we introduce some notation.

Let m_k be the number of sheets on the surface Φ_k.

Denote by Φ^ν, $\nu = 1, 2, \dots$, the part of the Riemann surface Φ consisting of $\bigcup_{k=1}^{\nu} \tilde{\Phi}_k$; moreover we include Γ_1^- into Φ^ν but exclude Γ_ν^+. Let $\bar{n}_\nu(a)$ be the number of points of Φ^ν lying over a, and $n_\nu(a)$ is the number of the same points, but with each ramification point of order m counted $m+1$ times. The number

$$n_\nu = \sum_{k=1}^{\nu} m_k$$

is called the number of sheets of the surface Φ^ν. Besides, we let $n_0 = n_0(a) = 0$ for all a. Obviously, $\bar{n}_\nu(a) \leq n_\nu(a) \leq n_\nu$, and $n_\nu(a) = n_\nu$ for all $a \neq b_1, b_2$. In the case where $b_1 = b_2 = a_1$, we denote by $n'_\nu(b_j)$, $j = 1, 2$, the sum over k, $1 \leq k \leq \nu$, of the numbers of points (in the case of algebraic ramification points we take into account their orders) of the Riemann surfaces $\Phi_k \backslash \{B_j^k\}$ lying over the point a_1. It is easy to see that

$$2n_\nu - n'_\nu(b_1) - n'_\nu(b_2) = n_\nu - n_\nu(a_1). \tag{6.1}$$

Later on, we write $n_\nu(b_j)$ instead of $n'_\nu(b_j)$, because this does not lead to any confusion but will allow us to combine the cases $b_1 = b_2$ and $b_1 \neq b_2$ in some formulas. By the Riemann–Hurwitz formula (4.1), the sum of all algebraic ramification points of the surface Φ_k is equal to $2m_k$ if we agree to count the orders of the points B_j^k, $j = 1, 2$, according to the number of sheets included in the cycle; that is, we add one to the order of a ramification point (or to zero, in the case of a simple point).

[10] In the case of periodic ends, the requirement that the curves $\Gamma_k \equiv \Gamma_1$ are J-curves is superfluous, since the equal bricks $\tilde{\Phi}_k \equiv \tilde{\Phi}_1$ can be pasted into a surface Φ if the cuts in $\Phi_k = \Phi_1$ are made along an arbitrary Jordan curve, the same for all Φ_k.

Then it is easy to check that

$$\begin{aligned}(6.2)\qquad &\sum_{j=1}^{2}(n_\nu - \bar n_\nu(a_j)) + \sum_{j=3}^{q}(n_\nu(a_j) - \bar n_\nu(a_j))\\ &\qquad = \sum_{j=1}^{2}(n_\nu - n_\nu(a_j)) + \sum_{j=1}^{q}(n_\nu(a_j) - \bar n_\nu(a_j)) = 2n_\nu\end{aligned}$$

for $b_1 \neq b_2$, and

$$\begin{aligned}(6.3)\qquad &n_\nu - \bar n_\nu(a_1) + \sum_{j=2}^{q}(n_\nu(a_j) - \bar n_\nu(a_j))\\ &\qquad = (n_\nu - n_\nu(a_1)) + \sum_{j=1}^{q}(n_\nu(a_j) - \bar n_\nu(a_j)) = 2n_\nu\end{aligned}$$

for $b_1 = b_2 = a_1$.

The quantities

$$\delta_g(a) = \delta_g(a, \Lambda_1) = 1 - \limsup_{\nu\to\infty} \frac{n_\nu(a)}{n_\nu} = \liminf_{\nu\to\infty} \frac{n_\nu - n_\nu(a)}{n_\nu},$$

$$\varepsilon_g(a) = \varepsilon_g(a, \Lambda_1) = \liminf_{\nu\to\infty} \frac{n_\nu(a) - \bar n_\nu(a)}{n_\nu}$$

will be called the *geometric deficiencies* and *indices*, respectively, of the end Λ_1, as well as of the end Λ_2, and of the surface Φ. It is clear that for $a \neq b_1, b_2$ we have $\delta_g(a) = 0$, and for $a \neq a_1, \dots, a_q$, we have $\varepsilon_g(a) = 0$. It is easy to see that $\varepsilon_g(a) + \delta_g(a) \leq 1$, $\varepsilon_g(a) \geq 0$, $\delta_g(a) \geq 0$. Besides, equalities (6.2) and (6.3) imply that

$$\sum_a (\delta_g(a) + \varepsilon_g(a)) \leq 2;$$

moreover, here the equality holds if and only if all geometric deficiencies and indices are obtained as exact limits; that is, for all a, $n_\nu(a) = n_\nu(1 - \delta_g(a)) + o(n_\nu)$ and $n_\nu(a) - \bar n_\nu(a) = \varepsilon_g(a) n_\nu + o(n_\nu)$.

For $b_1 = b_2 = a_1$, we will consider, in addition to $\delta_g(a_1)$, the quantity

$$\delta'_g(b_j) = \delta'_g(b_j, \Lambda_1) = 1 - \limsup_{\nu\to\infty} \frac{n'_\nu(b_j)}{n_\nu}, \qquad j = 1, 2.$$

Equality (6.1) implies that if $n'_\nu(b_j)/n_\nu = 1 - \delta'_q(b_j) + o(1)$ as $\nu \to \infty$, $j = 1, 2$, then $\delta_g(a_1) = \delta'_g(b_1) + \delta'_g(b_2)$. Later on, we write $\delta_g(b_j)$ and $\delta_g(b_j, \Lambda_1)$ instead of $\delta'_g(b_j)$ and $\delta'_g(b_j, \Lambda_1)$.

It is clear that the surface Φ belongs to the class F_q and can be represented by means of a line complex. It will be convenient for us to choose base curves in a special way. If Φ has ends of the first kind, we choose the base curve such that it includes the curve γ. If Φ has ends of the second kind, then γ is a closed Jordan curve, and we choose the base curve such that it intersects γ at two points only: at $a_1 = b_1 = b_2$ and at one more point b. In this case, for vertices we take one point on each of the arcs into which the curve γ is divided by the points a_1 and b. It is easy to see that with this choice of the base curve, the boundary between the parts of the line complex of the surface Φ corresponding to the neighboring bricks passes either through one edge ($b_1 \neq b_2$) or through two vertices joined by one edge ($b_1 = b_2$). A line complex of the surface Φ with ends of the first kind is shown in

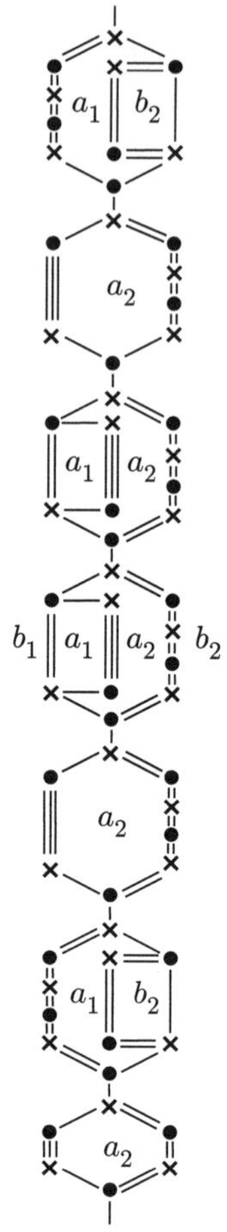

FIGURE 35

Figures 9 and 35, and a line complex of the surface with ends of the second kind is shown in Figures 13 and 36. In the case shown in Figure 9, the surface Φ has two logarithmic ends; in the case shown in Figure 13, the surface Φ has two periodic ends, which in this special case are called sine-ends.

It is easy to single out those parts of the line complex of Φ that correspond to surfaces Φ^ν and to verify that the geometric deficiencies and indices of the surface Φ can be computed directly from its line complex.

Now we assume that $\delta_g(b_1) > 0$ and $\delta_g(b_2) > 0$. We cut from Φ_k disjoint ε-neighborhoods $U(\varepsilon, B_j^k)$, $j = 1, 2$, of the points B_1^k and B_2^k, where $\varepsilon > 0$ is chosen to be so small that the 2ε-neighborhoods of the points B_1^k and B_2^k do not contain other ramification points of Φ_k, and the part of Γ_k contained in each of the neighborhoods $U(2\varepsilon, B_j^k)$ consists of one line segment. We find a quasiconformal mapping of the Riemann surface

$$\Phi_k' = \tilde{\Phi}_k \backslash (U(\varepsilon, B_1^k) \cup U(\varepsilon, B_2^k))$$

onto the rectangle $R_k = \{-1/\delta_g(b_2) \leq \xi \leq 1/\delta_g(b_1),\ 0 \leq \eta \leq 2\pi m_k\}$, $\zeta = \xi + i\eta$, such that the part of the boundary of Φ_k' lying over $\{|w - b_2| = \varepsilon\}$ and over

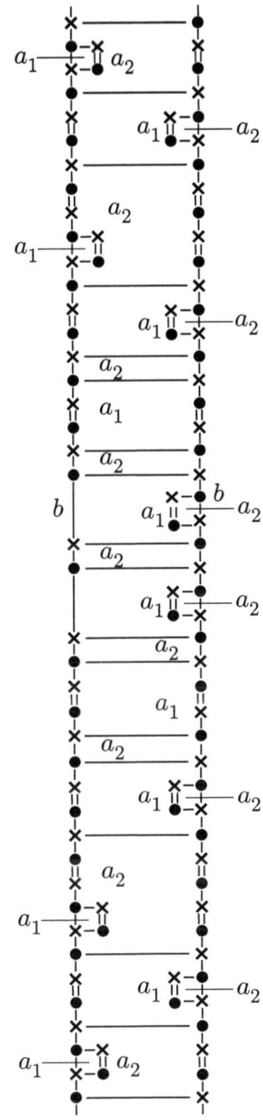

FIGURE 36

$\{|w - b_1| = \varepsilon\}$ is mapped onto the left and the right sides of the rectangle R_k, respectively, and the part of the boundary of Φ'_k lying on Γ_k^- and on Γ_k^+, is mapped onto the lower side and the upper side of R_k, respectively. Moreover the distortion of the quasiconformal mapping is constant on each side of the rectangle. The existence of the required quasiconformal mapping follows from Example 5 of Section 2. These quasiconformal mappings are not uniquely determined. Denote by p_k the infimum of the set of numbers $\sup_{w \in \Phi'_k} p(w, \zeta)$ over all quasiconformal mappings of Φ'_k onto R_k with the described correspondence between the boundaries. It is clear that $p_{-k} = p_k$. Since Φ'_k depends on ε, it follows that p_k depends on ε. We assume that ε is fixed.[11]

[11] Since, as we shall see, the sequence p_k is used in the definition of an almost periodic end, the almost periodicity of an end depends, generally speaking, not only on Λ_j, but also on the

Denote the function realizing the quasiconformal mapping of Φ'_k onto R_k with the indicated correspondence between the boundaries by

$$\zeta = \varphi_k(w);$$

we also assume that $p(w, \varphi_k(w)) < 2p_k$ for $w \in \Phi'_k$.

DEFINITION. Ends Λ_1 and Λ_2 of the Riemann surface Φ are called *almost periodic* if the following conditions are satisfied:

1°. $1 - \frac{n_\nu(b_j)}{n_\nu} + O(\frac{1}{n_\nu}) = \delta_g(b_j) > 0$, $j = 1, 2$;
2°. $\sum_{\nu=1}^{\infty} \frac{p_\nu m_\nu}{n_\nu^2} < \infty$;
3°. The following limit exists: $\lim_{\nu\to\infty} \frac{n_\nu(a) - \bar{n}_\nu(a)}{n_\nu} = \varepsilon_g(a) \geq 0$;
4°. $\lim_{\nu\to\infty} \frac{m_\nu}{n_\nu} = 0$.

If the surface Φ has periodic ends, then $m_k \equiv m_1$, $n_\nu = \nu m_1 = \nu n_1$, $n_\nu(a) = \nu n_1(a)$, $\bar{n}_\nu(a) = \nu \bar{n}_1(a)$, $n_\nu(b_j) = \nu n_1(b_j)$, $j = 1, 2$, and $p_\nu \equiv p_1$. It is clear that conditions 1°–4° hold; hence periodic ends are almost periodic. It is easy to see that all geometric deficiencies and indices of a surface Φ with periodic ends are rational numbers, and if $\varepsilon_g(a) > 0$, then $\varepsilon_g(a) + \delta_g(a) < 1$.

Conditions 1° and 3° together with (6.2) and (6.3) imply that for a surface Φ with almost periodic ends,

$$\sum_a (\delta_g(a) + \varepsilon_g(a)) = 2. \tag{6.4}$$

In what follows, we consider only surfaces Φ with almost periodic ends without explicitly mentioning this.

We construct a quasiconformal mapping of the surface Φ onto the complex ζ-plane, $\zeta = \xi + i\eta$. To this end we map the ε-neighborhoods $U(\varepsilon, \tilde{B}_1)$ and $U(\varepsilon, \tilde{B}_2)$ of the logarithmic ramification points, using the functions

$$\zeta = \frac{1}{\delta_g(b_2)} \ln \frac{w - b_2}{\varepsilon e} + i\alpha_2, \tag{6.5}$$

$$\zeta = \frac{1}{\delta_g(b_1)} \ln \frac{\varepsilon e}{w - b_1} + i\alpha_1, \tag{6.6}$$

onto the regions $G_2 = \{\xi < -1/\delta_g(b_2)\}$ and $G_1 = \{\xi > 1/\delta_g(b_1)\}$, respectively. We choose the real constants α_2 in (6.5) and α_1 in (6.6) in such a way that these functions map the line segment $\Gamma \cap U(\varepsilon, \tilde{B}_2)$ onto $\{\eta = 0, \xi < -1/\delta_g(b_2)\}$, and the line segment $\Gamma \cap U(\varepsilon, \tilde{B}_1)$ onto $\{\eta = 0, \xi > 1/\delta_g(b_1)\}$. We place the vertices A_1^k, A_2^k, A_3^k, A_4^k on the boundary Φ'_k in the order corresponding to the positive (counterclockwise) orientation of the boundary, so that $A_1^k A_2^k = \Gamma_k^-$, $A_3^k A_4^k = \Gamma_k^+$, the curve $A_2^k A_3^k$ lies over $\{|w - b_1| = \varepsilon\}$, and the curve $A_4^k A_1^k$ lies over $\{|w - b_2| = \varepsilon\}$. Under (6.5), the point A_1^k is mapped to the point

$$D_1^k = -\frac{1}{\delta_g(b_2)} + \frac{2\pi i}{\delta_g(b_2)}(n_{k-1} - n_{k-1}(b_2)) \quad \text{for } k \geq 1,$$

$$D_1^k = -\frac{1}{\delta_g(b_2)} - \frac{2\pi i}{\delta_g(b_2)}(n_{-k} - n_{-k}(b_2)) \quad \text{for } k \leq -1,$$

---

choice of ε. This flaw of the definition can be eliminated, but we are not going to do this because the present definition will not cause any inconvenience.

and the point A_4^k is mapped to the point $D_4^k = D_1^{k+1}$, $D_1^0 = D_1^1$. Under (6.6), the point A_2^k is mapped to the point

$$D_2^k = \frac{1}{\delta_g(b_1)} + \frac{2\pi i}{\delta_g(b_1)}(n_{k-1} - n_{k-1}(b_1)) \quad \text{for } k \geq 1,$$

$$D_2^k = \frac{1}{\delta_g(b_1)} - \frac{2\pi i}{\delta_g(b_1)}(n_{-k} - n_{-k}(b_1)) \quad \text{for } k \leq -1,$$

and the point A_3^k is mapped to the point $D_3^k = D_2^{k+1}$.

We consider a quasiconformal mapping of Φ_k' onto the trapezoid T_k with vertices at the points D_1^k, D_2^k, D_3^k, D_4^k such that the point A_j^k is mapped to the point D_j^k, $j = 1, 2, 3, 4$, and the distortion is constant on each side of the trapezoid. To this end, we first map Φ_k' onto the rectangle $R_k = \{-1/\delta_g(b_2) \leq \xi' \leq 1/\delta_g(b_1), 0 \leq \eta' \leq 2\pi m_k\}$ using the function $\zeta' = \varphi_k(w)$. Then we map the rectangle R_k onto the trapezoid T_k using a quasiconformal mapping $\zeta = \psi_k(\zeta')$ such that $\psi_k(-1/\delta_g(b_2)) = D_1^k$, $\psi_k(1/\delta_g(b_1)) = D_2^k$, $\psi_k(1/\delta_g(b_1) + i2\pi m_k) = D_3^k$, $\psi_k(-1/\delta_g(b_2) + i2\pi m_k) = D_4^k$, and the distortion is constant on each side of R_k. We show that the functions $\psi_k(\zeta')$ can be chosen in such a way that the characteristic $p(\zeta', \zeta)$ is bounded by a constant that does not depend on k. In fact, if we use inequality (2.12), we get the estimate ($k \geq 1$)

$$(6.7)\quad \begin{aligned} p(\zeta', \zeta) \leq{} & \frac{2\pi m_k}{\min(q_1^k, q_2^k)} + \frac{\max(q_1^k, q_2^k)}{2\pi m_k} \\ & + \frac{2\pi m_k}{h^2 \min(q_1^k, q_2^k)} \max\{\alpha_k^2, (q_2^k - q_1^k - q_1^k + \alpha_k)^2\}, \end{aligned}$$

where

$$h = \frac{1}{\delta_g(b_1)} + \frac{1}{\delta_g(b_2)},$$

$$q_j^k = 2\pi \frac{\{n_k - n_k(b_j)\} - \{n_{k-1} - n_{k-1}(b_j)\}}{\delta_g(b_j)}, \qquad j = 1, 2,$$

$$\alpha_k = 2\pi \left\{ \frac{n_{k-1} - n_{k-1}(b_1)}{\delta_g(b_1)} - \frac{n_{k-1} - n_{k-1}(b_2)}{\delta_g(b_2)} \right\}.$$

If $k \leq -1$, estimate (6.7) remains true with $\alpha_k = \alpha_{|k|}$, $q_j^k = q_j^{|k|}$, $j = 1, 2$. It is clear that it suffices to consider (6.7) for $k \geq 1$. Condition 1° implies that $n_\nu - n_\nu(b_j) = n_\nu \delta_g(b_j) + O(1)$, $j = 1, 2$. Therefore

$$\alpha_k = O(1), \qquad q_j^k = 2\pi(n_k - n_{k-1}) + O(1) = 2\pi m_k + O(1),$$

$$q_2^k - q_1^k + \alpha_k = O(1), \qquad \frac{q_j^k}{2\pi m_k} = 1 + O\left(\frac{1}{m_k}\right) = O(1).$$

Besides, since the quantities $n_k - n_k(b_j)$ form a non-decreasing sequence, $q_j^k \geq 2\pi/\delta_g(b_j)$, $j = 1, 2$, $k \geq 1$. Therefore

$$\frac{2\pi m_k}{q_j^k} = \frac{1}{1 + O(1/m_k)} = O(1).$$

These estimates imply that the right-hand side of (6.7) is uniformly bounded by some constant L. Then the characteristic of the quasiconformal mapping $\zeta = \Psi_k(w) = \psi_k(\varphi_k(w))$ of the surface Φ_k' onto the trapezoid T_k is bounded by $2Lp_k$.

Taking into account the fact that under the mapping of $U(\varepsilon, \tilde{B}_1)$, $U(\varepsilon, \tilde{B}_2)$, and Φ_k', $k = \pm 1, \pm 2, \dots$, onto G_1, G_2, and T_k, respectively, the distortion on the common part of the boundary of two of these regions is equal to the same constant, it is easy to see that we have constructed a quasiconformal mapping $\zeta = \zeta(w)$ of the whole Riemann surface Φ onto $\{|\zeta| < \infty\}$; moreover, this mapping is conformal in $U(\varepsilon, \tilde{B}_1)$ and $U(\varepsilon, \tilde{B}_2)$. Under this mapping the end Λ_1 is mapped into $\{\eta > 0\}$, the end Λ_2 is mapped into $\{\eta < 0\}$, and the curve Γ is mapped into $\{\eta = 0\}$.

We estimate the integral

$$\tag{6.8} \begin{aligned} \iint_{\zeta \bar{\in} T_1 \cup T_{-1}} p(\zeta, w)\, d\sigma(\ln \zeta) &= \sum_{|k|=2}^{\infty} \iint_{T_k} p(\zeta, w)\, d\sigma(\ln \zeta) \\ &\le \sum_{|k|=2}^{\infty} 2Lp_k \iint_{T_k} d\sigma(\ln \zeta) = 4L \sum_{k=2}^{\infty} p_k \iint_{T_k} d\sigma(\ln \zeta). \end{aligned}$$

But for sufficiently large k we have

$$\tag{6.9} \begin{aligned} \iint_{T_k} d\sigma(\ln \zeta) &= \iint_{T_k} \frac{d\sigma(\zeta)}{|\zeta|^2} \le \frac{1}{(\min_{\zeta \in T_k} |\zeta|)^2} \iint_{T_k} d\sigma(\zeta) \\ &\le \frac{h}{2\left\{\min\left(|D_1^k|, |D_2^k|\right)\right\}^2} \left\{|D_3^k - D_2^k| + |D_4^k - D_1^k|\right\} \\ &\le \frac{h(q_1^k + q_2^k)}{2\left\{\min_{\mu=1,2}\left(\frac{2\pi}{\delta_g(b_\mu)}[n_{k-1} - n_{k-1}(b_\mu)]\right)\right\}^2} \\ &= \frac{h}{2} \frac{4\pi m_k + O(1)}{\{2\pi n_{k-1} + O(1)\}^2} = \frac{h}{2} \frac{4\pi m_k + O(1)}{\{2\pi n_k - 2\pi m_k + O(1)\}^2} \\ &= \frac{h}{2} \cdot \frac{m_k}{n_k^2} \frac{4\pi + O(1/m_k)}{\left\{2\pi - 2\pi \frac{m_k}{n_k} + O(1/n_k)\right\}^2} = O\left(\frac{m_k}{n_k^2}\right). \end{aligned}$$

We used condition 4° here.

By condition 2°, inequality (6.9) implies that the series in the right-hand side of (6.8) converges. Hence

$$\tag{6.10} \begin{aligned} &\iint_{|\zeta| \ge 1} \{p(\zeta, w) - 1\}\, d\sigma(\ln \zeta) \\ &\qquad = \iint_{|\zeta| \ge 1, -1/\delta_g(b_2) \le \xi \le 1/\delta_g(b_1)} \{p(\zeta, w) - 1\}\, d\sigma(\ln \zeta) \\ &\qquad \le \iint_{|\zeta| \ge 1, -1/\delta_g(b_2) \le \xi \le 1/\delta_g(b_1)} p(\zeta, w)\, d\sigma(\ln \zeta) < \infty. \end{aligned}$$

Let $z = z(w)$ be a conformal mapping of the surface Φ onto $\{|z| < R\}$. Then the quasiconformal mapping $z = z(\zeta)$ obtained as a composition of the mappings $\{|\zeta| < \infty\} \to \Phi \to \{|z| < R\}$ with $p(\zeta, z) = p(\zeta, w)$, by (6.10), satisfies the assumptions of Theorem 3.1. Hence $R = \infty$, and with a suitable normalization of the mapping of Φ onto $\{|z| < \infty\}$ we have

$$\tag{6.11} \lim_{z \to \infty} \left|\frac{\zeta(z)}{z}\right| = \lim_{\zeta \to \infty} \left|\frac{\zeta}{z(\zeta)}\right| = 1.$$

Let us study the value distribution of the meromorphic function $w = \varphi(\zeta)$, which maps the complex z-plane onto Φ. By (6.11), it suffices to study the value distribution of the function $w = \varphi(\zeta)$ inverse to the function $\zeta = \zeta(w)$ constructed above. The function $w = \varphi(\zeta)$ establishes a quasiconformal mapping of $\{|\zeta| < \infty\}$ onto Φ. It is easy to see that for each a, the a-points of the function $\varphi(z)$ lie in some vertical strip $\{|\xi| < \xi_0\}$, $\xi_0 = \xi_0(a) > 0$. On the other hand, $\zeta = \zeta(w)$ maps the brick $\tilde{\Phi}_k$ into the region S_k consisting of the trapezoid T_k and two horizontal half-strips joined to its vertical sides. The region S_k is contained in the horizontal strip

$$\min_{\mu=1,2}\left\{\frac{2\pi}{\delta_g(b_\mu)}[n_{|k|-1} - n_{|k|-1}(b_\mu)]\right\} \\ \leq (-1)^{\operatorname{sgn} k}\eta \leq \max_{\mu=1,2}\left\{\frac{2\pi}{\delta_g(b_\mu)}[n_{|k|} - n_{|k|}(b_\mu)]\right\}.$$

Using conditions 1° and 4°, we see that the width of this strip is equal to $2\pi(n_{|k|} - n_{|k|-1}) + O(1) = 2\pi m_{|k|} + O(1) = o(n_{|k|})$. Since the set

$$\left(\bigcup_{j=1}^{k} S_j\right) \cup \left(\bigcup_{j=-1}^{-k} S_j\right), \qquad k \geq 1,$$

contains $2n_k(a)$ a-points of the function $\varphi(\zeta)$ ($2\bar{n}_k(a)$ a-points if we do not take into account the multiplicity of a-points), it is easy to show, using conditions 1°, 3°, and 4°, that

$$(6.12) \quad n(r, a, \varphi) = \begin{cases} \frac{r}{\pi} + o(r) & \text{for } a \neq b_1, b_2, \\ \frac{r}{\pi}(1 - \delta_g(a_j)) + o(r) & \text{for } a = a_j,\ j = 1, 2 \ (\text{if } b_1 \neq b_2), \\ \frac{r}{\pi}(1 - \delta_g(a_1)) + o(r) & \text{for } a = a_1 \ (\text{if } b_1 = b_2 = a_1), \end{cases}$$

$$(6.13) \qquad n_1(r, a, \varphi) = \varepsilon_g(a)\frac{r}{\pi} + o(r) \quad \text{for all} \quad a.$$

Taking into account (6.11), we see that relations (6.12) and (6.13) remain true if we replace $n(r, a, \varphi)$ by $n(r, a, f)$ and $n_1(r, a, \varphi)$ by $n_1(r, a, f)$ in them. Now we easily find that

$$(6.14) \quad N(r, a, f) = \begin{cases} \frac{r}{\pi} + o(r) & \text{for } a \neq b_1, b_2, \\ \frac{r}{\pi}(1 - \delta_g(a_j)) + o(r) & \text{for } a = a_j,\ j = 1, 2 \ (\text{if } b_1 \neq b_2), \\ \frac{r}{\pi}(1 - \delta_g(a_1)) + o(r) & \text{for } a = a_1 \ (\text{if } b_1 = b_2 = a_1), \end{cases}$$

$$(6.15) \qquad N_1(r, a, f) = \varepsilon_g(a)\frac{r}{\pi} + o(r) \quad \text{for all} \quad a.$$

Taking $a \neq b_1$, b_2 such that $\Delta(a, f) = 0$, we get

$$(6.16) \qquad T(r, f) = \frac{r}{\pi} + o(r).$$

Relation (6.16) implies that the type, and also deficiencies and indices are the same for all functions $f(Az + B)$, $A \neq 0$, that is, for all normalizations of functions that map the complex z-plane onto the surface Φ. Taking into account (6.14)–(6.16), we get the following theorem.

THEOREM 6.1. *If the meromorphic functions $w = f(z)$ maps the complex z-plane onto the Riemann surface Φ with almost periodic ends, then the function $f(z)$ has the normal type of the first order, and its deficiencies and indices coincide with the geometric deficiencies and indices of the surface Φ.*

Taking into account this theorem, we write simply $\delta(a, \Lambda_1)$ and $\varepsilon(a, \Lambda_1)$ instead of $\delta_g(a, \Lambda_1)$ and $\varepsilon_g(a, \Lambda_1)$. However, we cannot omit the subscript g in the notation $\delta_g(a)$ and $\varepsilon_g(a)$ because this can lead to a confusion.

EXERCISE. Suppose that the surface Φ has periodic ends and $\Phi_k \equiv \Phi_1$. Suppose that a rational function $w = R_1(z)$ maps the extended z-plane onto Φ_1, and, moreover, $R_1(\beta_j) = B_j^1 \in \Phi_1$, $j = 1, 2$, and $R(z) = R_1(L(z))$, where $L(z)$ is a linear-fractional transformation that maps the points 0 and ∞ to the points β_1 and β_2, respectively. Prove that the function $w = R(e^z)$ maps $\{|z| < \infty\}$ onto Φ. Conversely, each meromorphic function of the form $w = R(e^z)$, where $R(z)$ is a rational function, maps the complex z-plane onto some surface Φ with periodic ends.

We would like to make a few remarks in connection with the above study of the surfaces Φ. These remarks will be used later.

REMARK 1. When we constructed the quasiconformal mapping $\zeta = \zeta(w)$ of the Riemann surface Φ onto $\{|\zeta| < \infty\}$, the part of the Riemann surface which was mapped into the strip $\{-1/\delta_g(b_2) \le \xi \le 1/\delta_g(b_1)\}$ was $\Phi \backslash (U(\varepsilon, \tilde{B}_1) \cup U(\varepsilon, \tilde{B}_2))$, where ε is the fixed number from the definition of the quantity p_k entering condition 2°. Let us show that the quasiconformal mapping $\zeta_1 = \zeta_1(w)$ with similar properties can be constructed if we take instead of ε an arbitrary number ε_1 satisfying $0 < \varepsilon_1 < \varepsilon$. In fact, let us map $U(\varepsilon_1, \tilde{B}_2)$ and $U(\varepsilon_1, \tilde{B}_1)$ onto $\{\xi_1 < -1/\delta_g(b_2)\}$ and $\{\xi_1 > 1/\delta_g(b_1)\}$, respectively, using the functions (6.5) and (6.6), with ε_1 instead of ε. We map $\Phi \backslash (U(\varepsilon_1, \tilde{B}_1) \cup U(\varepsilon_1, \tilde{B}_2))$ onto the strip

$$\left\{-\frac{1}{\delta_g(b_2)} \ln \frac{\varepsilon e}{\varepsilon_1} \le \xi \le \frac{1}{\delta_g(b_1)} \ln \frac{\varepsilon e}{\varepsilon_1}\right\}$$

using the function $\zeta = \zeta(w)$, and then map this strip using the function

$$\xi_1 + i\eta_1 = \frac{\xi}{\ln \frac{\varepsilon e}{\varepsilon_1}} + i\eta \tag{6.17}$$

onto the strip $\{-1/\delta_g(b_2) \le \xi \le 1/\delta_g(b_1)\}$. The quasiconformal mapping (6.17) has characteristic $p(\zeta, \zeta_1) \equiv \text{const}$; therefore it is easy to verify that relations (6.10), (6.12), and (6.13) are valid also for the function $w = \varphi_1(\zeta_1)$ which is the inverse of $\zeta_1 = \zeta_1(w)$. Therefore in what follows, using these formulas, and also formulas (6.5) and (6.6), we regard ε as a sufficiently small number.

REMARK 2. The constructed quasiconformal mapping $w = \varphi(\zeta)$ maps the upper half-plane $\{\eta > 0\}$ onto the almost periodic end Λ_1. Let α and β be some positive numbers. Denote by $H_{\alpha\beta}$ the region in the ζ_1-plane, $\zeta_1 = \xi_1 + i\eta_1$, whose boundary is the polygonal line consisting of $\{\eta_1 = \alpha,\ \xi_1 \le -1/\delta_g(b_2)\}$, $\{\eta_1 = \beta,\ \xi_1 \ge 1/\delta_g(b_1)\}$, and the line segment joining the points $(-1/\delta_g(b_2)) + i\alpha$ and $(1/\delta_g(b_1)) + i\beta$, where the ray $\{\xi_1 = 0,\ \eta_1 > \max(\alpha, \beta)\}$ is contained in $H_{\alpha\beta}$.

We map $H_{\alpha\beta}$ onto $\{\eta > 0\}$ using the mapping

$$\zeta = \xi + i\eta = \begin{cases} \xi_1 + (\eta_1 - \alpha)i & \text{for } \xi_1 \le -1/\delta_g(b_2), \\ \xi_1 + (\eta_1 - \beta)i & \text{for } \xi_1 \ge 1/\delta_g(b_1), \\ \xi_1 + \left(\eta_1 - \frac{\beta-\alpha}{h}\xi_1 - \frac{\alpha/\delta_g(b_1)+\beta/\delta_g(b_2)}{h}\right) i & \text{for } \frac{-1}{\delta_g(b_2)} \le \xi_1 \le \frac{1}{\delta_g(b_1)}. \end{cases}$$

It is clear that $p(\zeta_1, \zeta) \equiv 1$ for $\xi_1 > 1/\delta_g(b_1)$ and $\xi_1 < -1/\delta_g(b_2)$, and $p(\zeta_1, \zeta) \equiv C = \text{const}$ for $-1/\delta_g(b_2) < \xi_1 < 1/\delta_g(b_1)$. Denote by $w = \varphi_{\alpha\beta}(\zeta_1) = \varphi(\zeta(\zeta_1))$ a function that gives a quasiconformal mapping of the region $H_{\alpha\beta}$ onto the almost periodic end Λ_1. It is easy to see that the function $\zeta = \zeta(w)$, inverse to $w = \varphi_{\alpha\beta}(\zeta)$, has the form (6.5) and (6.6) in $U(\varepsilon, \tilde{B}_2)$ and $U(\varepsilon, \tilde{B}_1)$, but with different constants α_1 and α_2. Next, by (6.10), we have

$$\begin{aligned} &\iint_{H_{\alpha\beta}} \{p(\zeta_1, w) - 1\}\, d\sigma(\ln \zeta_1) \\ &\quad = \iint_{H_{\alpha\beta}\cap\{-1/\delta_g(b_2)<\xi_1<1/\delta_g(b_1)\}} \{p(\zeta_1, w) - 1\}\, d\sigma(\ln \zeta_1) \\ &\quad \le C \iint_{\{-1/\delta_g(b_2)<\xi<1/\delta_g(b_1), |\zeta|>1\}} p(\zeta, w) \frac{d\sigma(\ln \zeta_1)}{d\sigma(\ln \zeta)}\, d\sigma(\ln \zeta) + C_1 \\ &\quad = C \iint_{\{-1/\delta_g(b_2)<\xi<1/\delta_g(b_1), |\zeta|>1\}} p(\zeta, w) \left|\frac{\zeta}{\zeta_1}\right|^2 d\sigma(\ln \zeta) + C_1 \\ &\quad \le C_2 \iint_{\{-1/\delta_g(b_2)<\xi<1/\delta_g(b_1), |\zeta|>1\}} p(\zeta, w)\, d\sigma(\ln \zeta) + C_1 < \infty, \end{aligned} \tag{6.18}$$

where C_1 and C_2 are some constants.

Denote by $n(r, a, \varphi_{\alpha\beta})$ or $\bar{n}(r, a, \varphi_{\alpha\beta})$ the number of a-points of the function $w - \varphi_{\alpha\beta}(\zeta)$ lying in $H_{\alpha\beta} \cap \{|\zeta| \le r\}$, depending on whether we count or not the order of the a-points. Let $n_1(r, a, \varphi_{\alpha\beta}) = n(r, a, \varphi_{\alpha\beta}) - \bar{n}(r, a, \varphi_{\alpha\beta})$. Using (6.12) and (6.13), it is easy to get

$$n(r, a, \varphi_{\alpha\beta}) = \begin{cases} \frac{r}{2\pi} + o(r) & \text{for } a \ne b_1, b_2, \\ \frac{r}{2\pi}(1 - \delta_g(a_j)) + o(r) & \text{for } a = a_j,\ j = 1, 2 \ \text{(if } b_1 \ne b_2\text{)}, \\ \frac{r}{2\pi}(1 - \delta_g(a_1)) + o(r) & \text{for } a = a_1 \ \text{(if } b_1 = b_2 = a_1\text{)}, \end{cases} \tag{6.19}$$

$$n_1(r, a, \varphi_{\alpha\beta}) = \varepsilon_g(a)\frac{r}{2\pi} + o(r) \text{ for all } a. \tag{6.20}$$

Now we prove the following theorem.

THEOREM 6.2. *Let $\varepsilon_1, \varepsilon_2, \dots, \varepsilon_q$ $(q \ge 2)$, and let δ_1, δ_2 be some numbers such that $\delta_j > 0$ for $j = 1, 2$, $0 \le \varepsilon_j \le 1$ for $j = 1, \dots, q$, $\delta_j + \varepsilon_j \le 1$ for $j = 1, 2$, and*

$$\delta_1 + \delta_2 + \sum_{j=1}^{q} \varepsilon_j = 2. \tag{6.21}$$

Let $a_1, \dots, a_q$ be distinct complex numbers, and γ a J-curve in the w-plane with ends at the points a_1 and a_2, which does not pass through the points $a_3, \dots, a_q$. Then there exists a Riemann surface Φ with almost periodic ends of the first kind, separated by a J-curve $\Gamma \subset \Phi$, such that Γ is projected into γ, $\delta_g(a_j) = \delta_j$, $j = 1, 2$; $\varepsilon_g(a_j) = \varepsilon_j$, $j = 1, \dots, q$; $\delta_g(a) = 0$ for $a \ne a_1, a_2$; and $\varepsilon_g(a) = 0$ for $a \ne a_1, \dots, a_q$.

If $\delta_1 + \delta_2 = 2$, then necessarily $\delta_1 = \delta_2 = 1$, $\varepsilon_j = 0$, $j = 1, \dots, q$. We can take for Φ the Riemann surface of the function $\ln \frac{w-a_2}{w-a_1}$, that is, a Riemann surface Φ with logarithmic ends. The logarithmic ends can be separated by an arbitrary J-curve on Φ, joining the logarithmic ramification points. In particular, the J-curve can be selected in such a way that it is projected into γ.

Now we can assume that $\delta_1 + \delta_2 < 2$. First we introduce an additional restriction.

A. If $\varepsilon_j > 0$, then $\varepsilon_j + \delta_j < 1$, $j = 1, 2$, and $\varepsilon_j < 1$, $j = 3, \dots, q$. From this it follows that $q \geq 3$, $\sum_{j=3}^{q} \varepsilon_j > 0$. We may assume without loss of generality that $\varepsilon_3 > 0$. Choose a positive integer m so large that, with the notation $D_j = m\delta_j - 1$, $j = 1, 2$, $E_j = m\varepsilon_j$, $j = 1, \dots, q$, the following conditions are satisfied:

(a) $D_j \geq 0$, $j = 1, 2$; moreover, if $\delta_j < 1$, then $D_j \leq m - 2$;
(b) If $E_j > 0$, then $D_j + E_j \leq m - 4$, $j = 1, 2$, and $E_j \leq m - 2$, $j = 3, \dots, q$;
(c) $q + 3 \leq E_3 \leq m - q - 3$.

To construct the surface Φ, we take surfaces Φ_k with the same number m of sheets, and with algebraic ramification points which are projected to the points $a_1, \dots, a_q$ only. There exists only a finite number of different surfaces Φ_k.[12] Since the curve γ for the construction of the surface Φ is fixed, Φ is composed of finitely many bricks $\tilde{\Phi}_k$ of different types. Choosing arbitrarily a sufficiently small ε, we define Φ'_k and find that $\max_k p_k = P < \infty$. It is clear that $n_\nu = m\nu$, $m_\nu = m$; therefore conditions 2° and 4° in the definition of an almost periodic end hold.

We take the surfaces Φ_k that have algebraic ramification points at the endpoints of the curve Γ_k of orders d_1^k and d_2^k, $d_j^k \geq 0$, $j = 1, 2$, over the points a_1 and a_2, respectively, and at most one algebraic ramification point of each of the orders e_1^k and e_2^k, $e_j^k \geq 0$, $j = 1, 2$. There is at most one algebraic ramification point of order $e_j^k \geq 0$, $4 \leq j \leq q$, over each point a_j, $j = 4, \dots, q$, and there are at most two algebraic ramification points of total order $e_3^k \geq 0$ over the point a_3. Theorem 4.2 implies that the only relations between the numbers e_j^k, $1 \leq j \leq q$, d_1^k, and d_2^k are the relations $d_j^k \leq m - 1$ and $d_j^k + e_j^k \leq m - 2$, if $e_j^k > 0$, $j = 1, 2$; $e_j^k \leq m - 1$, $3 \leq j \leq q$ (moreover, if there are two algebraic ramification points over a_3, we also have $e_3^k \leq m - 2$); and the Riemann–Hurwitz formula $d_1^k + d_2^k + \sum_{j=1}^{q} e_j^k = 2(m-1)$; that is, we can construct Riemann surfaces Φ_k with preassigned numbers d_j^k, e_j^k, provided they satisfy the relations listed above.

Choosing the sequence d_j^k, $j = 1, 2$, we shall distinguish two cases. In the first case $d_j^k = [kD_j] - [(k-1)D_j]$, $k = 1, 2, \dots$. Then

$$
\begin{aligned}
\frac{n_\nu - n_\nu(a_j)}{n_\nu} - \delta_j &= \frac{1}{\nu m} \sum_{k=1}^{k} (d_j^k + 1) - \delta_j = \frac{1}{\nu m} \sum_{k=1}^{\nu} d_j^k - \delta_j + \frac{1}{m} \\
&= \frac{1}{m} \left\{ \frac{1}{\nu} \sum_{k=1}^{\nu} d_j^k - D_j \right\} = \frac{1}{m} \left\{ \frac{[\nu D_j]}{\nu} - D_j \right\} = -\frac{1}{m\nu} \{\nu D_j - [\nu D_j]\}.
\end{aligned}
\tag{6.22}
$$

Hence

$$
-\frac{1}{n_\nu} = -\frac{1}{m\nu} < \frac{n_\nu - n_\nu(a_j)}{n_\nu} - \delta_j \leq 0. \tag{6.23$_1$}
$$

[12] Readers who are not acquainted with this fact, can replace this sentence by the following: "We shall use only a finite number of different surfaces Φ_k."

In the second case the choice of d_j^k differs from the first only in that we set $d_j^1 = [D_j] + 1$. The second case is applicable only if $\delta_j < 1$ and, by condition (a), $D_j \leq m - 2$. In the second choice of d_j^k we get, instead of (6.23_1), the inequality

$$(6.23_2) \qquad 0 < \frac{n_\nu - n_\nu(a_j)}{n_\nu} - \delta_j \leq \frac{1}{n_\nu}.$$

In both cases we have

$$(6.24) \qquad \left| 1 - \frac{n_\nu(a_j)}{n_\nu} - \delta_j \right| \leq \frac{1}{n_\nu}, \qquad j = 1, 2,$$

that is, $\delta_g(a_j) = \delta_j$, $j = 1, 2$, and the surface Φ satisfies condition 1° in the definition of an almost periodic end.

Now we can choose e_j^k, $1 \leq j \leq q$, $k \geq 1$, such that

$$(6.25) \qquad \lim_{\nu \to \infty} \frac{n_\nu(a_j) - \bar{n}_\nu(a_j)}{n_\nu} = \varepsilon_j, \qquad 1 \leq j \leq q.$$

For $1 \leq j \leq q$, $j \neq 3$, it suffices to take

$$e_j^k = [kE_j] - [(k-1)E_j] \leq E_j + 1.$$

By condition (b) such a choice of e_j^k is possible. Then $(j \neq 3)$

$$(6.26) \qquad \frac{n_\nu(a_j) - \bar{n}(a_j)}{n_\nu} = \frac{1}{\nu m} \sum_{k=1}^{\nu} e_j^k = \frac{[\nu E_j]}{\nu m} = \frac{[\nu m \varepsilon_j]}{\nu m} \to \varepsilon_j, \qquad \nu \to \infty.$$

Now for e_3^k we take

$$(6.27) \qquad e_3^k = 2(m-1) - d_1^k - d_2^k - \sum_{\substack{j=1 \\ j \neq 3}}^{q} e_j^k.$$

Since $m\delta_j - 2 \leq d_1^k \leq m\delta_j + 1$ for $j = 1, 2$, and $E_j - 1 \leq e_j^k \leq E_j + 1$ for $j \neq 3$, by condition (c) and (6.21) we have

$$\begin{aligned} e_3^k &\leq 2(m-1) - (m\delta_1 - 2) - (m\delta_2 - 2) - \sum_{j \neq 3} (m\varepsilon_j - 1) \\ &= 2m + q + 1 - m\left(\delta_1 + \delta_2 + \sum_{j \neq 3} \varepsilon_j \right) \\ &= 2m + q + 1 - m(2 - \varepsilon_3) = E_3 + q + 1 \leq m - 2, \end{aligned}$$

$$\begin{aligned} e_3^k &\geq 2(m-1) - (m\delta_1 + 1) - (m\delta_2 + 1) - \sum_{j \neq 3} (m\varepsilon_j + 1) \\ &= 2m - (q+3) - m(2 - \varepsilon_3) = E_3 - (q+3) \geq 0. \end{aligned}$$

Hence such a choice of e_3^k is admissible. Now, by (6.21), (6.24), (6.26), and (6.27) we get

$$\begin{aligned}\frac{n_\nu(a_3)-\bar{n}_\nu(a_3)}{n_\nu} &= \frac{1}{\nu m}\sum_{k=1}^{\nu} e_3^k \\ &= 2-\frac{1}{\nu m}\sum_{k=1}^{\nu}(d_1^k+1)-\frac{1}{\nu m}\sum_{k=1}^{\nu}(d_2^k+1)-\sum_{j\neq 3}\frac{1}{\nu m}\sum_{k=1}^{\nu} e_j^k \\ &\to 2-\delta_1-\delta_2-\sum_{j\neq 3}\varepsilon_j=\varepsilon_3 \qquad (\nu\to\infty).\end{aligned}$$

Thus, relation (6.25) has been proved for all j, $1\le j\le q$. We have proved that condition 3° in the definition of an almost periodic end is satisfied, and also that $\varepsilon_g(a_j)=\varepsilon_j$, $1\le j\le q$.

Note that we have not only proved (6.25) but also obtained the inequalities

$$\left|\frac{n_\nu(a_j)-\bar{n}_\nu(a_j)}{n_\nu}-\varepsilon_j\right|\le\frac{C_j}{n_\nu}, \tag{6.28}$$

where $C_j=1$ for $j\neq 3$ and $C_3=q+1$.

Now we eliminate the restriction A. We may assume that $q\ge 3$, letting $\varepsilon_3=0$ if necessary. We can choose numbers $\varepsilon_1^{(n)},\dots,\varepsilon_q^{(n)}\ge 0$, $n=1,2,\dots$, such that

(a) $\delta_1+\delta_2+\sum_{j=1}^q\varepsilon_j^{(n)}=2$;

(b) If $\varepsilon_j^{(n)}>0$, then $\varepsilon_j^{(n)}+\delta_j<1$, $j=1,2$, and $\varepsilon_j^{(n)}<1$, $j=3,\dots,q$;

(c) $\lim_{n\to\infty}\varepsilon_j^{(n)}=\varepsilon_j$, $1\le j\le q$.

We already know how to construct a surface $\Phi^{(n)}$ with almost periodic ends, such that $\delta_g(a_j)=\delta_j$, $j=1,2$, and $\varepsilon_g(a_j)=\varepsilon_j^{(n)}$, $1\le j\le q$. We denote the bricks of the surface $\Phi^{(n)}$ by $\tilde{\Phi}_k^{(n)}$. Now we construct a surface Φ, each of whose ends consists of the following sequence of bricks:

$$\tilde{\Phi}_1^{(1)},\dots,\tilde{\Phi}_{l_1}^{(1)},\tilde{\Phi}_1^{(2)}\dots,\tilde{\Phi}_{l_2}^{(2)},\dots,\tilde{\Phi}_1^{(n)},\dots,\dots,\tilde{\Phi}_{l_n}^{(n)},\dots;$$

the choice of the sequence $l_1,l_2,\dots$ will be specified later. We use the same notation for the surface $\Phi^{(n)}$ as we used for Φ but add the superscript n. If $\delta_j=1$ $(j=1,2)$, we take $d_j^{k,(n)}=m^{(n)}-1$ for all n. Then obviously $n_\nu(a_j)=0$, and condition 1° for the point a_j is satisfied for Φ. Now let $0<\delta_j<1$. We show that we can choose $d_j^{k,(n)}$ according to the first or second case, depending on n, in such a way that inequality (6.24) is satisfied for Φ. Let us choose $d_j^{k,(1)}$ as in the first case. Then, obviously, (6.23_1) and (6.24) are satisfied for $1\le\nu\le l_1$. We introduce the notation $\nu_\tau=l_1+l_2+\dots+l_\tau$. Suppose that (6.24) is satisfied for $1\le\nu\le\nu_\tau$. Consider the expression

$$Q_\tau=1-\frac{n_{\nu_\tau}(a_j)}{n_{\nu_\tau}}-\delta_j.$$

If $Q_\tau>0$, then we choose $d_j^{k,(\tau+1)}$ as in the second case, and if $Q_\tau\le 0$, we use the first case. Then inequality (6.24) is valid also for $\nu_\tau<\nu\le\nu_{\tau+1}$. In fact, let $Q_\tau>0$, for example, that is,

$$0<1-\frac{n_{\nu_\tau}(a_j)}{n_{\nu_\tau}}-\delta_j\le\frac{1}{n_{\nu_\tau}}. \tag{6.29}$$

Then we choose $d_j^{k,(\tau+1)}$ according to the first case and denote $N = \nu - \nu_\tau$. By (6.23$_1$), we have

$$-\frac{1}{n_N^{(\tau+1)}} < \frac{n_N^{(\tau+1)} - n_N^{(\tau+1)}(a_j)}{n_N^{(\tau+1)}} - \delta_j \leq 0. \tag{6.30}$$

By (6.29) and (6.30), respectively, we have

$$0 < n_{\nu_\tau} - n_{\nu_\tau}(a_j) - \delta_j n_{\nu_\tau} \leq 1,$$

$$-1 < n_N^{(\tau+1)} - n_N^{(\tau+1)}(a_j) - \delta_j n_N^{(\tau+1)} \leq 0.$$

Summing these inequalities and observing that

$$n_{\nu_\tau} + n_N^{(\tau+1)} = n_{\nu_\tau + N} = n_\nu, \qquad n_{\nu_\tau}(a_j) + n_N^{(\tau+1)}(a_j) = n_\nu(a_j),$$

we find that

$$-1 < n_\nu - n_\nu(a_j) - \delta_j n_\nu \leq 1,$$

and inequality (6.24) follows. The case $Q_\tau \leq 0$ is treated similarly. Thus, we may assume that (6.24) holds for all ν independently of the choice of the sequence $l_1, l_2, \dots$. Hence $\delta_g(a_j) = \delta_j$, and condition 1° holds for Φ.

Let $\nu_\tau < \nu \leq \nu_{\tau+1}$. By (6.28) we get

$$|n_{l_s}^{(s)}(a_j) - \bar{n}_{l_s}^{(s)}(a_j) - \varepsilon_j^{(s)} n_{l_s}^{(s)}| \leq C_j, \qquad 1 \leq s \leq \tau;$$

then

$$|n_{l_s}^{(s)}(a_j) - \bar{n}_{l_s}^{(s)}(a_j) - \varepsilon_j n_{l_s}^{(s)}| \leq C_j + |\varepsilon_j^{(s)} - \varepsilon_j| n_{l_s}^{(s)}, \qquad 1 \leq s \leq \tau. \tag{6.31}$$

Similarly, for $N = \nu - \nu_\tau$, we have

$$|n_N^{(\tau+1)}(a_j) - \bar{n}_N^{(\tau+1)}(a_j) - \varepsilon_j n_N^{(\tau+1)}| \leq C_j + |\varepsilon_j^{(\tau+1)} - \varepsilon_j| n_N^{(\tau+1)}. \tag{6.32}$$

Summing inequalities (6.31) and (6.32) and taking into account

$$n_\nu = \sum_{s=1}^{\tau} n_{l_s}^{(s)} + n_N^{(\tau+1)} \tag{6.33}$$

and similar inequalities for $n_\nu(a_j)$ and $\bar{n}_\nu(a_j)$, we get

$$\begin{aligned} &|n_\nu(a_j) - \bar{n}_\nu(a_j) - \varepsilon_j n_\nu| \\ &\leq (\tau+1)C_j + \sum_{s=1}^{\tau} |\varepsilon_j^{(s)} - \varepsilon_j| n_{l_s}^{(s)} + |\varepsilon_j^{(\tau+1)} - \varepsilon_j| n_N^{(\tau+1)}. \end{aligned} \tag{6.34}$$

Since $\varepsilon_j^{(n)} \to \varepsilon_j$ as $n \to \infty$, taking into account (6.33) we get ($\nu \to \infty$)

$$\frac{1}{n_\nu} \left\{ \sum_{s=1}^{\tau} |\varepsilon_j^{(s)} - \varepsilon_j| n_{l_s}^{(s)} + |\varepsilon_j^{(\tau+1)} - \varepsilon_j| n_N^{(\tau+1)} \right\} \to 0. \tag{6.35}$$

We assume that $l_n \geq n$; then $\tau = o(\nu_\tau)$, and for $\tau \to \infty$ we have

$$\frac{\tau+1}{n_\nu} \leq \frac{\tau+1}{n_{\nu_\tau}} \leq \frac{\tau+1}{\nu_\tau} \to 0. \tag{6.36}$$

Relations (6.34)–(6.36) imply that, for Φ, the relation $\varepsilon_g(a_j) = \varepsilon_j$, $1 \leq j \leq q$, holds, and condition 3° is satisfied.

It is easy to see that[13] ($\nu_0 = 0$)

$$\sum_{\nu=1}^{\infty} \frac{p_\nu m_\nu}{n_\nu^2} \leq \sum_{\tau=1}^{\infty} \sum_{\nu=\nu_{\tau-1}+1}^{\nu_\tau} \frac{P^{(\tau)} m^{(\tau)}}{n_\nu^2}$$

$$\leq \sum_{\tau=1}^{\infty} P^{(\tau)} m^{(\tau)} \sum_{\nu=\nu_{\tau-1}+1}^{\nu_\tau} \frac{1}{\nu^2} < \sum_{\tau=1}^{\infty} P^{(\tau)} m^{(\tau)} \sum_{\nu=\nu_{\tau-1}+1}^{\infty} \frac{1}{\nu^2} \tag{6.37}$$

$$< \sum_{\tau=2}^{\infty} \frac{P^{(\tau)} m^{(\tau)}}{\nu_{\tau-1}} + P^{(1)} m^{(1)} \sum_{\nu=1}^{\infty} \frac{1}{\nu^2}.$$

It is clear that for arbitrary given sequences $\{P^{(\tau)}\}$, $\{m^{(\tau)}\}$ we can, by means of a suitable choice of $l_1, l_2, \ldots$, make the sequence $\nu_{\tau-1}$ so rapidly increasing that

$$\sum_{\tau=2}^{\infty} \frac{P^{(\tau)} m^{(\tau)}}{\nu_{\tau-1}} < \infty. \tag{6.38}$$

Inequality (6.37) then implies that condition 2° holds for Φ. Condition (6.38) implies that $m^{(\tau)} = o(\nu_{\tau-1})$ as $\tau \to \infty$, that is, condition 4° is satisfied.

Theorem 6.2 has been proved.

THEOREM 6.3. *Let $\varepsilon_1, \varepsilon_2, \ldots, \varepsilon_q$ ($q \geq 2$), δ_1, δ_2 be some numbers such that $\delta_j > 0$, $j = 1, 2$, $1 \geq \varepsilon_j \geq 0$, $j = 1, \ldots, q$, $\delta_1 + \delta_2 + \varepsilon_1 \leq 1$, and*

$$\delta_1 + \delta_2 + \sum_{j=1}^{q} \varepsilon_j = 2.$$

Let $a_1, \ldots, a_q$ be distinct complex numbers and let γ be a J'-curve with ends at the points a_1 and with all points $a_2, \ldots, a_q$ on the same side of the curve γ with the possible exception of one point, which is lying on the other side of γ. Then there exists a Riemann surface Φ with almost periodic ends of the second kind, which are separated by a J-curve $\Gamma \subset \Phi$, such that Γ is projected into γ, $\delta_g(a_1) = \delta_1 + \delta_2$, $\varepsilon_g(a_j) = \varepsilon_j$, $j = 1, \ldots, q$, $\delta_g(a) = 0$ if $a \neq a_1$, $\varepsilon_g(a) = 0$ if $a \neq a_1, \ldots, a_q$, and $\delta_g(b_j) = \delta_j$, $j = 1, 2$.

We do not give a proof of Theorem 6.3 because it is very close to the proof of Theorem 6.2. First we construct a Riemann surface Φ under the following additional restrictions.

B. All points $a_3, \ldots, a_q$ lie on the same side of the curve γ, and the point a_2 is on the other side; $1 > \varepsilon(a_2) > 0$; if $\varepsilon_1 > 0$, then $\varepsilon_1 + \delta_1 + \delta_2 < 1$; $\varepsilon_j < 1$ for $3 \leq j \leq q$.

The construction is carried out in the same way as in the corresponding part of the proof of Theorem 6.2, but the reference to Theorem 4.2 is replaced by the reference to Theorem 4.3. Then we turn to the general case, as before, but if the given points $a_2, \ldots, a_q$ lie on the same side of the curve γ, we add a point a_{q+1}, which is separated from $a_2, \ldots, a_q$ by the curve γ in the construction of the auxiliary surfaces $\Phi^{(n)}$, and let $\varepsilon_{q+1}^{(n)} > 0$, $\varepsilon_{q+1}^{(n)} \to 0$ as $n \to \infty$.

REMARK. In the hypothesis of Theorem 6.3 we can require only that the curve γ be a J'-curve with endpoints at a_1, and $a_j \notin \gamma$, $2 \leq j \leq q$. To prove this

---

[13]Recall that $P^{(\tau)} = \max_k p_k^{(\tau)}$.

modification, we replace the reference to Theorem 4.3 in the proof by a reference to the Exercise at the end of Section 4.

EXERCISE. Suppose that we have added condition A to the assumptions of Theorem 6.2 or condition B to the assumptions of Theorem 6.3, and, besides, the numbers $\delta_1, \delta_2, \varepsilon_1, \dots, \varepsilon_q$ are rational. Show that in these theorems we can choose Φ to be Riemann surfaces with periodic ends.

Theorems 6.1, 6.2, and 6.3 immediately imply the following.

THEOREM 6.4. *Let $q \geq 2$ complex numbers $a_1, \dots, a_q$ be given, and let $\delta(a_j)$ and $\varepsilon(a_j)$, $1 \leq j \leq q$, be non-negative numbers such that $0 \leq \delta(a_j) + \varepsilon(a_j) \leq 1$ and*

$$\sum_{j=1}^{q} \{\delta(a_j) + \varepsilon(a_j)\} = 2,$$

and, moreover, only one or two of the numbers $\delta(a_j)$ are positive. Then there exists a meromorphic function $f(z)$ such that its deficiencies and indices at the point a_j are equal to $\delta(a_j)$ and $\varepsilon(a_j)$, respectively, $1 \leq j \leq q$.

REMARK. We can drop the requirement that the number of index values in Theorem 6.4 is finite, extending somewhat the notion of an almost periodic end. Without giving the details, we provide hints which will allow the reader to reconstruct the corresponding argument. We consider the case where $\delta(a_j) = \delta_j > 0$, $j = 1, 2$. Now we have sequences $\{a_1, a_2, \dots\}$ and $\{\varepsilon(a_1), \varepsilon(a_2), \dots\}$. If the points a_1 and a_2 are not accumulation points of the sequence $\{a_3, a_4, \dots\}$, there are no new substantial difficulties. We carry out the construction of the surface Φ in the same way as in the proof of Theorem 6.2, but the number $q^{(n)}$ of the base points for the surface $\Phi^{(n)}$ will depend on n, and $q^{(n)} \to \infty$ as $n \to \infty$, while before we had $q^{(n)} \equiv q$. In the general case we need to make more substantial changes in the construction of the surface Φ.

As before, we choose a J-curve γ with endpoints at a_1, a_2 such that it does not pass through the points a_j, $3 \leq j < \infty$. Now we cannot choose neighborhoods $U(\varepsilon, B_j^k)$, $j = 1, 2$, with fixed ε that does not depend on k, and we take $U(\varepsilon_k, B_j^k)$, where $\varepsilon_k = \varepsilon_{-k}$, $0 < \varepsilon_j \leq \varepsilon_k < 1$ for $|k| < |j|$. Now $\Phi'_k = \tilde{\Phi}_k \backslash (U(\varepsilon_k, B_j^k) \cup U(\varepsilon_k, B_2^k))$, and the role of $U(\varepsilon, \tilde{B}_j)$ is played by

$$U(\{\varepsilon_k\},\ \tilde{B}_j) = \bigcup_{|k|=1}^{\infty} \{U(\varepsilon_k,\ B_j^k) \backslash \Gamma_k\}.$$

We let $\varepsilon = \varepsilon_1$ in formulas (6.5) and (6.6); then the functions (6.5) and (6.6) are conformal mappings of $U(\{\varepsilon_k\}, \tilde{B}_2)$ and $U(\{\varepsilon_k\}, \tilde{B}_1)$ onto the regions

$$G_2 = \bigcup_{\nu=1}^{\infty} \left\{ \xi < \frac{1}{\delta_g(b_2)} \ln \frac{\varepsilon_\nu}{\varepsilon e},\ 2\pi \frac{n_{\nu-1} - n_{\nu-1}(b_2)}{\delta_g(b_2)} \leq |\eta| \leq 2\pi \frac{n_\nu - n_\nu(b_2)}{\delta_g(b_2)} \right\}$$

and

$$G_1 = \bigcup_{\nu=1}^{\infty} \left\{ \xi > \frac{1}{\delta_g(b_1)} \ln \frac{\varepsilon e}{\varepsilon_\nu},\ 2\pi \frac{n_{\nu-1} - n_{\nu-1}(b_1)}{\delta_g(b_1)} \leq |\eta| \leq 2\pi \frac{n_\nu - n_\nu(b_1)}{\delta_g(b_1)} \right\},$$

respectively.

The definition of the number p_k also has to be changed. Now we shall find a quasiconformal mapping of Φ'_k onto the rectangle

$$R_k = \left\{\frac{1}{\delta_g(b_2)}\ln\frac{\varepsilon_k}{\varepsilon e} \le \xi \le \frac{1}{\delta_g(b_1)}\ln\frac{\varepsilon e}{\varepsilon_k},\ 0 \le \eta \le 2\pi m_k\right\}$$

such that: (1) The part of the boundary of Φ'_k lying over the circle $\{|w-b_2| = \varepsilon_k\}$ (over $\{|w-b_1| = \varepsilon_k\}$) is mapped with constant distortion onto the left (right) side of the rectangle; (2) The part of the boundary of Φ'_k lying on Γ_k^- (on Γ_k^+) is mapped onto the lower (upper) side of R_k; moreover, the points on $\Gamma_k^\pm$ satisfying $\varepsilon_k \le |w-b_2| \le \varepsilon$ (satisfying $\varepsilon_k \le |w-b_1| \le \varepsilon$) are mapped to the points with the real part equal to

$$\xi = \frac{1}{\delta_g(b_2)}\ln\frac{|w-b_2|}{\varepsilon e}$$

(with the real part equal to $\xi = \frac{1}{\delta_g(b_1)}\ln\frac{\varepsilon e}{|w-b_1|}$), and the points of the remaining part of $\Gamma_k^\pm$ are mapped with constant distortion onto intervals with $-1/\delta_g(b_2) \le \xi \le 1/\delta_g(b_1)$. After that, the number p_k is defined as before.

We replace condition 2° in the definition of an almost periodic end by the following condition.

2° bis. $$\sum_{\nu=1}^{\infty}\frac{p_\nu m_\nu \ln\frac{1}{\varepsilon_\nu}}{n_\nu^2} < \infty.$$

Now, modifying somewhat the previous argument, we find a quasiconformal mapping of Φ onto $\{|\zeta| < \infty\}$ such that $\bigcup_{|k|=1}^{\infty}\Phi'_k$ is mapped into $\{|\zeta| < \infty\}\setminus\{G_1 \cup G_2\}$ by some function $\zeta = \zeta(w)$, and $U(\{\varepsilon_k\}, \tilde{B}_2)$ and $U(\{\varepsilon_k\}, \tilde{B}_1)$ are mapped by the functions (6.5) and (6.6).

Observe that now the mapping $\zeta = \psi_k(\zeta')$ is such that we still have $\xi = \xi'$, but only the set $R_k \cap \{-1/\delta_g(b_2) \le \xi' \le 1/\delta_g(b_1)\}$ is mapped into the trapezoid T_k, and the sets $R_k \cap \{\xi' \le -1/\delta_g(b_2)\}$ and $R_k \cap \{\xi' \ge 1/\delta_g(b_1)\}$ are mapped by affine transformations into rectangles joined to T_k from the left and from the right. Using condition 2° bis, we can show that

$$\iint_{|\zeta|\ge 1}\{p(\zeta,w)-1\}\,d\sigma(\ln\zeta) \le \iint_{\substack{\zeta\notin G_1\cup G_2\\ |\zeta|\ge 1}} p(\zeta,w)\,d\sigma(\ln\zeta) < \infty.$$

Let $w = \varphi(\zeta)$ be a function realizing the inverse mapping of $\{|\zeta| < \infty\}$ onto Φ. If we add the condition

5°. $\lim_{\nu\to\infty}\frac{\ln\varepsilon_\nu}{n_\nu} = 0$,

the arguments of a-points of $\varphi(\zeta)$ lying in $\{(-1)^j\eta > 0\}$, $j = 0, 1$, tend to $(-1)^j\frac{\pi}{2}$, and formulas (6.12) and (6.13) remain true. Now it is not difficult to prove analogs of Theorems 6.1, 6.2, and 6.3 for this wider class of generalized almost periodic ends. In the case of an analog of Theorem 6.3, we impose on the curve γ only the restrictions that $a_j \notin \gamma$, $j = 2, 3, \ldots$, and a_1 is an end of γ, and use the Exercise at the end of Section 4 (cf. Remark to Theorem 6.3). The fact is that the points $a_2, a_3, \ldots$ could be located in such a way (e.g., everywhere dense in the plane) that no matter how we draw a J'-curve γ with endpoints at a_1, there are infinitely many points of the sequence $a_2, a_3, \ldots$ on each side of γ.

7. Riemann surfaces with finitely many almost periodic ends

After the introduction of the notion of an almost periodic end, we can introduce the class of Riemann surfaces with finitely many almost periodic ends in the same way as we did for surfaces with finitely many logarithmic ends.

Let K be a simply connected finite-sheeted Riemann surface with boundary, bounded by a closed Jordan curve C that consists of p J-curves $\sigma_k = \widehat{B_k B_{k+1}}$, $k = 1, \dots, p$. As in Section 5, all indices k will be considered $\mod p$. Let b_k be the projection of the point B_k. For simplicity we assume that ∞ is not among the points b_k. Take p almost periodic ends Λ_k, $1 \leq k \leq p$, with the boundary of Λ_k being a J-curve $\sigma_k^- = \widehat{B_{k+1} B_k}$. Obviously, if $b_k \neq b_{k+1}$, then the end Λ_k is of the first kind, and if $b_k = b_{k+1}$, then the end Λ_k is of the second kind. We introduce the following notation:

$$\delta_g(b_k, \Lambda_k) = d_k^+, \qquad \delta_g(b_{k+1}, \Lambda_k) = d_k^-.$$

To the Riemann surface K with the points B_k, $1 \leq k \leq p$, excluded, we paste almost periodic ends Λ_k along the curves σ_k, $1 \leq k \leq p$. Then we get a Riemann surface F which is called a *Riemann surface with p almost periodic ends.* The part K of the Riemann surface F is called its *pseudokernel.* If we exclude from K sufficiently small neighborhoods of the boundary points B_k, $1 \leq k \leq p$, we get a Riemann surface with boundary K_0, which is called the *kernel* of the Riemann surface F. The Riemann surface F has logarithmic ramification points $\tilde{B}_1, \dots, \tilde{B}_p$, which are projected to $b_1, \dots, b_p$. A sufficiently small ε-neighborhood $U(\varepsilon, \tilde{B}_k)$ of a logarithmic ramification point $\tilde{B}_k$ contains the ε-neighborhood of the point B_k on K and intersects the almost periodic ends Λ_k and Λ_{k-1}. From now on, we assume that $\varepsilon > 0$ does not depend on k and does not exceed any ε in the definition of almost periodic ends Λ_k.

Obviously, the Riemann surface F belongs to some class F_q. Note that it is possible for a Riemann surface to have one almost periodic end ($p = 1$), which is, obviously, an end of the second kind. For example, the function $w = \cos\sqrt{z}$ (see Figure 15) maps the complex z-plane onto a Riemann surface F with one periodic end (sine-end).

Now we construct a homeomorphic mapping of the Riemann surface F onto the complex ζ-plane, which is quasiconformal outside a certain compact region on F.

As in Section 5, we assume that all regions on F which we consider contain all of their boundary points belonging to F. We make a cut in $U(\varepsilon, \tilde{B}_k) \cap K$ along a radius of $S_k = \{0 < |w - b_k| \leq \varepsilon, \ \arg w = \alpha_k'\}$ starting at B_k. This cut splits $U(\varepsilon, \tilde{B}_k)$ into two parts: $U^+(\varepsilon, \tilde{B}_k)$ and $U^-(\varepsilon, \tilde{B}_k)$, where $U^+(\varepsilon, \tilde{B}_k)$ has a non-empty intersection with Λ_k, and $U^-(\varepsilon, \tilde{B}_k)$ has a non-empty intersection with Λ_{k-1}. We denote the kernel $K \backslash \bigcup_{k=1}^p U(\varepsilon, \tilde{B}_k)$ by K_0. The regions $U_1^+(\varepsilon, \tilde{B}_k) = U^+(\varepsilon, \tilde{B}_k) \backslash \Lambda_k$ and $U_1^-(\varepsilon, \tilde{B}_k) = U^-(\varepsilon, \tilde{B}_k) \backslash \Lambda_{k-1}$ are circular sectors with the vertex at B_k and the angles $\tilde{\theta}_k^+$ and $\tilde{\theta}_k^-$, respectively, $\tilde{\theta}_k^+ + \tilde{\theta}_k^- \leq 2\pi$. Denote by Λ_k' the following part of the Riemann surface F:

$$\Lambda_k' = \Lambda_k \cup U^+(\varepsilon, \tilde{B}_k) \cup U^-(\varepsilon, \tilde{B}_{k+1}) = \Lambda_k \cup U_1^+(\varepsilon, \tilde{B}_k) \cup U_1^-(\varepsilon, \tilde{B}_{k+1}).$$

It is clear that

$$F = K_0 \cup \left(\bigcup_{k=1}^{p} \Lambda_k' \right).$$

We find a quasiconformal mapping of Λ'_k onto the region $H_k = \{\eta_k > 0\}\backslash T_k$, $\zeta_k = \xi_k + i\eta_k$, where T_k is a trapezoid with vertices at the points $-1/d_k^-$, $1/d_k^+$, $1/d_k^+ + i\theta_k^+$, $-1/d_k^- + i\theta_{k+1}^-$, where $\theta_k^+ = \tilde{\theta}_k^+/d_k^+$, $\theta_{k+1}^- = \tilde{\theta}_{k+1}^-/d_k^-$. We denote the region $\{\xi_k \leq 1/d_k^+,\ 0 \leq \eta_k \leq \theta_k^+\}$ by $H^{(k)}(\theta_k^+)$, the region $\{\xi_k \leq -1/d_k^-,\ 0 \leq \eta_k \leq \theta_{k+1}^-\}$ by $H^{(k)}(\theta_{k+1}^-)$, and the region $H^{(k)}(\theta_{k+1}^-, \theta_k^+) = H_k\backslash(H^{(k)}(\theta_{k+1}^-) \cup H^{(k)}(\theta_k^+))$ by $H^{(k)}(\theta_{k+1}^-, \theta_k^+)$. We map the region $H^{(k)}(\theta_{k+1}^-, \theta_k^+)$ onto Λ_k using the function $w = \tilde{\varphi}_k(\zeta_k)$, where $\tilde{\varphi}_k(\zeta_k)$ is the function $\varphi_{\alpha\beta}(\zeta_k)$ mentioned in Remark 2 to Theorem 6.1 (p. 394) with $\alpha = \theta_{k+1}^-$, $\beta = \theta_k^+$ $(H_{\alpha\beta} = H^{(k)}(\theta_{k+1}^-, \theta_k^+))$. This is a quasiconformal mapping. We map $H^{(k)}(\theta_{k+1}^-)$ onto $U_1^-(\varepsilon, \tilde{B}_{k+1})$ using the function

$$w = b_{k+1} + \varepsilon \exp\left\{d_k^- \zeta_k + 1 - id_k^- \alpha_{k+1}^-\right\} \tag{7.1}$$

(that is, $\zeta_k = \frac{1}{d_k^-} \ln \frac{w - b_{k+1}}{\varepsilon e} + i\alpha_{k+1}^-$), where the real constant α_{k+1}^- is chosen in such a way that $\{\xi \leq -1/d_k^-, \eta_k = 0\}$ is mapped into S_{k+1}. Similarly we map $H^{(k)}(\theta_k^+)$ onto $U_1^+(\varepsilon, \tilde{B}_k)$ using the function

$$w = b_k + \varepsilon \exp\{-d_k^+ \zeta_k + 1 + id_k^+ \alpha_k^+\} \tag{7.2}$$

(that is, $\zeta_k = \frac{1}{d_k^+} \ln \frac{\varepsilon e}{w - b_k} + i\alpha_k^+$), where the real constant α_k^+ is chosen in such a way that $\{\xi_k \geq 1/d_k^+,\ \eta_k = 0\}$ is mapped into S_k.

If we take into account the form of the function $\tilde{\varphi}_k(\zeta_k)$, which can be written explicitly in the regions $\{\xi_k \leq -1/d_k^-,\ \eta_k \geq \theta_{k+1}^-\}$ and $\{\xi_k \geq 1/d_k^+,\ \eta_k \geq \theta_k^+\}$, it is easy to see that the point $\zeta_k \in \partial H^{(k)}(\theta_{k+1}^-, \theta_k^+) \cap \partial H^{(k)}(\theta_{k+1}^-)$, that is, $\xi_k = \xi_k + i\theta_{k+1}^-$, $\xi_k \leq -1/d_k^-$ is mapped to the same point both under the function $\tilde{\varphi}_k(\zeta_k)$ and the function (7.1); namely it is mapped to the point on σ_k satisfying $|w - b_{k+1}| = \varepsilon \exp\{d_k^- \xi_k + 1\}$. Similarly, the point $\zeta_k = \xi_k + i\theta_k^+$, $\xi_k \geq 1/d_k^+$ is mapped to the point on σ_k such that

$$|w - b_k| = \varepsilon \exp\{-d_k^+ \xi_k + 1\},$$

both under the mapping $\tilde{\varphi}_k(\zeta_k)$ or under the function (7.2). In such a way we get a quasiconformal mapping $w = \varphi_k(\zeta_k)$ of the region H_k onto the Riemann surface Λ'_k, where $\varphi_k(\zeta_k) = \tilde{\varphi}_k(\zeta_k)$ for $\zeta_k \in H^{(k)}(\theta_{k+1}^-, \theta_k^+)$, the function $\varphi_k(\zeta_k)$ is given by formula (7.1) for $\zeta_k \in H^{(k)}(\theta_{k+1}^-)$, and by formula (7.2) for $\zeta_k \in H^{(k)}(\theta_k^+)$. Taking into account (6.18), we get

$$\begin{aligned} &\iint_{H_k} \{p(\zeta_k, w) - 1)\}\, d\sigma(\ln \zeta_k) \\ &\quad = \iint_{H^{(k)}(\theta_{k+1}^-, \theta_k^+)} \{p(\zeta_k, w) - 1\}\, d\sigma(\ln \zeta_k) < \infty. \end{aligned} \tag{7.3}$$

We denote the number of a-points of the function $w = \varphi_k(\zeta_k)$ in $H_k \cap \{|\zeta_k| \leq r\}$ by $n(r, a, \varphi_k)$; similarly we define $n_1(r, a, \varphi_k)$. By (6.19) and (6.20), we have

$$n(r, a, \varphi_k) = \begin{cases} \frac{r}{2\pi} + o(r) & \text{for } a \neq b_k, b_{k+1}, \\ \frac{r}{2\pi}(1 - \delta_g(b_j, \Lambda_k)) + o(r) & a = b_j,\ j = k, k+1 \ \ (\text{if } b_k \neq b_{k+1}), \\ \frac{r}{2\pi}(1 - \delta_g(b, \Lambda_k)) + o(r) & \text{for } a = b \ \ (\text{if } b_k = b_{k+1} = b), \end{cases} \tag{7.4}$$

$$n_1(r, a, \varphi_k) = \varepsilon_g(a, \Lambda_k)\frac{r}{2\pi} + o(r) \quad \text{for all} \quad a. \tag{7.5}$$

Let us observe that for fixed a all a-points of the function $\varphi_k(\zeta_k)$ lie in a strip $|\xi_k| \le K = K(a)$. The quantity

$$\chi = \chi(F) = \frac{d_1^+ d_2^+ \cdots d_p^+}{d_1^- d_2^- \cdots d_p^-}$$

will be called the *characteristic* of the surface F. If we change the way in which we count the points B_k on the curve C, the characteristic χ does not change. Let

$$\beta = 1 + \left(\frac{\ln \chi}{\pi p}\right)^2.$$

Let $\zeta_k = \rho_k e^{i\theta_k}$, $1 \le k \le p$. We map H_k using the function

$$\zeta' = w_k(\zeta_k) = \left(\frac{d_1^+ \cdots d_k^+}{d_1^- \cdots d_{k-1}^-}\zeta_k\right)^{\frac{2}{p}} e^{i\frac{2\pi}{p}(k-1)}, \qquad 1 \le k \le p,\ d_0^\pm = 1, \tag{7.6}$$

where the branch of $\zeta_k^{\frac{2}{p}}$ is chosen in such a way that ($\zeta' = \rho' e^{i\theta'}$)

$$\rho' = \left(\frac{d_1^+ \cdots d_k^+}{d_1^- \cdots d_k^-}\right)^{\frac{2}{p}} \rho_k^{\frac{2}{p}}, \qquad \theta' = \frac{2}{p}\theta_k + (k-1)\frac{2\pi}{p}, \quad 1 \le k \le p. \tag{7.6$'$}$$

The function (7.6) maps H_k onto the region

$$\left\{(k-1)\frac{2\pi}{p} < \arg \zeta' < k\frac{2\pi}{p}\right\} \setminus T_k',$$

where T_k' is the image of the trapezoid T_k under the mapping (7.6). The mapping $\zeta_k = \varphi_k^{-1}(w)$ sends the point $w \in S_k$ to the point

$$\zeta_k = \xi_k = \frac{1}{d_k^+} \ln \frac{\varepsilon e}{|w - b_k|},$$

and the mapping $\zeta' = \omega_k(\varphi_k^{-1}(w))$ sends the point $w \in S_k$ to the point

$$\zeta' = \left(\frac{d_1^+ \cdots d_{k-1}^+}{d_1^- \cdots d_{k-1}^-}\right)^{\frac{2}{p}} \left(\ln \frac{\varepsilon e}{|w - b_k|}\right)^{\frac{2}{p}} e^{i\frac{2\pi}{p}(k-1)}, \qquad k = 1, \dots, p. \tag{7.7}$$

The mapping $\zeta_{k-1} = \varphi_{k-1}^{-1}(w)$, $k = 2, \dots, p$, sends the same point $w \in S_k$ to the point

$$\zeta_{k-1} = \frac{1}{d_{k-1}^-} \ln \frac{|w - b_k|}{\varepsilon e} = e^{i\pi} \frac{1}{d_{k-1}^-} \ln \frac{e}{|w - b_k|},$$

and the mapping $\zeta' = \omega_{k-1}(\varphi_{k-1}^{-1}(w))$ sends it to the point

$$\begin{aligned}\zeta' &= \left(\frac{d_1^+ \cdots d_{k-1}^+}{d_1^- \cdots d_{k-2}^-} e^{i\pi} \frac{1}{d_{k-1}^-} \ln \frac{\varepsilon e}{|w - b_k|}\right)^{\frac{2}{p}} e^{i\frac{2\pi}{p}(k-2)} \\ &= \left(\frac{d_1^+ \cdots d_{k-1}^+}{d_{k-1}^- \cdots d_{k-1}^-}\right)^{\frac{2}{p}} \left(\ln \frac{\varepsilon e}{|w - b_k|}\right)^{\frac{2}{p}} e^{i\frac{2\pi}{p}(k-1)}, \qquad k = 2, \dots, p.\end{aligned} \tag{7.8}$$

Comparing (7.7) and (7.8), we see that the mapping $\zeta' = \omega(w) = \omega_k(\varphi_k^{-1}(w))$ remains single-valued and continuous on the radii $S_2, S_3, \dots, S_p$ for $w \in \Lambda_k' \cup S_k \cup S_{k+1}$, $k = 1, \dots, p$, and maps $(\bigcup_{k=1}^p \Lambda_k') \cup (\bigcup_{k=2}^p S_k)$ onto the region G' defined as the complex ζ'-plane with $\bigcup_{k=1}^p T_k'$ removed and with the cut along the positive real semi-axis. In the general case the function $\zeta' = \omega(w)$ is not single-valued on

S_1. In fact, under the mapping $\zeta' = \omega_1(\varphi_1^{-1}(w))$ the point $w \in S_1$, by (7.7), goes to the point

$$\zeta' = \left(\ln \frac{\varepsilon e}{|w - b_1|}\right)^{\frac{2}{p}} = \zeta^+,$$

and under the mapping $\zeta' = \omega_p(\varphi_p^{-1}(w))$, it goes to the point

$$\begin{aligned}\zeta' &= \left(\frac{d_1^+ \cdots d_p^+}{d_1^- \cdots d_{p-1}^-} e^{i\pi} \frac{1}{d_p^-} \ln \frac{\varepsilon e}{|w - b_1|}\right)^{\frac{2}{p}} e^{i\frac{2\pi}{p}(p-1)} \\ &= \chi^{\frac{2}{p}} \left(\ln \frac{\varepsilon e}{|w - b_1|}\right)^{\frac{2}{p}} = \chi^{\frac{2}{p}} \zeta^+ = \zeta^-.\end{aligned}$$

If $\chi = 1$, then $\zeta^+ = \zeta^-$, and the function $\zeta' = \omega(w)$ is continuous and single-valued also on S_1. In the general case we make an additional conformal mapping:

$$\zeta = (\zeta')^{1/(1 - i\frac{\ln\chi}{\pi p})}, \qquad \zeta' = \rho' e^{i\theta'}, \qquad \zeta = \rho e^{i\theta}, \tag{7.9}$$

where a single-valued branch is chosen in such a way that the region $\{0 < \theta' < 2\pi\}$ is mapped into the region

$$\left\{\frac{\ln\rho \ln\chi}{\pi p} < \theta < \frac{\ln\rho\ln\chi}{\pi p} + 2\pi\right\},$$

that is, into the complex ζ-plane with the cut along a logarithmic spiral $\theta = \frac{1}{\pi p} \ln\rho\ln\chi$, $0 < \rho < \infty$. Under this mapping, the point $\zeta^+ = \rho'$ on the upper boundary of the cut in the ζ'-plane goes to the point

$$\zeta = (\zeta^+)^{1/(1 - i\frac{\ln\chi}{\pi p})} = (\rho')^{1/\beta} \exp\left\{i \frac{\ln\chi\ln\rho'}{\pi p \beta}\right\},$$

and the point $\zeta^- = \chi^{\frac{2}{p}} \zeta^+ e^{2\pi i}$ on the lower boundary of the cut goes to the point

$$\begin{aligned}\zeta &= (\zeta^+)^{1/(1 - i\frac{\ln\chi}{\pi p})} \chi^{2/(p - i\frac{\ln\chi}{\pi})} e^{2\pi i/(1 - i\frac{\ln\chi}{\pi p})} \\ &= (\zeta^+)^{1/(1 - i\frac{\ln\chi}{\pi p})} \exp\left\{\frac{2\ln\chi + 2\pi i p}{p - i\frac{\ln\chi}{\pi}}\right\} = (\zeta^+)^{1/(1 - i\frac{\ln\chi}{\pi p})} e^{2\pi i}.\end{aligned}$$

That is, under the mapping (7.9), the points ζ^+ and ζ^- are sent to the points on the opposite boundaries of the cut in the ζ-plane that correspond to the same number in the complex plane. We denote the image of the region G' under the mapping (7.9) by G. It is easy to see that $G_0 = \{|\zeta| < \infty\} \setminus \bar{G}$ is a simply connected region bounded by a closed Jordan curve. The mapping $\zeta = \zeta(w)$, which is a composition of the mapping $\zeta' = \omega(w)$ and the mapping (7.9), is a homeomorphic mapping of $F \setminus K_0$ onto G. There exists a homeomorphic mapping of the kernel K_0 onto the region G_0 such that we get a homeomorphic mapping $\zeta = \zeta(w)$ of the whole Riemann surface F onto $\{|\zeta| < \infty\}$. Denote the inverse mapping to $\zeta = \zeta(w)$ by $w = \varphi(\zeta)$. It is clear that the mapping $w = \varphi(\zeta)$ is quasiconformal in G. Taking into account (7.3) we get

$$\begin{aligned}
\iint_G \{p(\zeta, w) - 1\}\, d\sigma(\ln \zeta) &= \iint_{G'} \{p(\zeta', w) - 1\} \left|\frac{d\ln\zeta}{d\ln\zeta'}\right|^2 d\sigma(\ln\zeta') \\
&= \frac{1}{\beta}\iint_{G'} \{p(\zeta', w) - 1\}\, d\sigma(\ln\zeta') \\
&= \frac{1}{\beta}\sum_{k=1}^{p}\iint_{H_k} \{p(\zeta_k, w) - 1\} \left|\frac{d\ln\zeta'}{d\ln\zeta_k}\right|^2 d\sigma(\ln\zeta_k) \\
&= \frac{4}{p^2\beta}\sum_{k=1}^{p}\iint_{H_k} \{p(\zeta_k, w) - 1\}\, d\sigma(\ln\zeta_k) < \infty.
\end{aligned} \tag{7.10}$$

If the function $w = f(z)$ is a conformal mapping of $\{|z| < R\}$ onto F, then the mapping $z = z(\zeta) = f^{-1}(\varphi(\zeta))$ satisfies, by (7.10), the assumptions of Teichmüller's Theorem 3.1; hence, $R = \infty$ and under the corresponding normalization of the function $w = f(z)$ meromorphic in $\{|z| < \infty\}$ we have

$$\lim_{z\to\infty}\left|\frac{\zeta(z)}{z}\right| = \lim_{\zeta\to\infty}\left|\frac{\zeta}{z(\zeta)}\right| = 1. \tag{7.11}$$

Let $d^k(a) = \delta_g(a, \Lambda_k)$, $e^k(a) = \varepsilon_g(a, \Lambda_k)$. If r is sufficiently large, then the region corresponding to $\{|\zeta| \le r\}\backslash G_0$ in the ζ'-plane is the intersection of G' with the region

$$\left\{\rho' \le r^\beta \exp\left(\theta'\frac{\ln\chi}{\pi p}\right)\right\}, \qquad 0 < \theta' < 2\pi,$$

and the corresponding regions in ζ_k-planes are the intersections of H_k with ($\zeta_k = \rho_k e^{i\theta_k}$)

$$\begin{aligned}
\rho_k &\le \frac{d_1^- \cdots d_{k-1}^-}{d_1^+ \cdots d_k^+} r^{\beta p/2} \exp\left\{\frac{\ln\chi}{2\pi}\left(\frac{2}{p}\theta_k + (k-1)\frac{2\pi}{p}\right)\right\} \\
&= \frac{d_1^- \cdots d_{k-1}^-}{d_1^+ \cdots d_k^+} r^{\beta p/2} \chi^{\frac{1}{\pi p}\{\theta_k + \pi(k-1)\}}, \qquad 0 < \theta_k < \pi, \quad 1 \le k \le p;
\end{aligned}$$

this follows from (7.6′) and (7.9). Taking into account relations (7.4) and (7.5), and also the fact that for each a the arguments of a-points in the ζ_k-plane tend to $\pi/2$, we find that

$$n(r, a, \varphi) = \frac{r^{\beta p/2}}{2\pi}\sum_{k=1}^{p}\{1 - d^k(a)\}\frac{d_1^- \cdots d_{k-1}^-}{d_1^+ \cdots d_k^+}\chi^{\frac{1}{p}(k-\frac{1}{2})} + o(r^{\beta p/2}), \tag{7.12}$$

$$n_1(r, a, \varphi) = \frac{r^{\beta p/2}}{2\pi}\sum_{k=1}^{p} e^k(a)\frac{d_1^- \cdots d_{k-1}^-}{d_1^+ \cdots d_k^+}\chi^{\frac{1}{p}(k-\frac{1}{2})} + o(r^{\beta p/2}). \tag{7.13}$$

By (7.11), relations (7.12) and (7.13) remain true if we replace $n(r, a, \varphi)$ by $n(r, a, f)$ and $n_1(r, a, \varphi)$ by $n_1(r, a, f)$. From this we get

$$N(r, a, f) = \frac{r^{\beta p/2}}{\beta p\pi}\sum_{k=1}^{p}\{1 - d^k(a)\}\frac{d_1^- \cdots d_{k-1}^-}{d_1^+ \cdots d_k^+}\chi^{\frac{1}{p}(k-\frac{1}{2})} + o(r^{\beta p/2}), \tag{7.14}$$

$$N_1(r, a, f) = \frac{r^{\beta p/2}}{\beta p\pi}\sum_{k=1}^{p} e^k(a)\frac{d_1^- \cdots d_{k-1}^-}{d_1^+ \cdots d_k^+}\chi^{\frac{1}{p}(k-\frac{1}{2})} + o(r^{\beta p/2}). \tag{7.15}$$

Since $d^k(a) = 0$, $1 \le k \le p$, for all a except at most p, we can argue as in the derivation of (6.16) to find that

$$T(r,f) = \frac{r^{\beta p/2}}{\beta p \pi} \sum_{k=1}^{p} \frac{d_1^- \cdots d_{k-1}^-}{d_1^+ \cdots d_k^+} \chi^{\frac{1}{p}(k-\frac{1}{2})} + o(r^{\beta p/2}), \tag{7.16}$$

and get the following theorem.

THEOREM 7.1. *Let $w = f(z)$ be a meromorphic function that maps the complex z-plane onto a Riemann surface F with p almost periodic ends and characteristic χ. Then the function $w = f(z)$ has a normal type of order*

$$\rho = \frac{\beta p}{2} = \frac{p}{2}\left\{1 + \left(\frac{\ln \chi}{\pi p}\right)^2\right\}; \tag{7.17}$$

moreover $T(r,f) \sim Kr^\rho$, $0 < K < \infty$, and the deficiencies and indices are equal to

$$\delta(a,f) = \frac{\sum_{k=1}^p d^k(a) \frac{d_1^- \cdots d_{k-1}^-}{d_1^+ \cdots d_k^+} \chi^{\frac{1}{p}(k-\frac{1}{2})}}{\sum_{k=1}^p \frac{d_1^- \cdots d_{k-1}^-}{d_1^+ \cdots d_k^+} \chi^{\frac{1}{p}(k-\frac{1}{2})}} \tag{7.18}$$

and

$$\varepsilon(a,f) = \frac{\sum_{k=1}^p e^k(a) \frac{d_1^- \cdots d_{k-1}^-}{d_1^+ \cdots d_k^+} \chi^{\frac{1}{p}(k-\frac{1}{2})}}{\sum_{k=1}^p \frac{d_1^- \cdots d_{k-1}^-}{d_1^+ \cdots d_k^+} \chi^{\frac{1}{p}(k-\frac{1}{2})}}, \tag{7.19}$$

respectively.

Since $\sum_a \{d^k(a) + e^k(a)\} = 2$ for $k = 1, \dots, p$, equalities (7.18) and (7.19) imply that

$$\sum_a \{\delta(a,f) + \varepsilon(a,f)\} = 2$$

for each function f satisfying the assumptions of Theorem 7.1.

For brevity, we speak about the deficiencies and indices of the surface F with almost periodic ends meaning the deficiencies and indices of the mapping meromorphic function $w = f(z)$, and define $\delta(a,F) = \delta(a,f)$ and $\varepsilon(a,F) = \varepsilon(a,f)$. In what follows, we consider surfaces with finitely many almost periodic ends for $\chi = 1$ only, without mentioning this explicitly. For convenience of references we rewrite formulas (7.17)–(7.19) for this case:

$$\rho = p/2, \tag{7.20}$$

$$\delta(a,f) = \frac{\frac{1}{d_1^+} d^1(a) + \frac{d_1^-}{d_1^+ d_2^+} d^2(a) + \cdots + \frac{d_1^- \cdots d_{p-1}^-}{d_1^+ \cdots d_p^+} d^p(a)}{\frac{1}{d_1^+} + \frac{d_1^-}{d_1^+ d_2^+} + \cdots + \frac{d_1^- \cdots d_{p-1}^-}{d_1^+ \cdots d_p^+}}, \tag{7.21}$$

$$\varepsilon(a,f) = \frac{\frac{1}{d_1^+} e^1(a) + \frac{d_1^-}{d_1^+ d_2^+} e^2(a) + \cdots + \frac{d_1^- \cdots d_{p-1}^-}{d_1^+ \cdots d_p^+} e^p(a)}{\frac{1}{d_1^+} + \frac{d_1^-}{d_1^+ d_2^+} + \cdots + \frac{d_1^- \cdots d_{p-1}^-}{d_1^+ \cdots d_p^+}}. \tag{7.22}$$

The quantity

$$L(F, B_1) = \frac{1}{d_1^+} + \frac{d_1^-}{d_1^+ d_2^+} + \cdots + \frac{d_1^- \cdots d_{p-1}^-}{d_1^+ \cdots d_p^+}$$

(B_1 is the point on the boundary of the pseudokernel at which we start counting), which is the denominator in both (7.21) and (7.22), is called the *weight* of the surface F at the point B_1. Taking into account that $\chi = 1$, it is easy to verify that for $1 \le j \le p$ we have

$$L(F, B_j) = \frac{1}{d_j^+} + \frac{d_j^-}{d_j^+ d_{j+1}^+} + \cdots + \frac{d_j^- \cdots d_p^- d_1^- \cdots d_{j-2}^-}{d_j^+ \cdots d_p^+ d_1^+ \cdots d_{j-1}^+} = \frac{d_1^+ \cdots d_{j-1}^+}{d_1^- \cdots d_{j-1}^-} L(F, B_1).$$

Thus, in general, the weight depends on a point at which it is taken. However, for brevity, we write $L(F)$ instead of $L(F, B_1)$.

It is clear that the expressions in the right-hand sides of (7.21) and (7.22) do not depend on the choice of the starting point B_1 since $\delta(a, f)$ and $\varepsilon(a, f)$ do not depend on this choice, but this can also be proved by a formal computation.

Let F_1 and F_2 be two surfaces with finitely many (p_1 and p_2, respectively) almost periodic ends; moreover, each of these surfaces has at least one logarithmic ramification point $\tilde{B}'$ and $\tilde{B}''$ which is projected to b. Let K_1 and K_2 be pseudokernels of the surfaces F_1 and F_2. For sufficiently small $\varepsilon > 0$, the regions $K_1 \cap U(\varepsilon, \tilde{B}')$ and $K_2 \cap U(\varepsilon, \tilde{B}'')$ are circular sectors with vertices at points $\tilde{B}' \in \partial K_1$ and $\tilde{B}'' \in \partial K_2$. Suppose that these two sectors contain two radii with the same projection. We make cuts along these radii and paste the surfaces F_1 and F_2 along them, joining each boundary of one cut with the opposite boundary of the other cut. We get a Riemann surface F, which also is a surface with p $(= p_1 + p_2)$ almost periodic ends.

We write $F = F_1 + F_2$ and say that the surface F is obtained by a junction of the surfaces F_1 and F_2 at the points $\tilde{B}'$ and $\tilde{B}''$.

It is easy to see that $\chi(F) = \chi(F_1)\chi(F_2)$. Therefore, if $\chi(F_1) = \chi(F_2) = 1$ (we agreed to consider only such surfaces), then $\chi(F) = 1$. On each of the joined surfaces F_1 and F_2, we start counting the logarithmic ramification points at $\tilde{B}'$ and $\tilde{B}''$, respectively, and on the junction $F = F_1 + F_2$, we start this counting at either the logarithmic ramification point corresponding to $\tilde{B}'$, or at the one corresponding to $\tilde{B}''$; the specific choice makes no difference. It is easy to verify that

$$L(F_1 + F_2) = L(F_1) + L(F_2),$$

and using formulas (7.21) and (7.22) we find that

$$\delta(a, F_1 + F_2) = \frac{L(F_1)\delta(a, F_1) + L(F_2)\delta(a, F_2)}{L(F_1) + L(F_2)}, \tag{7.23}$$

$$\varepsilon(a, F_1 + F_2) = \frac{L(F_1)\varepsilon(a, F_1) + L(F_2)\varepsilon(a, F_2)}{L(F_1) + L(F_2)}. \tag{7.24}$$

If we assume that we have already defined $F_1 + \cdots + F_{n-1}$, we define $F_1 + \cdots + F_n = \sum_{j=1}^n F_j$ as $(F_1 + \cdots + F_{n-1}) + F_n$.

Suppose that the surface $F_1 + \cdots + F_n$ is obtained in the following way. The surfaces F_1 and F_2 are joined at the points $\tilde{B}'$ and $\tilde{B}''$, and we start counting logarithmic ramification points of the surface $F_1 + F_2$ at one of the points corresponding to $\tilde{B}'$ or $\tilde{B}''$ (we denote it by $\tilde{\tilde{B}}''$). The surfaces $F_1 + F_2$ and F_3 are joined at the points $\tilde{\tilde{B}}''$ and $\tilde{B}'''$, and we start counting logarithmic ramification points of $F_1 + F_2 + F_3$ from one of the points corresponding to $\tilde{\tilde{B}}''$ and $\tilde{B}'''$ (we denote it by $\tilde{\tilde{B}}'''$), and continue in an obvious way till we get the surface $F_1 + \cdots + F_n$. It is not always possible to get the surface in this way, but if this is the case, we get, by induction, that

$$L\left(\sum_{j=1}^{n} F_j\right) = \sum_{j=1}^{n} L(F_j), \tag{7.25}$$

$$\delta\left(a, \sum_{j=1}^{n} F_j\right) = \frac{\sum_{j=1}^{n} L(F_j)\delta(a, F_j)}{\sum_{j=1}^{n} L(F_j)}, \tag{7.26}$$

$$\varepsilon\left(a, \sum_{j=1}^{n} F_j\right) = \frac{\sum_{j=1}^{n} L(F_j)\varepsilon(a, F_j)}{\sum_{j=1}^{n} L(F_j)}. \tag{7.27}$$

In the general case, for $n \geq 3$, these formulas are no longer true. Later on, we will need the following.

THEOREM 7.2. *Let $b_1, \dots, b_p$ be some complex numbers, not necessarily distinct, and let $a_1, \dots, a_q$ be distinct complex numbers such that all terms of the sequence $\{b_k\}$ are present among the terms of the sequence $\{a_j\}$. Then there exists a Riemann surface F with p almost periodic ends $\Lambda_1, \dots, \Lambda_p$, following after each other in the indicated order, where $\delta(a, \Lambda_k) = 0$ for $a \neq b_k, b_{k+1}$, and $\delta(b_k, \Lambda_k)$, $\delta(b_{k+1}, \Lambda_k)$, $\varepsilon(a_j, \Lambda_k)$, $1 \leq j \leq q$, $1 \leq k \leq p$, are arbitrary preassigned numbers satisfying $\delta(b_k, \Lambda_k) > 0$, $\delta(b_{k+1}, \Lambda_k) > 0$,*

$$\sum_{j=1}^{q} \{\delta(a_j, \Lambda_k) + \varepsilon(a_j, \Lambda_k)\} = 2, \quad \text{and} \quad 0 \leq \delta(a_j, \Lambda_k) + \varepsilon(a_j, \Lambda_k) \leq 1.$$

First we construct a pseudokernel K. If $b_k \neq b_{k+1}$, that is, all almost periodic ends Λ_k are of the first kind, the construction of the pseudokernel K is the same as in the proof of Theorem 5.1. In general, the construction is somewhat more complicated.

First we assume that not all of the numbers b_j, $j \geq 2$, are equal to b_1. Let $b'_1, \dots, b'_\nu$ and $b''_1, \dots, b''_\mu$ be subsequences of $b_1, \dots, b_p$ such that each of them has the same set of values as $b_1, \dots, b_p$, but: (a) $b'_k \neq b'_{k-1}, b'_{k+1}$ ($b'_0 = b'_\nu$, $b'_{\nu+1} = b'_1$), $k = 1, \dots, \nu$; (b) all numbers b''_k are distinct; (c) there is no subsequence with the same properties and more terms. Let $b'_k = b_{n_k}$, $1 = n_1 < n_2 < \cdots < n_\nu \leq p$; $b_{n_k} \neq b_{n_k - 1}$, $b''_k = b'_{m_k}$, $1 = m_1 < m_2 < \cdots < m_\mu \leq \nu$. We draw a closed Jordan curve c through the points $\{b''_1, \dots, b''_\mu\}$ that does not pass through the points $a_1, \dots, a_q$ different from $\{b_1, \dots, b_p\}$, has a continuous curvature, and consists of line segments in sufficiently small neighborhoods of the points $b''_1, \dots, b''_\mu$. After that, in the same way as at the beginning of the proof of Theorem 5.1, we construct a simply connected finite-sheeted Riemann surface K' with boundary, bounded by a closed Jordan curve C', consisting of ν J-curves $\sigma'_k = \widehat{B'_k B'}_{k+1}$, where the point B'_k is projected to b'_k. If $\nu = p$, we let $K = K'$ and $B_k = B'_k$. If $\nu < p$, we carry out an additional construction. Suppose that the sequence $b_1, \dots, b_p$ satisfies $b_{t-1} \neq b_t = b_{t+1} = \cdots = b_{t+s} \neq b_{t+s+1}$ (indices are taken $\mod p$). Let $b_t = b'_\tau = b''_\tau$. Take a J'-curve c_τ with endpoints at b''_τ such that the region G_τ it bounds is contained in the region bounded by the curve c, and $\bar{G}_\tau \backslash \{b''_\tau\}$ does not contain the points $a_1, \dots, a_q$, and moreover, for different t the regions G_τ are pairwise disjoint. We take s copies of the region G_τ and make in each of them the same cut along a line segment with one endpoint at b''_τ and the other endpoint at $\pi_\tau \in G_\tau$. We make the same cut on K' starting at B'_τ. We paste to K' s copies of the regions G_τ along this cut in such a way that the obtained Riemann surface remains simply connected, and an algebraic ramification point of order s appears over π_τ. Making such pasting for all t, we get a finite-sheeted simply connected Riemann surface K with boundary, bounded by a Jordan curve C consisting of p J-curves $\sigma_k = \widehat{B_k B_{k+1}}$, where the point B_k is projected to the point b_k, and the arc σ_k is projected into the arc $\widehat{b_k b_{k+1}}$ on the curve c if $b_{k+1} \neq b_k$, and into some arc c_τ if $b_{k+1} = b_k$. The surface K plays the role of the pseudokernel. Now, using Theorems 6.2 and 6.3, we construct almost periodic ends Λ_k with preassigned $\delta(a, \Lambda_k)$ and $\varepsilon(a, \Lambda_k)$, and the boundary σ_k^-. Joining them to the pseudokernel K, we get the desired surface F.

If $b_1 = \dots = b_p = b$, the construction of the pseudokernel K is simplified. Take a J'-curve c with endpoints at b, and let G be the interior of the curve c. Choose the curve c in such a way that $\bar{G}\backslash\{b\}$ does not contain the points $a_1, \dots, a_q$. Let $w_0 \in G$. We take for K the part of the surface of the function $(w - w_0)^{1/p}$ lying over G.

8. The inverse problem of the value distribution theory

First we prove the following theorem.

THEOREM 8.1. *Let δ_j, ε_j, $j = 1, \dots, q$, be non-negative numbers satisfying the conditions*

(1′) $\sum_{j=1}^q \{\delta_j + \varepsilon_j\} = 2$;
(2) $0 < \delta_j + \varepsilon_j \le 1$, $1 \le j \le q$;
(3) $0 < \sum_{j-1}^p \delta_j \le 2$; *moreover, if* $\sum_{j=1}^q \delta_j = 2$, *then all numbers* $\delta_1, \dots, \delta_q$ *are rational.*

Let $a_1, \dots, a_q$ be arbitrary distinct complex numbers. Then there exists a Riemann surface F with almost periodic ends such that $\delta(a_j, F) = \delta_j$, $\varepsilon(a, F) = \varepsilon_j$, $1 \le j \le q$.

In the case where $\sum_{j=1}^q \delta_j = 2$, the desired Riemann surface F was constructed in Theorem 5.1, namely, a surface with logarithmic ends. Note that for $\sum_{j=1}^q \delta_j = 2$ we cannot drop the restriction that the numbers δ_j, $1 \le j \le q$, are rational, even if we use Riemann surfaces with almost periodic ends. In fact, if $\varepsilon(a_0, \Lambda_k) > 0$ for at least one almost periodic end Λ_k and at least one point a_0, then, by (7.19), we have $\varepsilon(a_0, F) > 0$. But this contradicts the condition $\sum_{j=1}^q \delta_j = 2$. Hence, for each end Λ_k we have $d_k^+ = d_k^- = 1$, $1 \le k \le q$, and, necessarily, $\chi = 1$. Equality (7.21) implies that the deficiencies $\delta(a, F)$ are rational numbers.

In what follows we assume that $\sum_{j=1}^q \delta_j < 2$ and $\sum_{j=1}^q \varepsilon_j > 0$. The distribution of deficiencies and indices of a surface F with almost periodic ends will be described using a *scheme*

$$\begin{pmatrix} a_1 & \dots & a_q \\ \delta(a_1) & \dots & \delta(a_q) \\ \varepsilon(a_1) & \dots & \varepsilon(a_q) \end{pmatrix} = \begin{pmatrix} \delta_1 & \dots & \delta_q \\ \varepsilon_1 & \dots & \varepsilon_q \end{pmatrix},$$

and the construction of a surface with this distribution of deficiencies and indices will be called a *scheme solution.* Junction of Riemann surfaces will be symbolically represented as addition of the corresponding schemes using the following rule:

$$L_1\begin{pmatrix} \cdots & \delta_j^{(1)} & \cdots \\ \cdots & \varepsilon_j^{(1)} & \cdots \end{pmatrix} + L_2\begin{pmatrix} \cdots & \delta_j^{(2)} & \cdots \\ \cdots & \varepsilon_j^{(2)} & \cdots \end{pmatrix} = (L_1+L_2)\begin{pmatrix} \cdots & \frac{L_1\delta_j^{(1)}+L_2\delta_j^{(2)}}{L_1+L_2} & \cdots \\ \cdots & \frac{L_1\varepsilon_j^{(1)}+L_2\varepsilon_j^{(2)}}{L_1+L_2} & \cdots \end{pmatrix}, \tag{8.1}$$

where the factor in front of a scheme denotes the weight of the surface taken at the point at which the junction occurs. We mean that the joined surfaces both have positive deficiencies over one of the points a_{j_0}, and the junction takes place at logarithmic ramification points over a_{j_0}. Solving schemes-summands we, obviously, can always select pseudokernels in such a way that the junction is possible. We may assume without loss of generality that $a_{j_0} = a_1$. Thus, in all schemes which we are going to join, δ_1 should be > 0, and the weight must always be taken at a logarithmic ramification point over a_1. To solve a scheme with weight means to construct a Riemann surface with almost periodic ends, with preassigned value distribution, and with the given weight up to a rational factor. If we can solve the schemes in the left-hand side of equality (8.1) using surfaces F_1 and F_2 with weights $\frac{n_1}{m_1}L_1$ and $\frac{n_2}{m_2}L_2$, respectively, where n_1, n_2, m_1, m_2 are positive integers, then, joining n_2m_1 copies of the surface F_2 with n_1m_2 copies of the surface F_1, we, by (7.25), (7.26), and (7.27), get a surface F solving the scheme in the right-hand side of equality (8.1) with weight $n_1n_2(L_1+L_2)$.

We reduce solutions of complicated schemes to solutions of four main schemes (α-, β-, γ-, and δ-schemes), for which we provide a solution with a preassigned weight, up to a rational factor. We use the notation $\Theta_j = \delta_j + \varepsilon_j$, $1 \le j \le q$.

An *α-scheme* is a scheme with $\delta_1 > 0$, $\delta_2 = \delta_3 = \cdots = \delta_q = 0$:

$$L\begin{pmatrix} \delta_1 & 0 & \cdots & 0 \\ \varepsilon_1 & \varepsilon_2 & \cdots & \varepsilon_q \end{pmatrix}.$$

Let m be a large positive integer such that $2/(mL) < \delta_1$. We construct a surface Φ with almost periodic ends of the second kind solving a given α-scheme, and such that $d_1^+ = d_2^- = 2/(mL)$, $d_1^- = d_2^+ = \delta_1 - d_1^+$. Such a surface Φ exists by Theorem 6.3. The weight of such a surface Φ is equal to

$$\frac{1}{d_1^+} + \frac{d_1^-}{d_1^+ d_2^+} = \frac{2}{d_1^+} = mL.$$

A *β-scheme* is a scheme for which $\delta_j > 0$, $1 \le j \le q$, $\varepsilon_1 > 0$, and all numbers Θ_j are rational, $1 \le j \le q$:

$$L\begin{pmatrix} \delta_1 & \delta_2 & \cdots & \delta_q \\ \varepsilon_1 & \varepsilon_2 & \cdots & \varepsilon_q \end{pmatrix}.$$

Let $\Theta_j = n_j/n$, n_j, n be positive integers, $\varepsilon_1 > 1/n$. Then $\sum_{j=1}^q n_j = 2n$. We represent δ_j and ε_j as the sums

$$\delta_j = \sum_{k=1}^{n_j} \Delta_{jk}, \qquad \varepsilon_j = \sum_{k=1}^{n_j} E_{jk},$$

where

$$\Delta_{jk} > 0, \quad E_{jk} \geq 0, \quad \Delta_{jk} + E_{jk} = \frac{1}{n}, \quad 1 \leq k \leq n_j.$$

Let m be a positive integer satisfying $2/(mL) < 1/n$. Since $\varepsilon_1 > 0$, we can take $\Delta_{11} = 2/(mL)$, $E_{11} = n^{-1} - \Delta_{11} > 0$. We take a surface F with $2n$ almost periodic ends $\Lambda_1, \ldots, \Lambda_{2n}$ and with logarithmic ramification points $\tilde{B}_1, \ldots, \tilde{B}_{2n}$, where $\tilde{B}_1$ lies over a_1, and there are n_j logarithmic ramification points over a_j, $1 \leq j \leq q$. Since $n_j \leq n = \frac{1}{2}\sum_{j=1}^{q} n_j$, the sequence $\tilde{B}_\nu$ can be chosen in such a way that $b_\nu \neq b_{\nu\pm 1}$, $1 \leq \nu \leq 2n$, where b_ν is a projection of the point $\tilde{B}_\nu$. Here and later, all indices ν are taken $\mod 2n$. Suppose that $\tilde{B}_\nu$ is the logarithmic ramification point number s among those projected into a_j, $s = s(\nu)$, $j = j(\nu)$. Let $\Delta_\nu = \Delta_{js}$, $E_\nu = E_{js}$, $j = j(\nu)$, $s = s(\nu)$. Obviously, $\Delta_1 = \Delta_{11}$. We choose an almost periodic end Δ_ν in such a way that $\delta(a_{j(\nu)}, \Lambda_\nu) = \Delta_\nu n$, $\delta(a_{j(\nu+1)}, \Lambda_\nu) = \Delta_{\nu+1} n$, $\varepsilon(a_{j(\nu)}, \Lambda_\nu) = E_\nu n$, $\varepsilon(a_{j(\nu+1)}, \Lambda_\nu) = E_{\nu+1} n$. We have $d_\nu^- = d_{\nu+1}^+ = \Delta_{\nu+1} n$. The weight of the surface F is equal to

$$\frac{1}{d_1^+} 2n = \frac{1}{\Delta_{11} n} 2n = \frac{2}{\Delta_{11}} = mL.$$

Equality (7.21) implies

$$\begin{aligned}\delta(a_j, F) &= \frac{1}{2n}\sum_{\nu=1}^{2n} \delta(a_j, \Lambda_\nu) = \frac{1}{2n}\sum_j{}' \delta(a_{j(\nu)}, \Lambda_\nu) + \frac{1}{2n}\sum_j{}'' \delta(a_{j(\nu+1)}, \Lambda_\nu) \\ &= \frac{1}{2n}\sum_j{}' \{\delta(a_{j(\nu)}, \Lambda_\nu) + \delta(a_{j(\nu)}, \Lambda_{\nu-1})\},\end{aligned}$$

where $\sum_j'$ means that the sum is taken over those ν for which $j(\nu) = j$, and $\sum_j''$ means that the sum is taken over those ν for which $j(\nu+1) = j$. Therefore

$$\delta(a_j, F) = \frac{1}{2n}\sum_j{}' \{\Delta_\nu n + \Delta_\nu n\} = \sum_j{}' \Delta_\nu = \sum_{s=1}^{n_j} \Delta_{js} = \delta_j, \qquad 1 \leq j \leq q.$$

Similarly we get $\varepsilon(a_j, F) = \varepsilon_j$, $1 \leq j \leq q$.

REMARK. If we considered the "restricted" inverse problem only and did not impose any conditions on ε_j in the statement of Theorem 8.1, it would suffice to restrict our attention to α- and β-schemes. It is clear that if $\sum_{j=1}^{q} \delta_j < 2$, $q \geq 2$, we can always choose non-negative numbers ε_j such that $0 < \delta_j + \varepsilon_j < 1$ for $1 \leq j \leq q$, $\sum_{j=1}^{q}(\delta_j + \varepsilon_j) = 2$, and $\Theta_j = \delta_j + \varepsilon_j$ are rational numbers. We will not have to sum schemes; for this reason their weight does not matter for us, and we can disregard the requirement $\varepsilon_1 > 0$. Besides, if we refer to Theorem 6.4, we may restrict our attention by the case $q \geq 3$ and solve the "restricted" inverse problem using β-schemes only.

A *γ-scheme* is a scheme for which: (1) $\delta_j > 0$ for $1 \leq j \leq p$, $2 \leq p < q$ and $\delta_j = 0$ for $p+1 \leq j \leq q$, $1 < p < q$; (2) $2\Theta_j \leq \sum_{j=1}^{p} \Theta_j = \Theta$ for $1 \leq j \leq p$; (3) $\varepsilon_{p+1} > 0$:

$$L\begin{pmatrix} \delta_1 & \ldots & \delta_p & 0 & \ldots & 0 \\ \varepsilon_1 & \ldots & \varepsilon_p & \varepsilon_{p+1} & \ldots & \varepsilon_q \end{pmatrix}.$$

Since $\varepsilon_{p+1} > 0$, we have $\Theta < 2$. We show that all Θ_j, $1 \le j \le p$, can be represented as sums of n_j positive terms θ_{jk} in such a way that the following conditions are satisfied:

$$\Theta_j = \theta_{j1} + \cdots + \theta_{jn_j}, \qquad 1 \le j \le p, \tag{8.2}$$

$$\left|\frac{\theta_{jk}}{\theta_{il}} - 1\right| < 2 - \Theta, \qquad 1 \le j \le p, \quad 1 \le i \le p, \quad 1 \le k \le n_j, \quad 1 \le l \le n_j, \tag{8.3}$$

$$2n_j \le \sum_{j=1}^{p} n_j = n. \tag{8.4}$$

Let κ be a large positive integer, so that

$$\frac{2^{1-\kappa}}{1 - 2^{-\kappa}} < 2 - \Theta. \tag{8.5}$$

Let

$$\Theta_j = \sum_{k=1}^{\infty} \alpha_k^j 2^{-k}, \qquad \alpha_k^j = 0 \text{ or } 1, \quad 1 \le j \le p,$$

and let λ be a large positive integer so that for each j, $1 \le j \le p$, at least one of the numbers $\alpha_1^j, \alpha_2^j, \ldots, \alpha_\lambda^j$ is equal to 1. Let

$$n_j' = 2^{\lambda+\kappa} \sum_{k=1}^{\lambda+\kappa} \alpha_k^j 2^{-k}.$$

If at least two of the numbers n_j', $1 \le j \le p$, attain their maximum values, then $n_j = n_j'$, $1 \le j \le p$. If there is j_0 such that $n_{j_0}' > n_j'$ for $j \ne j_0$, we let $n_{j_0} = n_j'$ and $n_j = n_j' + 1$ for $j \ne j_0$. It is clear that (8.4) holds in the first case. In the second case, inequality (8.4) follows from the inequality

$$\Theta_{j_0} \le \sum_{\substack{1 \le j \le p,\\ j \ne j_0}} \Theta_j,$$

which holds by condition (2) in the definition of a γ-scheme. Now we let $\theta_{jk} = \Theta_j / n_j$, $1 \le k \le n_j$. It is easy to see that

$$\begin{aligned} \frac{\Theta_j}{n_j} &\le \frac{\Theta_j}{n_j'} = 2^{-\lambda-\kappa}\left(1 + \frac{\sum_{k=\lambda+\kappa+1}^{\infty} \alpha_k^j 2^{-k}}{\sum_{k=1}^{\lambda+\kappa} \alpha_k^j 2^{-k}}\right) \\ &\le 2^{-\lambda-\kappa}\left(1 + \frac{\sum_{k=\lambda+\kappa+1}^{\infty} 2^{-k}}{2^{-\lambda}}\right) = 2^{-\lambda-\kappa}(1 + 2^{-\kappa}). \end{aligned} \tag{8.6}$$

On the other hand,

$$\frac{\Theta_j}{n_j} \ge \frac{\Theta_j}{n_j' + 1} = \frac{\Theta_j}{n_j'}\left(1 - \frac{1}{n_j' + 1}\right) \ge 2^{-\lambda-\kappa}\left(1 - \frac{1}{n_j'}\right) \ge 2^{-\lambda-\kappa}(1 - 2^{-\kappa}). \tag{8.7}$$

Inequalities (8.6), (8.7), and (8.5) imply that

$$\left|\frac{\theta_{jk}}{\theta_{il}} - 1\right| \le \frac{1 + 2^{-\kappa}}{1 - 2^{-\kappa}} - 1 < 2 - \Theta,$$

that is, (8.3) holds.

We write δ_j and ε_j in the form

$$\delta_j = \sum_{k=1}^{n_j} \Delta_{jk}, \qquad \varepsilon_j = \sum_{k=1}^{n_j} E_{jk}, \qquad 1 \le j \le p,$$

where

$$\Delta_{jk} + E_{jk} = \theta_{jk}, \quad \Delta_{jk} > 0, \quad E_{jk} \ge 0, \quad 1 \le k \le n_j.$$

We can always do this in such a way that Δ_{11} is equal to rL, where r is a rational number. To achieve this in the case where $\varepsilon_1 = 0$, we possibly need to change somewhat θ_{11} and θ_{12} without changing the sum $\theta_{11} + \theta_{12}$; it is clear that this can be done without violating any of the strict inequalities in (8.3).

We take a surface F with n almost periodic ends $\Lambda_1, \dots, \Lambda_n$. We use the notation $\tilde{B}_\nu$, b_ν, $s = s(\nu)$, $j = j(\nu)$, Δ_ν, E_ν in the same way as in the case of a β-scheme, but now $1 \le \nu \le n$ and the indices ν are taken $\mod n$ rather than $\mod 2n$, and the set of values $\{b_1, \dots, b_n\}$ coincides with $\{a_j\}$, $1 \le j \le p$. In addition, we let $\theta_\nu = \theta_{js}$, $j = j(\nu)$, $s = s(\nu)$. The possibility to satisfy the condition $b_\nu \ne b_{\nu\pm 1}$, and hence the possibility to choose all ends to be ends of the first kind, follows from (8.4).

We choose a periodic end Λ_ν in such a way that the following conditions are satisfied:

$$\text{(8.8)} \qquad \begin{cases} \delta(a_{j(\nu)}, \Delta_\nu) = d_\nu^+ = \Delta_\nu \frac{\Theta}{\theta_\nu + \theta_{\nu+1}}, \ \varepsilon(a_{j(\nu)}, \Delta_\nu) = E_\nu \frac{\Theta}{\theta_\nu + \theta_{\nu+1}}, \\ \delta(a_{j(\nu+1)}, \Lambda_\nu) = d_\nu^- = \Delta_{\nu+1} \frac{\Theta}{\theta_\nu + \theta_{\nu+1}}, \\ \varepsilon(a_{j(\nu+1)}, \Lambda_\nu) = E_{\nu+1} \frac{\Theta}{\theta_\nu + \theta_{\nu+1}}, \\ \delta(a, \Lambda_\nu) = 0 \ \text{ for } \ a \ne a_{j(\nu)}, a_{j(\nu+1)}, \\ \varepsilon(a_j, \Lambda_\nu) - \varepsilon_j, \ p+1 \le j \le q, \\ \varepsilon(a, \Lambda_\nu) = 0 \ \text{ for } \ a \ne a_{j(\nu)}, a_{j(\nu+1)}, a_{p+1}, \dots, a_q. \end{cases}$$

Such an end can be constructed since by (8.3) we have

$$\text{(8.9)} \qquad \Delta_\nu \frac{\Theta}{\theta_\nu + \theta_{\nu+1}} + E_\nu \frac{\Theta}{\theta_\nu + \theta_{\nu+1}} = \theta_\nu \frac{\Theta}{\theta_\nu + \theta_{\nu+1}} = \frac{\Theta}{1 + \frac{\theta_{\nu+1}}{\theta_\nu}} < 1,$$

$$\text{(8.10)} \qquad \Delta_{\nu+1} \frac{\Theta}{\theta_\nu + \theta_{\nu+1}} + E_{\nu+1} \frac{\Theta}{\theta_\nu + \theta_{\nu+1}} = \theta_{\nu+1} \frac{\Theta}{\theta_\nu + \theta_{\nu+1}} = \frac{\Theta}{1 + \frac{\theta_\nu}{\theta_{\nu+1}}} < 1,$$

and also

$$\theta_\nu \frac{\Theta}{\theta_\nu + \theta_{\nu+1}} + \theta_{\nu+1} \frac{\Theta}{\theta_\nu + \theta_{\nu+1}} + \varepsilon_{p+1} + \dots + \varepsilon_q = \Theta + (2 - \Theta) = 2.$$

The weight of the surface F is equal to

$$\begin{aligned} &\frac{1}{\Delta_1 \frac{\Theta}{\theta_1 + \theta_2}} + \frac{\Delta_2 \frac{\Theta}{\theta_1 + \theta_2}}{\Delta_1 \frac{\Theta}{\theta_1 + \theta_2} \Delta_2 \frac{\Theta}{\theta_2 + \theta_3}} + \dots + \frac{\Delta_2 \frac{\Theta}{\theta_1 + \theta_2} \cdots \Delta_n \frac{\Theta}{\theta_{n-1} + \theta_n}}{\Delta_1 \frac{\Theta}{\theta_1 + \theta_2} \cdots \Delta_n \frac{\Theta}{\theta_n + \theta_1}} \\ &\quad = \frac{1}{\Delta_1} \frac{\sum_{j=1}^n (\theta_j + \theta_{j+1})}{\Theta} = \frac{2}{\Delta_1} = \frac{2}{\Delta_{11}} = \frac{2}{r} L. \end{aligned}$$

Using the notation $\sum_j', \sum_j''$ in the same way as in the solution of a β-scheme, by (7.21) we get

$$\begin{aligned}
\delta(a_j, F) &= \frac{\Delta_1}{2}\sum_{\nu=1}^{n}\frac{\theta_\nu+\theta_{\nu+1}}{\Delta_1\Theta}\delta(a_j,\Lambda_\nu)\\
&= \frac{1}{2\Theta}\left\{\sum_j{}'(\theta_\nu+\theta_{\nu+1})\delta(a_{j(\nu)},\Lambda_\nu)+\sum_j{}''(\theta_\nu+\theta_{\nu+1})\delta(a_{j(\nu+1)},\Lambda_\nu)\right\}\\
&= \frac{1}{2\Theta}\sum_j{}'\{(\theta_\nu+\theta_{\nu+1})\delta(a_{j(\nu)},\Lambda_\nu)+(\theta_{\nu-1}+\theta_\nu)\delta(a_{j(\nu)},\Lambda_{\nu-1})\}\\
&= \frac{1}{2\Theta}\sum_j{}'\left\{(\theta_\nu+\theta_{\nu+1})\Delta_\nu\frac{\Theta}{\theta_\nu+\theta_{\nu+1}}+(\theta_{\nu-1}+\theta_\nu)\Delta_\nu\frac{\Theta}{\theta_{\nu-1}+\theta_\nu}\right\}\\
&= \sum_j{}'\Delta_\nu=\sum_{s=1}^{n_j}\Delta_{js}=\delta_j, \qquad 1\le j\le p.
\end{aligned}$$

Similarly we get that $\varepsilon(a_j, F) = \varepsilon_j$, $1 \le j \le p$. For $p+1 \le j \le q$ the equality $\varepsilon(a_j, F) = \varepsilon_j$ is obvious.

A *δ-scheme* is a scheme satisfying the conditions (1) $\delta_j > 0$ for $1 \le j \le p$, $2 \le p < q$, and $\delta_j = 0$ for $p+1 \le j \le q$; (2′) $\Theta = \sum_{j=1}^{p}\Theta_j \le 1$; (3) $\varepsilon_{p+1} > 0$.

As we see, the only difference between the conditions for a δ-scheme and for a γ-scheme is that condition (2) is replaced by condition (2′). The scheme is solved almost the same way as a γ-scheme; for this reason we restrict ourselves to a description of differences between the solutions of γ- and δ-schemes. We used condition (2) in the proof of inequalities (8.4), (8.9), and (8.10). But inequalities (8.9) and (8.10) for a δ-scheme are obviously satisfied by condition (2′).

Condition (8.4) is not satisfied in general, and for this reason we cannot use only the ends of the first kind. If an end Λ_ν is of the first kind, formulas (8.8) remain true. If the end Λ_ν is of the second kind, that is, $a_{j(\nu)} = a_{j(\nu+1)}$, we choose it in such a way that

$$\begin{gathered}
\delta(a_{j(\nu)},\Lambda_\nu)=(\Delta_\nu+\Delta_{\nu+1})\frac{\Theta}{\theta_\nu+\theta_{\nu+1}},\\
d_\nu^+=\Delta_\nu\frac{\Theta}{\theta_\nu+\theta_{\nu+1}},\qquad d_\nu^-=\Delta_{\nu+1}\frac{\Theta}{\theta_\nu+\theta_{\nu+1}},\\
\delta(a,\Lambda_\nu)=0 \ \text{ for } \ a\ne a_{j(\nu)},\\
\varepsilon(a_{j(\nu)},\Lambda_\nu)=(E_\nu+E_{\nu+1})\frac{\Theta}{\theta_\nu+\theta_{\nu+1}},\\
\varepsilon(a_j,\Lambda_\nu)=\varepsilon_j,\qquad p+1\le j\le q,\\
\varepsilon(a,\Lambda_\nu)=0 \ \text{ for } \ a\ne a_{j(\nu)},a_{p+1},\dots,a_q.
\end{gathered}$$

Such an end can be chosen because

$$\delta(a_{j(\nu)},\Lambda_\nu)+\varepsilon(a_{j(\nu)},\Lambda_\nu)=(\Delta_\nu+\Delta_{\nu+1}+E_\nu+E_{\nu+1})\frac{\Theta}{\theta_\nu+\theta_{\nu+1}}=\Theta\le 1$$

by condition (2′). The remaining computations are carried out in the same way as for a γ-scheme.

Observe that in the solution of a δ-scheme we do not need to obtain (8.3); therefore we can take $\Theta_1 = \theta_{11} + \theta_{12}$, $\Theta_j = \theta_{j1}$, $2 \le j \le p$, that is, $n_1 = 2$,

$n_2 = \cdots = n_p = 1$. With this choice of n_j, condition (8.4) is not satisfied for $p = 2$ only, and only in this case it is necessary to use an end of the second kind. Although for $p = 2$ we can solve a δ-scheme with the help of a surface Φ with two almost periodic ends of the first kind (see Theorem 6.2), but, generally speaking, we will not be able to ensure the preassigned, up to a rational factor, weight of our scheme.

It is easy to see that a scheme can be a γ-scheme and a δ-scheme simultaneously.

Now we consider schemes of the general form. Let $\delta(a_j) > 0$ for $1 \le j \le p$, $p \le q$, and $\delta(a_j) = 0$ for $p+1 \le j \le q$ (if $p < q$). We may assume that $p \ge 2$, since for $p = 1$ we get an α-scheme. Set $\Theta = \sum_{j=1}^{p} \Theta_j$. We may assume without loss of generality that $\Theta_1 = \max_{1 \le j \le p} \Theta_j$. We consider two cases: Case A: $\Theta < 2$; Case B: $\Theta = 2$. We divide Case A into three subcases: (1) A$'$ if $\delta_1 = 1$, (2) A$''$ if $\Theta_1 = 1$ and $\varepsilon_1 > 0$, and (3) A$'''$ if $\Theta_1 < 1$.

SUBCASE A$'$. Set

$$\Theta' = \sum_{j=2}^{p} \Theta_j, \qquad \Theta'' = \sum_{j=p+1}^{q} \varepsilon_j.$$

It is clear that $\Theta' < 1$ and $\Theta'' > 0$. We write Θ_j, $2 \le j \le p$, as a sum of n_j positive summands θ_{jk} (see (8.2)) such that

$$\left| \frac{\theta_{jk}}{\theta_{il}} - 1 \right| < 2 - 2\Theta', \qquad 2 \le j \le p,\ 2 \le i \le p,\ 1 \le k \le n_j,\ 1 \le l \le n_i. \tag{8.11}$$

The possibility of such a representation can be proved similarly to the proof of representation (8.2) under condition (8.3). Now we write δ_j and ε_j in the form

$$\delta_j = \sum_{k=1}^{n_j} \Delta_{jk}, \qquad \varepsilon_j = \sum_{k=1}^{n_j} E_{jk}, \qquad 2 \le j \le p,$$

where

$$\Delta_{jk} + E_{jk} = \theta_{jk}, \qquad \Delta_{jk} > 0,\ E_{jk} \ge 0,\ 1 \le k \le n_j,\ 2 \le j \le p.$$

We take a surface F with $2n = 2\sum_{j=2}^{p} n_j$ almost periodic ends $\Lambda_1, \dots, \Lambda_{2n}$ and with logarithmic ramification points $\tilde{B}_1, \dots, \tilde{B}_{2n}$. The points $\tilde{B}_{2\nu}$, $1 \le \nu \le n$, are projected to a_1, and there are n_j logarithmic ramification points lying over a_j, $2 \le j \le p$. Suppose that $\tilde{B}_{2\nu-1}$, $1 \le \nu \le n$, is sth logarithmic ramification point among points projected to a_j, $s = s(\nu)$, $j = j(\nu)$. Let $\Delta_\nu = \Delta_{js}$, $E_\nu = E_{js}$, $\theta_\nu = \theta_{js}$, $j = j(\nu)$, $s = s(\nu)$. We take all numbers $\nu \mod n$. We take the ends Λ_k of the first kind such that for $\Lambda_{2\nu-1}$ the following is true:

$$\begin{aligned}
\delta(a_1, \Lambda_{2\nu-1}) &= d_{2\nu-1}^{-} = 1, \\
\delta(a_{j(\nu)}, \Lambda_{2\nu-1}) &= d_{2\nu-1}^{+} = \Delta_\nu \frac{2\Theta'}{\theta_\nu + \theta_{\nu+1}}, \\
\delta(a, \Lambda_{2\nu-1}) &= 0 \ \text{ for } \ a \ne a_1, a_{j(\nu)}, \\
\varepsilon(a_{j(\nu)}, \Lambda_{2\nu-1}) &= E_\nu \frac{2\Theta'}{\theta_\nu + \theta_{\nu+1}}, \\
\varepsilon(a_j, \Lambda_{2\nu-1}) &= k_\nu \varepsilon_j, \quad p+1 \le j \le q,
\end{aligned}$$

where

$$k_\nu = \frac{1}{\Theta''}\left(1 - \frac{2\theta_\nu \Theta'}{\theta_\nu + \theta_{\nu+1}}\right),$$

$$\varepsilon(a, \Lambda_{2\nu-1}) = 0 \quad \text{for} \quad a \neq a_{j(\nu)},\ a_{p+1}, \ldots, a_q,$$

and for $\Lambda_{2\nu}$ the following is true:

$$\delta(a_1, \Lambda_{2\nu}) = d_{2\nu}^+ = 1,$$

$$\delta(a_{j(\nu+1)}, \Lambda_{2\nu}) = d_{2\nu}^- = \Delta_{\nu+1}\frac{2\Theta'}{\theta_\nu + \theta_{\nu+1}},$$

$$\delta(a, \Lambda_{2\nu}) = 0 \quad \text{for} \quad a \neq a_1, a_{j(\nu+1)},$$

$$\varepsilon(a_{j(\nu+1)}, \Lambda_{2\nu}) = E_{\nu+1}\frac{2\Theta'}{\theta_\nu + \theta_{\nu+1}},$$

$$\varepsilon(a_j, \Lambda_{2\nu}) = l_\nu \varepsilon_j, \quad p+1 \le j \le q,$$

where

$$l_\nu = \frac{1}{\Theta''}\left(1 - \frac{2\theta_{\nu+1} \Theta'}{\theta_\nu + \theta_{\nu+1}}\right),$$

$$\varepsilon(a, \Lambda_{2\nu}) = 0 \quad \text{for} \quad a \neq a_{j(\nu+1)},\ a_{p+1}, \ldots, a_q.$$

Let us show that such almost periodic ends can be constructed. In fact, inequalities (8.11) imply that

$$\frac{2\theta_\nu \Theta'}{\theta_\nu + \theta_{\nu+1}} = \frac{2\Theta'}{1 + \frac{\theta_{\nu+1}}{\theta_\nu}} < \frac{2\Theta'}{1 + (-1 + 2\Theta')} = 1$$

and similarly

$$\frac{2\theta_{\nu+1}\Theta'}{\theta_\nu + \theta_{\nu+1}} < 1;$$

hence the numbers k_ν and l_ν are positive, and $\delta(a_{j(\nu)}, \Lambda_{2\nu-1}) + \varepsilon(a_{j(\nu)}, \Lambda_{2\nu-1}) < 1$, $\delta(a_{j(\nu+1)}, \Lambda_{2\nu}) + \varepsilon(a_{j(\nu+1)}, \Lambda_{2\nu}) < 1$. Besides, for $p+1 \le j \le q$ we have

$$k_\nu \varepsilon_j \le k_\nu \Theta'' < 1 \quad \text{and} \quad l_\nu \varepsilon_j \le l_\nu \Theta'' < 1.$$

Thus, by Theorem 7.2 we can construct a Riemann surface F with the indicated distribution of values of almost periodic ends.

Since $\delta(a_1, \Lambda_k) = 1$ for all Λ_k, we have $\delta(a_1, F) = 1$.[14]

Now let $2 \le j \le p$. Using the notation $\sum_j'$ and $\sum_j''$ in the same way as before, we get

$$(8.12)\qquad \begin{aligned} \delta(a_j, F) &= \frac{\sum_j' \frac{d_1^- \cdots d_{2\nu-2}^-}{d_1^+ \cdots d_{2\nu-1}^+}\delta(a_{j(\nu)}, \Lambda_{2\nu-1}) + \sum_j'' \frac{d_1^- \cdots d_{2\nu-1}^-}{d_1^+ \cdots d_{2\nu}^+}\delta(a_{j(\nu+1)}, \Lambda_{2\nu})}{\sum_{k=1}^{2n} \frac{d_1^- \cdots d_{k-1}^-}{d_1^+ \cdots d_k^+}} \\ &= \frac{2\sum_j' \frac{d_1^- \cdots d_{2\nu-2}^-}{d_1^+ \cdots d_{2\nu-1}^+}\delta(a_{j(\nu)}, \Lambda_{2\nu-1})}{\sum_{k=1}^{2n} \frac{d_1^- \cdots d_{k-1}^-}{d_1^+ \cdots d_k^+}}. \end{aligned}$$

It is easy to see that

$$\frac{d_1^- \cdots d_{k-1}^-}{d_1^+ \cdots d_k^+} = \frac{\theta_l + \theta_{l+1}}{2\Theta' \Delta_1}, \qquad k = 2l-1, 2l \ (1 \le l \le n).$$

[14]Obviously, the surface F can be constructed in such a way that it does not cover the point a_1, in which case the value a_1 is a Picard exceptional value for the mapping function $f(z)$.

Therefore

$$L(F) = \sum_{k=1}^{2n} \frac{d_1^- \cdots d_{k-1}^-}{d_1^+ \cdots d_k^+} = 2\sum_{l=1}^{n} \frac{\theta_l + \theta_{l+1}}{2\Theta' \Delta_1} = \frac{1}{\Theta' \Delta_1} \sum_{l=1}^{n} (\theta_l + \theta_{l+1}) = \frac{2}{\Delta_1}, \tag{8.13}$$

$$\begin{aligned} \sum_j{}' \frac{d_1^- \cdots d_{2\nu-2}^-}{d_1^+ \cdots d_{2\nu-1}^+} \delta(a_{j(\nu)}, \Lambda_{2\nu-1}) &= \sum_j{}' \frac{\theta_\nu + \theta_{\nu+1}}{2\Theta' \Delta_1} \Delta_\nu \frac{2\Theta'}{\theta_\nu + \theta_{\nu+1}} \\ &= \frac{1}{\Delta_1} \sum_j{}' \Delta_\nu = \frac{1}{\Delta_1} \sum_{s=1}^{n_j} \Delta_{js} = \frac{\delta_j}{\Delta_1}. \end{aligned} \tag{8.14}$$

From (8.12), (8.13), and (8.14) we get that $\delta(a_j, F) = \delta_j$. Similarly we find that $\varepsilon(a_j, F) = \varepsilon_j$, $2 \le j \le p$.

For $p + 1 \le j \le q$, taking into account (7.22) and the equality $k_\nu + l_\nu = \frac{2}{\Theta''}(1 - \Theta') = 2$, we find that

$$\begin{aligned} \varepsilon(a_j, F) &= \frac{\Delta_1}{2} \left\{ \sum_{\nu=1}^{n} \frac{d_1^- \cdots d_{2\nu-2}^-}{d_1^+ \cdots d_{2\nu-1}^+} k_\nu \varepsilon_j + \sum_{\nu=1}^{n} \frac{d_1^- \cdots d_{2\nu-1}^-}{d_1^+ \cdots d_{2\nu}^+} l_\nu \varepsilon_j \right\} \\ &= \frac{\Delta_1}{2} \varepsilon_j \left\{ \sum_{\nu=1}^{n} \frac{\theta_\nu + \theta_{\nu+1}}{2\Theta' \Delta_1} k_\nu + \sum_{\nu=1}^{n} \frac{\theta_\nu + \theta_{\nu+1}}{2\Theta' \Delta_1} l_\nu \right\} \\ &= \frac{\varepsilon_j}{4\Theta'} \sum_{\nu=1}^{n} (\theta_\nu + \theta_{\nu+1})(k_\nu + l_\nu) = \frac{\varepsilon_j}{2\Theta'} \sum_{\nu=1}^{n} (\theta_\nu + \theta_{\nu+1}) = \varepsilon_j. \end{aligned}$$

Thus, the constructed surface F has all desired properties.

SUBCASES A$''$ AND A$'''$. The corresponding schemes are solved in the following way. After finitely many operations of the form

$$\begin{aligned} &(\delta_\nu + \varepsilon_\nu) \begin{pmatrix} \delta_1 & \ldots & \delta_\nu & \ldots \\ \ldots & \ldots & \varepsilon_\nu & \ldots \end{pmatrix} \\ &= \delta_\nu \begin{pmatrix} \delta_1 & \ldots & \delta_\nu + \varepsilon_\nu & \ldots \\ \ldots & \ldots & 0 & \ldots \end{pmatrix} + \varepsilon_\nu \begin{pmatrix} \delta_1 & \ldots & 0 & \ldots \\ \ldots & \ldots & \delta_\nu + \varepsilon_\nu & \ldots \end{pmatrix}, \end{aligned} \tag{8.15}$$

where $\delta_1 > 0$, the problem is reduced to a solution with weight of schemes of the form

$$\begin{pmatrix} \delta_1 & 0 & 0 & \ldots \\ \varepsilon_1 & \varepsilon_2 & \varepsilon_3 & \ldots \end{pmatrix}, \begin{pmatrix} \delta_1 & \delta_2 & \ldots & \delta_n & 0 & \ldots & 0 \\ \varepsilon_1 & 0 & \ldots & 0 & \varepsilon_{n+1} & \ldots & \varepsilon_q \end{pmatrix},$$

where $\delta_1, \ldots, \delta_n > 0$. Moreover, the numbers δ_j, ε_j, $2 \le j \le q$, in these schemes are in general different from the original, and also the numbers a_j, $2 \le j \le q$, could be subjected to a permutation. The first of these schemes is an α-scheme. The second scheme is either a γ-scheme, or it satisfies the inequality $2\Theta_1 > \sum_{j=1}^n \Theta_j$ (we recall that for the original scheme we assumed that $\Theta_1 \ge \Theta_j$, $2 \le j \le p$, and in decompositions of the form (8.15) the quantities Θ_j do not change). In the latter case, taking into account $\sum_{j=2}^n \Theta_j < \Theta_1 \le 1$, we reduce the problem by a sequence of operations of the type

$$\begin{aligned}(\delta_{n-1}+\delta_n)&\begin{pmatrix}\delta_1 & \dots & \delta_{n-1} & \delta_n & \dots & 0\\ \varepsilon_1 & \dots & 0 & 0 & \dots & \varepsilon_q\end{pmatrix}\\ &=\delta_{n-1}\begin{pmatrix}\delta_1 & \dots & \delta_{n-1}+\delta_n & 0 & \dots & 0\\ \varepsilon_1 & \dots & 0 & 0 & \dots & \varepsilon_q\end{pmatrix}\\ &+\delta_n\begin{pmatrix}\delta_1 & \dots & 0 & \delta_{n-1}+\delta_n & \dots & 0\\ \varepsilon_1 & \dots & 0 & 0 & \dots & \varepsilon_q\end{pmatrix}\end{aligned}$$

to a solution with weight of a scheme of the form

$$\begin{pmatrix}\delta_1 & \delta_2 & 0 & \dots & 0\\ \varepsilon_1 & 0 & \varepsilon_3 & \dots & \varepsilon_q\end{pmatrix},$$

where $\delta_2=\Theta_2<\delta_1+\varepsilon_1+\Theta_1$. If we needed just to solve this scheme, we would be able to refer to Theorem 6.4; however, we need to solve this scheme with weight. Set $\Theta''=\sum_{j=3}^q\varepsilon_j$. If $\Theta''\geq 1$, then $\Theta_1+\Theta_2\leq 1$, and our scheme is a δ-scheme. Further consideration is needed in the case $\Theta''<1$.

In Subcase A$''$, the equalities $\delta_1+\varepsilon_1=1$, $\delta_2+\Theta''=1$ hold, and we decompose

$$\begin{aligned}&1\begin{pmatrix}\delta_1 & \delta_2 & 0 & \dots & 0\\ \varepsilon_1 & 0 & \varepsilon_3 & \dots & \varepsilon_q\end{pmatrix}\\ &=\delta_2\begin{pmatrix}\delta_1 & 1 & 0 & \dots & 0\\ \varepsilon_1 & 0 & 0 & \dots & 0\end{pmatrix}+\Theta''\begin{pmatrix}\delta_1 & 0 & 0 & \dots & 0\\ \varepsilon_1 & 0 & \frac{\varepsilon_3}{\Theta''} & \dots & \frac{\varepsilon_q}{\Theta''}\end{pmatrix}.\end{aligned}\tag{8.16}$$

The first scheme in the right-hand side of (8.16) is a β-scheme; the second is an α-scheme.

In Subcase A$'''$ we observe that $0<1-\Theta''=\Theta_1+\delta_2-1<2\Theta_1-1$. Then

$$\begin{aligned}(2\Theta_1-1)&\begin{pmatrix}\delta_1 & \delta_2 & 0 & \dots & 0\\ \varepsilon_1 & 0 & \varepsilon_3 & \dots & \varepsilon_q\end{pmatrix}\\ &=(\Theta_1+\delta_2-1)\begin{pmatrix}\delta_1 & \Theta_1 & 0 & \dots & 0\\ \varepsilon_1 & 0 & \varepsilon_3\frac{2(1-\Theta_1)}{\Theta''} & \dots & \varepsilon_q\frac{2(1-\Theta_1)}{\Theta''}\end{pmatrix}\\ &+(\Theta_1-\delta_2)\begin{pmatrix}\delta_1 & 1-\Theta_1 & 0 & \dots & 0\\ \varepsilon_1 & 0 & \varepsilon_3/\Theta'' & \dots & \varepsilon_q/\Theta''\end{pmatrix}.\end{aligned}\tag{8.17}$$

Here we take into account that

$$(2\Theta_1-1)\varepsilon_j=(\Theta_1+\delta_2-1)\varepsilon_j\frac{2(1-\Theta_1)}{\Theta''}+(\Theta_1-\delta_2)\frac{\varepsilon_j}{\Theta''},\qquad 3\leq j\leq q,$$

by the equality $(\Theta_1+\delta_2-1)2(1-\Theta_1)+(\Theta_1-\delta_2)=(1-\Theta'')2(1-\Theta_1)+(\Theta''+2\Theta_1-2)=\Theta''(2\Theta_1-1)$. The first scheme in the right-hand side of equality (8.17) is a γ-scheme; the second is a δ-scheme.

CASE B. We may assume that $\varepsilon_1>0$, but Θ_1 is not necessarily the largest number among Θ_j, $j=1,\dots,q$. The scheme is of the form

$$\begin{pmatrix}\delta_1 & \dots & \delta_q\\ \varepsilon_1 & \dots & \varepsilon_q\end{pmatrix},\qquad \delta_j>0,\ 1\leq j\leq q,\quad \varepsilon_1>0.\tag{8.18}$$

Let $\Theta_1<1$. We choose rational numbers r_j, $1\leq j\leq q$, such that (a) $0<r_1<\Theta_1<1$; (b) $1\geq r_j\geq\Theta_j$ for $2\leq j\leq q$; (c) $\sum_{j=1}^q r_j=2$; (d) $r_j-\Theta_j<\Theta_j\frac{\Theta_1-r_1}{2(1-\Theta_1)}$ for $2\leq j\leq q$. The fact that r_j can be chosen in such a way that (a), (b), and (c) are satisfied, is obvious. The fact that condition (d) can also be satisfied follows from the inequality

$$\sum_{j=2}^{q} \Theta_j \frac{\Theta_1 - r_1}{2(1-\Theta_1)} = \frac{2-\Theta_1}{2-2\Theta_1}(\Theta_1 - r_1) > \Theta_1 - r_1 = \sum_{j=2}^{q}(r_j - \Theta_j).$$

We decompose (8.18) in the following way:

$$(8.19)\qquad \begin{aligned}(1-r_1)\begin{pmatrix} \delta_1 & \cdots & \delta_q \\ \varepsilon_1 & \cdots & \varepsilon_q \end{pmatrix} &= (1-\Theta_1)\begin{pmatrix} \frac{\delta_1}{\Theta_1}r_1 & \cdots & \frac{\delta_q}{\Theta_q}r_q \\ \frac{\varepsilon_1}{\Theta_1}r_1 & \cdots & \frac{\varepsilon_q}{\Theta_q}r_q \end{pmatrix} \\ &\quad + (\Theta_1 - r_1)\begin{pmatrix} \delta_1' & \delta_2' & \cdots & \delta_q' \\ \varepsilon_1' & \varepsilon_2' & \cdots & \varepsilon_q' \end{pmatrix},\end{aligned}$$

where $\delta_1' = \delta_1/\Theta_1$, $\varepsilon_1' = \varepsilon_1/\Theta_1$, and for $2 \le j \le q$ we have

$$\delta_j' = \frac{\delta_j}{\Theta_j}\left\{\Theta_j - \frac{1-\Theta_1}{\Theta_1 - r_1}(r_j - \Theta_j)\right\} > \frac{\delta_j}{2} > 0,$$

$$\varepsilon_j' = \frac{\varepsilon_j}{\Theta_j}\left\{\Theta_j - \frac{1-\Theta_1}{\Theta_1 - r_1}(r_j - \Theta_j)\right\} > \frac{\varepsilon_j}{2} \ge 0,$$

by condition (d). The first scheme in the right-hand side of (8.19) is a β-scheme, and $\Theta_1' = \delta_1' + \varepsilon_1' = 1$ in the second scheme. Therefore we may assume that $\Theta_1 = 1$ in (8.18).

Let $\varepsilon_\nu > 0$, $2 \le \nu \le q$. We represent the scheme (8.18) as a sum of two schemes:

$$(8.20)\qquad \begin{aligned}(\delta_\nu + \varepsilon_\nu)&\begin{pmatrix} \delta_1 & \cdots & \delta_\nu & \cdots & \delta_q \\ \varepsilon_1 & \cdots & \varepsilon_\nu & \cdots & \cdots \end{pmatrix} \\ &= \delta_\nu \begin{pmatrix} \delta_1 & \cdots & \delta_\nu + \varepsilon_\nu & \cdots & \delta_q \\ \varepsilon_1 & \cdots & 0 & \cdots & \cdots \end{pmatrix} + \varepsilon_\nu \begin{pmatrix} \delta_1 & \cdots & 0 & \cdots \\ \varepsilon_1 & \cdots & \delta_\nu + \varepsilon_\nu & \cdots \end{pmatrix}.\end{aligned}$$

The second scheme (S) in the right-hand side of equality (8.19) corresponds to Subcase A″. As we have seen, it is possible to solve the scheme (S) constructing a surface F with weight L. The scheme (S) can be represented as a finite sum of basic schemes $(S_1), \dots, (S_t)$ with weights $L_1, \dots, L_t$, respectively, where, by (7.25), $L = L_1 + \cdots + L_t$. We shall write this junction in the form of the following equality: $L(S) = L_1(S_1) + \cdots + L_t(S_t)$. However, we need to get a scheme (S) with weight equal to ε_ν, up to a rational factor. We do the following. We solve a basic scheme (S_j) with weight equal to $r_j\varepsilon_\nu L_j/L$, $1 \le j \le t$, where r_j is a rational number, and denote the corresponding surface by F_j. Let $r_j = m_j/n$, where n is the least common denominator of the fractions $r_1, \dots, r_t$. Now we join $m_2m_3\cdots m_t$ copies of the surface F_1, $m_1m_3\cdots m_t$ copies of the surface F_2, ..., $m_1m_2\cdots m_{t-1}$ copies of the surface F_t. By (7.25), the weight of the obtained surface F' is equal to

$$\begin{aligned} m_2m_3\cdots m_t\frac{r_1\varepsilon_\nu L_1}{L} &+ m_1m_3\cdots m_t\frac{r_2\varepsilon_\nu L_2}{L} + \cdots + m_1m_2\cdots m_{t-1}\frac{r_t\varepsilon_\nu L_t}{L} \\ &= m_1m_2\cdots m_t\frac{\varepsilon_\nu}{nL}(L_1 + \cdots + L_t) \\ &= \frac{m_1\cdots m_t}{n}\varepsilon_\nu,\end{aligned}$$

that is, to a rational multiple of ε_ν.

Using formulas (7.26) and (7.27), it is easy to see that the surface F' solves the scheme (S), that is, the second scheme in the right-hand side of (8.20), with the desired weight.

As for the first scheme in the right-hand side of (8.20), we decompose it again according to formula (8.20), but now with another ν. Repeating this operation finitely many times, we get, besides the schemes corresponding to Subcase A$''$, a scheme of the form (8.18) with $\varepsilon_2 = \cdots = \varepsilon_q = 0$. Next, using successive operations

$$(\delta_{n-1}+\delta_n)\begin{pmatrix} \delta_1 & \dots & \delta_{n-1} & \delta_n & \dots \\ \varepsilon_1 & \dots & 0 & 0 & \dots \end{pmatrix}$$
$$= \delta_{n-1}\begin{pmatrix} \delta_1 & \dots & \delta_{n-1}+\delta_n & 0 & \dots \\ \varepsilon_1 & \dots & 0 & 0 & \dots \end{pmatrix} + \delta_n\begin{pmatrix} \delta_1 & \dots & 0 & \delta_{n-1}+\delta_n & \dots \\ \varepsilon_1 & \dots & 0 & 0 & \dots \end{pmatrix},$$

we reduce the scheme to schemes of the form

$$\begin{pmatrix} \delta_1 & 0 & \dots & 0 & 1 & 0 & \dots \\ \varepsilon_1 & 0 & \dots & 0 & 0 & 0 & \dots \end{pmatrix},$$

which are β-schemes.

This completes the proof of Theorem 8.1.

REMARK. We can drop the restriction that there are finitely many numbers ε_j and consider the case of infinite sequences $\varepsilon_1, \varepsilon_2, \dots, 0 \le \varepsilon_j \le 1$ and $a_1, a_2, \dots$. In such a case condition (1$'$) should be replaced by the condition

$$\sum_{j=1}^{q} \delta_j + \sum_{j=1}^{\infty} \varepsilon_j = 2,$$

and instead of Riemann surfaces with almost periodic ends, we consider surfaces with generalized almost periodic ends (see the remark at the end of Section 6), which are constructed in the same way as surfaces with almost periodic ends. The main results of Section 7 can be transferred onto this wider class of surfaces without substantial changes. In order to get the indicated generalization of Theorem 8.1 it will be necessary to solve schemes with infinitely many columns:

$$\begin{pmatrix} \delta_1 & \dots & \delta_q & 0 & 0 & \dots \\ \varepsilon_1 & \dots & \varepsilon_q & \varepsilon_{q+1} & \varepsilon_{q+2} & \dots \end{pmatrix},$$

where $\varepsilon_j > 0$ for $j \ge q+1$. These schemes correspond to Case A, and we solve them, as before, joining to each of the schemes the same set of columns of the form $\binom{0}{\varepsilon_j}$ with numbers $j \ge q+1$, including the case of α-, γ-, and δ-schemes. We cannot join any columns to β-schemes, but β-schemes occur only in Case B.

Now we prove the following theorem.

THEOREM 8.2. *Let δ_j, ε_j, $j = 1, \dots, q$, be non-negative numbers satisfying*

(1) $\sum_{j=1}^{q}\{\delta_j + \varepsilon_j\} \le 2$;
(2) $0 < \delta_j + \varepsilon_j \le 1$, $1 \le j \le q$;
(3) $0 < \sum_{j=1}^{q} \delta_j \le 2$.

Moreover, if $\sum_{j=1}^{q} \delta_j = 2$, then all the numbers $\delta_1, \dots, \delta_q$ are assumed to be rational. Let $a_1, \dots, a_q$ be arbitrary distinct complex numbers. Then there exists a meromorphic function $f(z)$ of finite order, such that $\delta(a_j, f) = \delta_j$, $\varepsilon(a_j, f) = \varepsilon_j$, $1 \le j \le q$; $\delta(a, f) = 0$, $\varepsilon(a, f) = 0$ for $a \ne a_j$, $1 \le j \le q$.

The theorem has been already proved in the case where (1) holds with equality: in this case we can take $f(z)$ to be a mapping function of the Riemann surface F

which was mentioned in Theorem 8.1. So now we can assume that

$$E = 2 - \sum_{j=1}^{q} \{\delta_j + \varepsilon_j\} > 0.$$

We take unequal numbers a_{q+1} and a_{q+2}, different from $a_1, \ldots, a_q$. Let $\varepsilon_{q+1} = \varepsilon_{q+2} = E/2$. Then there exists a meromorphic function $w = f_1(z)$ such that

$$\tag{8.21} \begin{cases} n(r, a, f_1) = Kr^\rho + o(r^\rho) \ \text{ for } \ a \neq a_1, \ldots, a_q, \\ n(r, a_j, f_1) = K(1 - \delta_j)r^\rho + o(r^\rho) \ \text{ for } \ 1 \leq j \leq q, \\ n_1(r, a_j, f_1) = K\varepsilon_j r^\rho + o(r^\rho) \ \text{ for } \ 1 \leq j \leq q+2, \\ n_1(r, a, f_1) = o(r^\rho) \ \text{ for } \ a \neq a_j, \ 1 \leq j \leq q+2, \end{cases}$$

where K and ρ are positive constants. This statement follows from Theorem 8.1 and formulas (7.12), (7.13), (7.18), (7.19) (using these formulas we assume, as we agreed on p. 408, that $\chi = 1$).

Conditions (8.21) imply that

$$\tag{8.22} n(r+1, a, f_1) - n(r-1, a, f_1) = o(r^\rho)$$

for each a. Let $\{z_{k\nu}\}_{k=1}^{\infty}$ be the sequence of all multiple $a_{q+\nu}$-points of the function $f_1(z)$, $\nu = 1, 2$. Take pairwise disjoint discs $C_{k\nu} = \{|z - z_{k\nu}| < r_{k\nu}\}$, $k = 1, 2, \ldots$, $\nu = 1, 2$, such that $r_{k\nu} < 1$ and

$$\tag{8.23} \sum_{\nu=1}^{2} \sum_{k=1}^{\infty} \iint_{C_{k\nu} \cap \{|z|>1\}} d\sigma(\ln z) < \infty.$$

Suppose that the function $w_1 = f_1(z)$ maps the complex z-plane onto some Riemann surface F_1. We denote the image of the disc $C_{k\nu}$ under this mapping by $D_{k\nu}$, $D_{k\nu} \subset F_1$. The region $D_{k\nu}$ contains one algebraic ramification point $W_{k\nu}^1$ which is projected to $a_{q+\nu}$, that is, to the image of the point $z_{k\nu}$. We take a small circular neighborhood $U_{k\nu}^1$ of this ramification point so that $U_{k\nu}^1$ does not contain other ramification points, and $U_{k\nu}^1 \subset D_{k\nu}$. The region $U_{k\nu}^1$ is the Riemann surface of the function $(w - a_{q+\nu})^{1/m_{k\nu}}$ lying over the disc $\{|w - a_{q+\nu}| < \varepsilon_{k\nu}\}$, where $m_{k\nu} - 1$ is the order of the ramification point $W_{k\nu}^1$, and $\varepsilon_{k\nu}$ is a sufficiently small positive number. Let $U_{k\nu}$ be the Riemann surface of the function $(w - a_{k\nu})^{1/m_{k\nu}}$, $|a_{k\nu} - a_{q+\nu}| < \frac{1}{2}\varepsilon_{k\nu}$, lying over the disc $\{|w - a_{q+\nu}| < \varepsilon_{k\nu}\}$. We choose the points $a_{k\nu}$ in such a way that all of them are distinct when $k = 1, 2, \ldots$, $\nu = 1, 2$, and all of them are different from the projections of the algebraic ramification points of the Riemann surface F_1.

Let $w = \varphi_{k\nu}(w_1)$ be a quasiconformal mapping of $U_{k\nu}^1$ onto $U_{k\nu}$, such that the points on the boundary of $U_{k\nu}^1$ are fixed and the ramification point $W_{k\nu}^1$ is mapped to the ramification point $W_{k\nu}$ of the same order $m_{k\nu} - 1$, which is projected to $a_{k\nu}$. The mapping $w = \varphi_{k\nu}(w_1)$ can be chosen in such a way that $p(w_1, w) < 3$ (see Example 8 in Section 2).

Let $F_0 = F_1 \backslash \bigcup_{k,\nu} U_{k\nu}^1$. Denote by F the Riemann surface which we get if we replace the regions $U_{k\nu}^1$ by the regions $U_{k\nu}$ on the surface F_1. It is clear that $F_0 = F \backslash \bigcup_{k,\nu} U_{k\nu}$. Denote the preimages of F_0 and $U_{k\nu}^1$ under the mapping $w = f_1(z)$ by G_0 and $G_{k\nu}$, respectively. Obviously, $G_{k\nu} \subset C_{k\nu}$. Let $w = \varphi(\zeta)$ be a quasiconformal mapping of the complex ζ-plane onto the Riemann surface F, given by the formulas

$$\varphi(\zeta) = \begin{cases} f_1(\zeta) & \text{for } \zeta \in G_0, \\ \varphi_{k\nu}(f_1(\zeta)) & \text{for } \zeta \in G_{k\nu},\ k = 1, 2, \dots,\ \nu = 1, 2. \end{cases}$$

It is clear that for each a the functions $f_1(\zeta)$ and $\varphi(\zeta)$ have the same number of a-points in $G_{k\nu}$. Since the diameter of each of the regions $G_{k\nu}$ is less than 2, relations (8.22) and (8.21) imply that

$$\text{(8.24)} \qquad \begin{cases} n(r, a, \varphi) = Kr^\rho + o(r^\rho) \ \text{ for } \ a \neq a_1, \dots, a_q, \\ n(r, a_j, \varphi) = K(1 - \delta_j)r^\rho + o(r^\rho) \ \text{ for } \ 1 \leq j \leq q, \\ n_1(r, a_j, \varphi) = K\varepsilon_j r^\rho + o(r^\rho) \ \text{ for } \ 1 \leq j \leq q, \\ n_1(r, a, \varphi) = o(r^\rho) \ \text{ for } \ a \neq a_j,\ 1 \leq j \leq q. \end{cases}$$

We can even claim that $n_1(r, a, \varphi) = n_1(r, a, f_1)$ for $a \neq a_1, \dots, a_{q+2}$, $k = 1, 2, \dots$, $\nu = 1, 2$, $n_1(r, a, \varphi) = 0$ for $a = a_{q+1}, a_{q+2}$, and $n_1(r, a, \varphi) = O(1)$ for $a = a_{k\nu}$.

Taking into account (8.23) we get that the mapping $w = \varphi(\zeta)$ satisfies

$$\iint_{|\zeta|>1} \{p(\zeta, w) - 1\}\, d\sigma(\ln \zeta) = \sum_{\nu=1}^{2} \sum_{k=1}^{\infty} \iint_{G_{k\nu} \cap \{|\zeta|>1\}} \{p(\zeta, w) - 1\}\, d\sigma(\ln \zeta)$$

$$\leq 3 \sum_{\nu=1}^{2} \sum_{k=1}^{\infty} \iint_{G_{k\nu} \cap \{|\zeta|>1\}} d\sigma(\ln \zeta) \leq 3 \sum_{\nu=1}^{2} \sum_{k=1}^{\infty} \iint_{C_{k\nu} \cap \{|\zeta|>1\}} d\sigma(\ln \zeta) < \infty.$$

Using Teichmüller's Theorem 3.1 the same way as we have repeatedly done before, we find that the surface F is of parabolic type, and relations (8.24) remain true for the meromorphic function $w = f(z)$ which maps $\{|z| < \infty\}$ onto F, after an appropriate normalization of $f(z)$. From this we see that

$$T(r, f) = \frac{K}{\rho} r^\rho + o(r^\rho),$$

$$N(r, a, f) = \frac{K}{\rho} r^\rho + o(r^\rho) \ \text{ for } \ a \neq a_1, \dots, a_q,$$

$$N(r, a_j, f) = \frac{K}{\rho}(1 - \delta_j) r^\rho + o(r^\rho) \ \text{ for } \ 1 \leq j \leq q,$$

$$N_1(r, a_j, f) = \frac{K}{\rho} \varepsilon_j r^\rho + o(r^\rho) \ \text{ for } \ 1 \leq j \leq q,$$

$$N_1(r, a, f) = o(r^\rho) \ \text{ for } \ a \neq a_j,\ 1 \leq j \leq q.$$

These relations imply that the function $w = f(z)$ is a desired function, whose existence was claimed in Theorem 8.2. It is clear that our remark to Theorem 8.1 is applicable to Theorem 8.2 also. It is clear from our argument that if $\delta_j = 1$, then the function $f(z)$ can be chosen in such a way that $f(z) \neq a_j$.

We can drop the restriction $\sum_{j=1}^{q} \delta_j > 0$ in Theorem 8.2, but in this case we have to avoid using Riemann surfaces with almost periodic ends because their mapping functions necessarily have deficient values.

We prove the following theorem.

THEOREM 8.3. *Let $\varepsilon_1, \varepsilon_2, \dots$ be a sequence of numbers satisfying $0 \leq \varepsilon_j \leq 1$, $\sum_{j=1}^{\infty} \varepsilon_j \leq 2$, and let $a_1, a_2, \dots$ be a sequence of pairwise distinct complex numbers. There exists a meromorphic function $w = f(z)$ such that $\varepsilon(a_j, f) = \varepsilon_j$ for $j = 1, 2, \dots$, $\varepsilon(a, f) = 0$ for $a \neq a_j$, $j \geq 1$; and $\delta(a, f) = 0$ for all a.*

Let $b_1, b_2, \dots$ be a sequence of distinct complex numbers, $b_j \neq a_k$, j, $k = 1, 2, \dots$. Let e_{jk}, $1 \le j \le k$, be integers satisfying

(1) $0 \le e_{jk} \le k-1$, $1 \le j \le k-1$;
(2) $0 \le \sum_{j=1}^{k} e_{jk} \le 2(k-1)$;
(3) $\lim_{k\to\infty} \frac{e_{jk}}{k} = \varepsilon_j$ for each j.

It is clear that such numbers e_{jk} can be selected. Let β'_k and β''_k be integers such that $0 \le \beta'_{jk}, \beta''_k \le k-1$ and

$$\sum_{j=1}^{k} e_{jk} + \beta'_k + \beta''_k = 2(k-1).$$

Let Φ_k be a Riemann surface from the class

$$\mathfrak{F}(k; a_1 : e_{1k}; a_2 : e_{2k}; \dots ; a_k : e_{kk}; b_{2k-1} : \beta'_k; b_{2k} : \beta''_k), \qquad k = 1, 2, \dots .$$

By Theorem 4.1 such a surface exists.

It is always possible to find a positive number u_k and a number r_k, $0 < r_k < 1/4$, such that $|u_k - k| < 1/4$ and there are no ramification points of the surface Φ_k over the disc $\{|w - u_k| \le r_k\}$. Denote by G_k a fixed one-sheeted region on Φ_k lying over $\{|w - u_k| < r_k\}$. Let $\Phi_k^0 = \Phi_k \backslash G_k$. Let $w_1 = \psi_k(w)$ be a quasiconformal mapping of the surface Φ_k^0 onto the region $\{|w_1 - u_k| \ge r_k\}$, under which the points on the boundary of Φ_k^0 remain fixed. The existence of such a quasiconformal mapping follows from Example 9 in Section 2. Let the characteristic of this quasiconformal mapping be bounded by a constant p_k. Let $r' = \min(r_k, p_k^{-1/2})$. We introduce the notation $G'_k = G_k \cap \{|w - u_k| < r'_k\}$, $\Phi'_k = \Phi_k \backslash G_k^1$.

It is easy to see that there exists a quasiconformal mapping of Φ'_k onto the disc $\{|\zeta - u_k| \le r'_k/4\}$ with constant distortion on the boundary and with characteristic at most p_k. In fact, first we map Φ'_k onto $\{|w - u_k| \ge r'_k\}$; to this end we map $\Phi_k^0 \subset \Phi'_k$ using the function $\psi_k(w)$, and leave the region $G_k \backslash G'_k$ fixed. Then we map the region $\{|w - u_k| \ge r'_k\}$ onto $\{|\zeta - u_k| \le r'_k/4\}$ using the linear-fractional transformation $\zeta = (r'^2_k/4)(w - u_k)^{-1} + u_k$. In this case the point on $\partial\Phi'_k$ which is projected to $w = u_k + r'_k$ is mapped to the point $w_1 = u_k + r'_k/4$. Denote by $\zeta = \omega_k(w)$ the constructed mapping of Φ'_k onto $\{|\zeta - u_k| \le r'_k/4\}$, and by $w = \Omega_k(\zeta)$, the inverse mapping.

We make a cut on Φ_k along the line segment $\Gamma_k \subset G'_k$, which is projected into the line segment

$$\gamma_k = \left\{u_k - \frac{r'_k}{2} \le u \le u_k + \frac{r'_k}{2},\ v = 0\right\}, \qquad w = u + iv.$$

Take the complex w-plane with cuts along γ_k, $k = 1, 2, \dots$, and denote it by P. We paste to P the surfaces Φ_k with cuts along Γ_k joining the upper (lower) side of the cut Γ_k to the lower (upper) side of the cut γ_k. We get a simply connected Riemann surface F.

Let $\tilde{G}'_k = G'_k \backslash \Gamma_k$, $G''_k = \{|w - u_k| < r'_k\} \cap P$. Denote the part of the Riemann surface F consisting of $\tilde{G}'_k$ and G''_k by W_k. It is a doubly connected Riemann surface with two algebraic ramification points of the first order over the points $u_k \pm r'_k/2$. We denote the boundary curves of W_k by C_{1k} and C_{2k}; both of them are projected into $\{|w - u_k| = r'_k\}$. Moreover, $C_{1k} \subset \partial G''_k$, $C_{2k} \subset \partial \tilde{G}'_k$. By Example 10 of Section 2, there exists a quasiconformal mapping of the surface W_k onto the annulus $\{r'_k/4 < |\zeta - u_k| < r'_k\}$, such that the point $w = u_k + r'_k e^{i\theta} \in C_{1k}$ is mapped to the point $\zeta = u_k + r'_k e^{i\theta}$, and the point $w = u_k + r'_k e^{-i\theta} \in C_{2k}$ is mapped to the point $\zeta = u_k + \frac{r'_k}{4} e^{i\theta}$. Denote this quasiconformal mapping by $\zeta = s_k(w)$, and its inverse by $w = S_k(\zeta)$. As was shown in Example 10, the characteristic of this mapping is bounded by a constant Q, which obviously does not depend on k.

Let $P_0 = P \backslash (\bigcup_{k=1}^{\infty} G''_k)$. The Riemann surface F can be partitioned into pairwise disjoint parts $P_0, W_1, W_2, \ldots, \Phi'_1, \Phi'_2, \ldots$. There exists a quasiconformal mapping of the surface F onto the complex ζ-plane which leaves P_0 fixed, maps the surfaces W_k by $\zeta = s_k(w)$, and the surfaces Φ'_k by $\zeta = \omega_k(w)$. It is easy to verify that such a mapping is one-to-one and continuous everywhere on the surface F. We denote the inverse quasiconformal mapping of $\{|\zeta| < \infty\}$ onto F by $w = \varphi(\zeta)$:

$$w = \varphi(\zeta) = \begin{cases} \zeta & \text{for } \zeta \in P_0, \\ S_k(\zeta) & \text{for } r'_k/4 < |\zeta - u_k| < r'_k, \ k = 1, 2, \ldots, \\ \Omega_k(\zeta) & \text{for } |\zeta - u_k| \leq r'_k/4, \ k = 1, 2 \ldots. \end{cases}$$

It is clear that $p(\zeta, w) \equiv 1$ for $\zeta \in P_0$, $p(\zeta, w) \leq Q$ for $r'_k/4 < |\zeta - u_k| < r'_k$, $k = 1, 2, \ldots$, and $p(\zeta, w) \leq p_k$ for $|\zeta - u_k| \leq r'_k/4$. Hence

$$\begin{aligned} \iint_{|\zeta|>1/2} \{p(\zeta, w) - 1\}\, d\sigma(\ln \zeta) &= \sum_{k=1}^{\infty} \iint_{|\zeta - u_k| \leq r'_k} \{p(\zeta, w) - 1\}\, d\sigma(\ln \zeta) \\ &\leq \sum_{k=1}^{\infty} \max(Q, p_k) \iint_{|\zeta - u_k| \leq r'_k} d\sigma(\ln \zeta) \\ &\leq \sum_{k=1}^{\infty} \max(Q, p_k) O\left(\frac{r'^2_k}{k^2}\right) = \sum_{k=1}^{\infty} O\left(\frac{1}{k^2}\right) < \infty. \end{aligned} \tag{8.25}$$

Here we have used the condition $r'^2_k \leq 1/p_k$.

Now we consider the distribution of the a-points of the function $w = \varphi(z)$. It is clear that all a-points except at most one lie in the discs $\{|\zeta - u_k| < r'_k < 1/4\}$. Moreover, the number of a-points in each of these discs is equal to k if $|a - u_k| \geq r'_k$, and to $k+1$ if $|a - u_k| < r'_k$. Therefore

$$n(r, a, \varphi) = \sum_{k=1}^{[r]} k + O(r) = \frac{r^2}{2} + O(r) \tag{8.26}$$

for each a. Similarly we get that

$$n_1(r, a, \varphi) = O(1) \tag{8.27}$$

for $a \neq a_j$, $j = 1, 2, \ldots$, and

$$\begin{aligned} n_1(r, a_j, \varphi) &= \sum_{k=1}^{[r]} e_{jk} + O(r) \\ &= \varepsilon_j \sum_{k=1}^{[r]} k + \sum_{k=1}^{[r]} o(k) + O(r) = \varepsilon_j \frac{r^2}{2} + o(r^2). \end{aligned} \tag{8.28}$$

for $a = a_j$.

Using Teichmüller's Theorem 3.1, which is applicable by (8.25), in a way standard for this chapter, we find that there exists a meromorphic function $w = f(z)$

for which relations (8.26)–(8.28) remain true if we replace φ by f and $O(r)$ by $o(r^2)$ in them. Now we get that

$$N(r, a, f) = \frac{r^2}{4} + o(r^2) \quad \text{for all} \quad a,$$

$$T(r, f) = \frac{r^2}{4} + o(r^2),$$

$$N_1(r, a_j, f) = \varepsilon_j \frac{r^2}{4} + o(r^2), \quad j = 1, 2, \dots,$$

$$N_1(r, a, f) = O(\ln r) \quad \text{for} \quad a \neq a_j, \ j = 1, 2, \dots .$$

It is clear that the function $w = f(z)$ satisfies the assumptions of Theorem 8.3.

APPENDIX A

On the Magnitude of Type for an Entire Function

It was proved in Theorem 1.3 of Chapter 2 that for an entire function $f(z)$, the characteristics $T(r, f)$ and $\ln M(r, f)$ have the same growth category. Let $\rho(r)$ be a proximate order such that $\rho(r) \to \rho$ as $r \to \infty$, where $\rho < \infty$ is the order of an entire function.

Set

$$\Delta_T = \limsup_{r\to\infty} \frac{T(r,f)}{r^{\rho(r)}}, \qquad \Delta_M = \limsup_{r\to\infty} \frac{\ln M(r,f)}{r^{\rho(r)}}.$$

Theorem 1.3 of Chapter 2 implies that Δ_T and Δ_M are equal to 0 or ∞ simultaneously, since $0 < \Delta_T < \infty$ implies $0 < \Delta_M < \infty$, and vice versa. Now we establish more precise relations between Δ_T and Δ_M. The inequality $\Delta_T \le \Delta_M$ is obvious, because $T(r, f) \le \ln^+ M(r, f)$. We now prove the following theorem.

THEOREM A.1. *The following inequalities hold:*

$$\Delta_M \le \pi\rho\csc(\pi\rho)\Delta_T \quad \text{for} \quad 0 \le \rho \le 1/2, \tag{A.1}$$

$$\Delta_M \le \pi\rho\Delta_T \quad \text{for} \quad 1/2 < \rho < \infty. \tag{A.2}$$

Observe that (A.1) and (A.2) hold with equalities for the Mittag-Leffler function (5.31) of Chapter 2 for $\rho(r) \equiv \rho$, $0 < \rho < \infty$.[1] If $\rho = 0$, the equality $\Delta_M = \Delta_T$ holds for each entire function (see Remark 1 to Theorem 4.5 of Chapter 2). Therefore we can restrict our attention to the case $\rho > 0$.

It is relatively easy to prove inequality (A.1). We may assume without loss of generality that $f(0) = 1$. Then $N(r, 0, f) \le T(r, f)$. Let $\{a_k\}$ denote a sequence of zeros. Then (cf. (4.3) in Chapter 2)

$$\begin{aligned} \ln M(r,f) &= \ln M\left(r, \prod_{k-1}^{\infty} E\left(\frac{z}{a_k}, 0\right)\right) \le \sum_{k=1}^{\infty} \ln\left(1 + \frac{r}{|a_k|}\right) \\ &= \int_0^\infty \ln\left(1 + \frac{r}{t}\right) dn(t, 0, f) = r\int_0^\infty \frac{n(t,0,f)}{t}\frac{dt}{r+t} \\ &= r\int_0^\infty N(t,0,f)\frac{dt}{(r+t)^2} \\ &\le r\int_0^\infty T(t,f)\frac{dt}{(r+t)^2} = \int_0^\infty T(\tau r, f)\frac{d\tau}{(1+\tau)^2}. \end{aligned} \tag{A.3}$$

---

[1]Similar examples can be constructed in the case where $\rho(r)$ is not identically equal to ρ. For example, if $0 < \rho < 1$, we can take the function (5.1) from Chapter 2 with $l(r) \equiv \rho(r)$ and use formulas (5.4′) and (5.11).

We may assume that $\Delta_T < \infty$, since for $\Delta_T = \infty$ inequality (A.1) is obvious. Let $T(r,f) \le (\Delta_T + \varepsilon) r^{\rho(r)}$ hold for $r \ge r_0(\varepsilon)$. Then (A.3) implies

$$\ln M(r,f) \le T(r_0,f) \int_0^{r_0/r} \frac{d\tau}{(1+\tau)^2} + (\Delta_T + \varepsilon) \int_{r_0/r}^{\infty} (r\tau)^{\rho(r\tau)} \frac{d\tau}{(1+\tau)^2},$$

$$\text{(A.4)}\quad \begin{aligned} \Delta_M = \limsup_{r\to\infty} \frac{\ln M(r,f)}{r^{\rho(r)}} &\le (\Delta_T + \varepsilon) \lim_{r\to\infty} \int_{r_0/r}^{\infty} \frac{(r\tau)^{\rho(r\tau)}}{r^{\rho(r)}} \frac{d\tau}{(1+\tau)^2} \\ &= (\Delta_T + \varepsilon) \int_0^{\infty} \frac{\tau^{\rho} d\tau}{(1+\tau)^2} = (\Delta_T + \varepsilon) \frac{\pi\rho}{\sin \pi\rho}. \end{aligned}$$

The possibility of passing to the limit under the integral sign is proved in the same way as in Exercise 1 in Section 2 of Chapter 2; the last integral can be easily computed by using the residue calculus. Since ε can be chosen to be arbitrarily small, inequality (A.4) implies (A.1).

The proof of inequality (A.2) is much more complicated; it was obtained only recently by N. V. Govorov [**Gov69**], although it was conjectured long time ago (see, for example, [**Hay64**, Chapter 1, 1.8]).

Let $f(z)$ be an entire function of order ρ, $1/2 < \rho < \infty$, and let $\rho(r)$ be a proximate order, $\rho(r) \to \rho$ as $r \to \infty$. We may restrict ourselves to the case where $0 < \Delta_M < \infty$. Besides, without loss of generality, we may assume that $f(0) = 1$.

Let ε_0 be an arbitrary number satisfying the condition

$$0 < \varepsilon_0 < \min\left(2\rho - 1, \frac{2}{3}\rho\right).$$

Choose ε, $0 < \varepsilon < \varepsilon_0$, such that the function $f(z)$ does not have zeros on the rays $\arg z = \pm\pi/(2\rho - \varepsilon)$. Set $\sigma = 1/(2\rho - \varepsilon)$, $0 < \sigma < 1$, and

$$T_1(r) = \frac{1}{2\pi} \int_{-\pi\sigma}^{\pi\sigma} \ln |f(re^{i\theta})|\, d\theta.$$

It is clear that

$$\text{(A.5)}\qquad T_1(r) \le m(r,f) = T(r,f).$$

The function $\zeta = -z^{2\rho - \varepsilon}$ is a conformal mapping of the angle $\{|\arg z| < \pi\sigma\}$ onto the region $D = \{0 < \arg \zeta < 2\pi\}$, that is, ζ-plane with a cut along the positive real semi-axis. We consider the function

$$\text{(A.6)}\qquad \Phi(\zeta) = f\{(-\zeta)^{\sigma}\}$$

in D, where the single-valued branch of the function $(-\zeta)^{\sigma}$ is chosen so that $(-\zeta)^{\sigma} > 0$ for $\zeta < 0$. Let $\zeta = \xi + i\eta$. We denote by ξ^+ and ξ^- the points $\zeta = \xi > 0$ lying on the upper and lower sides of the cut S, respectively. Obviously,

$$\Phi(\xi^{\pm}) = f(\xi^{\sigma} e^{\pm i\pi\sigma}).$$

Let

$$G(\xi) = \Phi(\xi^+)/\Phi(\xi^-) = f(\xi^{\sigma} e^{-i\pi\sigma})/f(\xi^{\sigma} e^{i\pi\sigma}), \quad \xi > 0, \qquad G(0) = 1.$$

It is clear that the function $G(\xi)$ is analytic on the ray $\{\xi > 0\}$, and the function $G(\xi^{1/\sigma})$ is analytic, also, at the point $\xi = 0$; moreover, $G(\xi) \ne 0$ for $\xi \ge 0$.

We choose a branch of $\ln G(\xi)$ which is continuous for $\xi \ge 0$ and satisfies the condition $\ln G(0) = 0$. Let

$$s(r) = \int_0^r \ln G(\xi) \frac{d\xi}{\xi}.$$

Let us show that

$$s(r) = o(r^{3/4}) \quad \text{as} \quad r \to \infty. \tag{A.7}$$

In fact, let $f(z)$ be without zeros on the ray $\{\arg z = \theta_0\}$. We choose a continuous branch of $\ln f(z)$ on this ray, such that $\ln f(0) = 0$. Applying Theorem 2.4 of Chapter 1 to $\ln f(z)$, we get $(R = 2r > r > t > 0)$

$$\begin{aligned} \ln f(te^{i\theta_0}) &= \frac{1}{2\pi} \int_0^{2\pi} \ln |f(Re^{i\theta})| \frac{Re^{i\theta} + te^{i\theta_0}}{Re^{i\theta} - te^{i\theta_0}} d\theta \\ &\quad - \sum_{|a_m|<R} \ln \left\{ \frac{|a_m|}{\bar{a}_m} \frac{R^2 - \bar{a}_m te^{i\theta_0}}{R(a_m - te^{i\theta_0})} \right\}, \end{aligned} \tag{A.8}$$

where a_m are the zeros of the function $f(z)$, and the branches of the logarithms in the sum are chosen so that $\operatorname{Im} \ln \frac{R}{|a_m|} = 0$. Then

$$\left| \operatorname{Im} \ln \left\{ \frac{|a_m|}{\bar{a}_m} \frac{R^2 - \bar{a}_m te^{i\theta_0}}{R(a_m - te^{i\theta_0})} \right\} \right| \le 2\pi$$

for each t, $0 \le t < \infty$. Since

$$0 < \operatorname{Re} \ln \left\{ \frac{|a_m|}{\bar{a}_m} \frac{R^2 - \bar{a}_m te^{i\theta_0}}{R(a_m - te^{i\theta_0})} \right\} = \ln \left| \frac{R^2 - \bar{a}_m te^{i\theta_0}}{R(a_m - te^{i\theta_0})} \right| \le \ln \frac{2R}{||a_m| - t|},$$

from (A.8) we get

$$\begin{aligned} |\ln f(te^{i\theta_0})| &\le \frac{R+t}{R-t} \cdot \frac{1}{2\pi} \int_0^{2\pi} |\ln |f(Re^{i\theta})|| \, d\theta + \sum_{|a_m|<R} \ln \frac{2R}{||a_m| - t|} + n(R, 0, f) 2\pi \\ &\le \frac{R+r}{R-r} \left\{ m(R, f) + m\left(R, \frac{1}{f}\right) \right\} + \sum_{|a_m|<R} \ln \frac{2R}{||a_m| - t|} + 2\pi n(R, 0, f) \\ &\le 6T(R, f) + 2\pi n(R, 0, f) + \sum_{|a_m|<R} \ln \frac{2R}{||a_m| - t|}. \end{aligned}$$

We integrate this inequality with respect to $d \ln t$ from $t = 1$ to $t = r$. We get

$$\int_1^r |\ln f(te^{i\theta_0})| \frac{dt}{t} \le 6T(R, f) \ln R + 2\pi n(R, 0, f) \ln R + \sum_{|a_m|<R} \int_1^r \ln \frac{2R}{||a_m| - t|} \frac{dt}{t}.$$

But

$$\begin{aligned} \int_1^r \ln \frac{2R}{||a_m| - t|} \frac{dt}{t} &\le \int_{|a_m|-1}^{|a_m|+1} \ln \frac{2R}{||a_m| - t|} dt + \ln 2R \int_1^r \frac{dt}{t} \\ &= 2 \int_0^1 \ln \frac{2R}{\tau} d\tau + \ln 2R \cdot \ln r \\ &= 2\{\ln(2R) + 1\} + \ln 2R \cdot \ln r < (\ln r)^2 + 4 \ln r + 6. \end{aligned}$$

Therefore

$$\int_1^r \frac{|\ln f(te^{i\theta_0})|}{t}dt \leq 6T(2r,f)\ln 2r + n(2r,0,f)\{2\pi\ln(2r) + (\ln r)^2 + 4\ln r + 6\}.$$

Since the order of the functions $T(r,f)$ and $n(r,0,f)$ does not exceed ρ, the relations

$$\int_0^r \frac{|\ln f(te^{i\theta_0})|}{t}dt = O(r^{\rho'})$$

and

$$\int_0^r \frac{|\ln f(t^\sigma e^{i\theta_0})|}{t}dt = O(r^{\rho'\sigma}) = o(r^{3/4})$$

hold for each ρ', $3/(4\sigma) > \rho' > \rho$. Relation (A.7) follows.

Denote the number of zeros of the function $\Phi(\zeta)$ in the region $\{0 < \arg\zeta < 2\pi,\ |\zeta| \leq r\}$ or, which is the same, the number of zeros of the function $f(z)$ in the sector $\{|\arg z| < \pi\sigma, |z| \leq r^\sigma\}$, by $n_2(r)$. Then $n_2(r) = O(r^{\rho'\sigma})$ and

$$n_2(r) = o(r^{3/4}). \tag{A.9}$$

Suppose that the arc $\{|\zeta| = t,\ 0 < \arg < 2\pi\}$ does not contain zeros of the function $\Phi(\zeta)$. Let

$$T_2(r) = \frac{1}{2\pi}\int_0^{2\pi} \ln|\Phi(re^{i\alpha})|\,d\alpha.$$

Then, applying the argument principle and the Cauchy–Riemann equations, we find that

$$\begin{aligned}
n_2(t) &= \frac{1}{2\pi}\int_0^t d\arg G(\xi) + \frac{1}{2\pi}\int_{\alpha=0}^{2\pi} d\arg\Phi(te^{i\alpha})\\
&= \frac{1}{2\pi}\arg G(t) + \frac{1}{2\pi}\int_0^{2\pi}\frac{\partial\arg\Phi(te^{i\alpha})}{\partial\alpha}d\alpha\\
&= \frac{1}{2\pi}\arg G(t) + t\frac{d}{dt}\left\{\frac{1}{2\pi}\int_0^{2\pi}\ln|\Phi(te^{i\alpha})|d\alpha\right\}\\
&= \frac{1}{2\pi}\arg G(t) + t\frac{dT_2(t)}{dt}.
\end{aligned}$$

We divide this equality by t and integrate from 0 to r. Taking into account $n_2(0) = T_2(0) = 0$, we get the equality

$$T_2(r) = \int_0^r \left\{n_2(t) - \frac{1}{2\pi}\arg G(t)\right\}\frac{dt}{t}. \tag{A.10}$$

Now we consider the analytic function

$$F(\zeta) = \exp\left\{\frac{\zeta}{2\pi i}\int_0^\infty \frac{s(t)}{(t-\zeta)^2}dt\right\} = \exp\{\zeta\psi(\zeta)\} \tag{A.11}$$

in the region D. Relation (A.7) implies the absolute convergence of the integral in (A.11). Integrating it by parts, we get

$$\psi(\zeta) = \frac{1}{2\pi i}\int_0^\infty \frac{s(t)}{(t-\zeta)^2}dt = \frac{1}{2\pi i}\int_0^\infty \frac{\ln G(t)}{t(t-\zeta)}dt. \tag{A.11$'$}$$

Let $0 < \xi_0 < \infty$. We show that

$$\psi(\xi_0^+) - \psi(\xi_0^-) = \frac{\ln G(\xi_0)}{\xi_0}. \tag{A.12}$$

The equality (A.12) is a special case of the well-known formulas due to Yu. V. Sokhotskii; we prove only the special case that we need.

Take a circle $\gamma = \{|\zeta - \xi_0| = r\}$, where $r < \xi_0$ is chosen so small that the function $\ln G(\zeta)$ is analytic in the disc $\{|\zeta - \xi_0| \le r\}$. If $\operatorname{Im}\zeta > 0$, then by the Cauchy theorem, we can write the function $\psi(\zeta)$ as

$$\psi(\zeta) = \frac{1}{2\pi i}\int_{\Gamma^-} \frac{\ln G(t)}{t(t-\zeta)}\,dt, \tag{A.13}$$

where the curve Γ^- consists of the positive real semi-axis with the interval $\{|\xi - \xi_0| < r\}$ removed and the lower semicircle $\{|\zeta - \xi_0| = r,\ \operatorname{Im}\zeta < 0\}$. It is clear that with the help of (A.13) the function $\psi(\zeta)$ can be continued analytically from $\{\operatorname{Im}\zeta > 0\}$ into the half-disc $\{|\zeta - \xi_0| < r,\ \operatorname{Im}\zeta < 0\}$ through the interval $\{\eta = 0, |\xi - \xi_0| < r\}$. Therefore $\psi(\xi_0^+)$ exists and

$$\psi(\xi_0^+) = \frac{1}{2\pi i}\int_{\Gamma^-} \frac{\ln G(t)}{t(t-\xi_0)}\,dt.$$

Similarly we get that

$$\psi(\xi_0^-) = \frac{1}{2\pi i}\int_{\Gamma^+} \frac{\ln G(t)}{t(t-\xi_0)}\,dt,$$

where the curve Γ^+ consists of the positive real semi-axis with the interval $\{|\xi - \xi_0| < r\}$ removed and the upper semicircle $\{|\zeta - \xi_0| = r,\ \operatorname{Im}\zeta > 0\}$. Since $\Gamma^- - \Gamma^+ = \gamma$, we have

$$\psi(\xi_0^+) - \psi(\xi_0^-) = \frac{1}{2\pi i}\int_{\gamma} \frac{\ln G(t)}{t(t-\xi_0)}\,dt = \frac{\ln G(\xi_0)}{\xi_0}.$$

It is also clear from the preceding argument that the functions $\psi(\xi^+)$ and $\psi(\xi^-)$ arc continuous for $0 < \xi < \infty$.

Let us show that

$$\zeta\psi(\zeta) \to 0 \quad \text{as} \quad \zeta \to 0,\ \zeta \in D. \tag{A.14}$$

The function $\ln G(\zeta^{1/\sigma})$ is analytic in a sufficiently small neighborhood of $\zeta = 0$ and

$$\ln G(\zeta^{1/\sigma}) = \sum_{k=1}^{\infty} c_k\zeta^k;$$

hence, the function $\ln G(\zeta)/\zeta$ is analytic for $\{|\arg\zeta| < \pi,\ |\zeta| \le r_0\}$, and

$$|\ln G(\zeta)/\zeta| = O(|\zeta|^{\sigma-1}) \quad \text{as} \quad \zeta \to 0. \tag{A.15}$$

Let $\operatorname{Im}\zeta \le 0$. We write

$$\begin{aligned}\psi(\zeta) &= \frac{1}{2\pi i}\int_{S_1} \frac{\ln G(t)}{t(t-\zeta)}\,dt + \frac{1}{2\pi i}\int \frac{\ln G(t)}{t(t-\zeta)} \\ &= \frac{1}{2\pi i}\int_{S_1} \frac{\ln G(t)}{t(t-\zeta)}\,dt + O(1), \qquad \zeta \to 0,\end{aligned} \tag{A.16}$$

where S_1 is the line segment joining 0 and ir_0, and S_2 is a curve consisting of the line segment joining the points ir_0 and r_0, and the ray $\{\xi \ge r_0,\ \eta = 0\}$. If $t \in S_1$ and $\operatorname{Im}\zeta \le 0$, then $|t - \zeta| > |\zeta|$. Taking into account (A.15) and (A.16) we obtain that

$$|\psi(\zeta)| \le \frac{1}{|\zeta|}\frac{1}{2\pi}\int_0^{r_0} O(t^{\sigma-1})\,dt + O(1) = \frac{1}{|\zeta|}O(r_0^\sigma) + O(1).$$

Hence

$$\limsup_{\zeta\to 0,\ \mathrm{Im}\le 0} |\zeta\psi(\zeta)| \le O(r_0^\sigma).$$

Since r_0 can be chosen to be arbitrarily small, relation (A.14) follows. A similar argument works in the case where $\mathrm{Im}\,\zeta \ge 0$. Thus, relation (A.14) has been proved.

Formulas (A.11), (A.12), and (A.14) imply that

$$\text{(A.17)}\qquad \frac{F(\xi^+)}{F(\xi^-)} = G(\xi) = \frac{\Phi(\xi^+)}{\Phi(\xi^-)}, \qquad \frac{\Phi(\xi^+)}{F(\xi^+)} = \frac{\Phi(\xi^-)}{F(\xi^-)},$$

and in a neighborhood of $\zeta = 0$,

$$\frac{1}{K} < |F(\zeta)| < K < \infty,$$

where K is some constant. Therefore the function

$$\Omega(\zeta) = \Phi(\zeta)/F(\zeta)$$

analytic in D admits analytic continuation into the whole complex ζ-plane and is an entire function. Taking into account (A.6) and (A.14), we see that $\Omega(0) = 1$.

Equality (A.11) implies that $(0 < \varphi < 2\pi)$

$$|\ln|F(re^{i\varphi})|| \le \frac{r}{2\pi}\int_0^\infty \frac{|s(t)|\,dt}{|t - re^{i\varphi}|^2} = \frac{1}{2\pi}\int_0^\infty \frac{|s(r\tau)|\,d\tau}{|\tau - e^{i\varphi}|^2}.$$

Since $|\tau - e^{i\varphi}| > (\tau + 1)\sin\frac{\varphi}{2}$ (see Section 5 of Chapter 2), taking into account (A.7), we get that

$$|\ln|F(re^{i\varphi})|| \le \frac{1}{\sin^2\frac{\varphi}{2}}\int_0^\infty \frac{o\{(rt)^{3/4}\}}{(t+1)^2}\,dt = o(r^{3/4})\csc^2\frac{\varphi}{2}.$$

Furthermore, it is obvious that $\ln|\Phi(re^{i\varphi})| \le o(r^{3/4})$. Therefore

$$\text{(A.18)}\qquad \ln|\Omega(re^{i\varphi})| \le C_1 r^{3/4}\csc^2\frac{\varphi}{2} + C_1,$$

where C_1 is a positive constant. Positive constants which do not depend on r and φ will be denoted by C_j.

Using the method due to V. I. Matsaev [**Mat60**] we prove the estimate

$$\text{(A.19)}\qquad \ln|\Omega(re^{i\varphi})| \le C_2 r^{3/4} + C_2.$$

Denote by G the region $\{|\arg(\zeta - 1) - \pi| < \frac{\pi}{16}\}\cap\{|\zeta| < 1\}$, by L the boundary of this region, and by L_1 the part $L \cap \{|\zeta| < 1\}$ of the boundary L. Let $d = \sin\frac{\pi}{16}$. It is clear that $\{|\zeta| < d\} \subset G$. The function $\mu(\zeta) = \exp\{-(\zeta-1)^{-4}\}$, as is easy to show, is analytic in G and continuous in $\bar{G}$. For $\zeta \in L_1$ we have

$$|\mu(\zeta)| = \left|\exp\left\{-\frac{e^{-4i\arg(\zeta-1)}}{|\zeta-1|^4}\right\}\right| = \exp\left\{-\frac{\cos\frac{\pi}{4}}{|\zeta-1|^4}\right\}.$$

If $\zeta \in L_1$, then, by the sine theorem,

$$\frac{1}{d} \ge \frac{|\zeta|}{d} = \frac{|\zeta - 1|}{|\sin\varphi|}, \qquad \varphi = \arg\zeta.$$

Therefore $(\zeta \in L_1)$

$$\text{(A.20)}\qquad |\mu(\zeta)| \le \exp\left\{-\frac{d^4}{\sqrt{2}\sin^4\varphi}\right\} \le \exp\left\{-\frac{d^4 2^{-\frac{9}{2}}}{\sin^4\frac{\varphi}{2}}\right\}.$$

For $\zeta \in L\backslash L_1$ we have

$$|\mu(\zeta)| \le \exp\left\{-\frac{\cos\frac{\pi}{4}}{|\zeta-1|^4}\right\} \le \exp\left\{-2^{-\frac{9}{2}}\right\}. \tag{A.21}$$

Combining estimates (A.20) and (A.21), we obtain the inequality

$$|\mu(\zeta)| \le \exp\left\{-\frac{C_3}{\sin^4\frac{\varphi}{2}}\right\} \tag{A.22}$$

on L. Let $|\zeta| < \frac{d}{2}r$, $r \ge 1$. Then $\zeta/r \in G$, and, by the Cauchy formula,

$$\Omega(\zeta) = \frac{1}{2\pi i\mu(\zeta/r)}\int_L \frac{\Omega(rt)\mu(t)}{t-\frac{\zeta}{r}}dt. \tag{A.23}$$

It is easy to see that

$$\max_{|\zeta|=\frac{d}{2}} \frac{1}{|\mu(\zeta)|} = e^{\left(1-\frac{d}{2}\right)^{-4}} = C_4,$$

$$\int_L \frac{|dt|}{\left|t-\frac{\zeta}{r}\right|} \le \frac{1}{\left(\frac{d}{2}\right)}\int_L |dt| = C_5,\ |\zeta| = \frac{d}{2}r.$$

Taking into account (A.18) and (A.22), we conclude that (A.23) implies

$$M\left(\frac{dr}{2},\Omega\right) \le \frac{1}{2\pi}C_4C_5e^{C_1}\max_{0<\varphi<2\pi}\exp\left\{\frac{C_1r^{3/4}}{\sin^2\frac{\varphi}{2}} - \frac{C_3}{\sin^4\frac{\varphi}{2}}\right\}. \tag{A.24}$$

We assume that r is so large that $2C_3 \le C_1r^{3/4}$. Than, finding the maximum in (A.24) by the standard methods of calculus, we get

$$M\left(\frac{dr}{2},\Omega\right) \le \frac{1}{2\pi}C_4C_5e^{C_1}\exp\left\{\frac{C_1^2}{4C_3}r^{3/2}\right\}$$

and

$$M(r,\Omega) \le C_6e^{C_6r^{3/2}}, \quad r > 0. \tag{A.25}$$

Now we consider the auxiliary function

$$\Omega_\eta(\zeta) = \Omega(\zeta)\exp\left\{-7C_1\zeta^{3/4}\sec\frac{3\pi}{16} - \eta\zeta^{5/3}\right\}$$

in the angle $W = \{|\arg\zeta| < \frac{\pi}{4}\}$, where η is an arbitrary positive number. If $|\theta| \le \frac{\pi}{4}$, then, taking into account (A.25), we obtain the estimate

$$\begin{aligned}|\Omega_\eta(re^{i\theta})| &\le C_6\exp\left\{C_6r^{3/2} - 7C_1\sec\frac{3\pi}{16}r^{3/4}\cos\frac{3\theta}{4} - \eta r^{5/3}\cos\frac{5\theta}{3}\right\}\\ &\le C_6\exp\left\{C_6r^{3/2} - 7C_1r^{3/4} - \eta r^{5/3}\cos\frac{5\pi}{12}\right\}.\end{aligned}$$

Hence $\Omega_\eta(re^{i\theta}) \to 0$ for $r \to \infty$, $|\theta| \le \frac{\pi}{4}$. The function $\Omega_\eta(\zeta)$ is analytic in the region W and is continuous in $\bar{W}$. Now we estimate $|\Omega_\eta(\zeta)|$ on the boundary of W using (A.18). We get

$$\left|\Omega_\eta\left(re^{\pm i\frac{\pi}{4}}\right)\right| \le e^{C_1}\exp\left\{7C_1r^{3/4} - 7C_1r^{3/4} - \eta r^{5/3}\cos\frac{5\pi}{12}\right\} \le e^{C_1}.$$

Hence, by the maximum modulus principle,

$$|\Omega_\eta(\zeta)| \le e^{C_1}$$

in W. Letting η tend to 0 in this inequality, we get the inequalities

$$\left|\Omega(\zeta)e^{-7C_1\zeta^{3/4}\sec\frac{3\pi}{16}}\right| \leq e^{C_1}$$

and

$$\text{(A.26)} \qquad |\Omega(\zeta)| \leq \exp\left\{C_1 + 7C_1|\zeta|^{3/4}\sec\frac{3\pi}{16}\right\}$$

in W. Inequality (A.18) implies the inequality

$$\ln|\Omega(\zeta)| \leq 7C_1|\zeta|^{3/4} + C_1$$

for $\frac{\pi}{4} \leq \arg\zeta \leq \frac{7\pi}{4}$. Thus, (A.26) holds for $|\zeta| < \infty$. Since the order of the entire function $\Omega(\zeta)$ does not exceed 3/4, $\Omega(\zeta) \neq 0$ for $\zeta \geq 0$, and $\Omega(0) = 1$, we have

$$\Omega(\zeta) = \prod_{n=1}^{\infty}\left(1 - \frac{\zeta}{\zeta_n}\right), \qquad \zeta_n \to \infty, \quad 0 < \arg\zeta_n < 2\pi.$$

Since $\Phi(\zeta) = F(\zeta)\Omega(\zeta)$ in D, the zeros of the function $\Phi(\zeta)$ in D coincide with the zeros of $\Omega(\zeta)$, and $n(r, 0, \Omega) = n_2(r)$. Therefore (cf. (A.3))

$$\text{(A.26}'\text{)} \qquad \ln M(r, \Omega) \leq r\int_0^{\infty}\frac{n_2(t)dt}{t(t+r)}.$$

By (A.11), (A.11′), (A.26′), and (A.10), we have that

$$\begin{aligned}
\ln|\Phi(-r)| &\leq \ln|F(-r)| + \ln M(r,\Omega) \\
&\leq \operatorname{Re}\left\{-\frac{r}{2\pi i}\int_0^{\infty}\frac{\ln G(t)}{t(t+r)}dt\right\} + r\int_0^{\infty}\frac{n_2(t)\,dt}{t(t+r)} \\
&= -r\int_0^{\infty}\frac{\frac{1}{2\pi}\arg G(t) - n_2(t)}{t}\frac{dt}{t+r} = r\int_0^{\infty}\frac{dT_2(t)}{t+r} \\
&= r\int_0^{\infty}T_2(t)\frac{dt}{(t+r)^2} = \int_0^{\infty}T_2(\tau r)\frac{d\tau}{(\tau+1)^2}.
\end{aligned} \qquad \text{(A.27)}$$

Taking into account (A.5) and (A.6), it is easy to see that

$$T_2(r) = (2\rho - \varepsilon)T_1(r^{\sigma}) < 2\rho T(r^{\sigma}, f).$$

Then (A.27) implies that

$$\ln|f(r^{\sigma})| \leq 2\rho\int_0^{\infty}T(\tau^{\sigma}r^{\sigma}, f)\frac{d\tau}{(\tau+1)^2},$$

whence

$$\text{(A.28)} \quad \ln|f(r)| \leq 2\rho\int_0^{\infty}T(\tau^{\sigma}r, f)\frac{d\tau}{(\tau+1)^2} = \frac{2\rho}{\sigma}\int_0^{\infty}T(tr, f)\frac{t^{1/\sigma}}{(t^{1/\sigma}+1)^2}\frac{dt}{t}.$$

Since the function $x(x+1)^{-2}$ increases on $[0,1]$ and decreases on $[1,\infty)$, we have

$$\frac{t^{1/\sigma}}{(t^{1/\sigma}+1)^2} \leq \frac{t^{1/\sigma_0}}{(t^{1/\sigma_0}+1)^2}, \qquad 0 < \sigma < \sigma_0 = \frac{1}{2\rho - \varepsilon_0}.$$

Therefore (A.28) implies that

$$\text{(A.29)} \qquad \ln|f(r)| \leq (2\rho)^2\int_0^{\infty}T(tr, f)\frac{t^{1/\sigma_0}}{(t^{1/\sigma_0}+1)^2}\frac{dt}{t}.$$

Applying this inequality to the function $f(ze^{i\theta})$, $0 \le \theta < 2\pi$, we get that $\ln|f(r)|$ in the left-hand side of (A.29) can be replaced by $\ln|f(re^{i\theta})|$ (we cannot do the same for (A.28) since σ, unlike σ_0, depends on θ). Therefore

$$\ln M(r,f) \le (2\rho)^2 \int_0^\infty T(tr,f)\frac{t^{1/\sigma_0}}{(t^{1/\sigma_0}+1)^2}\frac{dt}{t}.$$

In the same way as we obtained (A.4) from (A.3), we find that for an arbitrary small ε_1 the inequality

$$\begin{aligned}\Delta_M &\le (\Delta_T+\varepsilon_1)(2\rho)^2\int_0^\infty t^{\rho-1}\frac{t^{1/\sigma_0}}{(t^{1/\sigma_0}+1)^2}dt \\ &= (\Delta_T+\varepsilon_1)(2\rho)^2\sigma_0\int_0^\infty \frac{\tau^{\sigma_0\rho}}{(\tau+1)^2}d\tau = (\Delta_T+\varepsilon_1)(2\rho)^2\sigma_0\frac{\pi\sigma_0\rho}{\sin\pi\sigma_0\rho}.\end{aligned}$$

holds. Letting ε_1 and ε_0 tend to 0 in the right-hand side of this inequality, we get (A.2). This completes our proof of Theorem A.1.

Theorem A.1 immediately implies

COROLLARY. *Let $f(z)$ be an entire function of order ρ, $0 \le \rho \le \infty$. Then*

$$\limsup_{r\to\infty}\frac{T(r,f)}{\ln M(r,f)} \ge \frac{\sin\pi\rho}{\pi\rho} \quad \textit{if } 0 \le \rho \le \frac{1}{2}, \tag{A.30}$$

$$\limsup_{r\to\infty}\frac{T(r,f)}{\ln M(r,f)} \ge \frac{1}{\pi\rho} \quad \textit{if } \rho > \frac{1}{2}. \tag{A.31}$$

In fact, if $\rho = \infty$, the statement of the corollary is trivial. If $\rho < \infty$, then let $\rho(r)$ be a proximate order such that

$$\limsup_{r\to\infty}\frac{\ln M(r,f)}{r^{\rho(r)}} = 1.$$

Then

$$\limsup_{r\to\infty}\frac{T(r,f)}{\ln M(r,f)} \ge \limsup_{r\to\infty}\frac{T(r,f)}{r^{\rho(r)}},$$

and (A.30) and (A.31) follow from (A.1) and (A.2), respectively.

The fact that the estimates (A.30) and (A.31) are sharp follows from the same examples which were used to show that the estimates (A.1) and (A.2) are sharp. If $\rho = \infty$, we can use the function $f(z) = \exp(e^z)$, for which $T(r,f) = o(\ln M(r,f)) = o(e^r)$ as $r \to \infty$ (see (6.4) of Chapter 4).

Inequality (A.30) was first proved by G. Valiron [**Val30**] and A. Wahlund [**Wah29**]. Inequality (A.31) was conjectured by R. E. A. C. Paley [**Pal32**] and proved 35 years later by N. V. Govorov [**Gov69**]. V. P. Petrenko [**Pet69**] proved that (A.30) and (A.31) remain true for meromorphic functions even if we replace the order ρ by the lower order λ. The fact that it is possible to replace ρ by λ in (A.30) for entire functions and for $\lambda \le 1/2$ was proved in Section 3 of Chapter 5 (see inequality (3.30)).

Notes

Chapter 1

Theorem 2.1 was apparently first proved by F. Nevanlinna and R. Nevanlinna [**NN22**]. Formula (2.2) was first obtained by R. Nevanlinna [**NevR24**], who named it the Poisson–Jensen formula, since it is an immediate generalization of the classical formulas due to Poisson and Jensen [**Jen99**] (Theorem 2.5). Formula (2.4) was obtained in the same paper of R. Nevanlinna; it is often called the Schwarz–Jensen formula. Formula (2.3) was also discovered by R. Nevanlinna [**NevR25b**]. Important generalizations of formulas (2.2), (2.4), and (2.8) were found by M. M. Dzhrbashyan [**Dzh66**, Chapter IX]. Theorems equivalent to Theorem 2.6 were independently proved by T. Shimizu [**Shi29**] and L. Ahlfors [**Ahl29**].

Theorem 3.1 was first proved by F. Nevanlinna (see [**NevF22a**] and also [**NevF22b**]), and somewhat later by T. Carleman [**Carl23**]. Following the existing tradition, we use the name "Carleman formula" for (3.1). Theorem 3.4 was proved by B. Ya. Levin [**Lev41**], and later independently by M. Tsuji [**Tsu50**].

Theorem 4.1 is due to R. Nevanlinna [**NevR25a**]. Theorems 4.2 and 4.4 were obtained by T. Shimizu [**Shi29**] and L. Ahlfors [**Ahl29**]. Formulas (4.8) and (4.13) were obtained by H. Cartan [**Cart29b**], [**Cart29a**], [**Cart33**].

Theorems 5.1 and 5.2 were proved by R. Nevanlinna [**NevR25b**]. Theorem 5.3 was proved by M. Tsuji [**Tsu50**].

Relations (6.1)–(6.15) are well known. Inequalities (6.16)–(6.18) were proved by S. Hellerstein and L. Rubel [**HR64**]. G. Valiron [**Val31**] proved the following theorem containing Theorem 6.2 and equality (6.29).

Let $R(u,v)$ be a rational function of u and v, of degree d in v. If $f(z)$ is a meromorphic function, then $T(r,R(z,f(z))) = dT(r,f) + O(\ln r)$.

Theorem 6.4 is due to R. Nevanlinna [**NevR29**], [**NevR74**, Chapter X, Section 214].

Theorems 7.1 and 7.2 were proved by R. Nevanlinna [**NevR29**, pp. 24–27]. Theorem 7.3 was proved by A. Edrei and W. H. J. Fuchs [**EF62a**].

Chapter 2

Section 1 contains well-known facts presented in many texts. In using the term "growth category" we follow B. Ya. Levin [**Lev80**]. The term "magnitude of type" is not standard and apparently is introduced for the first time here. Usually they call the number $\sigma[\alpha]$ simply the "type". We find this inconvenient. We note also that the well-known Cauchy–Hadamard formulas allow one to determine the order and the type of the function $\ln M(r,f)$ in terms of Taylor coefficients of an entire function $f(z)$ (see, for example, Levin [**Lev80**, Chapter I, §2]). Sometimes the

Picard exceptional values are defined as values which are never taken by a function $f(z)$, and the Borel exceptional values are defined as values for which the order of $n(r,a)$ is less than the order of the function.

Various comparison functions for building the scale of growth, different from the function $\exp(r^\rho)$, were introduced at the beginning of the 20th century by E. Lindelöf, P. Boutroux, O. Blumenthal, G. Valiron, and others. Valiron in [**Val23**] calls the proximate order defined as in our Section 2 a *proximate order of Lindelöf* (although Lindelöf considered a much narrower class of comparison functions), and he calls the function $l(r)$ which we introduce in Section 5 a *proximate order of Boutroux*. Today the definition of a proximate order, as we introduce it in Section 2, is standard, and the results proved in Section 2 are well known (see, for example, [**Lev80**]). Theorem 2.1 is due to Valiron [**Val23**] (in a slightly weaker form it was proved already in [**Val14**, p. 213]), who, however, did not consider a lower proximate order. The presented proof of Theorem 2.1 is mostly due to S. M. Shah [**Sha46b**]; he also introduced a lower proximate order [**Sha48**]. In the definition of a slowly varying function one can omit the condition of the uniform convergence of $L(kr)/L(r)$ to 1. In fact, J. Karamata [**Kar30**] proved that if $\lim_{r\to\infty} L(kr)/L(r) = 1$ for all k, $0 < k < \infty$, then the convergence of $L(kr)/L(r)$ to 1 is uniform in k for $0 < a \leq k \leq b < \infty$. See also J. Korevaar, T. Aardenne-Ehrenfest and N. G. de Bruijn [**KAB49**].

Theorems 3.2 and 3.3 in this generality are apparently published here for the first time, although some of their corollaries, for example, inequality (3.6), are well known (see, for example, Levin [**Lev80**]). Theorem 3.4 is due to E. Borel [**Bor97**] and is contained in all standard courses of the theory of entire functions.

Theorem 4.1 was proved by J. Hadamard [**Had93**] for entire functions. Theorem 4.2 essentially goes back to E. Borel [**Bor00**], who, however, did not use proximate orders. The corollary of Theorem 4.2 was proved by R. Nevanlinna [**NevR29**] in a different way.

Theorem 4.4 was proved by E. Lindelöf [**Lin05**] for entire functions without using proximate orders. The generalization of the Lindelöf theorem consisting in the usage of proximate orders apparently was first done by B. Ya. Levin [**Lev80**, Chapter I, §13]. The Lindelöf theorem was carried over to meromorphic functions by Valiron [**Val33**]. Our proofs of Theorems 4.3, 4.4, and 4.5 are close to the presentation of Shah [**Sha60**]. Theorem 4.6 was proved by Valiron [**Val50**]. Our proof is different from the original one, although it uses the same idea.

The asymptotics of functions (5.1) was found by Valiron [**Val14**]; however in some special cases, for example, if $n(r,0) \sim \Delta r^\rho (\ln r)^\alpha$, it was known already to Lindelöf [**Lin02**]. Our presentation is different from Valiron's method. Stronger results than those presented here are known in the case where $l(r) = \rho(r)$ is a proximate order (Levin [**Lev80**]). Examples 1, 2 and 1′ are due to A. A. Goldberg [**Gol63**]. The asymptotic formulas for the functions considered in 2° in the case $n(r,0) \sim \Delta r^\rho (\ln r)^\alpha$, $\rho > 0$, were studied already by Lindelöf [**Lin02**]. For an arbitrary proximate order $\rho(r)$ the results in Section 5, 2° follow from more general results of Goldberg [**Gol64a**]. The function $E_\rho(z)$ in (5.31) was introduced (including the case of complex ρ) by G. Mittag-Leffler [**Mit05**], and estimates (5.40), (5.43) are due to him. Asymptotic properties of functions $E_\rho(z)$ have been studied in detail, including the case of imaginary values of the parameter ρ. A series of papers by different authors devoted to this function were published in Acta Mathematica, **29**

(1905); among recent works we would like to mention M. M. Dzhrbashyan [**Dzh66**] and M. N. Sheremeta [**She69**].

Chapter 3

The main result in Section 1, Theorem 1.3 (the lemma on the logarithmic derivative) was first proved by R. Nevanlinna [**NevR25a**], [**NevR29**]. Theorem 1.1, giving the sharpest known estimate of $m(r, f'/f)$, and inequality (1.17) following from it, are due to Vu Ngoan and I. V. Ostrovskii [**NO65**]. Theorem 1.2 was proved by R. Nevanlinna [**NevR31**], who generalized an earlier result (Exercise 1) of E. Borel [**Bor97**].

Theorem 2.1 — the second fundamental theorem — was proved by R. Nevanlinna [**NevR25a**], [**NevR29**]. We would like to mention that, as was discovered later, there are proofs of the second fundamental theorem which do not use the lemma on the logarithmic derivative. We refer only to the proofs due to L. Ahlfors, which can be found in M. A. Evgrafov [**Evg68**, Chapter X, §6], R. Nevanlinna [**NevR74**, Chapter XIII], L. Ahlfors [**Ahl35**], and S. Stoïlow [**Sto62**, Vol. II, Chapter X]. Remark 1 to Theorem 2.1 is due to K. Yosida [**Yos34**].

Theorem 2.4 belongs to H. Milloux [**Mil40**], [**Mil47**]. We present the proof given by K.-L. Hiong [**Hio55**]. He also proved Theorem 2.5 ([**Hio56**]). Theorems 2.6 and 2.7 are due to W. K. Hayman [**Hay59**], [**Hay64**, §3.3].

Theorem 3.1 for meromorphic functions of finite order was proved by R. Nevanlinna [**NevR25b**], and in the stated form it was proved by Ostrovskii [**Ost60c**], [**Ost61**]. Note that J. Dufresnoy [**Duf39**] gave an estimate of the sum $A_{\alpha\beta}(r, f'/f)$ $+B_{\alpha\beta}(r, f'/f)$ in terms of $S_{\alpha_1\beta_1}(r, f)$, $\alpha_1 < \alpha < \beta < \beta_1$; some estimates of different character were given by K. Habetha [**Hab61a**], [**Hab61b**].

Theorem 3.2 is due to M. Tsuji [**Tsu50**]. Our presentation follows the paper by Levin and Ostrovskii [**LO60**].

Chapter 4

The notions of deficient values were introduced by R. Nevanlinna [**NevR25a**] and G. Valiron [**Val26**] in 1925–1926. A stronger result than Theorem 2.1 was obtained by G. Valiron [**Val26**]. In our proof of Theorem 2.1 we follow [**Val60**]. The strongest result showing 'sparseness' of the set $E_V(f)$ can be found in R. Nevanlinna's book [**NevR74**, Chapter X, §3]. We state a weaker result: *For each* $\varepsilon > 0$ *and each continuous strictly monotone increasing function* $h(r)$, $0 \le r \le 1$, $h(0) = 0$, *such that* $\int_0^1 h(r)\, d\ln r < \infty$, *the set* $E_V(f)$ *can be covered by a sequence of discs with radii* r_ν *such that* $\sum_\nu h(r_\nu) < \varepsilon$. Theorem 2.1 corresponds to the case $h(r) = \pi r^2$. Example 1 of Section 2 is due to G. Valiron [**Val19**] (in our presentation we have changed some details). D. Drasin and D. F. Shea [**DS69**] constructed a similar example of an entire function $f(z)$ with $T(r, f) = O(\psi(r)\ln^2 r)$, where $\psi(r) \to \infty$ is an arbitrary preassigned function. On the other hand, if an entire function $f(z)$ satisfies $T(r, f) = O(\ln^2 r)$, then $E_V(f)\backslash\{\infty\} = \emptyset$ (G. Valiron [**Val14**]).

For $0 < \rho < \infty$ examples of meromorphic functions without deficient values were found by S. K. Singh [**Sin61**] (the idea of his construction can be easily extended to the cases $\rho = 0$ and $\rho = \infty$). Our construction uses a different idea.

Examples of meromorphic functions of order $\rho > 0$ with preassigned at most countable set of deficient values were first constructed by A. A. Goldberg [**Gol54a**],

[**Gol59a**]. W. K. Hayman [**Hay64**, §4.3], using A. A. Goldberg's idea, significantly simplified the construction. Our presentation in §4 is close to the presentation of W. K. Hayman; here we also used some of his observations made during our meeting in 1965. Lemma 4.1 (with a slightly different statement) was proved by W. H. J. Fuchs and W. K. Hayman [**FH62**] (see also [**Hay64**, §4.1.3]).

In our presentation of N. U. Arakelyan's example we, in addition to his paper [**Ara66**], used notes of his lectures at Kharkov University in April 1967. Theorems 5.2 and 5.3 are special cases of certain results of M. V. Keldysh presented in S. N. Mergelyan's papers [**Mer52**], [**Mer53**].

Example 1 of Section 6 is due to D. Dugué [**Dug47**]; Example 6 of Section 6 is due to W. K. Hayman [**Hay53**]. The result which we obtained as a corollary of Theorem 6.5 was first obtained (but under the somewhat stronger restriction $\rho - \lambda < 1$) by G. Valiron [**Val47**]. The remaining results of Section 6 are due to A. A. Goldberg (see §2 of the paper by P. P. Belinskiĭ and A. A. Goldberg [**BG54**]). Originally one of the essential steps of the construction of Example 2 was based on P. P. Belinskiĭ's theorem on univalent conformal mappings (see §1 of [**BG54**]), but later B. Ya. Levin pointed out that it is possible to construct such an example using the same idea, but not the theorem of P. P. Belinskiĭ. This construction was carried out by A. A. Goldberg [**Gol60b**]. In Section 6 these results are presented in more detail than in the paper, and some of them are published for the first time.

W. K. Hayman [**Hay52**] considered more general functions than in Example 6. His purpose was to establish the possibility of realization of (6.61) and to get more precise statements about the rate of convergence to $-\infty$ in (6.61).

Chapter 5

To Section 1. The notion of an asymptotic spot is essentially present already in the work of P. Boutroux [**Bou08**]. The term "asymptotic spot" was introduced by M. Heins [**Hei57**]. Another term used for the same notion is "asymptotic tract" (see, for example, G. R. MacLane [**Mac63**]). Theorem 1.1 is due to F. Iversen [**Ive14**] (see also [**NevR74**, n°.239]). Lemma 6.1 is due to E. Phragmén and E. Lindelöf (see, for example, B. Ya. Levin [**Lev80**, Chapter 1, §14]).

Theorem 1.2 is a weakened version of L. Ahlfors's theorem [**Ahl32a**] (see, also, R. Nevanlinna [**NevR74**, n° 258]). In our proof we use the method due to T. Carleman [**Carl33**] (see also A. Dinghas [**Din36**]).

Theorem 1.3 was proved by A. Wiman [**Wim05**] under the assumption that $\rho < \frac{1}{2}$. We state this theorem in the form in which it was first stated by M. Heins [**Hei48**]. Corollary 1 of Theorem 1.3 was first found by A. Edrei and W. H. J. Fuchs [**EF56**]; Corollary 3 (in which the order ρ appears instead of the lower order λ) was proved by J. E. Littlewood [**Lit08**]. W. K. Hayman [**Hay52**] proved that the quantity

$$C(\rho) = \inf \limsup_{r\to\infty} \frac{\ln \mu(r, f)}{\ln M(r, f)},$$

where the infimum is taken over all entire functions of order ρ, satisfies the inequality

$$-K_1(\ln \rho + 1) \le C(\rho) \le -K_2(\ln \rho + 1), \qquad \rho \ge 1,$$

where $0 < K_2 < K_1 < \infty$ are absolute constants.

Theorem 1.3′ was proved by K. Arima [**Ari52**]. V. P. Petrenko [**Pet67**] proved that if $f(z)$ is a meromorphic function of finite lower order λ, and l is a number satisfying the conditions

$$0 \le l \le 2\pi, \qquad l < K\lambda^{-1}\delta^{\frac{1}{2}}(a,f)/\ln\frac{e}{\delta(a,f)},$$

where K is an absolute constant, then there exists a sequence of arcs $\{|z| = r_k,\ \theta_k \le \arg z \le \theta_k + l\}$, $r_k \to \infty$, on which $f(z)$ tends to a uniformly in $\arg z$.

Theorem 1.4 was conjectured by A. Denjoy [**Den07**], who proved it under the additional assumption that the C_j are rays going out of $z = 0$. T. Carleman [**Carl21**] proved Theorem 1.4 with a weaker estimate than (1.1). Theorem 1.4 in its full extent was first proved by L. Ahlfors [**Ahl30**]. Proving Lindelöf's Lemma 1.2, we used the presentations of G. Valiron [**Val23**, Chapter V, §4] and of N. G. Chebotarev and N. N. Meĭman [**CM49**, Chapter II, §7].

To Section 2. The first example of an entire function (of order $\rho = \infty$) with deficient nonasymptotic value was constructed by W. K. Hayman [**Hay53**]; later A. A. Goldberg [**Gol57b**] constructed an analogous example in the case where $1 < \rho < \infty$, and finally, N. U. Arakelyan [**Ara66**] gave an example in the case where $1/2 < \rho < \infty$.

J. M. Anderson and J. Clunie [**AC66**] proved that if a meromorphic function $f(z)$ satisfies $T(r,f) = O(\ln^2 r)$, then each deficient value of $f(z)$ should be asymptotic (by Theorem 4.6 the function $f(z)$ has at most one deficient value).

Example 1 is due to O. Teichmüller [**Tei39**]. He considered the case $\rho = 1/2$ only, but his construction can be easily carried over to the case $0 < \rho < \infty$. Example 2 was constructed by A. A. Goldberg [**Gol66**]. Using the idea of Example 2, it is possible to construct an analogous example of a meromorphic function of the first order and of the maximal type (F. A. Baraket and A. A. Goldberg [**BG69**]). Such examples are not known for lower growth categories. The statement of Lemma 2.1 was first published in A. Wiman [**Wim15**]. Today much more general results are known (Levin [**Lev80**, Chapter II]).

Theorem 2.1 is well known; however, we have not succeeded in finding an appropriate reference. In any case, all estimates needed for its prove were known to E. Borel [**Bor00**].

Examples 3 and 4 are published here for the first time. The observation that there are Borel exceptional values which are not deficient is due to G. Valiron [**Val33**]. Example 5 is due to A. A. Goldberg [**Gol67**].

In examples of meromorphic functions of order $\rho > 0$ constructed in Section 3 of Chapter 4, each deficient value was asymptotic; in particular, these functions can have a countable set of asymptotic values. G. Valiron [**Val25**] proved that for each function $\psi(r) \to \infty$ as $r \to \infty$, there exists a meromorphic function $f(z)$ such that $T(r,f) < \psi(r)\ln^2 r$, $r > r_0$, and the set of asymptotic values of $f(z)$ has the cardinality of continuum. As was shown by G. Valiron [**Val35b**], if $T(r,f) = O(\ln^2 r)$, then $f(z)$ cannot have more than one asymptotic value. Y. Tumura [**Tum43a**] showed that the last condition can be replaced by the condition

$$\liminf_{r\to\infty} T(r,f)/\ln^2 r < \infty.$$

If we do not impose any restrictions on $T(r,f)$, the structure of the set of asymptotic values is known. Namely, M. Heins [**Hei53**], [**Hei55**] proved that if a set M, $\infty \in M$ is analytic (Suslin), then it is a set of asymptotic values of an entire function; the necessity of this condition was proved earlier by S. Mazurkiewicz

[**Maz31**]. For meromorphic functions, a result similar to Heins's was proved earlier by D. B. Potyagailo [**Pot51**], but his result was not published. We cannot present these interesting results here.

To Section 3. Theorem 3.1 sums up research of several authors. First, O. Teichmüller [**Tei39**] proved that under the condition $\delta(\infty, f) > 1-\cos\pi\rho$, where $\rho < 1/2$ is the order of the function $f(z)$, the equality $\limsup_{r\to\infty}|f(re^{i\theta})| = \infty$ holds for all θ, $0 \le \theta < 2\pi$, and he proved the statement of Theorem 3.1 under a stronger assumption $\delta(\infty, f) > \frac{1-\cos\pi\rho}{1-\varepsilon\cos\pi\rho}$, where $\varepsilon \ge 0$ is an absolute constant. Under the condition $\delta(\infty, f) > 1-\cos\pi\rho$, the statement of Theorem 3.1 was proved by A. A. Goldberg [**Gol54a**], [**Gol56**].

The mentioned works of O. Teichmüller and A. A. Goldberg relied on the Hadamard theorem on canonical representation of meromorphic functions of finite order. The Hadamard theorem does not apply to the study of functions of finite lower order. A generalization of the results to the case of functions of finite lower order became possible due to the idea of B. Kjellberg [**Kje60**], which can be described as follows: a function of finite lower order for some sequence $R = R_k \to \infty$ can be represented as a product of two factors, one of which is a quotient of two canonical products, and the second is very close to 1 in the disc $\{|z| \le R_k\}$. A precise statement, somewhat more general than the one due to B. Kjellberg, is contained in Lemma 3.1, which was essentially proved already in the paper by F. Nevanlinna [**NevF23**] (see also R. Nevanlinna [**NevR74**, Chapter VIII, §177]).

Theorem 3.1, its corollary, Theorems 3.2, 3.3, and inequality (3.29) were obtained by A. A. Goldberg and I. V. Ostrovskii (published in [**Ost63a**]). See also the preceding paper by A. A. Goldberg and I. V. Ostrovskii [**GO61a**], containing a somewhat weaker result. Theorem 3.1 and its corollary were independently obtained by A. Edrei [**Edr64**]. The statement of the corollary for $\lambda = 0$ was obtained significantly earlier by A. Edrei and W. H. J. Fuchs [**EF59b**].

The inequality

$$\limsup_{r\to\infty}\frac{N(r,0,f)}{\ln M(r,f)} \ge \frac{\sin\pi\rho}{\pi\rho}$$

for entire functions of order $\rho < 1$ is due to A. Wahlund [**Wah29**] and G. Valiron [**Val30**] (a simple proof was given by G. Valiron [**Val35a**]).

Theorem 3.4 is due to B. Kjellberg [**Kje60**] (see also Kjellberg [**Kje63**], where its refinement is presented). It is a generalization of the classical theorem proved independently by A. Wiman [**Wim15**] and G. Valiron [**Val14**]: each entire function $f(z)$ of order $\rho < 1$ satisfies

$$\limsup_{r\to\infty}\frac{\ln\mu(r,f)}{\ln M(r,f)} \ge \cos\pi\rho.$$

Different problems related to the Wiman–Valiron theorem are treated in many papers. A survey of these papers is given in Goldberg and Ostrovskii [**GO61b**].

The main idea of the method used in the proofs of Theorems 3.2–3.4 is due to A. Denjoy [**Den30**]. This method was later developed in G. Valiron [**Val35a**], B. Kjellberg [**Kje54**], [**Kje60**], and A. A. Goldberg and I. V. Ostrovskii [**GO61a**]. The last paper contains a general formulation of the method, using which one can immediately get Theorems 3.2, 3.3, and several other relations. We decided not to give this general formulation here because of its complexity.

The construction of the examples showing how sharp Theorems 3.2–3.4 are in the case $\lambda = \rho$ is due to A. Edrei and W. H. J. Fuchs [**EF60**], and the example showing how sharp Theorem 3.1 is belongs to O. Teichmüller [**Tei39**]. These authors considered functions with positive zeros and $n(r, 0) \sim \alpha r^\rho$ instead of the functions $f(z, \alpha)$.

To Section 4. The estimate (4.1) with $d(\rho) = |\sin \pi\rho|/C_1([\rho])$ was obtained by R. Nevanlinna [**NevR29**] and later sharpened by S. M. Shah [**Sha44**] and A. A. Goldberg [**Gol57a**]. The parameter $\kappa(f)$ was introduced by R. Nevanlinna [**NevR29**], who conjectured that the best lower estimate of this parameter in terms of ρ is $\kappa(f) \geq \nu_1(\rho)$.

The estimate (4.4) confirming R. Nevanlinna's conjecture for $\rho < 1$ and close to the best possible for $\rho > 1$ is due to A. Edrei and W. H. J. Fuchs [**EF59a**], [**EF60**]. The estimate (4.3) is due to I. V. Ostrovskii [**Ost63b**]; close but somewhat weaker estimate was obtained independently by A. Edrei [**Edr64**]. The method consists in extension of the method used by Edrei and Fuchs [**EF59a**], [**EF60**] with the help of the idea of B. Kjellberg mentioned in our notes to Section 3.

Lemmas 4.2 and 4.5 belong to Edrei and Fuchs [**EF59a**], [**EF60**].

Lemma 4.4 is due to Goldberg [**Gol54a**], [**Gol56**], who also found its generalizations [**Gol57c**]. The presented proof of Lemma 4.4 is taken from Edrei [**Edr64**]. Lemma 4.6, part B is an analog of G. Pólya's theorem on sequences (Pólya and Szegö [**PS98**, Vol. 1, Chapter 3, Problems 107–110]); it was obtained in a more general form by Shah [**Sha40**], [**Sha41**]. Lemma 4.6, part A is due to Ostrovskii [**Ost63b**].

A version of Theorem 4.4 with the lower order λ replaced by the order ρ is due to Edrei and Fuchs [**EF60**]. Theorem 4.4 was proved independently by Edrei [**Edr62**] and Ostrovskii [**Ost63a**]. The method is an extension of the method of Edrei and Fuchs [**EF60**].

To Section 5. Examples similar to Examples 1–5 can be constructed in a uniform manner using a result of Shah [**Sha46a**]. This paper deals with entire functions $f(z) = \sum_{k=0}^{\infty} a_k z^k$ such that the sequence $|a_k/a_{k+1}|$ is non-decreasing. Shah proves for them the formulas

$$\rho[f] = \limsup_{k\to\infty} \frac{\ln k}{\ln |a_k/a_{k+1}|}, \qquad \lambda[f] = \liminf_{k\to\infty} \frac{\ln k}{\ln |a_k/a_{k+1}|}.$$

Theorem 5.2 with a weaker estimate than (5.5) was obtained in a different way by Edrei and Fuchs [**EF59b**]. Our presentation follows Ostrovskii [**Ost63b**]. The corollary of Theorem 5.2 and Theorem 5.3 with somewhat weaker estimates were obtained by Edrei and Fuchs [**EF59b**]; they also proved Lemma 5.1. Much earlier, A. Pfluger [**Pfl46**] established that for an entire function $f(z)$ of finite order ρ with $\sum_{a\neq\infty} \delta(a, f) = 1$, the number ρ is a positive integer, and all deficiencies are of the form k/ρ, where k is an integer; thus the number of finite deficient values in this case does not exceed ρ. Edrei and Fuchs [**EF59a**] complemented this result of Pfluger by showing that all deficient values are asymptotic. They established an analogous result for "almost all" deficient values of meromorphic functions satisfying condition (5.7) with sufficiently small value of γ.

Example 8 is a representative of a large class of similar examples considered by F. Nevanlinna [**NevF29**].[2] He conjectured that if a meromorphic function $f(z)$ satisfies $\sum_a \delta(a,f) = 2$, then $\rho[f] = q/2$, where q is a positive integer. This conjecture is still unproved.[3]

Example 9 was constructed by Goldberg [**Gol64b**]; however his construction of the auxiliary function $g(z)$ is different. The construction of the function $g(z)$ presented here is due to Levin and Ostrovskii.

To Section 6. The relation $m(r,f) = O(1)$ for functions of the form (6.1) was proved by Keldysh [**Kel54**]. Theorem 6.1 representing an improvement of this result was obtained by Ostrovskii (published in Goldberg [**Gol60b**]). Lemma 6.1, which makes it possible to sharpen the method of Keldysh, is due to V.I. Smirnov [**Smi28**].

Theorem 6.2 is due to Keldysh [**Kel54**], relation (6.4) makes it possible to simplify the original proof.

Theorem 6.3 is due to Ostrovskii [**Ost61**].

Chapter 6

To Section 1. Functions with (ρ,η)-separated zeros and poles were introduced by Goldberg [**Gol60a**]; he also proved ([**Gol60a**], [**Gol61**]) Theorems 1.1, 1.2, 1.4 and the corollaries of Theorem 1.1 and constructed Examples 1 and 3. All other results of this section are due to A. Edrei, W. H. J. Fuchs, and S. Hellerstein [**EFH61**]. In the case where $f(z)$ is an entire function, Corollary 3 of Theorem 1.5 was proved by Edrei and Fuchs [**EF59b**]. The same paper contains the first proof of Corollary 1 under the a priori assumption $\rho[f] < \infty$.

The papers by Edrei, Fuchs, and Hellerstein [**EFH61**] and [**Hel63**] contain results showing that under a certain condition of arithmetic character imposed on arguments of the zeros and poles of a meromorphic function $f(z)$ for which $\rho[N(r,0,\infty)]$ is finite but sufficiently large, the inequalities $\delta(0,f) > 0$, $\delta(\infty,f) > 0$ hold.

To Section 2. The first results on connections between the growth of a meromorphic function and the distribution of its values with respect to the arguments are due to L. Bieberbach [**Bie19**] and R. Nevanlinna [**NevR25b**].

The paper of Bieberbach [**Bie19**] implicitly contains the following statement. Let $f(z)$ be an entire function of finite order such that all of its a- and b-points ($a \neq b$; $a, b \neq \infty$) belong to the system of rays $D = D(\alpha_1, \dots, \alpha_n)$. Then the growth of the function $f(z)$ is does not exceed the normal type of order $\omega(D)$. R. Nevanlinna [**NevR25b**] showed that this estimate is valid also in the case where the a- and b-points satisfy the condition

$$\sum_{j=1}^{n} \left\{ C_{\alpha_j \alpha_{j+1}}(r,a) + C_{\alpha_j \alpha_{j+1}}(r,b) \right\} = O(1).$$

[2]In fact, the function in Example 8 belongs to the class of functions constructed in Theorem 5.1 of Chapter 7 (Editor's Note).

[3]It is proved now in [**44**]; see the comments in Appendix B.

Besides, R. Nevanlinna [**NevR25b**] proved that this growth estimate holds for a meromorphic function $f(z)$ of finite order if

$$\sum_{j=1}^{n} \{C_{\alpha_j\alpha_{j+1}}(r,a) + C_{\alpha_j\alpha_{j+1}}(r,b) + C_{\alpha_j\alpha_{j+1}}(r,c)\} = O(1)$$

(a, b, and c are three different points from the extended complex plane) and the growth categories of at least two of the functions $N(r,a)$, $N(r,b)$, $N(r,c)$ do not exceed the convergence class of order $\omega(D)$.

Note that these results are contained in Corollary 3 from Theorem 2.1 and in Theorem 2.4.

In the results of L. Bieberbach and R. Nevanlinna cited above, finiteness of the order of $f(z)$ is an a priori assumption. M. G. Krein [**Kre47**] considered the entire functions $f(z)$ with zeros close to the real line, such that $1/f(z)$ can be represented in the form (2.29), and found growth estimates without an a priori assumption of finiteness of the order. We stated the result of Krein as Corollary 2 of Theorem 2.10. Theorems 2.9 and 2.10 can be viewed as generalizations of this result.

Edrei [**Edr55**] proved the following theorem. Let $D = D(\alpha_1, \dots, \alpha_n)$ be a system of rays, and $f(z)$ a meromorphic function such that for some $a \neq 0, \infty$ and an integer $l \geq 0$, all solutions of the equations

$$\text{(K.1)} \qquad f(z) = 0, \qquad f(z) = \infty, \qquad f^{(l)}(z) = a,$$

except, possibly, finitely many of them, lie on the rays of the system D. If $\delta(0,f) + \delta(\infty,f) + \delta(a,f^{(l)}) > 0$, then the order of the function $f(z)$ does not exceed $\omega(D)$. In the case where $\delta(a,f^{(l)}) > 0$, this theorem of A. Edrei is contained in the corollary of Theorem 2.5; the growth estimate in the Edrei theorem is less precise, and the argument constraints are more restrictive. It is essential that in the corollary of Theorem 2.5 the argument restrictions are imposed on solutions of only two of equations (K.1). In the case that $\delta(a,f^{(l)}) > 0$, Edrei's theorem cannot be derived from Theorem 2.5. However, using the method of the proof of Theorem 2.5, it is not difficult to prove (see Ostrovskii [**Ost61**]) a generalization of Theorem 2.5, also containing these cases of the Edrei theorem.

All results of the present section, except Theorems 2.7, 2.8, Corollary 2 of Theorem 2.10, and Examples 2 and 4, are due to Ostrovskii [**Ost57**], [**Ost58**], [**Ost60a**], [**Ost61**].

Using a method similar to the method of Ostrovskii [**Ost58**], [**Ost61**], Edrei and Fuchs [**EF62b**] proved the following theorem:

Let

$$L_j = \left\{z = re^{i\alpha_j(r)},\ r_0 \leq r < \infty\right\}, \qquad j = 1, 2, \dots, n,$$

be a system of n rectifiable curves in the plane, where $\alpha_j(r)$, $r \geq r_0$, $j = 1, \dots, n$, are continuous functions, and

$$\alpha_1(r) < \alpha_2(r) < \cdots < \alpha_n(r) < \alpha_1(r) + 2\pi = \alpha_{n+1}(r), \qquad r \geq r_0.$$

Suppose that the length $s_j(r_1, r_2)$ of the curve $L_j \cap \{r_1 \leq |z| \leq r_2\}$ admits the estimate

$$s_j(r_1, r_2) \leq B(r_2 - r_1),$$

where B is a constant. If all zeros and poles of a meromorphic function $f(z)$, except possibly finitely many, lie on the indicated system of curves, and for some $a \neq 0, \infty$

and integer $l \geq 0$, the inequality $\delta(a, f^{(l)}) > 0$ holds, then the order of the function $f(z)$ does not exceed

$$\max_{1\leq j\leq n} \limsup_{r\to\infty} \frac{9\pi B^2}{\alpha_{j+1}(r) - \alpha_j(r)}.$$

Theorem 2.7 is a simple corollary of the results of R. Nevanlinna [**NevR25b**] (Theorem 3.1 in Section 3) and W. K. Hayman [**Hay56**] (the corollary of Theorem 3.2 in Section 3).

Theorem 2.8 somewhat complements the following result due to Edrei and Fuchs [**EF62a**] (other results of the same character can be found in Edrei and Fuchs [**EF62b**] and Edrei [**Edr62**]).

Let $f(z)$ be a meromorphic function. Suppose that there exists a number δ, $0 \leq \delta < 1$, and a sequence $\{r_k\}$, $r_k \uparrow \infty$, such that the annuli

$$S_k = \left\{\frac{r_k}{\sigma_k} \leq |z| \leq \sigma_k r_k\right\}, \qquad k = 1, 2, \dots,$$

where $\sigma_k = 1 + \{\ln T(r_k, f)\}^{-\delta}$ have the following property: to each S_k there correspond $n \geq 1$ numbers

$$\alpha_{k1}, \alpha_{k2}, \dots, \alpha_{kn}; \quad \alpha_{k1} < \alpha_{k2} < \dots < \alpha_{kn} < \alpha_{k1} + 2\pi$$

so that the number of zeros and poles of the function $f(z)$ lying in

$$S_k \setminus \bigcup_{j=1}^{n} \{|\arg z - \alpha_{kj}| \leq \{\ln T(\sigma_k r_k, f)\}^{-1-\eta}\}$$

does not exceed

$$O(T^c(r_k))$$

($\eta > \delta$, $c < 1$ are constants).

Then the function $f(z)$ has at most $n + 1$ deficient values. If the function $f(z)$ has $n + 1$ deficient values, then 0 and ∞ are among them.

Example 2 is taken from Edrei [**Edr55**] and Example 4 is taken from Edrei and Fuchs [**EF62a**].

To Section 3. Theorem 3.1 is due to R. Nevanlinna [**NevR25b**]. We prove Theorem 3.2 using the Cartan–Ahlfors method (see R. Nevanlinna [**NevR74**, Chapter V, §5]), which Hayman [**Hay56**] used to prove the result which we state as a corollary of Theorem 3.2. Our presentation follows this paper of Hayman. Theorem 3.3 is published here for the first time. Theorem 2.3, which is an immediate corollary of Theorem 3.3, originally was obtained (by Ostrovskii [**Ost61**]) in a different way.

To Section 4. Theorem 4.1 (Ostrovskii [**Ost60b**]) is a generalization of the Bieberbach theorem [**Bie19**], which we stated as Corollary 2 of Theorem 4.1. Example 1 is due to Bieberbach [**Bie19**], Lemma 4.1 is due to O. Teichmüller [**Tei39**], and Example 2 is due to Ostrovskii [**Ost60b**].

To Section 5. Theorem 5.2 was obtained by Ostrovskii [**Ost60c**]. Theorems 5.3, 5.4, and 5.5 are due to Levin and Ostrovskii [**LO60**]. Lemma 5.3 is a classical result due to C. Carathéodory. Theorem 5.3 is an improvement of the following result of Edrei [**Edr55**]: If an entire function $f(z)$ is representable in the form $f(z) = e^{g(z)}h(z)$, where $g(z)$ is an entire function, $h(z)$ is an entire function of finite order, and all zeros of the function $f(z)f'(z)f''(z)$, except possibly finitely many, are real, then $\rho[g] \le 1$. Hellerstein [**Hel66**] found a generalization of Theorem 5.3. An extensive bibliography of works devoted to the question of Pólya and Wiman can be found in Levin and Ostrovskii [**LO60**].[4]

Chapter 7

We present Theorem 1.2 following S. Kakutani [**Kak35**], who stated it in a somewhat more general form and made a small error. Theorem 1.2 is contained in a stronger result of Y. Tumura [**Tum43b**] and H. L. Selberg [**Sel44**], [**Sel46**], who, in particular, proved that under the assumptions of Theorem 1.2, the relation $m(r, a) = O(\ln r)$ holds. Under the additional restriction that all algebraic branch points of P are projected to a (each region F_ν may have at most one branch point), Theorem 1.2 was proved by E. F. Collingwood [**Col24**] and H. Cartan [**Cart30**].

In Section 2 we present the necessary preliminaries from the theory of quasiconformal mappings. The theory of quasiconformal mappings is presented in L. I. Volkovyskiĭ [**Vol54**] and L. Ahlfors [**Ahl66**].

Most of the examples are contained (sometimes implicitly) in Volkovyskiĭ's monograph [**Vol50**]. Teichmüller [**Tei44**] considered a class of quasiconformal mappings solving the problem in Example 7, and found a mapping for which $\max_{|z|\le R} p(z, w)$ attains its minimum. Comparison with Teichmüller's result shows that the estimate (2.17) (see Volkovyskiĭ [**Vol50**, p. 110]) approaches the best possible estimate as $x_0 \to 0$.

Theorem 3.1 was obtained by Teichmüller ([**Tei38**]; see also H. Wittich [**Wit68**, Chapter VI]) as a corollary of his theorem on the moduli of ring domains. Wittich [**Wit48a**] gave another proof of the Teichmüller theorem, using a method of Ahlfors [**Ahl32b**] (Step III of our proof). We present Wittich's proof with significant simplifications. P. P. Belinskiĭ [**Bel53**], [**Bel54**] proved that under the assumptions of Theorem 3.1, the limit $\lim_{z\to\infty} w(z)/z$ exists.

The class F_q of Riemann surfaces, as well as the notion of a line complex, were introduced by R. Nevanlinna [**NevR32a**], [**NevR32b**]. A detailed theory of line complexes was developed by G. Elfving [**Elf34**]. Our presentation differs from that of Elfving. The dependence of the line complex on the choice of the base curve was studied by E. Drape [**Dra36**], H. Habsch [**Hab52**], and V. G. Tairova [**Tai62**], [**Tai64**]. F. Huckemann [**Huc56a**] extended the class F_q and the notion of a line complex.

Theorem 4.1 should follow from a more general result of H. Weyl [**Wey31**], but this result, in the author's own words, is stated so inefficiently that it is not clear how to get any specific corollaries of it.

Our proof, which is close to a certain argument of Tairova [**Tai64**], allows us to make a remark important for our purposes, at the end of the proof. The corollary of Theorem 4.1 was proved, using line complexes, by Le-Van-Thiem [**LeV50**]. In familiar sources (A. Hurwitz [**Hur91**], [**Hur02**], H. Weyl [**Wey31**], H. Röhrl

[4]This question is now completely solved. See the comments in Appendix B.

[**Röh63**]) we have not found anything indicating that the problem posed in Section 4 before Example 1 is not always solvable. Theorems 4.2 and 4.3 are published here for the first time.

The statement of the inverse value distribution problem is due to R. Nevanlinna (see, for example, [**NevR74**, Section 220]).

The hypothesis that the deficiencies of a meromorphic function of a finite order should satisfy some other relations besides $\sum_a \delta(a, f) \leq 2$ was stated by Teichmüller [**Tei39**]. Fuchs [**Fuc58**] proved that for functions of finite order, the relation $\sum_a \sqrt{\delta(a, f)} < \infty$ holds. This result was strengthened by Hayman [**Hay64**], who established that for functions of finite lower order and for $\alpha > 1/3$, the series $\sum_a \delta^\alpha(a, f)$ is convergent, and observed that for $\alpha < 1/3$ such a series can diverge. The behavior of this series for $\alpha = 1/3$ has not been studied,[5] but V. P. Petrenko [**Pet67**] proved that for $\beta > 1/3$ the series

$$\sum_a \delta^{1/3}(a, f) \ln^{-\beta} \frac{e}{\delta(a, f)}$$

converges.

Fuchs [**Fuc58**] and Hayman [**Hay64**] also obtained estimates for the sum of the series $\sum_a \delta^\alpha(a, f)$ in terms of $\lambda[f]$; the sharpest estimates are due to Petrenko [**Pet66**], [**Pet67**].

R. Nevanlinna [**NevR32b**] proved Theorem 5.1 using Riemann surfaces with finitely many logarithmic ends and without any algebraic branch points. To study the value distribution of the uniformizing functions he applied some results of the analytic theory of differential equations. Using the same method, Elfving [**Elf34**] studied uniformizing functions of surfaces with finitely many logarithmic ends, defined as in our Section 5. Ahlfors [**Ahl32b**] proved the results of R. Nevanlinna [**NevR32b**] without using differential equations.

The results of Sections 6, 7, and 8 are due to Goldberg [**Gol54b**] (this paper contains some errors). However, various special cases of these results were known before. A detailed survey of previous results can be found in Chapter VIII of the book by Wittich [**Wit68**]. Here we mention the most important papers. The class of Riemann surfaces with periodic ends was introduced by E. Ullrich [**Ull36**]. Wittich [**Wit47**], [**Wit48b**] used quasiconformal mappings for the study of the value distribution of the uniformizing functions of such surfaces, and obtained formulas (7.17), (7.18), (7.19) for surfaces with periodic ends. Note that for surfaces Φ with periodic ends it is not necessary to construct a quasiconformal mapping onto the complex plane, the problem with which we were concerned in Section 6, since the uniformizing meromorphic function can be easily written in an explicit form (see the exercise after Theorem 6.1). Le-Van-Thiem [**LeV49**], [**LeV50**] solved the inverse problem under the following restrictions: 1) The number of deficient and index values is finite; 2) The deficiencies and indexes are rational numbers; 3) If $\varepsilon(a) > 0$, then $\delta(a) + \varepsilon(a) < 1$. In the case of $\sum_a \delta(a) = 0$, Le-Van-Thiem solved the problem using elliptic functions; in the case $\sum_a \delta(a) > 0$ he used Riemann surfaces with finitely many periodic ends and characteristic $\chi = 1$. K. Pöschl [**Pös51**] noticed that if we drop the restriction $\chi = 1$, then, with the help of Riemann surfaces with finitely many periodic ends, the inverse problem can be solved, at best

[5] Weitsman [**159**] proved the convergence for $\alpha = 1/3$.

(for $0 < \sum_a \delta(a) < 2$), if the deficiencies and indexes belong to a very narrow subclass of the set of algebraic numbers. Somewhat later than Goldberg, F. Huckemann [**Huc56b**] solved the "restricted" inverse problem in the class of meromorphic functons of finite order under the additional assumption that the number of deficient values is finite, $\sum_a \delta(a) < 2$, and $\delta(a) < 1$.

Le-Van-Thiem [**LeV49**], [**LeV50**] introduced the notion of a conjunction of surfaces with periodic ends and obtained formulas (7.23)–(7.27) for them. However, he did not take into consideration the conditions under which conjunction is possible and the conditions under which the formulas (7.25)–(7.27) are valid; for this reason parts of his argument are not convincing. The idea to use the calculus of diagrams (see Section 8) is also due to Le-Van-Thiem.

It was conjectured that, if in Theorem 5.1 we restrict our attention to functions of finite order, then the requirements of rationality of deficiencies and of the finiteness of the number of deficient values become necessary.[6] For the entire functions this conjecture has been proved (Pfluger [**Pfl46**]).

In connection with Theorem 8.3 we note that the first example of a meromorphic function with infinitely many index values was constructed by G. af Hällström [**Häl52**], but in his construction severe restrictions were imposed both on the location of the index values and on the values of indexes. G. af Hällström [**Häl41**] also found the first example of a function with $\varepsilon(a) = 1$ (see, also, H. Wittich [**Wit68**, Chapter VIII, §5]).

The class of Riemann surfaces with almost periodic ends is, as was mentioned, a generalization of the class of Riemann surfaces with periodic ends. The class of Riemann surfaces with periodic ends was generalized also in other directions. We mention in this connection the papers of F. Huckemann [**Huc56b**], A. A. Goldberg [**Gol59b**], and H. P. Künzi and H. Wittich [**KW59**], where the reader can find further references.

[6] This is proved in [**44**].

APPENDIX B

A Survey of Some Results after 1970 (by A. Eremenko and J. K. Langley)

The literature on meromorphic functions[1] is very large. There is a comprehensive survey [**62**], which contains everything that was reviewed on the topic in the Soviet "Referativnyi Zhurnal" in 1953–1970, and a later large survey [**67**]. More recent surveys [**30**], [**80**], and [**48**] are shorter and have narrower scope.

Some books on specific topics in the theory of meromorphic functions published after 1970 are [**18**], [**20**], [**79**], [**101**], [**137**], [**134**], [**138**], [**164**]. A survey of the fast developing subject of iteration of meromorphic functions is [**7**].

Here we give a short survey of some results which are closely related to the problems considered in this book. Thus we do not include many important topics, such as geometric theory of meromorphic functions, iteration, composition, differential and functional equations, normal families, Borel and Julia directions, uniqueness theorems, and most regrettably, holomorphic curves and quasiregular mappings.

Chapter 1

Mokhon'ko [**122**] proved the following generalization of (6.29), Chapter 1. *Let $R(w,z)$ be a rational function of w whose coefficients are meromorphic functions $h(z)$ satisfying $T(r,h)=O(\phi(r))$, where ϕ is a fixed positive increasing function on $[0,\infty)$. Then for every meromorphic function f we have*

$$T(r,R(f(z),z))=\deg_w R\,T(r,f)+O(\phi(r)).$$

This is a purely algebraic result; its proof uses only properties (6.5), (6.6), (6.8′) and the property $T(r,f^2)=2T(r,f)$.

Another result in the same direction is due to Eremenko [**39**]. *Let $F(u,v,z)$ be a polynomial in u and v whose coefficients are meromorphic functions h of z satisfying $T(r,h)=O(\phi(r))$, where ϕ is a function as above, and assume that F is irreducible over the algebraic closure of the field of meromorphic functions. If f and g are meromorphic functions satisfying*

$$F(f(z),g(z),z)\equiv 0, \tag{B.1}$$

then

$$\deg_u F\,T(r,f)=(\deg_v F+o(1))T(r,g)+O(\phi(r)).$$

Both theorems have applications in the analytic theory of differential equations.

Vojta [**155**] noticed a formal analogy between the definition of the Nevanlinna characteristic and the definition of *height* in number theory. This analogy extends quite far, and it has been a source of many interesting results and conjectures in the

[1] If the domain is not specified explicitly, we mean meromorphic functions in **C**.

recent years. For example, under Vojta's analogy, Jensen's formula corresponds to the fundamental theorem of arithmetic, while the second main theorem without the ramification term corresponds to the Thue–Siegel–Roth theorem on Diophantine approximation. The full second main theorem corresponds to the famous unproved conjecture in number theory which is known as the "*abc*-conjecture". The analogy extends to the multi-dimensional generalizations of Nevanlinna theory, where it has been especially fruitful [**137, 102**]. As an example of the influence of Vojta's analogy on the one-dimensional value distribution theory, we mention the recent precise results on the error term in the second main theorem (see our comments to Chapter 3).

Analogies between function theory and number theory were noticed before Vojta. Osgood was the first to view the second main theorem as an analog of the Thue–Siegel Roth theorem [**129, 130, 131**]. See also [**69, 77**].

Formula (2.6) in Chapter 1 is the basis of the so-called "Fourier method" in the theory of entire and meromorphic functions, which was developed in the work of Rubel [**138**], Taylor, Miles, and others. The Fourier method is considered as a substitute for the Weierstrass representation, which is more effective in certain cases. One of the main results achieved with this technique is the theorem of Miles [**115**] on the quotient representation of meromorphic functions: *Every meromorphic function f can be written as $f = g_1/g_0$, where g_j are entire functions, possibly with common zeros, satisfying*

$$T(r, g_j) \leq AT(Br, f), \quad j = 0, 1,$$

where A and B are absolute constants (independent of f). Moreover, for every $B > 1$ there exists A such that the above statement holds for every f. Simple examples show that one cannot in general take $B = 1$. This result was improved by Khabibullin [**97**]: *Let f be a meromorphic function and $\epsilon > 0$ a non-increasing convex function on $[0, \infty)$. Then there esists a representation $f = g_1/g_0$ such that g_j are entire functions, and*

$$\ln(|g_1(z)| + |g_0(z)|) \leq \frac{A_\epsilon}{\epsilon(|z|)} T((1 + \epsilon(|z|))|z|, f) + B_\epsilon.$$

Khabibullin also extended Miles's theorem to meromorphic functions in $\mathbf{C}^n$. A survey of his results is [**98**].

The Fourier method is also one of the main tools for the study of functions which satisfy various restrictions on the arguments of a-points (see our comments to Chapter 6). A recent book on the Fourier method is [**99**].

Chapter 2

Various properties of increasing functions on $[0, \infty)$ play an important role in the theory of meromorphic functions. By 1970 it became clear that the lower order λ is at least as important as the order ρ. The main development since 1970 was "localizing" the notion of order.

Let Φ be an unbounded increasing function. A sequence $r_k \to \infty$ is called a sequence of *Pólya peaks* (of the first kind) of order μ if the inequalities

$$\Phi(rr_k) \leq (1 + \epsilon_k) r^\mu \Phi(r_k), \quad \epsilon_k < r < \epsilon_k^{-1}$$

hold with some $\epsilon_k \to 0$. Pólya peaks were formally introduced by Edrei [**33**], though they were used by Edrei and Fuchs already in 1963. Pólya peaks of order μ exist for all μ in a certain interval $[\lambda^*, \rho^*]$, which contains the interval $[\lambda, \rho]$. The endpoints λ^* and ρ^* of this interval are called the order and lower order in the sense of Pólya, respectively. The following formulas were given in [**31**]:

$$\rho^* = \sup\{p\colon \limsup_{x,A\to\infty} g(Ax)/A^p g(x) = \infty\}$$

and

$$\lambda^* = \inf\{p\colon \liminf_{x,A\to\infty} g(Ax)/A^p g(x) = 0\}.$$

Most contemporary results on functions of finite lower order use Pólya peaks.

Chapter 3

Milloux' inequality (Chapter 3, Theorem 2.4) has led to a rich vein of results developing the value distribution properties of meromorphic functions and their derivatives, in which a decisive role has been played by the paper [**71**] of Hayman.

Perhaps the most striking of the many results from [**71**] is *Hayman's alternative* (Chapter 3, Theorem 2.6): If a function f meromorphic in the plane omits a finite value a, and its kth derivative $f^{(k)}$, for some $k \geq 1$, omits a finite non-zero value b, then f is constant. Two principal questions arising in connection with Hayman's alternative are: (i) whether a version of Hayman's main inequality (Chapter 3, (2.23)) holds with $N(r, 1/(f-a))$ replaced by $\overline{N}(r, 1/(f-a))$; (ii) whether $f^{(k)}$ can be replaced by a more general term, such as a linear differential polynomial

$$F = L[f] = f^{(k)} + a_{k-1}f^{(k-1)} + \cdots + a_0 f, \tag{B.2}$$

with suitable coefficients a_j of small growth compared to f. A positive answer to (i) was given by Chen [**17**]. Question (ii) was answered affirmatively in [**103**], although there do exist exceptional functions f, which may be determined from the a_j, for which f and $F - 1$ have no zeros. A unified approach to the questions from [**17, 103**] may be found in [**14**]. It was shown further in [**19, 87**] that if P is a non-constant differential polynomial in f, all of whose terms have degree at least 2 in f and its derivatives, then a version of Hayman's inequality holds with $f^{(k)}$ replaced by P, and with $\overline{N}$ counting functions.

Question (i) is related to the issue of whether $f - a$ and $f^{(k)} - b$ $(k \geq 1, b \neq 0)$ can both fail to have simple zeros, in analogy with the sharp result that a non-constant meromorphic function cannot be completely branched over five distinct values. It has recently been shown [**128**] using normal family methods that if f is transcendental and meromorphic in the plane with only multiple zeros then f' takes every finite non-zero value infinitely often (see also [**14, 157**]).

The obvious example $f(z) = e^z$ shows that a transcendental entire function f may have the property that f and all its derivatives omit 0: thus the condition $b \neq 0$ is necessary in Hayman's alternative. However, Hayman showed in [**71**] (see Chapter 3, Theorem 2.7) that if f is an entire function such that ff'' has no zeros then $f(z) = \exp(Az + B)$ with A, B constants: this follows from applying (Chapter 3, Theorem 2.6) to f/f'. Clunie [**21**] established the corresponding result with f'' replaced by a higher derivative $f^{(k)}$, which on combination with Hayman's theorem on ff'' significantly improved earlier results of Pólya and others on the zeros of entire functions and their derivatives [**24, 135, 140, 141**]. The *Tumura–Clunie method*, as developed in [**21**], shows that if $\Psi_k[g]$ is a differential polynomial of the form $\Psi_k[g] = g^k + P_{k-1}[g]$, where $P_{k-1}[g]$ is a polynomial in g and its derivatives of total degree at most $k-1$, and with coefficients of small growth compared to g, and if g has few poles and $\Psi_k[g]$ has few zeros, then $\Psi_k[g]$ admits a simple factorisation. The application to zeros of $ff^{(k)}$ then follows by writing $f^{(k)}/f$ in the form $\Psi_k[f'/f]$. Variants and generalisations of the Tumura–Clunie method, allowing functions g with unrestricted poles, appear in a large number of papers including notably [**126, 152, 166**].

It was conjectured by Hayman in [**71**] that if f is meromorphic in the plane and $ff^{(k)}$ has no zeros, for some $k \geq 2$, then $f(z) = \exp(Az+B)$ or $f(z) = (Az+B)^{-m}$, with A, B constants and $m \in \mathbf{N}$. This was proved by Frank [**52**] for $k \geq 3$ and by Langley [**104**] for $k = 2$ (see also [**9, 106**]), while simple examples show that no such result holds for $k = 1$. Generalisations include replacing $ff^{(k)}$ by fF, where F is given by (B.2) with rational coefficients [**15, 54, 105, 148**], and by $ff'' - \alpha f'^2$, $\alpha \in \mathbf{C}$ [**8, 106, 124**]. Closely linked to Milloux' inequality and Hayman's alternative is the question of whether $G = f^n f'$ must take every finite non-zero value, when f is non-constant meromorphic and $n \in \mathbf{N}$, the connection being that $(n+1)G$ is the derivative of f^{n+1}, which has only multiple zeros and poles. Hayman proved in [**71**] that this is the case for $n \geq 2$ and f entire, and for $n \geq 3$ when f has poles. Following incremental results by a number of authors [**22, 86, 125**], the definitive theorem in this direction was proved in [**11**]: if f is a transcendental meromorphic function in the plane and $m > k \geq 1$, then $(f^m)^{(k)}$ has infinitely many zeros. Here the result is proved first for finite order, and the infinite order case is then deduced by applying a renormalisation method from normal families [**132, 133, 165**]. The closely related question of whether $f' + f^n$ may omit a finite value, which in turn is related to the Tumura–Clunie method, is resolved in [**11, 71, 125, 146**]. A more recent conjecture [**158**] asserts that if f is a transcendental meromorphic function, then $ff^{(k)}$ takes every finite, non-zero value infinitely often: this is known to be true for $k = 1$ [**11**], and for $k = 2$ and f entire [**108**].

A further development from [**71**] leads to two conjectures which remain open. Hayman observed in [**71**] that since the derivative f' of a transcendental meromorphic function f has only multiple poles, it follows that f' has at most one finite Picard value. It was subsequently conjectured by Mues [**123**] that the Nevanlinna deficiencies of $f^{(m)}$ satisfy, for $m \geq 1$,

$$\sum_{a \in \mathbf{C}} \delta(a, f^{(m)}) \leq 1. \tag{B.3}$$

This was proved by Mues [**123**] for $m \geq 2$, provided all poles of f are simple. In the general case the best known upper bound for the sum in (B.3) appears to be $4/3$ [**96, 163**].

The Mues conjecture (B.3) would follow from a positive resolution of the *Gol'dberg conjecture* that, for a transcendental meromorphic function f and $k \geq 2$,

$$\overline{N}(r, f) \leq N(r, 1/f^{(k)}) + o(T(r, f)) \tag{B.4}$$

as $r \to \infty$, possibly outside an exceptional set. This is in turn linked to a classical result of Pólya [**135**] that if f has at least two distinct poles, then $f^{(k)}$ has at least one zero, for all sufficiently large k. When f has poles of multiplicity at most $k-1$, the inequality (B.4) follows from a lemma of Frank and Weissenborn [**55**] (see also [**156**]), so that in particular if f has only simple poles, then (B.3) is true for every positive m. A related inequality was proved in [**53, 149**]: if $F = L[f]$ as in (B.2), where the a_j are small functions compared to f, then either

$$N(r,F) \leq N(r,1/F) + 2N(r,f) + o(T(r,f))$$

outside a set of finite measure or f is a rational function in solutions of the homogeneous equation $L[w] = 0$. This method is connected to Steinmetz' proof [**147**] of the second main theorem for small functions discussed below. No further results in the direction of (B.4) appear to be known, although it was proved by Langley in [**107**] that if f is meromorphic of finite order in the plane and $f^{(k)}$ has finitely many zeros, for some $k \geq 2$, then f has finitely many poles.

Examples abound of meromorphic functions with infinitely many poles such that the first derivative has no zeros, but it was proved in [**50**] (see also [**11**]) that if f is transcendental and meromorphic with $\limsup_{r\to\infty} T(r,f)/r = 0$, then f' has infinitely many zeros: the corresponding result with lim sup replaced by lim inf may be found in [**88**].

We remark that the above results have all been stated for functions meromorphic in the plane. Those which are proved only using properties of the Nevanlinna characteristic admit in some cases generalisation to functions of sufficiently rapid growth in a disc [**71**] or a half-plane [**111**]. Some related results for functions of slower growth in the disc appear in [**143, 145**].

An old conjecture of Nevanlinna was that one can replace constants a_k in the second main theorem by meromorphic functions $a_k(z)$ with the property $T(r, a_k) = o(T(r,f))$. Such functions a_k are usually called "small targets". For the case of an entire function f such a generalization was obtained by Chuang Chi Tai in 1964. Much later, the second main theorem without the ramification term was proved for meromorphic functions by Osgood, who used methods from number theory [**129, 130, 131**]. A substantial simplification of Osgood's proof was made by Steinmetz [**147**], who also used a beautiful idea of Frank and Weissenborn [**55**]. Osgood and Steinmetz proved that

$$\sum_{k=1}^{q} N(r,(f-a_k)^{-1}) \geq (q-2+o(1))T(r,f),$$

outside the usual exceptional set. The proof in [**147**] is simple and elegant; and uses only manipulations with Wronski determinants and the classical lemma on the logarithmic derivative. This makes it suitable for generalizations to holomorphic curves [**130, 131, 137**]. However, this version of the second main theorem does not take ramification into account. Simple examples like $f(z) = e^z + z$, where $\delta(\infty, f) = \delta(z, f) = 1$ and $N_1(r,f) \sim T(r,f)$, show that one cannot include the term N_1. However, the following form of the second main theorem holds with small targets:

$$\sum_{k=1}^{q} \overline{N}(r,(f-a_k)^{-1}) \geq (q-2+o(1))T(r,f),$$

where $\overline{N}(r,(f-a_k)^{-1})$ is the usual function counting zeros of $f-a$ disregarding multiplicity. This result was recently obtained by Yamanoi [**162**]. In [**161**] he separately treats the case of rational functions a_k when the proof is technically simpler. Yamanoi's proof is very complicated, and it will be hard to generalize to holomorphic curves. Surprisingly, it uses Ahlfors's theory of covering surfaces (and also algebraic geometry, moduli spaces of curves, and combinatorics). The idea to bring Ahlfors's theory to this context has its origin in the work of Sauer [**139**], who obtained a partial result for rational small targets. One application of Yamanoi's generalization of the second main theorem is the following. *Suppose that f and g are meromorphic functions in* $\mathbf{C}$ *satisfying a relation of the form* (B.1). *If the genus of the curve* $F(u,v)=0$ *is greater than* 1, *then* $T(r,f)+T(r,g)=O(\phi(r))$. This was conjectured by Eremenko in 1982, and the important special case that $\phi(r)=\ln r$, that is, F is a polynomial in all three variables, was proved by Zaidenberg in 1990.

Now we turn to the classic setting. The estimate in the lemma on the logarithmic derivative was improved by Gol'dberg and Grinštein [**66**]:

$$m(r,f'/f)\leq \ln^{+}\{T(\rho,f)(1-(r/\rho))^{-1}\}+\text{const},$$

where the constant depends on f. Vojta's analogy (see our comments to Chapter 1) stimulated new interest in refined estimates for the logarithmic derivative, as well as for the error term

$$S(r,f)=\sum_{j=1}^{q} m(r,a_j,f)+N_1(r,f)-2T(r,f)$$

in the second main theorem. Miles [**119**] derived from Gol'dberg's and Grinštein's estimate the following. *Let ψ be a continuous non-decreasing function such that*

$$\int_1^\infty \frac{dt}{t\psi(t)}<\infty.$$

Then for every meromorphic function f we have

$$m(r,f'/f)\leq \ln^{+}\psi(T(r,f))+O(1),$$

outside an exceptional set of finite logarithmic measure. The strongest results on the error term $S(r,f)$ belong to Hinkkanen [**89**], for example:

$$S(r,f)\leq \ln^{+}\psi(T(r,f))+O(1),$$

outside an exceptional set of finite logarithmic measure, where ϕ is as before. If one replaces ψ by $t\psi(t)$, then both results will hold outside an exceptional set of finite measure. All presently known results on the error terms in one-dimensional Nevanlinna theory are collected in the book [**18**].

No analog of the second main theorem holds without an exceptional set of r's. This can be seen from the result of Hayman [**72**]: *Let $\{E_k\}$ be closed sets of zero logarithmic capacity, and ϕ and ψ arbitrary unbounded increasing functions. Then there exist an entire function f and a sequence $r_k\to\infty$ such that $N(r_k,a,f)\leq \phi(r_k)\ln r_k$ for all $a\in E_k$, while $T(r_k,f)>\psi(r_k)$.*

Chapter 4

The result of Hayman we just cited shows that the set $E_V(f)$ of Valiron exceptional values, which always has zero capacity, can contain any F_σ set of zero capacity, but a complete description of possible sets $E_V(f)$ is not known.

For meromorphic functions f of finite order, Hyllengren [**95**] obtained a very precise description of the sets $E_V(f)$. Let us say that a set E satisfies the H-condition if there exist a sequence (a_n) of complex numbers and $\eta>0$ such that every point of E belongs to infinitely many discs $\{w:|w-a_n|<\exp(-\exp(\eta n))\}$. *For every meromorphic function f of finite order, and every $x\in(0,1)$, the set $E_V(x,f)$ of those $a\in\mathbf{C}$ for which $\Delta(a,f)>x$ satisfies the H-condition, and vice versa, for every set E satisfying the H-condition, there exist an entire function f*

and a number $x \in (0,1)$ such that $\Delta(a,f) > x$, for all $a \in E$. Notice that the H-condition is much stronger than the condition of zero capacity.

The first example of an entire function of finite order whose deficiency depends on the choice of the origin was constructed by Miles [**117**]. The order of this function was very large. Then Gol'dberg, Eremenko and Sodin [**65**] constructed such examples with any given order greater than 5. (For entire functions of order less than 3/2, deficiencies are independent of the choice of the origin.)

Chapter 5

Put $L(r,f) = \ln \mu(r,f)/\ln M(r,f)$, where $\mu(r,f) = \min\{|f(z)| : |z| = r\}$. Corollary 3 on p. 178 says that $\limsup_{r\to\infty} L(r,f) \geq -[2\lambda]$. Hayman [**70**] showed that the same holds with $-2[\lambda]$ replaced by $-2.19 \ln \rho$ when ρ is large enough. For functions of infinite order, he proved that

$$\limsup_{r\to\infty} \frac{L(r,f)}{\ln\ln\ln M(r,f)} \geq -2.19.$$

He also constructed examples of entire functions of large finite order for which $\limsup_{r\to\infty} L(r,f) < -1$. Then Fryntov [**56**], answering a question of Hayman, constructed entire functions of any given order $\rho > 1$ with the same property. Drasin [**29**] constructed entire functions of order one, maximal type, with the property $M(r,f)\mu(r,f) \to 0$. This may be contrasted with a remarkable theorem of Hayman [**73**], which says that *if f is an entire function of order one and normal type, and $M(r,f)\mu(r,f)$ is bounded, then $f(z) = c\exp(az)$ for some constants c and a.*

Thus for an entire function of order at least one, $\mu(r,f)$ can decrease at a higher rate than that of increase of $M(r,f)$. The situation changes dramatically if we consider the rate of decrease of $|f(z)|$ on an unbounded *connected* set. Hayman and Kjellberg [**78**] proved that *for every entire function f and every $K > 1$ all components of the set $\{z : \ln|f(z)| + K\ln M(|z|,f) < 0\}$ are bounded.*

Theorem 1.3′ has been the subject of many deep generalizations. First we mention the famous "spread relation" of Baernstein [**2, 4**] conjectured by Edrei in [**34**]: *If f is a meromorphic function of lower order λ, then for every $\epsilon > 0$ there are arbitrary large values of r such that the set of arguments θ where $|f(re^{i\theta})| > 1$ has measure at least*

$$\min\left\{\frac{4}{\lambda}\arcsin\sqrt{\frac{\delta(\infty,f)}{2}},\ 2\pi\right\} - \epsilon. \tag{B.5}$$

Similar sharp estimates of the measure of the set where

$$\ln|f(re^{i\theta})| > \alpha T(r,f)$$

were given in [**1**].

Fryntov, Rossi, and Weitsman [**57, 58**] proved that under the assumptions of the spread conjecture, the set $|f(re^{i\theta})| > 1$ must contain an arc of length (B.5). See also [**3**] for the sharp lower estimate of the length of the arcs in the set $\{\theta : \ln|f(re^{i\theta})| > \alpha \ln M(r,f)\}$.

Extremal functions for the spread relation and its generalizations were studied extensively, [**5, 36, 37, 142**].

The new methods introduced by Baernstein [**2, 4**] are based on the use of subharmonic functions, and especially, on a new type of maximal function, the so-called "star-function", which turned out to be very useful in solving a wide variety of extremal problems of function theory. An account of Baernstein's star function and its main applications is contained in the monograph [**79**].

One important application of the spread relation is the sharp estimate of the sum of deficiencies of a meromorphic function of lower order $\lambda \leq 1$ [**35**]. *If a meromorphic function f of lower order λ has at least two deficient values, then*

$$\sum_{a\in\overline{\mathbf{C}}} \delta(a,f) \leq \begin{cases} 1-\cos\pi\lambda, & 0<\lambda\leq 1/2, \\ 2-\sin\pi\lambda, & 1/2<\lambda\leq 1. \end{cases}$$

The sharp estimate of the sum of deficiencies of a meromorphic function in terms of its order or lower order λ is still not established for $\lambda > 1$. The conjectured extremal functions are described in [**32**].

The results of §2 show that neither Nevanlinna nor Borel exceptional values need be asymptotic values. On the other hand, Picard exceptional values are asymptotic. A natural question arises whether any condition of smallness of $N(r,a,f)$ in comparison with $T(r,f)$ will imply that a is an asymptotic value. The basic result belongs to the intersection of the papers [**16**], [**74**], and [**38**]. *Let f be a meromorphic function of lower order $\lambda \leq \infty$. If the order of $N(r,a,f)$ is strictly less than* $\min\{1/2,\lambda\}$, *then a is an asymptotic value.* Example 3 on p. 191 shows that this condition is sharp if only the lower order of f is taken into account. In [**38**], a weaker sufficient condition for a to be an asymptotic value is given, which uses both the order and lower order of f. Hayman [**74**] gives the following refined condition: *if*

$$T(r,f) - \frac{1}{2}r^{1/2}\int_r^\infty t^{-3/2}N(t,a,f)\,dt \to \infty,$$

then a is an asymptotic value.

The problem on p. 219 of optimal estimation of $\kappa(f)$ for functions of lower order greater than 1 is still open, even for entire functions. It has been solved only for entire functions with zeros on a ray [**84**].

The best estimates known at this time for entire and meromorphic functions with fixed $\lambda > 1$ are contained in [**120, 121**]. They are derived from the following sharp inequality, which is obtained by the Fourier method:

$$\limsup \frac{N(r,0)+N(r,\infty)}{m_2(r,f)} \geq \sup_{\lambda^*\leq\lambda\leq\rho^*} \sqrt{2}\frac{|\sin\pi\lambda|}{\pi\lambda}\left\{1+\frac{\sin 2\pi\lambda}{2\pi\lambda}\right\}^{-1/2},$$

where m_2 is the L^2-norm,

$$m_2^2(r,f) = \frac{1}{2\pi}\int_0^{2\pi} (\ln|f(re^{i\theta})|)^2 d\theta,$$

and λ^* and ρ^* are the order and lower order in the sense of Pólya, [**31**] (see also our comments to Chapter 2).

In 1929, F. Nevanlinna [**127**] found that meromorphic functions of finite order satisfying $N_1(r,f) \equiv 0$ have the following properties:

a) 2ρ is an integer, $2\rho \geq 2$,

b) all deficient values are asymptotic, and

c) all deficiencies are rational numbers with denominators at most 2ρ, and their sum equals 2.

It was natural to conjecture that one of the conditions

$$N_1(r,f) = o(T(r,f)) \tag{B.6}$$

or

$$\sum_{a\in\overline{\mathbf{C}}} \delta(a,f) = 2 \tag{B.7}$$

implies the properties a), b), and c). Notice that by the second main theorem, (B.7) implies (B.6) for functions of finite order.

It turns out that a strong form of this conjecture holds:

SMALL RAMIFICATION THEOREM. *If f is a meromorphic function of finite lower order with the property* (B.6), *then a), b), and c) hold, and*

$$T(r,f) = r^{\rho}\ell(r), \tag{B.8}$$

where ℓ is a slowly varying function in the sense of Karamata.[2]

As a corollary we obtain that conditions (B.6) and (B.7) for functions of finite lower order are equivalent.

This result has a long history, which begins with theorems of Pfluger and Edrei and Fuchs establishing the case of entire functions (see Corollary 2 on p. 241). Weitsman [**160**] proved that (B.7) implies that the number of deficient values is at most 2ρ. Then Drasin [**27, 28**] proved that for functions of finite order, (B.7) implies a), b), and c) and the regularity condition (B.8). Eremenko proposed a new potential-theoretic method (see, for example, [**48**]), which finally led to a simpler proof of Drasin's theorem. The small ramification theorem in its present form stated above is proved in [**44**].

These results show that besides the defect relation, there is an additional restriction on defects of functions of finite order: (B.7) implies that the number of deficient values is finite and all defects are rational.

There are other restrictions as well. Weitsman's theorem [**159**] says that for functions f of finite lower order,

$$\sum_{a\subset\overline{\mathbf{C}}} \delta(a,f)^{1/3} < \infty. \tag{B.9}$$

The story of this theorem is described in the notes to Chapter 7 (see p. 450). Weitsman's proof can actually be modified to produce an upper bound depending only on the lower order of f.

The second restriction concerns functions of finite order having a defect equal to 1. Lewis and Wu proved that such functions satisfy

$$\sum_{a\in\overline{\mathbf{C}}} \delta(a,f)^{1/3-\alpha} < \infty,$$

with some absolute constant $\alpha > 0$. Their proof gives $\alpha = 2^{-264}$, which is far from what is expected. (Lewis and Wu state their result for entire functions, but their proof applies to all functions with $\delta(a,f) = 1$ for some a.)

---

[2]That is, $\ell(cr)/\ell(r) \to 1$ as $r \to \infty$, uniformly with respect to $c \in [1,2]$.

Examples of entire functions of finite order with the property $\delta(a_n, f) \geq c^n$ for some $c \in (0,1)$ are constructed in [**45**], but a large gap remains between these examples and the result of Lewis and Wu.

Recent research on value distribution for meromorphic functions of the form

$$\sum \frac{c_k}{z - z_k}, \quad \sum \frac{|c_k|}{|z_k|} < \infty \tag{B.10}$$

was mainly concentrated on the functions with $c_k > 0$. Such functions are (complex conjugate to) gradients of subharmonic functions of genus zero with discrete mass. The main conjecture is that every function of the form (B.10) has zeros. This was proved in [**50**] under the additional assumption that $\inf c_k > 0$.

Chapter 6

Entire functions whose zeros lie on (or are close to) finitely many rays were intensively studied. Under certain conditions, one can estimate $\delta(0, f)$ from below, as in Corollary 4 on p. 270. The strongest results in this direction belong to Hellerstein and Shea [**81**] and Miles [**116**]. One of the results of [**81**] says that $\delta(0, f) > B_q(\theta_1, \dots, \theta_n)$ *for all entire functions of genus* q *with zeros on the rays* $\arg z \in \{\theta_1, \dots, \theta_q\}$*, and* $B_q \to 1$ *as* $q \to \infty$ *while the rays remain fixed.* In the case of one ray, they obtained $B_q(\theta) = 1 - (\pi^2 e^{-1} + o(1))/\ln q$, $q \to \infty$. For entire functions of infinite order with zeros on a ray, Miles [**116**] proved that $N(r, 0, f)/T(r, f) \to 0$ as $r \to \infty$ avoiding an exceptional set of zero logarithmic density. However, it may happen that $\delta(0, f) = 0$ for such functions, as Miles shows by an example constructed in the same paper.

Hellerstein and Shea [**81**] also considered meromorphic functions of finite order whose zeros $\{z_n\}$ and poles $\{w_n\}$ lie in the opposite sectors $|\arg z_n| \leq \eta$ and $|\arg w_n - \pi| \leq \eta$, where $0 \leq \eta < \pi/6$. For such functions, they obtained a sharp estimate of $\kappa(f)$ (definition on p. 219) from above.

For entire functions with zeros on finitely many rays, there are relations between the order and lower order (see p. 264). These relations were further investigated in [**118, 60**] and [**136**].

Miles [**118**] considers the class of meromorphic functions f whose zeros belong to a finite union of rays X and poles belong to a finite union of rays Y, where $X \cap Y = \emptyset$, and such that the exponent of convergence of the union of zeros and poles is a given number q. He then produces a non-negative integer $p = p(q, X, Y)$ such that

$$\lim_{r\to\infty} \frac{T(r, f)}{r^p} = \infty \quad \text{if} \quad p > 0 \quad \text{and} \quad \lim_{r\to\infty} \frac{T(r, f)}{\ln r} = \infty \quad \text{if} \quad p = 0,$$

and these growth estimates are sharp in the considered class. The integer p depends in a subtle way on the arithmetical properties of the arguments of the rays X and Y, and this integer is in general hard to compute or estimate.

Gleĭzer [**60**] considers entire functions with zeros on n rays. If $n = 1$ or $n = 2$, we have $\rho \leq [\lambda] + n$, where [ ] is the integer part. This follows from Theorem 1.1, Chapter 6. However, if $n = 3$, then the difference $\rho - \lambda$ can be arbitrarily large. In this case, Gleĭzer proved that $[\rho] \leq 3([\lambda] + 1)$. For arbitrary n, Qiao [**136**] proved that $\rho \leq 4^{q-1}([\lambda] + 1)$.

In [**59, 61**], Gleĭzer extended Theorem 4.1 by taking into account not only the order but the lower order in the sense of Pólya. He used Baernstein's star-function.

There has been remarkable progress in the problems considered in Section 5. The conjecture of Pólya and Wiman stated on p. 321 was proved by Hellerstein and Williamson [**82, 83**]. *If f is a real entire function such that all zeros of $ff'f''$ are real, then f belongs to the Laguerre–Pólya class.* In [**85**] the same authors with Shen classified all entire functions (not necessarily real) with the property that $ff'f''$ has only real zeros. The classification of meromorphic functions with the property that all their derivatives have only real zeros was achieved by Hinkkanen [**90, 91, 92, 93, 94**].

Sheil-Small [**144**] proved the conjecture of Wiman (1911) that every real entire function of finite order with the property that ff'' has only real zeros belongs to the Laguerre–Pólya class. Bergweiler, Eremenko, and Langley [**13**] extended Sheil-Small's result to functions of infinite order. Then Langley [**110**] extended this result to the derivatives of higher orders: *If f is a real entire function of infinite order, with finitely many non-real zeros, then $f^{(k)}$ has infinitely many non-real zeros for every $k \geq 2$.*

For real entire functions of finite order with finitely many non-real zeros, which do not belong to the Laguerre–Pólya class, Bergweiler and Eremenko [**12**] proved that the number of non-real zeros of $f^{(k)}$ tends to infinity as $k \to \infty$. Together with Langley's result, this confirms another conjecture of Pólya (1943).

Chapter 7

The inverse problem (as stated on p. 377) was completely solved by Drasin [**25**]. A simplified proof is given in [**26**]. The general idea is the same as in Chapter 7: quasiconformal surgery and a version of the theorem of Belinskii and Teichmüller are involved. However, unlike Chapter 7, Drasin does not construct the Riemann surface spread over the sphere explicitly but uses a more flexible technique.

Theorem 8.1 in Chapter 7 actually gives a complete solution of the inverse problem for finitely many deficient values in the class of meromorphic functions of finite order. (This was not known in 1970 when the book was written. That condition (3) of this Theorem 8.1 is necessary follows from the small ramification theorem above.)

On the narrow inverse problem in the class of meromorphic functions of finite order with infinitely many deficiencies, there is the following result [**42**]:

Let $\{a\}$ be an arbitrary infinite countable subset of $\overline{\mathbf{C}}$, and $\{\delta_n\}$ positive numbers satisfying the following conditions:
(i) $\delta_n < 1$,
(ii) $\sum_n \delta_n < 2$, *and*
(iii) $\sum_n \delta_n^{1/3} < \infty$.
Then there exists a meromorphic function f of (large) finite order such that $\delta(a_n, f) = \delta_n$, and f has no other deficient values.

The order of this function depends on the quantities in the right-hand sides of (i), (ii) and (iii).

Conditions (ii) and (iii) are necessary because of the small ramification theorem, and Weitsman's theorem (see our comments to Chapter 5 above). Condition (i) cannot be removed because of the Lewis and Wu theorem stated above, but it is not known what the precise condition on δ_n is, if $\delta_1 = 1$.

The class of meromorphic functions with finitely many critical and asymptotic values, which was used in Chapter 7 to investigate the inverse problem, is interesting independently of this application. Let us call this class S. The first general result on functions of this class belongs to Teichmüller [**150**], who proved that the second main theorem becomes an asymptotic equality for functions of this class:

$$\sum_{j=1}^{q} m(r, a_j, f) + N_1(r, f) = 2T(r, f) + Q(r, f),$$

where a_j are all critical and asymptotic values.

Langley [**109**] found that the growth of a function $f \in S$ cannot be arbitrary:

$$c(f) := \liminf_{r\to\infty} \frac{T(r, f)}{\ln^2 r} > 0.$$

This constant $c(f)$ can be arbitrarily small, but in the case that f has only three critical and asymptotic values, we have [**49**] $c(f) \geq \sqrt{3}/(2\pi)$ and this is best possible. On the other hand, there are no restrictions from above on the growth of functions of class S [**114**].

Class S plays an important role in holomorphic dynamics (iteration of entire and meromorphic functions); see, for example, the survey [**7**]. In [**23**] an application of almost periodic ends is given. In [**32**] the method of Chapter 7 is extended to a new class of Riemann surfaces, which the authors call "Lindelöfian ends". The corresponding functions have infinitely many critical values and thus do not belong to the class S.

Appendix A

Govorov's original proof of the Paley conjecture was a byproduct of his research on the Riemann boundary problem with infinite index [**68**]. His theorem was generalized to meromorphic functions by Petrenko, to subharmonic functions in $\mathbf{R}^n$ by Dahlberg, and to entire functions of several complex variables by Khabibullin; see his survey [**98**].

Petrenko introduced the following quantity, which he called the "deviation of f from the point a":

$$\beta(a, f) = \liminf_{r\to\infty} \frac{\ln^+ M(r, (f-a)^{-1})}{T(r, f)}. \tag{B.11}$$

This differs from the defect in one respect: the uniform norm of

$$\ln^+ |f(re^{i\theta}) - a|^{-1}$$

stands in the numerator instead of the L^1 norm. Petrenko's generalization of Govorov's theorem proved in Appendix A can be restated as

$$\beta(a, f) \leq \pi\lambda \tag{B.12}$$

for all meromorphic functions of lower order λ and all $a \in \overline{\mathbf{C}}$. This was the starting point of a study of deviations $\beta(a, f)$ by Petrenko and others. The results obtained before 1978 are summarized in his book [**134**]. The main difference between the theory of deviations and the theory of defects is the absence of a first main theorem: there is no simple relation between $\beta(a, f)$ and solutions of the equation $f(z) = a$.

We only present a sample of the results. By analogy with defects, one can expect that the set of exceptional values in the sense of Petrenko

$$P(f) := \{a \in \overline{\mathbf{C}} : \beta(a, f) > 0\}$$

is small. This is indeed the case: *for every meromorphic function f, the set $P(f)$ has zero logarithmic capacity; for functions of finite lower order it is at most countable (but may have the power of continuum for functions of infinite lower order).* The following analog of the defect relation for functions of finite lower order was established by Marchenko and Shcherba [**113**]:

$$\sum_{a\in\overline{\mathbf{C}}} \beta(a, f) \leq \begin{cases} 2\pi\lambda, & \lambda \geq 1/2, \\ \pi\lambda \csc \pi\lambda, & \lambda < 1/2. \end{cases}$$

Moreover, an analog of Weitsman's theorem (see our comments to Chapter 5) holds: *for functions f of finite lower order we have*

$$\sum_{a\in\overline{\mathbf{C}}} \beta(a, f)^{1/2} < \infty,$$

and the exponent $1/2$ is best possible. A version Baernstein's spread relation also holds with deviations instead of deficiencies [**134**]. It is worth mentioning here that, according to Baernstein [**4**], the idea of introducing the star-function that led to the proof of the spread relation occurred under the influence of Petrenko's proof of (B.12).

The inverse problem for deviations turned out to be simpler than the inverse problem for deficiencies. A complete solution for functions of finite order is given in [**42**]: *For every at most countable set $\{a_n\}$ of points and every sequence of positive numbers β_n satisfying the condition $\sum \beta_n^{1/2} < \infty$, there exists a meromorphic function f of finite order such that $\beta(a_n, f) = \beta_n$ and $\beta(a, f) = 0$ for $a \notin \{a_n\}$.*

In general, there is no relation between the sets $E_N(f)$ and $P(f)$: *For every pair (A, B) of at most countable subsets of $\overline{\mathbf{C}}$, there exists a meromorphic function f of any given non-zero order such that $E_N(f) = A$ and $P(f) = B$* [**64, 65**]. On the other hand, if $T(2r, f) = O(T(r, f))$, then $P(f) = E_V(f)$ [**41**].

Bergweiler and Bock [**10**] found an analog of (B.12) for functions of infinite lower order. The idea was to replace $T(r, f)$ in the denominator of (B.11) by $A(r, f)$. Notice that if one uses the Ahlfors definition of $T(r, f)$, then $A(r, f) = dT(r, f)/d\ln r$; for example, if $T(r, f) = r^\lambda$, then $A(r, f) = \lambda T(r, f)$. Bergweiler and Bock proved that *for every meromorphic function f of order at least $1/2$ and every $a \in \overline{\mathbf{C}}$ we have*

$$b(a, f) := \liminf_{r\to\infty} \frac{\ln^+ M(r, (f-a)^{-1})}{A(r, f)} \leq \pi,$$

and then Eremenko [**46**] established the following analog of the defect relation:

$$\sum_{a\in\overline{\mathbf{C}}} b(a, f) \leq \begin{cases} 2\pi, & \lambda > 1/2, \\ 2\pi \sin \pi\lambda, & \lambda \leq 1/2, \end{cases}$$

assuming that there are at least two values of a with $b(a, f) > 0$. It follows that for every meromorphic function the set $\{a \in \overline{\mathbf{C}} : b(a, f) > 0\}$ is at most countable.

Even Drasin's theorem on the extremal functions for the defect relation (see our comments to Chapter 5) has its analog for $b(a, f)$ [**47**]:

If f is of finite lower order and

$$\sum_{a \in \overline{\mathbf{C}}} b(a, f) = 2\pi,$$

then the following limit exists:

$$\lim_{r \to \infty} \frac{\ln T(r, f)}{\ln r} = \frac{n}{2},$$

where n is an integer, and $b(a, f) = \pi/n$ or 0 for every $a \in \overline{\mathbf{C}}$.

Bibliography

[Ahl29] Ahlfors, L.: Beiträge zur Theorie der meromorphen Funktionen. In: VII Congrés des mathematiciens scand., Oslo, 84–88 (1929)

[Ahl30] Ahlfors, L.: Untersuchungen zur Theorie der konformen Abbildung und der ganzen Funktionen. Acta Soc. Sci. Fenn., **1**, no. 9, 1–40 (1930)

[Ahl32a] Ahlfors, L.: Über die asymptotischen Werte der meromorphen Funktionen endlicher Ordnung. Acta Acad. Aboensis. Math. et Phys., **6**, no. 9, 1–8 (1932)

[Ahl32b] Ahlfors, L.: Über eine in der neueren Wertverteilungstheorie betrachtete Klasse transzendenter Funktionen. Acta math., **58**, 375–406 (1932)

[Ahl35] Ahlfors, L.: Zur Theorie der Überlagerungsflächen. Acta Math., **65**, 157–194 (1935)

[Ahl66] Ahlfors, L.: Lectures on quasiconformal mappings. Van Nostrand Mathematical Studies, Toronto (1966)

[Akh27] Akhiezer, N.I.: A new approach to necessary conditions under which an entire function of integer order has a given type (Ukrainian). Zap. fiz.-mat. otd. AN USSR, Kiev, **2**, no. 3, 29–33 (1927)

[AC66] Anderson, J. M., Clunie, J.: Slowly growing meromorphic functions. Comment. Math. Helv., **40**, 267–280 (1966)

[Ara66] Arakelyan, N. U.: Entire functions of finite order with an infinite set of deficient values. (Russian) Dokl. Akad. Nauk SSSR, **170**, 999–1002 (1966). English transl.: Sov. Math., Dokl. **7**, 1303–1306 (1966)

[Ari52] Arima, K.: On maximum modulus of integral functions. Journ. Math. Soc. Japan, **4**, 62–66 (1952)

[BG69] Baraket, F. A., Goldberg, A. A.: Deficient nonasymptotic values of first order meromorphic functions. (Russian) Sibirsk. Mat. Zh., **10**, 483–487 (1969). English transl.: Sib. Math. J. **10**, 345–348 (1969)

[Bel53] Belinskiĭ, P. P.: Behavior of a quasi-conformal mapping at an isolated point. (Russian) Doklady Akad. Nauk SSSR (N.S.), **91**, 709–710 (1953)

[Bel54] Belinskiĭ, P. P.: Behavior of a quasi-conformal mapping at an isolated singular point. (Russian) Lvov. Gos. Univ. Uch. Zap. Ser. Meh.-Mat., **29**, no. 6, 58–70 (1954)

[BG54] Belinskiĭ, P. P., Goldberg, A. A.: Application of a theorem on conformal mappings to questions of invariance of deficiencies of meromorphic functions. (Russian) Ukraïn. Mat. Zh., **6**, 263–269 (1954)

[Bie19] Bieberbach, L.: Über eine Vertiefung des Picardschen Satzes bei ganzen Funktionen endlicher Ordnung. Math. Z., **3**, 175–190 (1919)

[Bor97] Borel, É.: Sur les zeros des fonctions entieres. Acta math., **20**, 357–396 (1897)

[Bor00] Borel, É.: Leçons sur les fonctions entieres. Paris (1900); Second edition (1921)

[Bou08] Boutroux, P.: Sur l'indétermination d'une fonction uniforme au voisinage d'une singularité transcendente. Ann. Sci. Ecole Norm. Sup., **25**, 318–370 (1908)

[Carl21] Carleman, T.: Sur les fonctions inverses des fonctions entieres d'ordre fini. Ark. for Mat. Astr. och Fys., **15**, no. 10 (1921)

[Carl23] Carleman, T.: Über die Approximation analytischer Funktionen durch lineare Aggregate von vorgegebenen Potenzen. Ark. for Mat., Astr. och Fysik, **17**, no. 9, 1–30 (1923)

[Carl33] Carleman, T.: Sur une inégalité différentielle dans la théorie des fonctions analytiques. C. r. Acad. sci., **196**, 995–997 (1933)

[Cart29a] Cartan, H.: Sur la fonction de croissance attachée à une fonction méromorphe de deux variables et ses applications aux fonctions méromorphes d'une variable. C. r. Acad. sci., **189**, 521–523 (1929)

[Cart29b] Cartan, H.: Sur la croissance des fonctions méromorphes d'une ou de plusieurs variables complexes. C. r. Acad. sci., **188**, 1374–1376 (1929)

[Cart30] Cartan, H.: Sur les valeurs exceptionnelles d'une fonction méromorphe dans tout le plan. C. r. Acad. sci., **190**, 1003–1005 (1930)

[Cart33] Cartan, H.: Sur les zeros des combinaisons linéaires de p fonctions holomorphes données. Mathematica, **7**, 5–31 (1933)

[CM49] Chebotarev, N. G., Meĭman, N. N.: The Routh-Hurwitz problem for polynomials and entire functions. Real quasipolynomials with $r = 3$, $s = 1$. (Russian) Appendix by G. S. Barhin and A. N. Hovanskiĭ. Trudy Mat. Inst. Steklov., **26**, 1–331 (1949)

[Col24] Collingwood, E. F.: Sur les valeurs exceptionnelles des fonctions entieres d'ordre fini. C. r. Acad. sci., **179**, 1125–1127 (1924)

[Den07] Denjoy, A.: Sur les fonctions entieres de genre fini. C. r. Acad. sci., **145** (1907)

[Den30] Denjoy, A.: Sur une théorème de Wiman. C. r. Acad. sci., **193**, 828–830 (1930)

[Din36] Dinghas, A.: Bemerkung zu einer Differentialungleichung von Carleman. Math. Z., **41**, 713–716 (1936)

[Dra36] Drape, E.: Über die Darstellung Riemannscher Flächen durch Streckenkomplexe. Deutsche Math., **1**, 805–824 (1936)

[DS69] Drasin, D., Shea, D. F.: On the Valiron deficiencies of integral functions. Bull. London Math. Soc., **1**, 174–178 (1969)

[Duf39] Dufresnoy J.: Sur les fonctions méromorphes dans une angle. C. r. Acad. sci., **208**, 718–720 (1939)

[Dug47] Dugué, D.: Le défaut au sens de M. Nevanlinna depend de l'origine choisie. C. r. Acad. sci., **225**, 555–557 (1947)

[Dzh66] Dzhrbashyan, M. M.: Integral transforms and representations of functions in the complex domain. (Russian) Nauka, Moscow (1966)

[Edr55] Edrei, A.: Meromorphic functions with three radially distributed values. Trans. Amer. Math. Soc., **78**, no. 2, 276–293 (1955)

[Edr62] Edrei, A.: Meromorphic functions with values that are both deficient and asymptotic. Studies in mathematical analysis and related topics. Essays in honor of George Pólya. Stanford University Press, pp. 93–103 (1962)

[Edr64] Edrei, A.: The deficiencies of meromorphic functions of finite lower order. Duke Math. J., **31**, 1–21 (1964)

[EF56] Edrei, A., Fuchs, W. H. J.: On the regularity of the growth of meromorphic functions with highly deficient zeros and poles. Report no. 59, Cornell University, Ithaca, N.Y. (1956)

[EF59a] Edrei, A., Fuchs, W. H. J.: Valeurs deficientes et valeurs asymptotiques des fonctions méromorphes. Comment. Math. Helv., **33**, 258–295 (1959)

[EF59b] Edrei, A., Fuchs, W. H. J.: On the growth of meromorphic functions with several deficient values. Trans. Amer. Math. Soc., **93**, no. 2, 292–328 (1959)

[EF60] Edrei, A., Fuchs, W. H. J.: On the deficiencies of meromorphic functions of order less than one. Duke Math. J., **27**, no. 2, 233–249 (1960)

[EF62a] Edrei, A., Fuchs, W. H. J.: Bounds for the number of deficient values of certain classes of meromorphic functions. Proc. London Math. Soc., **12**, 315–344 (1962)

[EF62b] Edrei, A., Fuchs, W. H. J.: On meromorphic functions with regions free of poles and zeros. Acta math., **108**, 113–145 (1962)

[EFH61] Edrei, A., Fuchs, W. H. J., Hellerstein, S.: Radial distribution and deficiencies of the values of a meromorphic function. Pacif. J. Math., **2**, 135–151 (1961)

[Elf34] Elfving, G.: Über eine Klasse von Riemannschen Flächen and ihre Uniformisierung. Acta Soc. Sci. Fenn., N. S., **2**, no. 3, 1–60 (1934)

[Evg68] Evgrafov, M. A.: Analytic functions. Second revised and enlarged edition. Nauka, Moscow (1968). English transl.: Dover, New York, 1978.

[Fuc58] Fuchs, W. H. J.: A theorem on the Nevanlinnas deficiencies of meromorphic functions of finite order. Ann. of Math., **68**, 203–209 (1958)

[FH62] Fuchs, W. H. J., Hayman, W. K.: An entire function with assigned deficiencies. Studies in mathematical analysis and related topics. Essays in honor of George Pólya. Stanford University Press, 117–125 (1962)

[Gol54a] Goldberg, A. A.: On deficiencies of meromorphic functions. (Russian) Dokl. Akad. Nauk SSSR (N.S.), **98**, 893–895 (1954)

[Gol54b] Goldberg, A. A.: On the inverse problem of the theory of the distribution of the values of meromorphic functions. (Russian) Ukraïn. Mat. Zh., **6**, 385–397 (1954)

[Gol56] Goldberg, A. A.: Some asymptotic properties of meromorphic functions. (Russian) Uchen. Zap. Lvov. Univ., Ser. Mekh.-Mat., **38**, no. 7 (1956)

[Gol57a] Goldberg, A. A.: An estimate of the sum of the deficiencies of a meromorphic function of order less than unity. (Russian) Dokl. Akad. Nauk SSSR (N.S.), **114**, 245–248 (1957)

[Gol57b] Goldberg, A. A.: An example of an entire function of finite order with a non-asymptotic deficient value. (Russian) Uzhgorod. Gos. Univ. Nauchn. Zap., **18**, 191–194 (1957); English translation: Amer. Math. Soc. Transl. (2), **18**, 45–47 (1961)

[Gol57c] Goldberg, A. A.: On an inequality connected with logarithmically convex functions. (Ukrainian) Dopovidi Akad. Nauk Ukraïn. RSR, no. 3, 227–230 (1957)

[Gol59a] Goldberg, A. A.: On the set of deficient values of meromorphic functions of finite order. (Russian) Ukraïn. Mat. Zh., **11**, 438–443 (1959)

[Gol59b] Goldberg, A. A.: On a class of Riemann surfaces. (Russian) Mat. Sb. (N.S.), **49 (91)**, 447–458 (1959)

[Gol60a] Goldberg, A. A.: Meromorphic functions with separated zeros and poles. (Russian) Izv. Vyssh. Uchebn. Zaved. Matematika, no. 4 (17), 67–72 (1960)

[Gol60b] Goldberg, A. A.: Some problems of the value distribution theory (Russian). An appendix in the Russian translation of [**Wit68**], pp. 263–300.

[Gol61] Goldberg, A. A.: Distribution of values of meromorphic functions with separated zeroes and poles.(Russian) Dokl. Akad. Nauk SSSR, **137**, 1030–1033 (1961); English translation: Soviet Math. Dokl., **2**, 389–392 (1961)

[Gol63] Goldberg, A. A.: Three examples of entire functions. (Ukrainian) Dopovidi Akad. Nauk Ukraïn. RSR, no. 4, 443–446 (1963)

[Gol64a] Goldberg, A. A.: The integral with respect to a semi-additive measure and its application to the theory of entire functions. III. (Russian) Mat. Sb. (N.S.), **65 (107)**, 414–453 (1964)

[Gol64b] Goldberg, A. A.: On the possible value of the lower order of an entire function with a finite deficient value. (Russian) Dokl. Akad. Nauk SSSR, **159**, 968–970 (1964)

[Gol66] Goldberg, A. A.: Non-asymptotic deficient values of meromorphic functions. (Russian) Dokl. Akad. Nauk SSSR, **171**, 254–257 (1966)

[Gol67] Goldberg, A. A.: Non-asymptotic deficient values of entire functions of infinite order. (Russian) Mat. Zametki, **1**, no. 1, 81–86 (1967)

[GO61a] Goldberg, A. A., Ostrovskiĭ, I. V.: Some theorems on the growth of meromorphic functions. (Russian) Zap. Matem. Otd. Kharkov. Univ. and Kharkov. Mat. Ob., **27**, 3–37 (1961)

[GO61b] Goldberg, A. A., Ostrovskiĭ, I. V.: New investigations on the growth and distribution of values of entire and meromorphic functions of genus zero. (Russian) Uspekhi Mat. Nauk, **16**, no. 4, 51–62 (1961)

[Gol69] Goluzin, G. M.: Geometric theory of functions of a complex variable. American Mathematical Society, Providence, R.I. (1969)

[Gov69] Govorov, N. V.: The Paley conjecture. (Russian) Funktsional. Anal. i Prilozhen., **3**, no. 2, 41–45 (1969)

[Hab61a] Habetha, K.: Eine Bemerkung zur Werteverteilung meromorpher Funktionen in der Halbebene. Arch. Math., **12**, 43–50 (1961)

[Hab61b] Habetha, K.: Über die Werteverteilung in Winkelraumen. Math. Z., **77**, 453–467 (1961)

[Hab52] Habsch, H.: Die Theorie der Grundkurven und das Aquivalenzproblem bei der Darstellung Riemannscher Flächen. Mitt. Math. Seminar Giefien, no. 42, 1–51 (1952)

[Had93] Hadamard J.: Sur les proprietés des fonctions entieres et en particulier une fonction etudiée par Riemann. Journ. math, pure appl., **9**, 171–215 (1893)

[Häl41] Hällström, G. af: Zwei Beispiele ganzer Funktionen mil algebraischem Hochstindex einer Stellensorte. Math. Z., **47**, 161–174 (1941)

[Häl52] Hällström, G. af: Eine quasikonforme Abbildung mit Anwendungen auf die Wertverteilungsiehre. Acta Acad. Aboensis. Math. et Phys., **18**, no. 8, 1–16 (1952)

[Hay52] Hayman, W. K.: The minimum modulus of large integral functions. Proc. London Math. Soc., **2**, 469–512 (1952)

[Hay53] Hayman, W. K.: An integral function with a deficient value that is neither asymptotic nor invariant under change of origin. J. London Math. Soc., **28**, 369–376 (1953)

[Hay56] Hayman, W. K.: Questions of regularity connected with the Phragmen-Lindelöf principle. J. Math. pures et appl., **35**, 115–126 (1956)

[Hay59] Hayman, W. K.: Picard values of meromorphic functions and their derivatives. Ann. of Math., **70**, 9–42 (1959)

[Hay64] Hayman, W. K.: Meromorphic functions. Clarendon Press, Oxford (1964)

[Hay65] Hayman, W. K.: On the characteristic of functions meromorphic in the plane and of their integrals. Proc. London Math. Soc., **14A**, 93–128 (1965)

[Hei48] Heins, M.: Entire functions with bounded minimum modulus; subharmonic functions analogues. Ann. of Math., **49**, 200–213 (1948)

[Hei53] Heins, M.: The set of asymptotic values of an entire function. 12-te Skand. matematikerkongr., Lund, 56–60 (1953)

[Hei55] Heins, M.: Meromorphic functions with assigned asymptotic values. Duke Math. J., **22**, 353–356 (1955)

[Hei57] Heins, M.: Asymptotic spots of entire and meromorphic functions. Ann. of Math., **66**, 430–439 (1957)

[Hel63] Hellerstein, S.: On a class of meromorphic functions with deficient zeros and poles. Pacif. J. Math., **13**, 115–124 (1963)

[Hel66] Hellerstein, S.: On the zeros of an entire function and its second derivative. Ill. J. Math., **10**, 488–496 (1966)

[HR64] Hellerstein, S., Rubel, L. A.: Subfields that are algebraically closed in the field of all meromorphic functions. Journ. analyse math., **12**, 105–111 (1964)

[Hio55] Hiong, King-Lai: Nouvelle démonstration et amélioration d'une inégalité de M. Milloux. Bull. sci. math., **79**, 135–160 (1955)

[Hio56] Hiong, King-Lai: Sur la limitation de $T(r, f)$ sans intervention des poles. Bull. sci. math., **80**, 175–190 (1956)

[Huc56a] Huckemann, F.: Zur Darstellung von Riemannschen Flächen durch Streckenkomplexe. Math. Z., **65**, 215–239 (1956)

[Huc56b] Huckemann, F.: Über den Einfluß von Randstellen Riemannschen Flächen auf die Wertverteilung. Math. Z., **65**, 240–282 (1956)

[Hur91] Hurwitz, A.: Über Riemann'sche Flachen mit gegebenen Verzweigungspunkten. Math. Ann., **39**, 1–61 (1891)

[Hur02] Hurwitz, A.: Über die Anzahl der Riemann'schen Flachen mit gegebenen Verzweigungspunkten. Math. Ann., **55**, 53–66 (1902)

[Ive14] Iversen, F.: Recherches sur les fonctions inverses des fonctions méromorphes. Thèse, Helsingfors, 1–67 (1914)

[Jen99] Jensen, J. L. W. V.: Sur un nouvel et important théorème de la théorie des fonctions. Acta math., **22**, 359–364 (1899)

[Kak35] Kakutani, S.: On the exceptional value of meromorphic functions. Proc. Phys.-Math. Soc. Japan, **17**, 174–176 (1935)

[Kar30] Karamata, J.: Sur un mode de croissance reguliere des fonctions. Mathematica, **4**, 38–53 (1930)

[Kel54] Keldysh, M. V.: On series of rational fractions. (Russian) Doklady Akad. Nauk SSSR (N.S.), **94**, 377–380 (1954)

[Kje54] Kjellberg, B.: A relation between the maximum and minimum modulus of a class of entire functions. 12-te Scand. matematikerkongr., Lund, 135–138 (1954)

[Kje60] Kjellberg, B.: On the minimum modulus of entire functions of lower order less than one. Math. Scand., **8**, 189–197 (1960)

[Kje63] Kjellberg, B.: A theorem on the minimum modulus of entire functions. Math. Scand., **12**, 5–11 (1963)

[KAB49] Korevaar, J., Aardenne-Ehrenfest, T., van, Bruijn, N. Q., de: A note on slowly oscillating functions. Nieuw Archief voor Wiskunde, **23**, 77–86 (1949)

[Kre47] Krein, M.: A contribution to the theory of entire functions of exponential type. (Russian) Izvestiya Akad. Nauk SSSR. Ser. Math., **11**, 309–326 (1947)

[KW59] Künzi, H. P., Wittich, H.: The distribution of a-points of certain meromorphic functions. Michigan Math. J., **6**, 105–121 (1959)

[LeV49] Le-Van-Thiem: Über das Umkehrproblem der Wertverteilungslehre, Comment. Math. Helv., **23**, 26–49 (1949)

[LeV50] Le-Van-Thiem: Sur un problème d'inversion dans la théorie des fonctions méromorphes. Ann. Sci. Ecole Norm. Sup., **67**, 51–98 (1950)

[Lev41] Levin, B. Ya.: On functions holomorphic in a half-plane, Trudy Odess. Derzh. Univ., **3**, 5–14 (1941)

[Lev80] Levin, B. Ya.: Distribution of zeros of entire functions. American Mathematical Society, Providence, R.I. (1980)

[LO60] Levin, B. Ya., Ostrovskiĭ, I. V.: The dependence of the growth of an entire function on the distribution of zeros of its derivatives. (Russian) Sibirsk. Mat. Zh., **1**, 427–455 (1960) English transl.: AMS Translations (2) **32**, 323–357 (1963)

[Lin02] Lindelöf, E.: Mémoire sur la théorie des fonctions entières de genre fini. Acta Soc. Sci. Fenn., **31**, no. 1 (1902)

[Lin05] Lindelöf, E.: Sur les fonctions entières d'ordre entier. Ann. Sci. École Norm. Sup., **22**, 365–395 (1905)

[Lit08] Littlewood, J. E.: A general theorem on integral functions of finite order. Proc. London Math. Soc., **6**, 189–204 (1908)

[Mac63] MacLane, G. R.: Asymptotic values of holomorphic functions. Rice Univ. Studies, **49**, no. 1, 83 pp. (1963)

[Mar77] Markushevich, A. I.: Theory of functions of a complex variable. Vol. I, II, III. Chelsea Publishing Co., New York (1977)

[Mat60] Matsaev, V. I.: On the growth of entire functions that admit a certain estimate from below. (Russian) Dokl. Akad. Nauk SSSR, **132**, 283–286 (1960); English translation: Soviet Math. Dokl., **1**, 548–552 (1960)

[Maz31] Mazurkiewicz, S.: Sur les points singuliers d'une fonction analytique. Fund. math., **17**, 26–29 (1931)

[Mer52] Mergelyan, S. N.: Uniform approximations of functions of a complex variable. (Russian) Uspehi Matem. Nauk (N.S.), **7**, no. 2 (48), 31–122 (1952)

[Mer53] Mergelyan, S. N.: On completeness of systems of analytic functions. (Russian) Uspehi Matem. Nauk (N.S.), **8**, no. 4 (56), 3–63 (1953)

[Mil40] Milloux, H.: Les fonctions méromorphes et leurs dérivées. Extensions d'un théorème de M. R. Nevanlinna. Applications, Paris (1940)

[Mil47] Milloux, H.: Les derivées des fonctions meromorphes et la théorie des defauts. Ann Sci. Ecole Norm. Sup., **3**, 289–316 (1947)

[Mit05] Mittag-Leffler, G.: Sur la représentation analytique d'une branche uniforme d'une fonction monogène. Acta math., **29**, 101–181 (1905)

[NevF22a] Nevanlinna, F.: Über die Beziehungen zwischen dem Anwachsen einer analytischen Funktion und der Verteilung ihrer Nullstellen und Pole. V Congrès des mathématiciens scand., Helsingfors (1922)

[NevF22b] Nevanlinna, F.: Sur les relations qui existent entre la distribution des zéros et des pôles d'une fonction monogène et la croissance de son module. C. r. Acad. sci., **175**, 676–679 (1922)

[NevF23] Nevanlinna, F.: Bemerkungen zur Theorie der ganzen Funktionen endlicher Ordnung. Soc. Sci. Fenn. Comment. Phys.-math., **2**, no. 4, 1–7 (1923)

[NevF29] Nevanlinna, F.: Über eine Klasse meromorpher Funktionen. 7-te Congrés Math. Scand., Oslo, 81–83 (1929)

[NN22] Nevanlinna, F., Nevanlinna, R.: Über die Eigenschaften analytischer Funktionen in der Umgebung einer singulären Stelle oder Linie. Acta Soc. sci. Fenn., **50**, no. 5, 1–46 (1922)

[NevR24] Nevanlinna, R.: Untersuchungen über den Picard'schen Satz. Acta Soc. sci. Fenn., **50**, no. 6, 1–42 (1924)

[NevR25a] Nevanlinna, R.: Zur Theorie der meromorphen Funktionen. Acta math., **46**, 1–99 (1925)

[NevR25b] Nevanlinna, R.: Über die Eigenschaften meromorpher Funktionen in einem Winkelraum. Acta Soc. Sci. Fenn., **50**, no. 12, 1–45 (1925)

[NevR29] Nevanlinna, R.: Le théorème de Picard-Borel et la théorie des fonctions méromorphes. Paris (1929)

[NevR31] Nevanlinna, R.: Remarques sur les fonctions monotones. Bull. sci. math., **55**, 140–144 (1931)

[NevR32a] Nevanlinna, R.: Über die Riemannsche Fläche einer analytischen Funktion. Verhandl. Iinternat. Mathematiker-Kongr., Zurich, B. I. (1932)

[NevR32b] Nevanlinna, R.: Über Riemannsche Flächen mit endlich vielen Windungspunkten. Acta math., **58**, 295–373 (1932)

[NevR74] Nevanlinna, R.: Eindeutige analytische Funktionen. Springer-Verlag, Berlin (1974)

[NO65] Ngoan, Vu., Ostrovskiĭ, I. V.: The logarithmic derivative of a meromorphic function. (Russian) Akad. Nauk Armjan. SSR Dokl., **41**, 272–277 (1965)

[Ost57] Ostrovskiĭ, I. V.: A generalization of a theorem of M. G. Kreĭn. (Russian) Dokl. Akad. Nauk SSSR (N.S.), **116**, 742–745 (1957)

[Ost58] Ostrovskiĭ, I. V.: On meromorphic functions taking certain values at points lying near a finite system of rays. (Russian) Dokl. Akad. Nauk SSSR, **120**, 970–972 (1958)

[Ost60a] Ostrovskiĭ, I. V.: On the relation of the growth of a meromorphic function to the distribution of its values according to their arguments. (Russian) Dokl. Akad. Nauk SSSR, **132**, 48–51 (1960); English translation: Soviet Math. Dokl., **1**, 485–488 (1960)

[Ost60b] Ostrovskiĭ, I. V.: An estimate for the deficiency of a meromorphic function for which two values are distributed inside a certain angle. (Russian) Izv. Vyssh. Uchebn. Zaved. Matematika, no. 2 (15), 138–148 (1960)

[Ost60c] Ostrovskiĭ, I. V.: On the location of the zeros of the derivative of an entire function whose zeros lie close to the real axis. (Russian) Dokl. Akad. Nauk SSSR, **130**, 973–976 (1960); English translation: Soviet Math. Dokl., **1**, 114–117 (1960)

[Ost61] Ostrovskiĭ, I. V.: The connection between the growth of a meromorphic function and the distribution of the arguments of its values. (Russian) Izv. Akad. Nauk SSSR. Ser. Mat., **25**, 277–328 (1961)

[Ost63a] Ostrovskiĭ, I. V.: Deficiencies of meromorphic functions of order less than one. (Russian) Dokl. Akad. Nauk SSSR, **150**, 32–35 (1963)

[Ost63b] Ostrovskiĭ, I. V.: A problem from the theory of distribution of values. (Russian) Dokl. Akad. Nauk SSSR, **151**, 34–37 (1963)

[Pal32] Paley, R. E. A. C.: A note on integral functions. Proc. Cambridge Philos. Soc., **28**, 262–265 (1932)

[Pet66] Petrenko, V. P.: Certain estimates for the values of the deficiencies of a meromorphic function. (Russian) Sibirsk. Mat. Zh., **7**, 1319–1336 (1966); English transl.: Sib. Math. J. **7**, 1035–1047 (1966)

[Pet67] Petrenko, V. P.: On the growth of meromorphic functions of finite lower order and on the magnitudes of their deficiencies. (Russian) Sibirsk. Mat. Zh., **8**, 1156–1189 (1967); English transl.: Sib. Math. J. **8**, 883–909 (1967)

[Pet69] Petrenko, V. P.: Growth of meromorphic functions of finite lower order. (Russian) Izv. Akad. Nauk SSSR Ser. Mat., **33**, 414–454 (1969); English transl.: Math. USSR, Izv. **3**, 391–432 (1969)

[Pfl46] Pfluger, A.: Zur Defektrelation ganzer Funktionen endlicher Ordnung. Comment. Math. Helv., **19**, 91–104 (1946)

[PS98] Pólya, G., Szegö, G.: Problems and theorems in analysis. I. Series, integral calculus, theory of functions. Springer-Verlag, Berlin (1998)

[Pös51] Pöschl, K.: Über die Wertverteilung der erzeugenden Funktionen Riemannscher Flächen mit endlich vielen periodischen Enden. Math. Ann., **123**, 79–95 (1951)

[Pot51] Potyagailo, D. B.: Asymptotic and boundary properties of meromorphic functions. Master's Thesis, Lvov University (1951)

[Pri56] Priwalow, I. I.: Randeigenschaften analytischer Funktionen. VEB Deutscher Verlag der Wissenschaften, Berlin (1956)

[Röh63] Röhrl, H.: Unbounded coverings of Riemann surfaces and extensions of rings of meromorphic functions. Trans. Amer. Math. Soc., **107**, 320–346 (1963)

[Rub63] Rubel, L. A.: A Fourier series method for entire functions. Duke Math. J., **30**, 437–442 (1963)

[Sel44] Selberg, H. L.: Über einen Satz von Collingwood. Arch. Math. Naturv., **B 48**, no. 9, 119–126 (1944)

[Sel46] Selberg, H. L.: Eine Ungleichung der Potentialtheorie und ihre Anwendung in der Theorie der meromorphen Funktionen. Comment. Math. Helv., **18**, 309–326 (1946)

[Sha40] Shah, S. M.: A theorem on integral functions of integral order. J. London Math. Soc., **15**, 23–31 (1940)

[Sha41] Shah, S. M.: A theorem on integral functions of integral order, II. J. Indian Math. Soc., N.S., **5**, 179–188 (1941)

[Sha44] Shah, S. M.: A note on meromorphic functions. Math. Student, **12**, 67–70 (1944)

[Sha46a] Shah, S. M.: On the lower order of integral functions. Bull. Amer. Math. Soc., **52**, 1046–1052 (1946)

[Sha46b] Shah, S. M.: On proximate orders of integral functions. Bull. Amer. Math. Soc., **52**, 326–328 (1946)

[Sha48] Shah, S. M.: A note on lower proximate orders. J. Indian Math. Soc., **12**, no. 1-2, 31–32 (1948)

[Sha60] Shah, S. M.: An extension of Lindelöf's theorem to meromorphic functions. Tôhoku Math. J., **12**, 333–338 (1960)

[She69] Sheremeta, M. N.: Asymptotic behavior of functions of Mittag-Leffler type and their application. (Russian) Izv. AN Arm. SSR. Matematika, **4**, no. 2, 144–148 (1969)

[Shi29] Shimizu, T.: On the theory of meromorphic functions. Japanese J. Math., **6**, 119–171, (1929)

[Sin61] Singh, S. K.: On exceptional values of entire and meromorphic functions. Tôhoku Math. J., **13**, 373–380 (1961)

[Smi28] Smirnov, V. I.: Sur les valeurs limites des fonctions régulières à l'intérieur d'un cercle. Zhurnal Leningr. Fiz.-Mat. Ob-va, **2**, no. 2, 22–37 (1928)

[Sto62] Stoïlow, S.: Teoria funcţiilor de o variabilă complexă, vol. 1-2, Editura Academiei Republicii Populare Române, 1954. Russian transl.: Theory of functions of a complex variable. Inostrannaya Literatura, Moscow (1962)

[Tai62] Tairova, V. G.: On "Streckenkomplexe" of closed Riemann surfaces of genus zero. (Russian) Izv. Vyssh. Uchebn. Zaved. Matematika, no. 4 (29), 155–160 (1962)

[Tai64] Tairova, V. G.: "Streckenkomplexe" of a certain class of closed Riemann surfaces. (Russian) Sibirsk. Mat. Zh., **5**, 929–951 (1964)

[Tei38] Teichmüller, O.: Untersuchungen über konforme und quasikonforme Abbildung. Deutsche Math., **3**, 621–678 (1938)

[Tei39] Teichmüller, O.: Vermutungen und Sätze über die Wertverteilung gebrochener Funktionen endlicher Ordnung. Deutsche Math., **4**, 163–190 (1939)

[Tei44] Teichmüller, O.: Ein Verschiebungssatz der quasikonformen Abbildung. Deutsche Math., **7**, 336–343 (1944)

[Tsu50] Tsuji, M.: On Borel's directions of meromorphic functions of finite order. Tôhoku Math. J., **2**, no. 2, 97–112 (1950)

[Tum43a] Tumura, Y.: Recherches sur la distribution des valeurs des fonctions analytiques. Japanese J. Math., **18**, 797–876 (1943)

[Tum43b] Tumura, Y.: Sur une extension d'un théorème de M. Teichmüller. Proc. Imp. Acad. Tokyo, **19**, 55–59 (1943)

[Ull36] Ullrich, E.: Zum Umkehrproblem der Wertverteilungslehre. Nachr. Qes. Wiss. Göttingen, Math.-Phys. Kl., **1**, no. 9, 135–150 (1936)

[Val14] Valiron, G.: Sur les fonctions entières d'ordre nul et d'ordre fini et en particulier les fonctions à correspondance régulière. Ann. fac. sci. Univ. Toulouse, **5**, 117–257 (1914)

[Val19] Valiron, G.: Sur les zéros des fonctions entières d'ordre fini. Rend. Circ. mat. Palermo, **43**, 255–268 (1919)

[Val23] Valiron, G.: Lectures on the general theory of integral functions. Toulouse (1923)

[Val25] Valiron, G.: Sur les valeurs asymptotiques de quelques fonctions méromorphes. Rend. Circ. mat. Palermo, **49**, 415–421 (1925)

[Val26] Valiron, G.: Sur la distribution des valeurs des fonctions méromorphes. Acta math., **47**, 117–142 (1926)

[Val30] Valiron, G.: Sur un théorème de M. Wiman. Opuscula math. A. Wiman dedicata, 1–12 (1930)

[Val31] Valiron, G.: Sur la dérivée des fonctions algébroïdes. Bull. Soc. math. France, **59**, no. 1-2, 17–39 (1931)

[Val33] Valiron, G.: Remarques sur les valuers exceptionnelles des fonctions méromorphes. Rend. Circ. mat. Palermo, **57**, 71–86 (1933)

[Val35a] Valiron, G.: Sur le minimum du module des fonctions entières d'ordre inférieur à un. Mathematica, **11**, 264–269 (1935)

[Val35b] Valiron, G.: Sur le nombre des singularités transcendantes des fonctions inverses d'une classe d'algébroïdes. C. r. Acad. sci., **200**, 713–715 (1935)

[Val47] Valiron, G.: Valeurs exceptionnelles et valeurs déficientes des fonctions méromorphes. C. r. Acad. sci., **225**, 556–558 (1947)

[Val50] Valiron, G.: Sur les valeurs déficientes des fonctions méromorphes d'ordre nul. C. r. Acad. sci., **230**, 40–42 (1950)

[Val60] Valiron, G.: Fonctions entières d'ordre fini et fonctions méromorphes. Genève (1960)

[Vol50] Volkovyskiĭ, L. I.: Investigation of the type problem for a simply connected Riemann surface. (Russian) Trudy Mat. Inst. Steklov., **34**, 1–171 (1950)

[Vol54] Volkovyskiĭ, L. I.: Quasiconformal mappings (Russian), Lvov State University Publishers, Lvov (1954)

[VLA91] Volkovyskiĭ, L. I., Lunts, G. L., Aramanovich, I. G.: A collection of problems on complex analysis. Dover Publications, Inc., New York (1991)

[Wah29] Wahlund, A.: Über einen Zusammenhang zwischen dem Maximalbetrage der ganzen Funktion und seiner unteren Grenze nach dem Jensensche Theoreme. Arkiv Math., **21 A**, no. 23, 1–34 (1929)

[Wey31] Weyl, H.: Über das Hurwitzsche Problem der Bestimmung der Anzahl Riemannscher Flächen von gegebener Verzweigungsart. Comment. Math. Helv., **3**, 103–113 (1931)

[WW96] Whittaker, E. T., Watson, G. N.: A course of modern analysis. An introduction to the general theory of infinite processes and of analytic functions; with an account of the principal transcendental functions. Cambridge University Press, Cambridge (1996)

[Wim05] Wiman, A.: Sur une extension d'une théorème de M. Hadamard. Ark. för Math., Astr., Fys., **2**, no. 14, 1–5 (1905)

[Wim15] Wiman, A.: Über eine Eigenschaft der ganzen Funktionen von der Höhe Null. Math. Ann., **76**, 197–211 (1915)

[Wit47] Wittich, H.: Über die Wachstumsordnung einer ganzen transzendenten Funktion. Math. Z., **51**, 1–16 (1947)

[Wit48a] Wittich, H.: Zum Beweis eines Satzes über quasikonforme Abbildungen. Math. Z., **51**, 278–288 (1948)

[Wit48b] Wittich, H.: Über eine Klasse meromorpher Funktionen. Arch. Math., **1**, 160–166 (1948)

[Wit68] Wittich, H.: Neuere Untersuchungen über eindeutige analytische Funktionen. Springer-Verlag, Berlin (1968)

[Yos34] Yosida, K.: On a class of meromorphic functions. Proc. Phys.-Math. Soc. Japan, **16**, 227–235 (1934)

References Added to the English Edition

[1] Anderson, J. M.; Baernstein, A. The size of the set on which a meromorphic function is large. Proc. London Math. Soc. (3) 36 (1978), no. 3, 518–539.

[2] Baernstein, A. Proof of Edrei's spread conjecture. Proc. London Math. Soc. (3) 26 (1973), 418–434.

[3] Baernstein, A. A generalization of the $\cos \pi\rho$ theorem. Trans. Amer. Math. Soc. 193 (1974), 181–197.

[4] Baernstein, A. How the $*$-function solves extremal problems. Proceedings of the International Congress of Mathematicians (Helsinki, 1978), pp. 639–644, Acad. Sci. Fennica, Helsinki, 1980.

[5] Baernstein, A. Regularity theorems associated with the spread relation. J. Analyse Math. 31 (1977), 76–111.

[6] Barth, K. F.; Brannan, D. A.; Hayman, W. K. The growth of plane harmonic functions along an asymptotic path. Proc. London Math. Soc. (3) 37 (1978), no. 2, 363–384.

[7] Bergweiler, W. Iteration of meromorphic functions. Bull. Amer. Math. Soc. (N.S.) 29 (1993), no. 2, 151–188.

[8] Bergweiler, W. On the zeros of certain homogeneous differential polynomials. Archiv der Math. 64 (1995), 199–202.

[9] Bergweiler, W. Normality and exceptional values of derivatives, Proc. Amer. Math. Soc. 129 (2001), 121–129.

[10] Bergweiler, W.; Bock, H. On the growth of meromorphic functions of infinite order. J. Anal. Math. 64 (1994), 327–336.

[11] Bergweiler, W.; Eremenko, A. On the singularities of the inverse to a meromorphic function of finite order. Rev. Mat. Iberoamericana 11 (1995), 355–373.

[12] Bergweiler, W; Eremenko, A. Proof of a conjecture of Pólya on the zeros of successive derivatives of real entire functions. Acta Math. 197 (2006), 145–166.

[13] Bergweiler, W.; Eremenko, A.; Langley, J. K. Real entire functions of infinite order and a conjecture of Wiman. Geom. Funct. Anal. 13 (2003), no. 5, 975–991.

[14] Bergweiler, W.; Langley, J. K. Multiplicities in Hayman's alternative. J. Aust. Math. Soc. 78 (2005), no. 1, 37–57.

[15] Brüggemann, F. Proof of a conjecture of Frank and Langley concerning zeros of meromorphic functions and linear differential polynomials. Analysis 12 (1992), no. 1/2, 5–30.

[16] Chang, Kuan Heo. Asymptotic values of entire and meromorphic functions. Sci. Sinica 20 (1977), no. 6, 720–739.

[17] Chen Zhen Hua. Normality of families of meromorphic functions with multiple valued derivatives. Acta Math. Sinica 30 (1987), 97–105.

[18] Cherry, W.; Ye, Zhuan. Nevanlinna's theory of value distribution. The second main theorem and its error terms. Springer-Verlag, Berlin, 2001.

[19] Chuang, C. T. On differential polynomials. Analysis of One Complex Variable, World Sci. Publishing, Singapore 1987, pp. 12–32.

[20] Chuang, Chi-Tai; Yang, Chung-Chun. Fix-points and factorization of meromorphic functions. Translated from the Chinese. World Scientific Publishing Co., Inc., Teaneck, NJ, 1990.

[21] Clunie, J. G. On integral and meromorphic functions. J. London Math. Soc. 37 (1962), 17–27.

[22] Clunie, J. G. On a result of Hayman. J. London Math. Soc. 42 (1967), 389–392.

[23] Clunie, J.; Eremenko, A.; Rossi, J. On equilibrium points of logarithmic and Newtonian potentials. J. London Math. Soc. (2) 47 (1993), no. 2, 309–320.

[24] Csillag, P. Über ganze Funktionen, welche drei nicht verschwindende Ableitungen besitzen. Math. Ann. 110 (1935), 745–752.

[25] Drasin, D. The inverse problem of the Nevanlinna theory. Acta Math. 138 (1976), no. 1-2, 83–151.

[26] Drasin, D. On Nevanlinna's inverse problem. Complex Variables Theory Appl. 37 (1998), no. 1-4, 123–143.

[27] Drasin, D. Quasiconformal modifications of functions having deficiency sum two. Ann. of Math. (2) 114 (1981), no. 3, 493–518.

[28] Drasin, D. Proof of a conjecture of F. Nevanlinna concerning functions which have deficiency sum two. Acta Math. 158 (1987), no. 1-2, 1–94.

[29] Drasin, D. The minimum modulus of subharmonic functions of order one and a method of Fryntov. J. London Math. Soc. (2) 54 (1996), no. 2, 239–250.

[30] Drasin, D. Meromorphic functions: progress and problems, Proc. Intl. Congr. Math. (Zürich, 1994), vol. 2, 828–835. Birkhäuser, Basel, 1995.

[31] Drasin, D.; Shea, D. Pólya peaks and the oscillation of positive functions. Proc. Amer. Math. Soc. 34 (1972), 403–411.

[32] Drasin, D.; Weitsman, A. Meromorphic functions with large sums of deficiencies. Advances in Math. 15 (1975), 93–126.

[33] Edrei, A. Sums of deficiencies of meromorphic functions. J. Analyse Math. 14 (1965), 79–107.

[34] Edrei, A. Sum of deficiencies of meromorphic functions. J. d'Analyse Math. 19 (1967), 53–74.

[35] Edrei, A. Solution of the deficiency problem for functions of small lower order. Proc. London Math. Soc. (3) 26 (1973), 435–445.

[36] Edrei, A.; Fuchs, W. H. J. Asymptotic behavior of meromorphic functions with extremal spread. I. Ann. Acad. Sci. Fenn. Ser A I Math. 2 (1976), 67–111.

[37] Edrei, A.; Fuchs, W. H. J. Asymptotic behavior of meromorphic functions with extremal spread. II. Ann. Acad. Sci. Fenn. Ser A I Math. 3 (1977), 141–168.

[38] Eremenko, A. The growth of entire and subharmonic functions on asymptotic curves. Siberian Math. J. 21 (1981), 5, 573–683.

[39] Eremenko, A. Meromorphic solutions of algebraic differential equations. Russian Math. Surveys 37 (1982), 4, 61–95. Correction: 38 (1983), 6.

[40] Eremenko, A. Deviations of meromorphic functions of finite lower order. (Russian) Teor. Funktsiĭ Funktsional. Anal. i Prilozhen. No. 40 (1983), 56–64.

[41] Eremenko, A. Defects and deviations of finite-order meromorphic functions. (Russian) Dokl. Akad. Nauk Ukrain. SSR Ser. A 1985, no. 1, 18–21.

[42] Eremenko, A. Inverse problem of the theory of distribution of values for finite-order meromorphic functions. Siberian Math. J. 27 (1987), 3, 377–390.

[43] Eremenko, A. On the set of defect values of a finite-order entire function. Ukrainian Math. J. 39 (1987), 3, 225–228.

[44] Eremenko, A. Meromorphic functions with small ramification. Indiana Univ. Math. J. 42 (1993), no. 4, 1193–1218.

[45] Eremenko, A. A counterexample to the Arakelyan conjecture. Bull. Amer. Math. Soc. (N.S.) 27 (1992), no. 1, 159–164.

[46] Eremenko, A. An analogue of the defect relation for the uniform metric. Complex Variables Theory Appl. 34 (1997), no. 1-2, 83–97.

[47] Eremenko, A. An analogue of the defect relation for the uniform metric. II. Complex Variables Theory Appl. 37 (1998), no. 1-4, 145–159.

[48] Eremenko, A. Value distribution and potential theory. Proceedings of the International Congress of Mathematicians, Vol. II (Beijing, 2002), pp. 681–690, Higher Ed. Press, Beijing, 2002.

[49] Eremenko, A. Transcendental meromorphic functions with three singular values. Illinois J. Math. 48 (2004), no. 2, 701–709.

[50] Eremenko, A.; Langley, J.; Rossi, J. On the zeros of meromorphic functions of the form $f(z) = \sum_{k=1}^{\infty} a_k/(z - z_k)$. J. Anal. Math. 62 (1994), 271–286.

[51] Fedorov, M. A.; Grishin, A. F. Some questions of the Nevanlinna theory for the complex half-plane. Math. Physics, Analysis and Geometry 1 (1998), 223–271.

[52] Frank, G. Eine Vermutung von Hayman über Nullstellen meromorpher Funktionen. Math. Z. 149 (1976), no. 1, 29–36.

[53] Frank, G. Über die Nullstellen von linearen Differentialpolynomen mit meromorphen Koeffizienten. Complex Methods on Partial Differential Equations, pp. 39–48, Math. Res. 53, Akademie-Verlag, Berlin 1989.

[54] Frank, G.; Hellerstein, S. On the meromorphic solutions of nonhomogeneous linear differential equations with polynomial coefficients. Proc. London Math. Soc. (3) 53 (1986), 407–428.

[55] Frank, G.; Weissenborn, G. Rational deficient functions of meromorphic functions. Bull. London Math. Soc. 18 (1986), no. 1, 29–33.

[56] Fryntov, A. A counterexample concerning the maximum and minimum of a subharmonic function. Proc. Amer. Math. Soc. 122 (1994), no. 1, 97–103.

[57] Fryntov, A.; Rossi, J.; Weitsman, A. On the longest arc relation for δ-subharmonic functions. Complex Variables Theory Appl. 34 (1997), no. 1-2, 99–108.

[58] Fryntov, A.; Rossi, J.; Weitsman, A. Circular means of fine Green's functions and the longest arc relation. Complex Variables Theory Appl. 37 (1998), no. 1-4, 211–224.

[59] Gleĭzer, E. V. Meromorphic functions with zeros and poles in small angles. (Russian) Sibirsk. Mat. Zh. 26 (1985), no. 4, 22–37, 203.

[60] Gleĭzer, E. V. The growth of entire functions with zeros on a system of rays. (Russian) Ukrain. Mat. Zh. 38 (1986), no. 3, 297–302, 404.

[61] Gleĭzer, E. V. Meromorphic functions with zeros and poles in small angles. II. (Russian) Sibirsk. Mat. Zh. 31 (1990), no. 2, 9–20, 223; English transl., Siberian Math. J. 31 (1990), no. 2, 192–202.

[62] Goldberg, A. A. Meromorphic functions. Mathematical analysis, Vol. 10 (Russian), pp. 5–97. Akad. Nauk SSSR Vsesojuz. Inst. Naučn. i Tehn. Informacii, Moscow, 1973.

[63] Goldberg, A. A. Nevanlinna's lemma on the logarithmic derivative of a meromorphic function. (Russian) Mat. Zametki 17 (1975), no. 4, 525–529.

[64] Goldberg, A. A.; Eremenko, A.; Sodin, M. L. Exceptional values in the sense of R. Nevanlinna and in the sense of V. P. Petrenko. I. (Russian) Teor. Funktsiĭ Funktsional. Anal. i Prilozhen. No. 47 (1987), 41–51; English transl., J. Soviet Math. 48 (1990), no. 6, 648–655.

[65] Goldberg, A. A.; Eremenko, A.; Sodin, M. L. Exceptional values in the sense of R. Nevanlinna and in the sense of V. P. Petrenko. II. (Russian) Teor. Funktsiĭ Funktsional. Anal. i Prilozhen. No. 4 (1987), 58–70; English transl., J. Soviet Math. 49 (1990), no. 2, 891–899.

[66] Goldberg, A. A.; Grinštein, V. A. The logarithmic derivative of a meromorphic function. (Russian) Mat. Zametki 19 (1976), no. 4, 525–530.

[67] Goldberg, A. A.; Levin, B. Ya.; Ostrovskiĭ, I. V. Entire and meromorphic functions. Complex analysis, I, pp. 1–193, 254–261, Encyclopaedia Math. Sci., 85, Springer, Berlin, 1997.

[68] Govorov, N. V. Riemann's boundary problem with infinite index. Edited and with an introduction and an appendix by I. V. Ostrovskiĭ. Translated from the 1986 Russian original by Yu. I. Lyubarskiĭ. Birkhäuser Verlag, Basel, 1994. xii+252 pp.

[69] Gundersen, G.; Hayman, W. K. The strength of Cartan's version of Nevanlinna theory. Bull. London Math. Soc. 36 (2004), no. 4, 433–454.

[70] Hayman, W. The minimum modulus of large integral functions, Proc. London Math. Soc., 2 (1952), 469–512.

[71] Hayman, W. K. Picard values of meromorphic functions and their derivatives, Ann. of Math. 70 (1959), 9–42.

[72] Hayman, W. K. On the Valiron deficiencies of integral functions of infinite order. Ark. Mat. 10 (1972), 163–172.

[73] Hayman, W. K. The minimum modulus of integral functions of order one. J. Analyse Math. 28 (1975), 171–212.

[74] Hayman, W. K. On Iversen's theorem for meromorphic functions with few poles. Acta Math. 141 (1978), no. 1-2, 115–145.

[75] Hayman, W. K. An approximation theorem with applications to Iversen's theorem. Aspects of contemporary complex analysis (Proc. NATO Adv. Study Inst., Univ. Durham, Durham, 1979), pp. 209–223, Academic Press, London-New York, 1980.

[76] Hayman, W. K. On a meromorphic function having few poles but not tending to infinity along a path. Ann. Polon. Math. 39 (1981), 83–91.

[77] Hayman, W. K. Warings Problem für analytische Funktionen. Bayer. Akad. Wiss. Math.-Natur. Kl. Sitzungsber. 1984, pp. 1–13 (1985).

[78] Hayman, W. K.; Kjellberg, B. On the minimum of a subharmonic function on a connected set. Studies in pure mathematics, pp. 291–322, Birkhäuser, Basel, 1983.

[79] Hayman, W. K. Subharmonic functions. Vol. 2. London Mathematical Society Monographs, 20. Academic Press, Inc. [Harcourt Brace Jovanovich, Publishers], London, 1989.

[80] Hayman, W. K. Nevanlinna theory 1945–1995. XVIth Rolf Nevanlinna Colloquium, Eds.: Laine/Martio, Walter de Gryuyter and Co., Berlin, New York, 1996.

[81] Hellerstein, S.; Shea, D. Minimal deficiencies for entire functions with radially distributed zeros. Proc. London Math. Soc. (3) 37 (1978), no. 1, 35–55.

[82] Hellerstein, S.; Williamson, J. Derivatives of entire functions and a question of Pólya. Trans. Amer. Math. Soc. 227 (1977), 227–249.

[83] Hellerstein, S.; Williamson, J. Derivatives of entire functions and a question of Pólya. II. Trans. Amer. Math. Soc. 234 (1977), no. 2, 497–503.

[84] Hellerstein, S.; Williamson, J. Entire functions with negative zeros and a problem of R. Nevanlinna. J. Analyse Math. 22 (1969), 233–267.

[85] Hellerstein, S.; Shen, Li Chien; Williamson, J. Reality of the zeros of an entire function and its derivatives. Trans. Amer. Math. Soc. 275 (1983), no. 1, 319–331.

[86] Hennekemper, W. Über die Werteverteilung von $(f^{k+1})^{(k)}$. Math. Z. 177 (1981), 375–380.

[87] Hinchliffe, J. D. On a result of Chuang related to Hayman's alternative. Comput. Methods Funct. Theory 1 (2002), 293–297.

[88] Hinchliffe, J. D. The Bergweiler-Eremenko theorem for finite lower order. Result. Math. 43 (2003), 121–128.

[89] Hinkkanen, A. A sharp form of Nevanlinna's second fundamental theorem. Invent. Math. 108 (1992), no. 3, 549–574.

[90] Hinkkanen, A. Iteration and the zeros of the second derivative of a meromorphic function. Proc. London Math. Soc. (3) 65 (1992), no. 3, 629–650.

[91] Hinkkanen, A. Reality of zeros of derivatives of meromorphic functions. Ann. Acad. Sci. Fenn. Math. 22 (1997), no. 1, 21–38.

[92] Hinkkanen, A. Zeros of derivatives of strictly non-real meromorphic functions. Ann. Acad. Sci. Fenn. Math. 22 (1997), no. 1, 39–74.

[93] Hinkkanen, A. Iteration, level sets, and zeros of derivatives of meromorphic functions. Ann. Acad. Sci. Fenn. Math. 23 (1998), no. 2, 317–388

[94] Hinkkanen, A. Zeros of derivatives of meromorphic functions with one pole. Complex Variables Theory Appl. 37 (1998), no. 1-4, 279–369.

[95] Hyllengren, A. Valiron deficient values for meromorphic functions in the plane. Acta Math. 124 (1970), 1–8.

[96] Ishizaki, K. Some remarks on results of Mues about deficiency sums of derivatives. Arch. Math. (Basel) 55 (1990), 374–379.

[97] Khabibullin, B. N. Growth of entire functions with prescribed zeros and representation of meromorphic functions. (Russian) Mat. Zametki 73 (2003), no. 1, 120–134; English transl., Math. Notes 73 (2003), no. 1-2, 110–124.

[98] Khabibullin, B. N. The representation of a meromorphic function as the quotient of entire functions and Paley problem in $\overline{\mathbf{C}}^n$: survey of some results. Mat. Fiz. Anal. Geom. 9 (2002), no. 2, 146–167.

[99] Kondratyuk, A. A. Fourier series and meromorphic functions, Vyshcha shkola, L'viv, 1988.

[100] Kotman, L. J. *An entire function with irregular growth and more than one deficient value.* Complex analysis, Joensuu 1978 (Proc. Colloq., Univ. Joensuu, Joensuu, 1978), pp. 219–229, Lect. Notes Math., vol. 747, Springer, Berlin, 1979.

[101] Lang, S. Introduction to complex hyperbolic spaces. Springer-Verlag, New York, 1987.

[102] Lang, S. Number theory. III. Diophantine geometry. Encyclopaedia of Mathematical Sciences, 60. Springer-Verlag, Berlin, 1991. xiv+296 pp.

[103] Langley, J. K. On Hayman's alternative, Mathematika 32 (1985), 139–146.

[104] Langley, J. K. Proof of a conjecture of Hayman concerning f and f''. J. London Math. Soc. (2) 48 (1993), no. 3, 500–514.

[105] Langley, J. K. On second order linear differential polynomials. Results Math. 26 (1994), 51–82.

[106] Langley, J. K. A lower bound for the number of zeros of a meromorphic function and its second derivative. Proc. Edinburgh Math. Soc. 39 (1996), 171–185.

[107] Langley, J. K. The second derivative of a meromorphic function of finite order. Bulletin London Math. Soc. 35 (2003), 97–108.

[108] Langley, J. K. The zeros of $ff'' - b$. Results Math. 44 (2003), no. 1-2, 130–140.

[109] Langley, J. K. Critical values of slowly growing meromorphic functions. Comput. Methods Funct. Theory 2 (2002), 539–547.

[110] Langley, J. K. Nonreal zeros of higher derivatives of real entire functions of infinite order. J. d'Analyse 97 (2005), 357–396.

[111] Levin B. Ya.; Ostrovskii, I. V. The dependence of the growth of an entire function on the distribution of zeros of its derivatives. Sibirsk. Mat. Zh. 1 (1960), 427–455. English transl., AMS Transl. (2) 32 (1963), 323–357.

[112] Lewis, J.; Wu, Jang-Mei. On conjectures of Arakelyan and Littlewood. J. Analyse Math. 50 (1988), 259–283.

[113] Marchenko, I. I.; Shcherba, A. I. On the magnitudes of deviations of meromorphic functions. (Russian) Mat. Sb. 181 (1990), no. 1, 3–24; English transl., Math. USSR-Sb. 69 (1991), no. 1, 1–24.

[114] Merenkov, S. Rapidly growing entire functions with three singular values. To appear in Illinois J. Math.

[115] Miles, J. Quotient representations of meromorphic functions. J. Analyse Math. 25 (1972), 371–388.

[116] Miles, J. On entire functions of infinite order with radially distributed zeros. Pacific J. Math. 81 (1979), no. 1, 131–157.

[117] Miles, J. Some examples of the dependence of the Nevanlinna deficiency upon the choice of origin. Proc. London Math. Soc. (3) 47 (1983), no. 1, 145–176.

[118] Miles, J. On the growth of meromorphic functions with radially distributed zeros and poles. Pacific J. Math. 122 (1986), no. 1, 147–167.

[119] Miles, J. A sharp form of the lemma on the logarithmic derivative. J. London Math. Soc. (2) 45 (1992), no. 2, 243–254.

[120] Miles, J.; Shea, D. An extremal problem in value-distribution theory. Quart. J. Math. Oxford Ser. (2) 24 (1973), 377–383.

[121] Miles, J.; Shea, D. On the growth of meromorphic functions having at least one deficient value. Duke Math. J. 43 (1976), no. 1, 171–186.

[122] Mohonko A. Z. Characteristics of certain meromorphic functions. (Russian) Teor. Funkciĭ Funkcional. Anal. i Priložen. No. 14 (1971), 83–87.

[123] Mues, E. Über eine Defekt- und Verzweigungsrelation für die Ableitung meromorpher Funktionen, Manuscripta Math. 5 (1971), 275–297.

[124] Mues, E. Über die Nullstellen homogener Differentialpolynome, Manuscripta Math. 23 (1978), 325–341.

[125] Mues, E. Über ein Problem von Hayman, Math. Z. 164 (1979), 239–259.

[126] Mues E.; Steinmetz, N. The theorem of Tumura-Clunie for meromorphic functions, J. London Math. Soc. (2) 23 (1981), 113–122.

[127] F. Nevanlinna (see the old ref list).

[128] Nevo, S.; Pang, X.; Zalcman, L. Picard-Hayman behavior of derivatives of meromorphic functions with multiple zeros, Electron. Res. Announc. Amer. Math. Soc. 12 (2006), 37–43.

[129] Osgood, C. F. A number theoretic-differential equations approach to generalizing Nevanlinna theory. Indian J. Math. 23 (1981), no. 1-3, 1–15.

[130] Osgood, C. F. Sometimes effective Thue-Siegel-Roth-Schmidt-Nevanlinna bounds, or better. J. Number Theory 21 (1985), no. 3, 347–389.

[131] Osgood, C. F. Simplified proofs of generalizations of the n-small function theorem. Analysis of one complex variable (Laramie, Wyo., 1985), pp. 158–174. World Sci. Publishing, Singapore, 1987.

[132] Pang Xuecheng. Bloch's principle and normal criterion. Sci. China Ser. A 32 (1989), 782–791.

[133] Pang Xuecheng. On normal criterion of meromorphic functions. Sci. China Ser. A (33) (1990), 521–527.

[134] Petrenko, V. P. Growth of meromorphic functions. (Russian) Izdat. Kharkov. Gos. Univ., "Vyshcha shkola", Kharkov, 1978.

[135] Pólya, G. Über die Nullstellen sukzessiver Derivierten. Math. Z. 12 (1922), 36-60.

[136] Qiao, Jianyong. Two problems in the value distribution theory. Acta Math. Sinica 11 (1995), 4, 365–371.

[137] Ru, Min. Nevanlinna theory and its relation to Diophantine approximation. World Scientific Publishing Co., Inc., River Edge, NJ, 2001.

[138] Rubel, L. Entire and meromorphic functions. Springer-Verlag, New York, 1996.

[139] Sauer, A. Deficient rational functions and Ahlfors's theory of covering surfaces. Ark. Mat. 39 (2001), no. 1, 151–155.

[140] Saxer, W. Sur les valeurs exceptionelles des dérivées successives des fonctions méromorphes. C.R. Acad. Sci. Paris 182 (1926), 831–833.

[141] Saxer, W. Über die Picardschen Ausnahmewerte successiver Derivierten. Math. Z. 17 (1928), 206–227.

[142] Shea, D. On the set where a meromorphic function is large. Complex analysis and its applications (Hong Kong, 1993), pp. 95–106. Pitman Res. Notes Math. Ser., 305, Longman Sci. Tech., Harlow, 1994.

[143] Shea, D. F.; Sons, L. R. Value distribution theory for meromorphic functions of slow growth in the disk. Houston J. Math. 12 (1986), 249–266.

[144] Sheil-Small, T. On the zeros of the derivatives of real entire functions and Wiman's conjecture. Ann. of Math. (2) 129 (1989), no. 1, 179–193.

[145] Sons, L. R. Value distribution for unbounded functions in the unit disc. Complex Var. Theory Appl. 7 (1987), 337–431.

[146] Steinmetz, N. Über die Nullstellen von Differentialpolynomen. Math. Z. 176 (1981), 255–264.

[147] Steinmetz, N. Eine Verallgemeinerung des zweiten Nevanlinnaschen Hauptsatzes. J. für die reine u. angewandte Mathematik 368 (1986), 134–141.

[148] Steinmetz, N. On the zeros of $(f^{(p)} + a_{p-1}f^{(p-1)} + \cdots + a_0 f)f$. Analysis 7 (1987), 375–389.

[149] Steinmetz, N. On the zeros of a certain Wronskian. Bull. London Math. Soc. 20 (1988), 525–531.

[150] Teichmüller, O. Eine Umkehrung des zweiten Hauptsatzes der Wertverteilungslehre. Deutsche Math. 2 (1937), 96–107.

[151] Ter-Israeljan, L. A. A meromorphic function of zero order with a non-asymptotic defect value. (Russian) Mat. Zametki 13 (1973), 195–206. English translation: Math. Notes 13 (1973), 116–123.

[152] Tohge, K. Logarithmic derivatives of meromorphic or algebroid solutions of some homogeneous linear differential equation. Analysis (Munich) 19 (1999), 273–297.

[153] Toda, N. On a modified deficiency of meromorphic functions. Tôhoku Math. J. 22 (1970), 635–658.

[154] Tumura, Y. On the extensions of Borel's theorem and Saxer-Csillag's theorem. Proc. Phys. Math. Soc. Japan (3) 19 (1937), 29–35.

[155] Vojta, P. Diophantine approximations and value distribution theory. Lecture Notes in Mathematics, 1239. Springer-Verlag, Berlin, 1987.

[156] Wang, Y. F. On Mues' conjecture and Picard values, Sci. China Ser. A 36 (1993), 28-35.

[157] Wang Y. F.; Fang, M. Picard values and normal families of meromorphic functions with multiple zeros. Acta Math. Sinica New Ser. 14 (1998), 17–26.

[158] Wang, Y. F.; Yang, C. C.; Yang, Lo. On the zeros of $(f^{(k)})^n f - a$. Kexue Tongbao 38 (1993), 2215–2218.

[159] Weitsman, A. A theorem on Nevanlinna deficiencies. Acta Math. 128 (1972), no. 1-2, 41–52.

[160] Weitsman, A. Meromorphic functions with maximal deficiency sum and a conjecture of F. Nevanlinna. Acta Math. 123 (1969), 115–139.

[161] Yamanoi, K. Defect relation for rational functions as targets. With errata by the author. Forum Math. 17 (2005), no. 2, 169–189.

[162] Yamanoi, K. The second main theorem for small functions and related problems. Acta Math. 192 (2004), no. 2, 225–294.

[163] Yang, Lo. Precise estimate of total deficiency of meromorphic derivatives. J. d'Analyse Math. 55 (1990), 287–296.

[164] Yang, Lo. Value distribution theory. Translated and revised from the 1982 Chinese original. Springer-Verlag, Berlin; Science Press, Beijing, 1993.

[165] Zalcman, L. Normal families: new perspectives. Bull. Amer. Math. Soc. (N.S.) 35 (1998), 215–230.

[166] Zhang Z.-L.; Li, W. Tumura-Clunie's theorem for differential polynomials. Complex Variables 25 (1994), 97–105.

Author Index

Subject Index

Notation Index